BALANCE HYDROLOGICS, INC.
1760 SOLANO AVENUE, SUITE 209
BERKELEY, CALIFORNIA 94707

HANDBOOK of
Integrated Pest Management for Turf and Ornamentals

Edited by
ANNE R. LESLIE
U.S. Environmental Protection Agency
Washington, D.C.

LEWIS PUBLISHERS
Boca Raton   Ann Arbor   London   Tokyo

Photography credits

Front Cover: upper left, patio area at Sycamore Hospital, Miamisburg, OH, photo courtesy of *Grounds Maintenance* magazine; lower left, landscaping at Tysons Space Center, McLean, VA, photo courtesy of *Grounds Maintenance* magazine; right, hole #9 at Devil's Pulpit Golf Club, Toronto, photo by Doug Ball.

Back Cover, from top: hole #10 at Prairie Dunes Country Club, Hutchinson, KS, photo by P. Stan George; parking mall at Ambassador College, Pasadena, CA, photo courtesy of *Grounds Maintenance* magazine; hole #17 at Prairie Dunes Country Club, Hutchinson, KS, photo by P. Stan George; hole #9 at Devil's Paintbrush Golf Club, Toronto, photo by Doug Ball.

Photographs used on section divider pages were taken by the editor.

Library of Congress Cataloging-in-Publication Data

Handbook of integrated pest management for turfgrass and ornamentals /
    edited by Anne R. Leslie.
        p. cm.
     Rev. ed. of: Integrated pest management for turfgrass and
ornamentals. 1989.
     Includes bibliographical references and index.
     ISBN 0-87371-350-8
     1. Turfgrasses—Diseases and pests—Integrated control—Handbooks,
manuals, etc. 2. Landscape plants—Diseases and pests—Integrated
control—Handbooks, manuals, etc. 3. Turf management—Handbooks,
manuals, etc. I. Leslie, Anne R., 1931– . II. Title: Integrated
pest management for turfgrass and ornamentals.
SB608.T87H35   1994
635.9′29—dc20   93-43149
                                                                     CIP

DEDICATION

The beauty of a lovingly tended landscape around a home can be enjoyed by all who view it throughout the changing seasons. At the time this was written, winter had yielded its severity to a brilliant succession of bulbs, flowering trees and shrubs, wildflowers, and green lawns.

For Earth Day '93, I was one of a number of EPA scientists who elected to talk with Washington, D.C. public schoolchildren. I wanted to focus my talk on Integrated Pest Management (IPM), and I decided to share some of the bounty of my garden with several fourth- to sixth-grade classes, to illustrate that a "weed" is often just a plant out of place.

I knew that the children would enjoy what many gardeners would try to eliminate, but I was not prepared for the eagerness of the children to hold the plants and branches — not just the lilac and wisteria, but even the dandelions, violets, and henbit.

We talked about the Environmental Protection Agency, Earth Day, and my job of informing people about IPM, pests, and pest control. I gave them copies of EPA's brochure, "Healthy Lawn, Healthy Environment" — and I thought about this book and my hope it will help people be sensitive to the natural environment as they care for their own or someone else's patch of earth.

It is in this spirit that I dedicate this book to children everywhere, and especially to my own five sons, who make all my work worthwhile — Douglas, the landscaper and philosopher, Neal, the electronics engineer, Eric, the musician, Travis, the cabinetmaker, and Brian,who cared about everyone and everything during his nine years on earth.

Anne R. Leslie is a chemist in the Office of Pesticide Programs of the U.S. Environmental Protection Agency (EPA). She has been collecting and disseminating information on Integrated Pest Management (IPM), both urban and agricultural, since joining the IPM unit of EPA in 1985.

Anne completed her undergraduate degree in chemistry at the University of Arizona and did graduate studies at the University of Utah toward a Ph.D. degree in biochemistry. She received a Master of Science degree in Biochemistry at McGill University in Montreal. She joined the EPA in 1980, working on the guidelines for pesticide registration, and later on assessment of exposure to pesticides. Her work in the IPM unit includes editorship of several books and chapters on IPM, presentations to trade groups and professional societies, participation in the USGA Green Section Research Committee, management of IPM technology transfer grants, and development of EPA brochures on IPM for Lawns, IPM in Schools, and IPM for Homeowners.

This year the IPM functions in EPA are moving to a new one-year pilot division of the Office of Pesticide Programs (OPP), the Biopesticides and Pollution Prevention Division (BPPD). Anne will be one of 33 staff organized in multidisciplinary teams to accelerate the registration and reregistration of biological pesticides, encourage the development and use of safer pesticides, and promote reduce pesticide use. She will be continuing to develop educational materials to inform registrants and the public about pesticide pollution prevention and IPM.

PREFACE

PURPOSE

The U.S. Environmental Protection Agency (EPA) assisted in the development of this book as one of many projects designed to enhance pesticide pollution prevention. The Pollution Prevention Act of 1990 established as national policy a new hierarchy of environmental protection, placing pollution prevention first, followed by recycling and environmentally safe pollution treatment. Reduced use is seen as the one clear method for making progress toward pesticide pollution prevention and has become a stated Agency goal in relation to both agricultural and urban environments.

The EPA believes that sound IPM practices are integral to any progress toward real pesticide use reductions. For over a decade, the Agency has defined IPM as "...an effective and environmentally sensitive approach to pest management that relies on a combination of common-sense practices. IPM programs use current, comprehensive information on the life cycles of pests and their interactions with the environment. This information, in combination with available pest control methods, is used to manage pest damage by the most economical means, and with the least possible hazard to people, property, and the environment. IPM programs take advantage of all pest management options possible, including, but not limited to the judicious use of pesticides."

INTENDED AUDIENCE

This *Handbook of Integrated Pest Management for Turfgrass and Ornamental*s is intended for professionals who deal with urban landscaping and turf management of all kinds. Whether the area is a park, a highway right-of-way, a commercial landscape, a home lawn, a schoolground, or a golf course, there are choices to be made on management of weeds, insects, and microorganisms that can damage desirable plants.

SCOPE

This book is not the result of a symposium, as was a previous EPA publication on turfgrass IPM. Instead, it is the product of several years of work identifying cutting-edge IPM research and management for turf and ornamentals. Some 59 manuscripts covering an array of topics were solicited. While there were other areas that could have been included, limits had to be placed upon the total volume of the text.

This book focuses on various management problems and offers the manager the tools to implement an IPM approach. The opening section describes the urban landscape ecosystem with chapters detailing some of the concerns over possible pollution by pesticides and the potential effects on nontarget systems. Following a section on how to create a healthy landscape, ensuing sections detail studies of selected problem pests and effectiveness of IPM alternatives. Many chapters offer references for further reading, and some of the chapters are reprints of information that may not be available in all parts of the country.

A section on special considerations for golf courses addresses site selection and preparation, as viewed by the environmentalist and by the golf course architect. Much significant research designed to examine potential problems with highly managed turf areas has been supported by the U.S. Golf Association as a function of its Green Section Research Committee over the past ten years. Such research is directly applicable to turf sites other than golf courses, and some of the results appear as chapters in other sections.

Although synthetic chemical controls are not covered in most of the book, it would not be complete without a discussion of the calibration of delivery systems. Certain biological controls require application by spraying, and delivery of the correct amount is critical in achieving the desired efficacy, just as for chemical controls.

The last section is a collection of case studies and success stories by managers who have embraced the IPM concept. As a whole, this handbook could well be used as a text for courses on landscape management. The primary intent, however, is that it serve as a useful reference for those involved in management of turf and ornamental pests.

ACKNOWLEDGMENTS

One of the most striking aspects of compiling this book was the cooperative spirit of all who contributed their time, whether in writing, reviewing, or revising. There were more than 100 reviewers drawn from the turf and chemical industries, academia, and government.

The list is too long to acknowledge each one, but I would like to accord special mention to Dr. Mark Welterlen, editor of *Grounds Maintenance* magazine, who devoted a substantial amount of his time to the review of a number of the manuscripts. I would also like to thank Dr. James Watson, vice president of the Toro Company. Dr. Watson provided much helpful advice, critiques, and a long list of reviewers and potential authors for topics not addressed in the original outline for the book. He, along with Jim Snow, Dr. Mike Kenna, and ten other members of the U.S. Golf Association's Green Section Research Committee, contributed a number of excellent reviews, several manuscripts, and much encouragement.

DISCLAIMER

Each chapter reflects the viewpoint of its author(s), and the content of any particular contribution should not be construed to represent the policy or beliefs of the EPA. It is particularly important to note that certain chapters deal with pest control agents not specifically registered by the EPA for use as pesticides. Such uses would constitute a violation of the Federal Insecticide, Fungicide and Rodenticide Act and are included in this text solely as research findings. Mention of these uses must not be taken as either an endorsement or recommendation by the EPA.

CONTENTS

Section I: Dynamics of the Urban Landscape

Chapter 1
Toward Sustainable Lawn Turf 3
Richard J. Hull, Steven R. Alm, and Noel Jackson

Chapter 2
The Special Needs of Trees 17
James R. Clark and Nelda P. Matheny

Chapter 3
Fate of Pesticides in the Turfgrass Environment 29
A. Martin Petrovic, Nina Roth Borromeo, and Mark J. Carroll

Chapter 4
Sewage Sludge Compost for Establishment and Maintenance of Turfgrass 45
J. Scott Angle

Chapter 5
Integrated Resource Management 53
Ronald G. Dodson and Nancy P. Sadlon

Chapter 6
Effects of Pesticides on Beneficial Invertebrates in Turf 59
Daniel A. Potter

Section II: Preparation for a Healthy Landscape

Chapter 7
Determining the Health of the Soil 73
Charles H. Peacock

Chapter 8
Physical Problems of Fine-Textured Soils 79
Robert N. Carrow

Chapter 9
Physical Problems of Coarse-Textured Soils 85
Robert N. Carrow

Chapter 10
Principles of Turfgrass Growth and Development 91
John E. Kaufmann

Chapter 11
Choosing the Right Grass to Fit the Environment 99
Nick E. Christians and Milton C. Engelke

Chapter 12
Integrating Cultural and Pest Management Practices for Sod Production 115
Stephen T. Cockerham

Chapter 13
Advances in Implementing IPM for Woody Landscape Plants 125
M. J. Raupp, C. S. Koehler, and J. A. Davidson

Chapter 14
Calibrating Turfgrass Chemical Application Equipment 143
H. Erdal Ozkan

Chapter 14A
Simple Hand Sprayer Calibration 163
Philip Catron

Section III: Special Considerations for Golf Courses

Chapter 15
**Siting and Design Considerations to Enhance the
Environmental Benefits of Golf Courses** 167
Richard D. Klein

Chapter 16
**Design and Management of Constructed Ponds:
Minimizing Environmental Hazards** 173
Thomas Jon Smayda and Brenda L. Packard

Chapter 17
**Minimizing Environmental Impact by Golf Course Development: A Method and
Some Case Studies** 185
Michael J. Hurdzan

Chapter 18
Pest Management Strategies for Golf Courses 193
L. B. McCarty and Monica L. Elliott

Section IV: Blemishes on the Perfect Landscape

Chapter 19
Survival of Trees in Metropolitan Areas 205
Donald A. Rakow

Chapter 20
IPM: A Seattle Street Tree Case Study 213
Sharon J. Collman

Chapter 21
Major Insect Pests of Turf in the U.S. 219
R. L. Brandenburg and J. R. Baker

Chapter 22
Major Insect Pests of Ornamental Trees and Shrubs 237
Peter B. Schultz and David J. Shetlar

Chapter 23
**Symptomology and Management of Common Turfgrass Diseases in the
Transition Zone and Northern Regions** 249
Peter H. Dernoeden

Section V: Integrated Management of Weeds

Chapter 24
Understanding Turfgrass Growth Regulation 267
John E. Kaufmann

Chapter 25
Turfgrass Weed Management — An IPM Approach..275
Joseph C. Neal

Appendix A
Weed Illustrations 281

Chapter 26
**Plan Before You Plant: A Five-Step Process for
Developing a Landscape Weed Management Plan** 293
Joseph C. Neal

Chapter 27
Weed Mangement Guide for Herbaceous Ornamentals................................301
Andrew Senesac and Joseph Neal

Chapter 28
IPM of Wildflower-Grass Mixes in the Eastern U.S. 315
John M. Krouse

Section VI: Integrated Management of Insects

Chapter 29
Turfgrass Insect Detection and Sampling Techniques 331
Lee Hellman

Chapter 30
Use of Insect Attractants in Protection of Ornamental Plants 337
Whitney S. Cranshaw

Chapter 31
Life Cycles and Population Monitoring of Pest Mole Crickets 345
William G. Hudson

Chapter 32
**Decision-Making Factors for Management of Fire Ants and
White Grubs in Turfgrass** 351
Beverly Sparks and S. Kristine Braman

Chapter 33
Timing Controls for Insect Pests of Woody Ornamentals 361
James L. Hanula

Chapter 34
**A Case Study of the Impact of the Soil Environment on Insect/Pathogen Interactions:
Scarabs in Turfgrass** 369
Michael G. Villani, Stephen R. Krueger, and Jan Nyrop

Section VII: Integrated Management of Disease

Chapter 35
Disease Management for Warm-Season Turfgrasses 387
Monica L. Elliott

Chapter 36
Use of Disease Models for Turfgrass Management Decisions 397
William W. Shane

Chapter 37
Managing Cool-Season Lawn Grasses to Minimize Disease Severity 405
Peter H. Dernoeden

Chapter 38
Biological Control of Turfgrass Diseases 409
Eric B. Nelson, Lee L. Burpee and Mark B. Lawton

Chapter 39
IPM for Tree and Shrub Diseases 429
George W. Hudler

Section VIII: An Inventory of Biological Controls

Chapter 40
Beneficial Insects and Mites 443
Richard Weinzierl and Tess Henn

Chapter 41
Commercial Biological Controls for Insect and Mite Pests of Ornamentals 455
Carol S. Glenister

Chapter 42
Inoculative Biological Control of Mole Crickets 467
J. Howard Frank

Chapter 43
Nematodes as Bioinsecticides in Turf and Ornamentals 477
Ramon Georgis and George O. Poinar, Jr.

Chapter 44
Biological Control for Plant-Parasitic Nematodes Attacking Turf and Ornamentals 491
George O. Poinar, Jr. and Ramon Georgis

Chapter 45
Microbial Control of Insect Pests of Landscape Plants 503
Whitney S. Cranshaw and Michael G. Klein

Chapter 46
The Role of Endophytes in IPM for Turf 521
Melodee L. Fraser and Jane P. Breen

Chapter 47
Stress Tolerance of Endophyte-Infected Turfgrass 529
Michael D. Richardson and Charles W. Bacon

Section IX: An Inventory of New Generation Chemical Controls

Chapter 48
Botanical Insecticides and Insecticidal Soaps 541
Richard Weinzierl and Tess Henn

Chapter 49
Oils as Pesticides for Ornamental Plants 557
Warren T. Johnson

Section X: Putting It All Together

Chapter 50
Golf Course Turf Pest Monitoring Program: 1991 Final Report 585
James D. Willmott, Maher Tawadros, and Jennifer Grant

Chapter 51
Avoid IPM Implementation Pitfalls 591
Tim Rhay

Chapter 52
An Award-Winning Management Plan 597
Laurie R. Broccolo

Chapter 53
A Lawn Care Alternative Service 603
Philip Catron

Chapter 54
Maintenance of Infields and Other Bare Soil Areas 611
Tim Rhay

Chapter 55
IPM in Municipal Parks Maintenance — A Case Study 613
Tim Rhay

Chapter 56
The Turfgrass Information File and Turfbyte 615
Robert Emmons, Peter Cookingham, and Duane Patton

Chapter 57
Enhancing Technology Transfer to the Homeowner 621
Laura Pottorff and William M. Brown, Jr.

Index 627

Dynamics of the Urban Landscape

Historic oak tree at
Sea Island, GA

A garden wedding at the Mohonk Resort in New York State

Pitfall trap used to determine populations of arthropods in grass (photo courtesy of *Grounds Maintenance* magazine)

Research plots (pesticide leaching
from turf) at Cornell University

Chapter 1

Toward Sustainable Lawn Turf

Richard J. Hull, Steven R. Alm, and Noel Jackson, Plant Sciences Department, University of Rhode Island, Kingston, RI

CONTENTS

The Value of Turf to the Urban/Suburban Environment 3
Lawn Ecosystem Dynamics 4
Factors Contributing to Sustainable Turf..5
    Nutrient Use Efficiency..6
    Water Use Efficiency 6
    Resisting Pests 6
    Healing Wounds 6
Improving Grass Adaptability to Turf Use 6
    Energy Partitioning...6
    Shoot Morphology 7
    Nutrition and Water Use 7
    Insect Predation 8
    Pathogens and Diseases..9
    Weed Infestation..10
Managing Lawn Turf for Sustainability 11
References 13

Extrapolating from a 1989 survey conducted by the Pennsylvania Department of Agriculture,[1] the total land area in the U.S. devoted to turf is about 40 million acres, of which 75% is in home lawns. This is in reasonable agreement with a 25 to 30 million acre estimate for total lawn turf made by Roberts and Roberts in 1986.[2] Citizens of the U.S. spend more than $30 billion in the annual maintenance of turf. Given the magnitude of the nation's turf enterprise, it is not surprising that the impact of turf on the environment and the resources committed to its maintenance have received critical attention. This attention challenges both the turf manager and researcher to work toward a low input, sustainable turf ecosystem. A sustainable turf can be defined as one which requires few material inputs and has only positive environmental impacts. Such a turf will be largely self-sustaining because it is so structured and managed that it will incur minimal loss of resources and therefore require few inputs.

THE VALUE OF TURF TO THE URBAN/SUBURBAN ENVIRONMENT

Before considering the problems and opportunities of establishing a self-sustaining turf, we must first dispose of the widely held view that turf is a luxury item which deserves a low ranking in the priorities for human investment.

Increasingly, the human species lives under crowded conditions. An urban or suburban environment supports most people today. Relatively few of us reside or work in a rural setting. This intensively developed environment poses numerous problems which were largely unknown when the nation was mostly farms and small towns. As an increasing percentage of the land area is paved or otherwise rendered impervious to water, storm runoff has become a major problem. Disposing of large quantities of relatively poor quality stormwater often overtaxes sewage treatment systems and results in erosion and pollution of surface water bodies. Turf constitutes the most water permeable ground cover in urbanized areas. Research conducted over the past decade has shown that storm runoff from a well-maintained turf rarely occurs, and then only from steep slopes of frozen ground.[3] Enhanced water infiltration by turf not only reduces runoff and protects the quality of surface water, it also recharges groundwater supplies[2] and, in coastal regions, repels saltwater intrusion.

Turf and other landscape plants moderate the sometimes harsh climate of the urban landscape. Turf along with shrubs and trees planted in close proximity to homes can reduce ambient air temperatures by as much as 7 to 14°F due to the cooling effect of evapotranspiration and shading.[2] This translates into reduced need for summer air conditioning. The strategic planting of lawns along with other landscape plants can reduce air conditioning costs by as much as 50% in high temperature, low humidity areas. Such landscaping strategies could reduce by 25% the 200 billion kilowatt hours spent annually in the U.S. for air conditioning.[4]

In addition to cooling, turf and other landscape plantings improve air quality, reduce noise levels, and increase wildlife habitat in urban and suburban communities. Turf, shrubs, trees, and exposed soil absorb significant quantities of air pollutants such as sulfur dioxide, carbon monoxide, and ozone and they trap airborne particulate matter. Through photosynthetic activity, landscape plants reduce atmospheric carbon dioxide levels while contributing oxygen. Although the amount of carbon bound into the turf biomass turns over at a reasonable rate, its positive impacts on air quality are greatest during summer, when atmospheric pollution is most serious. Trees and other woody landscape plants have the capacity to sequester large amounts of carbon dioxide and retain that carbon for many years.

Home lawns and sod-covered playing fields help reduce injury to children and others engaged in athletics or any type of outdoor activity. The cushioning effects of turf reduce abrasions, sprains, and even fractures resulting from physical contact sports, falls, and many types of accidents.[2] Well-maintained lawns, free of flowering weeds, minimize the probability of bee stings and the sometimes fatally acute reaction to them suffered by some people. Contracting arthropod-borne diseases can also be minimized by appropriate landscaping. Maupin et al.[5] reported that, of the deer ticks capable of transmitting Lyme disease that were found on residential properties, 9% were present in ornamental vegetation and only 2% were present on lawns. Wooded areas contained 67% of the ticks and 22% were collected from unmaintained edge areas. Thus, the likelihood of contacting spirochete-infected ticks from turf areas is remote compared to the probability of such encounters in the nearby woods.

A growing body of evidence testifies to psychological and even physiological benefits to people from ornamental plantings including turf.[6] The presence of plants in the workplace and home environments contributes to a feeling of well-being and helps reduce emotional tension and the frequency and severity of depression. The rate of recovery from surgery was even increased and the amount of pain-killer taken was reduced when hospital patients were placed in rooms with a view of a landscaped park compared to a view of adjacent buildings.[7]

Most of the benefits derived from landscape plantings cited above are likely not even considered by people when they invest the $30 billion each year on lawn maintenance. People simply value the presence of a well-maintained lawn and attractive plantings around their homes and, when given a choice, their workplace. This affinity of people for living things in their environment should not be underestimated or ignored. A valid case can be made that much of the physical and social deterioration occurring in our larger cities can at least, in part, be attributed to the hostile and dehumanizing environment created by a landscape devoid of lawns and ornamental plantings. When these elements are introduced into the urban landscape, the physical environment is improved and the social fabric of the neighborhood is strengthened. This certainly has been the experience of the extensive Philadelphia Green Program of the Pennsylvania Horticultural Society, to name one example. In this context, it is difficult to understand how ornamental plantings and the resources committed to them can be seen as any less important than investments in other social or physical elements of our urban and suburban communities.

LAWN ECOSYSTEM DYNAMICS

Even the most casual landscaper can testify to the fundamental instability of lawns and ornamental plantings. Without regular attention frequently involving the use of fertilizers, pesticides, and water, lawns quickly deteriorate. Why is this so? The poor quality of soil, or what substitutes for soil in many residential developments, may not provide the proper aeration, water, or nutrition required for healthy turfgrass growth. The selection of turfgrasses poorly suited to a particular landscape site may also explain lawn decline. It is good to remember that most turfgrasses are introduced species, are often subjected to intensive use, and do not constitute a climax vegetation where turf is normally grown. All these factors contribute to ecological instability and a dynamic grass community. However, even lawns composed of appropriate grasses growing in a good soil will often lose vigor, become thin, and exhibit weed encroachment unless they regularly receive fertilizer, are irrigated, and are occasionally treated with an herbicide. Even if clippings are retained on the turf and no nutrients are removed to deplete soil resources, turf decline normally occurs unless offset by proper management. There is, however, no inherent reason why a carefully constructed lawn cannot be reasonably stable; the concept of a low-input sustainable turf or landscape should be the rule, not a goal for the future. This is the challenge for turfgrass science and landscape management.

Lawns are rarely sustainable without significant inputs because achieving self-sufficiency has not been an important goal of the turf manager or the turfgrass researcher. Until recently, turfgrasses generally were not selected for efficiency in their use of nutrients or water. These were always added as needed. Resistance or tolerance to insect predation and other environmental stresses were infrequently considered. Insecticides could eliminate insects and stresses were reduced by adding water, raising soil pH, or otherwise modifying the turf environment. Only disease resistance has been addressed systematically by turfgrass breeders over the past 20 years, and substantial progress has been made in developing pathogen-tolerant grasses, although much remains to be done. Weeds were considered a consequence of improper turf management and, when deemed undesirable, they could be controlled by herbicides. Only now, when many of these interventions are less available or have become more costly and environmental considerations have entered the management equation, are serious efforts being made to create a sustainable turf which, once established, will require few additions of basic plant resources.

Before progress can be made toward achieving more self-sustaining turf, the various factors which contribute toward its present instability must be identified and quantified. What follows is an analysis of these factors and some ideas as to how they might be addressed. Most of these subjects are treated in much greater detail and authority by the authors of other chapters in this book, but this is offered to provide a context in which to view the efforts toward developing low-input sustainable turf.

Annual nutrient losses often constitute a major contributor to decline in turf quality. Turf frequently loses from 1 to 3 lb of nitrogen per 1000 ft^2/year, so at least that much often must be added to maintain a quality lawn. Other nutrients are also lost, but nitrogen is the most significant and its deficiency certainly contributes to the yellowing and thinning of turf that occurs whenever nitrogen applications are omitted. Such turf suffers a nutrient imbalance and this contributes to its increased susceptibility to pathogens which cause such diseases as dollar spot, red thread, and necrotic ring spot, to name only a few. The stand thinning which results from disease attack opens the door for weed invasion.

Once this occurs, a fundamental defect in most cool-season turfgrasses becomes apparent.

Cool-season turfgrasses possess the C-3 photosynthetic pathway for carbon dioxide fixation.[8] This means that, under elevated temperatures, their photosynthetic capacity becomes limited because of increased competition between oxygen and carbon dioxide for the binding site on the enzyme responsible for fixing carbon dioxide. This competition can reduce net photosynthesis 50%, decreasing the energy available for shoot and root growth. Many summer weeds are warm-season plants which fix carbon dioxide via a different enzyme for which oxygen offers no competition. These C-4 weeds are much more efficient in fixing carbon dioxide under the high temperatures of summer and, thus, become more competitive with C-3 lawn grasses at that time. Most C-4 plants are not favored during the cool conditions of spring or mid-fall and can only become established if the turf is first thinned by disease injury, insect feeding, or damage resulting from human activity. During cooler seasons, when C-4 weeds become injured by cold temperatures, C-3 weeds will become established, unless the turfgrasses can recover rapidly. Many of our cool-season turfgrasses are not capable of rapid recolonization of areas vacated by summer weeds.

Warm-season turfgrasses grown in the southern states do not lose competitiveness during the hot summer months, and weed invasion depends upon turf injury or errors in management. During cooler seasons, these C-4 turfgrasses become less competitive with C-3 weeds, which will become established if the turf has been thinned during or after the growing season. In both cases, weeds are opportunistic and take advantage of weaknesses in the turf resulting from innate competitive deficiencies or improper management strategies. It should be evident that sustainability in turf depends upon utilizing turfgrasses which are efficient in retaining nutrients and which are capable of resisting or repairing injury so weed invasion can be prevented.

FACTORS CONTRIBUTING TO SUSTAINABLE TURF

Grasses have been bred for use in lawn turf for a relatively short time. The first truly improved turfgrass cultivar, "Merion" Kentucky bluegrass, was released in 1947. This means that serious turfgrass breeding has been under way for only 45 years. During that time much emphasis has been given to aesthetic qualities of the grass: color, leaf size, leaf angle, lack of flowers when grown as turf, sod density, and tolerance of close mowing. Features that would contribute to sustainability are only now being considered in turfgrass improvement programs. What are some of these sustainability characteristics?

NUTRIENT USE EFFICIENCY

Among the more important properties of a sustainable turfgrass must be improved efficiency in nutrient acquisition and utilization. Because nutrient loss from the soil must be minimized, the turfgrass should have the capacity to absorb nutrient ions efficiently and maintain a low concentration of those nutrients in the soil solution. In that way, water leaching through the soil will remove very few nutrient ions, especially nitrate. Once within the turfgrass plants, nutrients must be used efficiently. This is a complex and not very well understood area involving metabolic efficiencies and enzyme turnover rates which result in high dry matter production per unit of mineral nutrient within the plant. These nutrient efficiency ratios have been determined for some crops[9] and turfgrasses,[10] but we are unaware of any effort to screen for them in turfgrass breeding programs.

WATER USE EFFICIENCY

Also essential for a sustainable turfgrass is efficiency in water use. Two very different phenomena are involved here. Turfgrasses must maintain an evapotranspiration rate sufficient to provide leaf cooling and the acropetal transport of nutrient ions from the roots, but no more. A greater water use rate would result in poorer water use efficiency. A deep and extensive root system will enable the turfgrass to mine the entire soil profile for available water and thereby delay the time when water might become inadequate for proper growth. At the same time, turfgrasses must be drought tolerant because, even in the humid parts of the U.S., periods of insufficient rainfall do occur. Grasses efficient in acquiring and using water will forestall the onset of drought stress, but when such stress occurs, a sustainable turfgrass should be capable of shutting down and entering a stress-induced dormancy or quiescence. During this time, the grass must sustain minimal injury so it can recover quickly when moisture is again available. In that way, vulnerability to weed invasion will be minimized and periods of disease susceptibility will be brief. Drought-induced dormancy will not protect turf from injury caused by continued use and recovery from such mechanical damage may be slow if use is not curtailed during periods of excessive drought. Obviously, in parts of the country that are chronically dry, irrigation will be required, although even here water efficient grasses will reduce the quantity of water needed to maintain a quality turf.

RESISTING PESTS

Selecting turfgrasses capable of resisting the assault of pests is part of the sustainability strategy. There remains much to learn concerning mechanisms by which plants repel the attack of pathogens. In general, healthy, unstressed plants are more resistant to disease organisms, but there clearly are times when the combination of high inoculum levels and optimum climatic conditions all but guarantee a disease problem, especially on closely mowed turf. Research on disease resistance mechanisms is progressing. Improvement of field crops has demonstrated that disease resistance can be achieved through breeding. Modest progress has been achieved in selecting for resistance to some diseases in newer cultivars of the major turfgrasses.[11] Certainly this is an area where much can be done to enhance turf sustainability, however, it is a complex problem and progress will be slow.

Insect predation can devestate a turfgrass stand. This often occurs when the grass is weakened by poor management, excessive use, or through environmental stress. Most healthy vigorous turfgrasses can sustain a modest degree of insect predation. Breeding turfgrasses for insect repulsion has some potential, but it has not progressed very far. Establishing a turf environment which discourages insect feeding and reproduction appears to be more immediately encouraging as will be discussed later.

HEALING WOUNDS

The capability to heal wounds is important to a sustainable turf. Injury will occur and grasses which can recover and fill opened areas quickly will be less likely to experience weed invasion. Rhizomatous or stoloniferous grasses have an advantage over bunch-type grasses because they can generate organs designed to colonize open areas of soil. Maintaining a mixture of grass species including creeping types will, along with sound management, help provide a self-healing turf less susceptible to weed invasion.

IMPROVING GRASS ADAPTABILITY TO TURF USE

Before considering the ways by which turfgrasses can be made more sustainable by incorporating the features outlined above, it might be useful to consider some very basic characteristics which contribute to grass well-being when maintained as a closely mowed turf.

ENERGY PARTITIONING

Turfgrasses should allocate a large portion of their photosynthetic product to root growth throughout the growing season and beyond. Except for repair of injury and replacement of damaged or senescent leaves, a turfgrass does not need to partition most of its energy resources into shoot growth. Grass would acquire nutrients and water better and tolerate root feeding by insects if it maintained an extensive root system. Close mowing reduces photosynthetic leaf tissue capable of generating energy and resources for root growth, and normally results in a restricted root system. However, if partitioning of that energy between shoot and root growth could be shifted to favor the latter, a more sustainable turfgrass would result.

We know which management practices or conditions favor root growth: adequate, but not excessive nitrogen and elevated phosphorus concentrations in the soil; near neutral soil pH; higher mowing levels; and adequate moisture throughout the soil profile. Even under ideal management, cool-season turfgrasses partition less energy to root growth as the hot summer progresses.[8] This results in a turf vulnerable to injury from a number of agents, including normal wear and numerous environmental stresses. Of course, cool-season turfgrasses, because of their C-3 photosynthesis, have less energy available to them during the hot summer. This may be at the heart of summer growth decline normally observed in cool-season turfgrasses. Other factors, such as reduced protein synthesis or the production of heat shock proteins, may also contribute to high-temperature growth suppression.[12] Progress in altering the physiology of cool-season grasses to tolerate better elevated temperatures has been frustrating. In time, C-3 grasses metabolically tolerant of prolonged high-temperature conditions may be available, but in the meantime, attention might be addressed more constructively to grasses which already exhibit high-temperature adaptation, e.g., zoysiagrass and buffalograss.[12] Of course, the suitability of these warm-season grasses to northern conditions must be evaluated before they can be widely recommended.

SHOOT MORPHOLOGY

A prostrate growth habit which enables turfgrasses to retain more photosynthetic leaf surface when closely mowed, will contribute to a more sustainable turf. Many improved aggressive Kentucky bluegrass cultivars possess this characteristic. It is also being sought in new selections of tall fescue and perennial ryegrass. Prostrate shoot growth not only permits the retention of more leaf surface, but these leaves shade the soil better and retard germination and establishment of warm-season weeds. A lower-growth form and more compact leaf canopy tends to reduce air movement over leaf surfaces, thereby increasing the atmospheric boundary layer and reducing evapotranspiration. For this reason, many of the newer prostrate turfgrasses are more efficient users of water. All these conditions, associated with the prostrate growth habit, contribute to greater sustainability of turf.

NUTRITION AND WATER USE

We have already outlined some principles underlying water and nutrient use efficiency by turfgrasses. These are important because nutrients and water are resources most often limiting to grass growth. Reducing turf dependency on the application of these items as part of a turf management program is central to the concept of sustainable turf.

Root depth has already been identified as a factor in efficient water and nutrient acquisition from the soil. Kentucky bluegrass cultivars that perform well under low fertility and no irrigation (low maintenance) have been observed to have larger and deeper root systems and maintain a lower shoot-to-root ratio than cultivars which perform well only under high-maintenance conditions.[13] In this study, low-maintenance cultivars produced 56% more root mass which was 12% deeper in the soil. In another study, field measurements of ten Kentucky bluegrass cultivars revealed a five-fold range in root mass during late summer and early fall.[14] In general, those cultivars with the greatest root mass produced the highest quality turf under medium-low maintenance conditions. Thus, it appears that there is considerable genetic variation in plant characteristics important to a sustainable turfgrass. Breeding programs directed toward enhancing these characteristics should generate cultivars better able to thrive under natural levels of nutrients and water.

The capacity of roots to absorb nutrients has been studied in comparisons among food-crop cultivars. Substantial variation in both the capacity (maximum rate of nutrient uptake) and the intensity (affinity of nutrients for roots) of nutrient absorption have been observed.[15] Such research on turfgrasses is limited, but Cisar et al.[16] also observed substantial variation among turfgrass species for several parameters describing the kinetics of nitrate, phosphate, and potassium uptake by roots. Their field studies demonstrated that these kinetic values corresponded with nutrient accumulation in turf clippings, especially with respect to nitrogen. While these are preliminary observations, they are encouraging because the possibility of obtaining turfgrass genotypes inherently more efficient in recovering nutrients from the soil solution is clearly indicated. If enhanced effi-

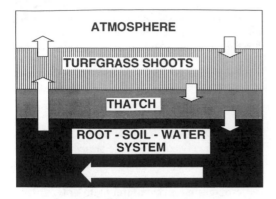

Figure 1 Model for a self-sustaining turf. A steady state ecosystem is visualized in which resources cycle efficiently within the plant-soil-atmosphere continuum with little net gain or loss.

ciency for nutrient absorption can be expressed throughout the year, turfgrasses so endowed would recover virtually all nutrients mineralized from soil organic matter, no true leaching or denitrification losses would occur, and little if any fertilizer would have to be applied (Figure 1). Such developments would constitute a major step toward achieving sustainable turf.

Mutually beneficial relationships between plants and microorganisms frequently occur in nature. The vast majority of higher plants, including over 80% of all grass species, form vesicular-arbuscular mycorrhizae (VAM) that, in some situations, may improve phosphorus uptake, stress tolerance, and resistance to root diseases. Research on mycorrhizae in turfgrasses has been limited to relatively few inconclusive studies, but investigations are under way to survey the species of VAM fungi associated with the various turfgrasses (Figure 2). Efforts to isolate and culture VAM fungi, if successful, will allow the development of techniques for introducing inoculum of the most effective VAM species. This basic work is needed to establish the conditions whereby the beneficial effects of VAM associations can be maximized in the turf environment. Practical application of this technology will greatly advance progress toward achieving sustainable turf.

INSECT PREDATION

Public concern over potential environmental and human health risks from urban pesticide usage will mandate much more limited and selective pesticide use on turf. Progress in this area will come through research on biological and bio-rational insect controls and better trained turfgrass managers. Training is needed in pest identification, biology, ecol-

ogy, thresholds, pesticides, application techniques, etc.

With the widespread availability of turfgrass seed containing high levels of a fungal endophyte, there is now an effective alternative to pesticides for managing most surface-feeding turfgrass insect pests.[17-19] The adverse effects of foliar-feeding insects, including chinch bugs, billbugs, and sod webworms, can be minimized by seeding grasses containing high levels of *Acremonium* endophytes. There are many endophytic cultivars of perennial ryegrass, tall fescue, and fine fescues (Chewings, hard, creeping red, and blue), but none within Kentucky bluegrass or bentgrasses at this time (Table 1). For many purposes, endophytic grasses seem to be the preferred means of controlling these foliar pests although this, combined with genetic resistance, will provide true sustainable control. This leaves the white grub complex (Japanese, Oriental, black turfgrass ataenius, June beetles, etc.) with few alternatives to chemical control in the Northeast.

Considerable research is being conducted worldwide on the use of entomopathogenic nematodes

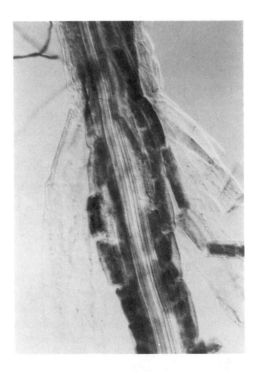

Figure 2 Vesicular-arbuscular mycorrhizal fungus *Glomus intraradices* in root of creeping bentgrass (*Agrostis palustris*). Fungal hyphae, arbuscules, and circular vesicles are stained dark blue. *(Courtesy of R. E. Koske, University of Rhode Island.)*

Table 1 **Examples of endophyte-containing turfgrass cultivars**

Grass Species	Cultivar	Producer
Perennial ryegrass	Advent	Jacklin Seed Co.
	Assure	LESCO, Inc.
	Citation II	Turf-Seed, Inc.
	Commander	LESCO, Inc.
	Gettysburg	Jonathan Green, Inc.
	Legacy	LESCO, Inc.
	Pinnacle	Normarc
	Repell II	Loft's Seed/Great Western
	Target	Medalist America
Tall fescue	Guardian	Pickseed West, Inc.
	Mesa	Jonathan Green, Inc.
	Shenandoah	Willamette Seed Co.
	Titan	Seed Research of Oregon
	Tribute	Loft's Seed, Inc.
Chewings fescue	Jamestown II	Loft's Seed, Inc.
	Longfellow	International Seeds, Inc.
	Shadow[E]	Turf-Seed, Inc.
	Victory	Pickseed West, Inc.
Hard fescue	Aurora (E)	Turf-Seed, Inc.
	Biljart	Van Der Have Oregon, Inc.
	Spartan	Pickseed West, Inc.
Blue fescue	SR 3200	Seed Research of Oregon

for control of turf insect pests (Figure 3). Certain nematode species appear to be better adapted for control of surface-feeding pests (e.g., *Steinernema carpocapsae* for surface-feeding Lepidoptera and *Steinernema scaperisci* for mole crickets[20]) while others (*Heterorhabditis bacteriophora* and *Steinernema glaseri*) appear to be better adapted to control grubs in the soil.[21,22] The production and storage of *Steinernema carpocapsae* has progressed to the point where this species is being distributed and sold much like a conventional insecticide. Considerable research still needs to be conducted on the production, storage, and efficacy of all nematode strains, but studies to date are encouraging.

There is also renewed interest in biological control of turf insects with other microorganisms. Fungal pathogens such as *Metarhizium anisopliae* and *Beauveria bassiana* have received considerable research attention in the past, and have recently gained renewed interest and support in the U.S. *Bacillus popilliae*, milky spore disease, is still being manufactured *in vivo* by Fairfax Biological Laboratory in Clinton Corners, N.Y. This bacterium is pathogenic to Japanese beetle larvae, if they ingest enough spores (approximately 7000 to 10,000).[32] Soil temperatures greater than 21°C are also required for infection. Milky disease bacteria apparently can provide general suppression of larval populations rather than acting as microbial insecticides.[23] Unfortunately, after nearly 60 years of

availability, data on efficacy of this product is insufficient to state what levels of control (if any) can be predicted and under what conditions. If infective spores could be produced economically *in vitro,* and spores could be applied to turf at infective dose levels over the entire turf surface, this product might gain widespread use.

There are also sod and soil characteristics which minimize insect injury. Turf can withstand considerable insect feeding if sufficient moisture is available. Under adequate moisture levels, Japanese beetle grub densities of 50/ft^2 have been observed with little apparent turf damage.[24] Other landscape plants, however, may be less tolerant of the adults emerging from such grub populations. Thatch, in addition to being a barrier to water and fertilizer applications, also harbors chinch bugs, sod webworms, and cutworms. Deep, infrequent irrigation should be encouraged over shallow, frequent irrigation to encourage extensive rooting and take full advantage of the grass' inherent ability to tolerate feeding damage.

PATHOGENS AND DISEASE

All turfgrasses have major disease problems caused by fungal pathogens, but the incidence and severity of particular diseases may vary widely among grass species and turf uses. There is also considerable variation within species as to their disease susceptibility. Thus, selection and/or breeding for disease

Figure 3 The entomopathogenic nematode *Steinernema carpocapsae* which has been used effectively for the control of lepidopterus turf insects. *(Courtesy of Randy Gaugler, Rutgers University.)*

resistance is feasible. This approach to disease management is being increasingly pursued in turfgrass improvement programs but progress tends to be slow, since resistance to several diseases is desired. At the same time, new disease-resistant cultivars must maintain desirable agronomic features. The development of such germplasm, incorporating all of these traits, inevitably is a time-consuming and painstaking task.

Some soils afford an environment hostile to particular soil-borne pathogens because they harbor populations of antagonistic microorganisms.[25] They are termed suppressive soils. Natural biological control agents present in these soils prevent or minimize (immediately or subsequently) establishment of a pathogen and/or development of disease. Fungi and bacteria involved in these processes are being isolated, identified, and cultivated with a view to their utilization as inoculum for seed treatment, seed bed amendment, and inclusion in topdressing mixtures. Organic amendments to soils (animal manures, plant residues, sewage sludge, composts, etc.) may enhance native populations of antagonistic microorganisms. Some of these materials may be used as carriers to introduce inoculum of beneficial microorganisms.

For a fungal disease outbreak to occur in turf, several requirements must be met.[26] These include, a susceptible host, the presence of inoculum of the pathogen, and suitable environmental conditions for the disease process. If host and pathogen are present, then environmental conditions, affecting both partners (host and pathogen) in disease development, are major determining factors as to whether turf remains healthy or succumbs. Temperature extremes, water availability extremes, fertility levels, soil pH, soil aeration and drainage, mowing regimes, amount of play or use, pest management routines, etc. are influenced (to a greater or lesser extent) by management decisions that can exacerbate or mitigate disease problems. Thus, cultural practices affect the turf environment and can be manipulated to have an impact on disease incidence and severity, particularly for less intensively managed turf.

As the demand for higher quality lawns and sports turf escalates, turf diseases become more of a problem. Even minor blemishes are a cause for concern in pristine turf, particularly in golf course situations where quality expectations are exceedingly high. Intensive turf management practices, necessary to meet these quality expectations, subject grasses to greater stress that is reflected in increased disease susceptibility. There is little leeway to modify cultural practices to combat all diseases, hence the reliance upon fungicides. Since a few turf diseases develop and damage turf in a matter of a few hours, routine preventive treatment with fungicides is the only practical means for the turf manager to protect his turf and his job.

As was noted at the outset, only a small percentage of managed turf is devoted to such high-intensity use or scrutiny. In most turf situations, management practices can incorporate many of the concepts mentioned above and described further in this book. If this is done, there is every reason to believe that most turf can be maintained with minimal injury from pathogens.

WEED INFESTATION

As was stated earlier, weeds invade turf as a consequence of opportunities provided by injury or failures in management although weed aggressiveness and sanitation practices can play a role. Turfgrasses which spread via rhizomes or stolons are most likely to heal turf injuries before weeds become established. This, however, will be a function of the time-of-year when injury occurs. During mid-summer in most locations, no cool-season turfgrass will spread rapidly into disturbed areas and C-4 weeds will gain a foothold before the stressed C-3 turfgrasses can fill in the injury. Later in the season, cool temperatures of early-fall will suppress or kill the warm-season invaders. If they are removed, the turfgrasses will recolonize exposed areas. Removal of existing weed carcasses is important because they will hold their place and prevent turfgrass growth into the sites occupied by weeds. Toxic allelochemicals released from some decomposing weeds may retard stem growth of turfgrasses and prevent closure of injured areas. Thus weed encroachment, when it is not massive, can be managed by stimulating cool-season turfgrass growth during the fall, while mechanically removing weeds killed by frost or by herbicides.

Warm-season grasses maintain vigorous growth throughout the summer and at that time normally are not subject to weed invasion unless injured. With the onset of cool temperatures in the fall, these grasses become dormant. In the deep South, overseeding warm-season grasses with perennial ryegrass or other fine-leaved C-3 grasses maintains a green turf and also helps resist weed invasion. If these turf areas are not overseeded and sustain injury, cool-season weeds can become established. However, some weeds likely to invade warm-season turf during the dormant season will often become stressed when hot temperatures return and the turfgrasses may then displace them. However, this spring transition period also provides an opportunity for summer weeds to invade warm-season turf. Thus, proper turf management is the key to effective weed control in warm-season turfgrasses as it is in those adapted to cooler climates.

A thick, vigorous turf will resist weed invasion. With proper management and the avoidance of injury, weeds can be kept below threshold levels with little if any specific efforts directed toward weed control. This is probably a matter of competition between an established turf and invading seedling weeds. The production of allelochemicals by turfgrass roots may contribute to weed suppression by producing a soil environment toxic to most germinating weed seeds. This subject has received little study in turf, but it likely is a factor in the sustainability of turf stands, and in the future may be enhanced through specific breeding or genetic transformation efforts.

Controlling turf weeds by specific biological agents offers little promise at this time. Some success in *Poa annua* control using the bacterial pathogen *Xanthomonas campestris* has been achieved under laboratory conditions.[27] This major weed of golf greens is difficult, if not impossible, to control through management practices, including application of herbicides, so biological control agents are attractive. In most cases, the weed complex in turf is sufficiently diverse that specific insects or pathogens are not likely to provide general weed control if conditions are favorable for weed invasion.

The best weed management in turf, therefore, rests in avoiding those conditions which permit weed emergence.

Can a sustainable turf be kept weed free without the use of herbicides? Probably not over the long term. Turf injury will happen and conditions favoring weed invasion will occur sometimes even under good management practices. On those occasions, herbicides may have to be employed to prepare the turf for overseeding or other corrective measures. What probably can be achieved in a sustainable turf is elimination of the programmed use of preventive herbicides. It may never be practical to obtain a weed-free seeding without using at least one herbicide. Once established, however, turf can be managed so as to avoid weed encroachment and herbicide use may be reduced to an infrequent as-needed measure. This depends upon successfully controlling other turf pests, using improved, efficient grass cultivars, and managing the turf so as to minimize injury. Under such circumstances, herbicide use may be reduced to a fraction of that commonly used today on well-maintained turf.

MANAGING LAWN TURF FOR SUSTAINABILITY

Earlier we emphasized the importance of minimizing losses as the basis for maintaining a sustainable turf area. Creating a self-contained turf ecosystem in a steady state relationship with its environment is the goal (Figure 1). That means, if little is to be added, then little must be lost.

Clipping removal is the most obvious source of nutrient loss from lawn turf. Starr and DeRoo[28] determined that one-third of the nitrogen required by turf will be supplied from clippings if they are retained on the lawn. Gradual decomposition of clippings from the current and past seasons constitutes a slow-release nitrogen supply that is most available when the grass is best situated to use it. Less than 5% of the nitrogen contained in clippings was lost from the turf. Consequently, the simple practice of retaining grass clippings on the turf will permit a 30% reduction in the amount of fertilizer nitrogen applied.

Nitrate leaching is a potential loss of nitrogen from turf. Most research indicates that, with proper fertilizer management, this loss is minimal.[28] Our research has indicated that cool-season turf is most vulnerable to nitrate leaching during late-summer and fall. During those seasons, when root activity is low and nutrient absorption rates are depressed, mineralization of soil organic matter can release nitrate to the soil solution more rapidly than it can

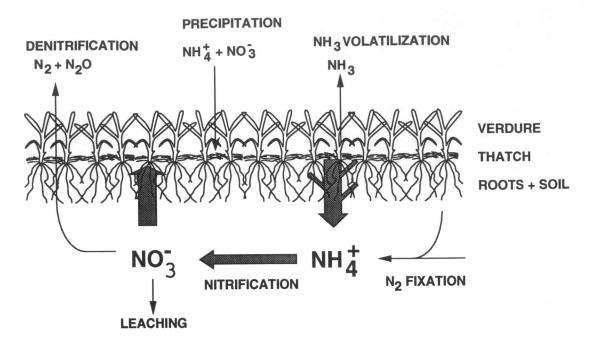

PRECIPITATION

DENITRIFICATION $NH_4^+ + NO_3^-$ NH_3 VOLATILIZATION
$N_2 + N_2O$ NH_3

VERDURE

THATCH

ROOTS + SOIL

NO_3^- ⟸ NH_4^+ ⟵

NITRIFICATION N_2 FIXATION

LEACHING

Figure 4 The nitrogen cycle in turf. In a self-sustaining turf, N losses through NH_3 volatilization, NO_3^- leaching and denitrification are balanced by influx of NH_3 and NO_3^- in precipitation and associative N_2 fixation.

be absorbed (Figure 4). Even unfertilized turf exhibits an increased nitrate concentration in soil water during the late-summer, which often does not decline again until March or April. Soluble nitrogen fertilizers applied at those times will be subject to significant nitrate leaching if heavy precipitation results in water percolation through the soil profile. Late-fall applications (mid- to late November) of organic nitrogen sources normally result in limited mineralization of the nitrogen and oxidation to nitrate, so leaching is minimal. By the time such nitrogen is oxidized to nitrate, the turfgrass roots have grown enough to absorb it efficiently and nitrate increases in the soil solution are not observed. If turf managers understand this relationship between nitrogen availability, turf root activity, and soil conditions, and fertilize accordingly, nitrate leaching from turf can be kept to levels comparable to undisturbed forest sites. Such areas are used as reference sites for estimating background or minimum nitrate leaching situations.

A recent study from Ohio[30] demonstrated that while seeded turf permitted more nitrate leaching than sodded turf during the first 3 months after establishment, the greater rooting from the seeded turf promoted more efficient nitrate uptake by the grass and less nitrate leaching thereafter. This study was conducted for less than 2 years but indicates the importance of managing sodded turf

to maximize root development and conserve nitrogen within the turf. Turf sod is often heavily fertilized prior to harvest and the high nitrogen status of the grass may depress root growth over that obtained from seedling grasses. These factors should be considered when establishing a lawn for sustainable management.

Seasonal nitrate leaching from warm-season turfgrasses has received less study, but a similar pattern was reported from work in Florida and Texas.[29] Bermudagrass turf managed as a lawn or golf course green was investigated. Again, the fall-winter dormant season was the time of greatest nitrate leaching and this extended further into the spring than was normally observed with cool-season grasses. Because warm-season grasses do not experience the heat stress of summer, their potential for nitrate leaching remains low and did not increase until fall. As with cool-season grasses, heavy irrigation or precipitation following an application of inorganic nitrogen sources promoted greater nitrate leaching from bermudagrass. The basic management strategies for conserving nitrogen apply to both grass types and center around using fertilizer when grass roots are best able to absorb nutrients from the soil solution.

Other nitrogen losses from turf, ammonia volatilization, and denitrification (Figure 4), have been less well quantified. They are not thought to account for more than a small percentage of fertilizer nitrogen applied to turf. The potential for such losses should not be neglected, however. Titko et al.[31] concluded that granular urea applied to turf under conditions of high relative humidity, elevated temperatures, and followed by periodic wetting of the soil surface could lose as much as 60% of its nitrogen to ammonia volatilization. The simple expedient of irrigating turf after urea application reduced ammonia losses by more than 95%. Other nitrogen sources, with the possible exception of ammonium salts which are rarely used on turf, offer less opportunity for ammonia volatilization losses. Denitrification losses from turf have received less attention and are not considered to be significant, except when soils are water logged and temperatures are high. Because nitrate is the substrate for denitrification, turf managed to maintain low nitrate levels in the soil solution will not incur significant denitrification losses, even when conditions for such losses are optimum.

Nutrient losses can be further reduced and hence turf sustainability increased by proper irrigation management, thatch reduction, and grass selection. As indicated above, nitrate leaching depends upon water percolating through the soil profile. That can be controlled in many areas by irrigation management. This was dramatically demonstrated by Morton et al.[3] who concluded that overwatering contributed more to nitrate leaching from lawn turf than any other factor. On the other hand, irrigation following fertilizer application will reduce ammonia volatilization and avoid turf burning when soluble nitrogen sources are applied during hot weather. Collecting leachate from turf via drainage tiles and reusing it for irrigation may conserve both water and nutrients and contribute to a more sustainable turf. Proper irrigation will also prevent drought stress and avoid turf injury which would promote weed invasion and aggravate injury from insects and diseases. While minimal water use is a characteristic of sustainable turf, proper water utilization is an essential component of turf management for sustainability.

Thatch impedes water and nutrient penetration from turf to soil and, thus, must be considered in sustainable turf management. A dry thatch layer will promote rainwater runoff and tends to promote ammonia volatilization following urea application. Thatch accumulates when dead crowns and stems are generated more rapidly than they will decompose. Excess fertilization often stimulates thatch accumulation, as does improper mowing and clipping management. Clipping retention on turf provides a slow release nitrogen source within the thatch which promotes thatch decomposition by microorganisms. In general, most turf management practices which promote sustainability will also tend to discourage thatch accumulation. Thatch often is more a symptom of poor turf management than a problem that must be managed.

The number of turfgrass cultivars currently available is bewildering even to the turf professional. Most of them exhibit superior traits and the future will surely provide grass genotypes even better suited to maintaining a sustainable turf. Many turfgrass evaluation trials are currently being conducted under minimal maintenance conditions and the cultivars emerging from them will be better suited to a sustainable lawn turf. It is important to recognize that a plant monoculture is inherently unstable and not consistent with the goals of sustainability. Consequently it follows that a mixture of grasses will provide a more resilient turf better able to respond favorably to stress and injury. Nutrient and water utilization will be more efficient in a turf composed of two or three grass species and several cultivars of each species. Injury caused by disease or insects will be less damaging to a turf composed of diverse grasses. Including rhizomatous grasses such as Kentucky bluegrass and creeping red fescues in a turf mix will ensure rapid recovery from injury and less weed invasion. Thus, sustainability in turf is more than a matter of management practices. It begins with landscape design and the selection of an appropriate grass mixture.

Turf and landscape sustainability is not a simple matter. It depends upon understanding and acting on many variables in formulating a management plan. Most of it is common sense if the concept of managing the landscape to minimize loss is accepted as the underlying cultural philosophy. What is not lost need not be replaced and a landscape managed on that principle will require little intervention and be environmentally benign.

REFERENCES

1. **Evans, W. C., Truckor, P. G., and Knopf, D. P.,** Pennsylvania Turfgrass Survey — 1989, Pennsylvania Turfgrass Council and Pennsylvania Department of Agriculture, Harrisburg, PA, 1991, 92.

2. **Roberts, E. C. and Roberts, B. C.,** *Lawn and Sports Turf Benefits,* The Lawn Institute, Pleasant Hill, TN, 1987, 31.

3. **Morton, T. G., Gold, A. J., and Sullivan, W. M.,** Influence of overwatering and fertilization on nitrogen losses from home lawns, *J. Environ. Qual.,* 17, 124, 1988.

4. **National Academy of Science,** Policy Implications of Greenhouse Warming, Report of the Mitigation Panel, National Academy Press, Washington, DC, 1991.

5. **Maupin, G. O., Fish, D., Zultowsky, J., Campos, E. G., and Piesman, J.,** Landscape ecology of Lyme disease in a residential area of Westchester County, New York, *Am. J. Epidemiol.,* 133, 1105, 1991.

6. **Kaplan, R. and Kaplan, S.,** *The Experience of Nature: A Psychological Perspective,* Cambridge University Press, Cambridge, 1989, 340.

7. **Ulrich, R. S.,** View through a window may influence recovery from surgery, *Science,* 224, 420, 1984.

8. **Hull, R. J.,** Energy relations and carbohydrate partitioning in turfgrasses, in *Turfgrass, Agronomy No. 32,* Waddington, D. V., Carrow, R. N., and Shearman, R. C., Eds., American Society of Agronomy, Madison, WI, 1992, chap. 5.

9. **Gerloff, G. C.,** Plant efficiencies in the use of nitrogen, phosphorus, and potassium, in *Plant Adaptation to Mineral Stress in Problem Soils,* Wright, M. J., Ed., Cornell University Agricultural Experimental Station, Ithaca, NY, 1976, 161.

10. **Mehall, B. J., Hull, R. J., and Skogley, R. C.,** Cultivar variation in Kentucky bluegrass: P and K nutritional factors, *Agron. J.,* 75, 767, 1983.

11. **Schumann, G. L. and Wilkinson, H. T.,** Research methods and approaches to the study of diseases in turfgrasses, in *Turfgrass, Agronomy No. 32,* Waddington, D. V., Carrow, R. N., and Shearman, R. C., Eds., American Society of Agronomy, Madison, WI, 1992, chap. 19.

12. **DiPaola, J. M. and Beard, J. B.,** Physiological effects of temperature stress, in *Turfgrass, Agronomy No. 32,* Waddington, D. V., Carrow, R. N., and Shearman, R. C., Eds., American Society of Agronomy, Madison, WI, 1992, chap. 7.

13. **Burt, M. G. and Christians, N. E.,** Morphological and growth characteristics of low- and high-maintenance Kentucky bluegrass cultivars, *Crop Sci.,* 30, 1239, 1990.

14. **Mehall, B. J., Hull, R. J., and Skogley, C. R.,** Turf quality of Kentucky bluegrass cultivars and energy relations, *Agron. J.,* 76, 47, 1984.

15. **Barber, S. A.,** *Soil Nutrient Bioavailability: A Mechanistic Approach,* Wiley-Interscience, New York, 1984, chap. 3.

16. **Cisar, J. L., Hull, R. J., and Duff, D. T.,** Ion uptake of cool season turfgrasses, in *Proceedings 6th International Turfgrass Research Conference,* Ide, H. and Takatoh, H., Eds., International Turfgrass Society and Japaneese Society of Turfgrass Science, Tokyo, 1989, 233.

17. **Funk, C. R., Clark, B. B., and Johnson-Cicalese, J. M.,** Role of endophytes in enhancing the performance of grasses used for conservation and turf, in *Integrated Pest Management for Turfgrass and Ornamentals,* Leslie, A. R. and Metcalf, R. L., Eds., U.S. Environmental Protection Agency, Washington, DC, 1989, chap. 18.

18. **Siegel, M. R., Dahlman, D. L., and Bush, L. P.,** The role of endophytic fungi in grasses: new approaches to biological control of pests, in *Integrated Pest Management for Turfgrass and Ornamentals,* Leslie, A. R. and Metcalf, R. L., Eds., U.S. Environmental Protection Agency, Washington, DC, 1989, chap. 16.

19. **Dahlman, D. L., Eichenseer, H., and Siegel, M. R.,** Chemical perspectives on endophyte-grass interactions and their implications to insect herbivory, in *Microbial Mediation of Plant-Herbivore Interaction,* Barbosa, P., Krischik, V., and Jones, C. G., Eds., John Wiley & Sons, New York, 1991, 227.

20. **Klein, M. G.,** Efficacy against soil-inhabiting insect pests, in *Entomopathogenic Nematodes in Biological Control,* Gaugler, R. and Kaya, H. K., Eds., CRC Press, Boca Raton, FL, 1990, 195.

21. **Alm, S. R., Yeh, T., Hanula, J. L., and Georgis, R.,** Biological control of Japanese, oriental and black turfgrass ataenius (Coleoptera: Scarabaeidae) larvae with entomopathogenic nematodes (Nematoda: Steinernematidae, Heterorhabditidae), *J. Econ. Entomol.,* 1992.

22. **Yeh, T. and Alm, S. R.,** Effects of entomopathogenic nematode species, rate, soil moisture, and bacteria on control of Japanese beetle (Coleoptera: Scarabaeidae) larvae in the laboratory, *J. Econ. Entomol.,* 2144, 1992.

23. **Klein, M. G.,** Use of *Bacillus popilliae* in Japanese beetle control, in *Use of Pathogens in Scarab Pest Management,* Glare, T. R. and Jackson, T. A., Eds., Intercept Press, Andover, UK, 1992, 179.

24. **Casagrande, R.,** personal communication, 1992.

25. **Baker, K. F. and Cook, R. J.,** *Biological Control of Plant Pathogens,* American Phytopathology Society, St. Paul, MN, 1982.

26. **Smith, J. D., Jackson, N., and Woolhouse, A. R.,** *Fungal Diseases of Amenity Turf Grasses,* 3rd ed., E. & F. N. Spon, London, 1989.

27. **Vargas, J. M., Jr., Roberts, D., Danneberger, T. K., Otto, M., and Detweller, R.,** Biological management of turfgrass pests and the use of prediction models for more accurate pesticide applications, in *Integrated Pest Management for Turfgrass and Ornamentals,* Leslie, A. R. and Metcalf, R. L., Eds., U.S. Environmental Protection Agency, Washington, DC, 1989, chap. 11.

28. **Starr, J. L. and DeRoo, H. C.,** The fate of nitrogen fertilizer applied to turfgrass, *Crop Sci.,* 21, 531, 1981.

29. **Petrovic, A. M.,** The fate of nitrogenous fertilizers applied to turfgrass, *J. Environ. Qual.,* 19, 1, 1990.

30. **Geron, C. A., Danneberger, T. K., Traina, S. J., Logan, T. J., and Street, J. R.,** The effects of establishment methods and fertilization practices on nitrate leaching from turfgrass, *J. Environ. Qual.,* 22, 119, 1993.

31. **Titko, S., III, Street, J. R., and Logan, T. J.,** Volatilization of ammonia from granular and dissolved urea applied to turfgrass, *Agron. J.,* 79, 535, 1987.

Contribution No. 2772 of the Rhode Island Agricultural Experiment Station and the Plant Sciences Department, University of Rhode Island, Kingston, RI 02881.

Chapter 2

The Special Needs of Trees

James R. Clark and Nelda P. Matheny, HortScience, Inc., Pleasanton, CA

CONTENTS

Introduction 17
    The Unique Nature of Landscape Settings 17
    Integrated Pest Management Programs for Landscape Trees 18
The Unique Nature of Trees 18
Tree Development and Pest Management 19
    Tree Age 19
    How Trees Die 20
    Mortality Spiral 21
    Cultural Practices, Maintenance of Vigor, and Pest Management 21
        Pruning as an Example of the Culture-Pest Relationship 21
Key Plants in the Landscape 22
    Taxonomic Level 23
    Location in the Landscape 24
    Severity of a Pest Problem and Capacity for Control 25
    Designating Key Plants 25
Summary — A Prescription for Tree Care 25
    Proper Plant Selection 25
    Use of Quality Stock 26
    Appropriate Installation Procedures 26
    Pruning 26
    Irrigation 26
    Mulch/Turf 26
    Nutrient Management 26
    Pest Management 26
Acknowledgments 26
References 26

INTRODUCTION

Trees play a pivotal role in the environmental, aesthetic, architectual, and engineering functions of a landscape. By virtue of their large size, longevity, and variety, trees have a diversity of character and form that is welcome in urban areas. Their care is a critical activity in a program of landscape management.

For trees, pest management cannot be separated from plant management. When we perform activities that enhance plant health, we are also performing pest management. The interrelated nature of pest management and day-to-day care is applied through programs such as Monitor® (Bartlett Tree Expert Co.), Plant Health Care (Davey Tree Expert Co.), or IPM.[1] Each of these integrates cultural practices with pest management.

Recognition of the relationship between pest management and tree care is a key element in successful maintenance programs and will be developed as the central theme in this chapter. For this discussion, pest management includes disorders caused by insects, fungi, bacterial, nematodes, and viruses as well as decay and abiotic problems.

THE UNIQUE NATURE OF LANDSCAPE SETTINGS

Landscapes are most frequently created in two ways. The first consists of planting a group of nursery-grown, exotic species onto a site (which by the nature of the construction process has been substantially disturbed). The second retains preexisting trees on a site that undergoes development. In either case, little of the "natural" system remains. The mix of plant and animal species is artificial, the environment is disturbed, and human influence is pervasive. Even those projects which go to great lengths to preserve trees are artificial creations. Therefore, the normal checks and balances between host and pest that exist in natural environments are disrupted and/or nonexistent.

Planted landscapes are frequently assemblages of exotic species and cultivars, requiring more intensive programs of care than those of natural systems. These collections of plants differ from natural forests in the wide variety of plant species that co-exist in an artificial environment. This species composition may provide a "potentially diverse resource base for phytophagous and sapophytic insects."[2] In these settings, plant and pest combinations not encountered in nature are commonplace.

In landscapes, trees are grown for their environmental (e.g., modification of climate, reduction of atmospheric contaminants) and aesthetic functions rather than as a timber product. They are considered to be relatively permanent and highly durable components of the environment. The general public seems to have little tolerance for the natural decline and regeneration process that sustains forests. They will not permit declining and senescent trees in parks, near homes, and along streets.

The care of plants in landscapes also differs from plants in nurseries. The level of culture is less intensive for landscape trees than nursery-grown trees, with lower inputs of irrigation, fertilizer, and pest control. Pesticide use in landscape settings is more restricted.

INTEGRATED PEST MANAGEMENT PROGRAMS FOR LANDSCAPE TREES

Integrated pest management (IPM) programs for landscape trees are a direct outgrowth of the nature of these plantings and character of the individual plants themselves. Landscapes lend themselves to IPM for several reasons. First, pest control based upon programmed or timed spray efforts are frequently ineffective and often result in resurgent pest populations. Second, the character of urban environments as a mosaic of small, distinctly managed spaces results in reinfestation following control efforts. Third, the public increasingly demands lower inputs of pesticides, resulting in more integrated programs of tree care.

Unlike traditional plant and pest management, IPM seeks to take advantage of the inherent strengths and weaknesses of a landscape. In a rather direct way, IPM incorporates principles of ecology into the management of artificial assemblages. IPM requires viewing landscapes as systems, where plants and animals (including pests) interact with the physical environment. It relies upon directed,

specific, low intensive treatment of problems and does so only after a period of monitoring for natural controls. IPM also sees the landscape in a holistic way, where pests may be but symptoms of more fundamental problems. It may be argued that IPM is most successful in those landscapes which most strongly mimic natural systems.

From the landscape management perspective, the goals of IPM are (1) to reduce aesthetic injury, (2) to control pest problems in an effective and environmentally sound manner, and (3) to prevent predisposition of host trees to other pests and noninfectious disorders. The physical damage that results from the growth, feeding, oviposition, and related activities of a pest may be significant. But the fact that damage by pests may predispose the tree to a more serious problem is just as important.

However, the concerns of the manager for plants and pests do not always touch upon a major concern of the general public, i.e., that the appearance of a landscape affects their use and enjoyment of outdoor spaces. In our view, pest management is as much driven by concerns over leaf litter, aphid honeydew, tent caterpillar tents, and pests entering their home as much as by plant health.

Pest managers must see programs of care as responding to the needs of both plant and client. In this view, even low-level infestations of some pests may be serious by exacerbating secondary problems, even though no aesthetic injury results. Cultural practices, although acceptable in terms of aesthetic quality, may predispose the tree to pest problems. Finally, even relatively minor problems may be seen as significant to the public.

THE UNIQUE NATURE OF TREES

Trees are long-lived organisms, distinguished from other landscape plants by their size, single axis, massive amounts of secondary growth, and other characteristics.[3,4] That the residential environment is a difficult one is seen in the longevity of trees in natural and nonnatural settings. As a general rule, trees growing in landscapes have a much shorter life-span than do those in their natural environments. For example, Monterey pine (*Pinus radiata*) lives to be 120 years in its native forest along California's central coast. When planted in the San Francisco Bay area, it may live only for 50 years; in the Central Valley of California, only 25 years.

Trees develop in response to their environment. There exists a balance among parts of the tree, as reflected in ratios of root:shoot mass and sapwood area:foliage mass.[5] This balance develops within a specific environment or cultural system.[6] For example, under conditions of good fertility root:shoot ratio is generally low. If fertility declines, the relative amount of root to shoot would gradually increase.

Since they are long-lived and oriented about a single axis, the large size and volume trees attain increases their value to a landscape. However, pest management of a such a plant becomes increasingly more difficult as it grows larger. Assessing tree health, monitoring pest populations, evaluating damage levels, establishing action thresholds, and applying treatments are more difficult for trees than for smaller landscape plants and turf.

TREE DEVELOPMENT AND PEST MANAGEMENT

The unique character and developmental patterns of trees requires that their IPM programs differ from those for turf and herbaceous plants. Since trees are long-lived, respond relatively slowly to change and can rarely be rescued when in decline, the goal of IPM for trees is to maintain vigor. Susceptibility to stress of all types is related to vigor: a healthy plant is less likely to be injured than a plant in poor condition. Maintaining the cultural (and environmental) conditions which provide the resources necessary for growth and defense will increase resistance to pests.

At the same time, environment and biotic stresses need not result in visible damage to have an adverse effect upon tree health. Even low-intensity drought, defoliation, or low temperature will predispose a plant to secondary stress.[7] Management of pest populations not only eliminates visible damage that may result; it also enhances overall plant vigor by reducing an important predisposing stress. As Harris[1] observed, "Although a few of the pathogens that cause stem cankers, dieback, decline and some root rots aggressively attack healthy plants, many attack only plants that are weakened or low in vigor." The development of dieback and decline problems in trees has also been linked to stress factors.[8]

Predisposition is "the tendency of nongenetic factors, acting prior to infection, to affect the susceptibility of plants to disease."[7] It has been discussed primarily in the context of plant diseases. However, there is evidence that plants of low vigor are also more susceptible to attack by some insects.[9] We also observe that plants of low vigor are more sensitive to site changes such as grading, soil compaction, and mechanical injury.

In summary, we suggest that trees which are low in vigor are more likely to be predisposed to pest problems. Moreover, factors which reduce the capability of the tree to grow and defend itself may

Table 1 Factors that predispose trees to stress

Drought[a]
Low temperature
Defoliation
Low soil aeration
Nutrient deficiency
Chemical injury
Mechanical damage
Insect and disease infestation
Competition with other vegetation
Pruning
Mineral element excess/toxicity
Air and soil pollution
High temperature
Exposure
Limited root volume
Allelopathy
Soil compaction

[a] First seven factors taken from Schoeneweiss.[26] Others added by authors.

be either biotic or abiotic in origin. We have expanded Schoenewiess's list of predisposing factors to include this idea (Table 1). In this view, such factors fall into three general groups:

1. biological (pest infestation, competition);
2. environmental (temperature, radiation);
3. cultural (irrigation, fertilization, pruning).

The nature of the predisposing factor seems to be less important than its overall effect on plant vigor. A tree may be impacted by a predisposing factor with little or no change in its outward appearance. Foliage color, density, and shoot elongation may be affected, but the effects are not usually obvious and are more difficult to assess in tall trees.

Recognition of the importance of predisposition to pest management can be a critical one, for predisposing factors may be linked to a number of significant tree problems (Table 2). In several cases, there is no practical control/cure of the problem which kills the tree. The primary manner in which such problems are managed is through keeping predisposing stress at a minimum.

TREE AGE

The overall effect of predisposing factors on tree health will vary as a function of age. Old trees are less able to respond to changes such as site development, soil compaction.[4] Manion[10] noted that tree age was involved in many tree decline complexes. It appears that old age in trees is associated with an increase in susceptibility to predisposing factors.

Table 2 **Relationship between predisposition and pest problems**

Predisposing Factor	Pest Problem
Abiotic	
Drought	Botryosphaeria canker[27]
	Branch dieback (Diplodia) on oak[28]
	Ips beetle on Monterey pine[29]
	14 genera of insects[18]
Overirrigation/ wet soils	Elm leaf beetle[18]
	Gypsy moth[18]
	Phytophthora root rot[30]
	Armillaria root rot[30]
Low temperature injury	Botryosphaeira canker[27]
Overfertilization	Hemlock adelgid[31]
	Brown rot on stone fruits[32]
Nutrient excess/ saline soils	Phytophthora root rot[30]
	Armillaria root rot[30]
	Elm leaf beetle[18]
	Gypsy moth[18]
Air pollution	Ips beetle on Monterey pine[19]
	Gypsy moth[33]
Defoliation	Botryosphaeria canker[27]
Rapid growth rates	Cypress bark moth[2]
	Fireblight[34]
Old age	Red turpentine beetle on Monterey pine[11]
Mechanical injury	Armillaria root rot[30]
	Sequoia pitch moth[29, 35]
	Red turpentine beetle[29]
	Ash borer[29]
Biotic	
Insect infestation	Cryptocline twig blight on oak (oak pit scale)[28]
	Nectria canker (scale)[36]
	Ips beetle on Monterey pine (red turpentine beetle, sequoia pitch moth)[29]
Disease	Dwarf mistletoe infestation (Ips beetle)[19]

For example, we observe root rots such as *Phytophthora* to be a significant cause of death in mature oaks (*Quercus* sp.) in California and madrone (*Arbutus menziesii*) in Washington. The presence of these pathogens seems intensified by frequent summer irrigation, generally for trees growing in association with turf. Yet young seedlings of both oak and madrone are frequently observed growing in heavily irrigated landscape beds. Under similar conditions, mature trees rapidly succumb to the disease. The ability of these species to survive infection by root rot appears to decrease with age.

We do not know if the death of mature trees is due to an increase in the intensity of infection or a decrease in host resistance.

A similar situation exists with Monterey pine (*Pinus radiata*) and attack by bark beetles, a frequent cause of death. "A factor contributing to susceptibility of Monterey pines to bark beetles is their age."[11] This is true regardless of the intensity of care and maintenance.

With age, the increase in susceptibility of trees to pest problems is based in part upon age-related changes in development. These involve both the diminishing supply of carbohydrates and mineral nutrients as well as the general patterns of resource partitioning in woody plants. As trees age, the net amount of available carbohydrates declines due to increased respiratory demands and complexity of transport.[12] Since "defense" (both physical and chemical components) has a relatively low priority for resources,[13] the ability to resist insect attack, form barriers to disease and decay, and related responses also declines.

HOW TREES DIE

Tree death is frequently a slow and complex process. Vascular diseases such as Dutch elm disease and oak wilt may kill a tree within a year. However, in nature a gradual death, involving a number of factors, is more common.[14] The concept of "tree decline" as lacking a specific cause reflects this view.[8,15] In general, premature debilitation associated with decline involves a wide range of causes, biotic and abiotic, that may be either chronic or acute.

When the tree dies, it is either as a result of structural failure, environmental degradation, or pest infestation.[4] Taken alone, a stress factor may not be severe enough to physically damage the tree, but the cumulative effect of many stresses is. At some point, the energy required for defense, maintenance of structural stability, and replacement of damaged parts exceeds the supply and the tree dies as a result.

Overall, tree defense systems (the chemical and mechanical barriers to pests and decay) have a low priority for resources such as carbohydrates and mineral elements.[13] As a result, when a tree is under stress or where resources are limited, the tree's ability to resist infection or repel insects may be reduced. Predisposing stresses such as defoliation and drought may reduce the annual carbon gain below basic maintenance needs for protection.

In addition, some species allocate few resources to defense processes.[15] For these species (such as the *Populus* sp.), maintenance of growth rates that are more rapid than the rate of tissue loss from either wood decay or direct pest injury are required for survival. As Shigo[17] observed, "Trees survive as long as they can form new parts in new positions faster than old parts are breaking down."

While environmental stresses act to retard and alter plant development, they may also play a role in enhancing pest development. Mattson and Haack[18] described the variety of ways in which drought stress may alter the host-insect relationship and increase insect problems. Several of the important impacts involved insect development (such as rate of growth), activity of insect detoxification systems, reduction in natural enemies, etc. It seems clear that the effect of some environmental stresses on pests may be beneficial at the same time that they are detrimental to the tree.

MORTALITY SPIRAL

Manion[10] described a disease spiral as a series of related events in the development of a disease. Franklin et al.[14] broadened this concept into a series of linked events which lead to death — a "mortality spiral." They observed that while the most frequent proximate cause of death in many Pacific Northwest forests was mechanical failure, the contributing factors which led to failure were suppression of the crown by competition and defoliation by insects.

The concept of a mortality spiral has been applied to landscape trees.[4] For coast live oak (*Quercus agrifolia*) growing in the San Francisco Bay area, we observe the proximate causes of death are structural failure and root rot (*Armillaria mellea*) (Figure 1). Both are aggravated by frequent summer irrigation, drought, changes in grade, mechanical injury to roots, defoliation by oak moth, etc. These lead to low vigor in the tree that reduces its ability to limit the spread of natural fungal associations. Over a period of years, decay and *Armillaria* affect more and more of the tree's wood, until either death or structural failure occurs.

A similar situation exists for Monterey pine (*Pinus radiata*). A frequent cause of death is the *Ips* bark beetle attack. In general, any condition that reduces tree vigor contributes to the likelihood of a bark beetle outbreak.[19] Wounding the tree (e.g., pruning) during the growing season attracts adult beetles. Drought stress and insects such as the red turpentine beetle reduce tree vigor. Healthy Monterey pines usually resist attack by the production of pitch which kills the beetles. Trees weakened by these predisposing factors are less able to do so. Chemical control is rarely effective without correcting those factors which predispose the tree to attack.

The conclusion we reach from an examination of mortality spiral for landscape trees is that tree care and pest management cannot be separated. The maintenance of healthy, vigorous trees relies upon pest management. In turn, effective IPM programs require a healthy, vigorous tree. Pest management and tree care must be viewed as integrated and interactive.

CULTURAL PRACTICES, MAINTENANCE OF VIGOR, AND PEST MANAGEMENT

Horticulturists and arborists have traditionally employed a number of cultural practices to enhance and maintain vigor. The practices of pruning, fertilization, irrigation, etc. are an important component of an integrated pest management program. But these cultural treatments can be harmful if either poorly timed or inappropriately applied and may reduce plant vigor and actually increase susceptibility to pest problems (Table 3). If maintenance of tree vigor is to be an integral part of a pest management program, then the implementation of tree care practices must be high quality and appropriately timed and applied. In short, they must reflect the long-term needs of the plant.

Pruning as an Example of the Culture-Pest Relationship

Arborists consider pruning one of the most important components of a quality tree care program (Table 4). Harris[1] (citing Brown[41]) observed that pruning was influential in the control or elimination of 20 diseases and pests. Pruning plays a role in pest management by removing the pest itself or its source of further infection/infestation (sanitation) and altering microclimate conditions around tissues to affect the host and pest.

Yet, pruning can negatively affect plant development and pest problems (Table 4). This may occur by either poor practice or inappropriate timing. Poor pruning practice involves activities such as use of flush cuts, overthinning, topping, and the use of spurs. These create large wounds, increase the potential for bark damage, reduce the overall amount of photosynthetically active tissue, and decrease structural stability.

Timing of pruning directly influences susceptibility to pests and may do so in at least four ways (Table 5). First, the potential for infestation by some insects increases if pruning occurs during periods of adult activity. Second, wounds created by pruning may permit infection by some fungi and bacteria, if environmental conditions are appropriate. Third, the tree response to pruning may involve release of volatile chemicals which attract pests. Finally, pruning may stimulate regrowth of tissues which are preferred by some organisms (e.g., aphids, powdery mildew).

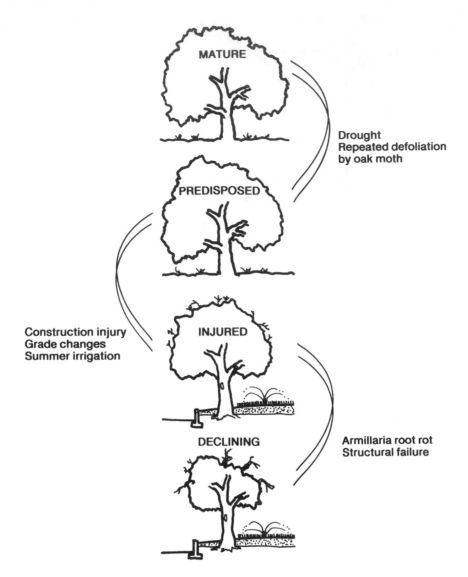

Figure 1 Biotic and abiotic factors, and stages of decline that comprise a mortality spiral for coast live oak (*Quercus agrifolia*) in the San Francisco Bay Area of California. *(Modified from Clark, J. and Matheny, N., J. Arboric., 17, 173, 1991.)*

The relationship between cultural practices and pest problems seems to be a rather general one. A diverse set of maintenance procedures (pruning, irrigation, fertilization, etc.) are linked to a range of pests (insects and disease) (Table 3). Inappropriate application of water, fertilizer, and/or mulch may alter the host-pest balance in favor of the pest.

A similar concern exists for pest management programs. When inappropriately implemented, they may actually increase pest problems. Programmed spray efforts are often ineffective in controlling the target pest, for reasons of timing, pesticide coverage, etc. In some cases, programmed sprays result in an increase in the populations of nontarget pests (scale, mites) and a reduction in beneficial insects.[20]

Application of chemicals to control organisms such as *Phytophthora* eliminates an important part of soil biology. Routine use of lawn care products that contain both fertilizers and herbicides may injure nearby tree roots.

KEY PLANTS IN THE LANDSCAPE

Integrated pest and plant management begins with knowledge of pest biology, host-plant ecology, and elements of tree maintenance that promote host vigor. Since there is a significant genetic component to host-plant resistance, plant selection is a central part of any IPM program. If IPM is to be successfully implemented in landscapes, plants must

Table 3 **Examples of the relationship between intensity of maintenance practice and the presence of pest problems**

Cultural Practice	Intensity of Application	
	Excessive/High	Inadequate/Low
Nitrogen fertility	Pitch canker on pines[37]	*Verticillium* wilt
	Brown rot[32]	Sycamore anthracnose
	Fire blight[34]	*Cytospora* canker
	Phytophthora root rot	
	Piercing-sucking insects[32]	
	(aphids, adelgids,	
	leafhoppers, scale)	
Irrigation	Root rots[7]	Bark beetles[22]
	(*Armillaria, Phytophthora*)	Borers[22]
	Sawflies[22]	Lepidopterous defoliators[22]
	Fire blight[34]	Sucking insects[22]
	Canker diseases[7]	Canker diseases[7]
	Borers[7]	(*Sieridium, Cytospora, Botryosphaeria*)
		Annosus root rot[7] (*Fomes annosus*)
Pruning	Aphids	Anthracnose
	Powdery mildew	Aphids (species that prefer interior
	Wilt disease[38]	canopy)[39]
	(Canary Island palm)	Dutch elm disease[40]
Degree of shade	Powdery mildew	*Hendersonula* on walnut[7]
	Black scale	Spider mites

Table 4 **Benefits and problems associated with pruning**

Benefits	Enhanced compartmentalization response
	Prevention of pest problems[a]
	Removal of pest/sources of inoculum
	Alteration of crown microclimate
	Enhanced structural stability
Problems	Reduced structural stability
	Damage to existing decay barriers
	Create entry points for pests and decay fungi
	Reduce photosynthetic capacity
	Damage to bark and cambium from increased exposure

[a] See Brown[41] (as cited by Harris[1]).

be selected for their adaptation to site condition and resistance to pest problems.

Some plant taxa have a disproportionate number of pest and cultural problems and are labelled these "key plants."[21] Key plants are "plants in the landscape that are most likely to incur problems year after year." Neilsen[22] observed that "relatively few tree and shrub species harbor most of the pest problems encountered in landscapes. These plants can be considered key for the purposes of IPM." In addition, "plants or groupings (that) contribute significantly to the value of the landscape may also be considered as key plants for monitoring."[22] A plant located in a prominent place in a landscape might be "key" while the same plant in a less critical location would not.

Three problems exist in current evaluation of key plants. The first involves how the designation of "key" is related to taxonomic level. The second concerns the relationship between key plant and landscape location. Finally, designation of a key plant must involve an evaluation of the severity of pest problems and the capacity for their control.

TAXONOMIC LEVEL

Raupp et al.[21] considered genera such as *Pyracantha, Malus, Cornus, Prunus,* and *Rosa* as being key. Yet, within each group exists substantial variation in susceptibility to insects and/or disease. Crabapple species and cultivars vary widely in their susceptibility to scab, powdery mildew, fireblight, and cedar apple rust. Such variation is common among landscape plants. Elm (*Ulmus*) species vary widely in their susceptibility to elm leaf beetle.[23] Within the *Cuppressaceae*, there is wide variation in resistance to the cypress tip miner.[24] There are a number of similar examples of variation in pest susceptibility/tolerance.

This variation in host resistance requires that

Table 5 **Relationship of pruning time (in California) to pest susceptibility for selected tree species**[42]

Plant Group	Period to Prune	Reason for Timing
Alnus rhombifolia	Nov. to Mar.	Pruning during growing season attracts flathead borer[43]
x *Cupressocyparis leylandii*	Summer or winter	Pruning during rainy or foggy weather increases potential for cypress canker disease[44]
Eucalyptus sp.	Winter to early spring	Pruning during growing season attracts eucalpytus long-horned borer[45]
Olea europaea	Summer	Pruning during rainy season allows transport of olive knot bacteria to wounds[46]
Pinus sp.	Nov. to mid-Dec.	Pruning from mid-Jan. to mid-Oct. increases susceptibility to bark beetles, sequoia pitch moth, western gall rust, and pitch canker disease[29,43]
Quercus agrifolia, Q. lobata	Nov. to Jan.	Pruning when infection by *Diploidia* is least likely[28]
Rosaceae	Summer to winter	Avoid pruning during spring rains when fire blight bacteria are active[34]
Ulmus sp.	Nov. to Jan.	Pruning during fall and winter when bark beetle (carrying Dutch elm disease) activity is reduced[47]

Table 6 **Variation in host susceptibility as a function of taxonomic level**

Taxonomic Level	Problem	Variation in Tolerance
Genus		
Betula	Aphids	All susceptible
Species		
Prunus	Brown rot, bacterial canker	*P. subhirtella* (susceptible) *P. serrulata* (resistant)
Cultivar		
Malus	Susceptibility to scab, fireblight, mildew, etc.	Wide range of resistance among cultivars
Pyrus calleryana	Branch attachment and crown stability	*P. c.* 'Bradford' (poor) *P. c.* 'Aristocrat' (good)
Platanus x *acerifolia*	Anthracnose	*P.* x *a.* 'Bloodgood' (resistant) *P.* x *a.* 'Yarwood' (susceptible)

identification of key plants occur on a taxonomic level that acknowledges such differences. It seems inappropriate to designate all elms as key plants due to elm leaf beetle when not all species within the genus are equally susceptible.

Key plants should be evaluated at the lowest taxonomic denominator. Where an entire genus is problematic, key plants are defined at that level. Where intrageneric and infraspecific variation occurs, plants are defined as key at that level (Table 6).

LOCATION IN THE LANDSCAPE

In our view, key plants must first be identified at an appropriate geographic level, whether it be local, regional, or national. Overall susceptibility to pests may vary by region of the country, local environment and culture, and pest. For example, the "Columbia" and "Liberty" cultivars of London plane (*Platanus* x *acerifolia*) are resistant to anthracnose in the eastern U.S. but not in the west.[25] Designation of these cultivars as key plants may occur in the west, but not eastern areas of the U.S.

The interaction between location and pest may be more dependent upon meso- and microclimate than geography. We observe the appearance of mildew on crape myrtle (*Lagerstroemia indica*) in the Bay Area of California to be largely dependent upon the presence of fog. In areas with regular summer fog, mildew is a large concern. Thus, crape myrtle might be a key plant in those areas. Where fog is not present, mildew is minor and crape myrtle is not a key plant. *Euonymus* shrubs develop more powdery mildew when planted in heavy shade than when found in open, sunny locations. Groundcover species of *Hypericum* develop more infestations of thrips in full sun than in partial shade.

An assessment of plants as key must also evaluate location in a particular landscape, in relation to other components. If cork oak (*Quercus suber*) is planted in locations where irrigation is heavy (turf, bottom of slopes), *Phytophthora* root rot is a predictable problem. Where irrigation is more limited (shrub beds, top of slopes), the root rot is not of consequence.

SEVERITY OF A PEST PROBLEM AND CAPACITY FOR CONTROL

Not all pest problems are equally significant in the landscape. Similarly, the ability of an arborist to control and manage pests varies widely. We believe designation of plants as key should consider these components as well.

The most serious pests are those for which there is no reasonable control. In the Bay Area of California, examples include the midge pod gall on honeylocust (*Gleditsia triacanthos* f. *inermis*) and verticillium wilt on pistache (*Pistacia chinensis*). The choices for a landscape manager are either to live with the problem or eliminate the plant from the landscape.

An equally serious management situation is one where a control procedure is possible but cannot be provided. In the Bay Area of California, an inability to provide drainage on irrigated sites will increase the likelihood of root rot. Similarly, being unable to provide irrigation to pines will increase their susceptibility to bark beetles by a large degree.

In summary, some pest problems are simply intractable; implementation of effective prevention or control procedures is not possible. We believe these situations are more serious than problems where either a control can be implemented or susceptibility can be minimized through culture.

DESIGNATING KEY PLANTS

An evaluation of key plants must consider the following:

(1) the regularity of pest infestation and its impact on plant health and appearance;
(2) location in the landscape;
(3) the capacity to control the pest.

At least four levels of "key plants" appear to exist (Table 7). At one end of the spectrum are plants without any significant pest or cultural problems

Table 7 **Ranking of key plants in the landscape**

Key Plant Group	Characteristics of Group
I	No pest or cultural problems
II	Irregular, primarily aesthetic problems
	Easily controlled
	No aesthetic concerns for host
III	Irregular problems, affecting both visual appeal and plant vigor
	Pest can be managed by spot control and cultural practice
	Plant prominant in landscape, aesthetic concerns significant
IV	Chronic problem, leading to tree decline or death
	Control procedures either lacking or involving extensive pesticide application
	Focal point of landscape; having special values

(Group I). In contrast, there are plants which have intractable, regularly occurring, predictable problems that are a concern for both vigor and appearance (Group IV).

As a long-term goal to reduce maintenance, plant health managers could strive to eliminate the "Group IV key plants" from the landscape and replace them with different taxa more resistant to the pest. Proper plant selection at the onset of design and grouping of plants with common and/or compatible cultural requirements is a key element to IPM programs.

SUMMARY — A PRESCRIPTION FOR TREE CARE

Trees are long-lived organisms that respond slowly to change and tree care should consist of routine, low-intensive practices implemented in a high-quality fashion, respecting the needs of individual taxa. The guiding principle for IPM programs involving trees should be the maintenance of vigor, providing pest control, and cultural practices that avoid predisposition. For trees, the cultural component of traditional IPM programs plays a highly significant role. The key elements in this program of care follow:

PROPER PLANT SELECTION

It is not enough to simply avoid using key plants in the landscape, for their evaluation as "key" involves localized site and design factors as well as genetic ones. Selections must be adapted to the site conditions. It is doubtful that modification of site conditions such as soil pH and texture can be practically implemented.

USE OF QUALITY STOCK

Whether retained on-site or grown in a nursery, plant material must be of the highest quality. Basic concerns about quality such as cleanliness and structural stability must be addressed at planting and/or retention. Nursery stock should conform to the American Standard for Nursery Stock.

APPROPRIATE INSTALLATION PROCEDURES

Planted trees must be installed at grade with baskets, wires, twine, and other materials removed.

PRUNING

Pruning to correct structural problems should occur at the time of planting/retention. Mature trees should be pruned on a 2- to 3-year cycle. Seasonal timing of pruning should occur to avoid pests (see Table 5). Pruning should be performed by International Society of Arboriculture Certified Arborists and follow professional pruning standards (available from both the International Society of Arboriculture and the National Arborist Association).

IRRIGATION

Irrigation should occur in response to species needs and seasonal climate. Aspects of this issue include:

(1) adequate irrigation following transplanting;
(2) for retained native trees in summer-dry climates, irrigate once per month;
(3) for species sensitive to root and crown rots, avoid irrigation within 10 ft of the trunk; irrigate once per month;
(4) for retained native trees in summer wet climates, irrigate during periods of drought;
(5) for established trees in cold and/or dry winter climates, irrigate in the fall.

MULCH/TURF

Newly planted trees should have a mulched area at least 1 ft from the trunk (with 2 ft preferred). Established trees should have as large a mulch area as possible, preferably to the dripline. Mulch should be organic materials such as wood chips, bark, etc.

NUTRIENT MANAGEMENT

Adequate soil fertility should be maintained. Soil tests and examination of plant growth should be used as indicators of the need to fertilize. In most cases, nitrogen will be the mineral element most needed.

PEST MANAGEMENT

A program of routine monitoring for pests should be present. Recommendations for control procedures should be made by a Licensed Pesticide Applicator.

The implementation of an IPM program for trees involves issues beyond pest biology, action thresholds, and monitoring. It must reflect the special place trees hold in the landscape and for the public. Further, it must consider the unique growth patterns of trees, their longevity, structural complexity, and exposure to decades of environmental and biological stresses.

The criteria for decision-making about treatment remains a significant stumbling block for implementation. On a physiological basis, decisions may involve considerations of species, age, vigor, history, and patterns of mortality. However, the guiding factor for IPM should be maintenance of vigor and management of mortality.

IPM for trees requires specific, comprehensive knowledge about pest management and tree care. Managers must have an intimate knowledge of both pest and tree biology, including the typical mortality patterns for trees. They must be able to identify predisposing stresses and pests which are potentially lethal and design long-term, low-intensity management programs to minimize stress and reduce debilitating pest problems.

ACKNOWLEDGMENTS

We appreciate the thoughtful comments and constructive review of Pavel Svihra, Dick Harris, and Larry Costello.

REFERENCES

1. **Harris, R.,** *Arboriculture — Integrated Management of Landscape Trees, Shrubs and Vines*, 2nd ed., Prentice-Hall, Englewood Cliffs, NJ, 1992, 674.
2. **Frankie, G. and L. Ehler,** Ecology of insects in urban environments, *Annu. Rev. Entomol.*, 23, 367, 1978.
3. **Clark, J.,** Age-related changes in trees, *J. Arboric.*, 9, 201, 1983.
4. **Clark, J. and Matheny, N.,** Management of mature trees, *J. Arboric.*, 17, 173, 1991.
5. **Waring, R.,** Estimating forest growth and efficiency in relation to canopy leaf area, in

Advanced Ecology Research, MacFadden, A. and Ford, E., Eds., Academic Press, New York, NY, 1983, 327.

6. **Cannell, M.,** Dry matter partitioning in tree crops, in *Attributes of Trees as Crop Plants*, Cannell, M. and Jackson, J., Eds., Institute for Terrestrial Ecology, Natural Environment Research Council, Huntingdon, England, 1985, 160.

7. **Schoeneweiss, D.,** Predisposition, stress and plant disease, *Annu. Rev. Phytopathol.*, 13, 193, 1975.

8. **Houston, D.,** Dieback and declines of urban trees, *J. Arboric.*, 11, 65, 1985.

9. **Waring, R. and Pitman, G.,** A Simple Model of Host Resistance to Bark Beetles, Research Note No. 65, Oregon State University, Forest Research Lab, Corvallis, OR, 1980.

10. **Manion, D.,** *Tree Disease Concepts*, Prentice-Hall, Englewood Cliffs, NJ, 1981.

11. **Barr, B., Hanson, D., and Koehler, C.,** Red Turpentine Beetle — A Pest of Pines, University California Cooperative Extension Leaflet 21055, University of California, 1978, 4.

12. **Nooden, L.,** Whole plant senescence, in *Senescence and Aging in Plants*, Nooden, L. and Leopold, A., Ed., Academic Press, New York, 1988, 391.

13. **Waring, R.,** Characteristics of trees prediposed to die, *BioScience*, 37, 569, 1987.

14. **Franklin, J., Shugart, H., and Harmon, M.,** Tree death as an ecological process, *BioScience*, 37, 550, 1987.

15. **Sinclair, W. and Hudler, G.,** Tree declines: four concepts of casuality, *J. Arboric.*, 14, 29, 1988.

16. **Loehle, C.,** Tree life history strategies: the role of defenses, *Can. J. For. Res.*, 18, 209, 1988.

17. **Shigo, A.,** *Tree Pruning: A Worldwide Photo Guide*, Shigo and Trees Associates, Durham, NH, 1990, 186.

18. **Mattson, W. and Haack, R.,** The role of drought in outbreaks of plant-eating insects, *BioScience,* 37, 110, 1987.

19. **Koehler, C., Wood, D., and Scarlett, A.,** Bark Beetles in California Forest Trees, University of California Cooperative Extension Leaflet 21034, University of California, 1978, 8.

20. **DeBach, P. and Rose, M.,** Environmental upsets caused by chemical eradication*, Calif. Agric.*, 31, 8, 1977.

21. **Raupp, M., Davidson, J., Holmes, J., and Hellman, J.,** The concept of key plants in integrated pest management for landscapes, *J. Arboric.*, 11, 317, 1985.

22. **Nielsen, D.,** Insect-tree relationships in an urban environment, *J. Arboric.*, 2, 158, 1976.

23. **Hall, R., Townsend, A., and Barger, J.,** Suitability of thirteen different host species for elm leaf beetle, *Xanthogaleruca luteola*, *J. Environ. Hort.*, 5, 143, 1987.

24. **Koehler, C. and Moore, W.,** Resistance of several members of the *Cupressaceae* to the cypress tip miner, *Argyresthia cupressella*, *J. Environ. Hort.*, 1, 87, 1983.

25. **McCain, A. and Svihra, P.,** Sycamore Anthracnose, University of California Cooperative Extension, Division of Natural Resources, Publ. 7070, University of California, 1989.

26. **Schoeneweiss, D.,** The role of environmental stress in diseases of woody plants, *Plant Dis.*, 65, 308, 1981.

27. **Sinclair, W., Lynn, H., and Johnson, W.,** *Diseases of Trees and Shrubs*, Cornell University Press, Ithaca, NY, 1987.

28. **Costello, L., Hecht-Poinar, E., and Parmeter, J.,** Twig and Branch Dieback of Oaks in California, University of California Cooperative Extension Leaflet 21462, University of California, 1989, 4.

29. **Koehler, C.,** Tree injuries attract borers, *Growing Points*, 27, 2, 1991.

30. **Svihra, P.,** A practical guide for diagnosing root rot in ornamentals, *J. Arboric.*, 17, 294, 1991.

31. **McClure, M.,** Nitrogen fertilization of hemlock increases susceptibility to hemlock woody adelgid, *J. Arboric.*, 6, 245, 1991.

32. **McHugh, J.,** Overfertilization increases brown rot, *Am. Fruit Grower*, 112, 9, 1992.

33. **Endress, A., Jeffords, M., Case, L., and Smith, L.,** Ozone-injured acceptability of yellow-poplar and black cherry to gypsy moth larvae, *J. Environ. Hort.*, 9, 221, 1991.

34. **McCain, A.,** Fire Blight of Ornamentals and Fruits, University of California Cooperative Extension Leaflet 2715, University of California, 1981, 3.

35. **Koehler, C., Frankie, G., Moore, W., and Landwehr, V.,** Relationship of infestation by the sequoia pitch moth (Lepidoptera: Sesiidae) to Monterey pine trunk injury, *Environ. Entomol.*, 12, 979, 1983.

36. **Shigo, A.,** *A New Tree Biology*, Shigo and Trees Associates, Durham, NH, 1986, 595.

37. **Svihra, P.,** Pitch canker: a new disease of Monterey pines, *Growing Points*, 25, 2, 1986.

38. **California Department of Food Agriculture,** Wilt Disease of Canary Island Palm, Division of Pest Management, IPM Info., 1981a.

39. **Olkowski, W. and Olkowski, H.,** Establishing an integrated pest control program for urban trees, *J. Arboric.*, 1, 167, 1975.

40. **California Department of Food Agriculture,** Dutch elm disease in CA, Division of Pest Management, IPM Info., 1981b.

41. **Brown, G.,** *The Pruning of Trees, Shrubs and Conifers*, Faber and Faber, London, 1972.

42. **Kontaxis, D.,** Contra Costa Pest-O-Gram and More, No. 153, University of California Cooperative Extension, Pleasant Hill, CA, 1992, 10.

43. **Svihra, P. and Koehler, C.,** Flatheaded Borer in White Alder, University of California Cooperative Extension Leaflet 7187, University of California, 1989, 2.

44. **McCain, A. and Hamilton, D.,** Cypress Canker, University of California Cooperative Extension Leaflet 2997, University of California, 1977, 2.

45. **California Department Forestry,** Eucalyptus Longhorn Borer — Stop the Spread, California Department Forestry, undated.

46. **Van Steenwyck, R., Teviotdale, B., and McHenry, M.,** Control Guide for Olive Pests and Diseases, University of California Cooperative Extension Leaflet 21370, University of California, 1983, 4.

47. **Byers, J., Svihra, P., and Koehler, C.,** Attraction of elm bark beetles to cut limbs of elm, *J. Arboric.*, 6, 245, 1980.

Chapter 3

Fate of Pesticides in the Turfgrass Environment

A. Martin Petrovic and Nina Roth Borromeo, Department of Floriculture and Ornamental Horticulture, Cornell University, Ithaca, NY

Mark J. Carroll, Department of Agronomy, University of Maryland, College Park, MD

CONTENTS

Introduction 29
Foliar Interception of Pesticides 30
    Pesticide Losses from Turf Foliage 30
    Fate of Pesticides in Thatch 35
Pesticide Leaching 35
    Groundwater and the Hydrologic Cycle 35
    Environmental Factors 36
    Soil Properties 37
    Pesticide Properties 39
    Leaching of Turf-Applied Pesticides 39
Pesticide Runoff 41
Summary 41
References 42

INTRODUCTION

As the public and scientific communities become more aware of the potential for environmental contamination by pesticides, the fate of turf-applied chemicals continues to receive growing attention. In particular, there is concern regarding the potential for surface-applied chemicals to leach and contaminate both groundwater and surface water.

Groundwater constitutes 96% of the freshwater supply in the U.S. and provides 40 to 50% of the U.S. population with its main source of drinking water.[1] Pesticides have been detected in groundwater monitoring studies throughout the U.S.[1,2] The U.S. Environmental Protection Agency (EPA) has recently completed a national survey of pesticides in drinking water wells.[2] The survey included 1300 community and rural domestic water wells that were sampled for 101 pesticides and 25 pesticide degradates. Twelve of the 126 pesticides and pesticide degradates were detected in one or more of the wells sampled. The two most frequently detected pesticides were the herbicides DCPA (acid metabolites) and atrazine which were found in 6.4 and 1.7% of the community water systems, respectively. It appears that the leaching of pesticides is

threatening the quality of our groundwater, but it is not clear to what extend turf-applied pesticides are contributing to this pollution.[3]

The demand for high-quality turfgrass in athletic facilities, home lawns, recreational areas, and other private and public settings often requires the use of chemicals (fertilizers and pesticides) as part of the management strategy. The three major groups of turfgrass pests include weeds, insects, and diseases, which can be chemically controlled by the use of herbicides, insecticides, and fungicides. Pest damage can reduce turfgrass density and uniformity, making it impossible to provide a safe, uniform playing surface for athletic uses.[4] The encroachment of pests can also reduce the aesthetic and functional value of turf by degrading color and inhibiting turfgrass growth. Therefore, it is often necessary to supplement cultural management practices with chemical inputs in order to sustain the desired level of turf quality.[4-6]

It is clear that the use of pesticides will continue to be a major tool in the management of high quality turfgrass areas for the foreseeable future. The fate of these pesticides is an important public issue that has only recently gained the widespread attention of scientists and public health officials as a research topic. Exposure of humans and nontarget organisms to foliar residues, runoff into surface waters and the potential contamination of groundwater are the current primary concerns of most state and federal agencies.

A multitude of interactive physical, chemical, and biological processes determine the fate of a pesticide in the environment. These include the processes of drift, volatilization, photodecomposition, runoff, adsorption, plant uptake, microbial degradation, and chemical degradation.[4,6-8] This chapter provides an overview of the scientific information available on the fate of pesticides applied to turfgrass. Table 1 contains a summary of important information from pesticide fate studies conducted with turfgrasses and was used to formulate the summary section of this chapter.

FOLIAR INTERCEPTION
OF PESTICIDES

In contrast to most agricultural crops, turfgrass is a permanent crop that fully covers the soil surface. As such, pesticide applications are made to turfgrass foliage. In mature turfgrass stands most pesticide enters the soil only after being washed off canopy foliage and moving through thatch. Direct pesticide contact with the soil surface at application is usually restricted to areas devoid of plants, or to areas where low aerial shoot density permits direct fall-through of the pesticide.

Foliar interception of pesticides applied to turf has not been intensively investigated. Research work has been limited in this area because of high cost and technical expertise required to analyze pesticides. Furthermore, the proportion of applied pesticide intercepted by the foliage simultaneously depends upon the application procedures used, as well as plant, chemical, and meteorological factors.

Liquid and dry pesticide applications are made to turf using a sprayer or granular spreader. With liquid applications the height of the nozzle above the canopy, nozzle design, sprayer operating pressure and the volume of liquid applied all influence the portion of pesticide intercepted by the foliage.[9] Increasing nozzle elevation above the canopy increases spray drift and evaporation losses thus lowers the proportion of applied pesticide intercepted by foliage. Similarly, any combination of spray volume, application pressure, and nozzle design that decreases spray droplet size will increase evaporation losses, thereby decreasing the proportion of applied pesticide intercepted by the foliage. Applying an excessive volume of spray will promote foliar droplet runoff, which will also lower the portion of pesticide intercepted by foliage. Foliar interception of granular pesticide formulations may be influenced by the size of the granule and the presence of moisture on the foliage at the time of application. Small diameter, low mass granules contacting wet foliage are more likely to fully dissolve on leaf surfaces than are larger diameter and heavier mass granules applied to dry foliage.

Pesticide chemical properties and environmental conditions at the time of application can also influence foliar interception of pesticides. Wind and low humidity favor evaporative losses of liquid pesticide applications, while high humidity favors increased foliar penetration of highly ionic pesticides.[10] High temperatures enhance pesticide volatility as do certain formulations of select pesticides.[9] Kuhr and Tashiro[11] reported that foliar interception is greater for emulsifiable concentrate formulations than for granular formulations of the same pesticide applied at identical rates of application to Kentucky bluegrass (*Poa pratensis* L.).

Work conducted on crop species other than turf suggests that crops that fully cover the soil surface typically intercept between 55 to 95% of a ground applied pesticide.[9] The few studies conducted on turfgrass indicate that the portion intercepted by the foliar portion of a turfgrass canopy is somewhat lower. Sears et al.[12] reported that 42% of the total diazinon applied as a liquid was intercepted by the foliage of Kentucky bluegrass grown in pots. Carroll et al.[13] collected Kentucky bluegrass foliage 18 to 48 hr after making a 0.6 kg ai/ha application of dicamba and found that 66% of the sprayer applied dicamba was present on the foliage. Prinster and Hurto[14] applied different formulations of five pesticides to Kentucky bluegrass and reported a mean foliar interception value of 21%. No published work on the effect of turfgrass canopy form and structure on pesticide interception exists. It is likely, however, that open, vertically oriented leaf blade canopies, such as those produced by tall fescue (*Festuca arundinacea* Schreb.) and Kentucky bluegrass, allow a higher percentage of pesticide to come into contact with the soil surface than do more prostrate-growing stoloniferous grasses such as bentgrass (*Agrostis stolonifera* L.) and hybrid bermudagrass (*Cynodon* spp.).

PESTICIDE LOSSES FROM
TURF FOLIAGE

Pesticides residing on foliage are subject to the same avenues of dissipation as pesticides in soil. Once on foliage, a pesticide can be absorbed into the leaf, volatilized back into the atmosphere, chemically or microbially degraded, or removed by water. The contribution of each process to total foliar dissipation has not been examined because of the difficulty in isolating each specific process. For example, in the case of volatilization, separating

Table 1 Summary of fate of field pesticide studies conducted on turfgrass

Nature of Study	Pesticide	Formulation	Rate (kg/ha)	Irrigation	Sampling Period	% of Applied	Pesticide Residue Detected Average Conc. (µg/L)	Range in Conc. (µg/L)	Ref.
Foliar	Dicamba	Amine salt	0.6	—	18–48 hr	66	189[a]	93–318[a]	13
Foliar	Diazinon	Liquid[b]	4.0	—	c	1.4[d]	—	—	12
	Chlorpyrifos	Liquid[b]	2.0	—	c	2.3[d]	—	—	
	Isofenphos	Liquid[b]	2.0	—	c	2.4[d]	—	—	
	Diazinon	Liquid[b]	4.5	—	c	1.3[d]	—	—	
	Diazinon	Granular	4.5	—	c	<0.1[d]	—	—	
Foliar	2,4–D	Amine salt	2.24	—	c	2.7[d]	—	—	25
	2,4–D	Granular	2.24	—	c	0.2[d]	—	—	
Leaching	Diazinon	2G	4.9	Daily	1 wk	0.2			60
					2 wk	0.6			
					3 wk	1.0			
Turf with thatch				4 d	1 wk	0.3			
					2 wk	0.4			
					3 wk	0.6			
Turf without thatch				Daily	1 wk	1.4			
					2 wk	2.6			
					3 wk	3.4			
				4 d	1 wk	0.5			
					2 wk	0.9			
					3 wk	1.2			
Leaching	2,4–D	Amine salt	1.1	1.25 cm/wk	17 mths	0.4	0.87		61
				3.75 cm/wk			0.62		
			3.3	1.25 cm/wk			0.72		
				3.75 cm/wk			0.55		
	Dicamba	Amine salt	0.10	1.25 cm/wk	17 mths	1.0	0.55		
				3.75 cm/wk			0.29		
			0.33	1.25 cm/wk			0.34		
				3.75 cm/wk			0.26		
Leaching	2,4–D	Not given	1.12	76 to 228 mm per event	0–7 d	0.7		0 to 312	3
					25–46 d	0.1			
					68–76 d	0			

Table 1 (Cont.)   **Summary of fate of field pesticide studies conducted on turfgrass**

Nature of Study	Pesticide	Formulation	Rate (kg/ha)	Irrigation	Sampling Period	% of Applied	Pesticide Residue Detected Average Conc. (µg/L)	Pesticide Residue Detected Range in Conc. (µg/L)	Ref.
Leaching	2,4DP	Not given	1.12		0–7 d	0.8		0 to 210	
					25–46 d	0.2			
					68–76 d	0			
	Dicamba	Not given	0.28		0–7 d	1.7		0 to 251	
					25–46 d	0.2			
					68–76 d	0			
	Pendimethalin		1.68			ND[e]			
	Chlorpyrifos		1.12			ND			
Runoff	2,4-D	Not given	1.12	76 to 228 mm per event	0–7 d	1.6			3
					25–46 d	0.1			
					68–76 d	0			
	2,4-DP	Not given	1.12		0–7 d	1.1			
					25–46 d	0.1			
					68–76 d	0			
	Dicamba	Not given	0.28		0–7 d	0.9			
					25–46 d	0.1			
					68–76 d	0			
	Pendimethalin		1.68			ND			
	Chlorpyrifos		1.12			ND			
Groundwater	Chlordane	Not given	NA[f]	Not given	18 mths	Not given	<0.125 to 2.59	<0.125 to 7.2062	62
	Chlorothalonil						<0.015 to 0.08	<0.015 to 0.38	
	Chlorpyrifos						<0.05 to 0.05	<0.05 to 0.1	
	2,4–D						<0.05 to 0.10	<0.05 to 0.24	
	Dacthal deacid[g]						<0.20 to 0.29	<0.20 to 1.07	
	Dicamba		NA				<0.05 to 0.03	<0.05 to 0.06	
	DCBA						<0.20 to 9.38	<0.20 to 298	
	Heptachlor epoxide[g]						<0.03 to 0.07	<0.03 to 0.16	
	Isofenphos		NA				<0.75 to 0.57	<0.75 to 1.17	
	Trichloro pyridinol						<0.10 to 0.24	<0.10 to 0.76	

Soil	Chemical	Formulation[b]	Rate	Rainfall	Time[c]	Foliar level[a]	Soil residue	% dislodgeable[d]
Leaching sand	Dacthal					ND[e]		
	Diazinon					ND		
	Anilazine					ND		
	Iprodione					ND		
	Siduron					ND		
	MCPP					ND		63
	2,4–D	Dimethyla-mine salt	1.12	57 to 115 mm	21 to 28 d	1.4 to 0.86	21 to 45	0 to 105
				34 to 91 mm		0.0003–0.12	0 to 8.1	0 to 18
	Dicamba	"	0.42	57 to 115 mm		1.8 to 4.3	22 to 34	0 to 56
				34 to 91 mm		0 to 0.82	<0.1 to 19	0 to 24
	Carbaryl	50W	7.94	57 to 115 mm		0.1 to 0.98	3 to 28	0 to 100
				34 to 91 mm		0 to 0.004	0 to 2	0 to 4
	Chlorothalonil	4F	13.43	57 to 115 mm		0	0	<1
Sandy loam	2,4–D	Same as above	Same as above	62 to 135 mm		0.003 to 0.0004	0.9 to <0.1	<1
				26 to 70 mm		0.004 to 0.0004	0.2 to 0.5	<1
	Dicamba	"	"	62 to 135 mm		0.02 to 0.004	<0.1 to 0.2	<1
				26 to 70 mm		0.02 to 0012	0.5 to 0.8	<1
	Carbaryl	"	"	62 to 135 mm		0.001 to 0	<0.1 to 0.3	<1
				26 to 70 mm		0.007 to 0	0 to 7.2	0 to 20
	Chlorothalonil	"	"	62 to 135 mm		0	0	<1
				26 to 70 mm		0.0004	0.7	0 to 1
Silt loam	2,4–D	"	"	65 to 136 mm		0.1 to 0.005	0.1 to 5.2	0 to 25
				23 to 74 mm		0.007 to 0.0002	<0.1 to 1.1	0 to 2
	Dicamba	"	"	65 to 136 mm		0.2 to 0.04	0.4 to 3.1	0 to 14
				23 to 74 mm		0.01 to 0.012	0.4 to 0.8	0 to 1
	Carbaryl	"	"	65 to 136 mm		0.04 to 0.0002	<0.1 to 11	0 to 32
	Chlorothalonil	"	"	23 to 74 mm		0.002 to 0	<0.40 to 2.8	0 to 4
				65 to 136 mm		0.002	0.6	0 to 4
				23 to 74 mm		0.0002	0.1	<1

[a] Units of foliar levels = mg dicamba/*kg dry foliage; [b] Formulation not specified; [c] Immediately after application; [d] Percent of applied that was foliar dislodgeable residue; [e] ND = Not detected above detection limit; [f] NA = Not applied in recent years or never applied; [g] Pesticide metabolites.

soil airborne losses from foliar losses would be difficult. In any event, Cooper et al.[15] observed that there was a 13% volatilization loss of pendimethalin over a 5-day period following an application to Kentucky bluegrass. In contrast, several studies have examined total pesticide loss from turf foliage over time and have shown that pesticide levels decline in an exponential manner with most pesticides having a foliar half-life of 4 to 10 days.[11,14,16,17] The relatively uniform dissipation of pesticides from turfgrass foliage is likely due, in part, to rapid production of new turfgrass foliage which uniformly dilutes foliar pesticide levels.

Photodegradation is another important process that can decrease the amount of pesticide in the environment. Photolysis of pesticide molecules can occur when the molecule is in the atmosphere, on foliage, or when present on the soil surface. Once the pesticide has moved into an underlying horizon, photolysis is no longer a significant path of degradation. This is because organic matter absorbs short wavelengths, which blocks UV-induced reactions.[4,18] Pesticides on the soil surface can be chemically transformed by UV light in the 290- to 450-nm range.[19]

Foliar dissipation of pesticides can be greatly accelerated by precipitation after application. Foliar washoff is dependent on the amount of precipitation, the chemical properties of the pesticide, and the type of formulation selected. Water-soluble compounds wash off more readily from foliage than water-insoluble compounds. Washoff of water-soluble pesticides is reduced markedly with increasing time after application.[10,11] Pesticides formulated as emulsifiable concentrates are more resistant to washoff than those formulated as wettable powders. Washoff of wettable powder and salt-concentrate formulations may be influenced by rainfall intensity, while the washoff of emulsifiable concentrate and flowable formulations is not usually influenced by different rainfall intensities.[13,20,21]

Pesticides also may be removed from foliage by nonaqueous physical dislodgement. Loss by this avenue is of interest to public officials concerned with human and animal exposure to pesticide-treated turf. The amount of pesticide susceptible to foliar dislodgement is determined by placing foliage samples in a water and detergent or mild organic solvent solution for 15 to 60 sec,[16] or by wiping grass blades with cheesecloth and extracting the residue collected with acetone.[12] Total foliar pesticide residues are typically determined by subjecting foliage to an organic solvent, such as methanol, for at least 30 min.[12,16]

Dissipation of pesticide dislodgeable residues is similar to total foliar pesticide loss in that dislodgeable residues are highest immediately following pesticide application and decline exponentially with time.[12,14,22,23] In addition, like total foliar pesticide washoff, pesticide formulation can affect both the initial foliar dislodgeable residue amount and the subsequent washoff of these residues. Foliar dislodgeable residues immediately after application were reported to be 20 times higher for a liquid formulation of diazinon than for a granular formulation of equal rate.[12] In contrast, dislodgeable residue dissipation for the initial 24-hr period following application appears to be far greater for liquid formulations. Diazinon foliar dislodgeable residues collected 24-hr after application were essentially the same for both the liquid and granular formulations in the Sears et al.[12] study, despite the fact that initial liquid formulation dislodgeable residues were 20-fold greater than granular formulation residues immediately following application. The dislodgeable residues of wettable powder, granular and flowable formulated pesticides are more readily washed off foliage than are emulsifiable concentrates.[14]

In field studies, no more than 5 to 6% of the 2,4-D applied to Kentucky bluegrass foliage could be dislodged immediately after spraying by vigorous scuffling of cheesecloth "booties" over the turf.[24,25] Disappearance of dislodgeable residues was very rapid. Less than 0.1% of applied amounts could be dislodged by scuffling after 5 to 7 days. Mowing caused only a small reduction in dislodgeable residues. A total of 18 mm of rainfall 1 hr after spraying reduced dislodgeable residues to less than 0.01% of the applied amounts. Granular formulations of 2,4-D (with fertilizer) were less dislodgeable than the same quantity of 2,4-D applied as a liquid for the first 2 days following application. After this time dislodgeable residues were similar.

The time required before an individual can safely reenter an area sprayed with a pesticide is called the "safe reentry interval." The safe reentry interval is determined by monitoring the dissipation of dislodgeable foliar residues after applying a pesticide to a specific crop. The standards for safe pesticide dislodgeable foliar residue levels are set by the EPA and may be superseded by more stringent state agency regulations. Goh et al.[22,23] determined the safe reentry intervals for chlorpyrifos and dichlorvos using the State of California estimated safe dislodgeable foliar residue standards for these two insecticides. These two insecticides are the active ingredients of Dichloron™, which was applied at the maximum rate of application to a lawn area comprised of tall fescue and clover. The treated areas were subdivided into two areas:

one receiving 13 mm of precipitation immediately after application, and the other receiving no irrigation after Dichloron application. Irrigation immediately after application of Dichloron markedly reduced the safe reentry times for both insecticides. The safe reentry interval for dichlorvos was 4 hr when water was applied to turf, and 16 hr when it was not. Chlorpyrifos never exceeded the estimated safe level when the turf was watered after Dichloron application. However, when no irrigation was applied, 6 hr as required to dissipate chlorpyrifos to a foliar dislodgeable residue level considered safe for reentry.

The reentry times determined for the aforementioned study are likely to be valid only for the conditions under which the study was conducted and should not be extended to other regions of the country. Willis and McDowell[16] statistically compared foliar dislodgeable residue half-lives within the southwestern and southeastern regions of the U.S. for a range of crops and insecticides and found that half-lives were consistently shorter in the warmer, drier Southwest. Apparently, the higher humidities and lower maximum temperatures in the Southeast slow foliar dislodgeable residue dissipation of most insecticides.

FATE OF PESTICIDES IN THATCH

Turfs that receive excessive amounts of fertilizer, or have low soil earthworm activity resulting from repeated applications of benomyl, ethoprop, carbaryl, bendiocarb, and other pesticides may develop a thatch layer immediately above the soil surface.[26] Thatch is an intermingled layer of both dead and living roots, shoots, and stems. The presence of thatch forms a second canopy layer through which pesticides must travel to enter the upper strata of soil. Thatch, with its high percentage of organic matter, tightly adsorbs water-insoluble pesticides, thus preventing the movement of these pesticides into the soil.[27-29] The effect of thatch on the transport of partially or completely water-soluble pesticides is not clear. Significant movement through thatch has been reported for fenamiphos. Conversely, 96 to 99% of the total amount of carbaryl recovered from the top 10 cm of soil plus thatch was present in the thatch.[30]

In addition to reducing the entry of many pesticides into the soil, thatch also may accelerate pesticide dissipation in turf.[31] Enhanced biological degradation of pesticides due to high microbial populations in thatch may be the process responsible for accelerating the dissipation of some pesticides applied to turf.[28] Enhanced biological degradation of pesticides is the topic of a separate chapter in this book.

PESTICIDE LEACHING

There are two basic types of processes that can affect the amount of pesticide that is available for transport past the root zone: transformation and retention processes. Retention processes (such as adsorption) do not actually decrease the amount of pesticide in the environment, but do reduce the amount available for leaching. Transformation processes (such as biodegradation) can decrease and completely eliminate the amount of pesticides in the soil, and thus the amount available for offsite transport.[8]

Pesticide leaching depends on various soil properties, environmental and climatic factors, management practices, and properties of the chemicals applied.[4,6-8] Soil characteristics that affect pesticide leaching potential include moisture content, organic matter content, hydraulic conductivity, porosity, and microbial activity. Climatic properties include precipitation, temperature, and rates of evapotranspiration. Management practices such as method of irrigation, drainage, development of thatch layers, and the type of chemical applied will also affect the potential for pesticide leaching.[4,7,32] Certain properties of the pesticide itself will affect the chance for leaching, such as rate of application, water solubility, persistence (half-life), and adsorption potential. Also, it may be possible that the turfgrass environment differs from other agricultural crop covers in certain properties that may reduce the potential for pesticide leaching. These properties include a relatively high shoot density and the presence of thatch and a high organic matter content in the surface soil.

GROUNDWATER AND
THE HYDROLOGIC CYCLE

Groundwater can be defined as "subsurface water that occurs beneath the water table in soils and geologic formations that are fully saturated".[33] As with any component of the hydrologic cycle, there is an input, transport, and discharge of water within the groundwater system.[34] Precipitation is the initial water input; some of this precipitation falls onto the ground and percolates through the root zone. At this point, water can be absorbed by plants and lost via evapotranspiration (ET), some water may drain laterally into streams, and the remaining water will move beyond the root zone via gravity.[35]

Downward-moving water must first percolate through an upper partially saturated soil zone known as the zone of aeration, the unsaturated zone, or the vadose zone. Eventually, this percolating water will reach the water table, which is the upper surface of the zone of saturation. In the zone of saturation, all pore spaces are filled with water; wells and boreholes extract water from this region. Groundwater in the saturated zone moves through permeable geologic formations called aquifers, which are capable of storing and transporting water.[1,34,35]

Aquifers may be described as either confined and unconfined. An unconfined (or water-table) aquifer contains water under atmospheric pressure; the upper surface of the water is the water table, and it moves up and down according to seasonal recharge and discharge. A confined or artesian aquifer is bounded on top and bottom by relatively impermeable geologic layers (called aquitards), such as clay. The pressure of the water in a confined aquifer is greater than atmospheric pressure. Perched aquifers are found where a lens of impermeable material exists above the water table, which forms a local water table (zone of saturation).[1,34,35]

The transport of groundwater is a very slow process; flow rates depend on hydraulic gradients and aquifer permeabilities.[1] Most aquifers consist of unconsolidated materials such as sand and gravel, and other high-permeability rocks such as limestone, basalt, sandstones, and conglomerates.[34,36] When the depth of the water table varies from one area to another in unconfined aquifers, these differences in elevation determine the hydraulic head. Groundwater moves from areas of recharge to points of discharge, following the hydraulic gradient (from regions of high pressure to areas of lesser pressure).[1,34,36] Movement of groundwater is much slower than surface bodies of water, such as rivers, and in most locations flow rate ranges from 1 m/day to 1 m/year.[34] In certain limestone regions possessing "Karst topography," groundwater movement may be much quicker (>300 m/day). Natural discharge zones include oceans, rivers, marshes, and artificial (man-made) discharge zones including wells and drains.[35,36] Thus, precipitation and melting snow are the inputs to the groundwater system, transport occurs through the vadose zone to the water table via gravity, and then through aquifers according to the hydraulic gradient, and discharge occurs at natural sites (streams, rivers) and artificial sites (wells).

Obviously, the depth of the water table is one of the factors which determines if an area is at high risk for groundwater contamination by leaching. Shallow water tables are at greater risk for contamination. The depth of the water table below the soil surface fluctuates throughout the year, depending on inputs from precipitation and melting snow, and outputs through wells, streams, springs, etc.[34] Also, the depth of the water table varies in different geographic locations, depending on climate and other factors. For example, the depth of the water table may be only a meter in certain humid regions, compared to hundreds of meters in dry areas such as deserts.[34]

Of the total amount of water in the earth's hydrologic cycle, only 2.7% is fresh water. In the U.S., 96% of this freshwater is supplied by groundwater; lakes, streams and rivers constitute the remaining 4%.[1] Groundwater is therefore an important natural resource in the U.S.; it provides 40 to 50% of the population with its main drinking water, and in rural areas that percentage is as high as 95%. Groundwater is also utilized for public and rural supplies, irrigation needs, and industrial uses, such as for generation of thermoelectric power.[1]

Concern over pesticides in groundwater became more acute in the late 1970s when dibromochloropropane (DBCP), aldicarb, and atrazine were detected in various monitoring studies across the U.S.[32,37]

ENVIRONMENTAL FACTORS

Environmental factors, including precipitation, temperature, and evapotranspiration, play a vital role in the leaching of pesticides past the root zone. Losses via leaching can occur when the amount of precipitation or irrigation is greater than the field capacity of the soil, and are affected by the rate and distribution of rainfall and runoff losses from the soil.[38]

The first step in the leaching process is water infiltration, which is equivalent to precipitation minus evapotranspiration, interception, and runoff. Infiltration of water is affected by several factors, including rate of application, initial soil water content, and soil physical characteristics. Initial infiltration is faster if the soil water content is low, but the rate of water movement once it penetrates the soil is lower in a dry soil.[39] The potential for a pesticide to leach is greater in a soil at or near saturation (i.e., at maximum hydraulic conductivity).[4] High water-application rates result in shallow distribution in the soil, whereas slower application rates result in deeper penetration. Gerstl and Yaron[40] examined the distribution of two herbicides, bromacil and napropamide, in soil following irrigation at a high application rate (4 L/hr) and a low application rate (1.5 L/hr) through a simulated drip irrigation system. They found that the high irrigation-application rate resulted in a wide, shallow (lateral) distribution of water and herbicides, and the low irrigation-application rate produced a narrow, deeper distribution.

The timing of precipitation (or irrigation) following pesticide application will also affect leaching. When large amounts of water infiltrate the soil immediately following a pesticide application (on a permeable soil), the potential for leaching increases.[4] Pesticides in solution initially diffuse into micropores within the soil, decreasing the chance for leaching past the root zone. However, if preferential macropores are present, then a high-intensity precipitation event that exceeds the soil infiltration rate and occurs immediately following pesticide treatment could result in transport of the chemical past the root zone or cause runoff.[41]

Water is converted from a liquid to vapor state by radiant solar energy. Evaporation is the loss of soil water from the soil surface as water vapor. Plant removal of water from the soil into the atmosphere by water vapor loss from plant leaves is called transpiration. Evapotranspiration (ET) describes the combined loss of water by these two processes.[38] Water lost by ET is not available for drainage past the root zone. ET affects the leaching potential relative to precipitation/irrigation; if water losses by ET are less than the addition of water by precipitation/irrigation, then there is sufficient water to transport chemicals past the root zone. However, if ET is greater than water infiltration, then there is usually not enough water to leach chemicals past the root zone.[4]

The wide range of climatic factors in different geographic areas results in varying leaching potentials. In a humid temperate climate, for example, infiltration often exceeds ET rate. When infiltrating water reaches the field capacity of the soil, water will begin to percolate into underlying horizons (substrata). The greatest percolation takes place when evaporation is lowest and when the ground is not frozen (late fall and early spring months, November to April). During summer, ET is greater than infiltration (based on precipitation), which results in soil water depletion. During this time, plants rely on stored moisture (from the winter/early spring) or irrigation for survival. It is clear that climatic factors greatly affect the potential for leaching; thus, areas with high water percolation, such as the humid eastern part of the U.S. and heavily irrigated regions in the West, have higher leaching potentials.[38]

Temperature is another environmental factor that affects leaching potential. Temperature will affect such processes as photolysis, volatilization, sorption, chemical and biological degradation, evapotranspiration, etc. As temperatures increase, degradation rates also increase, thereby reducing the amount of pesticide available for leaching.[38] A study by Bouchard et al.[42] showed that degradation of three pesticides (metribuzin, metolachlor, and fluometuron) in soil increased with higher temperatures. For example, the half-life for metolachlor in a silt loam soil was 26 weeks at 15°C, 10.1 weeks at 23°C, and 5.2 weeks at 37°C in soil of 10 to 20 cm depth.

Higher temperatures favor reduced adsorption due to increased solubility.[19,43] This may increase the amount of pesticide in soil solution available for transport.[4] Higher temperatures also increase ET which increases water use of crops (including turf) during the hot summer months.[38] Overwatering may actually contribute to increased potential for leaching by increasing the amount of pesticide in soil solution, though concentrations may be smaller.

SOIL PROPERTIES

Soil physical, chemical, and biological characteristics play an important role in the leaching of pesticides. These physical and chemical properties include soil water content, bulk density, porosity, structure, hydraulic conductivity, water retention capacity, texture, clay content, organic matter (OM) content, and pH. Biological properties include the number and types of microorganisms and macroorganisms in the soil, as well as the type of vegetative cover.[4,19,32]

The water content of a soil will affect the initial infiltration and rate of mobility and distribution once water penetrates the soil. High soil water content (near or at saturation) will facilitate movement of pesticides past the root zone. Adsorption of pesticides onto soil particles increases with drier soils, reducing the potential for leaching.[4] However, degradation of pesticides in soil (which also reduces the amount available for transport) is more rapid in moist soil than in dryer soils. For example, a study by Walker[44] showed that napropamide was more persistent in low moisture content soil. The half-life of this chemical (at 28°C) was 54, 63, and 90 days in soil with a water content of 10, 7.5, and 3.5%, respectively.

Many of the soil physical properties that determine leaching potential are interrelated. Bulk density, hydraulic conductivity, texture, and structure of a soil all affect chemical transport. As soil porosity decreases, mass transport of pesticides is reduced;[4] correspondingly, as the bulk density of a soil increases, chemical movement via diffusion also decreases.[39] Hydraulic conductivity describes " ... a soil's ability to transmit water and entrained solutes."[4] A soil with a high hydraulic conductivity relative to its initial infiltration rate will have a high potential for leaching.

Textural and structural properties of a soil affect the water retention, porosity, hydraulic conductivity, and potential number of adsorption sites. Sandy soils have a greater potential for leaching than finer-textured soils due to a greater percolation rate (larger pore size), as well as a poorer ability to adsorb nutrients/pesticides.[39] Soil structure is affected by texture and organic matter; highly aggregated soils may have large macropore channels that permit preferential flow of pesticides beyond the root zone.[4]

Adsorption plays a major role in the potential for leaching. Both chemicals and soils vary in adsorptivity. Any factor that increases the chemical's adsorption to soil particles will decrease the potential for leaching. Some soil characteristics which affect adsorptivity include OM content, the percentage and type of clay, pH, field moisture capacity, cation-exchange capacity, and temperature. Of these, OM content is the most important factor in determining adsorption[39] since OM can adsorb a greater amount of pesticide residue than clay particles.[46] As little as 1 to 1.5% OM can significantly reduce chemical movement. Soil organic carbon decreases with soil depth, so that a pesticide that has moved past the root zone has less opportunity for sorption and greater opportunity for leaching.[47] Organic matter also harbors soil micro-organisms, which degrade chemicals. Thus, the presence of OM increases adsorptivity and indirectly aids in biodegradation, thereby decreasing leaching potential.[46] For ionic pesticides, pH also affects adsorption. For cationic compounds, adsorption (via ion-exchange with ions on clay or organic matter in the soil) increases with lower pH.[43] Anionic pesticides will most likely show the greatest adsorption at a higher pH.[19]

Certain soil factors also affect nonbiological transformation processes that break down pesticides. Nonbiological degradation of pesticides includes hydrolysis, oxidation and reduction reactions, and photolysis. Organic matter and clay mineral surfaces are the most likely sites of catalyzing the hydrolysis of pesticides in the soil environment. Hydrolysis may be affected by soil temperature and pH. Higher temperatures will promote hydrolysis; for example, rates of hydrolysis of methylcarbamate insecticides increase two- to threefold with every 10°C increase in temperature.[48]

The hydrolysis of certain pesticides, such as carbaryl, may be affected by soil pH. Larkin and Day[49] found that in solutions with a pH of 7, the half-life of carbaryl was 16.5 days, but under more acidic conditions (pH 6), it was much more persistent (171.4 days).[49] Chapman and Cole[50] also studied the effects of pH on carbaryl persistence and found that as conditions became more alkaline, carbaryl's stability decreased. At a pH of 4.5 (in a sterile 99:1 water-ethanol phosphate buffer at 25°C), carbaryl's half-life was 300 weeks, at pH 6 it was 58 weeks, at pH 7 the half-life was only 2 weeks, and at a pH of 8, the half-life was 0.27 week.

Biological factors in the soil environment have a substantial impact on the fate of applied pesticides. These factors include the presence of a vegetative cover, and the numbers and types of micro- and macroorganisms. Soils that are covered by vegetation (agricultural crop, turfgrass, etc.) are less likely to promote leaching. Vegetation intercepts the applied chemical, and reduces the amount which enters the soil; the plant may also absorb the chemical from the soil into its roots to be translocated. The higher OM associated within the root zone of a plant system may also support a microbial population capable of biodegrading the pesticide. The denser the plant cover, such as turf, the more successful vegetation will be in intercepting chemicals before they reach the soil and have the potential to leach.[51]

Soil microorganisms play a vital role in pesticide degradation.[52,53] Microbe populations are greatest in soil horizons with high OM (i.e., the root zone).[19] Biodegradation of pesticides in the soil is very important, from both a management perspective and an environmental perspective. Biodegradation may dramatically decrease chemical efficacy by rapidly removing the pesticide from the soil. However, a benefit of microbial breakdown of pesticides, by shortening the time the pesticide is found in the soil, is a reduction in the potential for leaching and groundwater contamination. Microbial breakdown is the primary path of degradation for most pesticides in the soil.[53]

While microorganisms may reduce potential pesticide leaching, macroorganisms may favor movement. Earthworms, for example, are found in cool, moist soils high in OM and create large macropore channels. Earthworm burrows have an average diameter of 3 to 12 mm.[54] The species *Lumbricus terrestris* and *Allolobophora nocturna* have burrows which penetrate to a depth of 150 to 240 cm; these burrows are mostly vertical, but near the soil surface they often exhibit extensive branching.[54] The presence of these continuous macropore channels may increase the risk for leaching of pesticides if an irrigation or rainfall event occurs soon after pesticide application. This type of movement is called short-circuiting or preferential flow, and may conduct the pesticide into soil depths with minimal biodegradation and adsorption capacities.[41]

Many studies have examined the effects of soil structure (presence of micro- and macropores) on water and solute movement through the soil. Rao et al.[55] investigated the movement of the herbicide picloram through a highly structured Molokai soil. They found that the peak concentration of the chemical remained in the upper 40 cm of the soil profile, but that a significant fraction leached beyond 81 cm, following a single application of 24 cm water. Based on calculated adsorption figures, the herbicide was retained more than expected, which they believe was the result of retention of the picloram in soil micropores.

Bouma et al.[56] have reported that the addition of a 5-cm-thick layer of sand to the surface of leaching columns greatly reduced preferential flow through macropores by causing a "more effective wetting of the soil matrix". They proposed decreasing water infiltration with drip irrigation, and the addition of a sandy layer to the surface, could aid in reduction of losses of nutrients and pesticides via macropores.

PESTICIDE PROPERTIES

In addition to the soil characteristics that affect chemical movement, the properties of the chemicals themselves are also important in determining leaching potential. These properties include water solubility, sorption characteristics, volatility, persistence, formulation, and the rate and timing of application.[4,32]

Water solubility and sorption characteristics of pesticides are major factors which determine the mobility of a chemical. The water solubility of non-polar compounds is proportional to the mass carried by soil solution, which is available for transport to groundwater.[4] According to Deubert,[46] chemicals with a solubility >30 mg/L will be mobile under certain conditions such as permeable soil (sandy), low adsorptivity, and high material persistence.

The adsorption characteristics of pesticides can be described by the soil adsorption coefficient (K_{OC}), which is a measure of its adsorptive capacity onto organic carbon. According to guidelines established by the EPA, pesticides with K_{OC} values less than 300 are considered to have a high potential for leaching.[57] Most currently used turf pesticides are strongly adsorbed (having K_{OC} values >300).[46]

The water solubility and K_{OC} of a pesticide are in many cases inversely related. For example, Bilkert and Rao[58] showed aqueous solubilities for oxamyl, aldicarb, and fenamiphos of 280,000, 6000, and 700 mg/L, respectively, and the calculated K_{OC} values in a sandy loam soil were 6, 20, and 197, respectively.

The persistence of a pesticide in the soil will also affect its potential to leach; compounds which are resistant to degradation have a greater chance for leaching under various conditions. The half-lives of most current pesticides are significantly shorter (weeks to months) than those of some older materials such as chlordane, a cyclodiene with a soil persistence of greater than 4 years.[19] Guidelines by the EPA[57] suggest that pesticides with a half-life greater than 21 days may have a greater

potential for leaching than those with shorter half-lives.

The formulation, and timing and rate of application of a pesticide will help to determine its potential for leaching. Ester formulations (such as 2,4-D ester) have lower water solubilities than salts (2,4-D amine salt), and are therefore less likely to leach. As would be expected, higher initial chemical application rates increase leaching potential. In general, as the amount of pesticide (or fertilizer) applied to the surface increases, the amount of material that can move through the soil profile (and beyond) also increases. The timing of pesticide application and subsequent irrigation or rainfall are equally important in reducing the risk for leaching. If an applied pesticide is given adequate time for utilization within the soil and degradation before receiving irrigation, then the chances for leaching are minimized.[4]

LEACHING OF TURF-APPLIED PESTICIDES

It is apparent that the fate of turfgrass pesticides in the environment is a highly complex subject which warrants thorough investigation. Although there are numerous fate-of-pesticide studies in the agricultural literature, there are few leaching studies which pertain specifically to turfgrass systems.

The initial distribution of a pesticide applied to a turfgrass system will greatly affect the amount available for leaching. Various studies have examined the distribution of turf-applied chemicals, comparing the amounts left in the canopy, thatch layer, and underlying soil. For example, Sears and Chapman[59] studied the distribution of four insecticides (chlordane, diazinon, chlorpyrifos, and CGA 12223) applied to a golf course fairway (annual bluegrass, *Poa annua* L.) on a sandy loam soil. This site received both natural precipitation and irrigation treatments. Following the irrigation event just after pesticide application, 2 to 8% of the insecticides were found in the top 1 cm of roots and soil. After 56 days, 60% of the chlordane applied and 9% of the chlorpyrifos applied were detected in the grass-thatch layer, but <1% of these pesticides were found in the underlying 2.5 cm of soil. Diazinon and CGA 12223 disappeared within 14 days.

A study by Branham and Wehner[60] revealed the importance of thatch and microbial degradation in minimizing leaching potentials from turfgrass systems. This study investigated the fate of diazinon, a turf insecticide, on Kentucky bluegrass, under two irrigation regimes (daily and every 4 days) and two thatch variables (with/without). The greatest degradation occurred on turf with thatch that was irrigated daily (only 7% of the applied chemical remained after 3 weeks). For turf both with and without thatch, 96% of the diazinon was found in the upper part of the soil profile (top 10 mm). Degradation was measured by the release of labeled CO_2, and was found to be highest in turf with thatch. Interestingly, greater irrigation did not increase leaching, but did enhance breakdown. Thus, the authors conclude that pest control may be hindered by thatch in two ways: the retention of pesticides within the thatch layer, which would decrease control of soil-inhabiting insects, as well as higher rates of degradation in the thatch layer.

A study by Gold et al.[61] examined the leaching of two herbicides, 2,4-D and dicamba, from home lawn-type turf. They investigated two rates of pesticide application and two irrigation regimes: a low treatment to prevent drought stress, and an overwatering treatment. For both herbicides, the low rate of chemical application-minimal irrigation treatment resulted in higher leaching concentrations, but in most samples, the concentrations were very low (less than 1 µg/L). The authors[61] feel that the lack of significant leaching is due to high microbial degradation rates in the root zone during the summer (higher temperatures).

A study by Watschke and Mumma[3] examined the leaching of nutrients and pesticides from sodded and seeded turf lysimeters, grown on a clay soil with 9 to 14% turfed slopes. Pendimethalin and chlorpyrifos were not detected in any of the leachate samples, presumably due to their low water solubilities and high adsorptive capacity. 2,4-D and dicamba were found at the highest concentrations in leachate collected 2 days after pesticide treatment. Dicamba leachate concentrations ranged from 0 to 251 µg/L, 2,4-D was found at levels ranging from 0 to 312 µg/L, and 2,4-DP levels ranged from 0 to 210 µg/L. Dicamba had the greatest percent recovery in the leachate, close to 1.7% of applied pesticide, during the first posttreatment interval (0 to 7 DAT); by the second interval (25 to 46 DAT), recovery had decreased to less than 0.3% for dicamba in leachate, and by 68 to 76 DAT, it was no longer detected. Recovery of 2,4-D in leachate was about 0.7% during the first collection interval and 0.1% in the second interval; none was recovered in the final collection. The leachate recovery values for 2,4 DP were 0.8% during the first collection interval, 0.2% in the second, and nondetectable in the third collection period. Based on current federal drinking water standards the concentration of 2,4-D was in excess of the standard (70 µg/L) in four of the 29 leaching events. Dicamba concentrations were above the standard of 210 µg/L only once. It appeared that sufficient dissipation of the

pesticides occurred by the second collection period to greatly reduce residues detected in leachate samples, which is consistent with the findings of Gold et al.[61]

The results from the previous two studies,[3,61] which only represented a small portion of the possible pesticide, soil, environmental interaction, do suggest that pesticides applied to home lawns do not seriously threaten groundwater quality. However, the chemicals applied to golf courses are in need of investigation, since rates and frequency of pesticide application, as well as irrigation regimes, are often higher than for home lawns. A study funded and coauthored by the EPA[62] examined pesticide leaching from four golf courses on Cape Cod, MA. Samples were collected from monitoring wells located at tees, greens and fairways four times over 18 months. The pesticide and pesticide metabolites analysis included eight herbicides, four fungicides, and five insecticides. Pesticides never observed above detectable levels included siduron, DCPA, anilazine, iprodione, and diazinon. The range in concentration of pesticide or pesticide metabolites detected in sample wells were: <0.125 to 7.2 µg/L for technical chlordane; <0.015 to 0.38 µg/L for chlorothalonil; <0.05 to 0.1 µg/L for chlorpyrifos; <0.05 to 0.24 µg/L for 2,4-D; <0.20 to 1.07 µg/L for dacthal diacid metabolite; <0.05 to 0.06 µg/L for dicamba; <0.20 to 8.94 µg/L for DCBA; <0.03 to 0.16 µg/L for heptachlor epoxide metabolite of chlordane; <0.75 to 1.17 µg/L isofenphos and <0.10 to 0.76 µg/L trichloropyridinol metabolite of chlorpyrifos.

Most of the detected chemicals came from monitoring stations on greens and tees; this could be due to greater number of applications made to these areas. Although the authors concluded that turf pesticides applied to four golf courses with vulnerable hydrogeology were found to have minimal impact on groundwater quality, they still feel that more studies of this nature need to be conducted, and that golf course superintendents should be cautious with their management of turf chemicals.[62]

Few studies have compared multiple factors (soils, pesticides, and irrigation) important in the leaching of turf-applied pesticide. Borromeo[63] studied the effect of soil texture (sand, sandy loam, and silt loam), pesticide mobility (dicamba > 2,4-D > carbaryl » chlorothalonil), and two irrigation regimes on the leaching from a Kentucky bluegrass turf. As expected, there was a far greater degree of pesticide leaching from the more mobile pesticides (dicamba > 2,4-D > carbaryl » cholorthalonil) applied to sand. There was virtually no pesticide leaching from the sandy loam and limited leaching from the silt loam. When irrigation to produce a leaching event occurred three times per week starting two DAT, leaching increased on the sand soil only. However, leaching in all three soils was limited (regardless of pesticide) when irrigation to saturation did not occur until 7 days after treatment. Thus, applications of highly mobile pesticides to turf on permeable soils such as sand (with low organic matter), followed by a substantial precipitation event, may result in pesticide leaching.

PESTICIDE RUNOFF

The runoff of pesticides applied to turfgrass sites has not been studied extensively. However, numerous factors that have been shown to influence pesticide runoff from agricultural settings may relate to turfgrass situations. These include climatic, soil, pesticide, and management variables such as tillage practices and crop rotation. The discussion that follows summarizes the nature of these factors as reviewed by Leonard.[64]

Climatic factors can play a major role in the degree of pesticide runoff. The time interval between the application of a pesticide and the runoff event will influence the concentration of the pesticide in the runoff water. The highest concentration of a pesticide in runoff occurs following the first significant runoff event after application. As time increases between the pesticide application and the first rainfall event, the amount of pesticide available for transport in runoff decreases. The intensity, amount, and duration of precipitation determines the amount of pesticide that may be washed off the foliage and thus, be available for runoff. Runoff can only occur when the intensity of precipitation exceeds the infiltration rate of the soil. Increasing precipitation intensity increases the rate of runoff and reduces the time to runoff. Cooler weather results in periods of low ET. If precipitation is high, runoff can be significant, especially if the ground is frozen.

Soil properties can influence the extent of pesticide runoff. Generally, more runoff is generated from finer textured or compacted soils. Sandy soils have a greater time to runoff, which reduces the initial concentration of dissolved pesticides in the runoff. Soils with a high moisture content are more susceptible to runoff from high intensity, short duration storms. As slope increases, so does the potential for runoff. Soils that are well aggregated, i.e., highly structured, have faster water infiltration rates and thus a lower potential for runoff. In certain turfgrass situations, hydrophobic soils (localized dry spots) occur, which may increase runoff.

Pesticides that are more water soluble are more likely to be washed off the turfgrass foliage and

into runoff. Once a turfgrass site is established, then soil erosion is minimal. Therefore, pesticides that are strongly sorbed onto the turfgrass plant and/or soil are much less likely to run off, as compared to an agricultural setting where erosion can occur. Pesticides that are more persistent in the turfgrass environment, because of their resistance to volatilization, chemical, photochemical, and biological degradation, have a higher probability for runoff. Liquid formulations of pesticides may be more readily transported in runoff water than granular ones. The concentration of a pesticide in runoff is proportional to the amount of pesticide in the runoff zone. Therefore, pesticides that are applied at higher rates have a higher potential for runoff.

Aspects of turfgrass management which may contribute to runoff are somewhat limited. Irrigation in excess of the infiltration rate of the soil can lead to runoff.[3] Reducing surface soil compaction and controlling localized dry spots by the use of wetting agents can reduce the potential for runoff from turfgrass sites.

Research on the degree of pesticide runoff from turfgrass sites is primarily limited to work conducted by Watschke and Mumma.[3] The site had field plots with slopes ranging from 9 to 14% on a clay soil. Pesticide runoff, as influenced by the method of establishment (seed vs sod), was determined over a 2-year period from 29 sampling dates. Since natural precipitation only produced runoff once during the study, irrigation was supplied at a rate of 15 cm/hr for 60 to 90 min to produce runoff. The pesticides pendimethalin and chlorpyrifos were never found above the detection limit in the runoff water. During the first 7 DAT the amount of dicamba, 2,4-D and 2,4-DP recovered in runoff water was 0.9, 1.6, and 1.1% of the applied pesticide, respectively. From 25 to 46 DAT, the amount of each pesticide recovered in runoff water was 0.1 % of the amount of pesticide applied. By 68 DAT, none of the three pesticides were observed above detection limits. Sodded plots had consistently higher infiltration rates (29.5 cm/hr) than the seeded plots (infiltration rate of 12.9 cm/hr). Thus, runoff was substantially less from sodded plots (0.8% of the total water applied) than from the seeded plots (11.6 to 13.4% of the total water applied). In general, runoff from these turfed sites was considered minimal since rainfall events for central Pennsylvania of 5 cm/hr for 1 hr would have a return period of 100 years.

SUMMARY

It is well documented that pesticides have been detected in groundwater. However, research on turfgrass (Table 1) seems to indicate that a variety of factors in the turf environment may reduce leaching to levels below those seen in production agriculture. These factors include: high shoot density compared to most agricultural crops, a thatch layer that hinders pesticide mobility into the soil and encourages microbial degradation, and an active root zone microbial population capable of rapid pesticide degradation. There are certain management practices, such as late fall pesticide applications (when temperatures are cooler and microbial activity is decreased), that may increase the chances for pesticide leaching. Macropore flow may be responsible for some of the chemicals detected in the groundwater; however, once the pesticides become sorbed onto soil particles or trapped within micropore solution, further leaching is greatly reduced. Certain conditions may be considered "high risk" for pesticide leaching, such as the application of highly water-soluble pesticides onto permeable soils (i.e., sands) followed by intense irrigation/precipitation events. These conditions may exist on high sand-content soils like golf course greens, some other athletic turf, and a few lawns. Thus, it appears that the implementation of sound management practices in high-risk turfgrass areas can help to minimize the potential for pesticide leaching and the contamination of the groundwater system. Examples of sound management practices that have been applied are (1) pesticide selection based on pesticide properties and site-specific soil properties, (2) application timing to avoid periods of intense rainfall, and (3) careful irrigation when highly water-soluble pesticides have been applied. Further research is warranted on pesticide leaching from turfgrass since the scope of studies to date has been limited to a few soils, pesticides, climatic conditions, and turfgrass management practices.

The runoff of pesticides applied to turfgrass has not been studied extensively. The results from one study would suggest the runoff events from turfgrass may be unusual. However, until studies are done on many more sites with different soils, grasses, and climatic conditions, broad statements on pesticide runoff from turfgrass are premature.

REFERENCES

1. **Pye, V. I., Patrick, R., and Quarles, J.,** *Groundwater Contamination in the United States*, University of Pennsylvania Press, Philadelphia, 1983.
2. **U.S. Environmental Protection Agency,** National Pesticide Survey, Summary results of EPA's National Survey of pesticide in drinking water wells, Fall, U.S. Environmental Protection Agency, Washington, DC, 1990.
3. **Watschke, T. L. and Mumma, R. O.,** The Effect of Nutrients and Pesticides Applied to Turf on the Quality of Runoff and Percolating Water, U.S. Department of Interior, Geological Survey (ER8904), Pennsylvania State University, University Park, PA, 1989.
4. **Walker, W. J.,** Environmental issues related to golf course construction and management: A literature search and review, Final report to the USGA Green Section, Spectrum Research, Sacramento, CA, 1990.
5. **Shurtleff, M. C., Fermanian, T. W., and Randell, R.,** *Controlling Turfgrass Pests*, Prentice-Hall, Englewood Cliffs, NJ, 1987.
6. **Turgeon, A. J.,** *Turfgrass Management*, Prentice-Hall, Englewood Cliffs, NJ, 1985.
7. **Arnold, D. J. and Briggs, G. G.,** Fate of pesticides in soil: predictive and practical aspects, in *Environmental Fate of Pesticides. Progress in Pesticide Biochemistry and Toxicology*, Vol. 7, Hutson, D. H. and Roberts, T. R., Eds., John Wiley & Sons, Chichester, England, 1990, 101.
8. **Cheng, H. H. and Koskinen, W. C.,** Processes and factors affecting transport of pesticides to ground water, in *Evaluation of Pesticides in Groundwater*, Garner, W. Y., et al., Eds., ACS, No. 315, Washington, DC, 1986, 2.
9. **Willis, G. H., Spencer, W. F., and McDowell, L. L.,** The interception of applied pesticides by foliage and their persistence and washoff potential, in *CREAMS — a field scale model for chemicals, runoff, and erosion from agricultural management systems,* U.S. Department of Agriculture, Conservation Research Report No. 25, 1989, 595.
10. **Caseley, J. C.,** Variations in foliar pesticide performance attributable to humidity, dew and rain effects, *Aspects Appl. Biol.*, 21, 215, 1989.
11. **Kuhr, R. J. and Tashiro, H.,** Distribution and persistence of chlorpyrifos and dichlorvos on turf, *Bull. Environ. Contam. Toxicol.*, 37, 27, 1986.
12. **Sears, M. K., Bowhey, C., Braun, H., and Stephenson, G. R.,** Dislodgeable residues and persistence of diazinon, chlorpyrifos and isofenphos following their application to turfgrass, *Pestic. Sci.*, 20, 223, 1987.
13. **Carroll, M. J., Hill, R. L., Pfeil, E., and Herner, A. E.,** Washoff of dicamba and 3,6-dichlorosalicylic acid from turfgrass foliage. *Weed Tech.*, 7, 437, 1993.

14. **Prinster, M. G. and Hurto, K. A.,** Dislodgeable residues of pesticides applied to lawn turfs, in *Agronomy Abstracts*, ASA, Madison, WI, 1989, 163.

15. **Cooper, R. J., Jenkins, J. J., and Curtis, A. S.,** Pendimethalin volatility following application to turfgrass, *J. Environ. Qual.*, 19, 508, 1990.

16. **Willis, G. H. and McDowell, L. L.,** Pesticide persistence on foliage, *Rev. Environ. Contam. Toxicol.*, 100, 23, 1987.

17. **Nash, R. G. and Beal, M. L., Jr.,** Distribution of silvex, 2,4-D, and TCDD applied to turf in chambers and field plots, *J. Agric. Food Chem.*, 28, 614, 1980.

18. **Parlar, H.,** The role of photolysis in the fate of pesticides, in *Environmental Fate of Pesticides. Progress in Pesticide Biochemistry and Toxicology*, Vol. 7, Hutson, D. H. and Roberts, T. R., Eds., John Wiley & Sons, Chichester, England, 1990, 245.

19. **Ross. S.,** *Soil Processes*, Routledge, London, 1989.

20. **McDowell, L. L., Willis, G. H., Smith, S., and Southwick, L. M.,** Insecticide washoff from cotton plants as a function of time between application and rainfall, *Trans. ASAE*, 28, 1896, 1985.

21. **Willis, G. H., McDowell, L. L., Smith, S., and Southwick, L. M.,** Rainfall amount and intensity effects on carbaryl washoff from cotton plants, *Trans. ASAE*, 31, 86, 1988.

22. **Goh, K. S., Edmiston, S., Maddy, K. T., and Margetich, S.,** Dissipation of dislodgeable foliar residue for chlorpyrifos and dichlorvos on turf, *Bull. Environ. Contam. Toxicol.*, 37, 27, 1986.

23. **Goh, K. S., Edmiston, S., Maddy, K. T., and Margetich, S.,** Dissipation of dislodgeable foliar residue for chlorpyrifos and dichlorvos treated lawn: implication for safe reentry, *Bull. Environ. Contam. Toxicol.*, 37, 33, 1986.

24. **Bowley, C. S., McLeod, H., and Stephenson, G. R.,** Dislodgeable residues of 2,4-D on turf, *Proc. Br. Crop Prot. Conf. Weeds*, 8A-10, 799, 1987.

25. **Thompson, D. G., Stephenson, G. R., and Sears, M. K.,** Persistence, distribution and dislodgeable residues of 2,4-D following its application to turfgrass, *Pestic. Sci.*, 15, 353, 1984.

26. **Potter, D. A., Cockfield, S. D., and Morris, T. A.,** Ecological side effects of pesticide and fertilizer use on turfgrass, in *Integrated Pest Management for Turfgrass and Ornamentals*, Leslie, A. R. and Metcalf, R. L., Eds., U.S. Environmental Protection Agency, Washington, DC, 1989, 33.

27. **Niemczyk, H. D. and Krueger, H. R.,** Persistence and mobility of isazofos in turfgrass thatch and soil, *J. Econ. Entomol.*, 80, 950, 1987.

28. **Niemczyk, H. D. and Krause, A.,** Degradation and mobility of insecticides applied to turfgrasses, in *Agronomy Abstracts*, ASA, Madison, WI, 1989, 162.

29. **Dell, C. J., Throssell, C. S., and Turco, R. F.,** Estimation of sorption coefficients for fungicides applied to turf, in *Agronomy Abstracts*, ASA, Madison, WI, 1991, 173.

30. **Niemczyk, H. D. and Filary, Z.,** Vertical movement and accelerated degradation of insecticides applied to turfgrass, *Ohio Turfgrass Foundation Newslett.*, March, 1988.

31. **Branham, B. E. and Wehner, D. J.,** The fate of diazinon applied to thatched turf, *Agron. J.*, 77, 101, 1985.

32. **Helling, C. S. and Gish, T. J.,** Soil characteristics affecting pesticide movement into ground water, in *Evaluation of Pesticides in Groundwater*, Garner, W. Y., et al., Eds., ACS Symposium Series No. 315, ACS, Washington, DC, 1986, 14.

33. **Freeze, R. A. and Cherry, J. A.,** *Groundwater*, Prentice-Hall, Englewood Cliffs, NJ, 1979.

34. **Hamblin, W. K.,** *The Earth's Dynamic Systems: A Textbook in Physical Geology*, 4th ed., Burgess Publishing, Minneapolis, 1985.

35. **Smith, C. J.,** Hydrogeology with respect to underground contamination, in *Environmental Fate of Pesticides. Progress in Pesticide Biochemistry and Toxicology*, Vol. 7, Hutson, D. H. and Roberts, T. R., Eds., John Wiley & Sons, Chichester, England, 1990, 47.

36. **Everett, L. G.,** *Groundwater Monitoring*, General Electric Co., Schenectady, NY, 1980.

37. **Cohen, S. Z.,** Pesticides in ground water: an overview, in *Environmental Fate of Pesticides. Progress in Pesticide Biochemistry and Toxicology*, Vol. 7, Hutson, D. H. and Roberts, T. R., Eds., John Wiley & Sons, Chichester, England, 1990, 13.

38. **Brady, N. C.,** *The Nature and Properties of Soils*, MacMillan, New York, 1990.

39. **Helling, C. S. and Dragun, J.,** Soil leaching tests for toxic organic chemicals, in *Test Protocols for Environmental Fate and Movement of Toxicants. Proceedings of a Symposium*, Association of Official Analytical Chemists, Arlington, VA, 1981, 43.

40. **Gerstl, Z. and Yaron, B.,** Behavior of bromacil and napropamide in soils. II. Distribution after application from a point source, *Soil Sci. Soc. Am. J.*, 47, 478, 1983.

41. **Wagenet, R. J.,** Process influencing pesticide loss with water under conservation tillage, in *Effects of Conservation Tillage on Groundwater Quality*, Logan, T. J., et al., Eds., Lewis Publishers, Chelsea, MI, 1987, 190.

42. **Bouchard, D. C., Lavy, T. L., and Marx, D. B.,** Fate of metribuzin, metolachlor, and fluometuron in soil, *Weed Sci.*, 30, 629, 1982.

43. **Yaron, B., Gerstl, Z., and Spencer, W. F.,** Behavior of herbicides in irrigated soils, *Adv. Soil Sci.*, 3, 121, 1985.

44. **Walker, A.,** A simulation model for prediction of herbicide persistence, *J. Environ. Qual.*, 3, 396, 1974.

45. **Rawls, W. J., Brakensiek, D. L., and Saxton, K. E.,** Estimation of soil water properties, *Trans. of the ASAE,* 25, 1316, 1982.

46. **Deubert, K. H.,** Environmental fate of common turf pesticides — factors leading to leaching, *USGA Green Section Record 28*, 4, 5, 1990.

47. **Green, R. E. and Khan, M. A.,** Pesticide movement in soil: mass flow and molecular diffusion, in *Fate of Pesticides in the Environment: Proceedings of a Technical Seminar*, Biggar, J. W. and Seiber, J. N., Eds., Division of Agriculture and Natural Resources, Oakland, CA, 1987, 87.

48. **Fukuto, T. R.,** Organophosphorous and carbamate esters: the anticholinesterase insecticides, in *Fate of Pesticides in the Environment: Proceedings of a Technical Seminar*, Biggar, J. W. and Seiber, J. N., Eds., Division of Agriculture and Natural Resources, Oakland, CA, 1987, 5.

49. **Larkin, M. J. and Day, M. J.,** The effect of pH on the selection of carbaryl-degrading bacteria from garden soil, *J. Appl. Bacteriol.*, 58, 175, 1985.

50. **Chapman, R. A. and Cole, C. M.,** Observations on the influence of water and soil pH on the persistence of insecticides, *J. Environ. Sci. Health*, B17, 487, 1982.

51. **Himel, C. M., Loats, H., and Bailey, G. W.,** Pesticide sources to the soil and principles of spray physics, in *Pesticides in the Soil Environment: Processes, Impacts, and Modeling*, SSSA Book Series No. 2, Cheng, H. H., Ed., SSSA, Madison, WI, 1990, 7.

52. **Bollag, J. M. and Liu, S. Y.,** Biological transformation processes of pesticides, in *Pesticides in the Soil Environment: Processes,* *Impacts, and Modeling*, SSSA Book Series No. 2, Cheng, H. H., Ed., SSSA, Madison, WI, 1990, 169.

53. **Alexander, M.,** Biodegradation of chemicals of environmental concern, *Science*, 211, 132, 1981.

54. **Edwards, C. A. and Lofty, J. R.,** *Biology of Earthworms*, Chapman and Hall, London, 1977.

55. **Rao, P. S. C., Green, R. E., Balasubramanian, V., and Kanehiro, Y.,** Field study of solute movement in a highly aggregated oxisol with intermittent flooding. II. Picloram, *J. Environ. Qual.*, 3, 197, 1974.

56. **Bouma, J., Belmans, C. F. M., and Dekker, L. W.,** Water infiltration and redistribution in a silt loam subsoil with vertical worm channels, *Soil Sci. Soc. Am. J.*, 46, 917, 1982.

57. **U.S. Environmental Protection Agency,** Protecting Ground Water: Pesticides and Agricultural Practices, EPA-440/6-88-001, U.S. Environmental Protection Agency, Office of Ground Water Protection, Washington, DC, 1988.

58. **Bilkert, J. N. and Rao, P. S. C.,** Sorption and leaching of three nonfumigant nematicides in soils, *J. Environ. Sci. Health*, B20, 1985, 1.

59. **Sears, M. K. and Chapman, R. A.,** Persistence and movement of four insecticides applied to turfgrass, in *Advances in Turfgrass Entomology*, Niemczyk, H. D., and Joyner, B. G., Eds., ChemLawn Corp., Columbus, OH, 1982, 57.

60. **Branham, B. E. and Wehner, D. J.,** The fate of diazinon applied to thatched turf, *Agron. J.*, 77, 101, 1985.

61. **Gold, A. J., Morton, T. G., Sullivan, W. M., and McClory, J.,** Leaching of 2,4-D and dicamba from home lawns, *Water Air Soil Pollut.*, 37, 121, 1988.

62. **Cohen, S. Z., Nickerson, S., Maxey, R., Dupuy, A., Jr., and Senita, J. A.,** A groundwater monitoring study for pesticides and nitrates associated with golf courses on Cape Code, *Ground Water Monit. Rev.*, 160, 1990.

63. **Borromeo, N. A.,** Leaching of Turfgrass Pesticides, M.S. thesis, Cornell University, Ithaca, NY, 1992.

64. **Leonard, R. A.,** Herbicide in surface water, in *Environmental Chemistry of Herbicides*, Vol. 1, Grower, R., Ed., CRC Press, Boca Raton, FL, 1988, chap. 3.

Chapter 4

Sewage Sludge Compost for Establishment and Maintenance of Turfgrass

J. Scott Angle, *Department of Agronomy, University of Maryland, College Park, MD*

CONTENTS

Introduction 45
Sludge Characteristics and Composting 45
Compost Characteristics 46
Compost Utilization 47
     Establishment 47
     Maintenance 47
Benefits of Compost Utilization 48
Limitations of Compost Utilization 49
Conclusions 49
References 50

INTRODUCTION

Use of sewage sludge compost and other processed sludge products for the establishment and maintenance of turfgrass is an environmentally sound and cost effective option for utilizing sewage sludge-derived products. As current methods of sludge disposal become either economically or environmentally unfeasible, novel disposal techniques must be adopted. Recent legislation has eliminated ocean disposal of sludge wastes as of 1992. Prior to this time, a significant portion of sludge produced in densely populated coastal areas was discharged from outfalls or dumped from vessels into the ocean as a simple and inexpensive method of disposal. Further, alternative sludge disposal methods, such as incineration and landfilling, are facing serious opposition due to concerns with air and groundwater pollution, respectively. Land application of sludge is thus one of the few viable alternatives for sewage sludge use or disposal.

Application to agricultural land where crops such as corn and soybeans are grown often requires that sludge be transported to distant rural areas, a process that greatly increases disposal costs. Application of sludge to turfgrass, however, is an attractive alternative since the areas where turfgrass production and use is concentrated are also where sewage sludge is generated in significant quantities. Transportation costs may be relatively low due to the close proximity to areas maintained to turfgrass.

Application to land producing turfgrass also eliminates many of the considerations related to sludge application to food and feed crops. Heavy metal uptake and accumulation in edible portions of crops can be a major limitation affecting sludge utilization on land and is a primary concern that limits sludge application rates. Since turfgrass is not consumed by humans or animals, heavy metal uptake is not a significant factor. Finally, turfgrass is capable of benefiting from the nutrients contained in sludge more so than many other types of crops. The rapid growth rate, extended growing season, and dense, fibrous root system of turf allows it to efficiently utilize the sludge-borne nutrients that might escape the root zone of other crops and eventually move into the groundwater.

SLUDGE CHARACTERISTICS AND COMPOSTING

From an agronomic perspective, there are few reasons to restrict the use of aerobically or anaerobically digested sludge on land used for producing turf. A number of aesthetic and health problems, however, are related to the application of digested sludges to turfgrass. The most significant problem with the use of such sludges is the odor associated

0-87371-350-8/94/$0.00+$.50

with these materials. Since sludge applied to turfgrass remains on or near the surface, conspicuous odors may be released. A health concern related to the use of digested sludges is their pathogen content. Digested sewage sludge may contain detectable levels of a variety of organisms such as pathogenic bacteria, viruses, and parasitic eggs.[14,40] Some of these pathogenic organisms can survive for long periods of time following application of digested sludge to turf, and it is therefore necessary to restrict human activities on land recently amended with digested sludge. Hence, while few agronomic reasons exist to restrict the application of digested sludge to turfgrass, aesthetic and health concerns prevent the use of these materials.

A viable method for rendering sludge safe for use on turf is to compost the sludge prior to application. The composting of sewage sludge has been practiced for many years, although modern and efficient composting methods have only recently been available.[12,13,18,32] A common method of sludge composting is called the Beltsville aerated static pile method.[11] This method mixes sludge with woodchips or other bulking agents in an approximately 1:2 (volume:volume) ratio. The mixture is formed into piles that may be expanded into windrows and air is forced through the piles for several weeks. During this time, microbial action degrades much of the organic matter in the sludge including many of the volatile organic compounds responsible for producing odors. Microbial activity also generates significant quantities of heat, bringing the temperature of the composting mixture to over 55°C. This temperature, if sustained for a period of 3 days, kills most of the pathogenic organisms associated with the sludge.[31,35] Another important characteristic of the composting process is a loss of moisture. Sewage sludge contains a considerable quantity of water and the heat produced during composting causes much of the water to evaporate. The final compost product is a relatively dry, friable material. This facilitates application and allows the material to be easily incorporated into soil or used as a topdressing for existing turfgrass.

COMPOST CHARACTERISTICS

Sludge compost, when properly produced and screened, resembles high quality topsoil. Physically, the material contains from 30 to 40% moisture and is granular. Particles are typically less than 1 cm in diameter. Sludge composting results in an "earthy" smell due to aromatic compounds released by a group of microorganisms called actinomycetes. Because the overall appearance of compost is similar

to topsoil, complaints from neighbors who are not familiar with compost are avoided.

When sewage sludge and compost are applied to agricultural land, one of the most significant concerns is the heavy metal content of the sludge. Some metal contaminants in sludge can potentially cause health problems to humans (e.g., cadmium) while heavy metals such as zinc, copper, and nickel have the potential to be phytotoxic.[7,42] With high metal additions to soil, crop failure can occur; whereas with lower metal additions to soil, subclinical toxicity can result in decreased growth.[22] The majority of sewage sludges contain relatively low metal concentrations when compared to allowable loading rates. However, unless sound pretreatment programs are in place, sludge may occasionally be excessively contaminated with heavy metals.

Since turfgrass is not a part of the domestic animal or human food chain, considerations related to human and animal toxicity are not a major issue. Only with extremely high sludge additions to soil (where the heavy metal content of sludge is high and soil pH low) would phytotoxicity to turf be a concern.[38] Most often, however, chelation, precipitation, and complex formation of heavy metals renders most of the sludge-borne metals inactive, thus protecting plants from excessive uptake of potentially toxic metals.[37] Further, under normal circumstances, application rates resulting in phytotoxicity would be prohibited by current U.S. Environmental Protection Agency (EPA) and state regulations.

Metal concentrations in compost are generally lower than in the original sludge because many of the sludge-borne metals are diluted as well as bound to the woodchips or other bulking agents added to the sludge. In addition, the metals remaining in compost are generally less available than metals in untreated sludge. The composting process binds metals so as to reduce their effective availability in soil.[36]

Many of the inorganic components of compost combine to add to its soluble salt content. During sewage and sludge treatment processes, inorganic salts and organic polymers are often added to the reaction mixture to enhance flocculation and precipitation of solids, some of which are subsequently concentrated in the sludge. Composting of sludge does not reduce the salt content, and may increase the salt concentration of compost as the organic fraction of the sludge is reduced by microbial respiration.

One of the most significant components of compost which requires careful management is its nutrient content. Assuming that the heavy metal content of compost is not excessive, the nitrogen content and availability controls the rate of compost applied to turfgrass. Compost contains approximately 2.0% nitrogen on a dry weight basis. The other macronutrients, phosphorus and potassium, are found in compost at concentrations of approximately 2% and less than 1%, respectively. As with the heavy metal content of sludge, the nutrient content may be quite variable. When sludge compost is applied to turfgrass based upon its nitrogen content and the nitrogen requirement of the crop, phosphorus will usually be added at rates sufficient to meet the requirements of the crop, however, potassium will normally be deficient. Supplemental applications of potassium fertilizer are usually required when compost is used as a fertilizer for turfgrass production. In addition to the macronutrients supplied by compost, this material also contains a variety of micronutrients. Copper, iron, manganese, and zinc are required by turfgrass for growth and may be found in compost in significant quantities. Hence, in soils that are deficient in micronutrients, such as very low or high pH soils, very sandy soils, or mine spoils, sludge compost can be a valuable source of micronutrients.

COMPOST UTILIZATION

Sludge compost can be used on turfgrass either during seed or sod establishment or to maintain previously established turfgrass. Methods and rates of application vary depending upon the use of the compost as well as the desired quality of the turf. As with other sources of fertilizer and soil amendments, the compost must be used in an environmentally safe manner that will not impact the groundwater or the long-term productivity of the soil.

ESTABLISHMENT

The addition and incorporation of compost into soil may significantly enhance the germination and subsequent establishment of turfgrass. This is especially true on poor, low-nutrient content soils, that can benefit greatly by the addition of compost.

Compost should be applied to soil so as not to exceed the metal loadings established by individual states. In those states where individual metal loadings are not available, USEPA guidelines should be followed. A soil test is necessary to establish the base fertility of the soil prior to sludge application. Compost should be evenly spread over the soil surface and subsequently worked into the soil to a depth of 15 to 30 cm. The soil pH should be adjusted to at least 6.5 with lime. Additional inorganic fertilizers may also be required if the compost and soil

mixture are deficient in specific nutrients. Seed, sod, sprigs, and root cuttings can all be established on compost-amended soil. Results by Angle et al.[4] have shown that seed and sod of Kentucky bluegrass, tall fescue, and perennial ryegrass respond favorably to the incorporation of moderate to high rates of sludge compost into soil. Burns and Boswell[6] have further shown that vegetative materials from warm season grasses such as bermudagrass, centipedegrass, and zoysiagrass can be successfully propagated on sludge-amended soil.

Application rates vary by state recommendations and the type of material being established. The rate of compost addition to turfgrass is usually based upon the nitrogen requirement of the crop. Heavy metals are generally not an important issue since cumulative metal loadings are often quite low. Sludge compost is assumed to have a nitrogen mineralization rate of 10% during the first year after application.[1,10] Previous studies where application rates were not based upon mineralization rates failed to effectively compare sludge compost with other more readily available sources of nitrogen.[26]

For the determination of the appropriate application rate, attention must be given to the nitrogen mineralization rate and the water content of the compost. For example, if 150 kg/ha of plant available nitrogen is to be applied during establishment, compost should be added to supply an equivalent amount of inorganic nitrogen. If the compost contains 2% total nitrogen on a dry-weight basis and the nitrogen mineralization rate is 10%, 7.5 Mg/ha compost (dry weight) must be applied to meet the desired nitrogen requirement. Additionally, if the compost contains 40% moisture by weight, the application rate must be adjusted for the moisture content. For the conditions noted above, 12.5 Mg/ha compost (wet weight) should be applied to the turfgrass to supply 150 kg/ha available nitrogen.

A further point to consider is the mineralization rate of the residual nitrogen remaining in the soil following the initial application. Organic nitrogen will continue to mineralize for several years after the initial application; thus, subsequent nitrogen fertilizer applications should include an effort to account for this additional plant available nitrogen.

MAINTENANCE

Compost may also be applied to the surface of established turfgrass in place of inorganic fertilizers. For maintenance, application is more difficult since lower rates of compost are added to turfgrass and the compost is often too coarse to be spread with ordinary drop-type spreaders. Effective application can be accomplished with the use of specially adapted rotary spreaders or manure spreaders set on the highest speed with the rear door barely open. Another method is to fill a screened roller with compost and roll it over the surface.

The compost should be added to turfgrass according to the nitrogen application schedule recommendations for each state. Frequently, recommendations require that fertilizer be added on a monthly or semi-monthly basis during the time when the turfgrass is actively growing. Hence, compost should be added to the turfgrass to meet the nitrogen requirement of the turf at that particular time. As before, the mineralization rate for nitrogen must be factored into the calculation of the application rate.

BENEFITS OF COMPOST UTILIZATION

Numerous agronomic advantages are observed when turfgrass is amended with compost, the most significant of which is the addition of macro- and micronutrients to soil. As has been previously discussed, compost is a good source of nitrogen, phosphorus, and many micronutrients. Utilization of compost on turfgrass can, therefore, replace most of the inorganic fertilizer used.

Because bound nutrients within compost are released slowly through the process of mineralization, the potential for ground and surface water pollution is minimized. Unlike inorganic fertilizers where all nutrients are available immediately after application, nutrients are made available only as microbial activity decomposes the compost and releases the nutrients. Hence, large quantities of nutrients are never available at any one time to potentially leach into groundwater.[24] From an environmental perspective, the use of compost as a slow-release fertilizer is considered a best management practice.

The influence of compost on soil physical properties is another benefit that significantly enhances turfgrass quality and growth. Soils with an initially poor physical structure, such as mine soils or soils disturbed due to construction, are markedly improved by the addition of compost. Compost stabilizes the soil through the aggregation of soil particles, thereby reducing soil bulk density and increasing infiltration and percolation of water into soil.[12,20] The effect of compost on soil bulk density is especially important in high traffic areas that are subject to compaction abuse. Duggan and Wiles[9] have shown that the addition of 450 Mg/ha compost to soil reduced soil bulk density from 1.43 to 0.99 g/cm^3. A corollary to this observation is a corresponding improvement in the water-holding capacity of compost-amended soil. Compost has a water-holding capacity approximately five times that of most mineral soils, thus soil water-holding capacity will be increased in direct proportion to the amount of compost added to soil.[21,23,25] However, since water is tightly bound to the organic fraction of compost, available soil moisture is not increased to a corresponding degree.

Compost has a cation exchange capacity approximately ten times higher than most loam soils.[21,25] When compost is added to soil, the soil exchange capacity will be increased.[33,34] This allows the soil to retain greater quantities of nutrients which can be used throughout the growing season. Further, the binding of nutrients helps to protect groundwater from the downward leaching of nutrients. Most composts also have a pH that is near neutral to one or two units higher than neutral. Soils amended with compost thus exhibit a pH close to that of the compost, an attribute especially beneficial in the acidic soils of the Northeast. A compost-induced increase in soil pH reduces the availability of heavy metals, further reducing their potential impact on the ecosystem.[9]

An additional environmentally related benefit of compost utilization on turfgrass is that compost-amended soils break down pesticides and other toxic synthetic compounds at a faster rate than unamended soils. Soils amended with compost exhibit higher microbial populations and are metabolically more active than soils that have not been amended with compost. Thus, compounds such as DDT are decomposed more rapidly in compost-amended soil.[2] Recently, the addition of compost and sludge to contaminated soil has been practiced on a wide scale as a means to "bioremediate" polluted soils. This practice currently appears to be a cost-effective means to clean many polluted soils of the toxic organic contaminants.

Throughout the years, sporadic reports have indicated that the application of compost to turfgrass significantly reduces the incidence of specific turf diseases. While not all of these observations have been confirmed, it does appear that compost additions to turfgrass may occasionally reduce the incidence or severity of several types of important diseases.

The most important compost effect on pathogens has been shown with nematodes. In 1958, Nutter and Christie[30] reported that sludge application to turf significantly reduced soil populations of lance, ring, and sting nematodes. These observations were subsequently confirmed by other research, which indicated that the compost increased soil populations of nematode predators.[19,41] In particular, populations of actinomycetes that attack nematodes and utilize the chitin in the cell wall are greatly enhanced in compost-amended soil.

Additional reports have shown that organisms that cause dollar spot (*Sclerotinia homoecarpa*), *Pythium* sp., and *Fusarium* sp. are inhibited by the application of sludge to soil.[29,41] Soil populations of the pathogens were not significantly decreased by the application of sludge. It was suggested, however, that enhanced foliar metal uptake caused the plants to exhibit fungitoxic activity.

Although sludge and compost effects on bacterial pathogens have not been specifically examined, many studies have shown that sewage sludge significantly reduces numbers of beneficial bacteria in soil. Soil populations of rhizobia were reduced to undetectable levels by the application of high rates of sludge to soil.[3,28] It is therefore possible that sludge may also reduce populations of bacterial pathogens in soil.

Gowans and Johnson[15] examined the overall aesthetic quality of turf fertilized with compost and ammonium nitrate. Turfgrass fertilized with compost was consistently superior to ammonium nitrate when the turf was subjectively rated for quality. Angle et al.[4] amended a silt loam soil with up to 720 Mg/ha compost. Amended soils were sodded with either tall fescue or Kentucky bluegrass. Kentucky bluegrass responded to the compost to a greater extent than the tall fescue. The overall quality of both species, however, was significantly enhanced by the incorporation of compost into soil. This one-time application continued to enhance turf quality for 510 days after sodding. In addition, these same authors demonstrated that the turfgrass growth rate was slightly increased throughout this time.

The application of sludge and/or compost to turfgrass was shown to be particularly effective in maintaining the "greenness" of turfgrass during droughty, summer months.[15,29,39] Bryan and Lance[5] have recently shown that the growth rate and clipping yield of bahiagrass was significantly increased by moderate applications of municipal waste and sewage sludge. An additional advantage associated with any slow-release fertilizer is the lack of a surge in growth immediately following application, thus reducing costs associated with mowing. Most inorganic fertilizers stimulate growth for a brief time after application as a result of the high available concentration of nutrients.

LIMITATIONS OF COMPOST UTILIZATION

Most concerns with compost addition to turfgrass are the same questions related to the use of any nutrient source. The most significant concern is that the available nutrient content of the compost applied to turfgrass does not exceed the nutrient requirements of the turf. If excess nutrients are present, they may leach out of the root zone and into groundwater. As a result, compost additions should be based upon the nitrogen content of the compost as well as the rate of nitrogen mineralization.

Nutrient losses from turfgrass are minimal when the nitrogen application rate closely corresponds to the nitrogen requirement of the crop. Gross et al.[16,17] has shown that both leaching and runoff of nitrogen from turfgrass are extremely low when fertilized at appropriate agronomic rates. Leaching of heavy metals is not generally a problem from sludge-amended land.[8] Only when extremely high rates of sludge are applied to soil over a period of many years, and the soil pH is allowed to fall below acceptable agronomic levels, will heavy metals potentially leach through the soil profile and into groundwater.[27]

Heavy metal phytotoxicity should generally not be a problem unless the compost is very high in heavy metals, the soil pH is very low, and a metal sensitive species of turfgrass is being grown on the compost-amended site. Some species of turfgrass, such as red fescue, are actually heavy metal-tolerant and may be grown on metal-polluted soils. Red fescue is commonly grown on metal-polluted soils, where little else will grow, as a means to stabilize these soils.

Another significant concern that must be addressed is the soluble salt content of compost. Some turfgrass species, such as Kentucky bluegrass, are very sensitive to the toxic effects of soluble salts. As previously noted, salts are often added during the sewage treatment process to precipitate and flocculate organic particles. Thus, many sludges and composts have a relatively high soluble salt content. If high rates of compost are incorporated into soil and immediately seeded with a sensitive bluegrass species, seedlings may die as a result of the osmotic pressure. To prevent the potential occurrence of salt toxicity, it is recommended that the compost-amended soil be seeded only after several rainfall or irrigation events have flushed some of the salts from the seedling root zone. This is especially critical for poorly drained soils that exhibit restricted solute leaching.

CONCLUSIONS

Considering the many beneficial properties of composted sludge, the application of compost to turfgrass is a logical use of this material. Compost satisfactorily replaces inorganic fertilizers on turfgrass and may actually provide many cultural and physical benefits beyond those observed when inorganic materials are applied as the source of nutrients. The application of compost to establish or maintain turfgrass, if properly managed and with an understanding of its limitations, should be encouraged whenever possible.

REFERENCES

1. **Adkins, R., Angle, J. S., Aycock, M. K., Bandel, A., Brodie, H., Decker, M., Gouin, F., James, B. R., Taylor, A., and Weismiller, R.,** Guidelines for Application of Digested Sewage Sludge and Composted Sewage Sludge to Agricultural Land, Fact Sheet No. 336, University of Maryland Cooperative Extension Service, 1989.
2. **Albone, E. S.,** Formation of bis(p-chlorophenyl acetone) (p,p'DDCN) from p,p'DDT in anaerobic sewage sludge, *Nature*, 240, 420, 1972.
3. **Angle, J. S., McGrath, S. P., Chaudri, A. M., Chaney, R. L., and Giller, K. E.,** Inoculation effects on legumes grown on soil previously treated with sewage sludge, *Soil Biol. Biochem.*, 25, 575, 1993.
4. **Angle, J. S., Wolf, D. C., and Hall, J. R.,** Turfgrass growth aided by sludge compost, *Biocycle*, 22, 40, 1981.
5. **Bryan, H. H. and Lance, C. J.,** Compost trials on vegetables and tropical crops, *Biocycle*, 32, 36, 1991.
6. **Burns, R. E. and Boswell, F. C.,** Effect of municipal sewage sludge on rooting of grass cuttings, *Agron. J.*, 68, 382, 1976.
7. **Chaney, R. L., Hundemann, P. T., Palmer, W. T., Small, R. J., and Decker, A. M.,** Plant accumulation of heavy metals and phytotoxicity from utilization of sewage sludge composts on cropland, in Proceedings of the 1977 National Conference on the Composting of Municipal Residue and Sludges, Information Transfer, Rockville, MD, 1977, 86.
8. **Dowdy, R. H. and Volk, V. V.,** Movement of heavy metals in soils, in *Chemical Mobility and Reactivity in Soil Systems*, Nelson, D. W., et al., Eds., Soil Science Society of America, Madison, WI, 1983, 229.
9. **Duggan, J. C. and Wiles, C. C.,** Effect of municipal compost and nitrogen fertilizer on selected soils and plants, *Compost Sci.*, 17, 24, 1976.
10. **Epstein, L., Keane, D. B., Meisinger, J. J., and Legg, J. O.,** Mineralization of nitrogen from sewage sludge and sludge compost, *J. Environ. Qual.*, 7, 217, 1978.
11. **Epstein, E. G., Willson, G. B., Burge, W. D., Mullen, D. C., and Enkiri, N. K.,** A forced aeration system for composting wastewater sludge, *J. Water Pollut. Control Fed.*, 38, 688, 1976.
12. **Evans, J. O.,** Using sewage sludge on farmland, *Compost Sci.*, 9, 16, 1968.
13. **Eweson, E.,** Making compost work, *Pollut. Eng.*, 6, 38, 1973.
14. **Goluke, C. G. and Gotaas, H. B.,** Public health aspects of waste disposal by composting, *Am. J. Public Health*, 4, 339, 1954.
15. **Gowans, K. D. and Johnson, E. J.,** Nitrogen source in relation to turfgrass establishment, *Ca. Turfgrass Cul.*, 23, 13, 1973.
16. **Gross, C. M., Angle, J. S., Hill, R. L., and Welterlen, M. S.,** Runoff and sediment losses from tall fescue under simulated rainfall, *J. Environ. Qual.*, 20, 604, 1991.
17. **Gross, C. M., Angle, J. S., and Welterlen, M. S.,** Nutrient and sediment losses from turfgrass, *J. Environ. Qual.*, 19, 663, 1990.
18. **Harrison, E. Z. and Richard, T. L.,** Municipal solid waste composting: policy and regulation, *Biomass Bioenergy*, 3, 127, 1992.
19. **Heald, C. M. and Burton, G. W.,** Effect of organic and inorganic nitrogen on nematode populations on turf, *Nematologia,* 14, 8, 1968.
20. **Hinesly, T. D., Braids, O. C., and Molina, J. E.,** Agricultural benefits and environmental changes resulting from the use of digested sludge on field crops, in *An Interim Report on a Solid Waste Demonstration Project*, U.S. Environmental Protection Agency, Washington, DC, 1979, 9.
21. **Hortenstine, C. C. and Rothwell, D. F.,** Evaluation of composted municipal refuse as a plant nutrient source and soil amendment on a Leon fine sand, in Proceedings of the Soil and Crop Science Society of Florida, 1969, 29, 312.
22. **John, M. K. and VanLaerhoven, C. J.,** Effects of sewage sludge composition, application rate and lime regime on plant availability of heavy metals, *J. Environ. Qual.*, 5, 246, 1976.
23. **King, L. D. and Morris, H. D.,** Land disposal of liquid sewage sludge. IV, *J. Environ. Qual.*, 2, 411, 1973.
24. **King, L. D. and Morris, H. D.,** Nitrogen movement resulting from surface application of liquid sewage sludge, *J. Environ. Qual.*, 3, 238, 1974.

25. **Kirkham, M. B.,** Disposal of sewage sludge on land: effects on soil, plants and groundwater, *Compost Sci.* 15, 6, 1974.

26. **Landschoot, P. J. and Waddington, D. V.,** Response of turfgrass to various nitrogen sources, *Soil Sci. Soc. Am. J.,* 51, 225, 1987.

27. **McGrath, S. P.,** Metal concentrations in sludges and soil from a long-term field trail, *J. Agric. Sci. Cambridge,* 103, 25, 1984.

28. **McGrath, S. P.,** Effect of heavy metals in sewage sludge on soil microbes in agricultural ecosystems, in *Toxic Metals in Soil-Plant Ecosystems,* Ross, S. M., Ed., John Wiley & Sons, Chichester, England, 1991.

29. **McWhiriter, E. L. and Ward, C. Y.,** Turfgrass mowing and fertility tests: perennial ryegrass overseeding on Tifgreen Bermudagrass, *Mississippi Farm Res.,* 34, 7, 1971.

30. **Nutter, G. C. and Christie, J. R.,** Nematode investigations on putting green turf, in Proceedings of the Florida State Horticultural Society, 1958, 71, 445.

31. **Osterle, P. G., Rhode, and Rudat, K. D.,** Berlin studies show efficiency of composting, *Compost Sci.,* 4, 19, 1973.

32. **Poincelot, R. P.,** A scientific examination of the principles and practice of composting, *Compost Sci.,* 15, 24, 1974.

33. **Premi, P. R. and Cornfield, A. H.,** Incubation study of nitrification of digested sewage sludge added to soil, *Soil Biol. Biochem.,* 1, 1, 1969.

34. **Schatz, S. M.,** Humus research — east and west, *Compost Sci.,* 7, 19, 1966.

35. **Sikora, L. J. and Sowers, M. A.,** Effect of temperature control on the composting process, *J. Environ. Qual.,* 14, 434, 1985.

36. **Simeoni, L. A., Barbarick, K. A., and Sabey, B. R.,** Effects of small-scale composting of sewage sludge on heavy metal availability to plants, *J. Environ. Qual.,* 13, 264, 1984.

37. **Sommers, L. E.,** Chemical composition of sewage sludges and analysis of their potential use as fertilizers, *J. Environ. Qual.,* 6, 225, 1977.

38. **Stucky, D. J. and Newman, T. S.,** Effect of dried anaerobically digested sewage sludge on yield and elemental accumulation in tall fescue and alfalfa, *J. Environ. Qual.,* 6, 271, 1977.

39. **Waddington, D. V., Moberg, E. L., and Duich, J. M.,** Effect of nitrogen source, potassium source and potassium rate on soil nutrient levels and the growth and elemental composition of 'Penncross' Creeping Bentgrass, *Agron. J.,* 64, 562, 1972.

40. **Wiley, J. S.,** Pathogen survival in composting municipal wastes, *Sewage Indust. Wastes,* 34, 80, 1962.

41. **Wilson, C. G.,** Milorganite and Dollar Spot — Those Damnable Eelworms, mimeo, The Sewage Commission, Wisconsin, 1974.

42. **Woodbury, P. B.,** Trace elements in municipal solid waste compost: a review of potential detrimental effects in plants, soil biota and water quality, *Biomass Bioenergy,* 3, 239, 1992.

Chapter 5

Integrated Resource Management

Ronald G. Dodson, *Audubon Society of New York State, Selkirk, NY*
Nancy P. Sadlon, *United States Golf Association, Far Hills, NJ*

CONTENTS

Nature's Balance 53
The Insect Eaters 53
Integrated Resource Management 54
Balancing Nature with Sustainable Resources 54
Essential Elements: Food, Water, Cover, Space 55
Putting the Essential Elements Together — Habitat Creation and Enhancement 55
Getting Started in Integrated Resource Management 56
Programs for Conservation 56
Land Development Guidelines 57
References 57

NATURE'S BALANCE

One important way of controlling what many people consider pests is to allow nature to do it. The natural biological process that we call "life" will do wonders if allowed to function properly. Nature's solution is a teeter-totter balance between harmful species and beneficial species. Nature's components such as plants with natural repellant actions, insect-eating birds, toads, frogs, and other wildlife can represent important players in maintenance and pest control.

It is fairly common knowledge that many species of wildlife depend totally on insects for food. Even those that do not depend entirely on insects consume substantial quantities. The trick is to determine which species of wildlife consume which species of insect and entice those insect eaters to utilize your property and help in pest control.

Encourage nature's pest control components to work in your habitat.

THE INSECT EATERS

Three of the most widely recognized insect eaters include bluebirds, swallows (especially purple martins), and bats. All three of these species are cavity nesting species. Artificial nest boxes are often successful in enticing these insect eaters to a property. Bluebirds are admired for their striking blue color, purple martins for their purple color and colonizing habits, and bats unfortunately are shunned for unsubstantiated public fears of rabies. In truth, all three species, including bats, are not harmful and can be beneficial insect eaters. Table 1 shows interesting facts about these three insect eaters.

Many other species of wildlife consume only insects or consume mostly insects as part of their

Table 1 **Dietary data on three insect-eating species**

Species	Preferred Food	Quantities	% Diet of Insects
Bluebirds	Beetles, crickets, cutworms, grasshoppers, supplemental food of berries	Eat 7 to 8 times their weight/night during winter feeding months	60–90%
Purple martins	Flying insects: flies, moths, dragonflies, locusts, beetles, butterflies, weevils, and mosquitoes	Eat their weight/night	40–100%
Bats	Flying insects of mosquito size: flies, moths, flying ants, caddisflies, ground beetles, and bugs	Eat 1,000 to 3,000 insects per night	100%

0-87371-350-8/94/$0.00+$.50
© 1994 by CRC Press Inc.

diet. Below is a "short list" of wildlife that consume insects:

Plovers	Bluebirds
Killdeer	Kinglets
Roadrunner	Vireo
Whippoorwill	Warblers
Chuck-will's widow	Meadow Lark
Poor-will	Robin
Nighthawk	Oriole
Swifts	Tanagers
Hummingbirds	Cardinal
Woodpeckers	Grosbeaks
Flycatchers	Finches
Kingbird	Sparrows
Phoebe	
Swallows	Skunks
Chicadee	Grey Squirrel
Tufted titmouse	Bats
Nuthatches	White-footed mouse
Mockingbird	
Catbird	Fish
Brown Thrush	Snakes (many species)
Wood Thrush	Amphibians

Little work has been done to quantify the exact numbers of specific insects eaten by all of the above-listed species. However, we do know wildlife consume considerable quantities of insects and should not be forgotten when developing a pest control program. Many of the selected readings referenced at the end of this chapter provide specific information about foods preferred by wildlife.

INTEGRATED RESOURCE MANAGEMENT

Plants and animals are subject to certain laws of nature and each species has certain requirements and tolerance ranges for the essential elements of food, water, cover, and space. Many people think only of feathered and furred creatures when the term "wildlife" is mentioned. In fact, all living creatures not tamed or domesticated should be considered wildlife. This includes reptiles, amphibians, crustaceans, spiders, fish, birds, mammals, and, yes, insects. In focusing on one animal, we will learn about many because individual plants and animals are active, growing, multiplying organisms that interact with their surroundings. This ecological interactive process represents the foundation of Integrated Resource Management. Unfortunately, in too many cases this interactive ecology is overlooked and an all-out assault on the

"pest" in question results in destruction of other wildlife as well. In most cases, this approach will prove to be about as feasible as stopping an earthquake. All-out war against one species often results in temporary control for the target organism, yet it ends in permanent control for unintended species that have beneficial properties. What we must strive toward is a balance — a balance of nature, a balance of social activities, and a balance of management activities.

By focusing on the improvement of habitat to attract a number of insect-eating birds and mammals, we can integrate their pest control actions. Improving habitat for one species generally helps others as well as results in greater diversity and a more stable, balanced community of plants and animals.

BALANCING NATURE WITH SUSTAINABLE RESOURCES

In the past, our interest in wildlife focused on the forms that we pursued for food and clothing. Today we recognize that all wildlife have value. We know that wildlife can be found everywhere. We also know that human activities have tended to alter the basic biological process in many areas. In many cases this has affected population levels of some wildlife species. Some species have experienced significant population explosions.

Conflict normally occurs when a human social or economic activity is impacted by wildlife. For example, increasing gull populations in areas where people picnic or moor their expensive boats have caused conflicts. Large concentrations of mosquitoes or black flies, where people tend to recreate outside, have resulted in social and political conflict.

There is no simple answer to any of these conflicts. Understanding and managing the natural resources in an appropriate manner will play an important role in this matter. Aesop's fable of the goose and the golden egg illustrates by contrast the concept of good resource management.

This fable is the story of a poor farmer who one day discovered in the nest of his pet goose a glittering golden egg. At first, he thinks it must be some kind of trick. But as he starts to throw the egg aside, he has second thoughts and takes it in to be appraised instead.

The egg is pure gold! The farmer can't believe his good fortune. He becomes even more incredulous the following day when the experience is repeated. Day after day, he awakens to rush to the nest and find another golden egg. He becomes fabulously wealthy; it all seems too good to be true.

But with his increasing wealth comes greed and impatience. Unable to wait day after day for the golden eggs, the farmer decides he will kill the goose and get them all at once. But when he opens the goose, he finds it empty. There are no golden eggs — and now there is no way to get any more. The farmer has destroyed the goose that produced them.[6]

In managing natural resources, we must not be so greedy as to destroy them; we must focus on sustainable use of our resources. Sustainable use is, for example, harvesting only one tree a year from a forest that is producing one tree a year. It is the act of assuring that future generations have the opportunity to live at least as well as we do.

ESSENTIAL ELEMENTS

All wildlife species require four basic things to survive — food to eat, water to drink, cover or shelter to protect them, and the necessary space to carry on the basic requirements of life.

FOOD

When planning to attract wildlife, providing food is often the first activity that comes to mind. You can manipulate food resources by simply setting up a bird feeder, or adopting a landscape plan that includes plants of high food value to a variety of wildlife species (i.e., fruit, nut, or seed-bearing vegetation).

WATER

The availability of water is often the most important factor in attracting wildlife. Birds not only use water for drinking but need a water source to keep their feathers clean in order to contain body heat. Managing your water resources should be a full-year commitment to enhance the value of your property for wildlife.

COVER

Cover is a general term applied to the aspect of an animal's habitat that provides protection for the animal to carry out life functions such as breeding, nesting, sleeping, resting, feeding, and travel. Anticipating the need for cover is related to planning food sources because animals often will not come to food if there is not a protected place for them to eat it. Hedgerows and taller grasses will be used as safe travel corridors by wildlife seeking food or water. Strive to landscape with a variety of flowers, grasses, shrubs, and trees to accommodate a diversity of wildlife, from ground-dwelling species to those who prefer living in the treetops.

Dead trees, or snags, provide important shelter and nesting sites for many insect-eating mammals and birds. When snags pose no safety hazard, consider leaving them in place. If there are few snags on site, mounting nest boxes is an excellent way to increase the breeding success of cavity-nesting species of birds. Forest understory also provides cover for safe travel and nesting. Brush piles can be added to a woodlot understory to enhance cover for small animals.

SPACE

Space is the foundation of the balance of nature. An animal will not tolerate an overabundance of its own kind within its space. This area may be a few square feet for a mouse or a few thousand acres for a grizzly bear. For example, the space requirements for various woodpeckers range from 10 to 175 acres (Table 2).

Within the spatial restrictions for a species, all other basic requirements (food, water, shelter) must be met or the species will not exist in that area. Each species has different requirements. Additionally, at any given time, there is a fixed limit for the kind and number of animals that may live in a habitat. This is called the "carrying capacity."

Within the area of space, physical makeup and location will determine the numbers and types of wildlife to be present. This physical makeup is referred to as habitat. A complete habitat is an area which fills the four basic needs of any particular species. Some habitats are obvious. A running stream is habitat for fish and a woodlot is habitat for songbirds. But habitat requirements can be very specific. For example, the stream "fish habitat" can be further divided into "bass habitat" and "trout habitat." Habitat is the single most important influence on wildlife.

PUTTING THE ESSENTIAL ELEMENTS TOGETHER — HABITAT CREATION AND ENHANCEMENT

To encourage species of wildlife that consume insects to utilize an area, one must remember the four basic requirements, and one must foster the concept of interspersion to maximize use. To provide a simple example: even though there may be adequate water, living space, and shelter in an area, you wouldn't expect to find bluebirds if there was no food available. All elements must be interspersed!

Table 2 **Space requirements for various woodpeckers**

	When Using Territory	Territory Size (Acres)	Maximum Pairs/ 100 Acres
Downy woodpecker	All year	10	10
Pileated woodpecker	All year	175	0.6
Red-bellied woodpecker	All year	15	6.7
Red-headed woodpecker	Breeding	15	6.7
Black-backed three-toed woodpecker	All year	75	1.3
Yellow-bellied sapsucker	Breeding	10	10

Adapted from Kress, S. W., The Audubon Society Guide to Attracting Birds, Charles Scribner's Sons, New York, 1985.

Many species of wildlife have adopted special ways of meeting these four basic needs. In order to avoid a winter scarcity of food, some animals hibernate, others migrate, and still others store food. Some animals are not dependent upon ponds and streams for water because they get enough moisture from their food. To intersperse is to mix so that things that are alike are not all in the same place. Food, water, and shelter should be interspersed throughout the space so that an animal can get all the things it needs without traveling too far.

This is sometimes referred to as "edge effect" — where one vegetation type meets another type. The border between a marsh and a meadow, the border between a woods and a field, or the area between a golf fairway and a pond provide edge effect (Figure 1). This "edge" supports more wildlife than pure woods, pure meadow, or pure marsh. Certain species require "pure" habitat, but the greatest diversity is at the edge.

GETTING STARTED IN INTEGRATED RESOURCE MANAGEMENT

1. Habitat Inventory
The first step in utilizing an Integrated Resource Management technique is to determine what you presently have on your property as far as habitat is concerned.

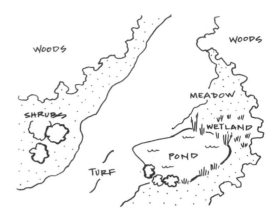

Figure 1 "Edge effect" in a landscape, which also illustrates interspersion of food, water, and shelter.

2. Animal Inventory
Next, a survey of animal types presently utilizing the property is important.

3. Interspersion Analysis
Last, an assessment of present "interspersion" is essential. More simply put, a review of where habitats are located and what elements are missing that are important to the desired species. Interspersion analysis can also help identify what is needed to discourage unwanted wildlife. Many people who have experienced an overabundance of "problem" wildlife, such as Canada geese, have found that their present habitat management techniques have resulted in optimum habitat being right where they don't want it (i.e., a golf green or tee area). Golf course ponds and municipal lakes, which are lacking in any border vegetation, are the perfect habitat for Canada geese. Adding plant material (shrubs) or other barrier devices such as a wire or fence, makes these water borders less desirable to Canada geese, which makes the pond more attractive to other species, such as the red-winged blackbird or kingbird. The USDA Soil Conservation Service has long recommended the planting of farm pond edges to help control goose populations. Understanding habitat preferences and needs not only helps one to increase beneficial wildlife, but aids in the understanding and control of problem species as well.

PROGRAMS FOR CONSERVATION

The New York Audubon Society has initiated an international conservation program called the Audubon Cooperative Sanctuary System. This system comprises individually owned properties such as homes, farms, and apartments, corporate and business lands, schools, and — with the sponsorship of the United States Golf Association — golf courses. The program is aimed at informing landowners about wildlife and environmental projects they can undertake on their own properties; encouraging landowners to become more involved in conservation actions, and knowledgeable about Integrated Resource Management. Landowners register their property in the Audubon Cooperative Sanctuary Program and Audubon provides continuous information and recognition, through educational fact sheets and news articles. The Audubon Cooperative Sanctuary Program for Golf Courses certification process provides a system for participating landowners to become certified in many different categories. They include the following:

Environmental Planning: Must show efforts to increase habitat types on the course, conserve water, and increase efforts at integrated pest management.

Public Involvement: Requires that at least one person outside the golf club serve on a Resource Committee to facilitate better communications with the general public about the conservation projects on the course.

Integrated Pest Management: Must show efforts to use turf management "scouting" and other IPM principles and application of minimum amounts of pesticides, as well as use of insect-eating birds, bats, or other biological controls as part of the management strategy.

Wildlife Cover Enhancement: Includes the use of native or naturalized vegetation, understory enhancement in woodlot areas, nesting boxes, wood piles, meadow or grassland establishment, or other efforts.

Wildlife Food Enhancement: Includes the use of bird feeders, and native or naturalized plants (grasses, wildflowers, shrubs) that provide food sources for wildlife species such as song birds.

Water Conservation: Includes use of conservation technology in the irrigation system, rainwater storage, recycling, and so forth, as well as use of drought-tolerant grass species, where applicable.

Water Enhancement: Includes efforts to enhance water quality and habitat for various wildlife species such as fish, amphibians, and birds, including efforts to document present water quality.

Several thousand acres of property are presently registered in the Audubon Cooperative Sanctuary Program in all 50 states and 18 additional countries.

Other assistance and advice on habitat enhancement and conservation projects is sometimes available through the U.S. Fish and Wildlife Service, a state Natural Resources department, or local environmental organizations.

LAND DEVELOPMENT GUIDELINES

As a guidance document for its programs, the New York Audubon Society has prepared a document entitled *Audubon Sustainable Resource Management Principles*. This document covers: site assessments, habitat sensitivity, native and naturalized plants, water conservation, waste management, renewable energy sources, transportation, recreation, chemical use, agriculture, building design, community design, durability, and community education.

REFERENCES

1. **Cox, J.,** *Landscaping With Nature — Using Nature's Designs to Plan Your Yard*, Rodale Press, Emmaus, PA, 1991.
2. **Emery, M.,** *Promoting Nature in Cities and Towns — A Practical Guide*, Croom Helm, Dover, NH, 1986.
3. **Martin, A., Zim, C., Herbert, S., and Nelson, A. L.,** *American Wildlife & Plants — A Guide to Wildlife Food Habits*, Dover Publications, New York, 1961.
4. **Schneck, M.,** *Your Backyard Wildlife Garden — How to Attract and Identify Wildlife in Your Backyard*, Rodale Press, Emmaus, PA, 1992.
5. **Kress, S. W.,** *The Audubon Society Guide to Attracting Birds*, Charles Scribner's Sons, New York, 1985.
6. **Convey, S. R.,** *The Seven Habits of Highly Effective People*, Simon & Schuster, New York, 1990.

Chapter 6

Effects of Pesticides on Beneficial Invertebrates in Turf

Daniel A. Potter, Department of Entomology, University of Kentucky, Lexington, KY

CONTENTS

Introduction 59
The Turfgrass Ecosystem 60
Earthworms and Thatch 60
Effects of Fertilizers and Pesticides on Earthworms 61
Effects of Pesticides on Natural Control of Pest Populations 63
High-Maintenance Lawn Care: Cumulative Effects 65
Acquired Resistance and Enhanced Microbial Degradation 66
Final Thoughts 68
Acknowledgments 68
References 68

INTRODUCTION

Turfgrasses are typically the most intensively managed plantings in the urban landscape. Use of turfgrass in the U.S. skyrocketed during the past 40 years as large tracts of land were developed to accommodate the growing urban population. As dense, uniform, dark green turf became increasingly valued for its aesthetic and recreational benefits, as an enhancement to property values, and as a symbol of social status and affluence, the use of pesticides and fertilizers also increased.[1-3] The commercial lawn care industry grew at an average annual rate of 22% between 1977 and 1984, with gross annual sales in 1983 exceeding $2.2 billion.[4] The growing popularity of golf has been accompanied by higher standards for playing conditions.[5] Turfgrass culture, in its many forms, has become a $25 billion per year industry in the U.S.,[5] a significant component of which involves the marketing and use of herbicides, insecticides, fungicides, growth regulators, and fertilizers.

Growing concerns about the hazards of pesticides, especially groundwater contamination and potential risks to human health,[6] are mandating that the turfgrass industry reduce its reliance on chemical pesticides. Nevertheless, the current lack of reliable alternatives suggests that chemical pesticides will remain important tools of the turf manager for the foreseeable future. Indeed, use of a pesticide is often the only practical way to avoid significant damage from a heavy or unexpected pest outbreak. The public's general intolerance of weeds or pest injury and the high replacement costs for damaged turf are factors that will continue to encourage preventative use of pesticides on lawns and golf courses.[3,7]

Pesticides and fertilizers may have profound effects on the structure and stability of the turfgrass system. Because some pesticides kill beneficial organisms as well as pests, their use may increase the risk of pest resurgences or secondary pest outbreaks. Pesticides and fertilizers may also affect decomposition and nutrient recycling in turfgrass by altering primary production or by directly or indirectly affecting soil organisms such as earthworms. Like many cultivated crops, turfgrass lacks the complexity of natural grassland and forest habitats and so would be expected to be relatively susceptible to pesticide-induced perturbations.

Greater consideration of the effects of pesticides and fertilizers on the environment will be important in the turfgrass industry's transition toward integrated pest management. This chapter summarizes recent research which has begun to clarify how pesticides, fertilizers, and other management practices affect beneficial invertebrates in turfgrass, and how this in turn affects key processes such as thatch degradation and natural regulation of pest populations. The goal of these studies is to identify effective management options, both chemical and nonchemical, that are compatible with long-term stability of the turfgrass ecosystem.

THE TURFGRASS ECOSYSTEM

Turfgrass consists of the roots, stems, and leaves of grass plants together with the underlying thatch and soil. Along with the familiar pest species,[8] this composite habitat supports a diverse assemblage of nonpest invertebrates. In one survey, 83 different groups of arthropods including numerous families of insects and mites, eleven types of nematodes, and many species of annelids, gastropods, and other invertebrates were collected from a bluegrass-red fescue turf in New Jersey.[9] More than 40 species of Staphylinidae (rove beetles), 30 species of Carabidae (ground beetles), ten species of Formicidae (ants), dozens of spider species, and other predominantly predaceous invertebrates have been collected from turfgrass sites in Kentucky.[10-13] Earthworms

(Oligochaeta: Lumbricidae), oribatid mites (Acari: Cryptostigmata), Collembola, and other soil-inhabiting invertebrates are important to decomposition of plant litter and nutrient recycling in pastures and forests.[9,12,13,15,16]

Plant-feeding, predatory, and soil-inhabiting invertebrates form a complex community that interacts with the living grass, thatch, and soil and contributes to the stability of the turfgrass system (Figure 1). It is noteworthy that excessive thatch accumulation and outbreaks of insects and diseases rarely occur in turfgrass that is under minimal maintenance. This implies that low-maintenance turf is a relatively stable habitat in which thatch accumulation is balanced by decomposition and in which pests are held in check by predators, parasites, or plant resistance.

EARTHWORMS AND THATCH

Aristotle, the Greek philosopher and scientist, called earthworms "the intestines of the earth." Earthworms

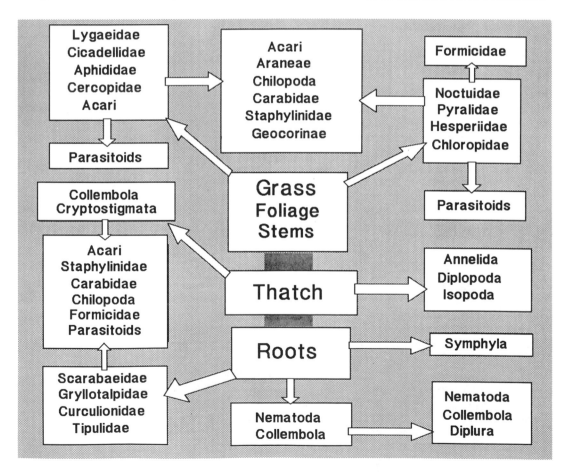

Figure 1 Proposed invertebrate food web in a typical turfgrass ecosystem. *(Adapted from Streu, H. T., Bull. Entomol. Soc. Am., 19, 89, 1973.)*

and other soil invertebrates aid the decomposition processes by fragmenting and conditioning plant debris in their guts before further breakdown by microorganisms.[17-19] Earthworms, in particular, enhance the chemical environment of soil by mixing organic matter into subsurface layers and by enriching the soil by their castings. Their burrowing activity increases aeration and water infiltration.[20] These processes are especially important in lawns and golf fairways that are cultivated mainly through the activity of earthworms. Cultural problems such as soil compaction or accumulation of thatch may become evident when earthworm populations are decimated by pesticides or through indirect effects of management.[21,22]

Thatch is a tightly intermingled layer of undecomposed roots, rhizomes, stolons, plant crowns, stems, and organic debris that accumulates between the soil surface and the green vegetation in turfgrass.[23] Thatch results from an imbalance between production and decomposition of organic matter at the soil surface.[24] Problems associated with excessive accumulation of thatch include reduced water infiltration,[25] shallow root growth with increased vulnerability to heat and drought stress,[24] and restricted penetration of fertilizers[26] and soil insecticides.[27] Thatch accumulation is common in managed turfgrass, especially when a high rate of nitrogen fertilizer is applied for several years.[14,24]

Recent experiments[15,16] confirmed the importance of earthworms to thatch degradation in Kentucky bluegrass (*Poa pratensis* L.) turf. Several hundred intact, preweighed pieces of thatch were buried in nylon bags having fine (53 μ), medium (1.2 mm) or coarse (5 mm) mesh. The mesh selectively admitted or excluded certain components of the soil fauna. In a companion experiment, thatch pieces were buried in identical, coarse mesh bags in either untreated plots with abundant earthworms, or in plots that had been treated with selected insecticides to eliminate the worms. Samples were recovered periodically for 23 months and extracted in Tullgren funnels to determine the invertebrate taxa present. The thatch was then dried, reweighed, incinerated, and analyzed for mineral soil content and for net loss of organic matter. Subsamples of thatch were also tested for microbial respiration, indicative of rate of microbial decomposition.

Dramatic differences were apparent in both experiments after only 3 months. Without earthworms the structure and composition of the thatch remained nearly the same, but when earthworms were present the pieces were broken apart and dispersed (Figure 2). The worms, mainly *Apporectodea* spp. and *Lumbricus terrestris*, incorporated large amounts of mineral soil into the thatch matrix (Figure 3). In both experiments, rates of net loss of organic matter and relative microbial respiration were much greater when earthworms were present than when worms were excluded. There was only a small difference in degradation rates of thatch from medium and fine mesh bags, suggesting that smaller soil invertebrates such as oribatid mites and Collembola are less important than earthworms in the initial breakdown of thatch.

Topdressing, i.e., the distribution of a thin layer of soil followed by physical incorporation of the soil into the thatch, is considered the best cultural method for reducing a thatch layer in turf.[24,28] This process increases the bulk density and moisture retention capacity of the thatch and creates a more favorable environment for microbial decomposition and turfgrass growth.[28] Topdressing, however, may be too expensive for large turf areas. Earthworms perform a function similar to topdressing by rapidly incorporating soil into the thatch matrix.[15] Preservation of earthworm populations is clearly important where thatch is a concern.

EFFECTS OF FERTILIZERS AND PESTICIDES ON EARTHWORMS

Excessive fertilization encourages accumulation of thatch by increasing vegetative production, but it may also contribute to the problem by affecting decomposition processes. Nitrogen fertilization commonly results in soil acidification[29] which may in turn inhibit microbial activity.[30] Furthermore, the repellent nature of NH_4^+ can adversely affect soil invertebrates.[31,32] Earthworms, in particular, are generally sparse in acidic soils.[32,33] Several studies have shown correlations between high rates of fertilization and thatch accumulation,[34,35] although this may not always occur.[36] Populations of earthworms and microarthropods were sampled in Kentucky bluegrass that had been fertilized with varying rates of ammonium nitrate for 7 years. Higher rates of fertilizer were correlated with soil acidification (Figure 4), decreased populations of earthworms and some oribatid mites, and a significant increase in thatch.[14]

Use of pesticides can also contribute to thatch development. For example, certain fungicides may reduce soil pH, which inhibits activity of microorganisms that are important to thatch decomposition.[37] Alternatively, fungicides may contribute to thatch accumulation by increasing rates of root and rhizome production.[38]

Another study compared the short-term and long-term impact of 17 commonly used pesticides on earthworm populations in Kentucky bluegrass turf.[16] A single application of the fungicide

Figure 2   (A) Thatch pieces recovered after 3 months from untreated plots (left) vs. plots that were treated with chlordane and carbofuran to eliminate earthworms. (B) Thatch pieces from fine-, medium-, and coarse-mesh bags recovered 23 months after burial. Note large amount of mineral soil incorporated by earthworms. (C) Cross sections of original thatch (top) and thatch recovered from medium- and fine-mesh bags 23 months after burial. Note relative lack of decomposition in the absence of earthworms. *(From Potter, D. A., Powell, A. J., and Smith, M. S., J. Econ. Entomol., 83, 205, 1990. With permission.)*

Figure 2 (Continued).

benomyl or the insecticides ethoprop, carbaryl, fonofos, or bendiocarb at labeled rates reduced earthworm populations by 60 to 99% (Table 1). Reductions were evident for at least 20 weeks. These pesticides also reduced the rate at which earthworms incorporated mineral soil into thatch pieces that had been buried in the treated plots. Other insecticides, including diazinon, isofenphos, trichlorfon, chlorpyrifos, and isazophos, caused less severe, but significant mortality in some tests (Table 1). None of the herbicides tested were significantly toxic to earthworms under field use conditions. The abundance of Cryptostigmata, Collembola, and ants may also be drastically reduced by some insecticides.[10,16,39]

EFFECTS OF PESTICIDES ON NATURAL CONTROL OF PEST POPULATIONS

Broad-spectrum insecticides that are applied for the control of turfgrass insect pests are generally also toxic to predaceous and parasitic arthropods. For example, a single, surface application of chlorpyrifos significantly reduced populations of spiders, rove beetles (Staphylinidae), and predatory mites in Kentucky bluegrass for at least 5 to 6 weeks (Figure 5). Other turfgrass insecticides, including isofenphos, trichlorfon, isazofos, carbaryl, and bendiocarb, also caused short-term reductions in abundance of some predator groups.[10a,b] Application of isofenphos, a relatively persistent organophosphate, to home lawns suppressed populations of nonoribatid Acari, Collembola, Diplopoda, Diplura, and Staphylinidae for as long as 43 weeks after treatment.[39] Not surprisingly, predatory arthropods are often less abundant and less diverse in high-maintenance lawns (i.e., those that receive scheduled applications of fertilizers and pesticides) than in lawns maintained without regular treatments.[12,13,40]

The fact that insect outbreaks are relatively uncommon in low-maintenance turfgrass suggests that many pests are normally held in check by

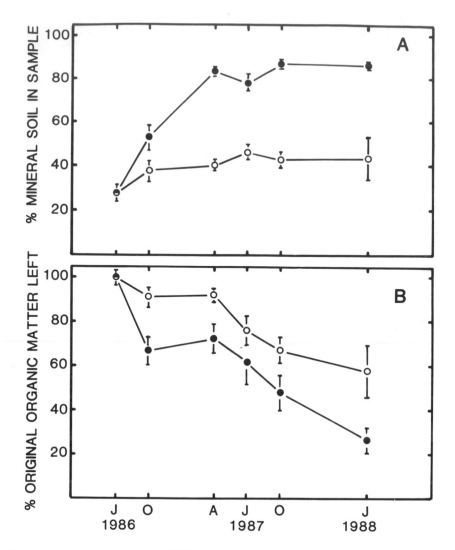

Figure 3 Incorporation of mineral soil (A) and net loss of organic matter (B) from thatch buried in coarse-mesh bags in untreated turf (•), or in plots treated with insecticides to eliminate earthworms and other soil invertebrates (o). *(From Potter, D. A., Powell, A. J., and Smith, M. S., J. Econ. Entomol., 83, 205, 1990. With permission.)*

indigenous natural enemies. Several authors[9,11,41] have cautioned that repeated or heavy pesticide usage could reduce the stability of the turfgrass ecosystem and lead to resurgences or secondary outbreaks of pests.

Southern chinch bug (*Blissus insularis* Barber) populations in Florida remained low in untreated St. Augustinegrass (*Stenotaphrum secundatum* (Walt.) Kutze) lawns where natural enemies were abundant, while at the same time reaching outbreak densities on lawns that were repeatedly treated with insecticides.[41] Resurgence of hairy chinch bug (*Blissus leucopterus hirtus* Montandon) populations following several years of chlordane use were at-

tributed to reduced populations of predators, including mites and possibly hemipterans.[9,43] In New Jersey, repeated application of carbaryl to turfgrass was associated with outbreaks of winter grain mite, *Penthaleus major* (Duges), apparently due to suppression of acarine predators.[42] Outbreaks of the greenbug, *Schizaphis graminum* (Rhondani) in Kentucky may be more common on high-maintenance lawns than on untreated turf.[44] Reinert[41] observed parasitism of eggs and predation on southern chinch bugs, and suggested that the combined activity of predators and parasites contributed to an observed collapse in chinch bug populations in late summer.

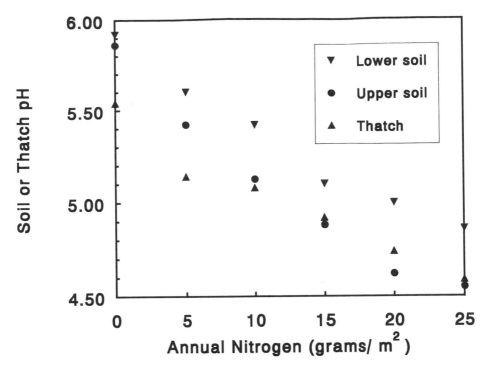

Figure 4 Decline in soil and thatch pH in Kentucky bluegrass associated with application of varying rates of ammonium nitrate fertilizer for 7 years. All regressions are significant at $p < 0.001$. *(Adapted from Potter, D. A., Bridges, B. L., and Gordon, F. C., Agron. J., 77, 367, 1985.)*

Rates of natural predation on sod webworm (*Crambus* and *Pediasia* spp.) eggs were compared in untreated Kentucky bluegrass and in turf that had received a single surface application of chlorpyrifos at the labeled rate.[12] Eggs were obtained from field-collected females and were placed on disks of filter paper in small dishes set level with the ground surface. Groups of 500 eggs were set out at 1, 3, or 5 weeks after the insecticide application and the numbers of eggs that were eaten or carried off by predators in 48 hours were compared between treated and untreated plots. Relative abundance of predators was measured with pitfall traps.

Predatory arthropods, especially ants, consumed or carried off as many as 75% of the eggs in the untreated plots within 48 hours. Chlorpyrifos reduced predator populations and suppressed predation on sod webworm eggs in the treated plots for at least 3 weeks after treatment (Table 2). Predator populations had begun to recover by 5 weeks after treatment, by which time rates of predation between treated and untreated plots were similar. Numerous predator species collected from the study site were found to consume sod webworm eggs in the laboratory.[12] More recent experiments revealed similarly high rates of predation on eggs and young

larvae of the Japanese beetle (*Popillia japonica* Newman) and on pupae of the fall armyworm (*Spodoptera frugiperda* J.E. Smith) in the laboratory and the field.[10b]

HIGH-MAINTENANCE LAWN CARE: CUMULATIVE EFFECTS

Although use of certain pesticides and fertilizers can clearly be harmful to beneficial invertebrates and to key processes such as thatch degradation and natural regulation of pest populations, another study[13] suggests that the cumulative effects of high-maintenance lawn care programs may not necessarily be so severe. Replicated plots of Kentucky bluegrass were maintained for 4 years on a schedule of treatments similar to that used by many lawn maintenance companies. This consists of four fertilizer applications (225 kg [AI]/ha total nitrogen annually from urea), broadleaf weed control (2,4-D, MCPP, and dicamba) in spring and fall, early spring application of bensulide for preemergent crabgrass control, two applications of chlorpyrifos for surface-feeding insects, and granular diazinon applied for control of white grubs in late August. Control plots were unmanaged except for mowing. Changes in soil and thatch pH and thatch

Table 1 **Relative toxicity of turfgrass pesticides to earthworms based upon the mean reduction in population density in two independent field tests**[a]

Treatment	Common Name	Rate kg (AI)/ha	Class[a]
Low toxicity (0–25% reduction)			
2,4-D	Dacamine 4D	2.2	H
Trichlopyr	Garlon 3 A	0.6	H
Dicamba	Banvel 4 E	0.6	H
Pendimethalin	Pre-M 60 WDG	3.4	H
Triodimefon	Bayleton 25 WDG	3.0	F
Fenarimol	Rubigan 50 WP	3.0	F
Propiconazole	Banner 1.1 EC	3.4	F
Chlorothalonil	Daconil 2787	12.7	F
Isofenphos	Oftanol 5 G	2.2	I-OP
Moderate toxicity (26–50% reduction)			
Trichlorfon	Proxol 80 WP	9.0	I-OP
Chlorpyrifos	Dursban 4 E	4.5	I-OP
Isazophos	Triumph 4 E	2.2	I-OP
Severe toxicity (51–75% reduction)			
Benomyl	Benlate 50 WP	12.2	F
Diazinon	Diazinon 14 G	4.5	I-OP
Very severe toxicity (76–99% reduction)			
Carbaryl	Sevin SL	9.0	I-C
Bendiocarb	Turcam 2.5 G	4.5	I-C
Ethoprop	Mocap 10 G	5.6	I-OP
Fonofos	Crusade 5G	4.5	I-OP

Note: Treatments were applied to Kentucky bluegrass in April and watered in, and earthworms were sampled using formalin drenches after 7 to 9 days. Adapted from Potter et al.[16]

[a] (H) Herbicide, (F) fungicide, (I-OP) organophosphate insecticide, and (I-C) carbamate insecticide

thickness were monitored, and impact on earthworms and beneficial and nontarget arthropods was determined from pit-fall trap collections, extraction of soil and thatch samples, and sweep-net and formalin-drench samples.

Even after 4 years on this relatively heavy treatment schedule, earthworm numbers were not significantly reduced in the treated plots. Numbers of oribatid mites actually increased. Moreover, there was only a modest decline in soil pH (6.2 to 5.9), possibly because the site was treated with agricultural limestone 2 years before the experiment began. Thatch accumulation was greater in the high maintenance turf (10.7 mm vs. 3.3 mm in control plots), but was still not excessive. Predator populations, specifically Araneae, Staphylinidae, and Carabidae, were suppressed by the insecticides, particularly the late summer soil treatment with diazinon, but predators repopulated the treated plots by the following spring. Effects of the program on nontarget herbivorous arthropods were variable; flea beetle (Chrysomelidae) populations were generally higher, and leafhopper (Cicadellidae)

populations were somewhat lower in the high-maintenance plots.

The apparent recovery of the turf ecosystem from disturbances caused by this relatively heavy schedule of treatments suggests that the impact of more moderate programs would be less severe and of shorter duration. However, the rate of thatch accumulation would probably have been greater had we used pesticides that are more toxic to earthworms, or had the applications been made in early spring or fall, when earthworms are more active at the soil surface. Furthermore, the rate at which predators, earthworms, and other beneficial organisms would repopulate larger turf areas such as home lawns and golf courses would undoubtedly be slower than occurred in our study.

ACQUIRED RESISTANCE AND ENHANCED MICROBIAL DEGRADATION

Repeated or heavy use of certain pesticides on turfgrass may encourage other problems, including

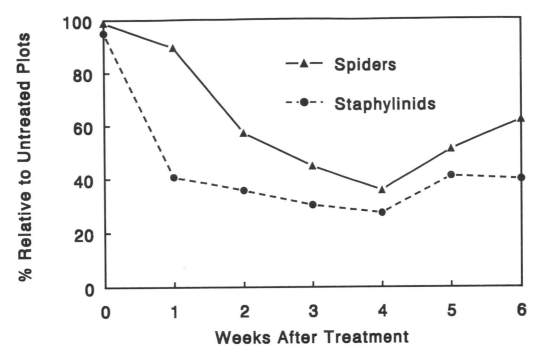

Figure 5 Impact of surface application of chlorpyrifos, an organophosphate insecticide, on spider (Araneae) and rove beetle (Staphylinidae) populations in Kentucky bluegrass turf. Data are based on pitfall trap captures, and are expressed as mean percent of numbers collected in untreated control plots. *(Adapted from Cockfield, S. D. and Potter, D. A., Environ. Entomol., 12, 1260, 1983.)*

Table 2 **Mean percentage of sod webworm eggs eaten or carried off by predators in replicated Kentucky bluegrass plots**

	Percent Eggs Missing from		
Weeks Posttreatment	Untreated Plots	Treated Plots	*p* Level[a]
1	37.7	0.9	$p < 0.001$
3	17.6	0.1	$p < 0.05$
5	75.4	61.3	$p < 0.08$

Note: Plots were untreated or received a single, surface application of chlorpyrifos (Dursban 4 E) at the labelled rate. Percentages are based upon groups of 500 total eggs placed in the turf for 48 hours at 1, 3, or 5 weeks after treatment.

[a] Probability of a greater *t* statistic, one-sided paired *t* test.

Reprinted with permission from Cockfield, S. D. and Potter, D. A., J. Econ. Entomol., 77, 1542, 1984.

acquired resistance of pests to insecticides or fungicides, and enhanced microbial degradation of pesticide residues.

Acquired resistance can become a problem when insecticides are applied repeatedly, or over a number of years. Resistance of insects to cyclodiene insecticides had become widespread by the early 1970s, even before environmental concerns resulted in cancellation of their registrations for turf. Resistance to organophosphates or carbamates has been documented for chinch bugs and greenbugs[45] and, in at least one instance, for white grubs.[46] Metcalf[47] listed general resistance management tactics directed at reducing the single-factor selection pressure on pest populations that occurs with conventional chemical control. These include (1) reducing the frequency and extent of treatments, (2) avoiding pesticides with long environmental persistence and slow release formulations, (3) reduced use of residual treatments, (4) avoidance of treatments that apply selection pressure on both immature and adult stages, and (5) increased use of cultural, biological, and other nonchemical methods in integrated pest management. Each of these tactics is applicable to the turfgrass system.

Enhanced microbial degradation occurs when pesticide residues are degraded more rapidly than usual by microorganisms. It occurs most often in soils that have been conditioned by prior exposure to a pesticide. Enhanced degradation has been reported for isofenphos, diazinon, ethoprop, and other insecticides used on turf,[48] and it has been implicated in reduced residual effectiveness of isofenphos on golf courses previously treated with that chemical.[49] Thatch from plots that had been treated with isofenphos were found to rapidly degrade residues of carbaryl and diazinon. Similarly, residues of diazinon, chlorpyrifos, carbaryl, and isazophos were rapidly degraded in plots that had previously been treated with the same chemicals.[50] The risk of inducing enhanced microbial degradation provides another strong argument against the indiscriminate or nonessential use of pesticides on lawns or golf courses.

FINAL THOUGHTS

The intent of this chapter is not to condemn the responsible use of pesticides and fertilizers on turf. Indeed, the use of a chemical pesticide is often the only practical way to prevent significant damage from heavy infestations or unexpected outbreaks of pests. However, pesticide applications, like human medicines, may have adverse side-effects which should be weighed against the overall benefit that the treatment will provide.

The accumulated evidence suggests that turfgrass is a complex ecosystem with many buffers. However, we are only beginning to understand the roles of microorganisms, earthworms, predators, parasites, and other invertebrates in maintaining this natural balance. Unnecessary or excessive use of chemical pesticides can aggravate thatch and pest problems by interfering with the activities of beneficial organisms, or by encouraging development of acquired resistance to pesticides or enhanced microbial degradation.

These potential problems should serve as an incentive for turfgrass managers to select pesticides or alternative tactics that are known to cause fewer adverse side effects (e.g., earthworm toxicity), to apply them at the proper time and rate, and to use them only as necessary to control specific problems. I believe that most homeowners and professional turfgrass managers are aware of and concerned about the perceived and real hazards of overuse of pesticides. Awareness of the potential negative side-effects to beneficial organisms can be an additional incentive for selective and responsible use of these tools. The challenge now is for turfgrass scientists to provide safe, effective, and reliable alternatives that are compatible with long-term stability of the turfgrass ecosystem.

ACKNOWLEDGMENTS

I am grateful to T. B. Arnold, M. C. Buxton, S. D. Cockfield, F. C. Gordon, C. G. Patterson, A. J. Powell, C. T. Redmond, and M. S. Smith for their contributions to this research. This work was supported by USDA Grants SR89-57-E-KY, SR91-31-E-KY, and 91-34103-5836, and by grants from the U.S. Golf Association and the O. J. Noer Research Foundation, Inc. This is Contribution No. 91-7-68 of the Kentucky Agricultural Experiment Station and is published with approval of the director.

REFERENCES

1. **Anon.,** National Household Pesticide Usage Study, 1976-1977, EPA/540/9-80-002, U.S. Environmental Protection Agency, Washington, DC, 1979.
2. **National Research Council**, *Urban Pest Management*, Committee on Urban Pest Management, Environmental Studies Board, Commission on Natural Resources, National Academy Press, Washington, DC, 1980.
3. **Potter, D. A. and Braman, S. K.,** Ecology and management of turfgrass insects, *Ann. Rev. Entomol.,* 36, 383, 1991.
4. **Anon.,** Lawn care receipts vault to more than $2 billion, *Lawn Care Ind.,* 8, 1, 1984.
5. **Roberts, E. C. and Roberts, B. C.,** *Lawn and Sports Turf Benefits*, The Lawn Institute, Pleasant Hill, TN, 1987.
6. **U.S. General Accounting Office,** Lawn Care Pesticides. Risks Remain Uncertain While Prohibited Safety Claims Continue, Report to the Committee on Environment and Public Works, U.S. Senate, GAO/RCED-90-134, 1990.
7. **Potter, D. A.,** Urban landscape pest management, in *Advances in Urban Pest Management*, Bennett, G. W. and Owens, J. M., Eds., Van Nostrand Reinhold, New York, 1986.

8. **Tashiro, H.,** *Turfgrass Insects of the United States and Canada*, Cornell University Press, Ithaca, NY, 1987.

9. **Streu, H. T.,** The turfgrass ecosystem: impact of pesticides, *Bull. Entomol. Soc. Am.,* 19, 89, 1973.

10. **Cockfield, S. D. and Potter, D. A.,** Short-term effects of insecticidal applications on predaceous arthropods and oribatid mites in Kentucky bluegrass turf, *Environ. Entomol.,* 12, 1260, 1983.

10a. **Terry, L. A., Potter, D. A., and Spicer, P. G.,** Effect of insecticides on predatory arthropods and predation on eggs of Japanese beetle (Coleoptera: Scarabaeidae) and pupae of fall armyworm (Lepidoptera: Noctuidae) in turfgrass, *J. Econ. Entomol.,* 86, 871,1993.

11. **Cockfield, S. D. and Potter, D. A.,** Predation on sod webworm (Lepidoptera: Pyralidae) eggs as affected by chlorpyrifos application to Kentucky bluegrass turf, *J. Econ. Entomol.,* 77, 1542, 1984.

12. **Cockfield, S.D. and Potter, D.A.,** Predatory arthropods in high- and low-maintenance turfgrass, *Can. Entomol.,* 117, 423, 1985.

13. **Arnold, T. B. and Potter, D. A.,** Impact of a high-maintenance lawn-care program on nontarget invertebrates in Kentucky bluegrass turf, *Environ. Entomol.,* 16, 100, 1987.

14. **Potter, D. A., Bridges, B. L., and Gordon, F. C.,** Effect of N fertilization on earthworm and microarthropod populations in Kentucky bluegrass turf, *Agron. J.,* 77, 367, 1985.

15. **Potter, D. A., Powell, A. J., and Smith, M. S.,** Decomposition of turfgrass thatch by earthworms and other soil invertebrates, *J. Econ. Entomol.,* 83, 205, 1990.

16. **Potter, D. A., Buxton, M. C., Redmond, C. T., Patterson, C. G., and Powell, A. J.,** Toxicity of pesticides to earthworms (Oligochaeta: Lumbricidae) and effect on thatch degradation in Kentucky bluegrass turf, *J. Econ. Entomol.,* 83, 2362, 1990.

17. **Swift, M. J., Heal, O. W., and Anderson, J. M.,** *Decomposition in Terrestrial Ecosystems. Studies in Ecology,* Vol. 5, Blackwell Scientific, Oxford, England.

18. **Curry, J. P.,** The invertebrate fauna of grassland and its influence on productivity. I. The composition of the fauna, *Grass Forage Sci.,* 42, 325, 1987.

19. **Curry, J. P.,** The invertebrate fauna of grassland and its influence on productivity. III. Effects on soil fertility and plant growth, *Grass Forage Sci.,* 42, 325, 1987.

20. **Lee, K. E.,** *Earthworms. Their Ecology and Relationships With Soil and Land Use,* Academic, New South Wales, Australia, 1985.

21. **Randell, R., Butler, J. D., and Hughes, T. D.,** The effect of pesticides on thatch accumulation and earthworm populations in Kentucky bluegrass turf, *HortScience,* 7, 64, 1972.

22. **Turgeon, A. J., Freeborg, R. P., and Bruce, W. N.,** Thatch development and other effects of preemergent herbicides in Kentucky bluegrass turf, *Agron. J.,* 67, 563, 1975.

23. **Ledeboer, F. B. and Skogley, C. R.,** Investigations into the nature of thatch and methods for its decomposition, *Agron. J.,* 59, 320, 1967.

24. **Beard, J. B.,** *Turfgrass: Science and Culture,* Prentice-Hall, Englewood Cliffs, NJ, 1973.

25. **Taylor, D. H. and Blake, G. R.,** The effect of turfgrass thatch on water infiltration rates, *Soil Sci. Soc. Am.,* 46, 616, 1982.

26. **Nelson, K. E., Turgeon, A. J., and Street, J. R.,** Thatch influence on mobility and transformation of nitrogen carriers applied to turf, *Agron. J.,* 72, 487, 1980.

27. **Niemczyk, H. D. and Krueger, H. R.,** Persistence and mobility of isazophos in turfgrass thatch and soil, *J. Econ. Entomol.,* 80, 950, 1987.

28. **Danneberger, T. K. and Turgeon, A. J.,** Soil cultivation and incorporation effects on edaphic properties of turfgrass thatch, *J. Am. Sci. Hort. Sci.,* 111, 184, 1986.

29. **Pierre, W. H.,** Nitrogenous fertilizers and soil acidity. I. Effect of various nitrogenous fertilizers on soil reaction, *J. Am. Soc. Agron.,* 20, 254, 1928.

30. **Martin, D. P. and Beard, J. B.,** Procedure for evaluating the biological degradation of turfgrass thatch, *Agron. J.,* 67, 835, 1975.

31. **Marshall, V. G.,** Effects of manures and fertilizers on soil fauna: a review, *Commonw. Bur. Soils Spec. Publ.,* 3, 1977.

32. **Edwards, C. A. and Lofty, J. R.,** *Biology of Earthworms,* 2nd ed., Chapman and Hall, London, 1977.

33. **Satchell, J. E.,** Lumbricidae, in *Soil Biology,* Burges, A. and Raw, F., Eds., Academic, London, 1967, 259.

34. **Engle, R. E. and Aldefer, R. B.,** The effect of cultivations, lime, nitrogen, and wetting agent on thatch development in one-quarter-inch bentgrass turf over a 10-year period, *N.J. Agric. Exp. Stn. Bull.,* 818, 32, 1967.

35. **Meinhold, J. H., Duble, R. L., Weaver, R. W., and Holt, E. C.,** Thatch accumulation in Bermudagrass turf in relation to management, *Agron. J.*, 65, 833, 1973.

36. **Shearman, R. C., Kinbacher, E. J., Riordan, T. P., and Steinegger, D. H.,** Thatch accumulation in Kentucky bluegrass as influenced by cultivar, mowing, and nitrogen, *HortScience*, 15, 312, 1980.

37. **Smiley, R. W. and Craven Fowler, M.,** Turfgrass thatch components and decomposition rates in long-term fungicide plots, *Agron. J.*, 78, 633, 1986.

38. **Smiley, R. W., Craven Fowler, M., Kane, R. T., Petrovic, A. M., and White, R. A.,** Fungicide effects on thatch depth, thatch decomposition rate, and growth of Kentucky bluegrass, *Agron. J.*, 77, 597, 1985.

39. **Vavrek, R. C. and Niemczyk, H. D.,** Effect of isofenphos on nontarget invertebrates in turfgrass, *Environ. Entomol.*, 19, 1572, 1990.

40. **Short, D. E., Reinert, J. A., and Atilano, R. A.,** Integrated pest management for urban turfgrass culture — Florida, in *Advances in Turfgrass Entomology*, Niemczyk, H. D. and Joyner, B. G., Eds., Hammer Graphics, Piqua, OH, 1982, 25.

41. **Reinert, J. A.,** Natural enemy complex of the southern chinch bug in Florida, *Ann. Entomol. Soc. Am.*, 71, 728, 1978.

42. **Streu, H. T. and Gingrich, J. B.,** Seasonal activity of the winter grain mite in turfgrass in New Jersey, *J. Econ. Entomol.*, 65, 427, 1972.

43. **Streu, H. T.,** Some cumulative effects of pesticides on the turfgrass ecosystem, Proc. Scotts Turfgrass Res. Conf. I. Entomology, O.M. Scott & Sons, Marysville, OH, 1969, 53.

44. **Potter, D. A.,** Greenbugs on turfgrass: an informative update, *Am. Lawn Applic.*, 3, 20, 1982.

45. **Reinert, J. A.,** Insecticide resistance in epigeal pests of turfgrass: 1. A review, in *Advances in Turfgrass Entomology*, Niemczyk, H. D. and Joyner, B. G., Eds., Hammer Graphics, Piqua, OH, 1982, 71.

46. **Ahmad, S. and Ng, Y. S.,** Further evidence for chlorpyrifos tolerance and partial resistance by the Japanese beetle (Coleoptera: Scarabaeidae), *J. N.Y. Entomol. Soc.*, 89, 34, 1981.

47. **Metcalf, R. L.,** Insect resistance to insecticides, in *Integrated Pest Management for Turfgrass and Ornamentals*, Leslie, A. and Metcalf, R. L., Eds., U.S. Environmental Protection Agency, Washington, DC, 1989.

48. **Felsot, A. S.,** Enhanced biodegradation of insecticides in soil, *Ann. Rev. Entomol.*, 34, 453, 1989.

49. **Niemczyk, H. D. and Chapman, R. A.,** Evidence of enhanced degradation of isofenphos in turfgrass thatch and soil, *J. Econ. Entomol.*, 80, 880, 1987.

50. **Niemczyk, H. D. and Filary,** *Ohio Turfgrass Foundation Newsletter*, March, 1988.

Preparation for a Healthy Landscape

A simple test for clay soil

Growth habit of bermudagrass

Trials of bluegrass/fescue mixtures
at Pennsylvania State University

Eugene, OR rose gardens on an IPM program,
using no fungicides (photo courtesy of Tim Rhay)

Growth of stoloniferous grass from plug

Chapter 7

Determining the Health of the Soil

Charles H. Peacock, North Carolina State University, Raleigh, NC

CONTENTS

Gathering Information from the Soil 73
    Careful Soil Sampling 73
    Analysis and Interpretation 74
Understanding Soil Test Reports 74
Using the Information from the Soil Test 75
    Soil pH..75
    Nutrient Levels 76
        Macronutrients 76
        Micronutrients 76
    Salinity 76
Recommendations for Fertilization 77
    Prior to Establishment 77
    Established Lawns 77

All turfgrass management programs focus on the primary cultural practices of mowing, fertilization, and irrigation. When focusing on turf fertilization it is necessary to have a basic understanding of how the soil provides the nutrients, what factors affect nutrient availability, and how these factors are managed under turf situations.

When turf cultural programs are developed as part of a management plan, soil testing provides important information regarding pH, available nutrient levels, and salt problems. While a wide range of soil chemical and physical characteristics can very accurately be determined, information from soil testing serves as a guideline to assist in developing a sound fertilization program.

GATHERING INFORMATION FROM THE SOIL

Any soil testing program depends on two factors. There is the *analytical* procedure for determining the levels of nutrients within the soil (a service function) and the *correlation and calibration* of data for basing fertilization recommendations (a research function). However, both of these functions depend on the sample being analyzed to truly represent the area from which it was taken. It has been estimated that 50% of the soil tests performed are of little value because they are not true representations of the area for which the results are intended to be used.

CAREFUL SOIL SAMPLING

In order for any soil test to be reliable, it must be *representative* of the area where the turf is growing. This means samples should be collected from as many different types of soil as there are on the site. Obvious changes in soil texture, drainage, and color should be considered, and separate samples taken for analysis from as many different soils as are present within a turf area. A composite of 15 to 20 samples from over the entire area of a given soil type should be taken at a depth of no more than 4 in. on established turf areas since this is where the majority of the turfgrass roots will be found. When an area is being prepared for seeding or sodding, samples should be taken to the depth to which fertilizer and lime will be incorporated during seedbed preparation, usually 4 to 6 in. Also, on previously fertilized turf, nutrient levels tend to be high near the surface and deep sampling will dilute the soil test level. Samples should be thoroughly mixed in a plastic bucket (to prevent metal contamination if possible). A portion of this will be submitted for analysis.

Sampling should be considered at a minimum of every 2 to 3 years. If the soil analysis shows acute deficiencies, then yearly testing may be necessary until these are corrected. If the soil pH is too low, the time for corrective applications of lime to completely react may be 5 to 6 months on a clay soil, so more frequent sampling than yearly is rarely necessary. Samples should be taken 3 to 6 months before the most active growing season begins,

0-87371-350-8/94/$0.00+$.50
© 1994 by CRC Press Inc.

thereby giving adequate time to apply liming materials and prepare the fertilization program schedule. Sampling should be done at about the same time of the year each time an area is tested because seasonal variation in nutrient levels does occur.

ANALYSIS AND INTERPRETATION

Soil testing programs now provide an extremely accurate analysis of the nutrients in the sample. Standardization of the nutrient extraction procedures and state-of-the-art analytical equipment have highly refined the analysis portion of the testing procedure. The shortcoming in any soil testing program is the correlation and calibration of fertilizer recommendations with research data. Fertilizer recommendations are based on responses expected for particular ranges of nutrients. This is particularly difficult for turfgrasses since turf response or performance is not being measured in pounds or bushels of yield per acre. Instead, factors such as turf cover, density, color, yield, and disease incidence have been used for soil test response calibration. Most soil test recommendations are based on ranges of extracted nutrients and an expected response from providing a certain amount of fertilizer to supplement nutrients already in the soil. While soil testing gains information about the nutrient levels present for a particular turf area, the interpretation transforms these numbers into an economically and environmentally sound fertilizer recommendation.

UNDERSTANDING SOIL TEST REPORTS

Soil test reports provide information on the status of certain nutrients in the soil. They also indicate the soil acidity (soil pH) and in certain cases the amount of salts present. Almost all soil test reports include the information found in Table 1. This information is used to develop fertilizer recommendations as to amounts, types of materials to use, and timing and rates of applications. While many more nutrients could be extracted from the soil and analyzed, it is extremely difficult, if not impossible, to correlate the levels extracted with a recommendation for fertilization. With many of the micronutrients (those required by the turf in the smallest amounts) and nitrogen, a plant tissue analysis is much more accurate to determine if an acceptable amount is being taken up and is therefore available for good nutrition.

Table 1 **Information commonly found in soil test reports and its importance in developing a fertilization program**

Test Result	Units Reported	Importance
Soil class	Mineral/Organic	Used in developing liming recommendations
Humic matter	Percent	Chemically active organic fraction related to CEC
Weight/volume or bulk density	g/cm^3	A broad classification of the soil density
Cation exchange capacity (CEC)	meq/100 cm^3	A relative measure of the nutrient holding capacity
Base saturation (BS)	Percent	That part of the CEC occupied by Ca, Mg, K (and Na in some soils)
Acidity	meq/100 cm^3	That part of the CEC occupied by H^+, and Al^{3+} — used for determining the lime requirement
pH	Standard units	Measure of active acidity — used with acidity for lime requirement
P, K	lb/acre, ppm, index values	Used with information from field research for fertilizer recommendations
Ca, Mg	Percent of BS	Used with information from field research for fertilizer recommendations
Mn, Zn, Cu, S	lb/acre, ppm, index values	Used with information from field research for fertilizer recommendations
Soluble salts	EC units, ppm	Relative content of salts in the soil and potential problems

Note: Units reported: Electrical Conductivity (EC) units = dS/m (dS = deciSiemens; or mmhos/cm); meq = milliequivalents, ppm = parts per million.

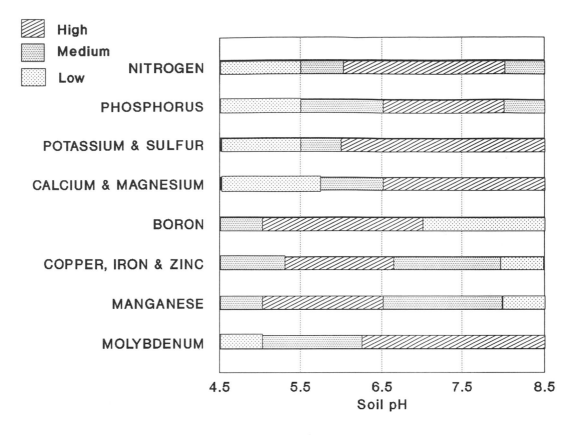

Figure 1 Soil pH and plant nutrient availability.

USING THE INFORMATION FROM THE SOIL TEST

SOIL pH

Soil pH measures the acidity or alkalinity of the soil. The soil pH and exchangeable acidity determine the need for lime and the amount is relative to these values. Soil pH is very important in turf nutrition because it affects the availability of all of the nutrients and also aluminum, which may be toxic to the plant roots (see Figure 1). It is desirable for most turfgrass species to have a soil pH in the 6.0 to 6.5 range. Exceptions to this are centipedegrass and bahiagrass with somewhat acidic optimum levels of pH 5.5.

A lime recommendation is made with each soil test report if the pH is too acidic. The lime requirement is normally expressed in tons/acre or for small turf areas in lb/1000 ft². If the amount required is in excess of 50 lb/1000 ft², the amount should be applied in multiple applications of 50 lb/1000 ft² at a 3- to 4-month interval. This will prevent applying too much material to the turf at one time and damaging the plants. The amount of ground limestone required to increase the soil toward a 6.5 to 7.0 pH

is best determined by a soil test. In lieu of this, for most soils the information in Table 2 can be used as a guideline.

At times the soil may be too alkaline or basic. If this condition exists because of natural amounts of lime within the soil, which can occur in coastal or other areas where naturally occurring limestone deposits exist, there is little which can be done to offset this problem. Where the alkalinity has been created by an over-application of lime several approaches can be taken including applying sulfur or

Table 2 **Approximate pounds of ground limestone needed per 1000 ft² of turf area to increase the soil pH toward 6.5 to 7.0 for three different soil types**

pH from Soil Test	Sandy	Loam	Clay
4.0	90	140	200
4.5	85	135	190
5.0	75	125	175
5.5	50	90	110
6.0	25	50	60

selecting acid-forming nitrogen fertilizers such as ammonium sulfate, ammonium nitrate, or urea.

Sulfur can be applied as a granular or wettable material to decrease soil pH. Sulfur is required at about $\frac{1}{3}$ the amount of lime to reduce the acidity the same number of units. It takes 3 to 5 lb of elemental sulfur to lower the pH $\frac{1}{2}$ a unit. In no case should more than 5 to 10 lb of sulfur be applied per 1000 ft^2 at any one application and it should be washed off the leaf surfaces and above ground plant parts immediately.

NUTRIENT LEVELS
Macronutrients

Of the macronutrients, also called major or primary nutrients, soil testing does not include nitrogen. Nitrogen levels change so rapidly in the soil that there is no reliable method for determining how to use the nitrogen analysis to make fertilization recommendations. All turfgrasses need nitrogen to maintain vigorous growth and sustain their perennial nature. Recommended nitrogen levels for different turf species under home lawn conditions are presented in Table 3.

Amounts of phosphorus, potassium, calcium, magnesium, and sulfur for fertilization are suggested based on nutrient levels extracted from the soil during the soil test. These levels are given within broad ranges and will vary somewhat for different turfgrasses. A soil range of 60 to 100 lb of phosphorus and 85 to 175 lb of potassium per acre is generally considered adequate for most turfgrasses. Fer-

Table 3 Suggested nitrogen fertilization rates for established lawns

Turf Species	Total lb N/1000 ft^2/year
Bahiagrass	1
Bermudagrass (common)	4.5
Bermudagrass (hybrid)	5 to 6
Centipedegrass	0.5 to 1
Fescue, tall	3 to 4
Kentucky bluegrass	3 to 4
Kentucky bluegrass/tall fescue	3 to 4
Kentucky bluegrass/fine fescue	3 to 4
Kentucky bluegrass/perennial ryegrass	3 to 4
Perennial ryegrass	3 to 4
St. Augustinegrass	3 to 4
Zoysiagrass	1.5 to 2

Note: All rates are per 1000 ft^2 based on an expected turf requirement for the year. If no soil test recommendations are available use a complete (N-P-K) turf-grade fertilizer in a 3-1-2 or 4-1-2 analysis in which $\frac{1}{3}$ to $\frac{1}{2}$ of the nitrogen is slowly available.

tilization programs indicate that within these levels only 20 lb of phosphorus or 20 to 45 lb of potassium would be needed annually per acre at the nitrogen levels listed in Table 3. Soil test levels below the minimums would suggest up to 90 lb of phosphorus or 90 to 135 lb of potassium annually.

Calcium and magnesium are seldom a problem in soils that require liming. Dolomitic limestone ($CaCO_3 + MgCO_3$) used to adjust the soil pH supplies both calcium and magnesium. If soil magnesium levels are adequate then calcitic limestone ($CaCO_3$) would be suggested to increase soil pH. Where soil pH is optimum and soil calcium is low, gypsum ($CaSO_4$) could be used to increase the soil calcium levels. If the soil is deficient in magnesium, but requires no pH adjustment then $MgSO_4$ (commonly called Epsom salt) could be used. A commonly used potassium source, sulfate of potash magnesia, is an excellent source of sulfur and magnesium as well as potassium. Sulfur is commonly supplied in the potassium or nitrogen fertilizers and naturally through atmospheric deposition. When it is needed it can be supplied in the sulfate form in the nitrogen or potassium fertilizers.

Micronutrients

This group includes boron, chlorine, cobalt, copper, iron, manganese, molybdenum, and zinc. Of these only copper, manganese, and zinc are routinely included in soil test reports. These are the ones most commonly deficient and in which there is the most reliable information on soil test levels on which to base fertilizer recommendations. Manganese availability is highly dependent on soil pH and recommendations for fertilization are based on soil pH conditions. While iron is probably the micronutrient most often deficient in turf, it is difficult to interpret a soil test level as to the availability and exact need for fertilization. A plant tissue analysis of chlorotic (yellow) tissue will give a good indication of iron deficiency and the need for corrective measures.

SALINITY

Salinity is the presence of excessive concentrations of soluble salts in the soil. Soluble salts can affect plants by suppressing growth through nonspecific salt effects related to osmotic problems of restricted water availability or by specific ion effects such as interference with plant uptake of nutrients or interactions with other elements within the soil. Many times a decrease in growth occurs without obvious signs of injury such as wilting, chlorosis, or leaf dieback.

Plants constantly respond to soil conditions by changing a number of internal physiological factors related to water content and salts by osmotic adjust-

Table 4 **Suggested monthly fertilization schedule for established lawns**

Turf	J	F	M	A	M	J	J	A	S	O	N	D
Bahiagrass					$^1/_2$		$^1/_2$					
Bermudagrass				$^1/_2$	1	$^1/_2$	1	1	$^1/_2$			
Centipedegrass						$^1/_2$						
Kentucky bluegrass	$^1/_2$–1		$^1/_2$–1						1		1	
Perennial ryegrass	$^1/_2$		$^1/_2$–1						1		1	
St. Augustinegrass					$^1/_2$–1	$^1/_2$–1	1	$^1/_2$–1				
Tall fescue	$^1/_2$–1		$^1/_2$–1						1		1	
Zoysiagrass				$^1/_2$		$^1/_2$–1		$^1/_2$				

Note: All rates are for nitrogen in pounds per 1000 ft^2 of turf area. These schedules were developed for the mid-Atlantic region and will vary for other locations. Contact the local Cooperative Extension Service office for more complete and detailed information.

ment and salt balance. Soil problems associated with salinity include both amounts and composition of salts, particularly sodium. Until the saturated extract of soils has an electrical conductivity (EC) >4 dS/m the soil problems are minimal. In regions where there is adequate rainfall, salts which are introduced into the soil from fertilizers, road deicing compounds, or irrigation water are periodically leached from the root zone. Even in arid or semiarid regions where rainfall is limited, irrigation can be used to leach salts from the root zone preventing damage to the plant. When the salt buildup exceeds an EC of 4 dS/m problems develop first with the root system and then physiologically within the plant.

Another aspect of salinity is the potential for impact on soil physical properties. If an excess of sodium is added in the irrigation water, the sodium can displace the calcium and cause dispersion of the soil structure if the base saturation due to Na exceeds 15%. This changes the permeability and affects water infiltration and percolation and oxygen exchange. Dealing with this problem means not only flushing the salts from the soil profile, but usually adding calcium (gypsum) to displace the sodium. If the total amount of salts is high and a large proportion of this is due to sodium then the problem is greatly compounded.

RECOMMENDATIONS FOR FERTILIZATION

PRIOR TO ESTABLISHMENT

Special consideration should be given to soil testing prior to establishing the turf. This is the opportune time to thoroughly incorporate lime, phosphorus, and potassium into the root zone. This allows for a quicker reaction and less time for pH adjustment. Liming materials and phosphorus fertilizers move very slowly downward in the soil profile. Applications made to the soil surface may move only $^1/_2$ to 1 in. downward in a year. Therefore,

applying and thoroughly incorporating these materials through rototilling or other tillage operations before the turf is planted will ensure good distribution throughout the root zone. While recommendations for soil pH adjustment, phosphorus, and potassium are best taken from soil test reports, information from Table 2 can be used for the lime requirement. In the absence of any information on soil pH or nutrition levels, for best seedling growth of all grasses except bahiagrass and centipedegrass, 50 lb of ground limestone per 1000 ft^2 and one of the following fertilizers should be applied:

(1) 40 lb of 5-10-5; 20 lb of 10-20-10; or
(2) 20 lb of 8-8-8 or 10-10-10 in combination with 4 lb of 0-46-0.

This should be thoroughly incorporated into the top 4 to 6 inches of the soil. Additional nitrogen should be provided at the first mowing after the grass is planted.

ESTABLISHED LAWNS

A soil test should be made every 2 to 3 years to determine the amounts of lime, phosphorus, and potassium needed by the lawn. A complete fertilizer with a ratio of nitrogen (N): phosphorus (P$_2$O$_5$): potash (K$_2$O) of 4:1:2 or 4:1:3 can be used in lieu of a soil test, but it is a poor substitute.

The proper amount of fertilizer and time of application vary with the grasses being grown. To help reduce turf loss, avoid high nitrogen fertilization of cool-season grasses in late spring or summer and to the warm-season grasses in the fall or winter. A suggested fertilization schedule for most of the turfgrasses commonly used for lawns is given in Table 4.

Proper turf maintenance must combine mowing, fertilization, and irrigation into a well-designed program. With turf fertilization, soil testing is a must to adjust and make the most efficient use of the materials being applied.

Chapter 8

Physical Problems of Fine-Textured Soils

Robert N. Carrow, *Crop and Soil Science Department, University of Georgia, Griffin, GA*

CONTENTS

Introduction 79
Common Problems 79
    Excess Quantities of Silt or Clay 79
    Soil Compaction 80
    Layers 81
    Salt-Affected Soils 81
    High Water Table 81
    Improper Contouring 81
Correcting Soil Physical Problems on Fine-Textured Soils 81
    Cultivation 82
    Soil Modification 82
    Drainage 82
Conclusions 83
References 83

INTRODUCTION

Soils are the major source of essential nutrients and water for turfgrass and ornamental growth. The nature of different soils arise from the diversity of components such as: (1) mineral matter — sand, silt, clay, gravel, stones; (2) organic matter — fresh, undecomposed organic matter as well as decomposed organic matter; (3) inorganic chemicals, such as iron oxides, gypsum, calcium carbonate, and many other chemicals; (4) biological entities including microorganisms, worms, mollusks, anthropods, vertebrates, and various plants; and (5) pore space which contains water, air, or both.

Soil properties are divided into three groups — physical, chemical, and biological. Examples of these properties are listed in Tables 1 and 2. A major challenge for turf and ornamental managers is to alter soil properties so as to obtain a better growing medium.

The objective of this article is to identify the major soil physical problems that confront turf and ornamental managers with fine textured soils (i.e., soils high in clay and/or silt content). Once a grower is aware of the specific problem(s) on his site, then appropriate management decisions can be made to alleviate the problem.

COMMON PROBLEMS

Certain soil physical problems are common on fine-textured soils. Identification of the primary problem is not always easy. What the grower often sees is *symptoms* or *secondary problems* such as: poor rooting, low soil aeration, black layer, low infiltration/percolation, waterlogged condition, on hard soils. These, however, can be caused by several primary soil physical problems. Unless the *primary or basic problem* is corrected, then dealing with the secondary problems is like treating a disease's symptoms but not the disease.

A basic principle of integrated pest management (IPM) is to start with a healthy plant that has the potential to persist through environmental or pest stresses. Since plants obtain water and nutrients from the soil, soil physical conditions have a dramatic influence on the physiological health of plants.[1]

EXCESS QUANTITIES OF SILT OR CLAY

A primary problem on fine-textured soil is simply excess quantities of silt or clay.[2] High silt/clay content results in a soil with too little macro (large) pore space, essential for water infiltration/percolation, root channels, and gas exchange (aeration).

0-87371-350-8/94/$0.00+$.50

Table 1　**Soil physical properties**

Aeration	The air exchange status of a soil. Often assessed in terms of oxygen diffusion rate or macro pore space
Bulk density	Density of a soil in terms of mass per unit volume. Is an indicator of soil strength and porosity
Color	Soil color is an indicator of aeration status, drainage conditions, and soil organic matter content
Pore-size distribution (porosity)	Three components are total pore space, macro pores, and micro pores. Macro pores are those that rapidly drain after a rainfall or irrigation event. These are important for water infiltration/percolation, air exchange, and root growth
Soil strength	Hardness of the soil. Soils with high soil strength often limit rooting. High soil strength is related to percent of clay, compaction level, and soil moisture status
Structure	Arrangement of primary soil separates (i.e., sand, silt, clay particles) into units held together by chemical and physical forces
Temperature	The temperature of a soil and its heat transfer ability
Texture	Soil particles are subdivided in texture classes based on particle size. The three primary classes are sand, silt, and clay. Texture has a major influence on all other soil properties
Water infiltration	Water movement into the soil
Water percolation	Water movement through the plant rootzone
Water drainage	Water movement beyond the rootzone and surface water movement

These soils can exhibit all of these symptoms discussed in the previous section.

Excessively fine-textured soils are generally poor growing media. An exception would be if the soil has a well-developed granular structure. Unfortunately, structural units are easily destroyed by vehicle and foot traffic, especially on recreational sites. Examples of areas of the country where excessive clay and silt content are present in soils include the Houston black clay areas, the very high clay content soils present within a mile of the mountains in Palm Springs, and the Piedmont red clay region of the Southeast. Many other sites also have this problem because the original topsoil has eroded, been mixed into the subsoil, or been sold. The remaining "topsoil" is often the high clay content B horizon.

SOIL COMPACTION

Soil compaction is a second, prominent problem on fine-textured soils.[1] Situations include: (1) compaction of the surface 2 to 8 cm, the most common type on established turf areas, (2) compaction of several inches (10 to 30 cm) of the surface due to heavy equipment traffic during construction or on established turf, (3) a buried zone of compacted soil several inches thick from heavy equipment operation during construction followed by application of a topsoil layer or tillage of the surface part of the compacted zone but not to the full depth, and (4) a compacted zone beneath the normal coring tine depth of 8 cm.

Table 2　**Soil chemical and biological properties**

Soil Chemical Properties
　Anaerobic produced chemical compounds. Compounds that appear under low soil O_2 conditions such as sulfides
　$CaCO_3$ levels. Acts as a buffering system in some alkaline soils
　Cation exchange capacity. Important for nutrient retention. Is a buffering system in all soils
　Nutrient levels and ratios with other nutrients
　Soil pH. Influences nutrient forms and plant availability
　Salt-affected soils. Sodic, saline, saline/sodic
　Toxic compounds. From heavy metals, naturally occurring compounds, herbicides, etc.
　Organic matter content. Influences many other chemical and physical properties
Biological Properties
　Insects. Beneficial and harmful
　Microorganism populations. Beneficial and pathogenetic
　Nematode populations. Beneficial and harmful
　Plant species present
　Small animals
　Worms

All of these problems increase in prevalence and seriousness as the clay content increases in a soil. For example, on a sandy loam soil a compacted zone beneath the normal core aeration zone may never form or take many years, but on a clay loam it may form in two or three coring operations.

Even soils with relatively high sand contents, if subjected to compaction, will exhibit the same symptoms as a much finer-textured noncompacted soil. Modification of golf greens and athletic fields to minimize the potential of soil compaction requires at least 85% sand content and often up to 95% sand. Compaction presses the soil particles together, increases soil density, and destroys the macropore space. The soil, therefore, performs physically as if it had a much higher clay content.

LAYERS

The third prominent soil physical problem is the presence of layers within the soil that impede drainage, rooting, and gas exchange.[3] Several types of layer situations can occur. Compacted zones, discussed above, are one type of layer. Deposition of clay and silt over a sandy soil creates a soil that has the characteristics of a fine-textured soil even if most of the profile is sand.

Another example is soil horizons that form under natural soil formation processes. Sometimes these can be distinctly different. Most often the B horizon (commonly the second horizon from the surface) will have a higher clay content than the surface A horizon. Thus, the surface several inches of a soil may have good physical properties but a subsurface horizon may limit deep rooting and water drainage. This is a typical occurrence on many Piedmont region soils.

SALT-AFFECTED SOILS

Salt-affected soils, especially those classified as *sodic*, may exhibit poor soil physical conditions.[4] High sodium content on the cation exchange sites of the clay fraction causes individual clay particles to be dispersed or deflocculated. This destroys granular structure and increases in seriousness as clay content increases. Sodic soils may arise through natural soil formation processes in semiarid and arid climates or from application of irrigation water containing excessive sodium. While sodic soils have several adverse effects on turfgrasses, the primary problem is due to loss of structure and poor soil physical properties.

Saline/sodic soils may revert to a sodic soil, thereby exhibiting a degradation of structure, if care is not exercised in managing the relative quantity of sodium vs. calcium, magnesium, and potassium. As long as the sodium ratio is not too high, the main problem on these soils is due to increased osmotic potential from high total salts; thereby, reducing plant water uptake.

Saline soils do not inherently exhibit poor soil physical conditions since the sodium content is low, but they do have high osmotic potentials from the total salt content. Also, salt-affected soils may exhibit any of the other physical problems discussed in this chapter.

HIGH WATER TABLE

High water tables on fine textured soils can result in many of the symptoms associated with poor structure or high clay content soils. Since the capillary fringe zone above the water table is considerably larger for fine-textured vs. sandy soils, even a water table 0.75 to 1.0 m deep can result in an excessively wet surface.

While a high natural water table is the most common situation, a perched water table can sometimes occur within a soil profile. Any distinctly different textural layer with an abrupt interface between layers may cause interference with water percolation and result in water perching above the interface for a period of time. Any turfgrass roots within this zone then experience low oxygen levels. Typically, a layer high in clay (even 3 to 8 mm) or a compacted zone is the cause of a perched water table.

IMPROPER CONTOURING

Improper contouring on many fine-textured soils contributes to soil physical problems. Fine-textured soils often have low infiltration and percolation rates. When excess water is channeled to a localized site, because of improper contouring, the symptoms (waterlogging, black layer, poor rooting, etc.) of adverse soil physical properties are enhanced. Examples of contouring problems include bowl-shaped greens, small depressions in greens or fairways, inadequate slopes for surface drainage of greens and tees, lack of diversion features to prevent excess surface water onto an area, and lack of landscape features to rapidly transport excess water from an area.

CORRECTING SOIL PHYSICAL PROBLEMS ON FINE-TEXTURED SOILS

It is very important to identify which of the above soil physical problems are present on a site in order to choose the best approach to correcting them, and to be able to evaluate the effectiveness of your choices. The spatial location of a problem area in a soil is also very important. For example, is the major problem surface compaction, a compacted zone at 10 cm, or a high water table at 1 m depth? The primary management approaches to dealing with the basic causes of soil physical problems follow.

CULTIVATION

Cultivation is often an option for improving soil physical properties. For ornamentals, cultivation would be conducted at establishment, but on recreational turfgrass sites, a routine cultivation program may be necessary. The basic benefit of cultivation is to create, at least on a temporary basis, macropores for water movement, aeration, and root growth. These macropores may be formed by loosening of the soil or by creation of holes or channels in the soil.

Many different turfgrass cultivation techniques are available with some penetrating up to 50 cm deep. Development of a sound cultivation program[1,5] depends upon (1) identifying the specific problem and its spatial location in the soil, (2) selecting the appropriate technique or techniques, (3) cultivating at the correct time of year and with the right frequency, (4) applying cultivation when the soil moisture is appropriate for the method, and (5) continually evaluating the program for effectiveness.

Effectiveness of a particular cultivation operation is dependent on a number of factors, including:

- size and number of holes made in the soil
- whether soil was removed. Generally holes made where soil is removed will remain open longer than where soil is compressed to form a hole. Hollow tine core units and Deep-Drill aerification remove soil.
- whether sand has been used to keep the cultivation hole open, as well as to dilute the layer with sand.
- clay type. Expanding clays when saturated can seal a hole, while nonexpanding clays do not.
- depth of cultivation. A deep cultivation hole should last longer than one at the surface
- factors that cause a layer to reform after cultivation. The most dramatic example is surface soil compaction created by frequent traffic. Other examples would be frequent flood deposited silt and layers with expanding clays that reseal. In such situations, a continuous cultivation program is essential.
- complete penetration through any existing layer. Cultivation with a procedure that will penetrate through the layer will have the best results.
- cultivation procedures that have a "loosening" action due to their cultivation tine action help to fracture the soil between holes and alleviate any compression of soil caused by tine penetration. Examples are Verti-Drain, Turf Conditioner, Aera-Vator, and Aerway slicer units.
- whether any layer is produced by the cultivation operation. Anytime a tool is pushed into the soil, some compaction results. However, if more compaction is alleviated than created the net result is beneficial. Sometimes cultivation units may create a compacted zone just below the depth of tine penetration on soils susceptible to compaction. Surface core aeration units with solid tines are most prone to this problem. Periodic deep cultivation to penetrate through this zone will correct this problem.
- cultivation methods that create pressure to the sides of the tines may create a more compacted soil between tine holes. As the holes seal or sand is added to the holes, the soil between holes may remain more compacted than originally. This is one reason why slicing and pin spiking are useful for short-term alleviation of surface compaction but not as the sole method of cultivation over time.

SOIL MODIFICATION

Either physical and/or chemical modification may be appropriate.[2,6,7] Physical modification on turf sites may be partial, such as core aeration and backfilling holes with sand, or complete, such as a U.S. Golf Association (UAGA) Green Section green.[6] During establishment of turf or ornamentals, various organic and inorganic physical amendments can be used to partially modify the soil.[2]

Chemical modification on fine-textured soils has included the use of gypsum on sodic soils, wetting agents, algae-based products, various "secret" formulations, and soil conditioners such as polyvinyl alcohols. The benefits of gypsum on sodic soils has sound research evidence. Research evidence, however, showing soil physical improvements of the other products, especially under traffic situations, is very limited. Sometimes materials are promoted to stimulate structure formation or stabilize structure on fine-textured soils. Since natural structural units (i.e., structural aggregates in a well-aggregated soil) cannot maintain their integrity under the compacting forces of traffic, it seems doubtful that these products will do better than nature. Sodic soils have no structure (i.e., a massive structure) throughout their profile. Even with gypsum, however, the surface structure is destroyed by traffic.

DRAINAGE

Subsurface and surface drainage techniques are very useful in eliminating many of the excess water problems on fine-textured soils.[6] Tile lines to drain wet areas and to drop a water table are sometimes necessary. Many of the surface drainage methods are essentially proper contouring and include: (1) providing adequate surface drainage; (2) use of grassed waterways, ditches, french drains, and mounds to divert water or to channel water across an area; (3) use of dry wells to drain small wet areas; and (4) use of deep probes to penetrate subsurface soil layers that pond water above a layer.

In addition to these three primary approaches to soil physical problems, other management techniques that are useful in some cases are (1) preventive measures to avoid any man-induced soil layers, including compacted zones from forming during construction or routine maintenance. (Man-induced layers can arise from improper topdressing mixes, a silt layer in sod, poor mixing of amendments, a thatch layer that becomes buried over time, and compacted zones during construction or cultivation operations); (2) avoiding excessive irrigation; (3) traffic control measures to spread traffic, such as larger tees and greens, a systematic cup rotation pattern, and control of golf car traffic on golf courses.

CONCLUSIONS

Many soil physical problems mentioned in this chapter may require a combination of cultivation, soil modification, drainage, and other approaches. For any procedure to be effective in alleviating soil physical problems, it must prevent/remove excess water from the site or alter the physical properties of the soil, primarily, pore-size distribution and bulk density. Unless pore-size distribution and soil density are significantly changed, very little change can be expected in the symptoms of water retention, infiltration, percolation, gas exchange, or rooting. To alleviate poor bulk density and pore-size distribution, cultivation or soil modification operations must reach the zone within the soil where these problems occur.

Remember that all soil physical properties (air, water, temperature) are controlled by physical laws of nature. To improve the physical characteristics of a soil as a growing medium requires an appreciation of how water and gases move (or are retained) in a soil in accordance to physical laws of nature.

There are no "magic bullet" compounds or procedures that are a cure-all for all soil physical problems. Once the primary problems are identified, the basic management tools to alleviate these problems are numerous cultivation procedures, a diversity of soil modification approaches, and many surface/subsurface drainage techniques.

On recreational turf sites, physical problems, especially compaction, are considered the most important factors adversely affecting turfgrass maintenance. Ornamental tree, shrub, and flower growth is also often limited due to physical conditions. Improvement of soil physical properties to promote healthy plants with deep, extensive root systems is a major component of an overall IPM program for turf and ornaments.

REFERENCES

1. **Carrow, R. N. and Petrovic, A. M.,** Effects of traffic on turfgrasses, in *Turfgrass*, Monograph No. 32, Waddington, D. V., Carrow, R. N., and Shearman, R. C., Eds., American Society of Agronomists, Madison, WI, 1992, chap. 9.

2. **Waddington, D. V.,** Soils, soil mixes, and soil amendments, in *Turfgrass*, Monograph No. 32, Waddington, D. V., Carrow, R. N., and Shearman, R. C., Eds., American Society of Agronomists, Madison, WI, 1992, chap. 10.

3. **Carrow, R. N.,** Understanding layered and compacted soils, *Golf Course Manage.*, 60(2), 52, 1992.

4. **Harivandi, M. A., Butler, J. D., and Wu, L.,** Salinity and turfgrass culture, in *Turfgrass*, Monograph No. 32, Waddington, D. V., Carrow, R. N., and Shearman, R. C., Eds., American Society of Agronomists, Madison, WI, 1992, chap. 6.

5. **Carrow, R. N.,** Developing turfgrass cultivation programs, *Golf Course Manage.*, 58(8), 14, 1990.

6. **Beard, J. B.,** *Turf Management For Golf Courses*, Burgess Publishing, Minneapolis, 1982, chap. 3.

7. **Daniel, W. H. and Freeborg, R. P.,** *Turf Managers' Handbook*, Harvest Publications, Cleveland, 1979.

Chapter 9

Physical Problems of Coarse-Textured Soils

Robert N. Carrow, Crop and Soil Science Department, University of Georgia, Griffin, GA

CONTENTS

Coarse-Textured Soils 85
Sand Classification 85
Common Physical Problems 86
    Low Water-Holding Capacity 86
    High Salt Content (Salinity) 88
    Layers 88
    Hydrophobic (Water Repellent) Sands 88
    High Water Table 88
    Improper Contouring 88
    "Hard" Sands 88
    "Soft" Sands 89
References 89

COARSE-TEXTURED SOILS

Soil scientists classify soils into 12 different texture classes based on their percent sand, silt, and clay contents. The three texture classes with the highest quantity of sand are sand, loamy sand, and sandy loam (Table 1). These are often termed "sandy" or "coarse-textured" soils. Obviously, a soil can be called "sandy" but contain considerable silt and clay. Physical, chemical, and biological properties can vary dramatically even between two soils within the same texture class.

Sandy or coarse-textured soils, whether natural or constructed, are used on many high-traffic sites because of their better soil physical properties relative to a soil with appreciable silt and/or clay. Certainly, a rootzone media developed to the specifications of the U.S. Golf Association (USGA) Green Section for golf greens and athletic fields would be expected to resist compaction and have ample macropores for water movement, gas exchange, and root penetration.[1] However, since sandy soils vary, certain physical problems may occur that adversely affect turfgrasses. The purpose of this article is to discuss major soil physical problems observed on coarse-textured soils. A very high percentage of physical problems observed on sandy soils would be caused by one or more of these factors.

Awareness of the specific problem(s) on a site is prerequisite to development of sound management programs. Any soil-based problem that adversely affects root growth soon will have a detrimental influence on overall plant vigor and physiological health. The beginning of any integrated pest management (IPM) program is to start with a healthy plant. In this paper the focus will be on physical properties, but problems can also arise on sandy soils due to adverse chemical and/or biological soil properties. Common soil chemical and biological problems are listed for reference in Table 2 but not discussed in detail.

Table 1 Sand, silt, and clay contents of coarse-textured (i.e., sandy) soils

Soil Texture Class[a]	Composition (% by Weight)		
	Sand	**Silt**	**Clay**
Sand	85–100	0–15	0–15
Loamy sand	70–90	0–30	10–30
Sandy loam	45–85	0–50	15–55

[a] U.S. Department of Agriculture texture classification.

SAND CLASSIFICATION

The separate class of "sand" is based on the diameter of the mineral particles in a soil (Table 3).

0-87371-350-8/94/$0.00+$.50

Table 2 **Primary chemical and biological problems observed on sandy soils**

Soil chemical problems
1. Nutrient deficiencies and imbalances affecting plant nutrition
2. Low CEC (cation exchange capacity) which contributes to low nutrient retention, low buffering capacity, and high leaching potential
3. Improper soil pH affecting nutrient availability
4. Salt-affected sand soil which influences nutrient uptake and nutrient balances
   - saline — most serious on sands
   - sodic
   - saline/sodic
5. Presence of free $CaCO_3$ that acts as a buffer system in alkaline sands and contributes to nutritional problems
6. Toxic compounds from heavy metals, allopathy substances, herbicides, etc. are potential problems due to low buffering capacity of most sands

Soil biological problems
1. All soils, fine-textured or coarse-textured, can contain weed seeds, disease organisms, and harmful insects, small animals, and worms
2. Low microorganism population. Sands are likely to contain fewer microorganisms than fine-textured soil
3. Less diversity of microorganisms than fine-textured soils. This creates the potential for microorganisms population balances to be easily shifted from beneficial to pathogenic
4. Nematodes. Many nematodes prefer sandier soils

Table 3 **Sand separate classes compared to silt and clay**

Separate	Particle Diameter (mm)	Number of Particles per Gram	Surface Area per 1 g[a] (cm^2)
Sand			
Very coarse sand	2.00–1.00	90	13
Coarse sand	1.00–0.50	720	26
Medium sand	0.50–0.25	5700	46
Fine sand	0.25–0.10	46,000	92
Very fine sand	0.10–0.05	722,000	229
Silt	0.05–0.002	5,776,000	460
Clay	Below 0.002	90,260,853,000	8,100,000

[a] 1 lb Soil = 454 g.

Sands are within the range of 2.00- to 0.05-mm diameter, a 40-fold range. A very coarse sand will not be the same as a very fine sand in its properties. Sometimes general names are used to designate a sand instead of the official separate classification (USDA system). Some common ones are (1) concrete sand — usually contains a wide particle size range of sands and fine gravel, (2) Mason's (mortar) sand — similar to a concrete sand but without the fine gravel, (3) dune sand — obtained from a wind or water deposited sand dune, often has a fairly narrow particle size range, and (4) river sand — can vary from very uniform sand to sand with considerable fines, derived from rivers. Thus, general sands are named not by well-defined particle ranges but by source or construction use. They may or may not be good for rootzone mixes.

For turfgrass sites that receive frequent traffic, rootzones are often constructed using sands that must meet well-defined particle size ranges.[1] However, on turfgrass or ornamental sites where sand is added to an existing soil and limited traffic is expected, sand size specifications are less important.

COMMON PHYSICAL PROBLEMS

LOW WATER-HOLDING CAPACITY

The most prevalent problem is low water-holding capacity. Available water for plants may range from 1.0 to 3.8 cm water per 30 cm depth of sand when organic matter content is less than 1% (by weight) and with no perched water table. Thus, drought stress occurs unless frequent irrigation is practiced.

A common approach to improving water holding capacity is by addition of 5 to 15% by volume of a well-decomposed organic matter. Generally, well-decomposed organic matter is recommended so that it will persist. Excessive moisture retention and low aeration can occur if too much organic matter is used. Many different organic sources can be used if they are well decomposed, have a good particle size for mixing, and do not contain excessive fines.[2] Common organic sources used on turf and ornamental sites are various peats, decomposed rice hulls, and composted sewage sludges.

Addition of silt and/or clay will enhance water retention but too much can easily seal the macropore channels. In high sand content media (>80% sand), particles are in direct contact with other sand particles. This produces a rigid matrix that resists compaction. However, relatively small quantities of silt and clay can accumulate at points of sand particle contact and start to seal off pore channels necessary for water movement. Silts are especially effective in sealing and decreasing pore continuity. For this reason, the USGA Green Section specifications for putting greens limit silt to less than 5% and clay to less than 3% by weight when severely compacted.[1] Sandy soils of between 65 and 85% sand may be prone to exhibiting poor water movement. These sandy soils have considerable silt/clay to seal many pores but too much sand to allow good structure formation. Structure develops when clay, silt, organic matter, and sand particles start to aggregate into structural units that open up new macropores, but high sand inhibits formation of strong aggregates. When aggregates form in a sand, they consist of primarily silt, clay, and organic-matter components.

Very fine sands will retain more water than coarser sands due to the presence of smaller pores. However, adding very fine sand to a medium to coarse sand is not recommended. Water retention would increase but infiltration and percolation rates would decline as the smaller particles fill the macropores.

Sometimes inorganic amendments (such as calcined clays, expanded shale, processed vermiculite, mica, porous ceramics, etc.) are used to enhance water retention.[2] To be effective, inorganic amendments must retain water in pores within the particles and the pores must be large enough to release the water to the turf plant, not accumulate salts in these pores, retain their physical structure and not deteriorate, and be cost competitive with various organic amendments at application rates to achieve comparable water retention.

Barriers that inhibit drainage will increase water retention in sands. In the USGA Green Section specifications for golf greens, a distinct interface is formed between the coarse sand (1- to 2-mm diameter) and pea gravel (6- to 10-mm diameter) layers that impedes drainage and, thereby, increases water content in the rootzone by creating a "perched" water table.[1] Water is retained around the coarse sand by adhesion-cohesion, while few interconnecting water films exist between the coarse sand and gravel because of distinct differences in particle sizes. Only after water "ponds" above the interface a few inches will drainage occur. Once drainage starts, it is rapid through the large pores. Thus, this "unique" type of barrier allows enhanced water retention while maintaining good drainage.

The Purr-Wick (golf greens) and PAT (athletic fields) construction systems use an enclosed cell method system to prevent drainage until water reaches a certain level, which can be adjusted as needed.[3] This is essentially a method of water table control. Other procedures to adjust the water table level to allow capillary rise of water to the roots have been used in flat sod fields but not on other turf areas.

In recent years, water-absorbing polymers have been promoted for use in turfgrass soils to enhance water-holding capacity.[4] Polymers are formed by combining two or more smaller molecules into a larger chain-like polymer. Examples include (1) natural polymers such as starches, and (2) synthetic polymers such as polyacrylamides (PAM), polyvinyl alcohols (PVA), and polyacrylates.

These are not new types of compounds (a particular chemical may be new since polymers can be formulated in many different lengths, co-linked, and other substances added). In fact, McGuire, Carrow, and Troll evaluated several PAMs and one PVA in a USGA-supported project for influence on moisture retention of sands in 1978 and found no influence.[5] Limited additional research has been conducted until recently. As more results are published, the potential for these materials will become clearer. Important questions are (1) effective application rates; (2) total and plant available water retained; (3) longevity of the materials; (4) influence of soil solutes on actual vs. potential water absorption; and (5) the potential for salt accumulation.[4]

Other procedures to influence rootzone water content or availability are (1) careful overhead irrigation which does not increase the ability of the sand to retain water but does allow for frequent addition, and (2) subsurface irrigation which has been attempted but is especially difficult on a sandy medium. Capillary rise on sands is less than if a soil has appreciable silt or clay; thereby requiring very careful placement of water emitters. Also, sands have slow unsaturated water flow through water films around sand particles which may limit water availability in high-demand periods; but they have very high saturated flow, so if output is increased by emitters to create saturated conditions, the water will rapidly drain downward out of the rootzone, and (3) any management practice to enhance rooting depth provides more water to the plant even though soil water-content is not altered.

HIGH SALT CONTENT (SALINITY)

High salt content in sandy soils dramatically influences plant water availability and is a second major soil physical problem. Solutes attract water films by adhesion forces (primarily H-bonding); thereby, increasing osmotic potential. Plant water availability is decreased even though total soil water content remains unaffected. In the case of sands, high sodium is detrimental primarily because it adds to the total salinity. While few structural aggregates are present in sands, high sodium will cause dispersion of any silt and clay which may migrate to the depth of routine irrigation water penetration to reform a mini-layer. In contrast, high sodium on a fine-textured soil has its major effect via destroying structure and secondary influence via reducing water availability.

Cultural means of alleviating salt buildup are primarily by leaching with excess water and/or the use of better quality irrigation water.[6] When leaching, the quantity of water depends upon the water quality, existing level of salts present in the soil, salt tolerance of the turf, and how much water goes toward evapotranspiration vs. leaching (i.e., arid climates require more total water).

LAYERS

Presence of fine-textured layers within the sand rootzone is another common soil physical problem.[7] As pointed out before, even small quantities of silt/clay can seal a zone within a sand if the particles accumulate at a micro-site. Layers arise from many sources but common ones are from sod, by use of a topdressing media with fines, incomplete on-site mixing of soil and amendments, and wind/water deposition. Sometimes clay lens (areas of silt/clay deposition) are present deeper in sand soil profiles due to water deposited fines as the soil developed.

Another type of layer is calcite formation at the surface due to irrigation water with high calcite (calcium carbonate) content. This can seal the surface and reduce infiltration. Acidification of the irrigation water to dissolve the layer is effective,

but the soil should be observed to make sure the layer does not form at the bottom of routine irrigation water penetration depth. Salts of various types can accumulate on the surface of soils in arid regions when sufficient capillary rise occurs. As the water moves to the soil surface it carries solutes which are precipitated out as the water evaporates.

Management of layers is primarily by two methods (1) prevention of layer formation, and (2) cultivation deep enough to penetrate the layer. Other methods may be appropriate in certain situations such as leaching of solutes or acidification of irrigation water.

HYDROPHOBIC (WATER REPELLENT) SANDS

Hydrophobic (water repellent) sands can be a major problem on very high sand content sites, where the surface area of the sands is limited. Organic coatings may form on sand particles that become highly water repellent.[8] Typically, the hydrophobic zone is 2 to 10 cm in depth and erratic over a site with affected and unaffected areas side-by-side.

Wetting agents are used to rewet the sands, often in conjunction with core aeration for creation of areas for water to collect and rewet adjacent soil. Normally, wetting agent treatment must be repeated since the organic coatings are not removed. Dr. Keith Karnok, University of Georgia, has been refining some treatments that may allow for dissolution of the coatings for a more permanent solution.[8]

HIGH WATER TABLE

As in any soil, a high water table can be present in a sand and result in excessive moisture. Each site should be evaluated to determine whether the water table is naturally high or whether a "perched" water table occurs due to a layer impeding drainage.

IMPROPER CONTOURING

Improper contouring that channels water to low spots can lead to two problems on sands: a wet area on which water stands and a dry site from which water came. Many times improper contouring enhances other problems such as layering or a high water table. One common example of poor contouring is golf greens with "pockets" that have no surface drainage. These pockets often collect fines over time and become susceptiable to scald and intracellular freezing. Proper contouring is best achieved prior to turf establishment.

"HARD" SANDS

"Hard" sands are a frequent complaint of turf managers for the first 1 to 3 years after turf establishment. The problem can also occur on older sites. These sand media do not have much resiliency, exhibit high ball bounce, and when on an athletic field have little "give." Thus, the term "hard" sands refers to the resiliency of the site and not to the strength or hardness of individual sand particles.

Hard sands can occur for several reasons. On new greens, the organic matter added will not result in the same degree of resiliency that occurs after turfgrass roots have grown. Bentgrasses, especially, develop a high mass of roots in the surface inch. Also, new greens will not have thatch, which contributes to resiliency.

Sands that are angular in shape, instead of more rounded, tend to fit together to form a rigid matrix. This matrix has less pore space than one formed with rounded sands. As a result, water movement is not as good and the sand is less resilient.

Construction specifications for high sand rootzone mixes normally indicate the use of sands with a narrow particle size distribution.[1] For example, sands with 75% of the particles in two adjacent particle size ranges are preferred. This ensures good pore space distribution as long as very fine sands are avoided that would plug many of the macropore size pores of a medium to coarse sand. Sands with wide particle size distributions will be "harder" since the particles fit together into a more dense media. This occurs whether the fine particles are very fine sand, silt, clay, or a combination. Many concrete and mortar sands are hard if used for growing turf because of their wide particle size ranges.

Organic content, whether it is achieved by organic matter addition or roots, will help soften any hard sand to some extent. Thatch development of 0.5 to 1.5 cm will also create greater resiliency.[9] However, hard sands, because of a wide particle size distribution or excessive fines, will still exhibit poor water movement and will remain hard when dry. Topdressing with a rounded sand in the medium to coarse range will improve conditions over time if done in conjunction with core aeration.

"SOFT" SANDS

"Soft" sands with a tendency to shift is another problem sometimes observed, especially on golf greens and athletic fields with high sand content rootzone mixes. Sands that are round and of a narrow particle size range, such as one with 70 to 80% of particles in one size range, will feel soft and not have good traction. These sands do not have enough fine particles to limit particle shifting. Other problems can contribute to soft soils but these are primarily on fine-textured soils; for example, water logged conditions and slippage zones. Obviously, high traffic and sharp turning creates more surface stability problems on any soil regardless of texture.

Solutions to "soft" sands are (1) avoiding sands that are too uniform, (2) adding organic matter to aid in stabilization, (3) promoting good root development, (4) in special situations using rootzone stabilization materials such as VHAF (vertical, horizontal, angular fibers) or Netlon mesh,[10] and (5) maintaining good moisture during times of high traffic. Moist, but not saturated, sands have appreciably more rigidity than a dry sand.

In conclusion, we often think of sandy soils as possessing good soil physical properties, which they do in comparison to fine-textured soils subjected to traffic. However, a surprising variety of soil physical problems can still occur on what we class as sand soils or sandy rootzone media. Management of sandy soils has its challenges. Turf and ornamental managers will continue to find that sands and sandy soils can be extremely variable growing media.

REFERENCES

1. **Beard, J. B.,** *Turf Management for Golf Courses*, Burgess Publishing, Minneapolis, 1982, chap. 3.
2. **Waddington, D. V.,** Soils, soil mixes, and soil amendments, in *Turfgrass*, Monograph No. 32, Waddington, D. V., Carrow, R. N., and Shearman, R. C., Eds., American Society of Agronomists, Madison, WI, 1992, chap. 10.
3. **Daniel, W. H. and Freeborg, R. P.,** *Turf Managers' Handbook*, Harvest Publishing, Cleveland, 1979.
4. **Koski, T.,** Polymers: can they work for you?, *Ornamental and Turf Newsletter*, Vol. 1(1), Iowa State University Extension Horticulture Publication, Ames, IA, April 1992, 2.
5. **McGuire, E., Carrow, R. N., and Troll, J.,** Chemical soil conditioner effects on sandy soils and turfgrass growth, *Agron. J.*, 70, 317, 1978.
6. **Harivandi, M. A., Butler, J. D., and Wu, L.,** Salinity and turfgrass culture, in *Turfgrass*, Monograph No. 32, Waddington, D. V., Carrow, R. N., and Shearman, R. C., Eds., American Society of Agronomists, Madison, WI, 1992, chap. 6.
7. **Carrow, R. N.,** Understanding layered and compacted soils, *Golf Course Manage.*, 60(2), 52, 1992.
8. **Tucker, K. A., Karnock, K. J., Radcliffe, D. E., Landry, G., Jr., Roncadori, R. W., and Tan, K. H.,** Localized dry spots as caused by hydrophobic sand on turfgrass greens, *Agron. J.*, 82, 549, 1990.

9. **Rogers, J. N., III. and Waddington, D. V.,** Impact absorption characteristics on turf and soil surfaces, *Agron. J.*, 84, 203, 1992.

10. **Baker, S. W.,** The effect of reinforcement materials on renovated turf on topsoil and sand rootzones, *J. Sports Turf Res. Instit.*, 66, 70, 1990.

Chapter 10

Principles of Turfgrass Growth and Development

John E. Kaufmann, Monsanto Company, St. Louis, MO

CONTENTS

Introduction 91
Turfgrass Morphology and Growth 91
    Shoots 92
    Roots 92
Understanding the Dynamics of the Turfgrass Ecosystem 93
    Building Turfgrass Biomass 93
    Turfgrass Regeneration and Rejuvenation 93
    Duration of Turfgrass Leaves 93
    Dormancy and Turfgrass Color 93
Key Environmental Factors Controlling Growth 94
    Cool-Season Grass Growth 94
    Warm-Season Grass Growth 94
Seasonal Growth Cycle of Perennial Cool-Season Grasses 95
Reproductive Variation Among Species and Varieties 97
Summary 97
References 97

INTRODUCTION

Within the past few years, many questions and concerns have been raised about the need for lawns. Conceptually, the lawn is a space between the home and the world, whether the world is desert, forest, or anything in between. While the origin of lawns is unknown, the principal need was most likely related to personal safety. The lawn or clearing was a buffer that provided visible exposure of intruders, in time for a home occupant to act in some fashion. The ability to quickly survey surroundings and clearly define when space has been violated continues to provide humankind a level of comfort that cannot be achieved in any other manner. Lawns continue to be the most cost-effective, practical and natural method of delineating and designating space around the home.

Questions have also been asked about the need for nonnative grasses in lawn culture. Clearly, space delineation requires height control of vegetation. Thus, to qualify as a turfgrass a plant species must, first and foremost, be able to withstand frequent mowing. The origin of most of the cool-season turfgrasses is in the same region of the world as the cradle of civilization, where it is thought that domesticated farm animals caused selection for grasses that tolerated intense grazing pressure. A turfgrass then, by definition, is a grass that maintains its growing point close to the soil surface. Most grasses native to the U.S. developed where the buffalo roamed freely and never were under intense grazing pressure. Therefore, lawn culture has primarily utilized the nonnative species that fulfill this basic requirement.

TURFGRASS MORPHOLOGY AND GROWTH

Turfgrass plants are made up of an aboveground portion called shoots and a belowground portion called roots (Figure 1). The crown is located at the juncture between roots and shoots near the soil surface, and is composed of compact internodes and the crown meristem or growing point. Turfgrass roots are fibrous, dense, pervasive, and generally shallow relative to roots of most other grasses and weeds. Shoots are made up of leaves and stems, and can be oriented vertically or horizontally. Horizontal underground shoots are called rhizomes and horizontal aboveground shoots are called stolons. Vertical shoots usually remain compacted except during flowering, when the stems elongate into seedstalks or culms.

0-87371-350-8/94/$0.00+$.50
© 1994 by CRC Press Inc.

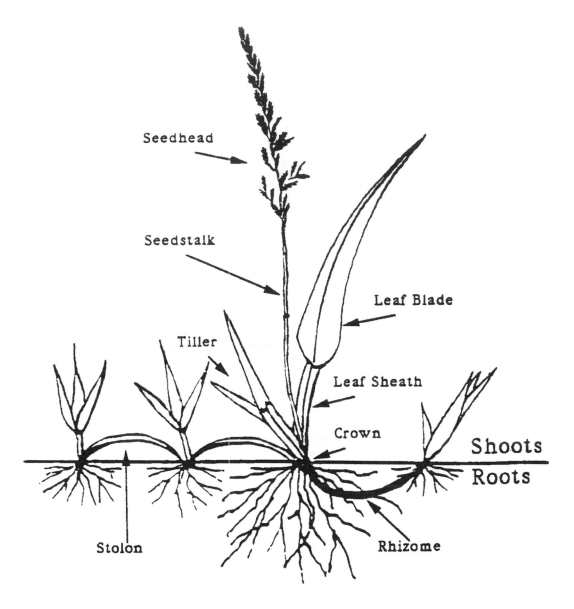

Figure 1 Diagram of a turfgrass plant.

SHOOTS

The crown gives rise to leaves, stolons, rhizomes and roots. Leaves form in sets, with each new leaf growing from within the encompassing sheath of the older leaf. The leaf proceeds through a vertical growth phase as a result of simultaneous cell division of two intercalary and the one basal meristem. Once a new leaf is well exposed above existing leaves, the final act of the intercalary meristems is the demarcation into leaf blade and leaf sheath. Leaf blades orient horizontally while sheaths remain vertical. Leaves appear in an alternating scheme on opposite sides of the leaf set. Crowns may also give rise to new leaf sets which are called tillers. Periodically, the tips of

rhizomes and stolons establish new crowns and new tillers. When a crown reaches sufficient maturity, and under certain environmental conditions, it may initiate a seedstalk in the place of a new leaf. In most turfgrass species, this reproductive organ elongates above the mowing height.

ROOTS

Turfgrass roots also initiate from crown. However, unlike leaves, roots have a growing point at the tip which is capable of branching into air spaces in the soil at every available opportunity. The term "grass-roots organization" undoubtedly evolved from observations of the root system of grasses.

UNDERSTANDING THE DYNAMICS OF THE TURFGRASS ECOSYSTEM

A comparison to trees illustrates the dynamic nature of the turfgrass ecosystem. For trees, it is immediately apparent by continuity of shape that the same tree exists from one year to the next. That is not true for lawns. Generally, lawns are built during the establishment phase and then are continuously regenerated every 6 weeks or so during those periods when the environment is conducive to growth. When the environment is not conducive to growth, the lawn deteriorates. As a result the turfgrass ecosystem is the most dynamic of the landscape and requires the most culture and management, even when compared to annual flowers.

BUILDING TURFGRASS BIOMASS

It is important to understand that the turf sward "builds" only during seedling establishment. Once the lawn is fully dense, the turfgrass ecosystem can only build three things: (1) organic matter in the soil through continuous death of leaves, stems and roots, (2) large quantities of clippings that may or may not be removed from the lawn, and (3) thatch. For the most part, leaves are highly biodegradable, do not contribute to thatch or diseases and do not need to be removed from the ecosystem. Turfgrass stems degrade more slowly. When the degradation rate of stems is further slowed for any reason (such as excessive use of certain antimicrobial pesticides) and/or growth is overstimulated (through excessive fertilization), the turfgrass ecosystem can build thatch. Turfgrass root systems are also dynamic. For some turfgrasses, two or more root systems can grow from the crown each year. Thus, growing roots is also a regenerative process where roots continuously grow and die, and thus build organic matter in the soil. The great prairie soils of the Midwestern U.S. were built as a result of this process.

TURFGRASS REGENERATION AND REJUVENATION

Tree maintenance is also much different from lawn maintenance. For turfgrasses, it is the health of the population rather than the health of the individual that is important. It is no secret that the easiest method to maintain the health of any population is to keep the average age very low through continuous introduction of new, young individuals.

Therefore, turf culture focuses on rejuvenation. Mowing stimulates production of new tillers and new leaves. Vertical mowing, dethatching and slicing are practices designed to rejuvenate turfs. Even the predominant effect of nitrogen fertilizer is not in greening up existing leaves, but stimulating growth of highly visible, new juvenile leaves that are greener than old aging leaves. As a result, there is a lag period between fertilization and the green color response. The length of the lag period is directly related to how favorable environmental conditions are for new growth. For example, cool-season grass response to summer fertilization is often not visible until cool weather in the fall. However, when environmental conditions are highly favorable for growth, fertilization actually speeds up the dynamics or turnover of the lawn ecosystem.

DURATION OF TURFGRASS LEAVES

The dynamic nature of turfgrass leaves can be calculated from the interval of time between the appearance of each new leaf, known as the plastrochron. For Kentucky bluegrass (*Poa pratensis* L.), grown under optimum conditions, the plastochron has been calculated to be about 5 to 7 days (Table 1). Additionally, the average tiller has about 5 to 7 leaves at any one time. As a result, the expected duration of any one leaf is about 25 to 49 days before the onset of senescence or dieback from the tip. Thus if the growth of a lawn is held back for a period of 6 weeks or so, it will begin to turn brown. The duration of leaves that subtend the seedstalk is considerably shorter than 6 weeks, as the energy needs of the rapidly elongating stalk drain leaf reserves quickly.

DORMANCY AND TURFGRASS COLOR

Lawns come in two basic colors, brown and green, as illustrated in Figure 2. When the lawn is brown, it may be dead, but more likely is merely dormant. Usually the crown meristems are awaiting conditions favorable to the initiation of new growth. All lawn grasses exhibit dormancy and off-color when conditions are unfavorable for growth. Superior turfgrass color occurs when the new leaf appearance rate is high and exceeds leaf senescence or death rate. Under these conditions turfgrasses build shoot density, exhibit a high amount and high visibility of juvenile tissue, and are green. The process of lawn dormancy begins with unfavorable conditions for initiation of new leaves and is followed by highly visible natural aging of existing leaves. Once natural aging occurs, dormant turfgrasses exhibit

Table 1 **Duration of turfgrass leaves**

New leaf appearance rate	5–7 days
Number of leaves per tiller	5–7
Expected leaf life	25–49 days

poor lawn quality. In aging grass leaves, much of the cell contents are translocated to other portions of the plant and the old leaves begin to lose weight. The leaf exhibits a dull grayish appearance, eventually discolors and dies back from the tip to the base. It is important to understand that tip dieback is a normal and inevitable event in the life of all turfgrass leaves and it will likely occur within 6 weeks after the leaf first appears.

KEY ENVIRONMENTAL FACTORS CONTROLLING GROWTH

The two major seasonal environmental factors controlling turfgrass growth are temperature and rainfall. The curves in Figures 3 and 4 are drawn to illustrate growth from the areas of the world where high or low temperatures are the principal factor controlling seasonal growth rates. In areas of the U.S. where annual temperatures are more uniform, such as southern Florida or the West Coast, the occurrence of rainfall is usually the principal factor controlling seasonal growth rate. The two basic types of turfgrasses are adapted to warm and cool climates and exhibit seasonal growth patterns related to that climate.

COOL-SEASON GRASS GROWTH

Cool-season grasses in the cool humid region of the U.S. exhibit a typical growth pattern shown in Figure 3. The long daylengths of spring orient growth vertically, resulting in a high amount of mowed vegetation. Often more than half of the vegetation grown for the year is removed during a 6-week period in the spring. During summer, heat and drought slow the rate of growth to a minimum.

Growth resumes in the fall, but short daylengths orient growth more horizontally and the amount of clippings is often only half of that in the spring.

Reproductive structures of cool-season grasses grow only once during spring. Elongation of the seedstalk usually occurs in May, but may extend into June. In mowed turfs, grasses that grow an extensive number of seedheads are described as being "stemmy" and are noted for poor turf quality during that time. Within the cool-season grasses one anomaly exists in annual bluegrass (*Poa annua),* which produces abundant seedheads during spring and then maintains a reduced level of production throughout the year.

WARM-SEASON GRASS GROWTH

Warm-season grasses exhibit peak vegetative growth during the summer, as illustrated in Figure 4. For the most part, seasonal growth patterns of warm-season grasses are influenced by soil moisture and total amount of solar radiation, and not by specific daylength responses. In temperate climates, cool temperatures in the fall cause growth cessation and the grasses remain dormant until late spring. In tropical climates, growth continues throughout the year.

Reproductive structures of warm-season grasses grow abundantly in early-summer and then are maintained at a reduced level throughout the growing season. Once again, there is an anomaly in the warm-season grasses in that zoysiagrass (*zoysia spp.*) exhibits seedhead development similar to cool-season grasses. In areas where zoysiagrass exhibits winter dormancy, seedhead appearance can occur before full vegetative green-up. Certain zoysiagrass varieties also exhibit some seedheads in the fall.

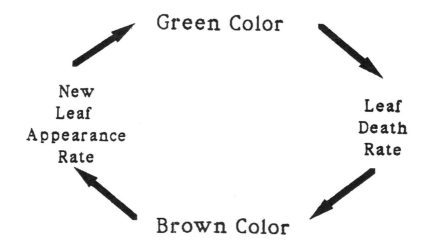

Figure 2 Life-cycle of lawn color.

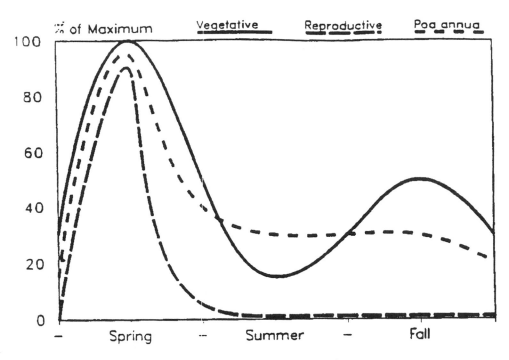

Figure 3 Cool-season grass growth rate.

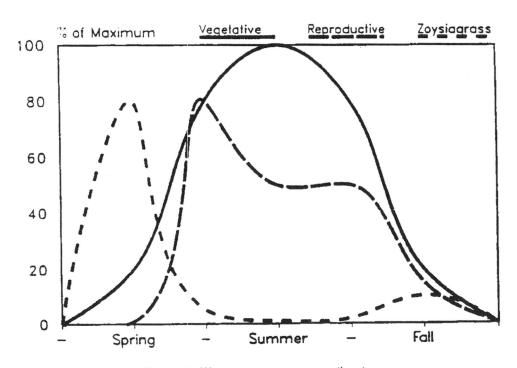

Figure 4 Warm-season grass growth rate.

SEASONAL GROWTH CYCLE OF PERENNIAL COOL-SEASON GRASSES

Growth of cool-season grasses is complicated by growth habit responses to long and short daylengths, and latent responses to cold temperatures (seedhead vernalization). With the possible exception of zoysiagrass, warm-season grasses do not exhibit these responses.

Figure 5 outlines the seasonal growth cycle of

cool-season grasses and identifies the following growth stages:

Stage I, Cold Dormancy or Pre-greenup: This is the appearance of the turf at the end of winter, often noted just after loss of snow cover. Appearance varies with kind of grass, quality or color of turf the previous fall, and severity of the winter. Under full sun, existing leaves that were not excessively damaged from winter conditions will green up through renewed chlorophyll synthesis. However, most leaves are damaged beyond repair and remain brown and fully visible until warmer temperatures hasten their degradation and they are hidden by new growth.

Stage II, Greenup and Initial Growth: As temperatures increase, new leaves grow from the crown apex within the encompassing leaf sheaths while old leaves degrade. Greenup may occur over a period of several weeks depending on rate of soil temperature increase. If this stage is prolonged by continued cool temperatures, the turf may reach 100 percent greenup while achieving only minimal vertical growth.

Stage III, Rapid Vertical Growth: The beginning of Stage III is most easily characterized by the need to mow. The grass begins to grow so fast that weekly mowings often remove much more than the recommended $1/3$ to $1/2$ of the existing leaf height. Long daylengths further stimulate vertical growth at this time. If spring temperatures warm rapidly and consistently, this stage can be entered before 100% greenup, and more than one mowing may be required before complete greenup has been achieved. Stage III ends when the first young, short seedheads appear in the turf area. The duration on Stage III varies with climate, but usually lasts 2 to 4 weeks.

Stage IV, Reproductive Physiology: In this stage, the seedstalk below the seedhead has begun to elongate. In many cool-season grass species, about the time the seedhead becomes visible in a mowed turf, the leaves on the tiller that bears the seedstalk stop growing and provide nutrients and energy to the developing seedstalk. Aging leaves on this tiller discolor and die very rapidly. At the same time lateral buds start developing into tillers, forming a new crown apex and a new vegetative plant.

Stage V, Revegetation: The dead tillers that bore the seedstalk degrade and fall into the thatch as the

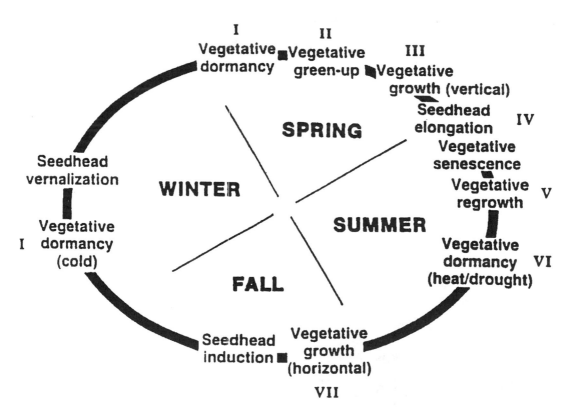

Figure 5 Seasonal growth habit of cool-season grasses.

lawn eventually replaces all the original leaf and stem tissue through initiation of new tillers. Thus, the green color of the lawn is maintained through development of new crown apices and new leaves. Occasionally this revegetation phase occurs at the onset of severe heat and drought. When that happens, the turf may not recover from the negative effects of the reproductive phase until the fall revegetation season.

Stage VI, Heat and Drought Dormancy: High heat loads on the turf result in excessive transpiration of water reserves from the soil. Occasional water deficits in the plant and above optimum temperatures cause growth to subside and subsequent senescence or death of existing leaves. For the most part, crown meristems merely remain quiescent and await more favorable growing conditions.

Stage VII, Fall Revegetation: The short daylengths of the fall encourage horizontal growth of turfgrasses and initiate the seedhead process that culminates the following spring. This is a period of frequent initiation of new tillers, rhizomes and stolons. Since vertical growth rate at this time of year is greatly reduced, traditional weekly mowing rarely removes excessive growth. Thus, both culture and environment are optimal for energy storage to facilitate winter survival.

REPRODUCTIVE VARIATION AMONG SPECIES AND VARIETIES

Normally the reproductive phase occurs in a lawn with minimal disruption of turfgrass quality. However, grass varieties or species that have difficulty maintaining quality during the reproductive phase are referred to as the "stemmy" types. In the region where cool-season grasses are grown, May and June are known as the stemmy months for the stemmy varieties. Certain cultivars of perennial ryegrass (*Lolium perenne L.*) and common Kentucky bluegrass exhibit poor quality at this time of the year while improved Kentucky bluegrass varieties maintain superior vegetative quality.

The relative ease or difficulty that species or varieties tolerate appears to be associated with two factors: (1) the overall tendency of the species or variety to produce seedheads (percentage of the crown apices with potential to flower) and (2) the tendency of those plants to follow through with flowering physiology in spite of frequent seedhead removal through mowing.

It is very interesting to observe that Kentucky bluegrass may be associated with the first factor while tall fescue (*Festuca arundinaceae Schreb.)*

may be related to the second. Compared to common, many improved Kentucky bluegrass varieties produce fewer reproductive apices both in the seed fields and in the lawn. For common types, rapid aging of a high number of subtending leaves and associated disease infection result in severe loss of turf quality.

On the other hand, tall fescue is often labelled the best summer adapted cool-season grass. Yet, unmowed tall fescue develops a seedhead, matures, and browns off while mowed tall fescue remains green. Thus a portion of the summer "tolerance" of tall fescue may be related to frequent mowing that removes the emerging seedhead before reproductive physiology kills the associated leaves.

SUMMARY

The inputs required to maintain a lawn are often maligned in today's society. Such a position is understandable if the lawn ecosystem is viewed merely as an extension of the general landscape. However, the fundamental precept of lawn culture is providing for the initiation of new plants and not in prolonging the life of the old. Lawn care is the management of a population, not individuals, and as such requires a unique and continuous management system. As examples, late-season fertilization is utilized to provide early spring growth to reduce the period of winter dormancy. Judicious watering is done during summer to reduce the period of summer dormancy. Vegetative demise of cool-season grasses associated with reproductive physiology has generally been reduced with the development of highly vegetative turfgrass varieties, or has been masked with fertilization. If humankind continues to require space delineation between the home and the outside world in order to achieve comfort, culture of turfgrasses in lawns will continue to be the best, most effective, and most enjoyable method of providing that space.

REFERENCES

1. **Beard, J. B.,** *Turfgrass Science and Culture*, Prentice Hall, Englewood Cliffs, NJ, 1973, 25.
2. **Etter, A. G.,** How Kentucky bluegrass grows, *Annals of the Missouri Botanical Garden*, 38(3), 293, 1951.
3. **Gould, F. W.,** *Grass Systematics*, McGraw Hill, New York, 1968, 20.

4. **Huffine, W. W. and Grau, F. V.,** History of turf usage, in *Turfgrass Science*, (American Society of Agronomy Monograph 14), Hanson, A. A. and Juska, F. V., Eds., 1969.

5. **Langer, R. H. M.,** *How Grasses Grow*, 2nd ed. (The Institute of Biology's Studies in Biology No. 34), Edward Arnold, London, 1979, 1.

6. **Turgeon, A. J.,** *Turfgrass Management*, Reston Publishing, Reston, VA, 1985, 15.

7. **Youngner, V. B.,** Physiology of growth and development, in *Turfgrass Science* (American Society of Agronomy Monograph 14), Hanson, A. A. and Juska, F. V., Eds., 1969.

Chapter 11

Choosing the Right Grass to Fit the Environment

Nick E. Christians, Department of Horticulture, Iowa State University, Ames, IA

Milton C. Engelke, Research and Extension Center, Texas A&M University, Dallas, TX

CONTENTS

Introduction 99
Zones of Adaptation 100
The Cool-Season Grasses 101
    Kentucky Bluegrass 102
    Tall Fescue 103
    Bentgrass 104
        Colonial Bentgrass 105
        Creeping Bentgrass 105
    Perennial Ryegrass 106
    Fine Fescues 106
The Warm-Season Grasses 107
    Bermudagrass 107
    Zoysiagrass 108
    St. Augustinegrass 109
    Buffalograss 111
References 112

INTRODUCTION

Grasses cover a greater proportion of the earth's land surface than any other plant species. According to Gould,[1] the grass family includes 600 genera and 7500 species. The U.S. is the natural habitat for approximately 1400 species.[2] Of this diverse family of plant material less than 200 species have any agronomic value beyond natural soil stabilization and land cover. As few as 30 to 35 are of major importance in turfgrass culture.[3] Each species has unique zones of adaptation. The center-of-origin of the grass and regions with similar climatic and edaphic conditions represent its optimum growth environments. Species and cultivars used outside their primary area of adaptation may require specific cultural inputs including the increased use of pesticides to overcome inherent biological limitations.

Amenity grasses (turfgrasses) serve a significant function in both agrarian and urban communities. Agronomically, grasses provide the primary basis for stabilization of the landscapes, and conservation of soil and water resources. Their extensive root systems hold soils in place during excessive rainfalls and dense protective canopies reduce water loss from soil surface evaporation during dry periods. This stability of the landscape results in improved foundations for factories, homes, and offices. Simultaneously, turf dissipates heat, reduces glare and dust, and aids in noise abatement.

Turf provides significant aesthetic value to the urban environment through enhancement of the mental, social, and physical health of our citizens. Agronomic and aesthetic attributes may be less obvious, but are equally as important as the recreational value of turf, where it is used extensively on athletic fields, parks, and golf courses. Actively growing, healthy turfgrass decreases the incidence of physical injury to athletes. Nationally, the turfgrass industry is estimated to exceed $27 billion, mostly for new construction, replacement and renovation cost, and routine maintenance.

The urban landscape receives 50% of the potable municipal water supply through irrigation during the summer months in the arid and semi-arid zones of the U.S.[4] Home owners routinely install irrigation systems. Industrial parks are designed and built with extensive and expensive landscape investments. The number of golf courses in the U.S. is projected to greatly increase by the year 2000. A number of golf courses in major metropolitan areas in the South consume an average of 2 million gallons of water daily to maintain actively growing, green turf. Our society insists on higher quality environments in which to live, work, and play. Urban sprawl and the demand for higher quality living and working conditions increase pressure on existing water supplies and on water treatment and distribution facilities. Predictions by the National Conference on Water (held in St. Louis, May 1978) suggested that U.S. water consumption rates will increase by 33% within the next 20 years.[5] The demand for quality turf will continue to increase as our quest for quality of life increases along with the demands for more leisuretime activities.

ZONES OF ADAPTATION

As shown in Figure 1, the U.S. is marked by considerable variability in climatic patterns and weather zones ranging from the Cool Humid (I) zone indicative of the North Central, Northeast, and the Pacific Northwest; the Warm Humid (II) zone of the South Central and South Eastern states; the Warm Arid (III) zone of the Southwest; and the Cool Arid (IV) zone of the West Central Northern states. The transition zone extends in a broad band across the central part of the U.S. and includes a section on either side of the edge between zones I and II, also reaching slightly into zones III and IV. Generally, this area is too cold for many of the warm-season grasses, and too warm for some of the cool-season grasses.

When selecting plant materials for use in zones with such diversity, it becomes evident that no one species will fit all niches. Furthermore, few turfgrass cultivars were developed or are adapted specifically for use under natural environmental conditions regardless of their zone of growth or utilization. Many early turfgrass varieties were developed during a period when water and energy (petroleum products) were abundant and inexpensive. They often require intensive cultural inputs including frequent watering, fertilization, and pesticide application to sustain acceptable turf quality. Consequently, many turfgrasses are being used in areas where they are not well suited or environmentally adapted. To compensate for biological inadequacies

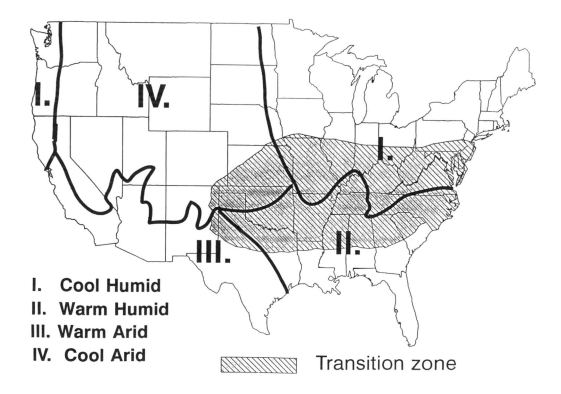

I. **Cool Humid**
II. **Warm Humid**
III. **Warm Arid**
IV. **Cool Arid**

Transition zone

Figure 1 Major areas of turfgrass adaptation in the U.S.

in the plant, turf managers often employ numerous cultural practices targeted toward modifying the environment.[6-9] Irrigation, fertilization, cultivation, subsurface soil remediation, and the use of numerous pesticides are examples of directed inputs to environmental modifications.

Cool-season grasses, such as creeping bentgrass, and perennial ryegrass may be used in warmer climates, but the additional stress on the grass during high temperature periods results in the need for more frequent irrigation, and timely fungicide applications to prevent diseases than would be required to manage these species in their respective zones of adaptation. On the other hand, some warm-season grasses can be used in Northern areas, outside of their zone of adaptation, but their inability to compete with cool-season weeds in the spring and fall generally results in the need for additional herbicides. Any grass forced into an environment to which it is not well suited will be more vulnerable to insect attacks, thereby increasing the need for insecticides. In general, technology has provided the turf manager with tools to successfully manipulate the environment, i.e., irrigation, cultivation equipment, pesticides, and fertilizers.

The turf industry is responding to the needs of the consumer through conscientious product development. Advances in irrigation technology provide more accurate irrigation through more precise delivery systems based on real-time weather monitoring, and consideration of the consumptive water needs of the grass. Chemical companies are gradually shifting to environmentally sensitive formulations for pest controls, including use of biological control systems such as nematodes for reducing mole crickets. Facets of the industry are likewise involved in breeding and development programs that utilize the biological strengths of the grass plant to produce cultivars with better adaptation to natural environmental conditions. The United States Golf Association (USGA), the Golf Course Superintendents Association of America (GCSAA), numerous private seed companies, and academic institutions are cooperatively and independently involved in genetic improvement of turfgrasses. As new cultivars are developed and tested, they likewise must be incorporated into the turfgrass community to provide ground covers that can be maintained with reduced management inputs.

Regardless of the intended use of the turf site, the primary consideration in selecting a grass is to identify those species/cultivars which are best adapted to the most limiting environmental condition. Grasses which are best suited to their environment will require the least use of inputs, including pesticides, and thereby provide a more environmentally sound approach to turfgrass management. Environmental and biotic factors most often limiting include temperature extremes, light levels, water quality and quantity, and traffic. In the 1980s, the turf industry saw a renewed emphasis placed on research to adapt grasses to the environment rather than to change the environment to meet the needs of the grass. A major factor influencing this was the changing regulatory climate that has placed new restrictions on how and where pesticides can be used. A second major factor has been the realization in much of the country that the once abundant supply of water that is needed for irrigation of turf may not always be available.

When choosing the right grass to fit an environment, it is necessary to have a thorough understanding of the strengths and weaknesses of each species and cultivar. It is also best to recognize the level of management to be used and specify the grasses that meet the requirements to sustain the desired turf quality. It must be recognized that no single grass will fulfill all the needs of the turf industry. Yet, we do have choices as to which grasses or combination of grasses to use. Good turfgrass management practices must always be used as periodic environmental fluctuations will impose excessive stress on even the best adapted varieties. Since the cultivar of grass once selected and properly maintained basically remains unchanged, the subsequent management involves manipulation of the environment to meet the requirements of the grass. It will be the intent of this chapter to discuss several of the more prominent turfgrass species and to identify accepted cultivars, and some of the cultural practices required for their use.

THE COOL-SEASON GRASSES

The cool-season grasses are generally adapted to the temperate and sub-arctic climates. Kentucky bluegrass, the bentgrasses, and fine fescues are widely used throughout much of northern U.S. and Canada. In contrast, tall fescue and perennial ryegrass are less winter hardy with their primary use in the mid-central regions of the U.S. including much of the transition zone. The cool-season grasses have a C-3 photosynthetic pathway. They are well adapted to cooler climates with ambient temperatures ranging from 15 to 22°C. Most cool-season grasses will experience considerable heat stress when temperatures exceed 40°C and may enter dormancy under prolonged stress.[10] Root growth is optimum when soil temperatures range from 12 to 20°C, and will cease to develop when soil temperatures reach 30°C.

KENTUCKY BLUEGRASS

Kentucky bluegrass (*Poa pratensis* L.) is the most widely used of the cool-season turfgrasses in the northern U.S. It spreads through the development of underground stems (rhizomes), asexually through apomictic (not-sexual) seed development, and to a limited extent, through sexual seed development. It can be found on lawns, golf course fairways and tees, cemeteries, parks, school grounds, athletic fields, and other areas where a dense grass cover is desired. The reason for its wide-spread use is that it has a number of advantages over alternative grasses.

Kentucky bluegrass has excellent recuperative and reproductive capacity. The rhizomatous nature of Kentucky bluegrass allows it to virtually repair itself. Perennial ryegrass, tall fescue, Chewing's fescue, and hard fescue, which are sometimes used in combination with, or as an alternative to Kentucky bluegrass, lack this extensive rhizome system and the capacity for rapid recovery. Even creeping red fescue, an alternative species that has rhizomes, lacks the extensive system that Kentucky bluegrass develops.

Kentucky bluegrass develops a dense turfgrass stand, has excellent color, and mows more cleanly than tougher bladed grasses such as perennial ryegrass. It also has a greater tolerance to cold temperatures than either perennial ryegrass or tall fescue. When mowed at the correct mowing height, 3.8 to 6.3 cm (1.5 to 2.5 in.), it is very competitive with weeds. Its tolerance of diseases is good when properly managed. Disadvantages of this species include its shallow root system, and its relatively high demand for water. However, it has the ability to survive during extended droughts through dormancy. A Kentucky bluegrass plant can lose its leaf tissue and part of its root system, but the crown (a compressed stem located near the soil surface from which the new leaves tillers, rhizomes, and adventitious roots arise) and the rhizomes can live for several weeks and regrowth will occur when water is available. Some of the other cool-season grasses will remain green longer under drought conditions, but none have the ability to recover that Kentucky bluegrass displays when the drought ends. Kentucky bluegrass is also poorly adapted to shade, except for certain cultivars, such as Glade, that have moderate shade tolerance.[11,12]

More than 300 cultivars (cultivated varieties) of Kentucky bluegrass have been developed over the past several decades. Each of the cultivars is genetically Kentucky bluegrass, but each has some unique, reproducible characteristic that sets it apart from the rest. These differences include variations in color, leaf angle, texture, and disease resistance, among others. These cultivars are generally divided into two categories: common and improved.

The "common" types, or "public varieties" as they are known in the turfgrass seed industry, are the older cultivars or selections from older cultivars, most of which have been in use for many decades. These "common" cultivars are characterized by an upright growth habit with a narrow leaf angle from vertical, and a relatively high susceptibility to the fungal disease "leaf spot" when they are intensely managed. Their positive attributes include early spring greenup and relatively good tolerance of environmental stress.

The improved types are newer releases, most of which have been selected or developed in the last few decades. The first of these improved types to be released was Merion, which was selected primarily for its tolerance to leaf spot. Since the release of Merion, many other improved cultivars of Kentucky bluegrass have been developed and today a wide variety of these improved types are available. As a group, the improved cultivars are known for their more prostrate growth, a slower growth rate, and improved tolerance to a number of grass diseases. These cultivars were generally selected under high moisture and high fertility regimes.

Research at Iowa State University comparing the adaptation of both common and improved cultivars to different environmental conditions over an 8-year period has demonstrated that the older "common" varieties are well adapted to lower maintenance conditions. Under these management regimes, the turf receives no supplemental irrigation and is allowed to go naturally into summer dormancy and to recover when temperatures cool and rainfall increases during the late summer.[13] Most of the common types were selected prior to World War II, or were selected in later years from grasses used in that time period, when low-intensity management systems were typically used. Less fertilizer was applied and irrigation was not widely practiced. Kentucky bluegrass was popular because of its ability to avoid drought through dormancy and to recover quickly in the late-summer and fall. Later work to investigate the morphological and physiological bases for these observations demonstrated that the low-maintenance varieties have several adaptations that allow them to better tolerate stressful environmental conditions.[14]

The observations from these trials have changed Kentucky bluegrass cultivar recommendations at Iowa State University. There was a time when the best cultivars from high-maintenance, irrigated trials were listed in extension bulletins and no distinctions were made as to the type of area on which the grasses were to be used. This is no longer the

Table 1 **Kentucky bluegrass cultivars recommended for high- and low-maintenance areas**

High-Maintenance

1.	Midnight	14.	Aspen	27.	Rugby
2.	Glade	15.	Escort	28.	Shasta
3.	Ram-1	16.	Mesa	29.	Bayside
4.	Majestic	17.	Sydsport	30.	Banff
5.	Enmundi	18.	Victa	31.	Dormie
6.	Bristol	19.	Baron	32.	Trenton
7.	Kimono	20.	Charlotte	33.	Nugget
8.	Merit	21.	Columbia	34.	Enoble
9.	Bonnieblue	22.	Mona	35.	Apart
10.	Eclipse	23.	Adelphi	36.	Touchdown
11.	Holiday	24.	Vanessa	37.	Parade
12.	Cheri	25.	Fylking	38.	Geronimo
13.	Barblue	26.	Admiral	39.	Plush

Low-Maintenance

1.	Kenblue	10.	Piedmont
2.	S.D. Common	11.	Fylking
3.	S-21	12.	Victa
4.	Vantage	13.	Monopoly
5.	Argyle	14.	Mesa
6.	Plush	15.	Ram-1
7.	Vanessa	16.	Harmony
8.	Parade	17.	Barblue
9.	Wabash	18.	Kimona

Note: The cultivars are ranked based on their performance over a 4-year period in high- and low-maintenance trials at Iowa State University. The grasses that are listed were chosen as the best from a group of 84 cultivars.

case, and cultivars are now recommended based on management regimes and environments to which they are best suited. This integrated pest management (IPM) approach results in lower inputs of pesticides and fertilizers.

The improved cultivars of Kentucky bluegrass are clearly the best choice when the area is to be irrigated, or when natural rainfall is sufficient to prevent summer dormancy. Under intense management regimes, the common cultivars prove to be disease prone and they will not perform as well as improved types. However, if the area is to receive a less intense management regime and is expected to go into summer dormancy during dry summers, the "common" cultivars will likely give more satisfactory results. Parks, school grounds, cemeteries, grassed areas along airport runways, low-maintenance home lawns, and other areas that may spend extended periods in dormancy will usually perform best when seeded to the "common" types. Cultivars currently recommended for high- and low-maintenance areas are listed in Table 1.

TALL FESCUE

Tall fescue (*Festuca arundinacea* Schreb.) is a cool-season bunch grass which is particularly well adapted to the transition zone of the U.S. The species originates from the European and Mediterranean areas, and was introduced into the U.S. in the mid-1800s and recognized for its forage value in the 1930s.[2] The leaves are rolled in the bud, and lack a mid-rib, in contrast to perennial ryegrass having a strong prominent mid-rib and folded in the bud. Both species have prominent veins on the upper leaf surface, with perennial ryegrass having a glossy undersurface.

"Rebel" was introduced as the first improved turf-type tall fescue. Since then over 125 cultivars have been developed with distinct improvements in growth habit, disease resistance, genetic color, heat and drought tolerance, and water use.

Tall fescue thrives best in fertile, well-drained, fine-textured soils, but will survive in a wide range of soils because of its extensive root system. It tolerates low fertility, yet responds to fertilization. This grass grows better on alkaline and saline soils than many other cool-season turfgrasses. While being highly drought resistant, tall fescue also withstands wet soils and survives compacted soils. This grass will remain green during adverse summer conditions when other grasses go dormant; however, tall fescue will also go dormant and turn brown when placed under severe stress. Its recovery from extended drought is inferior to that of Kentucky bluegrass in zone I (Figure 1), however, it is the primary cool-season turf for use in zone II, due to its excellent heat tolerance. Periodic reseeding is recommended throughout its area of use in order to maintain adequate turf density.

Tall fescue is recommended as a mowed turf on lawns, parks, cemeteries, and athletic fields. Tall fescue is grown in various climates from Canada to Florida and California to Georgia. Although best adapted to transition zone, tall fescue has been successfully used outside the areas where ample moisture is available and winters are not too severe. This grass will survive in arid climates when it is irrigated.

Shade tolerance is superior to most warm-season grasses. The two most severe diseases of tall fescue are leaf spot (*Drechslera dictyoides* Drechsl.) and brown patch (*Rhizoctonia solani* Kuhn). Nematodes may be more limiting to survival than high temperatures in the deep South. Another serious insect pest on tall fescue is the white grub (*Phyllophaga* spp.). The larva feed on actively growing root systems. Pesticide applications to control damage from white grubs must be timed according to the flight of the beetle for most effective control (usually 3 to 5 weeks following peak flight time). Since tall fescue is primarily a bunch grass, there is little opportunity for recovery from injury other than tiller proliferation. If damage occurs from insects, diseases, or utility, it may be necessary to reseed.

Tall fescue seedling vigor ranks high among turfgrasses. Strong vegetative buds usually ensure perpetuation of tall fescue turfs once the area has been established. Quality turf lawns will periodically (every 2 to 3 years) need overseeding during the optimum planting season to maintain density. Supplemental irrigation may be required to promote seed germination and establishment. Introduction of new seedlings into a thinning turf rejuvenates it and helps maintain stand density, quality, and competitiveness to weed invasion.

Tall fescue is the grass of choice in the southern part of Zone I where supplemental irrigation is not available. Regardless, nearly all grasses used for quality turf will require some supplemental irrigation to sustain active growth during high temperatures and low precipitation periods typical of summer months. Even though tall fescue is one of the most drought and heat tolerant cool-season grasses available for turf, it still requires considerable irrigation in the south. Tall fescue is considered a high water user when maintained in an active growth state. If the plant is allowed to enter into the summer stress period gradually (naturally), tall fescue is able to enter dormancy to avoid drought and heat stress. If irrigation is provided, sufficient water must be applied periodically to prevent wilt.

Several of the newly developed commercial cultivars of tall fescue are "enhanced" with endophytic organisms. Endophytes are organisms living inside grasses in a mutualistic (mutually beneficial) manner. These organisms have been reported to improve disease and insect tolerances of the grasses, as well as resulting in more vigorous plants compared to nonendophyte containing plants. White et al.[15] reports enhanced drought tolerance in plants containing endophytic organisms.

Considerable differences occur in the performance of the tall fescue varieties depending on the environment and level of management. Generally, no one variety is superior across all environments. Such contrasting performance has encouraged the use of blends containing two to four different varieties. This broadened genetic base provides some insurance against multi-use environments within a specific area. Examples include parks and other recreational areas where part may be irrigated routinely and heavily trafficked, while other areas may be marked by heavier shade with or without supplemental irrigation.

Cultivar recommendations and blends for specific locations are best obtained from local resource people, such as county extension agents. These professionals will be able to advise which cultivar(s) will perform best under your specific growing conditions and management level. A National Turfgrass Evaluation Program is conducted throughout the U.S. which provides information on the performance of the commercially available turfgrass cultivars under regional conditions. Assessing results from these trials requires careful examination for climate, soil, and management factors most similar to the area to be planted to tall fescue. These trials provide an excellent database for selecting those cultivars which are best adapted to the local conditions. This will ensure the fewest problems in establishing and maintaining a desirable tall fescue turf.

BENTGRASS

The genus *Agrostis* is composed of about 125 species, which are widespread in temperate and cold zones of the world and are present at high altitudes in the tropics and subtropics. The grasses that have turf potential include: Redtop (*Agrostis alba* L.), velvet bentgrass (*A. canina* L.), Colonial bentgrass (*A. tenuis* Sibth.,), and the most readily recognized species, creeping bentgrass (*A. palustris* Huds). All of these species are cool-season perennial grasses, which were introduced from Europe during the Colonial Period.[1,2,16] *A. tenuis* and *A. palustris* are of primary importance to the turf industry.

Colonial Bentgrass

Colonial bentgrass (*A. tenuis* Sibth.) is a fine-textured, cool-season, sod-forming perennial which spreads by short rhizomes and stolons to form a close tight turf. As with all bentgrasses, the leaves are rolled in the bud shoot. The ligule is membranous, truncate, and short. The above ground stems, although prostrate in growth, generally fail to root at the nodes. The grass is used to a limited extent for turf in some Northeast and Northwestern states (Zone I, Figure 1). Colonial bentgrasses are being improved genetically for use in golf course fairways and other intensely cultured turf areas.

Creeping Bentgrass

Creeping bentgrass (*A. palustris* Huds.) is a fine-textured, cool-season, stoloniferous perennial. The leaves are rolled in the bud shoot, and the ligule is membranous, rounded, or obtuse. The species lacks rhizomes; however, it spreads rapidly by stolons which root profusely at the nodes. Optimum growth occurs on moist, fertile soils with a pH of 5.5 to 6.5. The species tolerates a wide range of soil conditions, including salinity (variety Seaside) and frequent flooding.

Creeping bentgrass is the principle cool-season grass used for high quality golf greens in temperate and sub-arctic climates (Zones I and IV, Figure 1), and is used extensively throughout most of Zone III. It has excellent cold tolerance and is well adapted to frequent close mowing (0.5 to 0.75 cm).

Creeping bentgrass is marginally adapted to the southern zones of the U.S. (Zone II). However, it is used extensively throughout the transition zone and northern portions of the southern zone because of its superior turf quality on golf putting greens as compared to bermudagrass. Bentgrass has superior winter hardiness. Use of bentgrass as a greens surface in the southern states eliminates the necessity of winter overseeding and transitional management back to bermudagrass in the spring.

Utilization of intensive cultural practices with currently available cultivars of creeping bentgrass has extended their use throughout western and southern portions of the transition zone. Existing bentgrass cultivars generally lack sufficient heat tolerance to persist in the southern and southeastern U.S. (Zone II), unless compensatory management practices are employed to overcome this biological deficiency. Essentially, the lack of sufficient heat tolerance in both leaves and roots of bentgrass results in poor root development, severe wilt, poor traffic tolerance, and possible loss of stand during prolonged periods of heat. Syringing of greens has become a common practice in arid and semi-arid climates to enhance transpirational or evaporative cooling.

In the more humid southeastern U.S. (Zone II, Figure 1), syringing may not provide sufficient cooling to be effective. Additionally, the warm humid environment typical of the southeast, or which may be created through syringing, often results in a humid micro-environment conducive to severe disease activity, particularly *Pythium* blight.

Creeping bentgrass is less tolerant of wear than many other turfgrass species, has poor tolerance to soil compaction, and only average recuperative ability. Creeping bentgrass has a relatively shallow, dense, fibrous annual root system which requires constant replacement, especially under high soil temperatures. Typically, soil temperatures at 10 cm may reach 40°C or higher for several hours in mid-summer. No additional roots are formed and root senescence is accelerated. Close mowing and high temperatures result in a poorly developed, short root system with limited water absorption capabilities.

Creeping bentgrass is an aggressive stoloniferous species which can be propagated vegetatively or by seed. Currently available seeded cultivars include Penncross, Seaside, Carmen, Cobra, Putter, SR1020, Prominent, Providence, Lopez, Emerald, National, Penneagle, and Pennlinks. Penncross is the most widely used. Additional cultivars are being developed with specific emphasis on disease resistance, heat tolerance, and root persistence. Cultivar recommendations and blends for specific locations are best obtained from local resource people, such as county extension agents. These professionals will be able to advise which cultivar(s) will perform best under your specific growing conditions and management levels. A National Turfgrass Evaluation Program is conducted throughout the U.S. that provides information on the performance of the commercially available turfgrass cultivars under regional conditions. Assessing results from these trials requires careful examination for climate, soil, and management factors most similar to the area to be planted to creeping bentgrass. These trials provide an excellent database for selecting those cultivars which are best adapted to the local conditions. This will ensure the fewest problems in establishing and maintaining a desirable creeping bentgrass turf.

The seeded cultivars are utilized extensively in repairing damaged greens without major renovation. Numerous vegetative cultivars have been utilized throughout the northern states, but this has not occurred extensively in southern regions. Currently, most golf course greens in the U.S. are established with seeded cultivars.

PERENNIAL RYEGRASS

Perennial ryegrass (*Lolium perenne* L.) is thought to be native to the temperate zones of Asia and Africa,[3] but is widely used in both the cool-season and warm-season zones of the U.S. The species is recognized by its folded vernation, prominent midrib, the shiny backside of its leaves, and prominent veins on the upper side of its leaves. It maintains a tufted, bunch-type growth habit. It is known in the turf industry for its rapid germination and establishment, characteristics that make it a very useful species for rapidly reestablishing damaged areas on lawns, athletic fields, and golf courses. The primary limitation of perennial ryegrass is its poor tolerance of cold temperatures. It is often lost to winter-kill in the northern areas of the U.S. Its use in northern zones, where winter-kill is likely, expanded widely in the late 1980s, but it is often reestablished each spring.

The characteristics of its many cultivars vary widely. "Common" types are useful as forage grasses, but their rapid growth rate and poor mowing quality make them poorly adapted for use in intensively managed turf areas. Many "turf-type" cultivars have been developed in the past three decades. These grasses have shoot growth rates closer to that of Kentucky bluegrass, which is often mixed with perennial ryegrass for use in lawns and other turf areas. These turf type cultivars also mow more cleanly than the common types, although even the best is still inferior to the easily mowed Kentucky bluegrass in this regard. Like the tall fescues, endophyte-containing cultivars have been identified.

Perennial ryegrass is used in combination with Kentucky bluegrass on lawns and athletic fields, where it is known for its tolerance to traffic. It is rarely used alone on these areas because of its bunch-type growth habit and subsequent slow recovery from damage. Its use on golf course fairways and tees rapidly increased in zone II during the 1980s. It is generally used in monoculture on golf courses and is known for its rapid establishment from seed and its tolerance to lower mowing heights than Kentucky bluegrass.

In parts of zones II and III, perennial ryegrass is used for winter-overseeding of dormant (or nearly dormant) warm-season turf. Ryegrass is seeded at very heavy rates in September and October into a warm-season turf and managed as an overseeded turf through the winter. This is an expensive process and is generally used only on intensively managed lawns, athletic fields, and golf courses. This process has been found to be effective even at the low mowing heights used on golf course greens.

Perennial ryegrass is best adapted as a permanent turfgrass where winters and summers are moderate and where there is sufficient moisture. It thrives on fertile, well-drained soils with moderate fertilization. This species is also adapted for use on very poor soils, as it can be rapidly reestablished if turf dies.

Not as much is known about the adaptation of individual perennial ryegrass cultivars as is known about those of Kentucky bluegrass. But it is apparent that they vary in their adaptation to differing environmental conditions. The state extension service is the best source of information for the adaptation of specific cultivars to local environmental conditions.

FINE FESCUES

The term "fine fescue" is used in the turf industry to refer to a group of three similar species in the genus *Festuca*. The species include red fescue (*Festuca rubra* L.), Chewings fescue (*Festuca rubra* var. *commutata* Gaud.), and hard fescue (*Festuca longifolia* Thuill.). The fine fescues are identified by their extremely fine leaf texture. With a needle-like appearance, these grasses have the narrowest leaves of any of the turf species.

Red fescue is a complex species and has at least three recognized subspecies (ssp.). These include ssp. *rubra*, ssp. *tichophylla* Gaud., ssp. *litoralis* [Meyer] Auquir. All three have rhizomes, but only the *rubra* is considered to be a strong spreading type.[10] All of the subspecies lack the aggressive spreading habit of Kentucky bluegrass. The Chewings and hard fescue are bunch grasses that lack either rhizomes or stolons. Both Chewings and hard fescue are surprisingly good sod formers, however, and some cultivars of each species have been observed at the Iowa State University research area to maintain a high turf quality for a 10-year period at mowing heights as low as 2.54 cm (1 in.).[17]

The fine fescues are known primarily for their shade adaptation and low N requirement. They are often mixed with Kentucky bluegrass in commercial seed mixtures for lawns that have a combination of full sun and shade. Kentucky bluegrass will generally predominate in full sun and the fine fescues dominate in the shade.

The three species have a rather narrow moisture requirement. They do not do well in areas that are excessively wet and generally thin badly from disease infestation when maintained on saturated soils. They are susceptible to heat and drought and require frequent irrigation during extended drought periods, whereas in similar situations, Kentucky bluegrass will go into dormancy and readily recover in the fall.

The range of adaptation of the fine fescues is more limited than that of Kentucky bluegrass. The fescues are well adapted to the cool, moist conditions of zone I (Figure 1). In the upper great lakes region, the species are widely used in sun and shade. There is at least one golf course in central Michigan where greens, tees, and fairways have been established entirely with Chewings fescue. The area of adaptation is north of the transition zone (Figure 1) and east of Lincoln, Nebraska, plus zone I in the Pacific Northwest Region. To the west and south of zone I, summers are generally too hot and dry for fine fescue survival, and tall fescue is generally the grass of choice for shaded areas.

Those who would choose the species to fit the environment rather than modifying the environment to fit the grass will find the fine fescues to be very useful. Rather than trimming trees and ornamental plants to allow more sun for Kentucky bluegrass growth, or using fungicides to treat the diseases that often infest bluegrass in the shade, the well adapted fine fescues can be established in the shade where properly chosen cultivars will thrive with little additional care beyond mowing and moderate fertilization.

There are differences among the three species as to regional adaptation. Red fescue is better adapted to more humid regions and is often used in the cooler, moisture areas of the northeastern U.S. The Chewings fescues, and particularly the hard fescues, are well adapted to the drier regions of the midwest. The local state extension service will be the best source of information on the regional adaptation of species and cultivars.

THE WARM-SEASON GRASSES

The warm-season grasses are generally adapted to the tropical and subtropical climates. Bermudagrass (*Cynodon* spp.) and St. Augustinegrass (*Stenotaphrum secundatum* (Walt.) Kuntze) are widely used throughout much of southern U.S. and Mexico. Other species with excellent adaptation include: zoysiagrass (*Zoysia* spp.), buffalograss (*Buchloe dactyloides* (Nutt.) Engelm.), bahiagrass (*Paspalum*

notatum Flugge.), and centipedegrass (*Eremochloa ophiuroides* (Munro) Hack.). Zoysiagrass and buffalograss have considerable winter-hardiness and are used extensively in the mid-central zones of the U.S., including much of the transition zone.

The warm-season grasses have a C-4 photosynthetic pathway. They are well adapted to warmer climates with ambient temperatures ranging from 35 to 40°C. Most warm-season grasses will experience considerable cold stress when temperatures fall to 10°C and will enter dormancy when temperatures reach the freezing point.[10] Root growth is optimum when soil temperatures range from 30 to 35°C. Root growth ceases at temperatures below 12°C.

The major warm-season grasses in use in the U.S. include: bermudagrass, St. Augustinegrass, and zoysiagrass. Miscellaneous other species include: buffalograss, centipedegrass, bahiagrass, and kikuyugrass. The use of most of these grasses is limited to Zones II and III, although zoysiagrass and buffalograss have sufficient cold hardiness to extend into the northern boundaries of the transition zone as well as the southeastern portion of Zone IV, east of the Rocky Mountains.

The warm-season grasses are often vegetatively propagated, although several seeded varieties of bermudagrass are commonly used. Sodding, sprigging, plugging, or seeding are the alternative methods of establishing most warm-season grasses. The method selected is dependent on several factors including: the variety, availability of planting stock, budget, labor force, equipment inventory, and if renovating, how much down time will be necessary.

BERMUDAGRASS

Bermudagrass (*Cynodon* spp.) is a warm-season species native to Africa and introduced into the U.S. as early as 1751.[2] It is a sod-forming grass which is widely used throughout the southern U.S. for turf and ground covers.

The genus *Cynodon* comprises nine species, of which *C. dactylon* L. is the most common. Many of the fine textured low-growing commercial varieties such as Tifgreen, and Tifdwarf are interspecific hybrids between *C. dactylon* L. (tetraploid) and *C. transvaalensis* Burt-Davy (diploid). Consequently, the hybrids are sterile and must be produced vegetatively.

The species is readily identified by having both stolons and rhizomes, as well as to produce a distinctive three to five raceme seed head. With few exceptions, all commercial cultivars are folded in the emerging leaf bud — a major feature which distinguishes this grass from either zoysia (*Zoysia* spp.), or buffalograss (*Buchloe dactyloides*). Bermudagrass has a deep fibrous perennial root system, with rooting occurring at nodes of the stolons.

Bermudagrass is adapted to the tropical and sub-tropical climates of the U.S. (Zones II and III, Figure 1). Its use in the northern regions of the U.S. is generally limited by temperature. Ecotypes have been collected from East Lansing, MI, Minneapolis, MN, and Denver, CO, but most were found on south facing, protected micro-climates atypical of the zone. The practical northern limits of bermudagrass extend into the transition zone where low temperatures seldom exceed −12°C. Bermudagrass is sensitive to sub-freezing temperatures and will enter winter dormancy with a loss of leaf and stolon growth, and subsequent discoloration of the tissue. The plant remains dormant until soil temperatures (10-cm depth) rise and remain above 10°C. Root and rhizome growth will substantially increase when soil temperatures reach 15 to 20°C, with optimum soil temperatures of 25 to 30°C, and day-time ambient temperatures between 35 and 40°C. In warm, frost-free climates, bermudagrass will remain green and actively growing throughout the year. Generally, bermudagrass will enter winter dormancy in regions north of the 30°N latitude due to cool nights and low soil temperatures.[10]

Bermudagrass has no shade tolerance. At low-light intensities (50 to 60% of full sunlight) the plant becomes elongated and etiolated, with little or no root or rhizome development, resulting in weak sparse turf.

Water requirements of bermudagrass are cultivar dependent. The "common" type bermudagrasses tend to be more drought tolerant and have a lower water-use requirement than the hybrids. Supplemental irrigation is generally required throughout the low rain-fall months to ensure active green growth. If stressed, the plant will enter dormancy and may survive extended drought periods with little loss of stand. Generally the hybrids are more prone to loss of stand during prolonged drought periods.

Bermudagrass will survive on a wide range of soils from heavy, alkaline clays to deep sands. It tolerates a wide pH range, and saline soils. Bermudagrass has a high N requirement for good quality turf. The fine textured hybrid-type bermudagrasses are best suited for high-quality lawns and sports turf use. Bermudagrass is well suited to high traffic areas as it recovers rapidly from wear injury. Bermuda turf tolerates moderate levels of wear and compaction. Periodic cultivation and a high-fertility program is necessary on sports turfs (athletic fields, golf course tees and greens, etc.) where compaction and injury most often occur.

The "common" type bermudagrasses are well suited for industrial parks, golf course roughs, home lawns, and areas where stabilization and conservation of the soil resource is of primary importance. Bermudagrass is a good lawn grass because it responds well to management. With moderate fertilization, frequent mowing, and adequate moisture bermudagrass forms a dense fine textured turf. The grass performs well under low-maintenance conditions and has excellent resiliency under natural environmental conditions in zones II and III (Figure 1). As such it is well suited for use in stabilizing road sides, ditch banks, and industrial park areas.

Less breeding and development has occurred within the warm-season grasses (comparing cool-season grasses), consequently fewer cultivars are available (Table 2). Bermudagrasses can be established from seed, or vegetatively using sod, sprigs, or plugs. A select number of cultivars (Table 2) are available in seed form. When establishing from seed, a well prepared seed-bed is most desirable with the ground properly cultivated, contoured, and firmed prior to seeding. A starter fertilizer and/or organic amendments should be incorporated into the top 10 cm of the prepared seed bed.

Mowing requirements for bermudagrass turf are dependent on variety, use, and the level of maintenance. The hybrid bermudagrass will require more frequent mowing and a closer height-of-cut than the common types. The closer the mowing height, the more frequent mowing must occur. Golf greens mowed at 0.4 cm (5/32 in.) must be mowed daily, whereas roadside grasses mowed at 7.5 to 10 cm (3 to 4 in.) may need to be mowed monthly.

ZOYSIAGRASS

Zoysiagrass is indigenous to the Orient and was introduced into the U.S. as recently as 1895. It is a warm-season perennial sod former which spreads by both rhizomes and stolons. There are at least five species within the Zoysia genus: *Z. japonica* Steud. (Japanese lawngrass), *Z. matrella* (L.) Merr. (Manilagrass), and *Z. tenuifolia* Willd. (Mascarenegrass) are the three with the greatest turf potential.[3,10] Two additional species were obtained from Korea during an extensive collection trip in 1982. They include: *Z. macrostaycha,* Franch. et Sav. and *Z. sinica* Hance. Both species appear to have excellent salinity tolerance. Another species has been tentatively identified and designated *Z. korenia*.[27] Electrophoretic patterns suggest it may be a natural hybrid between *Z. japonica* and *Z. macrostaycha*. All the species are sexually compatible, although their flowering cycles are not always synchronized.

Table 2 **Partial listing of bermudagrass cultivars commercially available in 1992**

Cultivar	Texture	Type	Region
Arizona Common	Common	Seeded	Northern
NuMex Sahara	Common	Seeded	Southern
Gymond	Common	Seeded	Northern
Midiron	Common	Vegetative	Northern
Midway	Common	Vegetative	Northern
Santa Ana	Fine	Vegetative	Central S
U-3	Common	Vegetative	Northern
Tifgreen	Fine	Vegetative	Southern
Tifdwarf	Fine	Vegetative	Southern
Tifway	Fine	Vegetative	Southern
Tifway II	Fine	Vegetative	Southern
Ormond	Common	Vegetative	Southern
Texturf-10	Common	Vegetative	Southern

Generally, distinction among species is based on floral and morphological characteristics.

The Zoysia species have excellent low temperature hardiness. *Z. macrostaycha, Z. japonica,* and *Z. sinica* are broad-leaved, coarse-stemmed, and are found in the northern zones for Zoysia distribution in the Orient. *Z. matrella* is intermediate in leaf texture, appears to have more rhizomes, and is well adapted to the central zones of the Orient which closely correspond to the south-central U.S. and northern Mexico. *Z. tenuifolia* is the finest textured and generally the most cold susceptible, with distribution limited to tropical and subtropical climates.

Major biological attributes of the species include tolerance to shade, salinity, wear, and temperature extremes. The species have good drought tolerance and low nutritional requirements. Zoysiagrasses are adaptable to a broad range of soil conditions. Desirable turf characters include dense, fine-textured, slow-growing turfgrass with good color.

Few disease and insect problems are associated with zoysiagrass. The major problem in the southwest is reported to be rust, although leaf spot (*Drechslera dictyoides* Drechsl.), brown patch (*Rhizoctonia solani* Kuhn), and dollar spot (*Sclerotinia homoeocarpa* F.T. Bennett) have been observed. The primary insect damage occurs from billbugs (*Calendrinae* subfam.), white grubs, sod webworms, and mole crickets (*Gryllotalpidae* spp.).[3] An eriophyide mite (*Eriophyes zoysiae* (ACARI:ERIOPHYIDAE) has been identified on zoysia in the U.S. The mite causes leaf roll and a shepherd's crook appearance.[18]

Commercially available zoysiagrass cultivars and their date of release include: Meyer (1951), Midwest (1957), Emerald (1963), El Toro, which is patented (1984), Belair (1985), and Cashmere (1988). Cashmere is the first proprietary zoysiagrass cultivar released in the U.S. It is a *Z. matrella*, and thus is best adapted to the gulf coast states as are most finer textured zoysia species.

The geographic area of adaptation within the U.S. extends south from the coastal regions of New Jersey, west along the northern fringes of the transition zone (St. Louis, Kansas City) to the Great Plains. The primary utility of zoysiagrass is on home lawns, parks, sports fields, golf course roughs, fairways, and tees. More recently, it has also been used on collars around bentgrass greens and bunkers to restrict the spread of and contamination by bermudagrass (*Cynodon* spp.).[10]

Zoysias are used extensively on horse racing tracks and on golf course fairways throughout Japan. Several individual Japanese companies are also using conventional breeding programs to improve zoysia.

Zoysiagrass use in the U.S. is limited, due to slow recuperative ability and poor production traits contributing to the high initial cost of establishment. An improved seeded zoysiagrass cultivar will be readily accepted in the U.S. and Japan. In general, the zoysiagrasses are widely adapted geographically and hold considerable promise in reducing water use and cultural inputs for domestic and recreational turf. The primary effort of zoysiagrass improvement involves improving and developing cultivars which are more water use efficient, tolerant to saline and effluent water supplies, and are more economical to produce. Simultaneously, cultural practices must be defined to optimize use of new cultivars.

ST. AUGUSTINEGRASS

The genus *Stenotaphrum* includes six species distributed throughout the Old World in regions predominantly around the Indian Ocean.[19] *S. secundatum* (Walt.) Kuntze is indigenous to the West Indies and is commonly found in tropical Africa and Australia. According to Hanson et al.[2] the species was probably introduced into the U.S. during or before the Colonial Period.

St. Augustinegrass is widely distributed as a turfgrass throughout the tropical and subtropical regions of the U.S., and is also utilized as a forage grass in southern Florida. It ranks second only to bermudagrass for use as turf in the southern U.S. The geographic area of utilization is limited by its lack of cold and drought tolerance.[10] Its primary zone of adaptation is limited to southern zone II (Figure 1).

St. Augustinegrass spreads by stolons and forms a dense, coarse-textured turf.[20] It is vegetatively propagated by sprigs, plugs, or sod. Its extensive fibrous root system is conducive to sod production.[3] Seed head development does occur in all known accessions; however, viable seed formation has occurred only in diploid types (2n = 18).[21] Fertile florets are sparse and seed yields extremely low. The short thick seed heads are difficult to harvest and thresh.

St. Augustinegrass is distinct in appearance. It is coarse textured, highly stoloniferous, and a sod-forming perennial. The species generally lacks rhizomes. The leaves are folded in the bud shoot. The ligules are an inconspicuous fringe of hairs (0.3 mm long). Auricles are absent, the collar is continuous, glabrous, broad, and narrows to form a short stalk or petiole for the leaf blade. Its sheath is compressed, keeled, and overlapping. The inflorescence is short generally terminal, but occasionally axillary with fleshy raceme, with imbedded spikelets.

St Augustinegrass is used primarily in sandy, moist mucky, or clay soils extending from South Carolina to Florida and Texas (Zone II).[22] Its adaptation and use is generally limited by cold winter temperatures. It grows best on fertile well-drained soils, has relatively high water and nutritional requirements, does not tolerate poorly drained compacted soils and generally lacks traffic tolerance, and has excellent shade tolerance with adequate fertility and moisture. With proper cultural practices, St. Augustinegrass will tolerate a wide range in soil pH (5.5 to 8.5), however it may exhibit some chlorosis above pH 7.5.

St. Augustinegrass requires more water than most other warm-season grasses. There is limited genetic variability among existing St. Augustinegrass cultivars, although drought-resistant cultivars are being released by the University of Florida suggesting less supplemental irrigation may be necessary to maintain quality turf. Regardless, supplemental irrigation will be required throughout its zone of use. Under moderate fertility, St. Augustinegrass will develop thatch and if not properly maintained will require periodic dethatching. The large fibrous stolons are easily damaged with normal verticutting and aerification equipment, therefore the best time to "dethatch" is in the spring at the first sign of green-up.

St. Augustinegrass lacks winter hardiness, and will go dormant with the first fall frost. The species will remain dormant until soil temperatures rise above 12 to 15°C in the spring. The grass may remain active and green throughout the winter months along the gulf coast; however, along the northern limits of its range of adapation, it will enter winter dormancy. When temperatures drop below freezing for prolonged periods of time, considerable winter desiccation and death will occur. Total turf replacement may be required as often as every 3 to 4 years throughout its northern limits of use.

Recent advances in St. Augustinegrass breeding have seen additional cultivars being developed for use in zone II (Table 3).

St. Augustinegrass is vegetatively reproduced and marketed as sod or plugs. Vigorous stolon production is conducive to strip or plug planting as well as solid sodding. Sprigging of St. Augustinegrass is less desirable as sprig harvest equipment generally results in considerable mechanical damage to the stolon. Regardless, a level, firm, weed-free seed bed must be prepared whether plugging, sodding, or sprigging.

Properly maintained and established St. Augustine turf will require frequent fertilization, irrigation and mowing. St. Augustinegrass is known to be susceptible to numerous insects including the white grub and southern chinchbug and mole crickets. Sod webworms, army worms, and cutworms can be damaging when infestations are heavy. The most serious diseases include Panicum Mosaic Virus

Table 3 Cultivars of St. Augustinegrass commercially available in 1992

Cultivar	Source	Year Released
Seville	Pursley Proprietary	1980
Raleigh	NCSU	1989
Floratam	University of Florida Texas A&M	1975
FX10	University of Florida	Patent issued October 1991
FX33	University of Florida	1992 patent process

which causes SAD (St. Augustine Decline), brown patch (*Rhizoctonia solani* Kuhn), and grey leaf spot (*Pyricularia* (Cooke) Sacc).

St. Augustinegrass is second only to selections of zoysiagrass for its shade tolerance. It tolerates a wide range of soil types and a wide soil pH range. Wear tolerance is relatively poor compared with bermudagrass and zoysiagrass due to its thick stolons and coarse leaves. Common St. Augustinegrass is susceptible to numerous diseases and insects.[20,23] Devastation of lawns due to the southern chinch bug and rapid spread of St. Augustine decline virus (SADV) throughout Texas prompted the development and release of "Floratam" St. Augustinegrass. This cultivar is resistant to SAD. It was resistant to the Southern Chinch Bug; however, recent research indicates a population shift has occurred in the insect and a more virulent race is causing damage.[28] Floratam's area of adaptation is confined to the coastal plains of Texas due to insufficient cold tolerance.[20,25]

Plant Patent #4097 was issued for "Seville" St. Augustinegrass released in 1980 by The O. M. Scott and Sons Company. "Seville" has a rapid growth rate, but is more prostrate than other commercial St. Augustinegrass cultivars. It is recommended for use in Florida and South Texas.[24] Beard et al.[24] reported cold hardiness of "Seville" to be comparable to "Floratam," and therefore poorly adapted to the northern regions of the transition zone.

Another "improved" cultivar of St. Augustinegrass was released in the fall of 1980. "Raleigh" is resistant to SAD and is more winter hardy than Floratam. "Raleigh" ranked highest in winter survival of 31 St. Augustinegrass accessions, not including "Texas Common," in north Mississippi in 1976/77.[26] Over the past 11 years, Raleigh has become the dominant St. Augustinegrass used throughout the southern U.S. It is susceptible to brown patch, grey leaf spot, and to the southern chinch bug. Experimental St. Augustinegrasses from Texas and Florida, which are near release for commercial production, show either superior drought tolerance (Florida) or superior winter hardiness (Texas), as well has being improved for SAD resistance, salinity tolerance, chinch bug tolerance, and improved brown patch resistance.

BUFFALOGRASS

Buffalograss (*Buchloe dactyloides* (Nutt.) Engelm.) is a warm-season species native to the Great Plains extending from Montana to Central Mexico. The ploidy level ranges from hexaploids among the more northern accessions to diploids in Central Mexico. Buffalograss has been used as an integral component in range and road-side revegetation mixtures for highway stabilization and soil stabilization.

Buffalograsses natural area of adaptation extends from the northern Great Plains into the highlands of Central Mexico. The species has evolved under natural environmental stress conditions, and has the inherent biological characters to persist and perform under droughty conditions and a broad range of temperatures. The species is best adapted to heavy alkaline soils, however, the new turf-type cultivars have persisted and performed well on silt and sandy soils and at a pH range from 4.7 to 8.3. Buffalograss is one of the few native species which is used for turf, and it is used extensively for its soil stabilization on road sides and industrial park settings.

Buffalograss is a dioecious species. The male flower (staminate plant) has two to three flag-like, one-sided spikes on a seed stalk which extends above the turf canopy, whereas the female flower (pistillate plant) is a short spike, on which the glumes, lemma, and palea form a burr-like enclosure for the seed. The burr grows within the turf canopy and often goes undetected.

Both male and female plants are spread by stolons which may vary from a few inches to several feet in length. It is a fine leaved, warm-season, sod-forming perennial. The leaves are rolled in the bud, which aids in distinguishing it from most bermudagrasses. Internodes are 2 to 3 in. long with tufts of short leaves at each node. Plants will take root at the node and produce new stolons.

The turf-type buffalograsses were developed specifically for use as a low maintenance turf. Their aggressive vegetative growth habit makes them particularly well suited for use as an amenity turf for conservation and soil stabilization. It is well suited for home lawns, parks, recreational fields, and low-use areas such as roadsides and rough areas of golf courses. As a warm-season grass it is well adapted to the transition zones of semi-arid and sub-humid nonirrigated areas.

The agronomic strengths of the species lies in its ability to survive natural environmental conditions. The new buffalograsses will be useful in stabilization of the landscape, heat dissipation, and noise abatement. They are highly competitive under hot summer temperatures. With proper management, these turf-type buffalograsses will form a moderately dense, fine-leafed turf. Buffalograss has a lower water requirement than most other improved warm-season turfgrasses. (Engelke, et al. 1991) However, the buffalograss turf will enter summer dormancy with loss of green color under prolonged water stress. Regardless, it exhibits rapid recovery when water becomes available. It can tolerate long periods of flooding. Vegetative buffalograsses have good wear tolerance and excellent tolerance to heavily compacted soils. The species will persist under close mowing with an optimum mowing height of 3.6 to 5 cm (1.5 to 2.0 in.). It grows on a wide range of soil conditions, preferring the heavier alkaline soils. Buffalograss has poor to moderate shade tolerance only slightly better in shade tolerance than bermudagrass. It is recommended for use in highly lighted landscape areas.

Commercially available buffalograsses are propagated either vegetatively (Prairie — US Plant Patent #07539 and 609) or by seed (Texoka, Comanche, Sharps Improved, Plains and Top Gun). The early seed-producing cultivars were developed and released for forage production and soil conservation purposes whereas later cultivars were targeted more toward turf utility. Research efforts are being directed toward the development of seed-propagated turf-type buffalograsses. The seeded varieties are readily available on the commercial market. The seed is borne in a burr and must be treated to break seed dormancy prior to planting. A clean firm seed bed with supplemental irrigation is desirable. The optimum times to plant seed are in the late fall and early spring during periods of high natural rainfall. Fall-planted buffalograss burrs will imbibe moisture and initiate growth with optimum soil temperatures in the early spring. Seeding during the high-temperature dry summer months will require supplemental irrigation.

The vegetatively propagated cultivars (Prairie and 609) are best established by solid sodding, sprigging, or by using rooted or nonrooted plugs. The optimum time of year for establishment is dependent on the site of utilization. If irrigation and/or adequate moisture is available, planting can occur generally throughout the year provided the planting stock is grown in a similar climate. If however, the site is not serviced by or accessible for water application, establishment should be delayed until sufficient moisture is naturally available to favor establishment. Sprigging of buffalograss should be limited to the primary growing season to ensure adequate stands the first year. Sodding and plugging can be conducted throughout the year provided a sufficient soil base be retained around the crown (plugs) and the source of plant material is grown under similar environmental conditions as the target planting site.

Once established, the turf will require approximately 3 to 5 cm of water per month to persist. Additional irrigation and fertility may be necessary to provide high-quality, actively growing competitive turf.

Mowing frequency and height of cut are dependent on cultural conditions and objectives of the turfed site. High-visibility, high-traffic turf sites will require higher management levels. Regardless, the optimum level of management for the turf-type buffalograsses will be less than for most other warm-season grasses under the same conditions.

REFERENCES

1. **Gould, F. W. and Shaw, R. B.,** *Grass Systematics*, 2nd ed., McGraw-Hill, New York, 1969, 397.
2. **Hanson, A. A., Juska, F. V., and Burton, G. W.,** Species and varieties, in *Turfgrass Science Agronomy*, Hanson, A. A. and Juska, F. V., Eds., Madison, WI, 1969, 370.
3. Beard, J. B., *Turfgrass: Science and Culture*, Prentice-Hall, Engelwood Cliffs, NJ, 1973, chap. 17.
4. **Feldhake, C. M., Danielson, R. E., and Butler, J. D.,** Turfgrass evapotranspiration. I. Factors influencing rate in urban environments, *Agronomy J.,* 75, 824, 1983.
5. **Carter, L. J.,** Global 2000 report: vision of a gloomy world, *Science,* 209, 575, 1980.
6. **Daniel, W. H. and Freeborg, R. P.,** *Turf Managers Handbook*, Harvest Publishing Co., Cleveland, OH, 1979, chap. 13 and 14.
7. **Ledeboer, F. B., McKiel, C. G., and Skogley, C. R.,** Soil heating studies with cool-season turfgrasses. I. Effects of watt density, protection covers, and ambient environment on soil temperature, *Agron. J.,* 63, 677, 1971.
8. **McBee, G. G., McCune, W. E., and Beerwinkle, K. R.,** Effect of soil heating in winter growth and appearance of bermudagrass and St. Augustinegrass, *Agronomy,* 60, 228, 1968.
9. **Robey, M. L., Daniel, W. H., and Freeborg, R. P.,** The PAT (Prescription Athletic Turf) system — Purdue's installation, *Agron. Abst.,* 100, 1974.
10. **Turgeon, A. J.,** *Turfgrass Management*, Reston Publishing Co., Reston, VA, 1980.
11. **Taylor, L. H. and Schmidt, R. E.,** Performance of Kentucky bluegrass strains grown in sun and shade, *Agron. Abst.,* 113, 1977.
12. **Wood, G. M.,** Evaluating turfgrasses for shade tolerance, *Agron. J.,* 61, 347, 1969.
13. **Christians, N. E.,** Kentucky bluegrass for low-maintenance areas, *Grounds Maint.,* 24(8), 49, 1989.

14. **Burt, M. G. and Christians, N. E.,** Morphological and growth characteristics of low- and high-maintenance Kentucky bluegrass cultivars, *Crop Sci.*, 30, 1239, 1990.

15. **White, R. H., Engelke, M. C., Morton, S. J., Johnson-Cicalese, J. M., and Ruemmele, B. A.,** *Acremonium* endophyte effects of tall fescue drought tolerance, *Crop Sci.*, submitted.

16. **Ward, C. Y.,** Climate and adaptation, in *Turfgrass Science,* Hanson, A. A. and Juska, F. V., Eds., Agronomy. 14:27–29. Madison, Wisc.: Am. Soc. Agron. 1969.

17. **Christians, N. E.,** Fine Fescue Management Study, Iowa Turfgrass Research Report, Iowa State University Extension Publication FG-457, 1990, p. 12.

18. **Baker, E. W., Kono, T., and O'Neill, N. R.,** Eriophyes zoysiae (ACARI: ERIOPHYIDAE), a new species of Eriophyide mite on zoysiagrass, *Int. J. Acarol.*, 12(1), 3, 1986.

19. **Sauer, J. D.,** Revision of Stenotaphrum (Gramineae: Poniceae) with attention to its historical geography, *Brittonia*, 24, 202, 1972.

20. **Duble, R. L. and Novosad, A. C.,** Home Lawns, MP-1180 Texas Agricultural Extension Service, p. 19.

21. **Long, J. A.,** The Morphology, Reproduction, and Breeding Behavior of *Stenotaphrum secundatum* (Walt.) Kuntze, Ph.D. thesis, Texas A&M University, College Station, 1962, 61.

22. **Hitchcock, A. S. and Chase, A.,** Manual of the Grasses of the United States, USDA Misc. Publ. 200, U.S. Department of Agriculture, Washington, DC, 1950, 1051.

23. **Reinert, J. A.,** Antibiosis to the southern chinch bug in St. Augustinegrass accessions, *J. Econ. Entomol.*, 71, 21, 1978.

24. **Beard, J. B., Batten, S. M., and Pittman, G. M.,** St. Augustinegrass Cultivar Evaluation, Pr-3677 Texas Turfgrass Research 1978–1979, 1980, 44.

25. **Riordan, T. P., Meir, V. D., Long, J. A., and Gruis, J. T.,** Registration of Seville St. Augustinegrass, *Crop Sci.*, 20, 824, 1980.

26. **Wilson, C. A., Reinert, J. A., and Dudeck, A. E.,** Winter survival of St. Augustinegrass in North Mississippi, *Q. News Bull. South. Turfgrass Assoc.*, 12 (3), 20, 1977.

27. **Murray, J. J.,** personal communication.

28. **Busey,** personal communication.

Chapter 12

Integrating Cultural and Pest Management Practices for Sod Production

Stephen T. Cockerham, *University of California, Riverside, CA*

CONTENTS

Introduction 115
    A New Industry 116
Role and Value of the Sod Industry 116
    Role of the Sod Industry 116
    Value of the Sod Industry 116
    Environmental Applications 117
        Subsidence 117
General Pest Management Strategies 117
    Cultural Practices for Pest Management 118
Establishment 118
    Land Preparation 118
    Pre-Plant Pest Control 118
        Cultivation 118
        Fallowing 119
        Fumigation 119
    Establishment Nutrition 119
    Planting 120
        Cool-Season Grasses 120
        Netting 120
        Warm-Season Grasses 120
    Irrigation 120
Growing Turfgrass Sod 121
    Irrigation 121
        Drought Stress 121
    Mowing 121
    Turfgrass Sod Fertilization 122
    Pest Management 122
Market Preparation 123
    Inventory Maintenance 123
    Grooming 123
    Harvest 123
References 124

INTRODUCTION

Turfgrass sod production is often compared to turfgrass management, and just as frequently compared to general agriculture and to manufacturing. It is similar to, yet unlike, all three.

Whereas turfgrass management focuses upon the aesthetics and playability of a turf sward, sod production puts the priority on a product that can be marketed to a consumer. Where general agriculture focuses upon getting a crop to market in its harvest season, sod production harvests over a long period, even year-round in the sun belt. Manufacturing focuses upon efficiently turning raw materials

0-87371-350-8/94/$0.00+$.50

into a product, regardless of external influences, but sod production adds the variables of the environment and a living plant.

A NEW INDUSTRY

Turfgrass sod is a commercial product consisting of lawn-type grasses grown to the point of maturity so that they can be severed with a blade just below the soil level as a sod pad, physically picked up, carried to another site, laid on that site, and made to grow into a lawn. It is a perishable product to be installed within 36 hours from the time of harvest. After installation, the new lawn knits or roots into the soil in a few days becoming a high-quality turfgrass sward. The primary market for the sod product is the home lawn.

The turfgrass sod industry in the U.S. started on the East Coast in the early 1920s. Pastures of native grasses, mostly Kentucky bluegrass, were mowed short by entrepreneurial farmers and made to look somewhat like a lawn. The sod was cut and lifted with spades, then sold, at first for use on golf courses. Even though gardeners in Europe and the U.S. had long lifted and transplanted turfgrass sod, it was the Americans who developed it into a commercial product. As the business grew, turfgrass producers introduced horse-drawn sod cutters. By the 1930s, the home lawn was the predominant end product, and sod production had spread to the upper mid-west, especially in Michigan and around Chicago, Illinois. Today there is commercial sod production throughout the U.S. with the largest markets concentrated around the major urban areas.

During the late 1940s, the self-propelled sod cutter brought with it a more efficient way to harvest a sod pad cut to uniform thickness. Improved disease-resistant selections of Kentucky bluegrass were introduced during the 1950s, and so began the era of "cultured sod" as opposed to pasture sod.

The 1970s saw several major advances in sod production technology. By far the most important was the development of the sod harvester, which allowed two or three workers to perform the work of 20 to 30. Plastic netting laid at the soil line stabilized the mat and root structure, provided mechanical strength, and cut the crop maturity time by up to 75%. Developments in plant nutrition, weed, disease, and insect control have dramatically increased the uniformity, rooting ability, and general quality of the turfgrass sod. Marketing techniques have evolved or been adapted from other industries increasing sales and profitability. The availability of large self-propelled rotary mowers, wide vacuums and sweepers, totable forklifts, and hydraulically driven mower reels have all helped the rapid progress in the sod industry.[1]

ROLE AND VALUE OF THE SOD INDUSTRY

Turfgrass directly affects the way we live. It provides the play medium on many recreational facilities; it modifies the environment to make life easier and more pleasant; it provides opportunity for a pleasing and functional home landscape; and, in turn, the turfgrass industry has a significant direct impact on our economy as well as an indirect impact on the economy.[2]

ROLE OF THE SOD INDUSTRY

The turfgrass sod industry supplies high-quality, high-performance turf in a ready to use form. Environmental concerns and perceptions that focus on the turfgrass establishment wasting water, creating erosion, using fertilizer excessively, and pesticide application abuses are greatly reduced by the optimal, efficient, and safe resource input of professional growers. Each grower replaces thousands of those individual turf managers and homeowners that would be using the resources necessary to establish turf. The role of the turfgrass sod industry is to deliver a practical, immediate means to rapidly gain the many benefits of a turfgrass sward.

VALUE OF THE SOD INDUSTRY

The overall cost for the functional and amenity value of turf on a given site is reduced by the installation of the turfgrass as sod. Establishment of an equivalent quality lawn from seed requires substantial ownership input in time, labor, and financial resources — input that easily outweighs the initial higher cost of the sod product. Turfgrass sod creates a value-added significance to land-use and management. The instant-lawn result of sod installation has become important to the success of the housing industry — the single-family dwelling, the American dream.

In 1985, the turfgrass industry in the U.S. was estimated to be worth $25 billion with 371,000 employees. Turfgrass sod production was reported to have a significant part of the total industry value, employing 9400 with sales at $360 million.[3] Sod-farm size varies from under 10 acres to several thousand acres, averaging around 400 acres.

Sod has long been important for establishing turf where speed in developing ground cover was required. Installation of sod for erosion protection from wind and water has been of significant value in safeguarding banks, waterways, buildings, railroads, roads, and highways.

Sports fields and play areas often are so intensely scheduled for use that there is not time to grow turf from seed or stolons. Sod has been widely adapted for rapid establishment of sports surfaces. State-of-the-art sand construction of high-traffic sports facilities rely upon sod due to the complications in raising a uniform turf from seed in the pure sand.

ENVIRONMENTAL APPLICATIONS

Turfgrasses are important to the quality of life in the U.S. Environmental effects have been identified and are being further explored.[4] The use of sod assures a high quality turfgrass sward for the maximum environmental benefits. Conditions such as dust from bare soil, erosion protection, and fire protection are mitigated well before possible turf establishment from seed. Sodding has allowed cover to be established on poor soils, slopes, and in situations where seeding would be difficult or had failed.

Subsidence

Concern occasionally arises about soil depletion or subsidence from the potential mining of soil by sod harvesting. Several farms in various parts of the U.S. have long histories of production from certain fields with no noticeable loss of soil. After 20 or more years of continuous cropping, irrigation risers remain the same height and shallow soils have the same depth to bedrock.

Sod cut to an average $1/2$ in. soil thickness has shown that in 3-year-old sod approximately $1/16$ in. of soil is removed; in 2-year-old sod approximately $1/8$ in., and in 1-year-old sod $1/4$ in. At the same time clean, cultivated row crops, such as corn, soybeans, onions, peanuts, and cotton can lose from $1/2$ to $3/4$ in. of soil per year from wind and water erosion. The grass plant leaves more that it takes from the soil. There are tons of organic matter left in every acre of soil which helps build more soil. As soil is removed with a sod crop, the soil forming process goes forward with the building of new soil to replace some of that lost. Sod production usually improves soil productivity to the point that food and fiber crops planted after sod produce very high yields compared to the same crops grown before.[5]

GENERAL PEST MANAGEMENT STRATEGIES

Budgeting the sod operation involves forecasting sales and predicting product inventory availability. Cost control, product availability, and, ultimately, profit are dependent upon weather, and the timely and efficient use of sod production cultural practices and pest management. Mismanagement and pest infestation reduce the potential for getting the sod mature and marketable in keeping with the production schedule.

In a given market, most growers produce the same turfgrass species that perform well locally. For marketing, one of the simplest means of product differentiation, as well as one of the most difficult to sustain, is quality. Quality is the uniform appearance and density of the turfgrass, but the sod must appear fresh, be cleanly cut, be neatly stacked, and possess adequate sod strength. Weed, disease, and insect pests certainly impact product quality.

Pest management objectives may be achieved by preventing the establishment and spread of each pest, controlling those that become established, and maintaining infestations below economically damaging or annoying levels.[6] Sod production threshold considerations are the allowable levels for pest population buildup and distribution which begins to significantly reduce crop maturity, uniformity, and sod product quality; the ability of cultural practices to contain an infestation; and the point at which a pesticide should be applied. Sod producers must accomplish this at the lowest possible cost with minimum risk to employees, customers, and the environment.

In sod production pest management, as elsewhere, there has been difficulty with technology transfer and slow acceptance by growers. The conceptual complexity has limited implementation despite rising chemical costs, pest resistance, and societal concerns related to environmental impacts.[7]

Pest management strategies consider that turfgrasses grown for sod usually have fewer pest problems than established turf. Disease and insect populations tend to not build up, since the turf is nearly always new and young. The exception would be in market situations that force a long retention of mature inventory. Where permitted by law and economics, preplant soil fumigation may be used to effectively reduce or eradicate pests.

Nevertheless pests continue to be problems in production. The sod grower uses pest management strategies that are successful in other agricultural crops; pest identification, field monitoring, determination of control action guidelines, and application of effective methods for prevention and control. In some cases the symptoms caused by pest organisms closely resemble those caused by nutrient deficiencies or other soil problems. Monitoring the weather allows for prediction of the onset or disappearance of some pest problems.[8] Thorough knowledge of the complete culture of turfgrass by the sod grower is a business necessity.

Accurate monitoring of pests and natural enemies can determine the stage at which injury occurs and the threshold level at which pest management measures should be implemented. In turf and sod production, defining the economic injury level and action threshold is a major task.[9] The primary technical obstacle is the lack of simple monitoring techniques to establish action thresholds, and the few that do exist are only at the stage of being accepted by the early adopters. Disease identification kits and infrared photography are used with various levels of confidence and have not replaced the concept of management's footprints in the fields.

Thresholds tend to be site specific, tied to a number of variables such as weather, condition of turf, growth stage, as well as market demands. In a humid climate, such as Florida, diseases move very quickly, the action threshold will be low requiring immediate control from a fungicide. In an arid climate, such as southern California, disease would likely be allowed to develop to much higher levels, because of the improbability of the disease developing too quickly to stay on top of it and there would be more time possibly allowing the action taken to be cultural.

The decisions as to economic threshold are tied inexorably to the sod market demands reflecting the parameters of acceptable quality. While weeds in sod are totally unacceptable in some markets, consumers tolerate minimal weed presence in others. Insects and diseases are generally considered odious in all markets, and nematodes remain unacknowledged.

CULTURAL PRACTICES FOR PEST MANAGEMENT

The turfgrass seed industry has long recognized some of the problems faced by sod growers. Sod-quality seed is weed free and free of contaminant crop seeds. This is a particularly high-quality seed, at a premium price, that gives the sod grower a sense of security that when planting the crop no new problems are added to those already present. The grower, using information from various testing programs, has some assurance of selecting cultivars with potential resistance to major diseases.

Seeding rates, mowing schedules and techniques, clipping removal, soil fertility, irrigation practices, and weather affect the competitive relationship of turfgrasses and pests.[7] Healthy, growing young turfgrasses in sod production are able to significantly reduce the potential of injury from pest invasion. It is important in this intensely managed segment of the turfgrass industry to grow the crop quickly with avoidance of stresses, while vigorously curbing waste in labor, water, seed, fertilizers, and pesticides.

ESTABLISHMENT

In commercial sod production the operations are focused on bringing about optimum or even supra-optimum growth. Turfgrass is forced for maximum competitiveness of the grass with cultural practices intended to overcome environmental stresses, including pests.

It is not unusual for a sod farm to be located on a soil type that is less than ideal or with marginal water quality. As a result of the need to be close to a market, the availability and cost of arable land, and the availability and cost of water agronomic compromises are made that demand sound, efficient management for profitable output. Turfgrasses grown on poor soils with marginal water quality tend to be under stress and susceptible to pest infestation.

LAND PREPARATION

Land preparation is the first cultural practice in sod production and one of the most important. The quality of the work at this stage often determines whether or not a field will be profitable.[1]

To understand why preparation is so vital, just visualize how a sod harvester works. A sod cutter or harvester lifts the turf using a reciprocating blade that slices through the soil just under the grass. The blade is about $1/_2$-in. thick, 3-in. from front to back, and 18-in. wide. The working head of the machine rides on a roller, allowing the blade to follow most of the contours of the land with the intent of allowing a uniform depth of cut. As the unit moves forward, the roller and blades will bridge any undulation and holes that are narrower than the width of the blade.

One low spot a few inches wide can cause a hole in the sod pad, and the entire pad quality will be affected, resulting in it being thrown away. If the sod is harvested as square-yard pads (9 ft²), one rejected pad in 60 linear feet of harvest will be a 10% loss. Those bad spots can be eliminated during soil preparation.

PRE-PLANT PEST CONTROL

Disease and insect injury also cause sod pads to be rejected. Weeds in sod are not accepted in most markets and may also cause sod losses at harvest. Cultural practices have a significant impact on pest control at the soil-preparation stage of production.

It has been an accepted rule of thumb in sod production that any pesticide, particularly any herbicide, applied to the crop will increase plant stress. Thus exists a trade off in pesticide application for control of a pest to stop crop injury and the pesticide inducing it.

Cultivation

Although minimum tillage and no-tillage are widely used practices, plowing and disking are farming operations still common to many crops for soil preparation and as effective means of weed control. Weeds germinated by rainfall or pre-irrigation are mechanically killed with the cultivation equipment.

For cultivation to be effective in sod production, a minimum of soil should be moved to reduce bringing weed seeds to the surface and the majority of soil preparation must be completed.

Fallowing

In some markets, fallowing is an acceptable cultural practice. The field sits out of production for a period of time, a year or two, to cycle weeds and other pests. To be effective, the field is be kept clean of vegetation with cultivation. Nonselective herbicides, such as glyphosate, may be used if persistent or noxious perennial weeds such as nutsedge (*Cyperus* spp.) are present. Sometimes the field is worked to stir up the weeds and other pests. In many sod markets, the land is simply too valuable to be unproductive for the period of time needed to fallow.

Fumigation

Most weeds, nematodes, soil-borne diseases, and some insects can be eliminated by fumigating. Overall sod quality is improved and crop maturity time is reduced. Turfgrass forcing and fumigation require lower resource input, and pesticide use over several crop seasons is nearly eliminated due to the shortened time to sod maturity.

Three fumigants are commonly used on sod fields: *methyl bromide*, *chloropicrin*, and *metam-sodium* (Vapam or VPM). Methyl bromide is a highly toxic, odorless, colorless gas that is very effective for weed, insect, and nematode control. Chloropicrin is seldom used alone to fumigate sod fields because it does not control weeds, but it is mixed with methyl bromide to increase the effectiveness of disease control. Even when diseases are not targeted, a small percentage of chloropicrin is often mixed with odorless methyl bromide to provide a warning odor for safety. Chloropicrin is a form of tear gas, so it is easily detected.

Metam-sodium is effective against many of the same pests that methyl bromide and chloropicrin control. Although an excellent fumigant, it has not performed as consistently as the other materials.

Methyl bromide has been suspected to escape into the atmosphere in sufficient volume to cause environmental problems in the ozone layer, which is in violation of the 1990 U.S. Clean Air Act. International agreements have been formulated to reduce the future production and use of methyl bromide. Loss of the material will require an extensive reevaluation of pre-plant pest management for some growers.

Methyl bromide and chloropicrin are injected into the soil as gases, and a plastic tarp is laid on the soil immediately as the chemicals are injected. The tarp can be removed and seed planted 48 hours later. Workers must work upwind when applying methyl bromide and when removing the tarp to avoid excessive contact with the chemical.

Metam-sodium can be sprayed on the soil or it can be injected into the irrigation system and applied with the water. Application in the irrigation water is preferable because of the chemical's volatility. Backflow prevention is required with the injection of metam-sodium into an irrigation system and there is a point-source pollution potential at the site of the injector.

Surface application with a sprayer requires a large quantity of water as a carrier to get the metam-sodium into the soil surface. A follow-up irrigation is necessary to move it further into the soil and seal the soil surface to inhibit loss of the volatile chemical.

After application as a liquid, metam-sodium fumes in the soil and sealing the soil surface with a follow-up irrigation improves effectiveness. A waiting period of several days must pass for the metam-sodium to dissipate before planting. The big advantage of metam-sodium over the other fumigants is that even though a tarp does improve the effectiveness it is not required.

Recurrence of pest buildup after fumigation depends upon the sanitation of the sod operation. If irrigation water contains weed seed, as when the water source is an open canal, or if weeds are permitted to go to seed in the ditches or fence rows, the fumigation will have to be repeated in a short time.

Under normal conditions with good cultural practices the sod grower can expect a fumigation to be effective for 4 to 6 years, sometimes longer, before the pest population rebuilds to problem proportions. Excessive weed invasion is usually the indicator for repeat fumigation. The costs of fumigation can be amortized over several crops.

ESTABLISHMENT NUTRITION

A fertilizer application prior to planting assures that adequate nutrients, especially phosphorus, will be available to the emerging seedlings. Long experience has shown that most sod fields respond well and uniformly to a combination fertilizer product containing nitrogen and phosphorus, such as diammonium phosphate sulfate (16-20-0), applied preplant. The turfgrass competitiveness is maximized by having adequate levels always present in the soil beginning with germination through harvest. Excess fertilizer use, more than the amount required for optimum growth, is a waste of money, not environmentally sound management, and predisposes the plants to disease and stress susceptibility.

PLANTING

Turf species and variety selection is linked to local market and environmental conditions. Regional and local testing programs should be consulted for the optimum performing turfgrasses in a given market as well as the most vigorous sod formers.

Cool-Season Grasses

Kentucky bluegrass (*Poa pratensis*), perennial ryegrass (*Lolium perenne*), and tall fescue (*Festuca arundinaceae*) are the most frequently planted cool season grasses for sod production and are generally established from seed.

Sod growers can purchase sod-quality seed, premium seed that has been tested for a very high germination percentage and is exceptionally clean, free of weeds, and free of other crop seeds that might be contaminants.

Current theory in turfgrass science holds that two or more grasses should be planted together in blends or mixtures. Since each grass responds differently to various stresses, the more tolerant grass will dominate, increasing overall turf performance and competitiveness.

The seeding rates for sod production of cool season grasses vary greatly between climatic regions. In northern areas of the U.S., seeding rates are usually low, 20 to 40 lbs/acre of Kentucky bluegrass, where crops often take well over a year to mature. Seeding is usually done in the spring or the fall.

In the warm climates of the sun belt, a crop of cool-season grass can mature in as few as 3 months. Grasses grow almost year around and may be seeded at anytime during the year. There is a perception that grasses seeded out of their respective prime timing require significantly higher inputs for successful establishment. On the occasions when this is true, growers typically will postpone seeding as there is little advantage gained in the time to maturity vs. the cost input. Seeding rates of Kentucky bluegrass are 80 to 100 lbs/acre; perennial ryegrasses at 30 to 50 lbs/acre; and tall fescues at 200 to 250 lbs/acre. The higher seeding rates increase the competitiveness of the establishing grasses in these markets.

A stand established with poor uniformity due to seeding failures causes avoidable losses and excessive resource input. Mechanical seeders must be carefully calibrated and in good working condition.

Netting

In some markets netting has become important in the production of seeded turfgrass sod. This is a light polymer mesh netting laid over the soil surface at planting time. The turfgrasses grow through the netting which becomes an integral part of the sod product. Netted turf can be transplanted and rooted long before it would have the mechanical strength to be lifted as sod. Sod strength of cool-season grasses normally depends upon rhizomes and tillers — turfgrass roots contribute very little to tensile strength. Netting allows the harvest of much younger sod cutting 25% or more from crop maturity time. Netting use compresses the crop growing period making good management even more important. With the shorter growth period, there is a potential for reduced resource inputs including pesticide requirements.

Warm-Season Grasses

The warm-season turfgrasses that are grown from seed are planted like the cool-season grasses. Most of the warm-season grasses are planted vegetatively.

The most common method for large-scale vegetative planting of warm-season turfgrasses is stolonizing (sprigging). Stolon or sprig harvest machines do a pretty good job cutting sod into uniform length stolons and separating the soil from the plant material. Stolons can be and are often cut for planting using about any kind of flail or hammerknife including forage choppers.

Planting rates for stolons vary depending upon the grass and the speed of turf coverage desired. A practical and economical planting rate is 200 bushels per acre. Uniform stolon distribution and covering plus water management are necessary for an adequate stand of vigorous plants.

Many of the warm-season grasses used in sod production will regrow without being replanted. Those with rhizomes can be harvested cleanly as sod (e.g., bermudagrasses and zoysiagrasses) growing back from the rhizomes left below the harvester blade.

The sod of stoloniferous grasses that do not have rhizomes (e.g., St. Augustinegrass) is harvested to leave small strips or ribbons of turf in the field. These are allowed to spread and fill or are rotary tilled to distribute the rhizomes and stolons.

IRRIGATION

Water management at establishment is critical to the success of a sod crop with minimal losses.

Irrigation of a sod crop is more of an art than a science. Rarely are the irrigation systems as sophisticated as those seen on golf courses. Experienced irrigators understand the importance of uniform distribution, the vagaries of individual fields, environmental factors, and the needs of turfgrass species. Many growers use moisture sensors, computerized on-site weather stations, and consultants while relying on the judgment of an irrigator.

GROWING TURFGRASS SOD

The process of growing turfgrass sod includes irrigation, mowing, plant nutrition, and pest management. These are simultaneous operations, and poor execution of any one of them will have a direct impact upon profit.

IRRIGATION

Irrigation systems employed on sod farms are most often adapted from those developed for general agriculture, rather than the sophisticated, permanently installed equipment widely used on large turf installations. Permanently installed turfgrass management systems consist of controllers, buried lines, automatic valves, and pop-up sprinklers. They are expensive, but efficient systems.

On sod farms, one must consider the capital investment in the irrigation system. Arguments can be made for long-term returns on an investment in the automatic systems, but sod growers are often on leased land. Even on their own land, they are reluctant to commit the funds needed to install such a system on several hundred acres. Also, problems are likely when sod cutters and land preparation equipment work over and around the sprinklers.

Irrigation systems on sod farms use portable pipe or movable structures that allow work to go on in the fields. Even though automatic clocks and valves are available for use on portable systems, most require tending. Many require that an irrigator turn the water on and off at the desired running time.

A newly planted field will need water every day, at least, and may have to be irrigated more than once a day. A 6-week-old field may require irrigation every other day, and a mature field every third day. The water delivery must keep these individual water regimes in balance.

A harvest field must receive water the night before harvest, but not so much as to render it muddy or sloppy the next day. Fertilizer, iron, and pesticide applications may have to be watered in to prevent burn or to activate the material. Mowers can only work on fields that have dried enough to prevent rutting. Irrigation needs do not stop on the weekends or holidays, so someone must cover those periods. Irrigation scheduling impacts the entire production operation.

Turfgrass roots do not run deep. Established, mature turf has roots two or more feet deep, but the immature turf in sod production has most of its root growth near the soil surface. Irrigation only has to moisten a soil profile of about 4 to 8 in., or at most 1 foot. This means there is not a large reservoir from which the plant can draw moisture. The shallow profile must be replenished frequently.

On some sod farms, because of their unique situations, water is cheap. With the lower water cost, those companies can have a competitive advantage in pricing. Turfgrass sod is a good crop to irrigate with inexpensive, treated effluent water. However, the grower must be aware of the chemical content of the effluent water; dissolved salts, boron, certain metals, and other constituents can be in concentrations high enough to harm turf. For many growers, water is expensive and must be carefully managed to control production costs.

The uniform distribution of water is a key factor in producing quality sod. The design of the system and the condition of irrigation equipment are critical. If either is poor, there is little chance of success. If both are good, uniformity will depend upon the people operating the system.

Drought Stress

Wilting turf is the first symptom of drought stress. In the early stages of stress, the turf will recover with immediate water application. A severe drought area usually does not recover enough to be harvested with the rest of the field, and ultimately becomes waste. Seemingly minor turfgrass drought stress can result in heavy waste weeks or even months later at harvest.

The irrigation of turfgrasses for sod production remains more an art than a science. Weather, soil, species, and equipment make the application of water an operation unique to each sod farm.

MOWING

Mowing grass is about the most common, routine task in turf growing. It is also one of the biggest costs of sod production. Mowing can help develop and maintain turf quality, and can aid the sod crop to mature on time.

If the turf is allowed to grow too tall and is cut back excessively, the growth rate is significantly slower. The rule of thumb is to cut off no more than one-third of the leaf length at any one mowing. This is sometimes easier to say than to accomplish, especially in the spring. At that time, growth is so fast that it is difficult for a grower to keep up. Also, in many areas, spring rains can keep the ground too wet to support the mower's weight without damage, and this can disrupt mowing schedules and allow the grass to outgrow the mowing.

Mowing can be quite effective in weed control. Weeds that are often found as problems in agricultural crops and brought in by water or wind may not survive the frequent, close clipping of a sod field. This is an important advantage in evaluating potential weed control strategies.

TURFGRASS SOD FERTILIZATION

Turfgrasses are actually being forced to produce their maximum growth rate for sod production. Even though sod is sold on the basis of the quality and appearance of the verdure (the green foliage left on the plant after mowing) the grower must be sure it is balanced with root and rhizome development.

Nitrogen is extensively used by the plant. It is easily lost to the atmosphere and easily leached by irrigation water, so it must be applied regularly. Nitrogen recommendations for sod are based more upon general rules for the turf species, time of year, and soil type.

The standard recommendation for nitrogen fertilizer is 40 lbs of actual nitrogen per acre per month of growing season on turfgrasses grown for sod.

As soil temperatures move above the optimum range, nitrogen applications should stop. Cool-season grasses are extremely susceptible to stress in the presence of excess nitrogen at high temperatures.

Kentucky bluegrass is susceptible to rust disease (*Puccinia* spp.). Where it is a major problem, mowers can raise a roostertail of orange dust-like spores. Control of rust on Kentucky bluegrass is best done with an application of soluble nitrogen.

Phosphorus is required for various functions in the plant. Among the most important for the sod grower are the root growth and the rhizome development.

Potassium is seldom deficient on soils used for sod production. On sandy soils, sod growers have found that potassium has been beneficial in disease prevention and time to sod maturity.

Plants use iron in the formation of chlorophyll. When iron is deficient, grass plants become chlorotic. First the younger leaves turn yellow, showing up in irregular patches in the field. In contrast, nitrogen deficiency first becomes evident on the older leaves and is more uniform throughout a sod field.

Swards of thrifty turfgrasses that are actively growing with optimal nutrition are capable of avoiding or overcoming some pest problems. Weed invasion is minimized often allowing the grower to not use any herbicide. These healthy plants are not susceptible to disease development.

PEST MANAGEMENT

Pest management on sod farms has long followed the canons of integrated pest management (IPM). Even before it was politically correct to use IPM it was good business to do so. Cultural practices related to water and nutrition management and mowing are significant contributors to the avoidance of pest buildup.

If a particular pest, especially a disease such as *Pythium* blight or a weed such as *Poa annua*, begins to invade a field, care should be taken to reduce its spread. Equipment should be washed before moving between fields if the problem is a weed. For disease problems sanitation procedures, such as washing with an antiseptic solution (e.g., diluted bleach) or steam cleaning may be necessary. Personnel should avoid walking from the infected field to a clean field without cleaning shoes or boots. This is very difficult, since it disrupts daily operations.

Some weeds escape fumigation, in particular those with very hard seed coats. Fortunately, the weeds that do escape can usually be controlled selectively if they are not allowed to grow to maturity. The biggest problem with weed control after fumigation is keeping weeds out of the clean field. Good housekeeping — keeping ditches and fence rows clean — does pay off. Floodwater, irrigation from an open ditch, and run-off into reservoirs all may contain weed seed. Any of these can be a major source of contamination.

Most turf herbicides can be used on turfgrass grown for sod with only minimal effect on growth. When it is necessary to apply a herbicide, knowledge of the chemical activity is important since some inhibit root or rhizome development. Applied correctly, these herbicides would cause no turf management problems on a golf course — they just interfere with the timely production of sod. Several pre-emergent herbicides prevent hybrid bermudagrass stolons from pegging down. When applied to a bermuda sod field, they can have very disheartening results.

Any turf disease that is a local problem can occur in a sod field, but usually does not. Most pathogens build up in the debris in thatch. Under normal conditions, sod crops have little of this debris, and pathogens have little opportunity to increase. Unfortunately, a few diseases such as *Pythium* blight, southern blight, *Fusarium* blight, and spring deadspot can still devastate sod fields.

Sod fields damaged by a disease infestation may recover very slowly. One solution is to destroy the crop and replant, which is a difficult thing for a grower to do. Another solution is to scalp the turf and overseed. Kentucky bluegrass damaged with *Fusarium* blight does well overseeded with perennial ryegrass. Marketable sod can then be available in 30 to 60 days. The consumer will have no problem with the disease on the resulting Kentucky bluegrass/perennial ryegrass sod.

Grubs, cutworms, and armyworms are the main insect pests of sod. On occasion, sod webworm has caused problems. None of these pests seems to do much harm to sod in the field, but all can cause problems for the consumer. A large group of birds working in a field is a good indicator that some kind of worm is in the sod. Any irrigation pipe laying on the turf can be moved to show worms underneath. Once identified, the decision must be made as to whether the treatment threshold has been reached. Biological control materials, such as *Bacillus thuringiensis* (BT), are on the market and widely used for most of these worms, although there are some good insecticides still registered.

In recent years sod growers have made increasing use of scouts or pest control advisors. These are professionals, licensed in some states, who are very effective in identifying pests, in working with the grower to determine the treatment threshold, and in recommending economical, environmentally sound treatment strategies.

Typically, scouts walk the sod fields on a frequent, regular schedule. It is more important that the individual know what the normal crop is supposed to look like than be capable of instant diagnosis of problems. Once something unusual is observed, the scout looks for patterns to diagnostically separate operational problems from pest injury. Arcs, circles, ellipses, straight lines, triangles, and strips can be related to operation of the irrigation system, sprayer, spreader, or mower. Often correction requires a change in procedure and not a pesticide treatment.

Weeds in one or opposite field ends may indicate that equipment, such as a landplane, dragged the contamination into a clean area. This could indicate that a herbicide application would be necessary, but only as a spot treatment.

MARKET PREPARATION

The single most important factor that distinguishes turfgrass sod production from traditional turfgrass management is that it is sold, lifted, and transported from the site where it is grown to a site where it is installed as permanent turf. In most markets, commercial turfgrass sod is a high-quality, uniform product. It is also perishable, having a shelf-life of about 36 hours, and in the roll susceptible to a wide range of problems. Good quality sod is uniform with the appearance and density of the turfgrass (fresh, cleanly cut, and neatly stacked sod pads) and sod handling strength.[10]

INVENTORY MAINTENANCE

Turfgrass sod has a unique characteristic for an agricultural crop. It does not have to be harvested at maturity. Sod inventory can be held until the sales volume requires the product. Usually the crop is in pretty good condition and can be maintained at minimum levels even slightly stressed until needed. Only a short time is required to bring the quality up to standard. If the turf is allowed to stress excessively, pest management can become a problem which may require pesticide application.

GROOMING

In preparation for harvest, field operations for the final month focus on grooming the turf. The goal is to make the turf look like the finest lawn one can imagine. On the day of harvest, its height is to be uniform, its color as dark a green as the cultivar can be, and its density good, and it should be free of weeds, diseases, and insects. If pests are present at a scale that cannot be removed by hand, the product is not ready for harvest.

Fertilizer applications to mature sod are minimal and used only to maintain or enhance color. Nitrogen fertilizers should not be used on the turf in the last week before harvest due to the burn risk. Inside the sod roll or folded pad, ammonia and urea fertilizers can volatilize to act as fumigants, killing the grass.

Applications of iron materials are often used to improve color without endangering the sod in the roll or folded pad. Some iron products, such as ferrous sulfate, will readily burn the turf at application and great care must be used.

Risks are inherent in applying anything but water in the last couple of weeks. Many growers are reluctant to apply any material to sod that is ready for harvest. Weeds should be removed by hand. If there are too many for hand removal, the sod is not ready for harvest.

HARVEST

At maturity the turfgrass appearance is that of a magnificent green carpet of living plants, but the true test of a turfgrass sod, and ultimately the sod company, comes when the sod cutter goes into the field. If the turf holds together it is sod. If it does not, no matter how good it looks it is not a salable product. Conversely, if it is strong and tough and looks terrible, it still is not a salable product.

Sod is perishable, and should be harvested to order. The grower who delivers sod that has been on the pallet too long soon runs out of customers.

Harvest is the last opportunity for quality control. Poor cultural practices or chemical applications show up glaringly. Sod pads are accepted or rejected. Customers are satisfied or lost.

REFERENCES

1. **Cockerham, S. T.,** *Turfgrass Sod Production*, ANR Publications, University of California, Oakland, CA, 1988, 84.
2. **Gibeault, V. A., Cockerham, S. T., Henry, J. M., and Meyer, J.,** California turfgrass: its use, water requirement and irrigation, *Cal. Turfgrass Cult.*, 39(3,4), 1, 1989.
3. **Cockerham, S. T. and Gibeault, V. A.,** The size, scope, and importance of the turfgrass industry, in *Turfgrass Water Conservation*, Gibeault, V. A. and Cockerham, S. T., Eds., ANR Publications, University of California, Oakland, CA, 1985, 155.
4. **Gibeault, V. A.,** The effects of turf on the quality of life, *Golf Course Manage.,* 59(12), 59, 1991.
5. **Cockerham, S. T.,** Soil depletion, *West. Landscaping News*, 17(5), 19, 1977.
6. **Short, D. E., Reinert, J. A., and Atilano, R. A.,** Integrated pest management for urban turfgrass culture, in *Symposium on Turfgrass Insects: Advances in Turfgrass Entomology*, Niemczk, H. D. and Joyner, B. G., Eds., Hammer Graphics, Piqua, OH, 1982, 25.
7. **Balogh, J. C., Leslie, A. R., Walker, W. J., and Kenna, M. P.,** Development of integrated management systems for turfgrass, in *Golf Course Management and Construction: Environmental Issues*, Balogh, J. C. and Walker, W. J., Eds., Lewis Publishers, Boca Raton, FL, 1992, 355.
8. **Statewide Integrated Pest Management Project,** Division of Agriculture and Natural Resources, University of California, Integrated Pest Management for Citrus: Publication 3303, ANR Publications, University of California, Oakland, CA, 1991, 144 pp.
9. **Bruneau, A. H., Watkins, J. E., and Brandenberg, R. L.,** Integrated pest management, in *Turfgrass*, Waddington, D. V., Carrow, R. N., and Shearman, R. C., Eds., American Society of Agronomy, Madison, WI, 1992, 502.
10. **Cockerham, S. T.,** Buying sod for golf course use, *Golf Course Manage.*, 58(8), 48, 1990.

Chapter 13

Advances in Implementing Integrated Pest Management for Woody Landscape Plants

M. J. Raupp, *Department of Entomology, University of Maryland, College Park, MD*

C. S. Koehler, *Department of Entomological Sciences, University of California/Berkeley*

J. A. Davidson, *Department of Entomology, University of Maryland, College Park, MD*

CONTENTS

Perspectives and Overview 125
Reasons for the Development of IPM for Landscape Plants 126
Components of IPM for Landscape Plants 128
    Key Pests and Plants 128
    Monitoring 129
    Decision-Making 129
    Intervention 131
        Resistant Plant Materials 131
        Biological Control 131
        Biorational and Chemical Pesticides 133
        Cultural and Mechanical Tactics 134
Privatization of IPM Programs for Landscapes and Nurseries 134
Opportunities and Needs 135
Acknowledgments 135
References..135

PERSPECTIVES AND OVERVIEW

This review emphasizes advances in managing arthropod pests of woody ornamental plants in landscapes. We discuss a few examples from nurseries because of the similarity in pest complexes and management approaches. Several other reviews deal with the ecology and management of arthropod pests in turfgrasses[141] and other urban systems in general.[11,49-51,119] These are beyond the scope of this report.

With a few exceptions, the development of management procedures for landscape pests parallels procedures for agriculture. Although crop pests have been of concern to society for as long as agriculture has been practiced, pests in the landscape were mostly ignored until relatively recent times. The gypsy moth, *Lymantria dispar*, eradication campaign in Massachusetts between 1891 and 1900 was the first major attempt to manage a landscape pest,[48,195] although one might argue that the greater interest lay in protecting fruit crops than amenity plants. The gypsy moth campaign is significant also because of its contribution to the development of lead arsenate as an insecticide. Besides spraying for larval control, other intervention tactics included burning egg masses or treating them with creosote to prevent their hatch, burning wooded land to deny the larvae foliage, and banding tree trunks with burlap skirts for daily destruction of larvae beneath;[48] such practices might be considered components of an integrated pest management (IPM) program today. Lead arsenate was used widely for the next 50 years against many landscape and agricultural pests.[45,195] Spraying with successor inorganic and botanic products became standard practice until they were displaced by the synthetic organic insecticides following World War II.

Widespread concerns about hazards of insecticides to the public health and the environment, voiced in the late 1950s and early 1960s, were directed mainly at agriculture[18] but spread quickly to the landscape. The industry responded by offering insecticides of narrower spectrum and low persistence and with techniques for injecting and implanting systemic insecticides into ornamental trees.[11,83]

Integrated control, later termed integrated pest management, originated in agriculture[180] but by 1976 was being explored for use in the landscape.[137] Numerous refinements and advances have continued to the present.[42,75,89,128,149,150] Table 1 summarizes several events considered landmarks in pest management that are unique to the landscape situation. Although a few listings do not directly address management, their implications are nevertheless obvious.

REASONS FOR THE DEVELOPMENT OF IPM FOR LANDSCAPE PLANTS

The overriding justification for the development of IPM programs for landscape plants is the economic importance of the managed resources, which arises from several sources, not the least of which is the plants' aesthetic value.[66,162] The most readily identified economic indicator is the revenue derived from the production of ornamental plants. Nurseries and greenhouses generate the largest cash receipts of any crop commodity in 11 states. From 1983 to 1988, this income increased by more than $2.4 billion in the U.S.[185a] The sale of woody landscape plants represents an important component of this revenue.

Woody plants, especially trees, have enormous value once they are established in the landscape.

The urban forest is defined as those trees planted along streets and within municipal boundaries or reasonable distances beyond.[84] A 1990 evaluation of the urban forest in the U.S. estimated that it comprised 61 million street trees with an aggregate value of between $18 billion and $30 billion.[82] A 1991 report also indicated that about $425 million was spent each year on the management of these trees.[197] In addition to street trees, as many as 600 million additional trees live in yards and parks in the U.S.[82]

Not surprisingly, a great deal of pesticide use is associated with the management of pests in landscapes as well as in the production of landscape plants. Nationwide, about 27 million households used pesticides at least once in 1989 to control insect pests of lawns, trees, and gardens.[118] The cost of products associated with this control was in excess of $1 billion.[118] In addition to pest control implemented by private citizens, a large and thriving management control industry maintains plants in the landscape. In 1984, this industry generated an estimated $1 billion, approximately 18% of which was earned by spraying.[124]

A clear pattern of excessive pesticide use has emerged from the few instances in which pesticide use patterns have been examined. Several studies document dramatic reductions in the quantities of pesticides used and number of plants treated without sacrifice to the appearance or value of the plants following the implementation of IPM programs in landscapes. Reductions ranged from 4.9 to 99.8% of pre-IPM levels but typically exceeded 70%.[19,52,75,137,152,170,176] These studies serve as a strong

Table 1 **Landmarks in landscape pest management**

Event	Significance	References
Gypsy moth eradication campaign, Massachusetts (1891–1900)	First large-scale attempt to manage an insect pest of landscapes. Development of lead arsenate as an insecticide.	48
Dutch elm disease epidemics in Netherlands (1920s) and Ohio (1930s)	Recognition of the value of shade trees to urbanized societies.	182
Framework for implementing IPM in street-tree systems in California (1970s)	First demonstration of principles and practices of landscape IPM.	135–137
Publication of *Insects that Feed on Trees and Shrubs* (1976) and *Diseases of Trees and Shrubs* (1987)	Brought insect and disease identification to the level of the practitioner.	78, 173
Demonstration of feasibility of landscape IPM by private arboricultural firms in Maryland (1980s)	Led to privatization of landscape IPM services.	30, 75

indictment for the unnecessary use of pesticides in traditional pest-control programs for landscape plants.

Many studies have measured amounts of pesticides used in the production of woody landscape plants. A survey of 158 nurseries in Pennsylvania revealed that pesticides were applied at rates comparable to those found in agronomic systems.[171] As in the case of landscape plants, much of the pesticide use in nurseries appears to be unnecessary. By implementing an IPM program, Davidson et al.[28] reduced pesticide use by 87.6%. Pesticide reductions ranging from 45 to 96% were documented in the production of anthuriums in nursery situations using IPM.[65]

For several reasons, the heavy reliance on pesticides in managed landscapes and nurseries has created a climate favorable for the development of IPM programs. Heavy dependence on pesticides has resulted in the evolution of resistance in several key and secondary pests of agricultural systems. Although the magnitude of resistance problems in landscapes and nursery systems is unknown, several factors favor the development of resistance. First, when managed, these environments support numerous pests for which resistance to one or more pesticides has been documented. Recent reviews indicate that no fewer than 36 important insect and mite pests of landscape plants are known to be resistant to at least one kind of pesticide.[56,57] The taxon exhibiting the greatest number of resistant species is the Homoptera, particularly aphids. Coincidentally, aphids have been identified as the most important taxon of insect pests for municipal foresters[197] and rank high in the several lists of pests found in landscapes of homeowners[75,119,152] and in nurseries.[171]

Application procedures commonly used in landscapes and nurseries may also contribute to resistance. A traditional approach still widely used is the cover spray, in which all plants in a management area are treated.[19,75,125,136,176] In several eastern states, large geographic regions have been repeatedly treated aerially for control of the gypsy moth with a rather narrow array of insecticides, primarily *Bacillus thuringiensis* (Bt) and diflubenzuron, thereby increasing the likelihood for the evolution of resistance.

Although some circumstances favor the development of resistance in pests of landscape plants, at least four factors mitigate it. The managed landscape is a diverse and often highly heterogeneous management unit.[42,149,158] Pesticide applications usually occur on rather localized units often at the level of individual plants or landscapes. This practice provides treated areas that susceptible genotypes can readily colonize. A current trend away from the use of cover sprays in favor of spot treatments based on thresholds and scouting information further enhances the likelihood of preserving refuges for susceptible genotypes and beneficial organisms.[19,28,58,75,137,176] An emerging trend in the landscape maintenance industry is to substitute insecticides with short residual action, such as insecticidal soap and oil, for traditional materials with long residuals.[58,77] Lastly, an attribute of several IPM programs for landscape plants is that nonpesticidal control tactics are substituted for ones involving pesticides.[70,75,137,176] In their recent review, Leeper et al.[96] recommended all of these practices as tactics for delaying and managing resistance.

Along with the evolution of resistance, the disruption of natural enemy complexes by pesticides has often been cited as one of the primary reasons for pest outbreaks in agronomic systems.[25,177,180] Several cases of pesticide-associated pest outbreaks have been reported in landscape systems following pesticide applications. Luck and Dahlsten[100] found that mosquito fogging programs in a resort community greatly reduced the numbers of parasitoids attacking the pine needle scale, *Chionaspis pinifoliae*, thereby allowing populations of this pest to increase dramatically and injure trees. When the mosquito control program was modified, scale populations declined as mortality by predators and parasitoids increased.[100,157] A similar scenario was reported in a resort community in Michigan where the European fruit lecanium, *Parthenolecanium corni*, reached outbreak levels on street trees following weekly applications of dimethoate to control filth flies.[110] Following the implementation of an IPM program for control of the flies and the discontinuation of weekly sprays along streets, scale populations rapidly declined to innocuous levels, apparently because of increased activities of natural enemies.[110] Attempts to eradicate a localized infestation of the Japanese beetle, *Popillia japonica*, in Southern California resulted in serious outbreaks of citrus red mite, *Panonychus citri*, woolly whitefly, *Aleurothrixus floccosus*, purple scale, *Lepidosaphis beckii*, and citrus mealybug, *Planococcus citri* on landscape plants.[32] Pesticides used in the initial eradication program and in subsequent attempts to control secondary pest outbreaks were far more toxic to beneficial insects and mites than to the primary pests[32] and contributed to the outbreak of secondary pests. Eradication programs for the Mediterranean fruit fly, *Ceratitis capitata*, with insecticidal baits have resulted in outbreaks of several species of Homoptera and mites in urban areas of California.[38] Although the large scale application of pesticides clearly can result in pest outbreaks in landscapes, a study by McClure[107] revealed that incomplete application of pesticides to single trees can also result in dramatic increases in armored scale populations.

A final factor favoring the implementation of IPM programs in landscapes is the growing societal concern of safety with regard to the continued use of synthetic organic pesticides.[58,74] This concern is reflected in a recent dramatic increase in the number of jurisdictions within states, regulating pesticide use in the landscape.[34]

COMPONENTS OF IPM FOR LANDSCAPE PLANTS

Components such as monitoring, decision-making, and intervention constitute IPM programs for managed landscapes and nurseries.[28,75,128,135-137,152,176] We review progress in defining key pests and plants, in developing monitoring approaches and decision-making guidelines, and in reducing pest populations with resistant plant materials, biological, chemical, cultural, and mechanical control tactics.

KEY PESTS AND PLANTS

The biotic diversity of urban and suburban habitats has been well documented and tends to be much greater than typical agricultural systems in which most crops consist of a single species or cultivar.[42,138,185] Landscapes may be composed of several hundred species or cultivars. Frankie and Ehler[49] reported more than 330 species of woody landscape trees, shrubs, and vines found within the city limits of Austin, TX. In a survey of municipal foresters, Kielbaso and Kennedy[84] found more than 200 species or cultivars comprising the urban forest in western cities. A study of 26 homesites in Maryland revealed in excess of 133 species of landscape plants under the management of a single tree-care firm.[75]

The concept of key pests is widely used in traditional agronomic systems to define the focus for management activities.[177] Key pests of landscape plants have also been determined for a variety of geographic regions and management systems. Lists of key pests serve several important purposes. First, they clearly indicate that although the overall pest diversity is large in landscape systems, rather few insects and mites create the majority of problems. A national survey revealed that a group of 10 species or species groups accounted for 63% of the total insect problems encountered by municipal foresters.[197] Data gathered from scouting programs

in home landscapes and institutional settings disclosed that 10 species or functionally related groups accounted for 83 to 97% of the arthropod pests encountered annually.[152] Arthropod pests in nurseries appear similarly biased in that a group of 10 species or species groups accounted for 88% of the total problems encountered.[171]

Lists of key pests in a geographic region appear relatively stable temporally. Surveys of municipal arborists conducted six years apart yielded lists that were quite similar. In four geographic regions, the relative importance of individual pests or pest groups changed somewhat; however, lists shared from 70 to 90% of their pests between sample dates.[197] Not surprisingly, this pattern does not hold across geographic regions, where lists of the 10 most common pests have only 30 to 70% of their taxa in common.[197] Even within geographic regions, lists of key pests vary. The diversity of plants in the management unit will strongly influence the list of associated key pests. The managed landscape of one college campus was dominated by two genera of trees; as a result, a complex of only 10 arthropod pests accounted for 97% of all insect and mite pests found.[152] In contrast, homeowners in the same area generally had more diverse landscapes and the 10 most common arthropod pests accounted for only 83% of the total encountered.[152]

In most agronomic systems, crops are managed for production based on a unit area such as bushels or pounds per acre. The management unit is usually comprised of many hundreds or thousands of plants and the value of the crop is determined by the aggregate contribution of each plant to the yield. In most ornamental landscape systems, however, a single plant may be worth hundreds or thousands of dollars, and an individual plant may be the focus of management activities.[19,150] Furthermore, the floristic diversity of landscapes presents the inexperienced pest manager with a bewildering array of plants and associated pests. Key plants have been defined as those that provide aesthetic or functional attributes that contribute significantly to the landscape value.[127] An operational component was later added to this definition that identified key plants as those that were most likely to incur serious, perennial problems that dominate control practices.[149] The identification of key plants on a regional or local basis can assist in the design of pest-resistant landscapes by indicating which genera are pest prone and should be avoided. Also, for existing landscapes it identifies the genera that are pest prone and therefore will serve as the focus of monitoring and intervention activities.

This information is useful in planning the cost of IPM programs by landscape-management firms.[6,75]

An examination of more than 30,000 home landscape plants revealed that certain families and genera were much more pest prone than others.[149] For example, plants in the genus *Malus* represented 2.1 to 3.1% of the total plants monitored in four experimental IPM programs. However, the likelihood that any plant in this genus would have a pest or cultural problem ranged from 47 to 100%. Other plant genera such as *Viburnum* were relatively pest free. In general, rosaceous plants tended to be widely used and among the most pest prone in homeowner landscapes.[149] Similar lists of key plants have been generated by monitoring home landscapes in Minnesota[6] and by surveying municipal arborists and foresters in several midwestern cities.[132]

MONITORING

In most agronomic systems, control decisions are predicated on quantitative estimates of pest abundance determined by systematic sampling. Similar sampling approaches are not directly applicable to most managed landscape systems, primarily because of the diversity and heterogeneity of the managed resource. Exceptions include cases such as the gypsy moth in populated forests for which systematic sampling plans have been developed to make management decisions applicable to large acreage.[90,154] At present, decisions in landscape systems are based on information gathered using one or more monitoring techniques.[6,39,52,75,148]

Monitoring activities have been categorized according to their function as detection surveys, biological evaluations, loss or damage surveys, and evaluations of intervention activities.[22] With respect to landscape IPM programs, monitoring has been justified to provide information on the temporal and spatial abundance of pests and natural enemies to facilitate more effective selection, placement, and timing of control tactics.[148]

Several monitoring techniques are presently employed in landscape IPM programs. The most widely used is the visual inspection of plant material.[6,70,75,136,152] Correlative indicators of insect abundance such as frass counts for caterpillars[98,186] and honeydew droplet counts for aphids[39] have been used to monitor insects where direct pest observation is difficult. A variety of trapping approaches are also available for monitoring pest activity in landscapes. Pheromone traps have been used for several activities including survey and detection of pests such as the gypsy moth[44,154] and timing of developmental patterns to facilitate control actions for pests such as clearwing borers,[55,121,122,126,144] tip moths,[54,101] and scale insects.[79,156] Pheromones have also been used in conjunction with trap trees to depress localized populations of elm bark beetles in landscape settings.[93,133]

Monitoring devices such as colored sticky cards that do not rely on pheromones have been used to monitor beneficial insect activity in street trees and forests.[164,193] Investigators have used refugia such as burlap bands or plastic flaps to monitor gypsy moth activity.[154]

Forecasting models based on heat-unit accumulations have been developed for many pests of landscape plants including the elm leaf beetle, *Xanthogaleruca luteola*;[40] the bronze birch borer, *Agrilus anxius*;[1] the flatheaded apple tree borer, *Chrysobothris femorata*;[146] dogwood borer, *Synanthedon scitula*;[144] lilac borer, *Podosesia syringae*;[145] Nantucket pine tip moth, *Rhyacionia frustrana*;[54,101] bagworm, *Thyridopteryx ephemeraeformis*;[123] San Jose scale, *Quadraspidiotus perniciosus*;[79] walnut scale, *Quadraspidiotus juglansregiae*;[59] obscure scale, *Melanaspis obscura*;[142] and pine needle scale, *Chionaspis pinifoliae*.[15] The use of phenological models in landscape IPM programs was recently reviewed.[5]

DECISION-MAKING

The concept of decision-making as it pertains to the management of insect pests in ornamental plant systems is a complex one and begins with the correct identification of the pest causing the injury. At present, commercial pest managers, primarily arborists, can identify only about 50% of the key pests and beneficial arthropods commonly encountered in the landscape (M. J. Raupp and J. A. Davidson, unpublished data). Once a pest has been identified, the manager must decide if it is abundant enough to cause damage to the plant and to warrant intervention. Pedigo et al.[139] identified these central questions as the crux of decision making in their review of economic thresholds and injury levels. In a previous review, we argued that the same methodology used to generate economic injury levels for agronomic crops apply to woody landscape plants.[150] To avoid confusion between *aesthetic* injury levels and *economic* injury levels, we suggested that the goal of the resource manager be considered when constructing decision-making guidelines. For example, the goal of the retail or wholesale grower of nursery crops is clearly economic. Therefore, decision-making rules should be based on the cost of control, value of the crop, damage associated with the pest, and injury relative to the abundance of the pest. The value of the crop and the damage caused by the pest will be directly affected by the aesthetic perceptions of the consumer.

In other situations, such as home landscapes, plants are not grown or maintained for direct economic profit. Parameters such as the value of the crop or the economic damage relative to pest abundance do not clearly dictate management decisions in these settings. In such economically vague scenarios, we suggest that aesthetic injury levels should form the basis of management decisions.[150,151]

Although virtually no studies have examined economic injury levels or thresholds in ornamental plant systems, several have investigated one or more elements of decision-making. Empirical decision making rules have been proposed for the California oakworm, *Phryganidia californica*, and blue spruce aphid, *Elatobium abietinum*,[136] as well as for aphids, soft scales, pine needle scale, spider mites, spruce gall adelgids, pine shoot moths, defoliators on hardwoods and conifers, bronze birch borer, borers in lilac and ash (*Podosesia* spp.),[128] and the gypsy moth.[154]

At least six studies have attempted to quantify relationships between insect-related plant injury and aesthetic perceptions of damage as an aid to forming management decisions. Koehler and Moore[88] demonstrated that the abundance of the cypress tip miner, *Argyresthia cupressella*, was directly related to damage (unsightliness) on 13 cultivars of Cupressaceae. They used this relationship to recommend which cultivars would be the least pest-prone in landscape plantings. Dreistadt and Dahlsten[39] correlated tuliptree aphid, (*Macrosiphum liriodendri*), abundance with the production of honeydew and the complaints received by the city of Berkeley, CA. They used timed estimates of honeydew production and complaint records to define a treatment threshold. Buhyoff and Leuschner[14] photographed forested landscapes with varying levels of injury caused by southern pine beetle, *Dendroctonus frontalis*. The photographs were ranked according to their visual preference by a sample of students and nonstudents and a disutility function was generated. This function indicated that the perceived attractiveness of the landscape was greatly reduced even when injury levels were less than 10%. Based on the rapid decline in the attractiveness of the landscape when injured by southern pine beetle, Buhyoff and Leuschner[14] concluded that the goal of management programs should be to prevent new outbreaks of this pest.

Larew et al.[94] surveyed attendees of floriculture trade shows regarding their perceptions of injury caused by a serpentine leaf miner, *Liriomyza trifolii*

on chrysanthemums. In two independent surveys, one using illustrations of injury and the other using actual plants, a strong negative relationship was found between leafminer injury (percentage of leaves mined) and the willingness of respondents to accept injured plants for sale or purchase. In both surveys, the most rapid decrease in the willingness to buy injured plants occurred at injury levels less than 10%.[94]

The responses of customers in retail nurseries to photographs of arborvitae injured by the bagworm demonstrated a strong negative relationship between bagworm injury and plant damage as indicated by the consumers' perceptions of plant injury and willingness to purchase defoliated plants.[150,159,160] Consumers were also asked to select plants showing damage that warranted control. In both cases, more than half of those surveyed perceived damage and would initiate control at injury levels of less than 10% defoliation or discoloration.[150]

The most recent study to address the problem of decision-making in woody plant systems concerns the perception of injury caused by the orangestriped oakworm, *Anisota senatoria*, on street trees. Coffelt and Schultz[19] surveyed homeowners in the city of Norfolk, VA, regarding their perceptions of injury caused by the orangestriped oakworm. As in the previous studies, the majority of homeowners were not willing to accept defoliation exceeding about 10%. The authors determined, however, that starch reserves of these street trees were not affected until defoliation levels approached 25%.

The studies by Buyhoff and Leuschner,[14] Larew et al.,[94] Raupp et al.,[150] and Coffelt and Schultz[19] all revealed a strong negative association between injury and percent damage. In each case, damage was measured by quantifying people's perceptions of injury. A general pattern emerged indicating that the majority of people discriminate injury at or below 10% of the affected plant or landscape. This trend appeared in a variety of systems including defoliated landscapes, defoliated individual plants, illustrations of injured plants, and plants with mined leaves. The studies were also diverse in the types of audiences assayed. If this pattern is a general one in that people generally perceive and respond to plant damage at a level of about 10%, then the task of establishing aesthetic injury levels may be greatly simplified. Here, the determination of quantitative aesthetic injury levels is contingent upon establishing the relationship between pest abundance and plant injury.[150,151] However, we believe that further studies of the factors underlying individuals's perceptions of pest damage and their willingness to prevent that damage may reveal an economic basis for the association of damage, value, and perceptions that will permit the determination of economic injury levels in economically vague crop system such as managed landscapes.

INTERVENTION
Resistant Plant Materials

Selection of proper planting material is an important component of landscape IPM but is of value only where new or replacement plantings are required. In the past, plants, especially trees, were chosen on the basis of: availability; height; their deciduous or evergreen nature; flowering characteristics, foliage color, or other attribute of beauty; tolerance to temperature extremes; or tendency for roots to crack curbs or sidewalks or clog sewer or drainage pipes. We suggest that in the near future landscape plants will be selected more on the basis of what can, or cannot, be done to manage pests that threaten them.

For many years, agriculture has used and received the benefit of pest-resistant plant varieties. The majority are grains and forages, annual or short-lived perennials having low cash value per acre and developed through traditional plant-breeding efforts.[106,175] Breeding insect-resistant landscape ornamentals represents new challenges and opportunities,[161] yet the lack of consensus on aesthetic injury levels and thresholds and the perennial nature of these plants will both negate rapid progress in their development by plant breeding methods and discourage many investigators from even beginning the task.

As an alternative to plant breeding, the great variety and diversity of landscape material in the marketplace enables the identification of pest-resistant ornamentals from among already existing plant species and cultivars. Horticulturally desirable characteristics such as foliage and blossom color[23,143] and variegation[160] may also affect the resistance of woody ornamental plants to insect attack. An evaluation of insect resistance in some woody ornamental plants has already begun. Nearly every example illustrates that susceptibility to insects or mites varies widely among plant materials evaluated. Reluctance of the nursery industry to increase production of pest-resistant materials, and the failure of landscape architects to recognize the value of these plants and recommend their use, still impede more widespread propagation of pest-resistant varieties. Also, most insect-resistant ornamentals identified to date have not been evaluated for susceptibility to other important agents of unsatisfactory plant performance.

Weidhass[193] summarized other problems associated with the development or use of insect-resistant ornamentals.

Biological Control

Biological control tactics appeal to many landscape IPM program managers and clients alike because they epitomize the ecological, environmentally safe approach to pest management. Two recent reviews have focused on the use of classical biological control, conservation, and augmentation of natural enemies in urban settings and landscapes.[27,46] Despite the lack of applied research to show if, how, and when beneficials should be released in most landscape IPM programs, the use of augmentation continues to grow. This trend appears in the Suppliers of Beneficial Organisms in North America list.[12] The 1985 list recorded 45 parasites, predators, and pathogens sold by 53 companies, while the 1989 list shows 60 beneficials sold by 60 companies.

Dahlsten[26] suggested that diverse, discontinuous landscape plantings containing exotic plants may be well suited to the classical biological control approach. Dahlsten and Hall[27] discuss 28 insect pests that attack landscape plants found in urban areas around the world, for which control success by imported parasites and predators has been reported. They present a table showing 50 landscape-plant pest species, the country in which biological control was reported, and a reference. Of these, 34 species are primarily agricultural pests that occasionally attack landscape plants, and 16 are primarily landscape-plant pests. Of the 50 pests, 33 species are homopterans, 8 lepidopterans, and 1–3 each are found among the Dermaptera, Diptera, Hymenoptera, Thysanoptera, and Hemiptera. Most of the beneficials were imported to control the above pests in agricultural crops, and later followed their pest hosts to landscape settings. The majority of control successes have involved hymenopteran parasitoids against 28 species of pest scale insects. However, recently the coccinellid *Chilocorus kuwanae*[36] and the nitidulid *Cybocephalus nipponicus*[37] were imported to control the diaspidid *Unaspis euonymi*. Three years after releases were made in the Washington, D.C. area, both beetles had established and had effectively reduced scale pest populations.

The importance of conserving natural enemies in landscapes was clearly demonstrated in several cases of nontarget outbreaks following the application of pesticides to landscapes.[32,38,100,110,157] Several nursery and landscape studies have revealed large, diverse predator and parasitoid complexes that could produce significant management benefits

Table 2 Examples of evaluations of susceptibility of woody ornamentals to arthropod pests

Target Pest	Plants Evaluated	Results	References
Ceanothus stem gall moth, *Periploca ceanothiella*	40 species and cultivars of *Ceanothus*	More than half show no occurrence of insect. Severe infestations on *C. griseus* and its cultivars.	116
Acacia psyllid, *Acizzia uncatoides*	112 *Acacia* and *Albizia* species	Most showed low susceptibility. Several widely produced species showed high susceptibility.	117
Euonymus scale, *Unaspis euonymi*	8 species and cultivars of *Euonymus*	No occurrence on *E. kiautschovicus*	196
Nantucket pine tip moth, *Rhyacionia frustrana*	40 species of *Pinus*	Wide range in susceptibility in California. Generally consistent with findings in southeastern U.S.	168
5 root weevils in genera *Sciopithes, Otiorhynchyus, Nemocestes,* and *Dyslobus*	42 species and 27 hybrid rhododendrons	Species generally less susceptible than hybrids to leaf feeding. Dark red flowering hybrids or species generally susceptible to leaf feeding	3
Cypress tip miner *Argyresthia cupressella*	13 members of Cupressaceae	Low susceptibility of several *Juniperus chinensis* and *J. scopulorum* cultivars.	88
Fuchsia gall mite, *Aculops fuchsiae*	43 species and hybrids of *Fuchsia*	Wide range in susceptibility. Nine showed high resistance.	86
Elm leaf beetle, *Xanthogaleruca luteola*	12 elm species and hybrids and *Zelkova serrata*	European elms generally support greater adult survival and fecundity than American and Asian hosts.	63
Hawthorn lace bug, *Corythucha cydoniae*	14 species or cultivars of *Cotoneaster* and 5 of *Pyracantha*	Wide range in susceptibility to adult feeding or oviposition or nymphal feeding or survival. Leaf pubescence associated with reduced oviposition and nymphal survival.	163, 165, 166
Gypsy moth, *Lymantria dispar*	19 species or cultivars of shade and flowering trees	Wide range of defoliation with 7 cultivars showing no defoliation.	140

if appropriate conservation tactics were employed. Malaise trapping in a suburban English garden yielded 30, 25, and 16% of Britain's syrphid, ichneumonid, and coccinellid species, respectively.[138] A 3-year study of Maryland nurseries revealed 28 genera and 28 species of Coccinellidae.[178] The twice-stabbed lady beetle, *Chilochorus stigma*, reduced a large pine needle scale (*C. pinifolia*) population during a 3-year study of a Christmas tree plantation.[33] Parasitism of the Eastern tent caterpillar, *Malacosoma americanum*, increased manyfold when wildflowers were present in the vicinity of infested trees.[97] A 3-year study of first-generation prepupal parasitism of the mimosa webworm, *Homadula anisocentra*, in landscapes revealed rates of parasitism ranging from 44 to 47%.[9] Scale insects harbor large complexes of hymenopteran parasitoids in landscape settings, for example: *Q. juglansregia*, 4 to 6 spp.;[59,184] *Q. perniciosus*, 5 spp.;[184] *Melanaspis obscura*, 7 to 8 spp.;[142,183] and *Parthenolecanium quercifex*, 5 spp.[164] These studies revealed peak periods of parasitoid activity coincident with periods of crawler activity. Unfortunately, traditional recommendations for scale insect control often suggest the application of insecticides during periods of crawler activity. This practice may reduce the benefit of indigenous parasitoids in landscape settings.[142,164]

Several biotic factors complicate the implementation of biological control tactics in landscapes and nurseries. Not the least of these is the abundance of ants in the settings.[119] Ants are commonly associated with Homoptera in landscapes, and several studies suggest that ants may require management to implement biological control.[31,42]

Biorational and Chemical Pesticides

Biorational pesticides are microbial and biochemical agents, either naturally occurring or identical to natural products, that exert pesticidal activity.[174] Several materials in this growing group of products are receiving increasing study for landscape-plant pest control. They are dormant and summer oils, azadirachtin, soap, *Bacillus thuringiensis* (Bt), and entomogenous nematodes.

Johnson and Caldwell[77] surveyed 1225 landscape pesticide applicators and found that 43% used oil sprays during the dormant season, but only 8% used oil sprays during the growing season despite the fact that several products are formulated for this use. In contrast, a recent survey[125] of arborist firms practicing IPM showed nearly 50% using oil sprays in the summer as well as soap and Bt. Several workers[10,29,129] recently showed that oil sprays applied during the growing season are efficacious and safe to plants when used according to label directions. Grossman[61] reviewed the uses of horticultural oils on ornamental plant pests.

Products containing azadirachtin have extremely low mammalian toxicity but show repellency and/or toxicity to over 100 insect species when topically applied or fed to these species.[134] This compound was recently shown to have systemic activity against the birch leafminer, *Fenusa pusilla*, following injection.[102] An improved formulation of azadirachtin received expanded EPA registration in 1989 to control whiteflies, thrips, mealybugs, and several caterpillar pests of trees and shrubs.[71]

The insecticidal activity of fatty acids has long been known.[172] Miller[112] recently reviewed the advantages and disadvantages of insecticidal soaps and horticultural oil. Like foliar horticultural oil sprays, insecticidal soap is particularly effective against the soft-bodied sucking pests such as aphids,[188] adelgids,[69] scale insect crawlers,[130] and mites.[131]

The development of Bt as an alternative to synthetic organic insecticides for gypsy moth control has been reviewed.[43] The *kurstaki* and *thuringiensis* strains of this pathogen now have registration for more than twenty species of defoliating caterpillars. Recently, *B. thuringiensis* "San Diego" and "tenebrionis" have demonstrated effective control of the elm leaf beetle, *X. luteola*.[24]

The standard approach to controlling clearwing moth borers includes pheromone trapping of males and carefully timed bark treatments using residual insecticides to kill newly hatched larvae that must chew through the insecticide barrier to reach sapwood.[143,144] Inundative application of entomogenous nematodes appears to be an effective alternative to chemical sprays for sesiid and cossid larval control. Kaya and Brown[80] achieved 77 to 84% and 86 to 93% control of *Synanthedon culiciformis* in alder using the nematode *Steinernema feltiae* as bark sprays and gallery treatments respectively. In the same study, *S. feltiae* and *S. bibionis* were also used as spring and fall bark sprays to control *Synanthedon resplendens* in sycamore. *Steinernema feltiae* gave only 61% control in the fall, while *S. bibionis* was ineffective. Kaya and Lindegren[81] obtained 90% control of *Paranthene robiniae* in poplar and willow using *S. feltiae*. Lindegren and Barnett[99] controlled *Prionxystus robinae*, a serious cossid moth borer of large shade trees in fig tree orchards using nematodes. Application of *S. feltiae* and *Heterorhabdites heliothidis* to potted nursery plants controlled black vine weevil larvae. *S. feltiae* was ineffective at economical dosage levels, while *H. heliothidis* was effective at low dosages but difficult to store.[181]

In Czechoslovakia, Mracek and David[114] made soil applications of *Steinernema kraussei* to spruce plots containing diapausing sawfly (*Cephalcia abietis*) larvae. The mortality rate was 81 to 97%, and nematodes were found in the soil one year later. Mracek and Spitzer[115] showed that *S. kraussei* did not interfere with the major predators and parasitoids of *C. abietis*.

Where foliar sprays in landscapes are unacceptable, and tall, high-value trees require protection from potential defoliators such as the gypsy moth, trunk implants containing acephate or methamidophos have significantly reduced gypsy moth populations.[47,155,190]

Several tree-canopy pests traditionally controlled with foliar sprays migrate down the trunk to complete development on the ground. One such insect, the elm leaf beetle, *X. luteola*, has been controlled with carbaryl sprayed in a band around the trunk to intercept migrating larvae.[21,41] This technique should be tested against other migrators.

Cultural and Mechanical Tactics

Managed landscapes are fraught with both abiotic and biotic factors that predispose ornamental plants to insect attack. Many of these factors can be manipulated to aid in the reduction of pest problems. Problems often begin with the installation of the plant when edaphic conditions and ecological requirements receive less attention than aesthetic considerations.[66,187] Exposure to sunlight sometimes affects the frequency and severity of attack by insect pests. Understory plants or edge species such as azaleas or dogwoods are more likely to be attacked by the azalea lacebug, *Stephanitis pryiodes*, and dogwood borer, *Synanthedon scitula*, respectively, when planted in sunny locations.[143,147] However, other pests such as aphids may be more abundant on trees planted in shaded locations.[62]

Several recent reviews have examined the direct and indirect effects of drought on the suitability and susceptibility of plants to a wide variety of insects.[104,105,194] Many researchers have suggested that the effects of drought may be exacerbated in landscape settings where plants may have a reduced capacity to obtain water because of improper installation, restricted root zones, compaction of soil, and injury to roots during construction.[92,192] Drought may predispose many commonly utilized woody plants to attack by insects, mites, and diseases.[17,76,104,105,169]

Humans inadvertently and deliberately introduce agents into landscapes that alter the susceptibility of plants to pests. The effects of chemicals such as air pollutants on plant physiology and susceptibility to insect attack have been the focus of recent reviews.[2,91] As with other stressors, the effects of pollutants such as ozone may be idiosyncratic even within the same plant. For example, exposure to ozone reduced the suitability of *Populus deltoides* for the imported willow leaf beetle, *Plagiodera versicolora*, but did not affect the plant's suitability for an associated aphid, *Chaitophorus populicola*.[20]

Fertilization is a cultural practice frequently recommended to reduce the susceptibility of deciduous hardwoods and conifers to attack by boring Lepidoptera and Coleoptera.[53,64,143] Much literature indicates that herbivore performance and survivorship may be directly related to the moisture and nitrogen content of host tissues.[13,103,167] Not surprisingly, recent studies of performance of sucking pests such as the hemlock woolly adelgid, *Adelgis tsugae*, and the elongate hemlock scale, *Fiorinia externa*, showed that the performance of these insects improved on fertilized hemlock trees.[108,109] Conversely, fertilization of the white birch was found to have little effect on its ability to respond to attack by the bronze birch borer.[72] Clearly, much remains to be understood about the effects of fertilization on plant physiology and susceptibility to pests before this tactic can be generally recommended for managing pests in landscape situations.

Several mechanical methods are used to manage insect pests in landscapes. Sticky barrier bands and burlap bands have been used for the management of gypsy moth larvae on individual trees.[48,190] Pruning is a mechanical tactic commonly used in the landscape to remove pests from trees and shrubs. Sanitation pruning of infected branches serves as the cornerstone of management programs for Dutch elm disease.[60,73] However, freshly pruned elms are more attractive to the two principal scolytid vectors of this disease;[8,67] therefore, routine pruning of elms should be performed during periods of beetle inactivity.[8,16] A similar strategy is recommended for pruning oak trees where nitidulid beetles vector oak wilt[4] and for pines subject to attack by the Sequoia pitch moth, *Synanthedon sequoiae*.[87]

Wounding the boles of landscape trees during weed removal or mowing may increase the likelihood of attack by clearwing borers.[143] Therefore, mulching around the bole of the tree is recommended to reduce bark injury and thereby ameliorate borer problems.[143]

PRIVATIZATION OF IPM PROGRAMS FOR LANDSCAPES AND NURSERIES

A necessary step in the privatization of landscape IPM programs has been the demonstration of their feasibility in a variety of management systems. During the past two decades, at least 11 programs have been implemented for woody landscape plants, street trees, nursery crops, and the gypsy moth in forested urban areas. Universities have sponsored the majority of programs;[19,68,70,75,136,152,176] however, several were cooperative efforts between universities and municipalities,[19,136,176] institutions,[68,152] and private enterprise.[28,68,75] Other programs were implemented by state and federal agencies acting in concert[154] or individually.[170] In one instance, a commercial plant-care firm has reported results of its own program offered to clients.[52] Nine of these programs contained some type of evaluation that discussed changes related to the implementation of IPM. In several cases, the cost of managing the resource declined as a result of the implementation of IPM.[19,28,65,137,176] However, in at least two cases, costs increased, at least initially.[68,152] Nine of these studies reported pesticide reductions ranging from 5 to 99% associated with the implementation of IPM.[19,28,52,65,75,136,152,170,176]

The privatization of IPM is under way in several landscaping industries, but predominantly in the tree-care industries in the northeastern states and California.[125] Recent surveys of tree care firms indicated that a majority employed scouts that regularly and frequently inspected landscapes; some used monitoring techniques such as pheromone traps; and many intervened with biorational pesticides such as soap and oil and formulated biological control agents such as Bt.[58,125] Few firms utilized predators or parasitoids as control tactics.[58,125] Most disturbing was the finding that more than 30% of the firms continued to perform cover sprays, although more than 80% utilized some spot treatments as well.[58,125] Surveys of industry members in Maryland revealed a higher rate of adoption of IPM practices by individuals who have received formal IPM training compared to those who had not.[153] However, even those who had not participated in formal workshops had relatively high adoption rates of practices such as monitoring.[153] The tree care industry seems to be leaning toward reductions in the use of pesticides in landscapes primarily through the reduction or elimination of cover sprays in favor of well-timed, selective spot treatments.[52,58,125,153]

OPPORTUNITIES AND NEEDS

Landscape pests will not be managed in the twenty-first century in the same way as in the past. We predict that it will soon become unlawful or impractical to spray landscape vegetation in urban areas with the chemicals in common use today. We believe that IPM, with incrementally decreasing dependence on chemicals, represents the most realistic alternative course.

However, one should note several constraints that impede our ability to reach that IPM goal. First, the biology and ecology of many landscape pests and beneficial organisms, each having its complex of associated organisms and collectively interacting with several thousand important landscape plants, remain poorly known. Second, our knowledge of sampling and monitoring techniques and the consensus on management-action thresholds for pests that often cause aesthetic rather than economic losses remains incomplete. Third, appropriate control measures are not available in adequate or dependable supply. Finally, current commitment of the majority of the research and extension enterprises to the management of agricultural pests leaves too few resources available for landscape-pest management studies.

Despite these shortcomings, several pesticide-related incidents have influenced public opinion to be increasingly receptive to IPM.[74,85] Where the landscape is concerned, our hope is that this strong and growing public opinion will generate the resources required to develop the knowledge base and delivery systems to implement truly integrated approaches for managing landscape pests.

ACKNOWLEDGMENTS

We thank S. H. Dreistadt and L. P. Pedigo for commenting on previous drafts of this review. B. A. Barr and T. MacIntyre assisted in preparing the manuscript and M. A. Coffelt, D. A. Hermes, J. J. Kielbaso, M. S. McClure, and D. R. Smitley provided unpublished information. To these people we are indebted. This is scientific article A6155, contribution 8323 of the Maryland Agricultural Experiment Station.

REFERENCES

1. **Akers, R. C., Nielsen, D. G.,** 1984. Predicting *Agrilus anxius* Gory (Coleoptera: Buprestidae) adult emergence by heat unit accumulation. *J. Econ. Entomol.* 77:1459–1463.

2. **Alstadt, D. N., Edmunds, G. F. Jr., Weinstein, L. H.,** 1982. Effects of air pollutants on insect populations. *Annu. Rev. Entomol.* 27:369–384.

3. **Antonelli, A. L., Campbell, R. L.,** 1984. Root weevil control on rhododendrons. *Wash. State Univ. Ext. Bull.* 970. 4 pp.

4. **Appel, D. N., Peters, R., Lewis, R., Jr.,** 1987. Tree susceptibility, inoculum availability, and potential vectors in a Texas oak wilt center. *J. Arboric.* 13:169–173.

5. **Ascerno, M. E.,** 1991. Insect phenology and integrated pest management. *J. Arboric.* 17:13–15.

6. **Ball, J.,** 1987. Efficient monitoring for an urban IPM program. *J. Arboric.* 13:174–177.

7. **Barbosa, P., Schultz, J. C.,** Eds., 1987. *Insect Outbreaks.* New York: Academic. 578 pp.

8. **Barger, J. H., Cannon, W. N., Jr.,** 1987. Response of smaller European elm bark beetles to pruning wounds on American elm. *J. Arboric.* 13:102–104.

9. **Bastian, R. A., Hart, R. E.,** 1990. First-generation parasitism of the mimosa webworm (Lepidoptera: Plutellidae) by *Easmus albizziae* (Hymenoptera: Eulophidae) in an urban forest. *Environ. Entomol.* 19:409–414.

10. **Baxendale, R. W., Johnson, W. T.,** 1988. Evaluation of summer oil spray on amenity plants. *J. Arboric.* 14:220–225.

11. **Bennett, G. W., Owens, J. M.,** Eds., 1986. *Advances in Urban Pest Management.* New York: Van Nostrand Reinhold. 399 pp.

12. **Bezark, L. G.,** 1989. Suppliers list of beneficial organisms in North America. California Department of Food and Agriculture, Biological Control Services Program, Sacramento, CA.

13. **Brodbeck, B., Strong, D.,** 1987. See Ref. 7, pp. 347–364.

14. **Buhyoff, G. J., Leuschner, W. A.,** 1978. Estimating psychological disutility from damaged forest stands. *Forest Sci.* 24: 424–432.

15. **Burden, D., Hart, E.,** 1989. Degree-day model for egg eclosion of pine needle scale (Hemiptera: Diaspididae). *Environ. Entomol.* 18:223–227.

16. **Byers, J. A., Svihra, P., Koehler, C. S.,** 1980. Attraction of elm bark beetles to cut limbs on elm. *J. Arboric.* 6:245–246.

17. **Campana, R. J.,** 1983. See Ref. 51, pp. 459–480.

18. **Carson, R.,** 1962. *Silent Spring.* Boston: Houghton Mifflin Co. 368 pp.

19. **Coffelt, M. A., Schultz, P. B.,** 1990. Development of an aesthetic injury level to decrease pesticide use against orangestriped oakworm (Lepidoptera: Saturniidae) in an urban pest management project. *J. Econ. Entomol.* 83:2044–2049.

20. **Coleman, J. S., Jones, C. G.,** 1988. Acute ozone stress on Eastern cottonwood (*Populus deltoides* Bartr.) and the pest potential of the aphid *Chaitophorus populicola* Thomas (Homoptera: Aphididae). *Environ. Entomol.* 17:207–212.

21. **Costello, L. R., Scott, S. R., Peterson, J. D., Adams, C. J.,** 1990. Trunk banding to control elm leaf beetle. *J. Arboric.* 16:225–230.

22. **Coulson, R. N., Witter, J. A.,** 1984. *Forest Entomology.* New York: John Wiley and Sons. 669 pp.

23. **Cranshaw, W. S.,** 1989. Patterns of gall formation by the cooley spruce gall adelgid on Colorado blue spruce. *J. Arboric.* 15:277–280.

24. **Cranshaw, W. S., Day, S. J., Gritzmacher, T. J., Zimmerman, R. J.,** 1989. Field and laboratory evaluations of *Bacillus thuringiensis* strains for control of elm leaf beetle. *J. Arboric.* 15:31–34.

25. **Croft, B.,** 1990. *Arthropod Biological Control Agents and Pesticides.* New York: John Wiley and Sons. 723 pp.

26. **Dahlsten, D. L.,** 1986. Control of invaders. In *Ecology of biological invasions of North America and Hawaii.* H. A. Mooney, J. A. Drake, Eds. New York: Springer-Verlag, 275–302.

27. **Dahlsten, D. L., Hall, R. W.,** 1991. Biological control of insects in outdoor urban environments. In *Principles and Application of Biological Control* . T. Fisher, Ed. Berkeley: University of California Press. In press.

28. **Davidson, J. A., Cornell, C. F., Alban, D. C.,** 1988. The untapped alternative. *Am. Nurser.* 167(11):99–109.

29. **Davidson, J. A., Gill, S. A., Raupp, M. J.,** 1990. Foliar and growth effects of repetitive summer horticultural oil sprays on trees and shrubs under drought stress. *J. Arboric.* 16:77–81.

30. **Davidson, J. A., Holmes, J. J.,** 1984. A diversifying option for PCO companies. *Pest Management* 3(2):10–14, 53.

31. **DeBach, P., Huffaker, C. B., MacPhee, A. W.,** 1976. Evaluation of the impact of natural enemies. In *Theory and Practice of Biological Control*, C.B. Huffaker, P.S. Messenger, Eds. New York: Academic. 255–285.

32. **DeBach, P., Rose, M.,** 1977. Environmental upsets caused by chemical eradication. *Calif. Agric.* 31:8–10.

33. **DeBoo, R. F. , Weidhaas, J. A.,** 1976. Plantation research: XIV. Studies on the predation of pine needle scale, *Phenacaspis pinifolia* (Fitch), by the coccinellid, *Chilochorus stigma* (Say). Canadian Forest Service Report CC-X-119.

34. **Delaney, T. J.,** 1990. Posting and Notification Summary. Professional Lawn Care Association of America.

35. **Doane, C.C., McManus, M. L.,** Eds., 1981. *The Gypsy Moth: Research Toward Integrated Pest Management.* Forest Serv. Tech. Bull. 1584. USDA, Washington, DC. 757 pp.

36. **Drea, J. J., Carlson, R. W.,** 1987. The establishment of *Chilocorus kuwanae* (Coleoptera: Coccinellidae) in eastern United States. *Proc. Entomol. Soc. Wash.* 89:821–824.

37. **Drea, J. J., Carlson, R. W.,** 1988. Establishment of *Cybocephalus* sp. (Coleoptera: Nitidulidae) from Korea on *Unaspis euonymi* (Homoptera: Diaspididae) in the Eastern United States. *Proc. Entomol. Soc. Wash.* 90:307–309.

38. **Dreistadt, S. H., Dahlsten, D. L.,** 1986. Medfly eradication in California: lessons from the field. *Environment* 28(6):18–20,40–44.

39. **Dreistadt, S. H., Dahlsten, D. L.,** 1988. Tuliptree aphid honeydew management. *J. Arboric.* 14:209–214.

40. **Dreistadt, S. H., Dahlsten, D. L.,** 1990. Relationship of temperature to elm leaf beetle (Coleoptera: Chyrsomelidae) development and damage in the field. *J. Econ. Entomol.* 83:837–841.

41. **Dreistadt, S. H., Dahlsten, D. L.,** 1990. Insecticide bark bands and control of the elm leaf beetle (Coleoptera: Chrysomelidae) in northern California. *J. Econ. Entomol.* 83:1495–1498.

42. **Dreistadt, S. H., Dahlsten, D. L., Frankie, G. W.,** 1990. Urban forests and insect ecology. *BioScience.* 40:192–198.

43. **Dubois, N. R.,** 1981. See Ref. 35, pp. 445–453.

44. **Elkinton, J. S., Carde, R. T.,** 1981. See Ref. 113, pp. 41–55.

45. **Essig, E. O.,** 1931. *A History of Entomology.* New York: MacMillan. 1029 pp.

46. **Flanders, R.V.,** 1986. See Ref. 11, pp. 95–127.

47. **Fleischer, S. J., Delorme, D., Ravlin, F. W., Stipes, R. J.,** 1989. Implantation of acephate and injection of microbial insecticides into pin oaks for control of gypsy moth: time and efficacy comparisons. *J. Arboric.* 15:153–158.

48. **Forbush, E. H., Fernald, C. H.** 1896. *The Gypsy Moth, Porthetria dispar (Linn.).* Boston: Wright & Potter. 495 pp.

49. **Frankie, G. W., Ehler, L. E.,** 1978. Ecology of insects in urban environments. *Annu. Rev. Entomol.* 23:367–387.

50. **Frankie, G. W., Koehler, C. S.,** Eds., 1978. *Perspectives in Urban Entomology.* New York: Academic. 417 pp.

51. **Frankie, G. W., Koehler, C. S.,** Eds., 1983. *Urban Entomology: Interdisciplinary Perspectives.* New York: Praeger. 493 pp.

52. **Funk, R.,** 1988. Davey's plant health care. *J. Arboric.* 14:285–287.

53. **Funk, R., Rathjens, R.,** 1983. Fertilizer guide for trees. *Weeds, Trees, and Turf* 22:32–37.

54. **Gargiullo, P. M., Berisford, C. W., Canalos, C. G., Richmond, J. A.,** 1983. How to time dimethoate sprays against the Nantucket pine tip moth. *Georgia Forest Research Paper. No. 44.*

55. **Gentry, C. C., Holloway, R. L., Polet, D. K.,** 1978. Pheromone monitoring of peachtree borers and lesser peach tree borers in South Carolina. *J. Econ. Entomol.* 71:247–253.

56. **Georghiou, G. P.,** 1986. See Ref. 120, pp. 14–43.

57. **Georghiou, G. P., Mellon, R.,** 1983. Pesticide resistance in time and space. In *Pest Resistance to Pesticides.* G.P. Geoughiou, T. Saito, Eds. New York: Plenum. 1–46.

58. **Gerstenberger, P.,** 1991. IPM strategies. *Tree Care Indust.* 2:14–15.

59. **Gordon, F. C., Potter, D. A.,** 1988. Seasonal biology of the walnut scale, *Quadraspidiotus juglansregiae* (Homoptera: Diaspididae) and associated parasites on red maple in Kentucky. *J. Econ. Entomol.* 81:1181–1185.

60. **Gregory, G. F., Allison, J. R.,** 1979. The comparative effectiveness of pruning versus pruning plus injection of trunk and/or limb for therapy of Dutch elm disease in American elms. *J. Arboric.* 5:1–4.

61. **Grossman, J.,** 1990. Horticultural oils: new summer uses on ornamental plant pests. *The IPM Practitioner* (12):1–10.

62. **Hajek, A. E, Dahlsten, D. L.,** 1986. Discriminating patterns of variation in aphid (Homoptera: Drepanosiphidae) distribution on *Betula pendula. Environ. Entomol.* 15:1145–1148.

63. **Hall, R. W., Townsend, A. M., Barger, J. H.,** 1987. Suitability of thirteen different host species for elm leaf beetle, *Xanthogaleruca luteola* (Coleoptera: Chrysomelidae). *J. Environ. Hort.* 5:143–145.

64. **Ham, D. L., Hertel, G. D.,** 1984. Integrated pest management of the southern pine beetle in the urban setting. *J. Arboric.* 10:279–282.

65. **Hara, A. H., Nishijima, W. T., Hansen, J. D., Burke, B. C., Hata, T. Y.,** 1990. Reduced pesticide use in an IPM program for anthuriums. *J. Econ. Entomol.* 83:1531–1534.

66. **Harris, R.W.,** 1983. *Arboriculture: Care of Trees, Shrubs, and Vines in the Landscape.* Englewood Cliffs, NJ: Prentice-Hall. 688 pp.

67. **Hart, J. H., Wallner, W. E., Caris, M. R., Dennis, G. K.,** 1967. Increase in Dutch elm disease associated with summer trimming. *Plant Disease Reporter* 51:476–479.

68. **Hartman, J. R., Gerstle, J. L., Timmons, M., Raney, H.,** 1986. Urban integrated pest management in Kentucky — a case study. *J. Environ. Hort.* 4:120–124.

69. **Heller, P. R., Kellogg, S.,** 1990. Spring control of Cooley spruce gall adelgid with Safer's soap and conventional insecticides on Douglas fir in Berks County, PA, 1989. *Insecticide and Acaricide Tests: 1990.* 15:345.

70. **Hellman, J. L., Davidson, J. A., Holmes, J. J.,** 1982. Urban ornamental and turfgrass integrated pest management in Maryland. In *Advances in Turfgrass Entomology.* H. D. Niemczyk, B. G. Joyner, Eds. Piqua, OH: Hammer Graphics. 31–38.

71. **Henn, T., Weinzierl, R.,** 1989. Botanical insecticides and insecticidal soaps. University of Illinois, *Coop. Ext. Serv. Circ. 1296.*

72. **Hermes, D. A., Mattson, W. J.,** 1991. Does reproduction compromise defense in woody plants? In *Tree Insect Interactions: Variation Among Insect Feeding Guilds.* Y. N. Baranchikov, W. J. Mattson, F. Hain, T. L. Payne, Eds. USDA Tech. Rep. N.E. Forest Experimental Station. In press.

73. **Himelick, E. B., Ceplecha, D. W.,** 1976. Dutch elm disease eradication by pruning. *J. Arboric.* 2:81–84.

74. **Hock, W. K.,** 1984. IPM — is it for the arborist? *J. Arboric.* 10:1–4.

75. **Holmes, J. J., Davidson, J. A.,** 1984. Integrated pest management for arborists: implementation of a pilot program. *J. Arboric.* 10:65–70.

76. **Houston, D. H.,** 1985. Dieback and declines of urban trees. *J. Arboric.* 11:65–72.

77. **Johnson, W. T., Caldwell, D. L.,** 1987. Horticultural oil sprays to control pests of landscape plants: an industry survey. *J. Arboric.* 13:121–125.

78. **Johnson, W. T., Lyon, H. H.,** 1976. *Insects that Feed on Trees and Shrubs.* Ithaca, NY: Cornell University Press. 465 pp.

79. **Jorgensen, C. D., Rice, R. E., Hoyt, S. C., Westigard, P. H.,** 1981. Phenology of the San Jose scale (Homoptera: Diaspididae) *Can. Entomol.* 113:149–159.

80. **Kaya, H. K., Brown, L. R.,** 1986. Field application of entomogenous nematodes for biological control of clear-wing moth borers in alder and sycamore trees. *J. Arboric.* 12:150–154.

81. **Kaya, H. K., Lindegren, J. E.,** 1983. Parasitic nematode controls western poplar clear-wing moth. *Calif. Agric.* 37:31–32.

82. **Kielbaso, J. J.,** 1990. Trends and issues in city forests. *J. Arboric.* 16: 69–75.

83. **Kielbaso, J. J., Davidson, H., Hart, J., Jones, A., Kennedy, M. K.,** Eds., 1979. *Proceedings, Symposium on Systemic Chemical Treatments in Tree Culture.* Ann Arbor, Michigan: Braun-Brumfield, Inc. 357 pp.

84. **Kielbaso, J. J., Kennedy, M. K.,** 1983. See Ref. 51, pp. 423–440.

85. **Koehler, C. S.,** 1989. Prospects of implementation of IPM programs for ornamental plants. *Fla. Entomol.* 72:391–394.

86. **Koehler, C. S., Allen, W. W., Costello, L. R.,** 1985. Fuchsia gall mite management. *Calif. Agric.* 39(7,8):10–12.

87. **Koehler, C. S., Frankie, G. W., Moore, W. S., Landwehr, V. R.,** 1983. Relationship of infestation by the Sequoia pitch moth, (Lepidoptera: Sesiidae) to Monterey pine trunk injury. *Environ. Entomol.* 12:979–981.

88. **Koehler, C. S., Moore, W. S.,** 1983. Resistance of several members of the Cupressaceae to the cypress tip miner, *Argyresthia cupressella. J. Environ. Hort.* 1:87–88.

89. **Koehler, C. S., Raupp, M. J., Dutky, E., Davidson, J. A.,** 1985. Standards for a commercial arboricultural IPM program. *J. Arboric.* 10:293–295.

90. **Kolodny-Hirsch, D. M.,** 1986. Evaluation of methods for sampling gypsy moth (Lepidoptera: Lymantriidae) egg mass populations and development of a sequential sampling plans. *Environ. Entomol.* 15:122–127.

91. **Kozlowski, T. T.,** 1980. Responses of shade trees to pollution. *J. Arboric.* 6:29–41.

92. **Kozlowski, T. T.,** 1985. Tree growth in response to environmental stresses. *J. Arboric.* 11:97–111.

93. **Lanier, G. N.,** 1981. See Ref. 113, pp. 115–132.

94. **Larew, H. G., Knodel-Montz, J., Poe, S. L.,** 1984. Leaf miner damage. *Greenhouse Manager* 3:53–55.

95. **Lasota, J. A., Dybas, R. A.,** 1991. Avermectins, a novel class of compounds: implications for use in arthropod pest control. *Annu. Rev. Entomol.* 36:91–117.

96. **Leeper, J. R., Roush, R. T., Reynolds, H. T.,** 1986. See Ref. 120, pp. 335–346.

97. **Leius, K.,** 1967. Influence of wild flowers on parasitism of tent caterpillar and codling moth. *Can. Entomol.* 99:444–446.

98. **Liebold, A. M., Elkinton, J. S.,** 1988. Techniques for estimating the density of late-instar gypsy moth, *Lymantria dispar* (Lepidoptera: Lymantriidae), populations using frass drop and frass production measurements. *Environ. Entomol.* 17:381–384.

99. **Lindegren, J. E., and Barnett, W. W.,** 1982. Applying parasitic nematodes to control carpenterworms in fig orchards. *Calif. Agric.* 36(11–12):7–8.

100. **Luck, R. F., Dahlsten, D. L.,** 1975. Natural decline of a pine needle scale (*Chionaspis pinifoliae* (Fitch) outbreak at South Lake Tahoe, California, following cessation of adult mosquito control with malathion. *Ecology* 56:893–894.

101. **Malinoski, M. K., Paine, T. D.,** 1988. A degree-day model to predict Nantucket pine tip moth, *Rhyacionia frustrana* (Comstock) (Lepidoptera: Tortricidae), flights in southern California. *Environ. Entomol.* 17:75–79.

102. **Marion, D. F., Larew, H. G., Knodel, J. J., Natoli, W.,** 1990. Systemic activity of neem extract against the birch leafminer. *J. Arboric.* 16:12–16.

103. **Mattson, W.,** 1980. Herbivory in relation to plant nitrogen content. *Annu. Rev. Ecol. Syst.* 11:119–161.

104. **Mattson, W. J., Haack, R. A.,** 1987. The role of drought in outbreaks of plant-eating insects. *BioScience* 37:110–118.

105. **Mattson, W. J., Haack, R. A.,** 1987. See Ref. 7, pp. 365–407.

106. **Maxwell, F. G., Jennings, P. R.,** Eds., 1980. *Breeding Plants Resistant to Insects.* New York: John Wiley & Sons. 683 pp.

107. **McClure, M. S.,** 1977. Resurgence of the scale, *Fiorinia externa* (Homoptera: Diaspididae) on hemlock following insecticide application. *Environ. Entomol.* 6:480–484.

108. **McClure, M. S.,** 1977. Dispersal of the scale *Fiorinia externa* (Homoptera: Diaspididae) and effects of edaphic factors on its establishment on hemlock. *Environ. Entomol.* 6:539–544.

109. **McClure, M. S.,** 1991. Nitrogen fertilization increases susceptibility to hemlock woolly adelgid. *J. Arboric.* 17:227–230.

110. **Merritt, R. W., Kennedy, M. K., Gersabeck, E. F.,** 1983. See Ref. 51, pp. 277–299.

111. **Metcalf, R. L., McKelvey, J. J., Jr.,** Eds., 1976. *The Future for Insecticides.* New York: John Wiley & Sons. 524 pp.

112. **Miller, F. D.,** 1989. The use of horticultural oils and insecticidal soaps for control of insect pests of amenity plants. *J. Arboric.* 15:257–262.

113. **Mitchell, E. R.,** Ed., 1981. *Management of Insect Pests with Semiochemicals.* New York: Plenum. 514 pp.

114. **Mracek, Z., David, L.,** 1986. Preliminary field control of *Cephalcia abietis* L. (Hymenoptera: Pamphiliidae) larvae with steinernemated nematodes in Czechoslovakia. *J. Appl. Entomol.* 102:260–263.

115. **Mracek, Z., Spitzer, K.,** 1983. Interaction of the predators and parasitoids of the sawfly *Cephalcia abietis* (Pamphilidae: Hymenoptera) with its nematode *Steinernema kraussei. J. Invert. Path.* 42:397–399.

116. **Munro, J. A.,** 1963. Biology of the ceanothus stem-gall moth, *Periploca ceanothiella* (Cosens). *J. Res. Lepidoptera* 1:183–190.

117. **Munro, J. A.,** 1965. Occurrence of *Psylla uncatoides* on *Acacia* and *Albizia* with notes on control. *J. Econ. Entomol.* 58:1171–1172.

118. **National Gardening Survey,** 1990. Burlington, VT: National Gardening Assn. 8 pp.

119. **National Research Council.,** 1980. *Urban Pest Management.* Washington, DC: National Academy Press. 273 pp.

120. **National Research Council,** 1986. *Pesticide Resistance: Strategies and Tactics for Management.* Washington, DC: National Academy Press. 471 pp.

121. **Neal, J. W.,** 1981. Timing insecticide control of rhododendron borer with pheromone trap catches of males. *Environ. Entomol.* 10:264–266.

122. **Neal, J. W., Jr., Eichlen, T. D.,** 1983. Seasonal response of six male Sesiidae of woody ornamentals to clearwing borer (Lepidoptera: Sesiidae) lure. *Environ. Entomol.* 12:206–209.

123. **Neal, J. W. Jr., Raupp, M. J., Douglas, L. W.,** 1987. Temperature-dependent model for predicting larval emergence of the bagworm, *Thyidopteryx ephemeraeformis* (Haworth) (Lepidoptera: Psychidae). *Environ. Entomol.* 16:1141–1144.

124. **Neely, D., Himelick, E. B., Cline, S.,** 1984. Assessment of pesticide usages by commercial and municipal arborists. *J. Arboric.* 10:143–147.

125. **Neely, D., Smith, G. R.,** 1991. IPM strategies used by arborists. *J. Arboric.* 17:8–12.

126. **Nielsen, D. G.,** 1978. Sex pheromone traps: A breakthrough in controlling borers of ornamental trees and shrubs. *J. Arboric.* 4:181–183.

127. **Nielsen, D. G.,** 1983. Integrated pest management (IPM). *Brooklyn Botanic Garden Record* 40(1):70–72.

128. **Nielsen, D. G.,** 1989. Integrated pest management in arboriculture: From theory to practice. *J. Arboric.* 15:25–30.

129. **Nielsen, D. G.,** 1990. Evaluation of biorational pesticides for use in arboriculture. *J. Arboric.* 16:82–88.

130. **Nielsen, D. G., Dunlap, M. J.,** 1990. Scotch pine, pine needle scale control, Wayne Co., OH, 1988. *Insecticide and Acaricide Tests: 1990.* 15:350.

131. **Nielsen, D. G., Dunlap, M. J.,** 1990. Juniper, spruce spider mite control, Knox Co., OH, 1988. *Insecticide and Acaricide Tests: 1990.* 15:347.

132. **Nielsen, D. G., Hart, E. R., Dix, M. E., Linit, M. J., Appleby, J. E., Ascerno, M., Mahr, D. L., Potter, D. A., Jones, J. A.,** 1985. Common street trees and their pest problems in the North Central United States. *J. Arboric.* 11:225–32.

133. **O'Callaghan, D. P., Gallagher, E. M., Lanier, G. N.,** 1980. Field evaluation of pheromone-baited trap trees to control elm bark beetles, vectors of Dutch elm disease. *Environ. Entomol.* 9:181–85.

134. **Olkowski, W.,** 1987. Update: Neem — a new era in pest control products? *The IPM Practitioner* 10:1–8.

135. **Olkowski, W., Olkowski, H.,** 1978. Urban integrated pest management. *J. Arboric.* 4:241–46.

136. **Olkowski, W., Olkowski, H., Drlik, T., Heidler, N., Minter, M., Zuparko, R., Laub, L., Orthel, L.,** 1978. Pest control strategies: Urban integrated pest management. In *Pest Control Strategies.* E. H. Smith, D. Pimentel, Eds. New York: Academic. 215–233.

137. **Olkowski, W., Olkowski, H., van den Bosch, R., Hom, R.,** 1976. Ecosystem management: a framework for urban pest control. *BioScience* 26:384–89.

138. **Owen, D. F.,** 1978. See Ref. 50, pp. 13–29.

139. **Pedigo, L. P., Hutchins, S. H., Higley, L. G.,** 1986. Economic injury levels in theory and practice. *Annu. Rev. Entomol.* 31:341–68.

140. **Peterson, N. C., Smitley, D. R.,** 1991. Susceptibility of selected shade and flowering trees to gypsy moth (Lepidoptera: Lymantriidae) defoliation. *J. Econ. Entomol.* 84:587–592.

141. **Potter, D. A., Braman, K. S.,** 1991. Ecology and management of turfgrass insects. *Annu. Rev. Entomol.* 36:383–06.

142. **Potter, D. A., Jensen, M. P., Gordon, F. C.,** 1989. Phenology and degree-day relationships of the obscure scale (Homoptera: Diaspididae) and associated parasites in pin oak in Kentucky. *J. Econ. Entomol.* 82:551–55.

143. **Potter, D. A., Timmons, G. M.,** 1983. Biology and management of clearwing borers in woody plants. *J. Arboric.* 9:145–50.

144. **Potter, D. A., Timmons, G. M.,** 1983. Flight phenology of the dogwood borer (Lepidoptera: Sesiidae) and implications for control in *Cornus florida* L. *J. Econ. Entomol.* 76:1069–74.

145. **Potter, D. A., Timmons, G. M.,** 1983. Forecasting emergence and flight activity of the lilac borer (Lepidoptera: Sesiidae) based on pheromone trapping and degree-day accumulations. *Environ. Entomol.* 12:400–03.

146. **Potter, D. A., Timmons, G. M., Gordon, F. C.,** 1988. Flatheaded apple tree borer (Coleoptera: Buprestidae) in nursery-grown red maples: phenology of emergence, treatment timing, and response to stressed trees. *J. Environ. Hort.* 6:18–22.

147. **Raupp, M. J.,** 1984. Effects of exposure to sun on the frequency of attack by the azalea lacebug. *J. Amer. Rhodo. Soc.* 38(4):189–90.

148. **Raupp, M. J.,** 1985. Monitoring: an essential factor to managing pests of landscape trees and shrubs. *J. Arboric.* 11:349–54.

149. **Raupp, M. J., Davidson, J. A., Holmes, J. J., Hellman, J. L.,** 1985. The concept of key plants in integrated pest management for landscapes. *J. Arboric.* 11:317–22.

150. **Raupp, M. J., Davidson, J. A., Koehler, C. S., Sadof, C. S., Reichelderfer, K.,** 1988. Decision-making considerations for aesthetic damage caused by pests. *Bull. Entomol. Soc. Amer.* 34:27–32.

151. **Raupp, M. J., Davidson, J. A., Koehler, C. S., Sadof, C. S., Reichelderfer, K.,** 1989. Economic and aesthetic injury levels and thresholds for insect pests of ornamental plants. *Fla. Entomol.* 72:403–07.

152. **Raupp, M. J., Noland, R. M.,** 1984. Implementing plant management programs in residential and institutional settings. *J. Arboric.* 10:161–69.

153. **Raupp, M. J., Smith, M. F., Davidson, J. A.,** 1989. Educational, environmental, and economic impacts of integrated pest management programs for landscape plants. In *Integrated Pest Management for Turfgrass*

and Ornamentals. A. R. Leslie, R. L. Metcalf, Eds. Washington, DC: U.S. Environmental Protection Agency. 77–83.

154. **Reardon, R., McManus, M., Kolodny-Hirsch, D., Tichenor, R., Raupp, M., Schwalbe, C., Webb, R., Meckley, P.,** 1987. Development and implementation of a gypsy moth integrated pest management program. *J. Arboric.* 13:209–216.

155. **Reardon, R. C., Webb, R. E.,** 1990. Systemic treatment with acephate for gypsy moth management: population suppression and wound response. *J. Arboric.* 16:174–178.

156. **Rice, R. E., Flaherty, D. L., Jones, R. A.,** 1982. Monitoring and modeling San Jose scale. *Calif. Agric.* 36:13–14.

157. **Roberts, F. C., Luck, R. F., Dahlsten, D. L.,** 1973. Natural decline of a pine needle scale population at South Lake Tahoe. *Calif. Agric.* 27(10):10–12.

158. **Rowntree, R. A.,** 1984. Ecology of the urban forest-introduction to part I. *Urban Ecol.* 9:229–243.

159. **Sadof, C. S., Raupp, M. J.,** 1987. Consumer attitudes toward the defoliation of American arborvitae, *Thuja occidentalis,* by bagworm, *Thyridopteryx ephemeraeformis. J. Environ. Hort.* 5:164–166.

160. **Sadof, C. S., Raupp, M. J.,** 1991. Effect of variegation in *Euonymus japonica* var. *aureus* on two phloem feeding insects, *Unaspis euonymi* (Homoptera:Diaspididae) and *Aphis fabae* (Homoptera:Aphididae). *Environ. Entomol.* 20:83–89.

161. **Santamour, F. S., Jr.,** 1977. The selection and breeding of pest-resistant landscape trees. *J. Arboric.* 3:146–152.

162. **Schroeder, H. W., Cannon, W. N., Jr.,** 1987. Visual quality of residential streets: both street and yard trees make a difference. *J. Arboric.* 13: 236–239

163. **Schultz, P. B.,** 1983. Evaluation of hawthorn lace bug (Hemiptera: Tingidae) feeding preference on *Cotoneaster* and *Pyracantha. Environ. Entomol.* 12:1808–1810.

164. **Schultz, P. B.,** 1985. Monitoring parasites of the oak lecanium scale with yellow sticky traps. *J. Arboric.* 11:182–184.

165. **Schultz, P. B.,** 1985. Evaluation of selected *Cotoneaster* spp. for resistance to hawthorn lace bug. *J. Environ. Hort.* 3:156–157.

166. **Schultz, P. B., Coffelt, M. A.,** 1987. Oviposition and nymphal survival of the hawthorn lace bug (Hemiptera: Tingidae) on selected species of *Cotoneaster* (Rosaceae). *Environ. Entomol.* 16:365–367.

167. **Scriber, J. M., Slansky, F., Jr.,** 1981. The nutritional ecology of immature insects. *Annu. Rev. Entomol.* 26:183–211.

168. **Scriven, G. T., Luck, R. F.,** 1980. Susceptibility of pines to attack by the Nantucket pine tip moth in southern California. *J. Econ. Entomol.* 73:318–320.

169. **Scriven, G. T., Reeves, E. L., Luck, R. F.,** 1986. Beetle from Australia threatens Eucalyptus. *Calif. Agric.* 40(7–8):4–6.

170. **Sherald, J. L., DiSalvo, C. L. J.,** 1987. Integrated pest management in the National Capitol Region of the National Park Service. *J. Arboric.* 13:229–235.

171. **Shetlar, D.J., Heller, P.R.,** 1984. Survey of insecticide and miticide usage by 158 nurseries in Pennsylvania. *J. Environ. Hort.* 2:16–24.

172. **Siegler, E. H., Popenoe, C. H.,** 1925. The fatty acids as contact insecticides. *J. Econ. Entomol.* 18:292–299.

173. **Sinclair, W. A., Lyon, H. H., Johnson, W. T.,** 1987. *Diseases of Trees and Shrubs.* Ithaca, NY: Cornell University Press. 574 pp.

174. **Sine, C.,** 1990. *Farm Chemicals Handbook.* Willoughby, OH: Meister. 700 pp.

175. **Smith, C. M.,** 1989. *Plant Resistance to Insects: a Fundamental Approach.* New York: John Wiley & Sons. 286 pp.

176. **Smith, D. C., Raupp, M. J.,** 1986. Economic and environmental assessment of an integrated pest management program for community owned landscape plants. *J. Econ. Entomol.* 79:162–165.

177. **Smith, R. F., van den Bosch, R.,** 1967. Integrated control. In *Pest Control: Biological, Physical, and Selected Chemical Methods.* W. W. Kilgore, R. L. Doutt, Eds. New York: Academic. 295–342.

178. **Staines, C. L. Jr., Rothschild, M. J., Trumble, R. B.,** 1990. A survey of the Coccinellidae (Coleoptera) associated with nursery stock in Maryland. *Proc. Entomolog. Soc. Wash.* 92:310–313.

179. **Statistical Abstract of the United States,** 1990. Washington, DC: U.S. Department of Commerce. 991 pp.

180. **Stern, V. M., Smith, R. F., van den Bosch, R., Hagen, K. S.,** 1959. The integrated control concept. *Hilgardia* 29:81–101.

181. **Stimmann, M. W., Kaya, H. K., Burlando, T. M., Studdert, J. P.,** 1985. Black vine weevil management in nursery plants. *Calif. Agric.* 39(1–2):25–26.

182. **Stipes, J. R., Campana, R. J.,** Eds., 1981. Compendium of elm diseases. *Amer. Phytopath. Soc.* 96 pp.

183. **Stoetzal, M. B., Davidson, J. A.,** 1971. Biology of the obscure scale, *Melanaspis obscura* (Homoptera: Diaspididae) on pin oak in Maryland. *Ann. Entomol. Soc. Am.* 64:45–50.

184. **Stoetzal, M. B., Davidson, J. A.,** 1975. Seasonal history of seven species of armored scale insects of the Aspidiotini (Homoptera: Diaspididae). *Ann. Entomol. Soc. Am.* 68:489–492.

185. **Taylor, L. R., French, R. A., Woiwood, I. P.,** 1978. See Ref. 50, pp. 31–66.

186. **Volney, W. J. A., Koehler, C. S., Browne, L. E., Barclay, L. W., Milstead, J. E., Lewis, V. R.,** 1983. Sampling for California oakworm on landscape oaks. *Calif. Agric.* 37(9):6–9.

187. **Ware, G. H.,** 1984. Coping with clay: trees to suit sites, sites to suit trees. *J. Arboric.* 10:100–112.

188. **Warkentin, D.,** 1989. Control of green peach aphid, 1988. *Insecticide and Acaricide Tests: 1989.* 14:322.

190. **Webb, R. E., Boyd, V. K.,** 1983. Evaluation of barrier bands and insecticidal strips for impeding movement of gypsy moth caterpillars. *Melsh. Entomol. Ser.* 33:15–20.

191. **Webb, R. E., Reardon, R. C., Wieber, A. M., Boyd, V. K., Larew, H. G., Argauer, R. J.,** 1988. Suppression of gypsy moth (Lepidoptera: Lymantriidae) populations on oak using implants or injections of acephate and methamidophos. *J. Econ. Entomol.* 81:573–577.

192. **Weever, M. J., Stipes, R. J.,** 1988. White pine decline: a case study from Virginia landscapes. *J. Arboric.* 14:109–120.

193. **Weidhaas, J. A.,** 1976. Is host resistance a practical goal for control of shade-tree insects? In *Better Trees for Metropolitan Landscapes.* F. S. Santamour, Jr., H. D. Gerhold, S. Little, Eds. Washington, DC: USDA. Forest Service Tech. Rep. NE-22. 127–157.

194. **Weseloh, R. M.,** 1981. Relationship between colored sticky panel catches and reproductive behavior of forest tachinid parasites. *Environ. Entomol.* 10:131–135.

194. **White, T. C. R.,** 1984. The abundance of invertebrate herbivores in relation to the availability of nitrogen in stressed food plants. *Oecologia.* 63:90–105.

195. **White, W. B., McLane, W. H., Schneeberger, N. F.,** 1981. See Ref. 35, pp. 423–442.

196. **Williams, M. L., Ray, C. H., Daniels, I. E.,** 1977. The euonymus scale in Alabama. *Highlights of Agric. Res.* 24(2):15.

197. **Wu, Z., Jamieson, S., Kielbaso, J. J.,** 1991. Urban forest pest management. *J. Arboric.* 17:150–158.

Chapter 14

Calibrating Turfgrass Chemical Application Equipment

H. Erdal Ozkan, Department of Agricultural Engineering, The Ohio State University, Columbus, OH

(This chapter was reprinted from the author's Extension Bulletin 817 of the same title, published in 1991 by the Ohio Cooperative Extension Service in cooperation with the U.S. Department of Agriculture.)

CONTENTS

Introduction 143
    Why Calibrate? 144
    How Often Should You Calibrate? 144
Types of Equipment and Their Use 144
Factors That Influence Application Accuracy with All Equipment 144
Mixing and Loading Chemicals 145
    Calculating the Amount of Chemical Needed per Tank 146
Calibrating Liquid Application Equipment 146
    Low-Pressure Boom Sprayers 146
    How to Eliminate the Application Error 150
    Manual Hand Pump and Backpack Sprayers 151
    Hand Gun Sprayers 152
Dry Granular Spreaders and Their Calibration 153
    Drop Spreaders 153
    Rotary Spreaders 154
    Calibration 155
    Checking the Delivery Rate 155
    Checking the Distribution Pattern 158
    Other Considerations 158
Safety Guidelines 159
Useful Formulas 160
Conversion Tables 161

INTRODUCTION

You've been getting complaints about weed or insect control, or there is turf damage on your golf course or a customer's lawn. Your actual pesticide material usage doesn't match the total that should have been used for all the applications you and other workers have made. What can you do to get things back on track?

Chances are your equipment needs fine-tuning, or *calibration*. This chapter discusses:

- The importance of calibration (see Figure 1).
- The need for frequent calibration checks to ensure proper application.
- The types of application equipment and their use.

- The factors that influence the rate applied.
- How to calibrate the material needed to cover a given area of turf.
- How to troubleshoot for application problems and errors.
- The actual steps in calibrating the type of equipment you use in your operation.

Calibration refers to **all** the operations that ensure the correct amount of pesticide will be applied to the treated area. These operations include:

1. Calculating the amount of product.
2. Mixing properly and safely.
3. Adjusting equipment to deliver the desired rate uniformly across the spray swath.

0-87371-350-8/94/$0.00+$.50
© 1994 by CRC Press Inc.

Figure 1 Poorly calibrated equipment causes nonuniform spray coverage.

4. Determining effective swath/overlap.
5. Checking accuracy during operation.
6. Detecting and correcting errors.

WHY CALIBRATE?

Equipment used for applying chemicals to turfgrass must be calibrated periodically to achieve effective pest control and to reduce the potential for excessive pesticide residues remaining on sprayed surfaces. While applying too little pesticide may result in ineffective pest control, too much pesticide wastes money, may damage the turf and increases the potential risk of contaminating groundwater and environment.

A survey of 53 private and municipal golf course pesticide applicators in Nebraska revealed that only one of every six applicators applied pesticide carrier volumes within the acceptable error margin of ±5%. Application errors ranged from 83% underapplication to 177% overapplication. Thirty-six percent of the applicators overapplied pesticides by more than 5% of intended application rates; on average, they overapplied pesticide by 19%. The cost of overapplication for one group using Daconil 2787 ranged from 5 cents per acre to $16 per acre. Faulty pressure gauges and improperly sized hoses and plumbing components were contributing factors to inaccurate tank-mix application rates and nonuniform applications.

HOW OFTEN SHOULD YOU CALIBRATE?

Sprayers should be calibrated several times a year. Changes in operating conditions and the type of chemical used require a new calibration. Frequent calibration is even more important with liquid application because nozzles wear out with use, increasing the flow rate. For example, tests conducted at the Ohio State University indicated that the flow rate of a 0.2 gallon-per-minute capacity brass nozzle increased by 10% after only three hours of use with a wettable powder material.

The same University of Nebraska survey indicated that 67% of the operators who calibrated before every spray operation had application errors below 5%, while only 5% of those operators who calibrated their equipment less than once a year (once every two, three, four years) achieved the same degree of application accuracy.

Safety is extremely important when working with chemicals. Always wear gloves and protective clothing and follow label directions when handling chemicals and calibrating sprayers.

TYPES OF EQUIPMENT AND THEIR USE

Low-pressure boom sprayers, hand cans, backpack sprayers, hand gun nozzles and dry spreaders are the most common types of equipment used for applying chemicals on turfgrass and home lawns (Figure 2). Table 1 lists typical usage sites for the equipment, along with major advantages and drawbacks.

FACTORS THAT INFLUENCE APPLICATION ACCURACY WITH ALL EQUIPMENT

Certain factors influence application accuracy with all equipment, including:

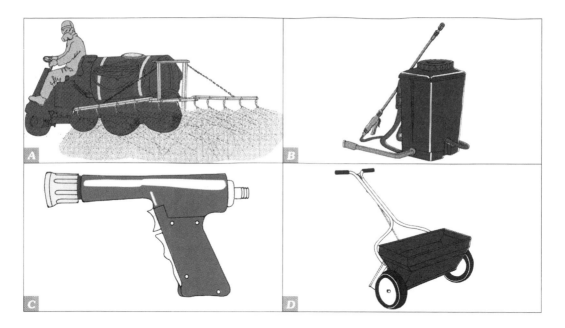

Figure 2 Four types of turf application equipment, as described in Table 1.

Table 1 **Common turf application equipment**

Equipment	Dry or Liquid	Use	Comments
Low-pressure Boom Sprayer (Fig. 2A)	L	Large areas Golf courses Commercial sites	Can cover larger areas in a short time. Usually mounted on a small tractor. The pressure should not exceed 60 psi. Requires frequent cleaning and servicing. Requires an external power source or tractor power.
Backpack/Hand Can (Fig. 2B)	L	Spot treatment	Durable, portable, easy to use. Difficult to keep the pressure up. Difficult to control a uniform application rate.
Hand Gun (Fig. 2C)	L	Small lawns and larger turf areas	Covers large areas quickly. Very little maintenance needed. Requires own motor or external power source. Application uniformity is highly dependent on walking pace and arm motion of the operator.
Spreaders (Fig. 2D)	D	Lawns and turf areas	Can cover larger area more quickly with rotary than drop spreader.

- Speed of travel — walking pace or tractor speed.
- Shape and size of nozzle orifice.
- Operating pressure.
- Application height.
- Effective swath width or the pattern in which the material is deposited.
- Uniform distribution of active ingredients in the applied material.

Each of these topics is discussed in the Calibrating Liquid Application Equipment section.

MIXING AND LOADING CHEMICALS

Mixing the pesticide thoroughly and carefully is an important step in good sprayer operation. Incomplete mixing results in varied application rates — too high at times, too low otherwise. Some chemicals, when mixed improperly with others, form a thick, mayonnaise-like mixture that will not spray properly and is difficult to clean out of the sprayer. Always read the label for proper mixing sequence, and test the compatibility of chemicals in a small container before mixing.

Generally, when preparing the spray mixture, fill the tank half full of water and pour in the correct amount of chemical while the pump is running. Then finish filling the tank. When using wettable powders, again fill the tank half full, make a slurry in a separate container or through an inductor system, and then add the slurry to the tank to ensure good mixing. Rinse the slurry container three times, pouring the rinsate into the tank. Finally, finish filling the tank with water.

CALCULATING THE AMOUNT OF CHEMICAL NEEDED PER TANK

You can spend hours in the field calibrating your sprayer to achieve the desired accuracy, but your efforts are wasted if you don't know how much chemical to put in the tank. In a survey conducted by agricultural engineers at the University of Nebraska, 38% of the applicators failed to add the correct amount of chemical to the tank.

The amount of chemical needed per tankful depends on the recommended rate and the size of area you can treat per tank of spray. Calculations are the same whether you are using a manual backpack sprayer or a powered boom sprayer. The only difference is in the units. For small manual sprayers, the rate is in gallons or quarts per 1,000 sq ft, and for boom sprayers the application rate is usually given in gallons per acre. So make sure you use the right units in tank mix calculations. Let's assume now that you have a boom sprayer.

- To find the area (acres) treated per tank, divide the tank capacity by the GPA application rate.
- To find the amount of pesticide needed per tank, multiply the number of acres treated per tank by the amount of actual chemical to be applied per acre.

Example 1

A sprayer has a 100-gallon tank, and it has been calibrated to apply 40 gallons per acre. The pesticide label recommends 2 quarts of commercial product per acre for broadcast application. Determine the quantity of pesticide to add to the tank:

- First, find the acres (A) each tank will spray:

$$\frac{100 \text{ gal per tank}}{40 \text{ gal per A}} = 2.5 \text{ A/tank}$$

- Next, calculate the number of quarts to be added to the tank: 2.5A/tank × 2 qts/A = 5 qts/tank

You must add 5 quarts of chemical to the tank.

Sometimes, chemical manufacturers give recommended rates in terms of active ingredient (A.I.) to be used per acre rather than the amount of total product per acre. In these cases, calculate the amount of material to be applied as shown in examples 2 and 3 for dry chemicals and liquid chemicals, respectively.

Example 2

The application rate of the chemical applied with the sprayer in Example 1 is 2 pounds A.I. applied per acre. The material to be used is a wettable powder containing 50% A.I. Determine the quantity of chemical to add to the tank:

Lbs. of formulated product per A =

$$2 \text{lbs} \times \frac{100}{50} = 4 \text{ lbs/A}$$

Lbs. of formulated product per tankful =

$$4 \text{lbs/A} \times 2.5 \text{ A/tank} = 10 \text{ lbs/tank}$$

Example 3

The chemical you are applying with the sprayer in Example 1 is liquid. The recommendation calls for 2 pounds A.I. per acre. You have purchased a formulation that contains 4 pounds A.I. per gallon. Determine the amount of product needed per acre:

Gal. of formulated product needed per A =

$$\frac{2 \text{ lbs A.I./A}}{4 \text{ lbs A.I./gal}} = 0.5 \text{ gal/A}$$

- Next, calculate the amount of chemical to add to each tankful:

$$0.5 \text{ gal/A} \times 2.5 \text{ A/tank} = 1.25 \text{ gal/tank}$$

CALIBRATING LIQUID APPLICATION EQUIPMENT

LOW-PRESSURE BOOM SPRAYERS

Low-pressure boom sprayers are used frequently for applying chemicals on large areas such as golf courses and recreational areas. Calibrating a boom sprayer is not as difficult as it sounds. Although there are many methods to use, the two methods described here are the most common. Choose a method that you are comfortable with and calibrate your sprayer often, but at least once a year.

For safety, use water instead of actual chemical mixtures when calibrating. However, some carriers, such as liquid fertilizers, are much denser than water and may cause the nozzle flow rate to vary from the rate obtained with water. In this case, determine the average nozzle output with water and use a conversion factor that is usually listed in the nozzle manufacturer's catalog.

Method I

This method is simple and requires few calculations. It is based on the spraying distance and the time needed for a nozzle to spray 1/128 of an acre. Because there are 128 ounces of liquid in one gallon, the ounces of liquid caught from one nozzle is directly equal to the application rate in gallons per acre, or GPA.

For example: If you catch an average of 80 ounces from a set of nozzles, the actual application rate of the sprayer is equal to 80 GPA. With this method, make sure that the time used to catch output from nozzles is the same as the time it takes to cover 1/128 acre. Table 2 shows the distance you must travel to cover 1/128 acre for different nozzle spacings.

To calibrate your sprayer, you need a measuring tape, a stopwatch or an ordinary watch that displays seconds, and a measuring jar graduated in ounces (Figure 3). A pocket calculator also will be handy. Follow these steps when calibrating boom sprayers for broadcast application:

1. Fill the sprayer tank with water.
2. Run the sprayer, inspect it for leaks and make sure all vital parts function properly.
3. Measure the distance in inches between the

Table 2 **Calibration distance for each nozzle to spray 1/128 acre**

Nozzle Spacing (in.)	Travel Distance (ft)	Nozzle Spacing (in.)	Travel Distance (ft)
18	227	30	136
20	204	32	127
22	185	34	120
24	170	36	113
26	157	38	107
28	146	40	102

nozzles (Figure 4). Then measure an appropriate distance in the field based on this nozzle spacing, as shown in Table 1.

4. Drive through the measured distance in the field (Figure 5) at your normal spraying speed, and record the travel time in seconds. You do not need to spray while traveling. Repeat this procedure and average the two measurements.
5. With the sprayer parked, run the sprayer at the same pressure level and catch the output from each nozzle in a measuring jar (Figure 6) for the travel time required in Step 4.
6. Calculate the average nozzle output by adding the individual outputs and then dividing by the number of nozzles tested. If an individual sample collected is more than 10% higher or lower than the average nozzle output rate, check for clogs and clean the tip, or replace the nozzle with a new tip.
7. Repeat steps 5 and 6 until the variation in discharge rate for all nozzles is within 10% of the average.

Figure 3 Equipment needed to calibrate a sprayer.

Figure 4 Measuring the distance between nozzles.

8. Then, the final average output in ounces is equal to the application rate in gallons per acre:

    Average output (ounces) = Application rate (GPA)

9. If needed, convert GPA to gallons per 1,000 square feet or ounces per 1,000 square feet.

    $$0.023 \times GPA = Gal/1,000 \text{ sq ft}$$

    $$2.94 \times GPA = 0z/1,000 \text{ sq ft}$$

10. Compare the actual application rate with the recommended or intended rate. If the actual rate is more than 5% higher or lower than the recommended or intended rate, you must make adjustments in pressure, travel speed and/or nozzle size.

11. You can start the adjustments by changing the pressure. Lowering the spray pressure will reduce the spray delivered; higher pressure means more spray is delivered. Don't vary from the pressure range recommended for your nozzles. (Refer to "Useful Formulas" to determine the new pressure rate.)

Figure 5 A distance of at least 200 ft is recommended to determine travel speed.

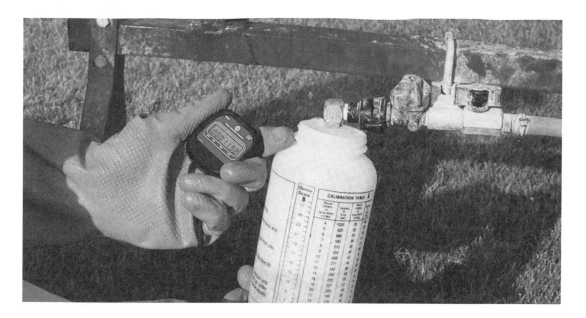

Figure 6 Use a cup graduated in ounces to measure nozzle flow rate.

12. You also can correct the application error by changing the actual travel speed. Slower speeds mean more spray is delivered; faster speeds mean less spray is delivered. (Refer to "Useful Formulas" to determine the new travel speed.)

13. If these changes don't bring the application rate to the desired rate, you may have to select a new set of nozzles with smaller or larger orifices.

14. Recalibrate the sprayer (repeat steps 5 through 12) after any adjustment.

Method II

As with any calibration procedure, this second method requires you to check the nozzle flow rate and the actual ground speed of the sprayer. Follow these steps:

Step 1. Determine Effective Swath Width (W) per Nozzle

For boom spraying, the effective spray width of each nozzle (W) is equal to the distance in inches between two nozzles.

Step 2. Determine Travel Speed (MPH)

To determine the travel speed, measure a known distance (Figure 5). Use fence posts or flags to identify this distance. A distance over 200 ft and a tank at least half full are recommended. Travel the distance determined at your normal spraying speed and record the lapsed time in seconds. Repeat this step and take the average of the two measurements. Use the following equation to determine the travel speed in miles per hour:

$$\text{Travel Speed (MPH)} = \frac{\text{Distance (feet)} \times 0.68}{\text{Time (seconds)}}$$

(0.68 is a constant to convert ft/sec to miles/hr)

Example:

After measuring a 250-foot distance, you traveled this course and it took 24 seconds for the first pass and 26 seconds for the second pass. Find the travel speed in miles per hour.

$$\text{Average time} = \frac{24 + 26}{2} = 25 \text{ seconds}$$

$$\text{MPH} = \frac{250 \text{ ft}}{25 \text{ sec}} \times 0.68 = 6.8 \text{ miles/hr}$$

Step 3. Determine Nozzle Flow Rate (GPM)

With the sprayer parked, run the sprayer at the same pressure level and catch the output from each nozzle (Figure 6) in a measuring jar for one minute (or collect output for half a minute and then double the ounces collected) to determine the nozzle flow rate in ounces per minute (OPM). Check for clogs and clean the tip or replace any nozzle tips having flow rates 10% more or less than the average of the other nozzles checked, and/or having obviously different fan angles or patterns. Repeat the above process until the variation in discharge rate for all nozzles is within 10% of the average.

Then, convert the final average output in OPM to gallons per minute (GPM) using the following equation:

$$GPM = \frac{OPM}{128} \quad (1 \text{ gallon} = 128 \text{ ounces})$$

Step 4. Determine Actual Application Rate in Gallons per Acre (GPA)

Use the following equation to determine the gallons per acre application rate.

$$GPA = \frac{GPM \times 5,940}{MPH \times W}$$

GPM: average nozzle flow rate in gallons per minute

MPH: travel speed in miles per hour

W: distance between two nozzles in inches

5,940: a constant to convert units to gallons/acre

Step 5. Compare Actual Application Rate with Recommended or Intended Rate

If the actual rate is more than 5% higher or lower than the recommended or intended rate, you must make adjustments in either the pressure or the travel speed or in both. If these changes don't bring the application rate to the desired range, then you may have to select a new set of nozzles with smaller or larger openings. (The intended application rate may be given in gallons or ounces per 1,000 sq ft. Refer to "Useful Formulas" to convert these rates to gallons per acre and to determine the new travel speed and/or the pressure rate.

Step 6. Recalibrate the Sprayer

Repeat steps 2 through 5 above until the application error is within 5% of the intended application rate.

Example:

You want to spray a golf course fairway at a rate of 1.25 gallons per 1,000 sq ft with a boom sprayer. The sprayer is traveling at 5 mph and has a nozzle spacing of 20 inches. The average flow rate of the nozzles is 100 ounces per minute at a pressure of 30 psi. What is the actual application error of this sprayer?

$$GPA = \frac{GPM \times 5,940}{MPH \times W}$$

First, you need to convert the flow rate of nozzles from ounces per minute to gallons per minute:

$$\frac{100 \text{ ounces}}{\text{minute}} \times \frac{1 \text{ gallon}}{128 \text{ ounces}} = 0.78 \text{ GPM}$$

$$GPA = \frac{0.78 \times 5,940}{5 \times 20} = 46.3 \text{ GPA (the actual application rate)}$$

To find the application error, you need to convert 46.3 GPA to gallons per 1,000 sq ft.

$$\frac{46.3 \text{ gallons}}{\text{acre}} \times \frac{1 \text{ acre}}{43.56 \times 1,000 \text{ sq ft}} = \frac{1.06 \text{ gallons}}{1,000 \text{ sq ft}}$$

Intended rate: 1.25 gal per 1,000 sq ft
Actual rate: 1.06 gal per 1,000 sq ft

$$\% \text{ application error} = \frac{1.25 - 1.06}{1.25} \times 100 = 15.2\%$$

HOW TO ELIMINATE THE APPLICATION ERROR

The sprayer is applying 15.2% less than the intended rate. Adjustments are necessary because errors over 5% are not acceptable. You can reduce the application error below 5% by trial and error. However, to save time, use the following equations to determine the new travel speed and the pressure required to increase the flow rate 1.06 to 1.25 gal. per 1,000 sq ft.

$$\text{Desired MPH} = \frac{\text{Measured rate} \times \text{Measured MPH}}{\text{Desired rate}}$$

$$\text{Desired MPH} = \frac{1.06 \times 5}{1.25} = 4.24$$

$$\text{Desired PSI} = \text{Measured PSI} \left(\frac{\text{Desired rate}}{\text{Measured rate}} \right)^2$$

$$= 30 \times \left(\frac{1.25}{1.06} \right)^2$$

$$= 41.7 \text{ psi}$$

To correct your application error, either reduce your travel speed from 5 mph to 4.24 mph, or increase the sprayer operating pressure from 30 psi to 41.7 psi.

Other Considerations

Here are some other things you should keep in mind when using a boom sprayer:

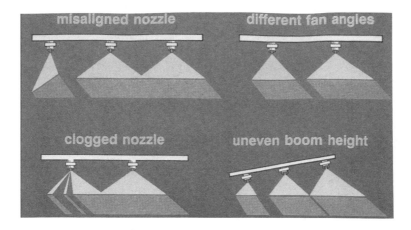

Figure 7 Common application errors that cause nonuniform coverage.

- Effective pest control is more than getting the right amount of chemicals on the ground. How the chemical is deposited on the spray target is as important as the amount deposited (Figure 7). Make sure all nozzle tips are properly aligned. Some nozzles require overlapping adjacent spray patterns. Check the nozzle catalog to determine the overlap required for a given type of nozzle, and adjust the boom height accordingly. There are some simple and practical ways to examine the spray pattern of the sprayer boom. Special portable spray patternator tables can be used to see the uniformity of spray distribution. If a spray table is not available, add nonstaining colorant to water in the tank and spray over an appropriate test area. You should see the color on the ground the same from one end of the boom to the other.
- A common cause of nonuniform coverage is clogged nozzles. To detect clogging, watch the nozzles periodically while spraying. Always carry extra nozzles in your tool box, and immediately replace bad nozzles with good ones. Never blow through nozzles to clear a clog.
- In most cases, the pressure gauges on sprayers do not represent the actual pressure at the nozzle tip. Therefore, check the pressure at the nozzle tip when calibrating your sprayer.

The flow rate of nozzles, especially those made from brass, increases as they become worn. Therefore, calibrate your sprayer as often as possible.

Always look for obvious things that may indicate your equipment is not working properly. For instance, if it appears that you have obviously too much or too little spray mixture remaining after spraying an area, stop immediately. Check your measurement of the area and make sure all vital parts of the sprayer are functioning properly.

Use quick bucket checks periodically. Always carry a marked container that will collect the correct amount of spray in 10 to 30 seconds. This quick check can also be used for other liquid application equipment such as hand guns and hand pumps.

MANUAL HAND PUMP AND BACKPACK SPRAYERS

Manual sprayers — hand pumps, hand cans and backpacks (Figure 8) — are designed for spot treatment and for spraying other areas not suitable for power sprayers. Most of these sprayers do not have pressure gauges or pressure controls. The pressure drops continuously as you spray. Therefore, you need to repressurize the tank at frequent intervals. When spraying, either hold the nozzle steady at a constant height and walk back and forth, or swing the nozzle in a steady, sweeping, overlapping motion. Maintaining a uniform walking speed is essential to keeping the application rate relatively uniform throughout the operation. Here is how you can calibrate manual sprayers.

Figure 8 Backpack sprayer.

Figure 9 Hand gun sprayer.

Step 1. Measure and mark off an area equal to 1,000 sq ft (such as 20 ft × 50 ft).

Step 2. Add a measured amount of water to the tank, spray the area and then measure the amount of water remaining in the tank. The difference between the amount in the tank before and after spraying is the amount used per 1,000 sq ft.

Step 3. Compare the measured rate with the intended or recommended rate, make necessary adjustments, and recalibrate the sprayer.

An alternative method is to record the time required to spray 1,000 sq ft and later catch and measure the spray from the nozzle (or nozzles) used for the same time period.

Hints for Improving the Uniformity of Application

Here are some hints for obtaining a more uniform application when using manual sprayers:

- Tie a weighted cord or chain to the wand near the nozzle. This can serve as a height gauge to maintain a set distance from the nozzle to the turf.
- Walk a known space that should be covered by the sprayer. Practice until you can consistently spray this area with the correct amount of material. A ticking stop watch or timer will improve your pacing.
- For hand pump sprayers, attach a pressure gauge to the spray wand and check how fast the pressure drops. Count the number of seconds needed for the pressure to drop 10 psi. Count the number of pump strokes needed to return the pressure to the proper level.
- For backpack sprayers, attach a gauge to the spray wand and determine how fast you need to pump to keep the gauge pressure constant while spraying.

HAND GUN SPRAYERS

When using a hand gun nozzle (Figure 9), four factors are critical to delivering the correct rate uniformly over the application area:

- The exact pressure
- Proper walking pace.
- A uniform hand/arm motion.
- Relatively constant nozzle height and angle in reference to the ground.

To calibrate a hand gun nozzle, spray a known area and record the time required to cover this area. Then use a bucket to catch the spray from the hand gun for the time period elapsed during spraying of the marked area. Determine the application rate by measuring the amount of liquid in the bucket.

Example:

You would like to spray at a rate of 1.5 gallons per 1,000 sq ft using a hand gun. To determine the actual application rate, you marked off an area 20 ft × 25 ft (500 sq ft). It took 2 minutes for you to spray this area. Then you sprayed into a bucket for 2 minutes and measured the amount of liquid in the bucket. It was 5.5 pints. What is the application rate in gallons per 1,000 sq ft and gallons per acre? What is the percent application error?

$$\frac{5.5 \text{ pints}}{500 \text{ sq ft}} \times 1,000 = \frac{11 \text{ pints}}{1,000 \text{ sq ft}}$$

$$(1 \text{ gallon} = 8 \text{ pints})$$

$$\frac{11 \text{ pints}}{1,000 \text{ sq ft}} \times \frac{1 \text{ gallon}}{8 \text{ pints}} = \frac{1.38 \text{ gallon}}{1,000 \text{ sq ft}}$$

OR

$$\frac{1.38 \text{ gallon}}{1,000 \text{ sq ft}} \times \frac{43,560 \text{ sq ft}}{\text{acre}} = \frac{60.11 \text{ gallons}}{\text{acre}}$$

(1 acre = 43,560 sq ft)

Percent Application Error =

$$\frac{1.5 - 1.38}{1.5} \times 100 = 8\%$$

Your actual application rate is 1.38 gal. per 1,000 sq ft. It should have been 1.5. This is an 8% application error, which is unacceptable (more than 5% of the intended rate). You should slightly reduce your walking pace, and repeat the calibration steps outlined above until the variation between the intended application rate and the actual measured rate is not more than 5% of the intended rate.

DRY GRANULAR SPREADERS AND THEIR CALIBRATION

Many lawn care companies and applicators select granular products as part of their program or even as a complete program. Proper selection, maintenance, calibration and use of granular applicators can reduce costs and maximize the results expected from the chemical being applied. Most granular spreaders are of two types: drop (gravity) and rotary (centrifugal). A discussion of each follows:

DROP SPREADERS

The typical drop-type spreader (Figure 10) has a full-length agitator mounted on the spreader axle. The agitator rotates over a series of openings in the bottom of the spreader hopper, and granules exit the hopper through these openings. The size of the openings can be adjusted to obtain different rates of application. Because all the fertilizer falls within the wheel base, slight overlapping of the wheels with each pass is necessary to ensure uniform coverage. Common problems associated with drop spreaders include:

- Skips
- Excessive overlap
- Nonuniform spread when turning corners
- Clogging of openings in the bottom of the hopper when the turfgrass is moist during the application.

However, drop spreaders are generally more precise than rotary spreaders and deliver a better pattern. In addition, because the granules drop straight down a short distance, there is less chemical drift and pattern distortion. To avoid nonuniform spread when turning corners, always turn off the spreader while making turns. This means that you should treat the strips at both ends of the area first. This allows you to start in this strip and turn the spreader on as you cross the start line.

ROTARY SPREADERS

With rotary spreaders (Figure 11), granules drop out of one or more adjustable openings at the bottom

Figure 10 Drop-type spreader.

Figure 11 Rotary spreader.

of the hopper onto a rotating plate and are spread in a semicircular arc. Rotary spreaders cover a wider swath than drop spreaders. Therefore, they are preferred by applicators servicing large lawns and golf courses.

Common problems associated with rotary spreaders include:

- Dry products with different granule sizes and weights do not spread uniformly.
- Overlapping to obtain a uniform pattern changes with each product.
- The drop holes often get clogged.
- Turning changes the rotating plate speed.

Although rotary spreaders are generally less precise than drop spreaders in terms of uniformity of distribution, with proper calibration and operation they too provide satisfactory results. The problem with uniform distribution of granules can be corrected by proper overlapping of adjacent swaths. Always try to minimize drift of small particles when using rotary spreaders.

Rotary spreaders throw material three to four feet in front of the spreader. The operator is another three to four feet behind the spreader hopper. This means that when the material finally reaches the turf, it is six to eight feet in front of the operator. It is essential then that the operator run a border along the edges of the area being treated so that there is enough room to maneuver the spreader for the next pass. This technique also eliminates the possibility of pesticides reaching nontarget areas.

Much attention is currently being given to nontarget application of fertilizer and pesticides. Because rotary spreaders throw material six to eight

feet wide, materials will likely be spread on sidewalks, driveways and patios. To prevent such nontarget applications, operators must use extreme caution around nontarget areas. Some spreaders have shut-off valves to limit the spread of materials to the right of the operator. The use of this shut-off valve along with some commercially available reflectors used to divert any other material will prevent nontarget applications to the right of the applicator.

Remember to open and close the spreader at the beginning and ending of each pass. Start walking before opening the lever and close it before stopping. Keeping the spreader open while turning around will cause a misapplication because the impeller plate is driven by one wheel. Therefore, the impeller stops when turning one direction, but turns faster than normal in the other direction. Always hold the spreader handle at the proper height to keep the impeller level.

Do not operate spreaders traveling backward. The pattern with rotary spreaders will be skewed and the delivery rate of drop spreaders will be difficult.

CALIBRATION

Labels on lawn care products list the rate at which the product should be applied. Some labels will also list the correct setting for some name-brand spreaders. However, these settings are just recommendating to provide a good reference point when calibrating spreaders (Figure 12). Seldom does a setting give the same application rate in all operating conditions and application sites. Variables such as particle size, particle density, particle shape, relative humidity, drop point of granules on

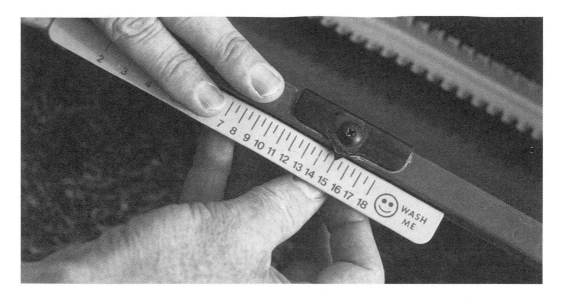

Figure 12 Spreader settings listed on product labels provide a good starting point when calibrating.

the impeller plate, impeller speed and ground speed all affect the application rate and the distribution pattern. Therefore, granular applicators should be calibrated for each type of granular product being applied and for change in operator, weather and field conditions.

Before beginning to calibrate a granular applicator, be sure that it is clean and all parts are working properly. Always wear rubber gloves and other protective clothing to prevent contact with chemicals during cleaning and calibration of equipment and actual application of chemicals.

The primary purpose of calibration is to determine the actual delivery rate of a spreader regardless of its type. However, with rotary spreaders, it is also necessary to check and correct the distribution pattern.

CHECKING THE DELIVERY RATE

"Delivery rate" refers to the average amount of product applied over a known area, and it usually is expressed in pounds per 1,000 square feet. The easiest way for an operator to check the actual output rate is to collect and weigh the granules actually applied to a known area. You can do this by following one of the two methods explained in Methods I and II. Both methods are acceptable, but the first one produces a more precise calibration.

Method I

a) Spread out a plastic tarp of known size on the ground (Figure 13).

b) Operate the spreader at a known speed across the tarp.

c) Determine the area covered. If the swath width is wider than the tarp (Figure 14), the area covered is equal to the area of the tarp. If the swath is narrower than the tarp (Figure 15), multiply the swath width by the tarp length to determine the area covered. It is highly recommended you keep the tarp width wider than the swath width for a more accurate measurement of the delivery rate.

d) Place all the granules on the tarp into a container and weigh them in ounces or pounds.

e) Divide the weight of granules collected by the area covered to determine the actual delivery rate in ounces per square foot or pounds per square foot. You may have to do some calculations to express the delivery rate in terms of lb/1,000 sq ft or lb/acre. Use the following equations when converting units.

$$(\text{oz/sq ft}) / 16 = \text{lb/sq ft}$$

$$\text{lb/sq ft} \times 1{,}000 = \text{lb/1,000 sq ft}$$

$$\text{lb/1,000 sq ft} \times 43.565 = \text{lb/acre}$$

$$\text{lb/sq ft} \times 43{,}565 = \text{lb/acre}$$

f) Compare the measured (actual) delivery rate with the rate recommended on the label. Make proper adjustments and recalibrate the spreader until you reach the desired accuracy (a variation within ±5% of the label recommendation is acceptable).

156

Figure 13 A tarp is used to determine the spreader application rate.

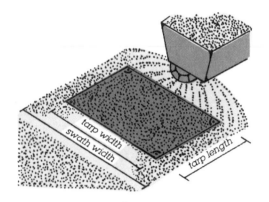

Figure 14 Area covered = tarp width × tarp length.

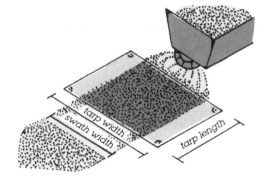

Figure 15 Area covered = swath width × tarp length.

Example:

You are calibrating a drop-type spreader with a swath width of 2 feet. You spread a 3-foot-by-6-foot piece of tarp on the ground. Then you filled the spreader hopper with granular fertilizer and traveled through the entire 6-foot length of the tarp at your normal application speed. Next, you collected all the granules on the tarp and weighed them. They weighed 2 ounces. What is the actual delivery rate of the spreader?

1. Determine area covered

Area covered = Swath width × travel distance marked

       Swath width = 2 ft
       Travel distance = 6 ft
       Area covered = 2 ft × 6 ft = 12 square feet

2. Determine delivery rate in ounces per square feet

$$\text{Delivery rate} = \frac{\text{weight of granules}}{\text{area covered}}$$

$$= \frac{2 \text{ oz}}{12 \text{ sq ft}}$$

$$= 0.166 \text{ oz/sq ft}$$

3. Determine the delivery rate in lb/1,000 sq ft

$$\frac{0.166 \text{ oz}}{\text{sq ft}} \times \frac{1 \text{ lb}}{16 \text{ oz}} \times 1{,}000 = 10.4 \text{ lb/1,000 sq ft}$$

4. Determine the delivery rate in lb/acre

$$\frac{10.4 \text{ lb}}{1{,}000 \text{ sq ft}} \times \frac{43{,}560 \text{ sq ft}}{\text{acre}} = 453 \text{ lb/acre}$$

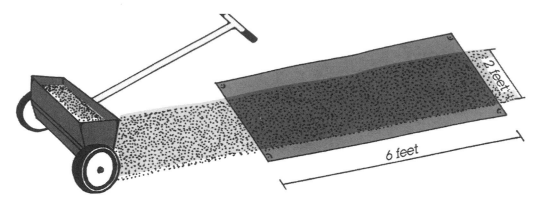

Figure 16 Determining the spreader delivery rate.

In this example, the spreader is broadcasting 10.4 pounds of fertilizer per 1,000 square feet or 453 pounds per acre. Let's assume the label on the 40-pound bag specifies a total coverage of 5,000 square feet. What is the application error?

Rate recommended = 40 lb/5,000 sq ft

= 8 lb/1,000 sq ft

Actual rate = 10.4 lb/1,000 sq ft

Percent error = $\dfrac{10.4 - 8}{8} \times 100 = 30\%$

This spreader is overapplying fertilizer by 30%. Reduce the size of openings at the bottom of the hopper and recalibrate the spreader until your error is less than 5%.

Spreader Swath Width (ft)	Distance Traveled (ft)	Area Covered (sq ft)
6	167	1,000
6	83	500
8	125	1,000
8	62	500
10	100	1,000
10	50	500
12	83	1,000
12	42	5000
14	71	1,000
14	36	500
16	63	1,000
16	31	500

Method II

The second method is not as precise as the first one, but it is preferred by many operators because it does not require a tarp. However, the principle is the same. You are trying to determine the amount of product applied over a known area. For calibrating, this area should be at least 500 sq ft for drop spreaders and 1,000 sq ft for rotary spreaders. The following table shows distances you need to travel to cover an area equal to 500 or 1,000 square feet for different swath widths.

Once you set up the known area that will be covered, follow the steps below to complete calibration.

1. Fill the spreader to a known level marked on the inside of the hopper (Figure 17).
2. Operate the spreader over the known area already set up.

Figure 17 Mark the initial product level in the hopper.

Figure 18 Weigh the amount needed to refill the spreader to the mark.

3. Weigh the amount of material needed to refill the spreader to the mark (Figure 18). Record the weight in pounds.
4. Pounds of product actually applied per square foot can be calculated as follows:

$$\frac{\text{Amount needed to refill (lb)}}{\text{Area covered (sq ft)}} = \text{Product applied}$$

If you need to convert this calculation to other units, use similar calculations explained in the previous method.

CHECKING THE DISTRIBUTION PATTERN

Drop-type spreaders provide fairly uniform distribution of granules across the swath width. The distribution is not affected much by product physical characteristics, weather conditions, etc. However, rotary spreaders are very sensitive to these variables, and patterns can be severely skewed if these variables are neglected. Therefore, always check the distribution pattern of granules when using rotary spreaders.

The best method to check the distribution pattern is to lay out a row of shallow cardboard boxes on a line perpendicular to the direction of travel (Figure 19). Boxes with partitions will prevent chemical particles from rebounding. All boxes should have the same area and height (1 sq ft area, 1 or 2 inches tall). Space the boxes 1 foot apart, and make sure the row is 1.5 to 2 times the width of the anticipated effective swath.

To conduct the test, fill the spreader at least half full, set it at the label setting for rate and pattern, and make three passes in the **same** direction over the boxes. Be sure to turn the spreader on and obtain proper walking speed before crossing the boxes. Put the material for each box into a test tube, vial or small narrow bottle. When the bottles are placed side by side in order, a plot of the pattern is visible. The effective swath width is twice the distance out from the center of the spreader to the point where the rate is one-half the average rate at the center of the pattern. For example, as illustrated in Figure 20, if the first three to four bottles from the center have material 4 inches deep and the bottles from the 6-foot positions (6 feet left and right of the spreader centerline) have material 2 inches deep, the effective swath width is 12 feet. Therefore, you should be making passes 12 feet apart to provide a uniform coverage throughout the area covered.

OTHER CONSIDERATIONS

In some cases, a rotary spreader's distribution pattern may not be uniform enough, even after all spreader adjustments have been made. A common recommendation in such cases has been to cut the rate of application in half and cover the field twice

Figure 19 Checking the distribution pattern of a rotary spreader.

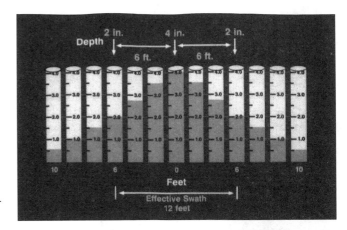

Figure 20 Determining the effective swath width.

at right angles. However, recent research by agricultural engineers at Louisiana State University indicates that cutting the rate in half and cutting the swath width in half while maintaining parallel swaths is more effective than the "right angle" method. The "half width" method is practical mainly for correcting severe pattern problems that remain after all mechanical adjustments to the spreader have been made.

There is always the question of which of the two modes of spreader operation — continuous, back-and-forth mode (Figure 21) or the circuitous, round-and-round mode (Figure 22) — gives a more uniform spreader distribution. Researchers at Louisiana State University analyzed the distribution pattern data from 90 spreader test runs involving four spreaders with several products and pattern settings. They found that when the continuous spreader pattern is generally acceptable, there is statistically no advantage to using the more cumbersome circular mode of operation. However, if the basic spreader pattern remains badly skewed after all the spreader adjustments are made, the circular mode may help improve the overlapping pattern. However, the half-rate, half-width method with the continuous mode is probably more effective in such cases. It is also easier to use and requires little additional travel over the turf area.

SAFETY GUIDELINES

Safety is critical when working with chemicals, especially pesticides. Minimize oral, dermal or inhalation exposure to chemicals, and always observe the following safety tips. Review the sprayer operator's manual and chemical labels for recommended procedures regarding safe use of equipment and chemicals.

Figure 21 Continuous, back-and-forth mode.

Figure 22 Circular, round-and-round mode.

- Wear protective clothing when calibrating, spraying and cleaning equipment (Figure 23). Goggles, rubber gloves and respirators or masks are standard equipment when handling pesticides.
- To avoid overflowing and spilling chemicals on the ground, never leave sprayer tanks unattended when filling.
- Pick up spilled materials promptly.
- Avoid contaminating drinking or other water supplies with direct dumping, runoff, drift or rinse water used to clean spraying equipment.
- Promptly dispose of bags and other containers.
- Never place nozzles in your mouth to blow through the orifice.
- Always wash hands thoroughly with soap and water before eating, drinking, or smoking.

Figure 23 Protective clothing is standard equipment when handling chemicals.

- Take a shower when pesticide handling or application is completed for the day.
- Store pesticides in properly labeled containers, and keep them locked up.

USEFUL FORMULAS

- **To convert ounces per minute (OPM) to gallons per minute (GPM):**

$$OPM/128 = GPM$$

- **To convert gallons per acre (GPA) to gallons per 1,000 square feet:**

$$0.023 \times GPA = \text{Gal per 1,000 square feet}$$

- **To convert gallons per 1,000 square feet to gallons per acre (GPA):**

$$\frac{(\text{Gallons per 1,000 sq ft})}{0.023} = GPA$$

- **To convert GPA to ounces per 1,000 square feet:**

$$2.94 \times GPA = \text{Ounces per 1,000 sq ft}$$

- **To determine travel speed:**

$$\frac{\text{Distance (ft)}}{\text{Travel time (sec)}} \times 0.68 = \text{Travel speed (mph)}$$

- **For any change in travel speed (MPH), calculate the resulting GPA using:**

$$GPA_2 = \frac{GPA_1 \times MPH_1}{MPH_2}$$

- **To determine the desired travel speed (MPH) when changing application rate (GPA):**

$$MPH_2 = \frac{GPA_1 \times MPH_1}{GPA_2}$$

- **To determine the resulting application rate (GPA) when changing nozzle pressure (psi):**

$$GPA_2 = GPA_1 \times \sqrt{\frac{PSI_2}{PSI_1}}$$

- **To determine the desired pressure (psi) that will produce a desired application rate (GPA):**

$$PSI_2 = PSI_1 \times \left(\frac{GPA_2}{GPA_1}\right)^2$$

GPA_2, GPA_1: Desired and measured application rates, respectively (gal/A).

MPH_2, MPH_1: Desired and measured travel speeds, respectively (miles/hours).

PSI_2, PSI_1: Desired and measured spraying pressures, respectively (lb/sq in.).

CONVERSION TABLES

Square Measure

144 square inches .... 1 square foot
9 square feet .... 1 square yard
30^1/$_4$ square yards .... 1 square rod .... 274^1/$_4$ square feet
43,560 square feet .... 1 acre
4,840 square feet .... 1 acre
4,480 square yards .... 1 acre
160 square rods .... 1 acre
640 acres .... 1 square mile

Linear Measure

1 inch .... 2^1/$_2$ centimeters .... 25^1/$_2$ millimeters
1 foot .... 12 inches
1 yard .... 3 feet
1 rod .... 5^1/$_2$ yards .... 16^1/$_2$ feet
1 mile .... 320 rods .... 1,760 yards .... 5,280 feet

Fluid Measure

^{1}/$_6$ fluid ounce .... 1 teaspoon (tsp)
^{1}/$_2$ fluid ounce .... 1 tablespoon (tbsp) .... 3 teaspoons
1 fluid ounce .... 2 tablespoons .... ^{1}/$_8$ cup
8 fluid ounces .... 1 cup .... ^{1}/$_2$ pint
16 fluid ounces .... 2 cups .... 1 pint
32 fluid ounces .... 4 cups .... 1 quart
128 fluid ounces .... 16 cups .... 1 gallon

Weights

1 ounce .... 23^1/$_3$ grams
1 pound .... 16 ounces .... 453^1/$_2$ grams
2^1/$_5$ pounds .... 1 kilogram .... 1,000 grams
1 ton .... 2,000 pounds .... 907 kilograms
1 metric ton .... 1,000 kilograms .... 2,205 pounds

Approximate Rates of Application Equivalents

1 ounce per square foot .... 2,722.5 pounds per acre
1 ounce per square yard .... 302.5 pounds per acre
1 ounce per 100 square feet .... 27.2 pounds per acre
1 pound per 1,000 square feet .... 43.56 ounces per acre .... 2.72 pounds per acre
1 pound per acre .... 1 ounce per 2,733 sq ft .... 8^1/$_2$ grams per 1,000 square feet
100 pounds per acre .... 2.5 pounds per 1,000 square feet
5 gallons per acre .... 1 pint per 1,000 square feet
100 gallons per acre .... 2.5 gallons per 1,000 sq ft .... 1 quart per 100 square feet

Chapter 14A

Simple Hand Sprayer Calibration

Philip Catron, *NaturaLawn of America, Inc., Damascus, MD*

CONTENTS

Getting Ready 163
Part I 163
    Calibration 163
Part II 163
    How Much Material Do I Put into My Hand Sprayer? 163

Hand cans and backpack sprayers have become a very popular and versatile tool for the turf/landscape manager. Successful use of this inexpensive equipment can be assured through simple calibration. Remember, when a pesticide fails to do the job, the fault lies most times with the applicator and/or the equipment — not the chemical.

GETTING READY

When you calibrate, I suggest you utilize a spray tip that gives you a flat spray fan, rather than a cone-shape spray. Many equipment dealers carry these spray tips. My preference is to use a TK-3 flood jet nozzle adapted to the wand of a hand can. By holding the spray wand steady in front, approximately 14 in. off the surface, you can spray a line approximately 4 ft wide.

PART I

The typical hand sprayer commonly in use has a usable capacity of $1^1/_2$ gal, or 192 oz (1 gal = 4 qt = 128 oz). You must first figure out how many square feet the sprayer will cover. This will vary from person to person, since height and walking speed for each of us is a bit different.

First: Try to find a comfortable walking pace (not too fast or too slow). Pretend you are walking hand-in-hand with your first date — that will be the right pace.

Second: Hold the hand sprayer and nozzle in a comfortable position with the nozzle about 14 in. above the ground.

CALIBRATION

Step 1:   Put about 1 gal of clean water into the sprayer and pump it up.

Step 2:   Holding your arm extended as described above, spray onto dry pavement without moving. Measure the width of your spray line and write it down.

Step 3:   Now, hold a measuring wheel by your side and walk for 15 seconds at that leisurely pace described previously, and measure how far you walked. Write down that number.

Step 4:   Multiply the number found in Step 2 by that found in Step 3 above. This gives you the known area (length × width) of what you could spray in 15 seconds. Write this number down.

Step 5:   Now, spray for 15 seconds into an empty container, then use a measuring device to see how many ounces you sprayed. Write it down. (This is the amount of solution you would have sprayed in the area calculated above.)

Step 6:   Now, multiply your answer found in Step 4 by 192, then divide that answer by the number of ounces you found in Step 5. This number now tells you how many square feet your $1^1/_2$-gal hand sprayer will cover.

PART II

HOW MUCH MATERIAL DO I PUT INTO MY HAND SPRAYER?

To calculate this figure, you must first do Part I! If you are using a pesticide that contains 4 lbs active ingredient per gallon and you wish to spray at a 1 lb/acre rate, then:

Multiply your answer from Step 6
in Part I by 32 (ounces) then
divide that by 43,560 (square feet).

This calculates the amount of pesticide (in ounces) you add to $1\frac{1}{2}$ gal of water in your hand sprayer. The basic formula is simply

$$\frac{\text{Rate desired to spray (in lbs ai/acre)}}{\text{Lbs or \% active ingredient}}$$

equals

Amount of pesticide needed per acre

Example: Using the formula with Dursban would look like this:

$$\frac{1.0 \text{ (lbs ai/acre)}}{4.0 \text{ (lbs ai/acre)}} = 0.25 \text{ (gal of Dursban/acre)}$$

Since your hand sprayer only will cover a small portion of an acre (you figured that out in Part I), simply take a proportionate part of the 0.25 gal of Dursban to put into the sprayer:

Example: A $1\frac{1}{2}$-gal hand sprayer covers 1300 ft² (which is close to 3% of an acre). You would then measure 3% of 0.25 gal to put into the sprayer (i.e., .03 × 32 oz = 1 oz).

Sounds simple, right? No! It probably sounds long and tedious, but it really takes longer to read through this than to actually do the calibration.

What's in it for you?

- Better use of pesticides (savings).
- Better control of weeds, etc., that you didn't seem to get before (less service calls).
- Happier customers (fewer cancellations).
- Employees that will "trust" how materials work (improved morale).

Try it — I think you will like the results, and when you have done the math a few times, it becomes second nature. Your hand sprayers and backpack sprayers are like any other pieces of equipment. When calibration and normal maintenance are done and common sense is used, they will give you many hours of failure-free operation. Abuse them and they will not work.

Special Considerations for Golf Courses: Site Selection and Preparation — A Critical Task

Aeration plugs on golf course

Sample showing roots following aeration holes

Royal Woodbine Golf Course near
Toronto, Canada — showing preservation of
environmentally sensitive areas
(photo courtesy of Michael Hurdzan)

Core sample showing layering of topdressing

Spray equipment on golf course

Chapter 15

Siting and Design Considerations to Enhance the Environmental Benefits of Golf Courses

Richard D. Klein, Community & Environmental Defense Services, Baltimore, MD

CONTENTS

Screening Potential Sites for New Golf Courses 168
Golf Course Layout and Design 168
Reducing the Impact of an Existing Golf Course 170
References 171

A golf course that is carefully sited, designed, and managed can provide many benefits to the aquatic environment. The recommendations presented in this chapter are derived from a study of 11 golf courses located in Maryland and Pennsylvania,[6] in addition to a review of the literature.

The results of the study indicated a general relationship between the quality of a waterway and the percentage of the drainage area which is in use as a golf course. As the percentage increases, the quality of the waterway declines. When a golf course accounts for more than 50% of the drainage area, then the receiving waters usually exhibit a moderate to severe level of degradation. Such a waterway would be unfit for most beneficial uses. This relationship applies to watersheds where the only other land uses are woodland, pasture, or cropland. Intense residential or commercial development can cause environmental impacts which obscure effects associated with a golf course.

The potential causes of the degradation revealed by our study include:

- stream channelization,
- destruction of wetlands,
- lack of a wooded buffer along waterways,
- elevated water temperature due to;
    - lack of shading vegetation,
    - reduction of groundwater inflow,
    - release of heated water from the surface of ponds,
    - entry of heated stormwater runoff from impervious surfaces,

- reduction of base flow due to ground or surface water withdrawals,
- release of toxic substances and oxygen deficient water from ponds,
- intermittent pollution incidents such as spills of pesticides, fertilizers, or fuel,
- loss of pesticides or fertilizers by way of ground or surface water runoff,
- entry of stormwater pollutants washed from parking lots and the other impervious surfaces associated with a golf course,
- accelerated channel erosion due to increased stormwater runoff velocity or prolonging the amount of time channels are exposed to erosive velocities,
- elimination of the scouring benefits of flooding by altering the frequency and/or magnitude of flooding,
- poor erosion and sediment control during the construction phase, and
- inadequate treatment of sewage and other wastewaters generated on the golf course.

After a list was compiled of the potential causes of degradation, an extensive search of the scientific literature was conducted. Experts throughout the nation were consulted to develop a better understanding of the relationships between a golf course and the aquatic environment. Finally, we analyzed the results of all this research and concluded it is possible to construct and operate a golf course without damaging the aquatic environment. The design and management recommendations presented in this chapter, however, must be closely followed in order to reduce the likelihood of environmental degradation.

The purpose of this chapter is not how to design a golf course to maximize the quality of the game, but how to minimize the impact upon the aquatic environment. Our study revealed that the typical golf course has the potential to negatively affect the quality of aquatic resources. Yet, if designed correctly, a golf course can be a method of clean water generation through an actively growing biological filter. If this potential is to be realized, then environmental protection must become a primary factor in all aspects of golf course design and management. The best place to begin is the site selection process.

SCREENING POTENTIAL SITES FOR NEW GOLF COURSES

When screening a number of tracts as a possible location for a golf course, it is suggested that priority be given to sites with the following characteristics.

(1) Pesticide and fertilizer movement will be lowest on sites with soils which are medium-textured, have a high organic matter content, a high cation exchange capacity, a low erosion and runoff potential, and where the water table and bedrock lie at least 4 ft below the surface.[1,3,4,7,13,15]

(2) The layout of the course should permit a vegetated buffer of 75 to 150 ft in width along all streams, wetlands, lakes, ponds, or other waterways. Ideally the buffer should be composed of trees and shrubs. A buffer of this width will serve to retard floodwaters, slow channel erosion, shade the waterway from the heating effects of the sun, filter a portion of the pollutants entrained in surface runoff, and contribute leaves and other plant parts to the food web of small, headwater streams.[9,12]

(3) Waterway crossings should not be needed, or the site selected should require the fewest number of crossings. By minimizing crossings, the disturbance of wetlands, stream channels, and the buffer will be diminished.

(4) Sites should be avoided where extremely sensitive species of aquatic life, such as trout, or rare, threatened, or endangered species may occur in downstream waters.

(5) Constructing a pond on a flowing stream can create a number of impacts, which include: barriers to fish migrations, thermal pollution, reduced stream flow, and the proliferation of algae with accompanying dissolved oxygen

deficiencies.[7,14] If a pond is needed at all, then site conditions should allow for construction of the pond without impounding an intermittent or perennial stream.

(6) Sufficient water must be available to meet irrigation needs without either causing a decrease of more than 5% in the low-flow (7-day, 10-year) of any waterway in the vicinity of the site or substantially reducing the yield of existing wells in the area.[7]

(7) Infiltration is the most effective means for controlling the effects of stormwater runoff.[14] Preferential consideration should be given to sites where parking lots, buildings, and other impervious surfaces can be placed near soils which are suitable for the infiltration of stormwater.[14]

(8) To minimize soil erosion and sediment pollution during the construction phase, sites with highly erodible soils or steep slopes (>15%) should be avoided, particularly if the course cannot be built without extensive disturbance on these soils and slopes.[5,7]

(9) Of all land uses, forest produces the least impact upon the aquatic environment.[5,8,20] Therefore preferential consideration should be given to sites with little forest cover or where opportunities to plant trees will be greatest.

(10) During a literature review a number of references were found to poisonings of waterfowl, raptors, and other birds as a result of ingestion of pesticides applied to turfgrass.[16-19] Sites should be avoided that are near congregating areas for waterfowl, raptors, or other birds, particularly if the golf course can only be built by bringing pesticide-treated turfgrass to the edge of a pond, lake, river, or other open body of water. But measures are also available for minimizing the threat to wildlife if a golf course is sited in the vicinity of congregation areas. These measures are described in the next section.

GOLF COURSE LAYOUT AND DESIGN

If the protection of aquatic resources were the sole consideration in selecting a site for a golf course, then the criteria listed above would render the choice an easy one. Of course, many other considerations must enter into the site selection process. As these considerations shift the selection in favor of sites which deviate from the criteria listed above, then the potential for impact increases. Fortunately mitigation measures are available which will reduce some of the impacts associated with golf courses.

Traditionally, one might look at a piece of raw land and lay out each fairway, green, tee, and rough in a way which will maximize the value of the course to the golfer. Once the site layout is "locked-in" one then looks for ways of mitigating the resulting impacts upon the environment. In some cases mitigation is successful, but in others the measures only partially offset the impact upon aquatic resources.

Instead, why not first identify all of the sensitive environmental features and avoid sites where a reliable means of mitigating impacts may not be available? Once these features are identified, the course can then be built around the sensitive areas. For example, the designer might identify all steep slopes, highly erodible soils, coarse-textured or shallow soils, and sensitive aquatic resource areas (streams, wetlands, etc.), then lay out the course so intrusion upon these features is minimized. By taking this approach, the dependence upon mitigation measures (particularly those with variable effectiveness) will be reduced. In other words, the most likely causes of environmental degradation will be "designed-out" of the course from the beginning. The golf course will reflect a "we care" attitude by protecting valuable environmental resources.

The following recommendations will aid the golf course architect in avoiding sensitive features and creating impacts which cannot be readily mitigated.

(1) An underdrain system should be installed beneath any portion of the fairways, greens, or tees which are sited on coarse-textured soils or where the depth to bedrock or the water table is less than 4 ft. The purpose of the drainage system is to collect water which may be contaminated with fertilizers or pesticides. The leachate should be treated by allowing it to soak into medium-textured soils lying 4 ft or more above bedrock and the water table. Or the leachate may be treated with a system such as a peat/sand filter.[2]

(2) If a waterway crossing must be used, then it should be designed to minimize the removal of trees and other shading vegetation. Cart paths should be constructed of permeable material, no wider than 8 ft, and placed on pilings from the edges of the floodplain. All streams should be bridged, not placed in a culvert.

(3) If a site is selected which may affect sensitive species, then a detailed analysis must be made of each impact associated with the golf course to determine if the degree of impact will exceed the level of this species' tolerance. Procedures for conducting such an analysis can be obtained from the local office of the U.S. Fish & Wildlife Service or the state natural heritage program.

(4) A pond should not be located on an intermittent or perennial stream. Upland ponds must not expose stream channels to an increase in either the rate or duration of floodwater velocity.[9a,9b] Upland ponds must not reduce flood-scour to a degree that silt and other fine material will accumulate within downstream channels. Ponds should be designed to minimize use by waterfowl, particularly geese. This can be done by keeping the pond small enough to interfere with the long take-off distance required by geese. Strips of tape or netting can be placed across the surface of the pond to reduce waterfowl use. Also, pesticide-treated turfgrass should not extend to the edge of the pond. If a buffer of tall grass or other dense growth is maintained between the pond and treated grass, then the potential for waterfowl grazing upon the turfgrass will be reduced. Local waterfowl biologists should be consulted when considering measures for a specific pond.

(5) If irrigation withdrawals will reduce the low-flow (7-day, 10-year) by more than 5%, then the Instream Flow Incremental Methodology, developed by the U.S. Fish & Wildlife Service,[10] should be used to determine if the reduction will significantly impact aquatic communities. Options to consider if the impact is deemed significant may include: installing runoff collection ponds in upland areas, extending an appropriation well to a deeper aquifer, using several wells located in different groundwater drainage areas to lessen the impact on any single waterway, or using treated wastewater to offset raw water needs.

(6) If impervious surfaces cannot be sited on soils suitable for infiltration, then a peat/sand filter should be used to control stormwater pollutant loadings.

(7) If grading or filling must occur on slopes steeper than 15%, then clearance should be timed to occur during that portion of the year when the potential for erosion is lowest (generally late summer and early fall). Work on steep slopes should be staged so denuded soils can be stabilized within a maximum of 14 days following initial exposure. The use of a silt fence and establishing turf by sod will reduce potential erosion on sensitive slopes.

(8) Similar tree species should be planted for each native tree removed during site development. Replacement trees should be planted within the same watershed where the loss occurs. The survival rate for each tree species should be taken into consideration. If the survival rate

averages 25%, then four trees should be planted for each tree felled.

(9) Monitoring should begin 1 year prior to the construction of a golf course and continue throughout the construction phase and the first 5 years the course is used. Ground and surface water should be analyzed quarterly for ammonia, nitrate, phosphorus, and pesticides.

Biological sampling should be performed quarterly, then, beginning in the third year, once annually, in August. Fish tissues should be examined once a year for any pesticides used on the course which have the potential to bioaccumulate. A groundwater monitoring program should also be established to detect effects upon existing wells or wetlands. Baseflow and water temperature should be monitored in any streams or rivers in the vicinity of the course.

REDUCING THE IMPACT OF AN EXISTING GOLF COURSE

(1) A combination of physical, chemical, and biological monitoring techniques should be employed to determine if the course is causing an impact and, if so, to identify the probable causes.

(2) The maintenance personnel responsible for identifying and controlling pests should become proficient in the use of integrated pest management (IPM), but IPM alone will not eliminate the potential for contamination of ground and surface waters with pesticides.

(3) If any area of a fairway, green, or tee is located on coarse-textured soils, or if the depth to bedrock or the water table is less than 4 ft, then one or more of the following measures should be employed.

    (a) The area should be fitted with an underdrain system to collect leachate so it can be treated through application to suitable soils or with a sand-peat filter.

    (b) The area should be filled with material which will increase the clay and organic matter content, reduce soil permeability, and increase the depth to groundwater, or

    (c) The area should be replanted with a grass species requiring minimal fertilizers, pesticides, and irrigation.

    (d) The area should be converted to a use requiring minimal maintenance, such as a rough.

(4) Fertilizers with a low leaching potential should be applied at the lowest acceptable rate and applied during periods when grass is actively growing.

(5) Irrigation should be performed on the basis of evaporative demand (evapotranspiration rates), rather then on a set schedule. If irrigation water is drawn from a well, a stream, or a river, then an analysis of the impact upon low-flows and aquatic organisms should be conducted. An analysis should also be conducted of the effects upon well-yields in the area. If either analysis indicates a problem, then the following options should be considered.

    (a) The construction of an upland pond to capture and store stormwater runoff. If this option is used then the ponds must be designed and sited to avoid either a significant increase or decrease in floodflows.

    (b) Relocate production wells to several groundwater drainage basins to reduce the impact upon individual streams or rivers and to lower the impact upon other groundwater users.

    (c) Relocate a surface water intake to utilize a stream, river, or lake which can meet irrigation needs without a negative impact upon aquatic communities.

    (d) Reduce or terminate water withdraws during critical periods.

    (e) Replant the course with grass species having a higher drought-tolerance.

(6) The first inch of stormwater runoff from all impervious surfaces should be delivered to an infiltration device[14] or a peat/sand filter.[2]

(7) A 75- to 150-foot buffer should be established along all wetlands, streams, rivers, tidal waters, ponds, or lakes.

(8) The use of chemical measures for managing ponds and lakes should be reduced or eliminated. Rather than using chemical substances to control algae, techniques with fewer long-term impacts should be used, such as reducing nutrient inputs, dredging, and so forth.[11]

(9) Wherever possible, the number of trees and shrubs on the course should be increased.

(10) Pesticides, fertilizers, fuels, and other toxic substances should be stored in a location where a spill will not result in rapid, uncontrollable entry into ground or surface waters.

(11) If the golf course existed prior to 1980, then soils on the greens, tees, and fairways should be analyzed for organochlorine and metallic pesticide residues. If residues are present, then measures should be taken to minimize movement to ground or surface waters, such as increasing the organic matter content of soil.

In summary, a golf course can be a tremendous asset to the aquatic environment. Past approaches to the siting, design, construction, and operation of a golf course have resulted in damage to streams, ground water, wetlands, and other aquatic systems. The recommendations presented in this paper will allow society to continue to enjoy the benefits associated with golf, while maintaining and even enhancing the quality of our water resources. The basis for the recommendations presented above are provided in "Protecting the Aquatic Environment from the Effects of Golf Courses."[7]

REFERENCES

1. **Frere, M. H., Ross, J. D., and Lane, L. J.,** The nutrient submodel, in *CREAMS — A Field Scale Model for Chemicals, Runoff, and Erosion from Agricultural Management Systems*, Southeast Watershed Research Laboratory, U.S. Department of Agriculture, Tifton, GA, 1980, chap. 4.
2. **Galli, J.,** Peat-Sand Filters: A Proposed Stormwater Management Practice for Urbanized Areas, Department of Environmental Programs, Metropolitan Washington Council of Governments, Washington, DC, 1990.
3. **Harrison, S. A.,** Effects of Turfgrass Method and Management on the Quantity and Nutrient and Pesticide Content of Runoff and Leachate, College of Agriculture, Pennsylvania State University, State College, PA, 1989.
4. **Jury, W. A., Focht, D. D., and Farmer, W. J.,** Evaluation of pesticide groundwater pollution potential from standard indices of soil-chemical adsorption and biodegradation, *J. Environ. Qual.,* 16(4), 422, 1987.
5. **Klein, R. D.,** Sediment Pollution: A Literature Review, Tidewater Administration, Annapolis, MD, 1984.
6. **Klein, R. D.,** The Relationship Between Stream Quality and Golf Courses, Community & Environmental Defense Services, Maryland Line, MD, 1989.
7. **Klein, R. D.,** Protecting the Aquatic Environment from the Effects of Golf Courses, Community & Environmental Defense Services, Maryland Line, MD, 1990.
8. **Lugbill, J.,** Potomac River Basin Nutrient Inventory, Metropolitan Washington Council of Governments, Washington, DC, 1990.
9. **Magette, W. L., Brinsfield, R. B., Palmer, R. E., Wood, J. D., and Dillaha, T. A.,** Vegetated Filter Strips for Agricultural Runoff Treatment, U.S. Environmental Protection Agency, Washington, DC, 1987.
9a. **Malcolm, H. R. and Sappington, J. M.,** Urban Stormwater Management Needs in North Carolina, Water Resources Research Institute, University of North Carolina, Raleigh, NC, 1983.
9b. **McCuen, R. H., Moglen, G. E., Kistler, E. W., and Simpson, P. C.,** Policy guidelines for controlling stream channel erosion with detention basins, University of Maryland, Department of Civil Engineering, College Park, MD, 1987.
10. **Milhous, R. T., Updike, M. A., and Schneider, D. M.,** Physical Habitat Simulation System Reference Manual — Version II, National Ecology Research Center, U.S. Fish & Wildlife Service, Fort Collins, CO, 1989.
11. **Olem, H. and Flock, G., Eds.,** *Lake and Reservoir Restoration Guidance Manual*, 2nd ed., EPA 440/4-90-006, North American Lake Management Society, U.S. Environmental Protection Agency, Washington, DC, 1990.
12. **Palfrey, R. and Bradley, E.,** The Buffer Area Study, Maryland Department of Natural Resources, Tidewater Administration, Annapolis, MD, 1981.
13. **Petrovic, A. M.,** The fate of nitrogenous fertilizers applied to turfgrass, *J. Environ. Qual.,* 19, 1, 1990.
14. **Schueler, T. R.,** Controlling Urban Runoff: A Practical Manual for Planning and Designing Urban BMP's, Metropolitan Washington Council of Governments, Washington, DC, 1987.
15. **Shoemaker, L. L., Magette, W. L., and Shirmahammadi, A.,** Modeling management practice effects on pesticide movement to ground water, *Ground Water Monit. Rev., Winter*, 1990.
16. **Stone, W. B.,** Poisoning of wild birds by pesticides, *N.Y. Fish Game J.,* 26(1), 37, 1979.
17. **Stone, W. B. and Gradoni, P. B.,** Recent poisonings of wild birds by diazinon and carbofuran, *Northeast. Env. Sci.,* 4(3,4), 160, 1985.
18. **Stone, W. B. and Gradoni, P. B.,** Poisoning of Birds by Cholinesterase Inhibitor Pesticides, New York Department of Environmental Conservation, Wildlife Pathology Unit, Delmar, NY, 1987.
19. **Stone, W. B. and Knoch, H.,** American brant killed on golf course by diazinon, *N.Y. Fish Game J.,* 29, 95, 1982.
20. **Troeh, F. R., Hobbs, J. A., and Donahue, R. L.,** *Soil and Water Conservation for Productivity and Environmental Protection*, Prentice-Hall, Englewood Cliffs, NJ, 1980.

Chapter 16

Design and Management of Constructed Ponds: Minimizing Environmental Hazards

Thomas Jon Smayda, *Vasey Engineering, Seattle, WA*

Brenda L. Packard, *Harding Lawson Associates, Seattle, WA*

CONTENTS

Introduction 173
Site Drainage Plan 173
Pond Design Considerations 174
    Pond Shape 174
    Pond Liners for Watertightness and Macrophyte Control 175
    Water Residence Time 175
    Aeration 177
    Pond Safety 178
Integrated Pest Management Plan Components 178
    Pest Identification 178
    Establishment of Pest Tolerance Levels 179
    Plant Establishment 179
    Pest Monitoring Program 180
    Mechanical and Physical Controls 181
    Biological Controls 181
    Chemical Controls 182
    Washwater Management 182
Conclusions 183
References 183

INTRODUCTION

Large developments such as golf course complexes are often designed to fit the natural environment. The design process begins with an identification of sensitive areas such as wetlands, steep slopes, and fish-bearing waters. A site drainage plan may then be conceptually designed to fit logically into the terrain. The development is then designed to fit within this mosaic of sensitive areas and drainage features.

Constructed ponds are a typical part of the drainage system, and are designed for flood control, irrigation reservoirs, aesthetics, and/or wildlife enhancement. Regardless of the primary purpose, pond design can readily be adjusted to improve water quality and to minimize pest problems. Given a good design, maintenance is simplified, and the potential need for herbicide application to control aquatic plants is reduced.

Also presented is an operations plan for golf course water management. The operations plan employs integrated pest management (IPM) principles to minimize the use of chemical pest controls and to provide improved management of these chemicals. The design and management strategies presented herein may be combined with a turfgrass management system to produce a more holistic approach to golf course IPM implementation than has often been employed in the past.

SITE DRAINAGE PLAN

Water quality onsite and in downstream receiving waters is impacted by the drainage plan. Development often leads to increased overland flow rates and requires construction of ponds, streams, wetlands, swales, or pipe conveyances. These features are first identified in a drainage plan which is formulated to route water through the site, provide for detention of peak flows, and ensure that good quality water is discharged to wetlands and other receiving waters.

Minimum requirements for the drainage system should include:

- Maintenance of discharge at natural locations in order to preserve existing downstream ecosystems. However, in some cases, a self-contained drainage system with no outlet from the site may be a feasible alternative.
- Erosion and sediment control program for the construction phase of the project.
- Flood and water quality control for site runoff.
- Off-site analysis to determine flow handling capacities and special water quality concerns of downstream locations. For instance, increased flows from the site may cause streambed erosion or exceed the capacity of existing downstream culverts; discharge to marine waters may require careful control of fecal contamination if shellfish beds are present; nutrient control may be important if eutrophication of marine or freshwaters is probable; and control of toxins such as pesticides and heavy metals is important especially if the receiving area has low assimilative capacity, such as oligotrophic lakes, coral and seagrass beds, or is used as a food or drinking water supply. Constructed ponds, wetlands, vegetated filter areas, infiltration areas, and sand/peat soils in greens are example methods used for runoff control. Minimization of substance application (source control) enables these features to operate most effectively.

These requirements serve to protect off-site receiving waters and are generally satisfied by a sequence of on-site water quality controls such as constructed ponds, swales, and wetlands, for example, as described by Schueler.[1] Onsite water quality is also important. Good water quality in onsite wetlands is protected by law and should also be maintained in constructed onsite ponds to prevent use of herbicides, alum, or other agrichemicals. Direct hydraulic connection which generally exists between onsite ponds and receiving waters is reason to avoid application of pesticides to onsite water features.

POND DESIGN CONSIDERATIONS

Constructed ponds can be designed for water quality control without compromising flood control, wildlife habitat, or aesthetic functions. Two general factors are considered for water quality control. First, good water quality is needed within the pond. On golf courses, the green committee will perceive "good water quality" to be a lack of objectionable odor (due to low oxygen), open water (not covered by vegetation), and blue (not brown with mud, or green with phytoplankton blooms). Essentially, water quality is generally dependent upon plant control. Algae should not be permitted to get to "pea soup" concentrations because of potential for clogged irrigation lines, low dissolved oxygen which could kill fish, and aesthetic nuisance including odor. Pond weeds such as cattails should not be permitted to colonize the entire lake if open water is desired. A good design, then, will minimize the potential for algae blooms and macrophyte encroachment, and will permit operation without the addition of herbicides or other chemical products.

The second factor of pond design is to protect beneficial uses of downstream waters. Minimization of the need for herbicide application by providing proper pond configuration is one method; provision for in-pond water quality improvement and construction of appropriate outlet structures are others. This section presents various pond design considerations.

POND SHAPE

The best shape for water quality improvement is to have the inlet and outlet separated by a fairly large distance, construct two or more deep portions separated by shallow berms or wetlands (so called multicelled ponds), and to have sufficient depth so that macrophytes do not fill the pond (Figures 1, 2, and 3). The necessary pond volume is often dictated by requirements for stormwater detention/retention, irrigation storage needs, or by the need for fill soils obtained from a pond location for use elsewhere on the site. Once the volume of the pond is established, considerations for water quality control, wildlife and fishery enhancement, and aesthetics govern the ultimate pond shape.

Macrophytes which grow to the surface are known to provide habitat for mosquito larvae, so that construction of open water areas rather than wetlands may be desired. A good way to control macrophyte abundance is to provide adequate water depth. The depth to which rooted plants may grow in a pond is closely correlated with summertime water clarity.[2]

$$Y = 0.83 + 1.22X, \qquad (1)$$

where $Y =$ maximum depth of plant growth (meters)

$X =$ average summer water clarity measured by secchi disk (meters).

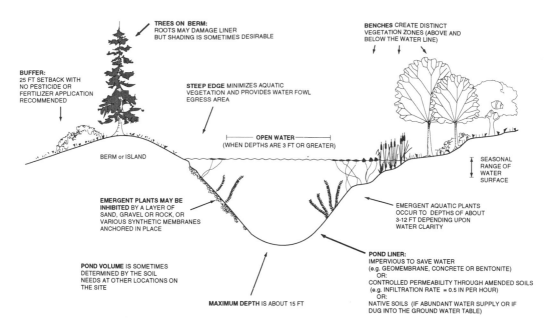

Figure 1 Cross section through a constructed pond to show how pond shape influences vegetation habitat. The amount of perimeter wetland vegetation should be controlled by pond design rather than by herbicide application.

So, if the clarity is 1 m, then macrophytes may be expected to exist to 2 m deep, and with 2 m clarity the plants may colonize the pond to a depth of about 3.3 m. In the Pacific Northwest a depth of 3 m is generally sufficient to inhibit macrophytes.

Banks may be steep, to minimize the width of the macrophyte zone, or gradual, supporting a more broad expanse of wetland vegetation (Figures 1 and 2). The amount of wetland vegetation and open water is controlled by proper design rather than by application of herbicides after construction. Above the water line the shoreline may be steep or gradual, which also exerts strong control on terrestrial vegetation zones. When designing to have an active water surface elevation for stormwater control (one that oscillates in response to storm flows), the selected vegetation should be tolerant of periodic inundation and drying. Native species of fruiting plants are endorsed because they supply food which can support increased numbers of songbirds.

POND LINERS FOR WATERTIGHTNESS AND MACROPHYTE CONTROL

A variety of geosynthetic membranes, expansive clays, and concrete products are available to create water-tight liners for the pond bottom. In excessively drained soils, except in cases where infiltration is deemed necessary, the pond may be lined with loamy sands to reduce the infiltration rate. An 18-in. layer may be compacted in a single lift to reduce infiltration to about 0.5 in./hr. A variety of geosynthetics, sands, and gravels are used to cover the bottom to reduce aquatic plant growth. The ideal sediment covering for macrophyte control[3] is opaque, remains in place (gas formation can float a plastic liner if not vented and anchored), and is durable. Retrofitting synthetic liners is difficult, perhaps easiest if set up on ice, and requires firm anchoring with top dressed gravel or the like.

WATER RESIDENCE TIME

Pond volume divided by the outflow rate defines water residence time. When residence time is short, 3 days for example, then the water quality in the pond is controlled by the influent water quality. Long residence time, a year for example, permits internal lake processes such as sediment-water interactions and biologically mediated changes to exert control on lake water quality. A target water residence time for constructed ponds is on the order of 5 to 50 days.

Fecal coliform bacteria have a die-off rate defined by the first order equation:[4]

$$N_t = N_o \, e^{-(1.1)(t)} \qquad (2)$$

where N_t = concentration of bacteria at time t
N_o = concentration of bacteria in the inflow
e = base of the natural logarithm system
t = elapsed time after inflow in days.

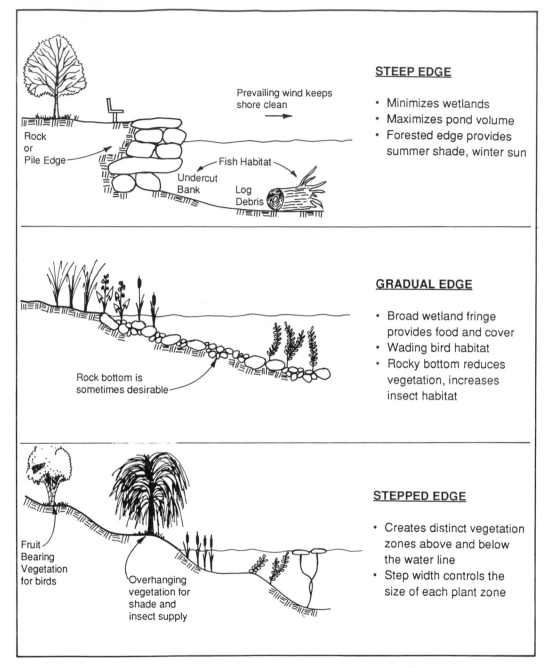

Figure 2 Pond edge designs can be altered to provide a range of wildlife habitat and aesthetic conditions. Generally, a variety of edge types is desirable for any one pond.

Therefore, with an inflow concentration of 1000 organisms per 100 ml, which is not uncommon in storm runoff on golf course sites, and a 5-day retention time, the expected outflow concentration should be on the order of 4 organisms per 100 ml, which is fairly low with respect to most beneficial uses in receiving waters, possibly excepting shellfish harvesting.

Water residence time also affects the rate of algae washout from a pond. Phytoplankton biomass is reduced as cells are flushed from the lake through the outlet. If the washout rate matches the growth rate, then no increase in biomass occurs. By design, a certain algae biomass is needed to keep the lake clarity moderate. A target secchi depth of 1 m may be achieved by encouraging some phytoplankton biomass to make the water a bit turbid, which serves to reduce the depth of macrophyte

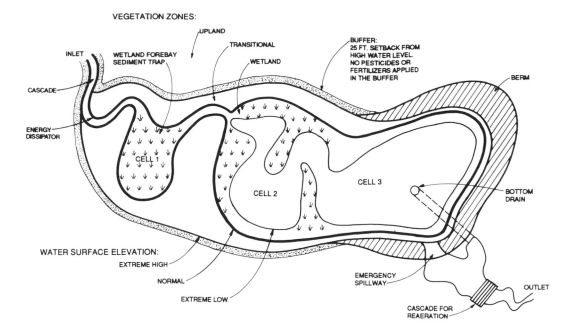

Figure 3 Plan view of a three-cell pond designed to have a fluctuating water surface for storm water control, and a controlled outlet located in the deep hole (hypolimnetic withdrawal).

colonization as described in Equation 1. Accrual of algae biomass depends upon regional influences such as temperature, sunlight, nutrient levels, and zooplankton grazing in addition to washout, and should be considered during the pond design process. However, unlike bacteria which die off with extended residence times, algae control can be achieved by shortening water residence time which washes the cells from the pond. Other proven methods for control of algal biomass include nutrient limitation by carefully controlling phosphorous in the watershed, stocking with insectivorous fish such as pumpkinseed or mosquito fish which have the indirect effect of promoting zooplankton biomass which graze on the phytoplankton, or ozonation to kill phytoplankton cells.

AERATION

Oxygen is needed in lake water to support fish life, prevent hydrogen sulfide odors, and reduce phosphorus solubility. Thermal stratification within a lake routinely occurs in temperate zones and exerts strong control on the distribution of oxygen. Stratification develops when warm weather causes an increase in surface water temperature. This warmed water effectively floats on top of cooler bottom water creating two distinct layers. Wind mixes the top layer which remains oxygenated because aeration occurs across the lake surface. The bottom water layer is isolated from the atmosphere and due to oxygen demand by the sediments and in the

water column, oxygen levels tend to be reduced, sometimes eliminated.

Excessive growth of phytoplankton and macrophytes can lead to anoxia each night when respiration occurs in the absence of photosynthesis. Additionally, oxygen is consumed by plants which die, accumulate on the bottom, and decay. Thus, excessive plant growth results in poor water quality by adding to the oxygen demand, and this situation may be exacerbated by thermal stratification.

Many methods exist to provide ample aeration. Creating a shallow lake which is mixed to the bottom is not endorsed because of the potential for colonization by macrophytes, which could then require the use of copper sulfate or other herbicides for their control. One method for aeration is to provide cascades in the inlet creek. Aeration in flowing waters and cascades can be readily calculated[5] and often natural flow rates can be augmented by pumping. An example design is to have a chain of lakes connected by swales, creeks, and cascades. Water from the lowest pond is pumped to the top pond to provide circulation and aeration (and possibly algae removal by filtration in vegetated swales). Such a system provides aesthetic appeal as well. If no outlet exists from the lowest pond and the system is a closed loop, then the design provides the ultimate in surface water quality protection.

Other aeration methods, paddles, fountains, bubbling, and ozonation have proven successful, but a good drainage plan can provide adequate oxygen without these devices. These systems, however, can readily be retrofitted to a pond should they prove necessary and may not be needed as part of the original design.

POND SAFETY

Since drowning may occur in any water depth if a person panics, limitations on pond depth for safety may not be necessary. A safe shoreline, however, is considered to be an important design feature.[6] A relatively flat shoreline for sure-footedness, a gradually sloped pond bottom, and compacted sediments provide key elements which minimize panic situations (Figure 4). Additionally, a thick band of vegetation around the pond perimeter is an effective deterrent to keep people out of the water. Reduction of deeper water aquatic plants such as lily ponds, pond weed, and milfoil may reduce anxiety in swimmers, prevent entrapment, and reduce any possibility of drowning.

Large storm flows through outlet structures can trap people. Therefore, properly designed outlet grate structures are an additional safety consideration.[7]

INTEGRATED PEST MANAGEMENT PLAN COMPONENTS

After a pond is constructed, the implementation of proper maintenance and pest control practices is necessary to preserve the intended function of the pond. Often, these practices include the use of chemical pesticides to reduce pest presence.

Integrated pest management concepts may be applied to golf course ponds in the same manner that they are applied to turfgrass management. The main principles of IPM[8,9] are reiterated here for use in golf course pond management as discussed in this paper:

(1) Identification of pest
(2) Establishment of injury and action levels for pest population
(3) Development and implementation of a reliable monitoring and record keeping program
(4) Use of good, regular maintenance practices to minimize pest problems
(5) Identification of a number of treatment options if a pest problem occurs. Pesticide use should be considered only as a last resort
(6) Evaluation of the effectiveness of the IPM system.

PEST IDENTIFICATION

Aquatic pests may be classified as vegetation, insects, and fish. Particular species of each category of pest will vary depending on geographic location and site-specific conditions.

Vegetation

Aquatic plants that may require control measures include the following:[10]

- Emergent aquatic plants including cattails, bulrush, spikerush, and others
- Submerged plants including watermilfoil, *Elodea*, *Potamogeton*, and others

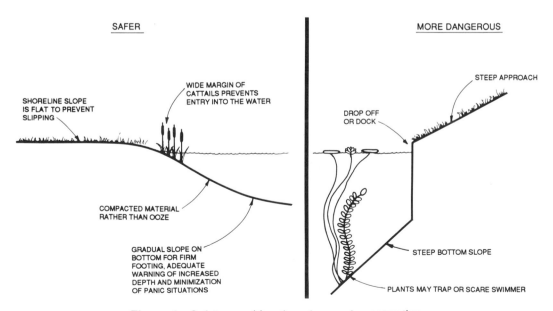

Figure 4 Safety considerations for pond construction.

Table 1   **Example action level determination**

Type of Problem	Basis for Action Level	Example Action Level
Pond turbidity	Turbidity has a number of causes (e.g., algae growth, bottom-feeding fish, sediment) and adversely affects site appearance.	Secchi disk depth 1 ft **or** Ability to see boots when standing in water 1-ft deep.
Macrophyte encroachment	Extensive macrophyte coverage of ponds may negatively impact site appearance. In addition, vegetation protects mosquito larvae from predators, thereby increasing the population.[12] Excessive vegetation may also deplete dissolved oxygen and can clog irrigation systems.	60% coverage of pond area. **or** Encroachment over irrigation withdrawal.
Mosquito nuisance	Determination of a significant mosquito nuisance problem will likely be based on the number of complaints the golf course management receives from golfers or offsite residents neighboring the course. In some areas mosquito-transmitted disease may also be an issue.	Daily average of 20 adults caught per trap night[13] or >1 bite per 15 min.[16]

- Floating plants including duckweed, water lilies, and others
- Aquatic algae including a variety of green and bluegreen forms, and fungal mats
- Invasive species such as purple or garden loosestrife.

Insects
Probably the most notorious and pervasive of all aquatic insect pests is the mosquito. Mosquito presence may be perceived as a nuisance, or in some situations it may pose a public health risk. Other insects such as midges can reach nuisance levels for golfers, grounds employees, and residents living adjacent to the golf course.

Fish
A pond inhabited by fish may become overpopulated by certain species and adversely affect the environment or other desirable aquatic organisms. For example, carp and bullheads tend to produce turbidity in ponds.[11]

ESTABLISHMENT OF PEST TOLERANCE LEVELS
Pest tolerance levels, commonly termed "injury" and "action" levels, must be established for pest populations in and around the ponds. In a golf course setting, aesthetics and nuisance conditions will dictate what levels of weed growth are tolerable at each particular site. Pest insect and fish levels will be based on any damage or nuisance created by the species. In any case, it is important to quantify pest populations for comparison to tolerance levels. These levels will be set as a result of experience with site conditions and careful monitoring of pest populations. When a pest problem occurs, the pest density should be recorded, and the tolerance level set at or below this level. Because these levels will be site specific, no attempt is made here to set typical tolerance levels. Instead, an approach for determination of action levels for several common pond-related problems is presented (Table 1).

PLANT ESTABLISHMENT
An initial step in preventing unwanted plants from becoming established in a pond is to plant desired species after site construction. As with turfgrass, it is advantageous to plant species native to the area and species that have a competitive advantage in the local environment. Species selected would depend upon the geographic area, the native species present, and the aesthetic preferences at the particular golf course. It is important to avoid monocultures and to instead diversify the pond ecosystem as much as possible. However, it should be noted that even desirable species may reach unacceptable levels and must be reduced in number.

Wetland vegetation consists of diverse species in distinct zones based on water depth. The vegetation surrounding a pond may be divided into planting or landscaping zones (Figure 3). These zones and vegetation found within each zone are summarized as follows.[1,14]

(1) **Deep Water Areas** contain submerged plants, such as pondweed.

(2) **Mid to Shallow Water Areas** contain emergent plants, such as arrowhead and cattail.

(3) **Transitional or Shoreline Areas** contain grasses and emergent plants, such as switchgrass, rushes, and sedges. These areas are regularly inundated.

(4) **Riparian Fringe Areas** contain predominantly woody plants such as swamp oak and dogwood. These areas are periodically inundated and have wet soils.

(5) **Floodplain Terrace Areas** contain trees, shrubs, and grasses such as willow, oak, spicebush, and fescue grass. Moist soils are present with infrequent inundations.

(6) **Upland Slopes** have moist to dry soils and are never or rarely inundated. Upland vegetation consists of a variety of trees, shrubs, and grasses. Species present vary depending upon the climate and geographic area.

After site preparation, desirable vegetation may be established by planting whole plants or dormant rhizomes and tubers. Seeding has not typically been successful for wetland plants, but wetland soils which have been imported for construction may contain a substantial seed bank of desired species. Appropriate timing in planting must also be considered. For example, early spring plantings are generally the most successful.[14]

PEST MONITORING PROGRAM

A monitoring program for aquatic pests including careful recordkeeping procedures is necessary to quantify pest abundance and to evaluate treatment effectiveness. As with other IPM programs, the results of monitoring are compared to the tolerance levels. Controls would be implemented, if necessary, when pest tolerance levels are approached. Further monitoring results collected after control methods are employed should be used to evaluate the effectiveness of the pest controls.[10] No attempt is made here to specify acceptable numbers of pests present in a pond. In practice, however, it is essential that pest levels be quantified and recorded (Table 1).

In addition to pest monitoring, the pond water level may be measured on a weekly or monthly basis by reading permanently installed staff gauges. This information will assist in selecting pest control options and determining results of treatment.

Vegetation

Excessive development of macrophytes and the presence of algae blooms detract from the aesthetics of the golf course and may indicate the presence of nutrients in site runoff.[15] Some problems caused by excessive growth are summarized in Table 1.

Macrophyte populations may be quantified as a percentage of pond area covered, total areal coverage, or ratio of emergent vegetation to open water. The golf course superintendent may use a scaled as-built construction drawing to aid in estimating areas of aquatic plant coverage. The coverage may be estimated visually, or actual measurements may be taken on the site. Another option is to install markers in the pond at locations designating the maximum tolerable extent of emergent vegetation encroachment into the pond.

Insects

Insects associated with aquatic environments, such as mosquitoes and midges, may become nuisance pests as numbers reach the tolerance level. Aquatic insect surveys may be done for larval or adult stages, depending upon the particular pest species, and type of information required. Surveys provide information on species identification, numbers, and location. Survey frequency may be based on the number and occurrence of complaints received about mosquito presence.

Larvae are identified by species and instar (developmental stage) and counted per unit area by collecting pond water samples with a dipping ladle.

Adults may be counted per unit time period by performing landing counts or setting light or CO_2 traps. Landing or biting counts are performed only after dark by recording the number of mosquitoes that land to bite exposed skin during a specified period of time, usually 2 to 15 min.[16] For species identification, capture the mosquitoes with a vial or aspirator and place them in a kill jar containing a cotton swab with ethyl acetate. Most mosquitoes are attracted to light, and some have a greater attraction to CO_2. Light traps should be set 6 ft above ground level in an open area near trees or shrubs. The presence of strong winds, competing light sources, and smoke should be avoided. Mosquitoes are typically collected from the traps on a daily basis for determining numbers and identifying species.[11] The use of traps involves more elaborate methods, inconvenience, and expense than conducting landing counts, and is less selective in the species collected.[16]

Fish

Pestiferous fish species may be detected as a result of fishing a pond, or from adverse effects caused by the fish (e.g., turbidity caused by carp and bullheads). Abundance may be quantified by keeping records of fish stocking and performing creel censuses of fish caught.[15] A fisheries biologist may be consulted if more accurate determinations are necessary.

MECHANICAL AND PHYSICAL CONTROLS

Mechanical/physical pest controls involve the manipulation of the environment to the disadvantage of the specific pest.

Vegetation

Mechanical/physical weed controls include the following:[17]

- An underwater mower, a mechanical weed harvester, rake, or a chain may be used to cut submersed and emersed weeds. The severed weeds must be removed from the water to prevent them from spreading and reproducing. Mechanical harvesting is not effective for controlling algae or duckweed.
- Drain and dry the pond. Lower the water level enough to expose weeds. In some cases several months may be required to control weeds by drying. In other cases seepage will prevent drying and would prohibit the use of this technique. Fish habitat must be considered before this control option is implemented.
- Shading of the water or sediment inhibits growth of some aquatic weeds. Plastic sheeting, pea gravel, soluble dyes, or bank vegetation may be used.
- Dredging ponds is effective (but expensive) in removing rooted plants and nutrient-rich sediments as well as increasing pond depth. Small ponds may be dredged by backhoe, dragline with clam bucket, or bulldozed after water level drawdown.
- Pond aeration reduces chance of odor and helps trap nutrients in the sediments which minimizes potential for algae blooms. Oxygen may be added by aeration, oxygenation, or ozonation. Rapid water exchange in the pond may maintain aerobic conditions.
- Cattails may be controlled by water drawdown in early spring, clipping of the new shoots, then flooding for a period of a month or so.

Insects

Control of insects may be accomplished by proper site-water management as described in a previous section of this paper. If a golf course is located in a residential area, the local residents may mitigate a mosquito problem by installing window, door, and porch screens, and by using personal insect repellent.

Fish

Mechanical/physical control methods include:[11]

- Seines, nets, traps, or other barriers to reduce movement of undesirable species into the particular environment.
- Draining the pond and then restocking with desirable species.
- A partial drawdown of the water level to force small forage fish out of protected areas, making them more vulnerable to predators.
- Removal of vegetation to achieve the same result as above.

BIOLOGICAL CONTROLS

Biological controls generally consist of the introduction of pathogens, predators (herbivores and carnivores), or competitive species into the aquatic ecosystem. Although biological control is not a new approach to pest control, practical and safe techniques are limited. In addition, unknown, long-term ecological consequences are possible. In spite of the current paucity of effective biological control agents, this approach shows promise in that it may bring long-term control at moderate cost.[2] However, more basic research into aquatic ecosystem ecology is required in order to develop easily available and more effective biological controls. Controls described here have been used successfully in some situations, although many more are still in the experimental stage. This method of control has not yet achieved the operational level of mechanical and chemical controls.

Vegetation

The following biological controls have been implemented in at least some areas:[2]

- Plant-eating fish: Sterile grass carp (white amur) may be used where they are legal. A typical stocking rate is approximately 15 fish per acre. *Tilapia zillii* has also been used successfully for macrophyte and algae control. This species does not tolerate extended periods of temperature below 10°C and requires annual restocking in less temperate regions.
- Algae-eating fish: *Tilapia mossambica* (Mozambique mouthbrooder), *Hypothalmicthys molitnx* (silver carp), and *Menidia ardens* (Mississippi silversides) are suggested for use in managed habitats where the fish cannot escape outside of the habitat. All species require warm-water habitats (above 10°C) and will likely need to be restocked annually in less temperate areas.
- Insects: Two beetles (*Neocletina bruchi* and *Neocletina eichhorniae*) and a moth (*Sameodes*

albiguttalis) have been effective in controlling water hyacinth. The flea beetle *Agasicles hygrophila* has been effective in alligator weed control.

- Competitive plant species (these will be present if initially planted after pond construction).
- Plant diseases: A fungus (*Cercuspora rodmanii*) has been effective in reducing water hyacinth populations.
- Other animals (e.g., muskrats, ducks, manatees) have also been used with limited success.[11]

Insects

The biological control of mosquitoes and other closely related pestiferous species of flies may be achieved with mosquito-eating fish, the bacterium *Bacillus thuringiensis israelensis* (BTI), or nematodes.[16]

Types of mosquito-eating fish include the top-minnow *Gambusia affinis* (mosquito fish), goldfish, some *Tilapia* species, and guppies (*Precilia reticulata*). Small ponds not connected with natural waters may be stocked with some kind of mosquito-eating fish. The fish may be protected from predatory cats and raccoons by providing an overhang around the edge of the pond or an escape area. Placing 1-ft sections of 4-in. or 6-in. diameter clay drainage pipe at the pond bottom provides protection from fish predators.[16]

Of the mosquito-eating fish species, goldfish are generally hardier, live longer, and are more easily available. *Tilapia* and guppies may die out during winter due to their intolerance of cold temperatures (guppies cannot survive below 11°C). Mosquito fish (*Gambosia*) can thrive in reduced-oxygen environments such as sewage treatment ponds.[16] State and local regulations must be considered before stocking any water body with non-native fish species.

Bacillus thuringiensis israelensis (BTI) is a bacterium that is commercially available for mosquito control as a larvicide. It is effective against several mosquito species in varying conditions and is highly selective, killing only mosquitoes and some related aquatic insects. BTI application is performed in the same manner as for chemical pesticides by using conventional spray equipment.[16]

The nematode *Romanomermis culcivorax* was commercially available for a number of years as a mosquito larvicide. Poor sales forced the withdrawal of this product from the market and it is no longer available. However, with increasing public concerns regarding the use of chemical pesticides, the potential exists for a resurrection of commercially available nematode pesticides.[16]

CHEMICAL CONTROLS

The use of registered, aquatically approved pesticides may be necessary as a last resort in some cases after other control options have been evaluated and found to be infeasible or ineffective. However, it is not the intent of this paper to endorse specific chemical pesticides or to describe the preparation or application of these pesticides. The focus of this paper is on techniques aimed at minimizing the need for chemical controls, and managing those chemicals that are used.

WASHWATER MANAGEMENT

The golf course wash area, used for cleaning mowers and chemical sprayers, poses a potential environmental hazard. The washwater generated in these areas may be a point source of nutrient, petroleum product, and/or pesticide contamination to the environment. The most common type of wash area consists of one dedicated site with no permanent surface and no containment of water and residues. Discharge of the washwater and residues is to surface water and soil. Another common type of wash area consists of a permanent asphalt or concrete pad that collects and directs the water, clippings, and residues into a catch basin. The drainage often is diverted into surface water and soil.[18]

It is strongly recommended that all washwater be completely contained in one or more dedicated areas and receive proper disposal in an environmentally safe manner. An improved design for washrack facilities may consist of an enclosed area such as a building or a roof over a concrete containment pad to prevent contaminated runoff from leaving the area after a storm event. The area should provide containment for all vehicle and equipment washwater generated onsite. A sump may be installed to collect the washwater for future reuse on the golf course.[19] No untreated washwater should be allowed to discharge to natural water courses. Reuse of washwater (and collected pesticide spills, if possible) on turf is highly recommended over disposal because pesticide waste may be regulated as a hazardous waste. However, reuse of pesticide-contaminated washwater on the golf course is permitted only when conducted in accordance with label requirements. Therefore, it is important that the collection sump be pumped regularly to prevent mixing of incompatible materials in the sump, and that the surrounding area be kept clean. Reuse of wastewater from pesticide spills requires careful consideration of dilution, application rates, and location of application.

Gilhuly[18] describes improvements that may be made to golf course washwater management facilities to provide for greater protection of water resources.

CONCLUSIONS

Golf course ponds can be designed to achieve a predetermined level of plant habitat, and minimize the need for aquatic pesticide application. IPM principals should be applied to pond maintenance to further reduce the introduction of agricultural chemicals into the water. It is recommended that golf courses include these aquatic controls during construction and in their IPM plans to complement their grounds management practices.

REFERENCES

1. **Schueler, T. R.,** Vegetative BMPs, in *Controlling Urban Runoff: A Practical Manual for Planning and Designing Urban BMPs*, Washington Metropolitan Water Resources Planning Board, Washington, DC, 1987, 1.

2. **Cooke, G. D., Welch, E. B., Peterson, S. A., and Newroth, P. R.,** *Lake and Reservoir Restoration*, Butterworth, Boston, 1986, 1.

3. **Perkins, M. A.,** An evaluation of pigmented nylon film for use in aquatic plant management, in *Lake and Reservoir Management*, EPA 440/5-84-001, U.S. Environmental Protection Agency, Washington, DC, 1984, 467.

4. **Horner, R., Brenner, M., and Jones, C.,** Design of Monitoring Programs for Determining Sources of Shellfish Bed Bacterial Contamination Problems, Report to Washington Department of Ecology, June 1989.

5. **Peavy, H., Rowe, D., and Tchobanoglous, G.,** *Environmental Engineering*, McGraw-Hill, New York, 1985, 1.

6. **Glenn, J., Harlan and Associates,** Thoughtful design is a prime factor in water safety, *Lake Line*, 2, 2, 1991.

7. **Kropp, R. H.,** Water quality enhancement design techniques, in *Stormwater Detention Facilities*, DeGroot, W., Ed., ASCE, New York, 1982.

8. **Daar, S.,** Least-toxic pest management for lawns, in *Least-Toxic Pest Management: Lawns*, The Bio-Integral Resource Center, Berkeley, CA, 1991, I-1.

9. **Leslie, A. R.,** Development of an IPM program for turfgrass, in *Integrated Pest Management for Turfgrass and Ornamentals*, Leslie, A. R. and Metcalf, R. L. Eds., U.S. Environmental Protection Agency, Washington, DC, 1989, 315.

10. **Packard, B. L. and Smayda, T. J.,** *Site Management Manual for the South Langley Golf Course and Residential Community*, Harding Lawson Associates, Seattle, 1992, chap. 5.

11. **Parker, R., Ramsey, C., and Thomasson, G., Eds.,** Aquatic Pest Control MISC 0134, Washington State University Cooperative Extension, 1990, 1.

12. **Martin, C. V. and Eldridge, B. F.,** California's experience with mosquitoes in aquatic wastewater treatment systems, in *Constructed Wetlands for Wastewater Treatment: Municipal, Industrial and Agricultural*, Hammer, D. A., Ed., Lewis Publishers, Boca Raton, FL 1990, chap. 31.

13. **Dill, C. H.,** Wastewater Wetlands: User friendly mosquito habitats in aquatic wastewater treatment systems, in *Constructed Wetlands for Wastewater Treatment: Municipal, Industrial and Agricultural*, Hammer, D. A., Ed., Lewis Publishers, Boca Raton, FL, 1990, chap. 39.

14. **Allen, H. H., Pierce, G. J., and Wormer, R. V.,** Considerations and techniques for vegetation establishment in constructed wetlands in aquatic wastewater treatment systems, in *Constructed Wetlands for Wastewater Treatment: Municipal, Industrial and Agricultural*, Hammer, D. A., Ed., Lewis Publishers, Boca Raton, FL, 1990, chap. 33.

15. **Schmidt, J. C., Ed.,** *How to Identify and Control Water Weeds and Algae*, 4th ed., Applied Biochemists, Mequon, WI, 1990, 1.

16. **Olkowski, W., Daar, S., and Olkowski, H.,** *Common-Sense Pest Control*, 2nd printing, Taunton Press, Newtown, CT, 1991, chap. 36.

17. **Packard, B. L. and Smayda, T. J.,** *Site Management Plan for the South Langley Golf Course and Residential Community*, Harding Lawson Associates, Seattle, 1992, chap. 4.

18. **Gilhuly, L. W.,** *Wash Rack Blues*, USGA Green Section Records, January-February 1993.

19. **Packard, B. L. and Smayda, T. J.,** *Water Quality Management Plan for the Whitehorse PUD Golf Course*, Harding Lawson Associates, Seattle, 1993, chap. 13.

Chapter 17

Minimizing Environmental Impact by Golf Course Development: A Method and Some Case Studies

Michael J. Hurdzan, *Hurdzan Golf Course Design, Columbus, OH*

CONTENTS

A Planning Method that Never Fails: The Team Approach 185
Selecting a Golf Course Site to Fit Your Objectives and Budget 187
Case Studies of Successful Environmental Golf Course Design 189

Golf courses can be good neighbors to all but the most fragile of environments or ecosystems **if** they are logically and scientifically designed, constructed, and maintained. As proof that a methodical planning and management approach works well, examples will be provided in which a golf course coexists and is integrated with a highly regulated agriculture crop, has protected and improved wetland habitats, reclaimed a prized ocean beach, fits harmoniously into a desert, and is blended upon a unique and protected glacial landscape. In each case a similar planning method was followed. The method was intensively reviewed by trained and concerned permitting agencies before construction began, scrutinized during construction, and once in operation, the courses have been monitored and found not to be deleterious to the site or surroundings.

The planning method is really quite simple, but it does require commitment by the would-be golf course developer to be sensitive to all related issues, patient with the process, willing to risk "upfront" money, and open to making concessions and compromises.

The degree of commitment and perseverance that the potential golf course developer is willing to give must be both a personal and business decision because a difficult site might take several years and large amounts of money before the dream becomes a reality.

A PLANNING METHOD THAT NEVER FAILS: THE TEAM APPROACH

The first step is to put together a compatible team of experts who are experienced and knowledgeable about the potential social, political, and technical issues involved in golf course development. Do not make the mistake of thinking that the "name or marquee" designer is going to impress anyone in the review process ahead. Instead, engage professionals of proven ability and expertise. The minimum expertise represented on the planning team should reside with the following professionals:

(1) golf course architect
(2) clubhouse architect
(3) landplanner (if housing or a resort is involved)
(4) lawyer
(5) accountant or financial expert
(6) golf course superintendent
(7) ecologist or biologist
(8) engineer
(9) owner or designated representative
(10) planning team coordinator

This list represents the **minimum** number of required professionals and to leave out one of them is often "penny wise but pound foolish."

First the owner/developer must clearly and specifically articulate the project's goals and objectives, and then place them in priority. For example, one objective might be 400 house lots, another could be a par-72, 7000-yard golf course, and others may include a large golf practice area, eight tennis courts, Olympic-size pool, 20,000-ft^2 clubhouse and parking for 300 cars. If space will not allow all of the objectives then their priority must be decided, and this prioritizing should be done only by the owner. Once all goals and objectives have been specified and prioritized, then the team, under the leadership of an experienced team coordinator, must develop a probable space or acreage requirement for each facility. This sets the scope and size of the project, as well as establishes a very approximate project cost.

Ideally, after the team has been selected it should be charged with helping the developer select the best site to achieve his objectives and goals. This is the ideal situation and can save much time and money for the developer by inexpensive problem avoidance instead of expensive problem solving. However, typically the golf course developer already has selected the site so now the team must forget the ideal and begin problem solving.

The team should compile a list of potential problems or threats that could impact the project, such as wetlands, water supply, historically significant sites, etc. A complete list of such concerns is not possible, for many are site or regionally specific. For discussion purposes, however, the following sample list is offered:

(1) Physical factors
   (a) property size
   (b) property shape
   (c) topography
   (d) on- and off-site drainage
   (e) soils and subsoil layers characteristics
   (f) type, health, age, and distribution of existing vegetation
   (g) sources of irrigation water
   (h) sources of power
   (i) size, location, and sensitivity of existing ecosystems
(2) Social factors
   (a) existing zoning restrictions
   (b) previous land use
   (c) historical significance
   (d) attitudes of neighbors
   (e) adjoining land use
   (f) buried utilities
   (g) rights-of-way or easements
   (h) mineral rights, restricted covenants, life estates
(3) Planning requirements and restraints
   (a) wetlands and required buffers
   (b) potential existing habitats or nesting/breeding areas
   (c) endangered species
   (d) soil engineering characteristics
   (e) seismic or disaster planning
   (f) open space
   (g) irrigation restrictions.

Developing a list of concerns initiates the research phase of planning, needed before any intelligent decisions can be made. The team coordinator usually assigns research responsibility to various team members as appropriate to their individual expertise, and sets a time line for completion and correlation of this information. The team coordinator then distributes the information with some generalized planning criteria resulting from the research phase.

Examples of these general planning criteria might be:

(1) ecological areas that must be avoided and/or list of buffer zone restrictions
(2) ecological areas that can be marginally impacted with specific instructions on limits and procedures
(3) possible main routes of ingress and egress and main arterial traffic patterns
(4) possible clubhouse location(s)
(5) areas of zoned or restricted use
(6) areas requiring extreme engineering or planning
(7) any possible phase plan for construction or development
(8) possible location of utilities or service centers.

With these broad guidelines the golf course architect and landplanner should begin to independently plan possible site plans using very simplistic and fast graphic techniques. This is called schematic design and the products are usually "stick and ball" centerlines of golf holes with proposed elevations at tee, landing zone, and green site; and housing or other areas shown as blocks or bubbles. The intent of this phase is to produce quick studies that can be presented to the rest of the team for critique and revision. Each revision session should refine the possible land use plan until finally the team agrees on a plan which represents the best package of compromises of all concerns and restrictions.

The next phase is called design development, and detail is added to each of the features and facilities of the plan, such as golf tees, greens, bunkers, fairways, street, roads, and building footprints, etc. This produces further smaller refinements of the concept and the end result becomes a master plan which can go forward for permitting and approvals.

Following the master plan, each professional or discipline now begins producing construction documents and specifications that will not only be used for competitive bidding between contractors, as well as guide construction, but also to ensure all construction is coordinated between each discipline. It is particularly important that topics such as drainage, utilities or buried services, massive earthmoving, and major construction activities are fully coordinated before **any** construction begins.

The more detailed the construction documents, the less time, money, and energy will be wasted during construction.

Throughout the planning process, possible and perceived environmental impacts are identified, reviewed, and responsibly addressed. Many times creative solutions are required for both construction and long-term use or maintenance, which might significantly add to the cost and time of construction. So, the most intelligent planners simply avoid areas or sites of environmental concern unless there is absolutely no other solution. Avoiding wetlands or sensitive habitats or endangered species, or whatever the concern, should be the first planning approach and not the last one. This appears to be common sense and not worth mentioning, but frequently great projects become stuck in permitting or litigation simply because of environmental ignorance or inattention.

SELECTING A GOLF COURSE SITE TO FIT YOUR OBJECTIVES AND BUDGET

This is a tale of two golf courses built within 20 miles of each other in Toronto, Canada, both designed by Hurdzan Golf Course Design, both built by Gateman-Milloy contractors in 1990, using identical materials. To the golfer the two look similar, but one course cost $2 million and the other cost almost four times that amount. Understanding why such cost disparities exist may help golf course developers in selecting sites and preparing reasonable budgets.

The expensive project, named Devil's Pulpit after a nearby rock formation, was conceived to be an ultraexclusive, members-only course, on 315 acres of dramatic and rolling sand and gravel glacial moraines, for a client who wanted a world-renowned golf course that would host major professional events. This site was on a highly regulated and protected geologic formation that required an extensive permit and approval process by town, regional, federal, and environmental agencies (a total of seven major agencies and several lesser ones).

The lesser-cost project, named Royal Woodbine after a famous horse racing track nearby, was conceived as a semiprivate club (members and fee players) on 165 acres of city-owned floodway leased to the developer, who wanted to operate a profit-producing golf course. The site had been an illegal dumping ground for many years along an environmentally dead stream that everyone wanted cleaned up, thus the permit process was less arduous and restrictive than for the Devil's Pulpit site.

I assisted in the selection of each site after reviewing several comparable properties. Site selection was based upon the program objectives of the client, the available construction budget, and intrinsic site factors such as topography, vegetation, soils, environmentally sensitive areas, on- and off-site views, and intuitive sense about which site would yield the best golf course considering all of the above factors.

Both golf courses consisted of 18 holes, with five sets of tees on each hole, with total yardage of approximately 5000 to 6800 yards; with 6000 ft^2 putting surfaces constructed of high sand rootzones; automatic run irrigation systems; landscaping; cart paths; bentgrass tees, fairway, and greens; stone bridges and walls.

On paper these projects sound very similar and thus one could expect their costs to be comparable. So then how can one cost four times more than the other? The difference in cost between these two projects (Table 1) focused on the environmental sensitivity of each site, the scale and scope of each golf course, and the construction techniques required on each site.

One should immediately notice the enormous difference for item 3, environmental protection — Devil's Pulpit spent $2 million to satisfy environmental concerns posed by the multitude of reviewing agencies. Specific expenditures included $1,300,000 to install artificial membranes beneath all ponds to protect groundwater quality, $250,000 to pipe springs through the property to ensure no change in water purity or temperature, $125,000 for silt control devices, $150,000 to build an environmentally isolated green near a wetland, and $175,000 for revegetation work.

At Royal Woodbine, only erosion control measures were required, and this cost only 1% of the environmental protection budget at the Pulpit. It took over a year and hundreds of thousands of dollars in professional fees and testing to get permits to construct Devil's Pulpit. At Royal Woodbine, it took only a few weeks and almost no cost. The lesson here is to carefully investigate environmental protection cost before site acquisition and decide if the nature of the site is worth the cost in time and money.

Items 4 and 5 (Table 1), earthmoving and shaping, are also vastly different owing to the excavation required to modify the huge and abruptly tumbling topography of the Devil's Pulpit site, which encompassed over 360 ft of contour change.

Earthmoving was estimated to be approximately 1 million cubic yards of material for Devil's Pulpit, while only 220,000 cubic yards of earthmoving was necessary on the Royal Woodbine site because

Table 1 **Comparative budgets for the two similar courses**

Item	Devil's Pulpit	Royal Woodbine
1. Clearing	$115,000	$39,000
2. Major drainage	$363,879	$62,000
3. Environmental protection	$1,989,000	$20,000
4. Earthmoving	$2,978,000	$750,000
5. Shaping	$325,000	$100,000
6. Green construction	$391,346	$285,000
7. Minor drainage	$120,451	$15,000
8. Irrigation	$1,200,000	$600,000
9. Walls	$354,000	$125,000
10. Bridges	$4,000	$100,000
11. Bunkers	$140,000	$100,000
12. Planting		
a. Sod	$600,000	$75,000
b. Seed and mulch	$355,200	$172,000
13. Maturation	$300,000	$200,000
14. Miscellaneous	$250,000	$100,000
Total	$9,485,894	$2,743,000

it had only 30 ft of contour change within the site. Similarly, the expense to shape and mold the earth into golf features once it was properly placed was proportional to the volume of earthmoving. The key point is that selecting a gently rolling site, as opposed to an excessively steep or flat site, will help keep golf course development costs modest.

Greens construction was nearly identical for both courses because exactly the same materials and methods were used in both cases. Greens were constructed of a laboratory-selected sand, blended off-site with a soil blender, and hauled to the greens in trucks after tile installation.

Minor drainage (item 7, Table 1, pipes of 12 in. or less) was much higher on the hilly Devil's Pulpit site because of the requirement to catch and contain surface runoff from a larger area, traveling at a higher velocity, over a longer distance. In short, gently rolling sites require far less underground drainage than do severely sloping sites, and small sites need less than large sites.

Royal Woodbine was built in a narrow stream valley that was protected from drying winds, had silt- and clay-laden soils that retain soil moisture, and fairways corridors were separated by lines of isolating and separating mounds and trees. This meant that the irrigation system at Royal Woodbine could be quite simple, requiring multi-row heads only in landing and major play areas, thus resulting in smaller and fewer irrigation pipes, fewer heads, controllers, valves, wires, etc., and a smaller pumping plant. Therefore, the cost to irrigate Royal Woodbine was $600,000, about half that of Devil's Pulpit, which is very windy, has sandy soils, few windbreaks, and large changes in elevation.

The combined cost for walls and bridges was 50% higher at Devil's Pulpit because Devil's Pulpit had 18,000 ft² of wall as compared to Royal Woodbine, which had 6250 ft², but eight bridges instead of one as at the Pulpit. Similarly, cost for bunkers at the Pulpit was higher because of the total area of sand bunkers used on both courses (unit costs were the same for both).

Almost as dramatic as the difference in cost due to environmental protection measures (item 3), was the amount spent to plant the golf course (item 12, Table 1). Although approximately the same area was planted on both courses, the difference between the projects is the amount of sod used. As my design associate, Dana Fry, who had primary responsibility for both projects, said, "Sod is addictive." Once the Devil's Pulpit client saw the first hillsides sodded, he was "hooked," he had the budget to indulge his "addiction" to sod, and environmentalists applauded. Before it was all said and done, over 80 acres of bluegrass and 20 acres of bentgrass sod were used on Devil's Pulpit. The only areas seeded were most of the greens and tees and 12 fairways to bentgrass, and deep rough areas to fine-bladed fescues. At Royal Woodbine the budget only allowed for 5 acres of sod with the other 145 acres being seeded — a savings of $600,000.

In retrospective defense of our benevolent Devil's Pulpit client, the decision to extensively use sod probably saved hundreds of thousands of dollars in the long run. The reason is that the Pulpit course received several torrential rainstorms during the planting season, which would have devastated seeded areas, eroded vast areas, and jeopardized the entire project. As it was, the course received minimal damage and matured quickly despite a very cold grow-in period. Sod is not only addictive, but it is also good insurance that the golf course will open on time which can be important if promises have been made to members, homeowners, or resort guests. Sod also slightly reduces maturation costs as shown in item 13 (Table 1).

It would be wrong to assume that Royal Woodbine is of the same stature, prestige, or golf appeal and impact as Devil's Pulpit. Devil's Pulpit and its new sister course, Devil's Paintbrush, will earn worldwide acclaim as the best 36 holes at one club anywhere in the world. That was the client's intention and he was willing to spend the money necessary to play in the big leagues of golf.

Both Devil's Pulpit and Royal Woodbine will meet the objectives of the owners and exceed the expectations of golfers, but at vastly different costs. However, neither site would have met the objectives of the other client, so each golf course purpose was perfectly suited to the site and the goals. An experienced golf course architect can greatly assist clients in site selection based on the objectives and budget of the club. This investing of professional fees in site selection can reap enormous savings in mistakes, time, and money later.

CASE STUDIES OF SUCCESSFUL ENVIRONMENTAL GOLF COURSE DESIGN

In the time of Old Tom Morris (mid 1800s to early 1900s), golf courses were laid out on the land with little or no modification to the ground except the clearing of brush or trees and seeding turfgrasses. Similarly, golf course maintenance was simple, requiring only that the greenskeeper occasionally mow the grass, topdress liberally, fertilize sparingly with natural fertilizer sources, and resod as needed. Golf had virtually no impact on the land.

During the era of Donald Ross (early to mid-1900s), golf courses required some modification to the land to permit enjoyable golf, but this "building" involved only crude earthmoving equipment powered mostly by horses and mules. Likewise, golf course maintenance was only slightly more sophisticated but still relied heavily on mowing, topdressing, and natural chemicals. Golf still had little impact on the land.

The golf courses built by Robert Trent Jones (mid-1900s to present) were the first to be "architected" and with the invention of powerful earthmovers, even the most inhospitable sites could be molded into pleasurable golf experiences. Since World War II, the proliferation of synthetic chemicals has ballooned, and these at times were excessively applied or used to move unadapted grasses into environments where they did not belong. That, coupled with the ever-increasing sophistication of science to assay or detect minute concentrations of chemicals, has resulted in public concern and reaction such as the U.S. cranberry scare of the late 1950s, DDT in food chains in the 1960s, agent orange in the 1970s, alar on apples in the 1980s, and recently, the holes in the ozone layer.

Sometimes these concerns were ill-founded, sometimes well-founded, but overall there was a public reaction to insist that human beings become better stewards of the earth. This rather sensible request has affected golf courses and brought some noticeable, often profound, legislated changes in golf course design and construction.

The intelligent professional has reacted positively to most of these required changes, and implemented suggestions that protect environments and recognize their long-term benefit as being in everyone's best interest. However, practicing professionals have on occasion rebutted the frivolous or unfounded concerns, by citing or sponsoring creditable scientific research to disprove or defuse emotional rather than rational concerns. Such adaptations in professional practice have not been easy, inexpensive, or enjoyable, for they have extended the planning process of a golf course by at least 6 months, raised its construction costs by an average of about 15%, and reduced the number of projects completed as a result of delayed approvals that took months or years.

But the demand for golf is high and the industry is resilient, and as painful as it may be, it is adapting to this renewed wave of environmental sensitivity. So this message is positive and it is worthwhile to evaluate some examples of how creative designers have satisfied environmental restructuring.

Obviously, environmental protection costs money which can often mount up into the millions of dollars, but I am not aware of any objection to golf courses that cannot be answered or mitigated with existing scientific data, except in extremely fragile or rare sites, upon which nothing should be built anyway.

The main environmental issues facing golf courses can be broadly categorized as:

(1) Wetlands
(2) Water quality
(3) Open space preservation
(4) Pesticides and fertilizers
(5) Habitat

not necessarily in that order.

It would be naive to think that these issues could be clearly separated, for they are intricately interwoven. It would be equally naive, however, not to recognize that effort to lessen or remove negative impacts in one area is not in some way beneficial to the others. Therefore, our intent is to show that if the main environmental concerns are addressed, there are ways to satisfy those concerns.

What follows are some examples of the innovative ways that golf course architects and contractors have devised to allow a golf course to be built and still protect or improve the environmental qualities of the site.

PROJECT:	Spanish Bay Golf Links
LOCATION:	Monterey, California
ARCHITECT:	Robert Trent Jones, Jr.
CONCERNS:	1. Water conservation
	2. Dunes protection
	3. Native vegetation
	4. Open space

The Spanish Bay site was previously mined for sand along some of the most beautiful coastline of California. It was privately owned and was slowly revegetating, but had been treated more as a dumping ground than as a recreational asset.

The permit process for this landmark project reportedly took 8 years because of the issues of coastal access by the public, groundwater issues, and isolated but sparse habitat. To satisfy those issues, coastal walkways were planned that were integrated into an overall golf course plan. Then, millions of cubic yards of sand were moved back to the site to rebuild a sand dunes landscape into which the golf course was developed.

During construction, subsurface drainage systems were built to capture and replenish groundwater, and the final product was planted to turf varieties that required low water, low fertilizer, minimum pesticides, and infrequent mowing. This was one of the first of the modern golf courses in North America to plant exclusively to British-type turf such as fine fescues (*Festuca rubra*).

PROJECT:	Ventana Canyon
LOCATION:	Tucson, Arizona
ARCHITECT:	Tom Fazio
CONCERNS:	1. Irrigation areas
	2. Native vegetation
	3. Habitat

Perhaps no large-scale development is more the antithesis of a desert environment than is a golf course. By definition the land is a desert because it lacks adequate water to support lush vegetation, while a golf course is commonly thought to be all lush vegetation. The obvious concerns voiced by environmentalists were impacts on habitat, water quality, and native vegetation. Mr. Fazio set a new standard for sensitive golf course planning and construction at Ventana Canyon by not forcing the course onto the site, but rather letting it melt into the landscape. Then, during construction, endangered plants were either moved to a nursery or protected, so that when later replanted they became assets of the site and not liabilities. Once built, the golf course used zonal maintenance to limit water and routine procedures to only selected areas. Tees and greens are intensively maintained and many areas are left in their natural desert state.

PROJECT:	Old Marsh Golf Club
LOCATION:	Palm Beach Gardens, Florida
ARCHITECT:	Pete Dye
CONCERNS:	1. Wetlands
	2. Surface drainage
	3. Endangered species

To the untrained eye, it would seem that Florida has all the wetlands, marshes, and estuaries that it needs. However, wetland preservation is a serious concern of environmentalists and conservationists, and preservation of these areas has become a deterrent to golf course permitting. However, at Old Marsh, and more recently at the Kiawah Island Ocean Course, Pete Dye worked closely with engineers to develop a construction system that allowed the golf course to be built in harmony with its surroundings. In fact, this has proven to be a positive impact during recent droughts in that area.

The system consists of drainage network which allows no surface water to enter the marshes without first going through a series of filtration ponds that remove all potential negative waterborne impacts such as pesticides, silt, or fertilizer. During droughts, the irrigation water used on the course, which is effluent water, has actually helped preserve the marshes by supplementing them daily with recycled irrigation water. Naturally, wildlife has prospered near the golf course and artificial habitats and nesting areas have been installed to further support the indigenous species.

PROJECT:	Willowbend Country Club
LOCATION:	Cape Cod, Massachusetts
ARCHITECT:	Michael Hurdzan
CONCERNS:	1. Wetlands
	2. Water quality
	3. Pesticides and fertilizers

Few other places in the U.S. are more concerned about or protective of their groundwater supplies than Cape Cod, simply because groundwater is the Cape's only source of water supply. Cape Cod is a sandy peninsula about 60 miles long by 10 miles wide, formed by a glacier about 10,000 years ago. Since it is mostly composed of sandy soils, whatever is applied to the surface eventually ends up in the drinking water. The Willowbend site was further complicated by the fact that much of the site was being, and was to remain, actively used for the highly regulated commercial growth of cranberries. The problem was to plan a golf course, in and among the cranberry bogs, that would not endanger the crop through golf course maintenance, nor would it have any negative impact on groundwater.

To achieve those environmental objectives, ideas were borrowed from Mr. Jones, Mr. Fazio, and Mr. Dye. The result was a golf course which utilized low-maintenance fescue-type turfgrasses, used zonal maintenance, and controlled drainage near sensitive areas. Fertilizer and pesticides used for golf course maintenance were matched to the cultural requirements and procedures of the cranberries so the entire complex would be treated similarly. However, a further step was taken to develop some fallow areas adjacent to the bogs that would serve as an environmental sink in the event of the misapplication of pesticides. This project remains significant in the U.S. because it represents the total integration of food production and recreation with positive effects on both.

In nearly every situation described above, the trend has been to less artificial golf course maintenance and greater utilization of natural materials, cultivation practices, and improved turfgrass varieties to reduce maintenance impact. Resolving environmental concerns about golf courses is global and universal. We find the same concerns all around the world. It is our contention that a golf course development can be a good friend to the environment with a positive effect on the quality of our lives.

I would prefer not to end on a negative note, but I believe that there is one issue that must be addressed by us as a society, and soon. That is the enormous power vested in environmental review boards, many of whose members at best have only a lay knowledge of environmental and developmental issues. As a result they often resort to making decisions based on emotion instead of information, and such decision-making then takes on a political or private bias. At some point in the near future, it is incumbent upon us as citizens of the world to insist that environmentalists be registered or certified based upon minimum technical skills, training, and background, if they are to render opinions or decisions which can so dramatically impact a proposed project. This same minimum competency, as established for persons serving on review boards, should be expected of golf course designers as well. Similarly, these environmental experts and golf course designers should be made to scientifically defend their decisions. There are opportunities for abuse, as in the case of an environmentalist reporting a wetlands violation to government offices and then approaching the same golf club offering his rather costly consulting services to satisfy the ensuing investigation. Such "make-work projects" tend to abuse the public's sympathy for environmental protection.

Properly designed, built, and maintained golf courses can protect and improve the environmental quality of their locations. But, it requires a scientific approach to this process, continuous research and monitoring, and a well-informed society. This end is one of the goals of this book, and I appreciate this opportunity to make my contribution.

Chapter 18

Pest Management Strategies for Golf Courses

L. B. McCarty and Monica L. Elliott, *University of Florida, Gainesville, FL*

(Contribution from the Florida Agric. Exp. Stn. J. Ser. No. R-03589.)

CONTENTS

Introduction 193
Background Information 193
    IPM Beginnings 193
    Previous Turf IPM Use 194
Strategies of IPM 194
    Cultural Controls 195
    Biological Controls 196
    Chemical Controls 196
Starting an IPM Program 197
Conclusions 198
References 202

INTRODUCTION

One of the most appealing aspects of the game of golf is the beauty of the golf course. The responsibility for maintaining this beauty and "acceptable" playing conditions falls on the superintendent and his staff. One method for meeting these objectives is the incorporation of a common sense approach to protecting the turf by information gathering, analysis, and knowledgeable decision making. Integrated pest management (IPM) utilizes the most appropriate cultural, biological, and chemical strategies for managing plant pests. Since the foundation of an effective IPM program is a sound agronomic program, Integrated *Plant* Management might be a better term for turf managers.

Unfortunately, the pressures to maintain tournament conditions throughout the year have often forced superintendents to abandon sound agronomic practices for "quick-fixes" to get them through the current crisis. For example, $\frac{1}{8}$-in. mowing heights, requests for "soft" greens with no scars or disruptions in consistency, and growing grasses outside their natural range of adaptability have forced superintendents to increase their inputs in terms of fertilizer, water, and pesticides. At the same time, public concerns over these inputs and restrictions on the availability of traditionally used resources, such as water, will require that golf courses consider incorporating IPM programs into their total management scheme. However, until playing condition expectations from the golfers themselves are changed, the superintendent will continue to feel pressured to do what is asked from the player and not necessarily what is best for the turfgrass or environment.

BACKGROUND INFORMATION

IPM BEGINNINGS

Modern IPM concepts and practices began to evolve in the late 1950s with apple production and was vastly expanded with cotton production in the 1960s. This evolved from the mid-1940s when the modern use of pesticides began to explode. Many felt at that time that pesticides were the "silver bullet" or ultimate specific weapon needed to control all pest problems. Many traditional pest and plant ecological studies were abandoned, as were nonchemical control alternatives. This led to a new generation of producers and scientists who had little experience with nonchemical approaches to pest or plant management. However, resistance to pesticides, especially insecticides, forced researchers and growers to seek alternative methods of pest control, thus, the birth of IPM.

In recent years, turf managers have begun to realize that their escalating dependence on pesticides and the lack of research and training in the pest management arena are now affecting their industry.

0-87371-350-8/94/$0.00+$.50

For example, in the early 1980s several very effective and relatively cheap pesticides were banned from the turf market. Two such pesticides were EDB (ethylene dibromide), a soil-injected nematicide, and chlordane, an insecticide. Managing the turf to withstand higher populations of nematodes or insects, particularly mole crickets and grubs, was not followed as long as EDB and chlordane were available. However, since the loss of these materials, nematodes, grubs, and mole crickets have become the most serious turf pest problems in many areas. Researchers are currently trying to find alternative methods of management and control for these pests based on pest life cycle and the use of biological control agents. Obviously, additional time and research will be necessary to solve the problems that were basically ignored for more than 40 years.

PREVIOUS TURF IPM USE

Only a handful of reports are available dealing with golf course turf IPM programs, primarily because food and fiber crops have received the majority of research funds for IPM projects. A country club in Massachusetts used IPM practices to control Japanese beetle grubs. Sex tabs and floral lures were used to attract these insects into traps instead of utilizing the traditional pesticides treatment approach. During the month of August, for example, 47 traps were placed in the rough and 160 gal of beetles were collected. Grub counts of 50 to 75 per sq ft were reduced to only 1 or 2 per sq ft.

A golf club in Idaho traditionally had disease problems on its bentgrass greens during the summer. The superintendent, with some major reservations from club officials, decided that continued and frequent fungicide use only provided temporary masking of disease symptoms and began searching for the underlying causes of the diseases. Poor drainage and soil layering were discovered on the most troublesome greens. Excessive nitrogen and water use rates had been applied in recent years and a distinct buildup of black layer had occurred.

The club began implementing IPM practices first by informing and soliciting support from the club members. After approval was obtained, greens were aggressively and frequently aerified deep enough to allow better and healthier root penetration. Natural organic fertilizers and a biostimulant were used to supplement and eventually replace a portion of the synthetic materials. Synthetic fungicide use was partially replaced by natural sea plant extract disease suppressants and a "compost soup" consisting of digested sewage sludge and wood wastes.

Benefits of the program were significant. Disease severity was reduced as were clipping yields with no loss of shoot density or color. The resulting balanced top growth produced reliable and predictable putting surfaces. Synthetic fertilizer use was reduced almost 50%. Other benefits included the return of earthworms to the soils and a reduction in thatch accumulation.

A pilot project, recently completed in South Carolina, incorporated traditional IPM strategies into golf turf management. Turfgrass Information and Pest Scouting (TIPS) was administered to seven golf courses in which scouting was performed and recommendations made to the superintendent on agronomic practices and judicious use of pesticides. Among the noted accomplishments of this project were (1) a 30% reduction of fungicides by monitoring weather parameters and not applying chemicals until conditions were favorable for disease development and (2) 35% reduction in nitrogen use without sacrificing the quality of the golf courses.

STRATEGIES OF IPM

Developing IPM strategies for a golf course requires reliable information about the following:

(1) Knowledge concerning all normal inputs required for growing the turf — not only what they are but also why they are required. This is supplemented by knowing pest life cycles and which management practices disrupt or influence these to reduce pest numbers. Understanding the logic behind a management practice, rather than just doing it "because that is the way we have always done it," allows the manager to make decisions to alter these practices to reduce pest problems or encourage turf growth to overcome or tolerate the pest.

(2) Use of a monitoring system to carefully follow pest trends to determine if a pesticide will be necessary and, if so, when it would be most effectively applied. Ideally, monitoring systems are based on known economical or aesthetic thresholds levels. Unfortunately, in many cases, these thresholds are not specifically known, thus are determined by the superintendent to reflect local conditions and threshold levels tolerated by club clientele.

A professional scout, who may be employed by several nearby courses, also may be used. Since these scouts visit several courses, pest trends are more easily recognized and information useful from one course can more easily be used to assist others. Typically a scout should hold a degree in agronomy, horticulture, entomology, or plant pathology with emphasis in pest management.

Tools required for scouting vary with pest problems, scout training, and golf course budget. A good set of eyes and an inquisitive mind are essential. These are supported by a standard 10× hand or pocket lens, soil probe, soil profile probe, spade, cup cutter, pocket knife, tweezers, scalpel, collection vials and paper bags, and field identification guides. Soap and water also are necessary for insect monitoring. More expensive but precise instruments may be used in a room designated as a diagnostic laboratory. Included are stereo- and compound-microscopes, soil sieves, pH meter, conductivity meter, and elementary soil analysis kits. These need to be supplemented by ongoing scout training at short courses, formal classes, appropriate diagnostic guides, and opportunities to visit similar set-ups to exchange ideas. Monitoring intensively maintained golf courses includes scouting greens, tees, fairways, roughs, ornamental plantings, and trees. Greens and tees generally require the greatest amount of attention and are monitored daily or every other day. Remaining areas are monitored less frequently, usually weekly. Monitoring frequency may require adjustment depending on climatic conditions and reports of nearby pest problems.

Greens and tees are scouted by simply walking around the area to observe insect and disease activity as well as other pest and noninfectious symptoms. Fairways and roughs are usually scouted from a golf cart or utility vehicle. This allows the scout closer examination if symptoms are observed.

The superintendent should provide guidance to the scout for traditional pest problem areas or hot spots. This allows the scout to concentrate their efforts in these areas when conditions favor those pests. In order to minimize play disruptions and to better recognize specific pest damage such as disease symptoms and nocturnal insect feeding, early morning scouting is suggested.

(3) Maintenance of careful records to measure the effectiveness of the IPM strategies. Generally, it is important to stress to club members that elimination of pests is not ecologically or economically desirable. However, if necessary, the decision to apply a pesticide will be supported by maintaining careful records should it be questioned by regulatory officials or the

public. In addition, an IPM program will constantly evolve as new control strategies, monitoring techniques, and threshold information become available.

Tactics involved with these IPM control strategies can be divided into cultural controls, chemical control, and biological control strategies. All are equally important in implementing a successful IPM program.

CULTURAL CONTROLS

The following tactics are integral features of a good cultural control strategy for pest management:

Host-Plant Resistance

Until recently, turfgrass breeders have been concerned primarily with improving the appearance and playability characteristics of grasses, including texture, density, and growth habit. Breeding for pest resistance has been a secondary concern. However, in other crops, one of the oldest means of pest control has been through careful breeding and selection of resistant or tolerant plants. If pest-resistant cultivars are available, they should be used on golf courses whenever it is practical. When building or renovating a course, cultivar selection is one of the first things to consider and can be the most important means of pest control once the grass is established. As turfgrass breeders put more emphasis on breeding for pest resistance, and as genetic engineering technology evolves for use in turf breeding, there should be more resistant or tolerant cultivars to select from in the future.

Pest-Free Propagating Material

Another easy but often overlooked means of preventing pest establishment in turf is the use of planting materials (seed or vegetative sprigs/sod) that are pest-free. Each state has a certification program to provide pest-free propagation material. Each bag of certified seed must provide information on purity and germination percentages. In addition, a weed seed listing must be provided, and no noxious weed seeds are allowed. If vegetative material such as sprigs or sod is being planted, inspect the turf for weeds, fire ants, and other pests. If applicable, ask to see results from a nematode assay of the planting material before purchasing. Remember, these steps will help to prevent or reduce pest problems during and after establishment.

Site Preparation

Proper preparation of a planting site is an important step in pest management, primarily due to its effects

196

on the health and viability of the turf. For turf managers this includes planning and constructing highly utilized areas, such as putting greens and tees, with irrigation and drainage systems capable of providing precise water management. If the soil remains saturated for too long, diseases and soil compaction eventually occur. Putting greens should provide adequate surface and subsurface drainage so play can resume quickly after heavy rain without soil compaction occurring. However, greens also need to be able to maintain adequate amounts of water and nutrients to promote healthy turf growth.

Another important aspect of site preparation, especially greens and tees, is the planting of the turf in pest-free soil. Soil fumigation controls soilborne pests such as nematodes, weeds, and plant pathogens. Likewise, pests can be added after turf establishment via infested topdressing. Do not accept unclean substitutes just because of price. Other considerations during site preparation include providing adequate sunlight to the turf area, proper ventilation around golf greens, proper drainage in fairways, greens large enough so traffic can be distributed over the surface, and an adequate supply of good quality irrigation water.

Basic Agronomic Practices
Probably the best defense against pest invasion is providing a dense, healthy, competitive turf. This is achieved after establishment by providing cultural practices which favor turf growth over pest occurrence. Important cultural practices in IPM programs include proper irrigation, fertilization, mowing, aerification, verticutting, and topdressing. Prolonged use of incorrect cultural practices and lack of understanding concerning interrelationships of these practices weakens the turf, encourages pest activity or invasion, and quite often encourages excessive thatch development. Thatch harbors many insects and disease pathogens. It also binds pesticides and reduces the efficiency of an irrigation program.

BIOLOGICAL CONTROLS
Pests in their native areas are usually regulated by predators and parasites that help keep populations at a constant level. Problems occur when pests, but not their natural enemies, are introduced into new areas and the pest populations increase unchecked. Biological pest control involves the use of natural enemies to reduce pest populations, indigenous and introduced, to aesthetically acceptable levels. Criteria for a successful biological control agent include: (1) not harmful to desirable plants and/or other nontarget organisms; (2) ability to reproduce quickly enough to prevent the pest from building to

major infestation levels; (3) ability to survive and maintain a population equilibrium between itself and the pest; (4) adapted to the environment of the host; and (5) free of its own predators or pathogens.

One inherent principle concerning classic biological pest control, where the agent is introduced only once or on a limited basis for permanent establishment, is accepting that a minimum level of the target pest will always be present. This low pest population is necessary for the biological control agent to have a continual food source after the target pest has been reduced to an acceptable level. Thus, complete elimination of the pest is not feasible when integrating biological control measures into the overall pest management scheme. Golfers must be educated to this fact and be willing to accept minor levels of pest pressure.

Various success stories have occurred using biological control agents involving parasites, predators, or diseases to control another organism. However, only a few examples of biological control measures are currently being used in commercial turf production. *Bacillus popilliae*, a bacterium that causes the milky spore disease, has been used with variable success in the control of Japanese beetle grubs. The white amur carp, a fish native to southeast Asia, has recently been introduced for submerged aquatic weed control in golf course ponds. Other potential agents for biological control of turf pests include endophytic fungi for insect control, a bacterium (*Xanthomonas* spp.) for annual bluegrass control, various rust fungi (*Puccinia* spp.) for nutsedge control, and several predacious nematodes for mole crickets and possible parasitic fungi and bacteria for turf nematode control. Research on antagonistic fungi and bacteria for biocontrol of diseases also shows promise.

Biological control agents are complex, not totally effective, and not always predictable. The concept of biological control has been so widely publicized that the general public views it as a viable and readily available alternative to all pesticides. Unfortunately, this is not the case, but this area is currently receiving much needed attention and hopefully will provide additional control agents in the future. The public must be informed that biological controls are not the answer to all pest problems, but may be a useful component of a good IPM program.

CHEMICAL CONTROLS
Not all pest problems can be solved by manipulating cultural practices in the plant environment or by the use of biological control agents. In these cases, pesticides become the second or third line of defense. In the IPM scheme, indiscriminate spraying

is eliminated and only judicious use of pesticides is employed, minimizing damage to biological control agents and the environment. This requires knowledge of the ecological interrelationships between the pest, the host plant, and the biological control agent. Judicious pesticide use involves making management decisions about the following:

(1) The pest must be properly identified and monitored with reliable techniques to establish aesthetic thresholds. A determination then must be made on when or whether further action is necessary. These threshold levels have been referred to in other IPM programs as economic, damage, or action thresholds. However, in golf course turf management, economic and related threshold level terms mean little since an acceptable aesthetic level, not crop yield, is the ultimate goal. An aesthetic threshold level deals with the amount of visual damage a particular turf area can withstand before action is required. Obviously, highly maintained areas such as putting greens have a low aesthetic threshold level before action is warranted, while areas such as roughs can withstand higher aesthetic threshold levels or more damage before action is needed. These threshold levels also vary with the expectations of the particular golf course, financial resources available, and available alternative control measures.

(2) The best control of many insects and weeds occurs at a particular stage in its life-cycle, which is usually during the early stages of development. For example, mole crickets are most susceptible to chemical control when they are small, usually during the months of May or June. Chemical applications at other times are less effective. This principle is more difficult to apply with plant pathogens and the diseases they cause. For some diseases, such as the root rot diseases caused by *Gaeumannomyces*, *Leptosphaeria,* and *Magnaporthe,* the turf must be treated months in advance of symptom development, usually based on historical experience with the disease at that site. Other diseases, such as dollar spot or Pythium blight, are not treated until the initial symptoms have developed or until weather conditions are conducive for disease development.

(3) If use of a pesticide is necessary, select the one that is most effective but least toxic to nontarget organisms or least persistent in the environment, whichever is more important in that location. Read the label completely and thoroughly. Spot treat, if possible, instead of applying "blanket" or "wall-to-wall" treatments. This requires effective scouting techniques and proper recording or mapping of pest outbreaks.

STARTING AN IPM PROGRAM

Pest problems are going to occur on a golf course, and even the best management program cannot guarantee that pest damage will not occur. The very nature of managing golf course turf predisposes it to stress since the turf is often maintained at its very edge of existence. For example, putting greens are often mowed below the height recommendation of the turf species in use, or turf species are used because of play characteristics even though they are not well adapted to the local environment. Each golfing facility differs and will require an IPM program tailored to its interest, level of expectation, and available budget. However, all IPM programs involve the same basic scheme. The following steps have proven successful in developing IPM programs and should provide a good starting point for golf course superintendents who are innovative enough to try such an approach.

(1) Define the role and responsibility of all people who will be involved in the IPM program. This includes establishing good communication between club officials, players, and maintenance staff. Emphasize and explain to these individuals exactly what will be involved and why. Do not lead anyone to expect perfection — either on the course or in the IPM program — as there will probably be as many problems as successes, especially during the development stages of the program.

Scouts who are conscientious and trained to recognize turf pest problems provide the base of a successful monitoring program. The superintendent will probably want to begin as the primary scout until a feel for IPM strategies is attained. Once this occurs, the responsibility might be delegated to an assistant. However, it should be emphasized that **all** employees should play an important role in recognizing pests and/or damage symptoms. The spray technician should also be familiar with identifying pests and understanding their life cycles.

(2) Determine management objectives for specific areas of the course and correct all practices which favor pest development or place undue stress on the turf. Obviously, putting greens and tees require a priority on pest control compared to roughs. Before implementing the IPM program, inspect and map each site on the golf

course. This will provide the foundation on which all management decisions can be based. For each hole, information that should be obtained includes: (1) turf species for each specific area on the hole, (2) mowing height and schedule, (3) irrigation amount and frequency, (4) soil drainage, (5) complete soil analysis, (6) fertilizer program, (7) traffic patterns, and (8) shade and air circulation concerns. A field history form similar to Table 1 can be used to record such foundation data.

Be prepared to improve the existing problems that weaken the turf, such as a poor irrigation or drainage system or severe tree shade and root effects. Solicit funds for these improvements before the IPM program is implemented. Otherwise, the potential success of the IPM program will be greatly reduced. Again, communicate openly with club officials and golfers.

(3) If at all possible, install a weather monitoring system. This will provide detailed, localized data on rainfall, soil temperatures, humidity and sunlight indexes, and evapotranspiration rates. Climatic conditions usually play the most important role in specific turf growth patterns and pest problems.

(4) Establish aesthetic or action threshold levels and begin monitoring and recording pest levels. A form similar to Table 2 may be used to record such levels. Threshold levels will vary according to location of the course, the specific pest being scouted, use of the turf area, expectations of the golfers, and budget constraints of the course. In addition to recording pest levels on forms, pinpoint pest problem areas on a map for each hole (Figure 1). For example, mole crickets usually lay eggs in the same locations each year. This may allow for spot treatment rather than a blanket pesticide application. Over time, these maps can indicate where pest problems annually occur and possibly allow superintendents to correct management or environmental variables influencing this occurrence. These maps also allow new crew members a visual aid in examining and treating problem areas. Computer programs are available that allow one to "draw" or "paint" such maps.

(5) Use pesticides correctly and only when threshold limits are reached. One of the goals of IPM is intelligent and prudent pesticide use. Pesticide use may not necessarily be reduced by an IPM program, although it often is, but it will allow for more efficient and effective use of pesticides. For example, by monitoring pest development, the pesticide can be used during the most susceptible stage of its life cycle. Utilize the safest, most effective pesticide available for the particular pest. Spot treat whenever possible.

(6) Evaluate the results of the cultural modifications and pesticide treatments by periodically monitoring the site environment and pest populations. Keep written records of site pest management objectives, monitoring methods, data collected, actions taken, and the results obtained. This will provide additional information for course officials and golfers who do not necessarily understand the IPM program but do understand desirable results. It will also aid in demonstrating that golf course superintendents are striving to reduce the inputs in maintaining the course and to obtain an ecological balan.ce between man and nature.

CONCLUSIONS

Golf course pest management strategies that do not rely solely on pesticides are in their infancy but are being developed and utilized. It is not an easy task, and it requires a commitment from **everyone** associated with the golf course. Strategies necessary for a successful IPM program have been outlined and should provide an initial starting point for golf course superintendents. It should be emphasized that IPM methods will take time to implement and develop to their full potential. An early goal for an IPM program would be to reduce pesticide applications 40 to 50% without sacrificing turf quality, even though initially this may not be attainable. Open communication between all maintenance staff, club officials, players, and the superintendent are necessary for such a program to be understood and eventually succeed.

Table 1 **Field history report form used for golf courses**

TURF IPM FIELD HISTORY REPORT FORM

Club ———— Superintendent ———— Date ————

Hole Number ———— Scout ————

Phone Number ————

Phone Number ————

Site	Turf Species	Area	Mowing Schedule	Soil Analysis			Soil Drainage	Fertilization				Frequency	Irrigation Scheduling
				pH	P	K		Amount (N/1000 sq.ft.)					
								Spring	Summer	Fall	Winter		
Green													
Tee													
Fairway													
Rough													
Driving Range													
Nursery Green													
Practice Green													

Comments on specific topics such as shade, overseeding blend, nitrogen carrier, top-dressing mix, weather, irrigation salinity levels, etc:

———————————————————————————————

———————————————————————————————

———————————————————————————————

Table 2 Field infestation report form used for golf courses

TURF IPM FIELD INFESTATION REPORT FORM

Club _____   Superintendent _____   Phone Number _____   Date _____

Hole Number _____   Scout _____   Phone Number _____

Site (turf species)	Mowing Height	Soil Moisture	Weeds Species	Weeds No. or %	Diseases Species	Diseases No. or %	Insects Species	Insects No.	Nematodes Species	Nematodes No.
Green										
Tee										
Fairway										
Rough										

Notes:

Weeds
1. Goosegrass
2. Crabgrass
3. Dallisgrass/Thin Paspalum
4. Torpedograss
5. Broadleaves
6. Nutsedge (Yellow, Globe, Purple, Annual, Kyllinga)
7. Poa annua
8. Kikuyugrass
9. Other

Diseases
1. Dollar Spot
2. Leaf Spot
3. Pythium Blight
4. Pythium Root Rot
5. Fairy Ring
6. Brown Patch (R. solani)
7. Rhizoctonia Leaf and Sheath Blight (R. zeae)
8. Bermudagrass Decline
9. Snow mold
10. Red thread
11. Algae/Moss
12. Other

Insects
1. Mole Crickets
2. Sod Webworms
3. Armyworms
4. Cutworms
5. White Grubs
6. Fire Ants
7. Mites
8. Grass Scales
9. Other

Nematodes
1. Sting
2. Lance
3. Stubby-Root
4. Root-Knot
5. Cyst
6. Ring
7. Spiral
8. Sheath
9. Other

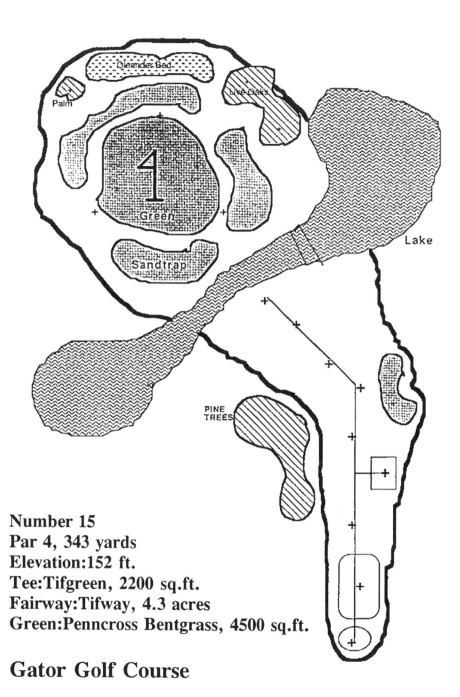

Number 15
Par 4, 343 yards
Elevation:152 ft.
Tee:Tifgreen, 2200 sq.ft.
Fairway:Tifway, 4.3 acres
Green:Penncross Bentgrass, 4500 sq.ft.

Gator Golf Course

Figure 1 Computerized drawn map of a golf hole enabling a scout to identify and mark pest problem areas.

REFERENCES

1. **Bowen, W. R., Gibeault, V. A., Ohr, H. D., and Thomason, I. J.,** IPM, an alternate approach to solving pest problems, *Golf Course Manage.*, 47, 16, 1979.

2. **Bruneau, A. H., Watkins, J. E., and Brandenburg, R. L.,** Integrated pest management, in *Turfgrass. Agronomy Monograph 32*, Waddington, D. V., Carrow, R. N., and Shearman, R. C., Eds., ASA, CSSA, SSSA, Madison, WI, 1992, 653.

3. **Hellman, J. L., Davidson, J. A., and Holmes, J.,** Urban ornamental and turfgrass integrated pest management in Maryland, in *Advances in Turfgrass Entomology*, Niemczyk, H. D. and Joyner, B. J., Eds., Chemlawn Corp., Columbus, OH, 1982, 31.

4. **Eskelson, D.,** Implementing IPM strategies, *Golf Course Manage.*, 60, 68, 1992.

5. **Leslie, A. R.,** An IPM program for turf, *Grounds Maint.*, 26, 84, 1991.

6. **McCarty, L. B., Robert, D. W., Miller, L. C., and Brittain, J. A.,** TIPS: an integrated plant management project for turfgrass managers, *J. Agron. Ed.*, 19, 155, 1991.

7. **Petraitis, J.,** Controlling the Japanese beetle without spraying, *Golf Course Manage.*, 49, 48, 1981.

8. **Pimentel, D.,** Perspectives of integrated pest management, *Crop Prot.,* 1, 5, 1982.

9. **Roberts, D. A.,** *Fundamentals of Plant-Pest Control*, W.H. Freeman and Co., San Francisco, 1978, 242 pp.

10. **Short, D. E., Reinert, J. A., and Atilano, R. A.,** Integrated pest management for urban turfgrass culture — Florida, in *Advances in Turfgrass Entomology*, Niemczyk, H. D. and Joyner, B. J., Eds., Chemlawn Corp., Columbus, OH, 1982, 25.

11. **Skorulski, J. E.,** Monitoring for improved golf course pest management results, *USGA Green Sect. Rec.*, 29, 1, 1991.

Blemishes on the Perfect Landscape:
What Can Go Wrong and Why

Distressed tree — burlap and wire not
removed from root ball during planting

Microscopic examination of grass
for disease diagnosis

Pheromone trap for monitoring pest insects
on trees (photo courtesy of *Grounds
Maintenance* magazine)

White grubs exposed when turf is pulled up

Fire ants, a major landscape pest in the South
(photo courtesy of Beverly Sparks)

Chapter 19

Survival of Trees in Metropolitan Areas

Donald A. Rakow, *Department of Floriculture and Ornamental Horticulture, Cornell University, Ithaca, NY*

CONTENTS

The Role of Trees in the Urban Environment 205
Belowground Site Characteristics 206
Restrictions in Rooting Space 208
Aboveground Site Characteristics 209
Site Limitations 209
Physical Damage to Trees 210
References 211

THE ROLE OF TREES IN THE URBAN ENVIRONMENT

To one degree or another, everyone appreciates the presence of trees in metropolitan settings. On a casual level, people know that trees will shade them on hot, sunny days, and that a well-placed row of trees breaks up the stark monotony of modern office buildings. Those with a deeper interest in urban trees may also be aware of their roles in environmental enhancement: air purification and oxygen regeneration, groundwater recharge, and reduction of stormwater runoff.[2]

Few people consider, however, that cities and suburbs are largely foreign habitats for trees, and that most trees are not well adapted to such sites. A stressed tree which will respond with renewed vigor to a change in its environment, can be assumed to be ill-adapted to its growing conditions.[23] While individual species may tolerate one or more of the stresses common to urban areas, there is no single, super species that adapts to all possible stresses and meets requirements for ornamental features.

In natural settings, trees commonly grow in groups or clusters, with mutual shading from tree to tree. But the typical way in which trees are planted in urban areas — in single rows with 30 to 40 ft between trees — leaves them exposed to reflected and reradiated heat from building facades, car tops, concrete, and asphalt.[2] This heat stress can combine with drought conditions to make it difficult for all but the hardiest of species to survive.

Trees that have been weakened by environmental stresses can be attacked by pathogens that are normally not aggressive. For example, severely defoliated plants seem to be more susceptible to attack by root rots, cankers, and dieback pathogens. In a 1981 study, Schoenweiss demonstrated that there are threshold levels of drought stress, freezing injury, and length of defoliation beyond which plants are susceptible to colonization by certain canker fungi. In each case, stem cankers stopped growing when the stress was removed. But, if the stress progressed beyond a second level of magnitude, i.e., longer period of drought stress or defoliation, or more severely cold temperatures, the progress of the disease became irreversible and the trees declined.[20] Thus, abiotic stresses can cause previously healthy trees to be attacked by pests that would otherwise not be sufficiently virulent (efficient) to colonize. Because of the prevalence of abiotic stresses in metropolitan locales, rates of pest occurrence tend to be much higher than in rural or park settings, leading to decreased life spans of urban trees.

Evidence of this can be found in cities throughout the U.S. and Canada. Of the approximately 1000 trees planted in New York City each spring, roughly half die within 1 year.[3] A study in Boston also found that the average life of trees planted in sidewalk cutouts is less than 10 years,[10] while a second study reported that the average life span of a downtown city tree is 7 years.[2] Because each tree represents an investment of approximately $300 plus associated costs of planting and post-plant care, this premature death translates into both an aesthetic and an economic loss to the cities.

As noted previously, tree species differ in their tolerance to abiotic stresses. Thus, tree selection should start with a thorough analysis of the site, rather than with the desired ornamental traits of particular species. The prevailing environmental conditions of sun/shade patterns, winds, soil limitations, and temperature extremes, as well as likely stresses, should be noted. Based in Schoenweiss' model, both the types and severity of the stresses should be determined. If a desired species is known to tolerate all expected conditions but one, then a modification of the site can be considered. For example, a desirable tree species may be adapted to expected extremes of winter cold and summer heat as well as to periodic droughts, but may be intolerant of high soluble salt levels in the soil. A concrete lip or form can be constructed around the base of the planting hole to prevent salt-laden run-off from flowing into the soil mass.

Modifying conditions and matching species to identified conditions, should be applied on as individualized a basis as possible. Avoid generalizing about entire blocks or even from one side of the street to the other. Solar radiation, average wind speeds, and soil physical features are just a few of the factors than can vary from one planting site to the next.

BELOWGROUND SITE CHARACTERISTICS

The methodical classification of soils based on the relative percentages of sand, silt, and clay particles has little application to most urban and disturbed suburban soils. Human activity, rather than the natural weathering of parent material, is the predominant factor in urban soil formation.[7] Examples of human activities that modify urban soils are the scraping away of topsoil layers, contaminating soil with concrete, metal, glass and other debris, filling with indigenous and/or foreign soil, and compaction by equipment, vehicles, and pedestrians.[7]

Craul[7] has indicated that there are seven characteristics that are common to urban soils. Each of these will be described in detail.

(1a) **Vertical variability:** unlike natural soil profiles, in which the texture and structure tend to change gradually from one profile to the next, urban soils exhibit abrupt changes from one layer to the one below. In grading a housing site or tall building, for example, the topsoil may be removed. Once the construction is complete, a foreign top layer may be added that bears no relation to what now subtends it. The interface between layers may occur at a depth from 2.5 to 14 in. below the surface.[8] Each layer may have very different characteristics of texture, structure, and fertility. As a result, it is not uncommon to have a very shallow zone in which tree roots can spread, under which is a compacted sub-layer. If shallow soil is combined with restricted rooting space (described in a later section), trees may lack sufficient root stability to remain standing during storms.

(1b) **Spatial variability:** from one planting pit to the next, the organic content of the soil, the concentration of dissolved salts, the degree of compaction, or the presence of foreign materials may all vary widely. Because of this potential for drastic contrast, soil tests must be conducted at every site prior to planting.[7]

(2) **Structural modification and compaction:** The soil *texture*, or relative percentages of sand, silt, and clay, cannot be changed without additions or subtractions to the soil itself. But the *structure* or relative degree of aggregation of the soil particles, is heavily influenced by human activity.

In natural settings, the annual drop of leaves and twigs provides a source of organic matter for the soil. The decomposed organic matter serves as a "glue" to bind individual soil particles and to increase aggregation. In urban settings, the presence of sidewalks and tree grates, as well as the solitary setting of trees, results in a greatly reduced input of raw organic material. Consequently, aggregation is not encouraged.

Once a soil has become well aggregated, it is very susceptible to compaction. Foot and vehicular traffic exert a force on the surface that is transferred into the soil, resulting in the breaking apart of soil aggregates and elimination of pore space. Such compactibility is most extreme for soils with high sand and/or silt or very low organic matter content; for soils under moist but not saturated conditions; and for soils with great variability of particle sizes. Compaction can occur very rapidly after a single extreme event (such as a bulldozer driving over the site), or can occur gradually as a result of chronic surface compression from pedestrians. Even vehicular vibrations can compact a soil.

Compacted soils have greatly elevated bulk densities relative to undisturbed soils of the same texture. Rooting is highly restricted at bulk densities above 1.7 g/cc, and urban soils frequently have bulk densities of 1.85 or above. As soils become more compacted, other properties, such as water infiltration and permeability, water-holding capacity, and oxygen diffusion, are also negatively affected. Trees grown in heavily compacted soils may decline or become more susceptible to attack by canker or root-rot pathogens.

(3) **Surface crusting.** Bare soil around the base of urban or suburban trees often becomes hard and crusted. This can be caused by physical compaction, or by the lack of organic matter or living roots in the top 4 in. of the soil. Also, as raindrops splash onto bare soil, they tend to break apart surface aggregates and force fine particles downward, filling more spaces and increasing compaction.[7]

Because individual planting pits are commonly surrounded by impervious sidewalks and asphalt, the surface of the pit may be the only point of ingress for water and oxygen.

Due to the frequent presence of hydrophobic (water-repelling) substances, it can be very difficult to re-wet soils once they have dried and crusted. The most effective solution may be to cultivate the top few inches, adding compost or other organic matter to the soil, and a surface treatment of wood chips or another mulch.

(4) **Restricted aeration and water drainage.** The combined effect of compacted subsoil and a crusted surface can severely restrict the infiltration and movement of water and oxygen into and through the soil. Both of these processes result in a decrease in the total pore space and particularly in the percent of macropores.

Macropores are essential for soil water drainage to occur under gravitational flow. Once gravitational flow has ceased, and the soil is between field capacity and permanent wilting point, the macropores serve as a network for the movement of oxygen and carbon dioxide. When the macropores are compressed, all of these processes are restricted. If the compression is at or near the surface, the soil becomes inhospitable to root development, even if the soil at a lower depth is well aggregated, and has a high percentage of large pores.[7]

(5) **Modified pH levels.** Urban soils tend to exhibit soil pH levels significantly higher than similar soil in rural areas. Cities as geographically diverse as Syracuse, New York, Philadelphia, Pennsylvania, and Berlin, Germany have been tested as having mean pH reactions

of 8.0, 7.6, and 8.0.[7] And in Ithaca, New York, where native soils are generally quite acidic, a survey of street tree pits revealed pH levels on the average of 8.1.[2]

There are three factors that are primarily responsible for this trend toward elevated pH levels. The first is the application of calcium chloride or sodium chloride as a sidewalk or roadway de-icing salt. Second is the release of concrete and limestone building rubble into the soil at the time of construction or demolition, or as a result of weathering. The third is the use of irrigation water with a naturally high pH.[7] The absence of annual depositions of organic matter which, as they decompose have an acidifying effect on soils, further drives this alkalinization of soils.

While pH levels near neutral (7.0) favor the availability of most essential elements, alkaline pH levels, such as have been reported in many cities, cause iron, manganese, and zinc to be tied up in the soil. These essential elements are necessary for chlorophyll production and thus for photosynthesis. When they are not present in sufficient concentration in the leaves, the tree's food-making capacity is reduced, and the leaves appear yellowish, particularly between the veins.[2]

(6) **Interrupted nutrient cycling.** By human standards, a natural forest ecosystem is a constantly messy affair: twigs and dead branches break from trees, deciduous leaves are deposited on the forest floor each autumn, and animals die and accumulate both on and under the forest soil.

As all of this organic material decomposes, it provides the soil with a regular supply of essential elements (especially nitrogen, phosphorus, and sulfur), as well as energy for soil-inhabiting organisms. Soil macrofauna, such as earthworms and beetles, are essential to aerate the soil and to contribute to the organic fraction. Microfauna, including aerobic bacterial species, are responsible for the conversion of the nitrogen in the decomposed organic material into a form that can be taken up by roots.[7]

In most urban, and many suburban sites, this regular cycling of essential elements is severely reduced. Plant debris are removed as soon as they fall, and surface crusting and/or subsoil compaction causes oxygen concentrations to fall below the level needed to sustain needed microorganisms.

Frequently, the result is greatly depressed levels of available nitrogen and phosphorus and, if the pH is high enough, iron, zinc, and manganese. The need for supplemental fertilization is thus increased. If the nutrient-deprived tree is not fertilized, the rate of growth or overall vigor of the tree will be affected.

(7a) **Presence of foreign materials.** A great variety of foreign materials, both solid and liquid, can be found in urban soils. Among the more common solid waste products are masonry, bricks, wood and paper, glass, metal, asphalt, and organic garbage. At their most benign, such materials can impede root penetration or movement, and disrupt the flow of water or nutrients. As certain organic solids decompose, anaerobic conditions in the soil can lead to the development of toxic materials including gases, alcohols, and organic acids that can damage tree roots and soil microflora.[7]

(7b) **High soluble salt levels.** The use of sodium chloride as a de-icing salt on sidewalks and roadways, though essential to ensure public safety, can do tremendous damage to street trees. As salt-laden snow from sidewalk de-icing operations melts, the solution tends to run into street tree pits. Once it infiltrates into the soil, the sodium and chloride ions dissociate. Chloride ions are readily taken up by the roots, where they collect in buds or leaves and result in marginal leaf necrosis or wilt. Analysis of chloride concentrations of *Tilia cordata* "Greenspire" trees in New York City revealed chloride levels of 1.8% of leaf dry weight, well above the threshold for damage in such trees.[3] The sodium which remains in the soil can have a detrimental effect on soil structure. Just as decomposed organic matter tends to aggregate soil particles, high concentrations of sodium ions cause a de-aggregation or loss of structure. This condition leaves the soil in a condition which is much more susceptible to being compacted.[13]

If trees in heavily salted areas do not receive sufficient rainfall or irrigation in the spring, the accumulated salts are not leached and the problem may persist. In Montreal, for example, spring soluble salts levels of 2000 ppm have been detected in certain street tree pits, well above the acceptable level of 500 ppm.[6]

There are other ways in which salt content in the soil is raised to dangerous levels. The affection that dogs have for tree trunks causes a buildup of urine, and thus sodium and potassium in the soil.[4] People frequently spill spent or gray water into tree pits as a means of watering the trees. But if this water contains nonhousehold concentrations of bleach or other toxic chemicals, considerable damage can result.

To reduce problems associated with road salts, many municipalities mix sand, light gravel, or cinders with the salt to reduce the concentration of sodium chloride. Other important measures are to heavily irrigate street tree pits in early spring to encourage leaching and to avoid piling salt-laden snow around the base of trees. Heavy irrigation does require closer essential element management.

(8) **Elevated soil temperatures.** The likelihood of an atmospheric heat island in urban areas, in which average air temperatures are measurably higher than in surrounding parklands, is the primary reason for elevated soil temperatures in such locales. This heat is especially likely to be transferred into the soil in sites in which the soil is bare and exposed.[7]

Elevated soil temperatures can be beneficial by accelerating the rate of root growth and organic matter decomposition. But warmer soils can also result in more rapid evaporative soil water losses and drying out, and in extreme cases can be lethal to roots. For most plants, soil temperatures above 85°F severely restrict or kill roots.[2]

RESTRICTIONS IN ROOTING SPACE

Considerable misunderstanding exists regarding the nature of tree root growth. A commonly held image is that trees develop an enormous carrot-like taproot, and that support roots extend a great distance below the soil surface. In fact, taproots are rare in mature trees, and most tree roots occur in the top 12 to 18 in. of the soil. Nearly 100% are found in the top 3 ft.[17]

Root growth is opportunistic, and occurs whenever the environment is favorable; that is, wherever sufficient oxygen, water, nutrients, and warmth can be found. Roots do **not** grow preferentially toward a source of increased oxygen or water, despite many published claims to the contrary. Tree roots typically spread outward over an area two to three times greater than the canopy. Such a wide-spreading root system allows the tree to exploit soil resources, and provides a firm anchoring for the trunk and limbs.

In urban sites, tree root development is often extremely restricted. Among the factors that can cause such restriction are: a textural or structural interface between planting hole soil and the soil surrounding the hole; compaction of the soil in the planting hole itself, resulting in depressed levels of oxygen movement; the presence of underground utilities, building foundations, or other human-made structures or systems; the establishment of the tree in an aboveground or partially aboveground container;[9] or the use of a vertical root barrier below the soil. Any of these scenarios can restrict both the area covered by the roots as well as the volume of the total root system. Since roots are essential for the uptake of water and essential elements and are the sites for synthesis of certain hormones, any factor that limits overall root development will limit the extent of shoot growth.[17]

The containerizing of trees can be especially damaging to tree root systems. Most free standing containers are both small in total volume and shallow in depth. Even with adequate drainage holes, a container can develop a nondraining zone of completely water-filled pores at the base of the soil. Roots cannot grow in this saturated zone, which further restricts total rooting volume.[18]

In streetside plantings, an interface can develop between the planting site and the surrounding soil as a result of glazing of the sides of the hole by mechanical planting equipment, or because of a distinct textural contrast between the soil types. If this interface is severe enough, it can act like a wall, restricting the movement of roots, water, or oxygen beyond its confines. More typically in urban street tree plantings, lack of lateral water movement results in periods of spring or fall flooding, interspersed with periodic summer droughts. In one study conducted in New York City, the presence of saturated planting holes was the most important environmental factor leading to decline in street trees.[4] But in a second study,[24] high atmospheric moisture demand, which can lead to drought stress, was found to be common, though infrequent, in trees in Manhattan. Still others have reported that water stress, particularly of newly planted trees, is among the most important factors in tree decline.[10]

Vertical root barriers, which direct roots away from growth under sidewalks, curbs, sewer lines, or foundations, can also reduce the total amount of rooting. This effect is less severe if the barrier is placed only on one side of the root system and if soil conditions are ideal for expansion on all other sides. Chemical barriers, which are made up of a geotextile fabric that has regularly spaced nodules impregnated with a slow-release herbicide, cause the redirection of root tips but still allow for some diffusion of water and oxygen across the fabric.

ABOVEGROUND SITE CHARACTERISTICS

Urban microclimates can differ greatly from contiguous rural or parklike settings in terms of radiation, humidity, and temperature levels,[16] due to the density and size of buildings and the intense development of infrastructural elements (roadways, plazas, sidewalks, curbing). Reflected and reradiated heat from cartops, sidewalks, asphalt, and building windows and surfaces can raise temperatures on leaf surfaces several degrees over the ambient atmosphere. In one large scale urban ecophysiological study, leaf temperatures of trees growing on a north-south running boulevard were contrasted with air temperatures in nearby Central Park. On one particular August day, average temperatures in the lower tree canopies were 106°F compared to the park maximum of 90°F.[3,24] The pattern of sunlight would lower photosynthetic rates, thus reducing the tree's ability to manufacture carbohydrates and maintain a vigorous state.

The significance of air pollution as an abiotic stress factor varies greatly with location. Those cities which are recognized as having the worst overall air pollution (Los Angeles, Mexico City, Tokyo, cities in eastern Europe, etc.) are also those most likely to have evidence of foliar damage to street trees from one or more of the most pervasive pollutants: ozone, fluorides, sulfur dioxide, or possibly nitric oxide or PAN. Of these pollutants, sulfur dioxide and the fluorides are generally point source contaminants. That is, they are derived from industrial processes at specific locations. Thus, modifying industrial or filtration processes could reduce the concentrations of these materials.

In contrast, ozone, PAN, and other oxidants are general atmospheric pollutants, formed by photochemical reactions between hydrocarbons and nitrogen oxides in the presence of sunlight. These materials are more difficult to control and may, in part, have precursors that are released by trees themselves.[11]

Symptoms of acute pollution damage to broadleaves is quite variable, but more typical symptoms include interveinal necrotic areas, marginal or tip necrosis, and stippling of the upper surface.[21] If foliar symptoms can be matched to one of the specific atmospheric pollutants, then replacement with a tolerant species may be the only viable solution. See Reference 21 for recommended species.

SITE LIMITATIONS

We all know that urban areas are crowded, and that space is often at a premium. Space restrictions commonly affect street trees which are unable to develop their full crown potential. The average distance between a street tree planting pit and a building is only 6 ft. This is barely enough crown space for broad spreading canopies. The tight fit can be further compromised when building awnings are added, or temporary scaffolding is erected. Frequently, tree branches are bent or even broken in the battle for aboveground space between the building and trees.

Limbs that venture in the other direction, over the street, may have an equally difficult time. Drivers of vans and delivery trucks typically show little awareness of trees and park their vehicles without regard for branches they may be breaking. Trees with broken limbs that are not remedially pruned may develop an increase in internal decay. By far the most serious problem that trees face with aboveground space, however, is with overhead power lines. Because of the potential danger of an interruption of utility service, trees must regularly be pruned to maintain clearance. If trees are pruned properly by the principles of crown reduction advanced by the National Arbortists Association and the International Society of Arboriculture, they can remain healthy and strong. But in some sections of the country, utility clearance crews may still practice topping or heading, the cutting back of major limbs to stubs, without regard for their location. In response to a topping, a tree produces masses of thin watersprouts. These are poorly attached to the limbs from which they emerge, they shade the interior and destroy the natural shape of the tree. Topping is a practice that must be stopped to ensure healthy trees in the future.

PHYSICAL DAMAGE TO TREES

Limb breakage from trucks is one form of casual damage that regularly occurs to trees in urban sites. Several other forms of casual human interference involve the integrity of the tree trunk and particularly the bark. Automobile bumpers banging into the base of tree trunks can knock off large patches of bark. Bicycle chains or dog leashes wrapped around the trunk can gradually abrade away a strip of bark. And mowers and string trimmers cause considerable amounts of damage as they come in contact with thin-barked trees.

The most effective way of reducing damage to the trunk is to develop a turf-free mulched ring around the base of the tree, at least 2 ft in radius. Where that is not possible, a plastic tree guard can be placed around the trunk. Other forms of casual tree damage are initiated under the guise of protecting or supporting the tree. In a Boston study, the staking of trees was found to cause more physical damage than either vehicles or vandalism.[10] Since that study was conducted, Harris has shown that a staked tree, when compared to one that can move

freely: grows taller; produces a decreased or even reverse stem taper; develops a smaller root system; offers more wind resistance; and is more subject to rubbing or girdling from stakes and ties. The combined impact of these effects makes a tree less able to stand after supports are removed, or more subject to injury if they are left on.[11] On the basis of these findings, it is now recommended that trees only be staked if they are in an especially windy or vandal-prone area, or if the crown is oversized relative to the roots. Trees that are staked should have guy wires left loose enough to allow for trunk sway. The support system should ideally be removed after one full growing season.

Attempts at protecting trunks from sunscald and physical injury have also been shown to be potentially damaging. Litzow and Pellett evaluated a dozen materials for their effectiveness as tree wraps. They found that the most commonly used material, Kraft tree wrap, did not provide the thermal insulation to the trunk needed to reduce the incidence of sunscald. They further reported that excess moisture retained behind the Kraft paper wrap, plastic bubble packing, and plastic tree guards that they evaluated might encourage fungal and bacterial growth, especially if the trunk had been wounded.[15]

Members of the International Society of Arboriculture (ISA), responding to a survey on trunk wrapping practices, indicated that the types of damage that they most frequently encounter as a result of this practice are: trunk girdling or constriction, insect injury, disease invasion, excess moisture, bark damage, and cambial death. The majority of respondents to the survey felt that the decision to use a trunk protective material should be made on an individual tree basis, considering the tree species and site considerations. Whenever a trunk wrap is used, it should be removed after the prescribed period of time, usually 1 or 2 years.[1]

Vandalism can also be a major contributor to tree loss in urban areas. Reports in both Montreal and Seattle indicate that vandalism can represent 5% of all tree losses.[5,6] Although a tree may not be killed by an act of vandalism, its ornamental value is normally destroyed. To reduce the incidence of vandalism in high-risk areas, Black suggests starting with larger caliper trees, attaching each tree to a single steel reinforcing bar with three covered wire ties, and working with community groups and schools to increase local support for tree plantings.[5] The second of these three suggestions must be exercised with caution — studies have shown that trees that remain tightly secured to supports for more than 1 year do not develop proper taper and are unable to stand independently when supports are removed.

This chapter has focused primarily on the environmental stresses that street trees are likely to encounter in their usually too brief life in metropolitan areas. Minimal emphasis has been placed on remediation of these problems. To improve the survival and health of urban trees, the greatest impact can be made by analyzing sites more carefully,[2,3] selecting species that can tolerate particular stresses,[19,22,23] and redesigning the ways in which trees are planted.[2,6,9,12,18] These concepts are expanded in another chapter.

REFERENCES

1. **Appleton, B. L. and French, S.,** Current attitudes toward and use of protective wraps, paints and devices, *J. Arbor.,* 18(1), 15, 1992.
2. **Bassuk, N.,** Urban trees, *Public Garden,* 6(1), 10, 1991.
3. **Bassuk, N. and Whitlow, T.,** Environmental stress in street trees, *Arbor. J.,* 12, 195, 1988.
4. **Berrang, P., Karnosky, D. F., and Stanton, B. J.,** Environmental factors affecting tree health in New York City, *J. Arbor.,* 11(6), 185, 1985.
5. **Black, M. E.,** Tree vandalism: some solutions, *J. Arbor.,* 4(5), 114, 1978.
6. **Bourque, P.,** Tree management in Montreal, *J. Arbor.,* 11(7), 200, 1985.
7. **Craul, P. J.,** A description of urban soils and their desired characteristics, *J. Arbor.,* 11(11), 330, 1985.
8. **Craul, P. J. and Klein, C. J.,** Characteristics of streetside soils in Syracuse, *N.Y. Proc. Metro. Tree Improvement Alliance (METRIA),* 3, 88, 1980.
9. **Flemer, W., III.,** How to cope with tree stress in urban environments, *Am. Nurseryman,* 156, 42, 1982.
10. **Foster, R. S. and Blaine, J.,** Urban tree survival: trees in the sidewalk, *J. Arbor.,* 4(1), 14, 1978.
11. **Harris, R. W.,** *Arboriculture: Integrated Management of Trees, Shrubs, and Vines,* Prentice-Hall, Englewood Cliffs, NJ, 1991, chap. 20.
12. **Hensley, D. L., Weist, S. C., and Gibbons, F. D.,** What to consider when planting trees in urban areas, *Am. Nurseryman,* 156, 81, 1982.
13. **Hudler, G. W.,** Salt Injury to Roadside Plants, Cornell Information Bulletin #169, New York State College of Agriculture and Life Sciences, Ithaca, NY, 1980.
14. **Kjelgren, R. K. and Clark, J. R.,** Microclimates and tree growth in three urban spaces, *J. Environ. Hortic.,* 10, 139, 1992.
15. **Litzow, M. and Pellett, H.,** Materials for potential use in sunscald protection, *J. Arbor.,* 9(2), 35, 1983.
16. **Oke, R. T., Kalanda, B. D., and Steyn, D. G. S.,** Parameterization of heat storage in urban areas, *Urban Ecol.,* 5, 45, 1980.
17. **Perry, T. O.,** The ecology of tree roots and the practical significance thereof, *J. Arbor.,* 8(8), 197, 1982.
18. **Rakow, D. A.,** Containerized trees in urban environments, *J. Arbor.,* 13(12), 294, 1987.
19. **Rhoads, A. F., Meyer, P. W., and Sanfelippo, R.,** Performance of urban street trees evaluated, *J. Arbor.,* 7(8), 127, 1981.
20. **Schoeneweiss, D. F.,** The role of environmental stress in diseases of woody plants, *Plant Dis.,* 65, 308, 1981.
21. **Sinclair, W. A., Lyon, H. L., and Johnson, W. T.,** *Diseases of Trees and Shrubs,* Cornell University Press, Ithaca, NY, 1987.
22. **Shurtleff, M. C.,** "People-pressure" tree problems, *Grounds Maint.,* 22, 1987.
23. **Steiner, K. C.,** Developing tree varieties for urban soil stresses, *Proc. Metro. Tree Improvement Alliance (METRIA),* 3, 57, 1980.
24. **Whitlow, T. H. and Bassuk, N. L.,** Ecophysiology of urban trees and their management — the North American experience, *HortScience,* 23(3), 542, 1988.

Chapter 20

Integrated Pest Management:
A Seattle Street Tree Case Study

Sharon J. Collman, *Cooperative Extension Liaison, U.S. EPA Region 10, Seattle, WA*

(This chapter was adapted from an article first published in *Forestry on the Frontier*, Proceedings of the 1989 National Convention of the Society of American Foresters [SAF Publication 89-02]. The original paper was presented at the Urban Forestry Technical Session at the 1989 SAF National/Convention held at Spokane, WA on September 24–27, 1989.)

CONTENTS

Abstract 213
Introduction 213
Integrated Pest Management (IPM) 214
The Ash Trees 214
Oystershell Scale 214
Scale Development 215
Management Options 215
Natural Enemies 215
Pesticides 216
Site Considerations 216
IPM as a Process 216
Toward the Future 217
References 217

ABSTRACT

Public concern about pesticide safety and environmental contamination requires that urban foresters move from reacting to apparent pest problems to a proactive mode of examining all aspects of a problem. A study of oystershell scale (*Lepidosaphes ulmi* Linne) which was "causing the decline of ash (*Fraxinus* sp.) cultivars along Seattle's streets" serves as a case study. Had we reacted to the oystershell scale without examining the site, trees, and scales, we would have developed a control program that would have missed the "target" altogether. The majority (59.7%) of the declining trees examined had only light or no scale infestation. Also, there were many trees with heavy scale infestations which looked quite healthy. In Seattle, the scale hatched 2 to 8 weeks later than any of the dates reported in the literature. In checking the original research reports, conflicting results of the effectiveness of dormant "sprays" were found. And finally, there was good evidence that this insect would be a good candidate for parasite or predator augmentation.

INTRODUCTION

The streets of Seattle have been enhanced by an ambitious community-based street tree program. Since 1968, new trees have been planted and established trees have been vigorously protected. The harshness of the urban environment and constraints of design, space, and soil limit the number of tree species and cultivars which can be planted.

To avoid problems associated with extensive plantings of only a few tree species and the potential buildup of insect and disease problems, Seattle wisely has chosen to use a mix of tree species and cultivars. Ash (and its cultivars) is an important component of the diversified plantings. It also is one of the relatively few trees reported to be tough enough to stand the abuse of city life.

In 1984, the late Marvin Black, then arborist for the city of Seattle, requested that I provide an integrated pest management (IPM) plan for oystershell scale attacking ash trees along the city streets. Healthy ash trees had begun to lose vitality and young trees failed to establish or they performed poorly. Dieback of twigs, branches, or large sections of trees was increasing. Concurrently, heavy infestations of oystershell scale, a sucking scale insect, were noticed encrusting the bark of the trunk and branches of unhealthy trees.

It is the process of developing the IPM recommendations that I want to focus on here. Had I accepted the information in the reference texts as valid without verification, I would have developed an IPM strategy that would have been ineffective in our region. In fact, the basic assumption that scale was killing the trees did not hold up once the trees were carefully inspected.

INTEGRATED PEST MANAGEMENT (IPM)

IPM has its origins in large-scale crop production systems and is based on volumes of scientific research, population modeling, and computerized data bases. By considering the information gained from monitoring the beneficial and pest insects, the development of the crop, temperature, capabilities of the staff, and other factors, a blend of strategies can be selected and timed to manage a pest with minimal negative impact on the surrounding environment. For example, one might augment natural controls, add or withdraw planned pesticide applications, change irrigation timing, control frost, etc., in order to manage the pest while keeping crop damage to an acceptable minimum.

For pests of the urban forest, we can borrow and extrapolate from crop or forestry research and draw from the growing research on pests of ornamentals. Instead of a single crop, we have a diverse inventory of native and introduced plant species and cultivars, and numerous environmental conditions (e.g., heat intensified by concrete surroundings, dense shading by buildings, or disturbed and shallow soils). The Olkowskis' definition of IPM as "a decision-making process" is more useful for the urban environment.

THE ASH TREES

The first step in approaching the problem was to examine the trees to determine possible causes for their decline rather than presume that oystershell scale was the cause ("guilt by association"). After considerable trial and error, a rapid method of surveying 36 linear miles of trees was devised. The results suggested that although oystershell scale was abundant on some declining trees, other causes should also be investigated.

There was no question that scales were abundant on dead branches and in some declining trees.

Of all the trees with dieback and thin canopies, only 40.3% had moderate to severe scale infestations. The remaining trees with dieback or thin canopies (59.7% total), had no scales at all or had only a very minor infestation of a few patches or scattered scales. In addition, there were vigorous, apparently healthy trees which supported moderate to severe scale infestations. Thus oystershell scale was not the sole cause of ash decline. In fact, one might equally postulate that the scale infestations were a result of declining ash rather than the cause.

Review of selected references revealed that, as a group, ash are fairly tolerant of diverse cultural conditions, generally preferring good soil and abundant moisture. Yet among the cultivars there are some preferred specific site conditions (e.g., some cultivars will tolerate drier soils and others will tolerate wetter soils with poor drainage). These specific cultural requirements may be a factor in the observed decline of a cultivar. For example, *F. quadrangulata* (blue ash) frequents dry limestone upland soils.[1] On the other hand, *F. pennsylvanica* (Marshall and Summit ash) "is usually found in wetter soils,"[2] though "once established it tolerates drought and sterile soils."[1] McArdle and Santamour stated that *F. excelsior* (Kimberly ash) "has not been popular lately because of a 'virus' or 'decline.'"[3] We wonder if this is not simply a case of latent graft incompatibility."

Seattle soils are generally mineral glacial till with low summer rainfall, and low pH which may not be ideal for some of the ash varieties. It was also evident that the root crown was planted well below the soil line in many instances. (I dug down and found the root crowns to be 7, 8, and 11 in. deep on three trees.) Before declaring the final cause of the ash decline, the root systems and soil/pit environment (moisture, pH, and possible herbicide contamination) should be investigated more thoroughly for other possible contributors to the decline of the trees.

Since a virus-like symptom was also present on the leaves of many trees, samples should be sent for laboratory diagnosis in the spring when the virus particles are more detectable.

OYSTERSHELL SCALE

In order to propose an IPM program for any insect one must understand its life cycle and natural enemies. A review of the research literature was necessary to capture the knowledge of past researchers. Additionally, it was necessary to conduct some field studies to correlate the "written word" from other regions of the U.S. with the realities of the Pacific Northwest.

Most of the reference books on pests of ornamentals gave conflicting reports on hatching dates for the scale and were vague on specific details of its biology. The original research papers in journals were more specific. Even so, the hatching dates and results of control methods were often conflicting. Collectively they provided pieces of a most interesting and fairly complex puzzle. By combining field studies of the scale in Seattle with reports from previous studies elsewhere in the U.S., it was possible to compare the local scale development with results from other areas of the country.

A heavily infected willow (*Salix* sp.) was selected for sampling purposes. Small branches were removed at irregular intervals and examined in the laboratory under a binocular dissecting scope. The results were consistent with other observations and samples taken from ash trees throughout the Seattle area. Each scale was overturned and the number of dead scales, or number of scales in each stage of development, as well as any evidence of parasites or predators, was noted. The same size sample was taken at each interval.

In reporting the results here, it is more important to focus on the implications for an IPM program than it is to focus on the numerical results, especially since this was not designed as a proper scientific study. The key point to remember is that any IPM program must be tied to the local conditions and observations.

SCALE DEVELOPMENT

Scales found encrusting a branch might have been alive or merely remnants of a past, but dead, infestation, so it was necessary to examine the infestation carefully with a hand lens. The newly hatched scales had a bright, white waxy cover (not to be confused with normal lenticels in bark); immature scales were pinkish to cream color; and adult scales were gray to gray-brown.

Hatching Date. Oystershell scale on the willow (this was consistent with activity on the ash throughout the city) began hatching on June 26. (The following year, followup samples again hatched at the end of June.) Yet in the literature, hatching dates varied from mid-March to mid-June, with no report of hatching in late June.

The *Pacific Northwest Insect Control Handbook*[4] recommends that pesticides be applied when scales hatch in mid- to late-May with a repeat application 10 days later. Since the scales actually hatched in late June a pesticide applied by calendar dates would have missed the target completely. The scales still would have been in the overwintering egg stage beneath the protective scale covering in May.

Crawler Stage. The scale crawlers have been reported to settle permanently and begin secreting their first scale covering within 4 hours of hatching. This accounts for the very rapid build up of scales on a patch of trunk or section of branch. It had been presumed there would be a very narrow window if pesticides were to be used, since scales would begin to secrete the waxy scale covering immediately. By observing the scales throughout the summer, it was determined that the white cap stage (that first scale covering) is very porous and that the first solid scale covering did not appear for 2 to 4 weeks. Thus, the "window" for application of any pesticides could be expanded from a few hours to several weeks.

Adult Scales. The adult scales were the largest in size and were brown to ash-gray in color. The adults had to be overturned to verify if they were alive, filled with eggs, or if they were remnant scale coverings from previous years. Live females underneath the scale cover were yellowish in color and the milky white eggs became yellowish prior to hatching. A fluffy white material under the scale indicated that the eggs had hatched, or had been consumed by predator mites.

MANAGEMENT OPTIONS

Although we often respond to pest problems by automatically spraying pesticides, there are many other management choices available. An assessment of the situation must include the life stage, and presence or absence of parasites and predators, and the conditions at the site.

NATURAL ENEMIES

In the literature,[5-9] weather, lady beetles, seven species of birds, a parasitic wasp, and several predatory mites are listed as natural enemies of oystershell scale. Each of these in turn must be evaluated for its potential in controlling oystershell scale in a region.

Birds. Seven bird species are reported to feed on oystershell scale, but only three of these are found in the Seattle area. Of these three, the brown creeper is a resident of wooded areas and would not likely feed in street trees; and the white breasted nuthatch and chickadee are both canopy feeders and would not likely move onto the lower trunk areas. Thus, birds in our area could only partially control an infestation if it occurred on the trunk of the tree, or on trees in exposed areas.

Parasitoids. Evidence of parasitization by the chalcid wasp *Aphelinus mytilaspidis* ranged from 0 to 71.4%. Researchers have successfully transferred this parasitoid to other areas by moving branches with parasitized scales into areas lacking parasites. In early studies, scale outbreaks in infested areas declined significantly as the parasite became established.[5,7]

Mites. Several mite species were observed among the samples.[9] The most common mite was *Hemisarcoptes malus*. This mite historically has proven to be an effective predator on scale eggs. Because it must find its way to a scale infestation by hitching a ride on lady beetles or by riding wind currents, this predator is most effective when prey is available in large numbers. It takes time to multiply to an effective eradication force. However, it may be possible to augment existing bio-control by introducing mite-infested scales from one area where they are plentiful to an area where mites are lacking.

Although this mite is reported to feed only on the scale eggs, it was frequently found on the female scales. In a sampling of 82 females, there were 45 live females. One half of these (22) had mites. The females with mites produced fewer eggs (averaging 2.5 eggs per female) than females without mites (averaging 6.3 eggs per female).

Lady beetles. The twice-stabbed lady beetle (*Chilochorus stigmata*) is recorded as a pest of oystershell scale, however, none were observed during this study.

PESTICIDES

Understanding the complexities of pest populations takes a bit of effort and detective work. From the discussions above one can see why the easier path of using pesticides is so often taken. However, in many commercial spray applications there is rarely follow-up monitoring to determine if, in fact, the scales have been killed. This leads to the possibly false assumption that the method is effective.

Most of the research on pesticides to control this insect was done from 1897 to 1947. The horticultural oils, insecticidal soaps, and pesticide formulations used then were vastly different from the oils, soaps, and pesticide formulations in use today. In examining all the research reports, it was clear that the effectiveness of lime sulfur or dormant oils was not consistent.

It is possible that the success or failure of dormant application of oil or lime sulfur might lie in the timing (early dormant application, coupled with late hatching date), in the strength of the application or in thoroughness of coverage (to infested bark), or combinations of the above. In fact, Lord[10] concluded: "In light of present knowledge, it would seem that many of the materials must have been successful owing to their ineffectiveness against the natural controls rather than any direct effect on the scales." Before deciding to use a pesticide, one must consider:

(1) the stage of development of the pest,
(2) whether significant numbers of predators or parasites are present,
(3) the management or control options available, and
(4) the site itself.

SITE CONSIDERATIONS

A street tree planting poses special considerations and challenges because there are parked cars, buildings, painted surfaces, cement, sidewalk eating areas, storm drains, and people, children, or animals. Any pesticide must be evaluated in terms of its potential effectiveness against the pest at the current stage of development and at that site.

For example, while lime sulfur is considered to be a less toxic pesticide and might be favored in many situations, it is highly caustic and can cause discoloration or spotting of the painted surfaces of buildings or automobiles. This makes it a risky choice along most streets.

Pyrethrins or rotenone are fairly acceptable to the public because of the short "life" and low mammalian toxicity, but these would not be effective once eggs were laid beneath the protection of the scale covering. Both of these pesticides are listed as extremely toxic to fish and should not be used near ponds with fish (as may occur in some parks).

The presence of storm drains beneath street trees should also be considered (especially if rain is predicted). Other considerations might include the presence of nearby children's play equipment, open office or apartment windows, or any limitations of application equipment and staff skills.

IPM AS A PROCESS

Although it appears that IPM is a complex process, it is really no more complex than hunting. A good hunter learns about the quarry and the weapons available. There is a time (season) and place (hunting grounds) for each. A good hunter often will track the quarry for several miles to verify the results, and determine if follow-up action is needed. Few hunters spend time in the lowlands when the quarry is in the alpine meadows, although it is not uncommon for some pest managers (especially homeowners) to spray damaged plants long after the pest has moved on. Few hunters would go out for tigers or lions in the Pacific Northwest. Similarly, there is no need to "shoot" at pests common in other areas (Japanese beetle or gypsy moth) but which have not yet become established in an area.

Perhaps IPM should be termed "Target" pest management. Of course one must first determine if there is a target: There is a 70% chance in Western Washington that the cause of a plant problem is cultural or environmental; most often the trees simply need deep watering. If there truly is an insect or disease to target, then one must understand the quarry in order to select a weapon (management option) and the time (season) when the target is vulnerable. The objective is to match the weapon to the target and the season, and to hit the target with a minimum negative impact on the surrounding environment.

It also just may be that the target should be the plant: that we should look beyond the pest to the plant itself and ask why the plant is unable to activate its own defense system. There is a good (70%) chance that the plant is struggling with a less than optimum site for that species or cultivar.

TOWARD THE FUTURE

I look forward to exciting times ahead. We need to look beyond our traditional noses and create new approaches to managing plants in our urban environments. It is time to find innovative new strategies and toss out some of the reactive, outmoded methods of pest "control."

As we gain experience in using computer technologies, many of the lessons learned from research in cropping systems can be adapted to urban forest. Once scientific research and pest management texts are available on computers or laser disc, the tedious and time consuming tasks of searching through volumes of books and scientific reports will be reduced substantially. A rapid access to the volumes of research will free us to develop innovative new approaches to managing

out-of-control pest populations or managing for healthy plants.

I wish us all well as we explore new frontiers in integrated pest management. Let us boldly go forth in creating new approaches to managing pests in the urban environment.

REFERENCES

1. **Dirr, M. A.,** *Manual of Woody Landscape Plants: Their Identification, Ornamental Characteristics, Culture, Propagation and Uses*, Stipes Publishing, Champaign, IL, 1983, 294.
2. **Bean, W. J.,** *Trees and Shrubs Hardy in the British Isles*, Vol. 11, 1978, 208.
3. **McArdle, A. J. and Santamour, F. S., Jr.,** Checklists of cultivars of european ash (*Fraxinus*) species, *J. Arboric.*, 10(1), 21, 1984.
4. *Pacific Northwest Insect Control Handbook*, Oregon State University and Washington State University, Pullman, WA, 1984.
5. **DeBach, P.,** *Biological Control by Natural Enemies*, Cambridge University Press, Cambridge, England, 1974, 323 pp.
6. **Turnbull, A. L. and Chant, D. A.,** The practice and theory of biological control of insects in Canada, *Can. J. Zool.*, 39, 697, 1961.
7. **Griswold, G. H.,** A study of oystershell scale, *Lepidosaphes ulmi* (L.), and one of its parasites, *Aphelinus mytilaspidis* LeB, I, II, *Cornell Univ. Agric. Exp. Sta. Mem.*, 93, 67, 1925.
8. **Siegler, E. H. and Baker, H.,** Parasitism of scales — San Jose and oyster shell, *J. Econ. Entomol.*, 17, 497, 1924.
9. **Ewing, H. E. and Webster, R. R.,** Mites associated with the oystershell scale (*Lepidosaphes ulmi* Linne), *Psyche*, August, 121, 1912.
10. **Lord, F. T.,** The influence of spray programs on the fauna of apple orchards in Nova Scotia. II. Oystershell scale, *Can. Entomol.*, 79, 196, 1947.

Chapter 21

Major Insect Pests of Turf in the U.S.

R. L. Brandenburg and J. R. Baker, *Department of Entomology, North Carolina State University, Raleigh, NC*

CONTENTS

Nuisance Pests — Rarely Cause Any Real Turf Damage 220
     Ants 220
     Cicada Killer Wasp 220
     Fire Ants 221
     Frit Fly 222
     Harvester Ants 222
     Scoliid Wasps 222
     Velvet Ants 223
     Wild Bees 223
Surface and/or Thatch Pests — Feed on Grass Above the Ground 224
    Chew on Leaves and Stems 224
     Armyworms 224
     Burrowing Sod Webworm 224
     Cutworms 225
     Fall Armyworm 225
     Sod Webworm 226
    Discolor Leaves and Stems 226
     Banks Grass Mite 226
     Bermudagrass Mite 227
     Bermudagrass Scale 227
     Chinch Bug 227
     Greenbug 228
     Leafhoppers 228
     Rhodesgrass Mealybug 228
     Twolined Spittlebug 229
     Winter Grain Mite 229
    Burrow into Leaves and Stems 229
     Billbugs 229
     Annual Bluegrass Weevil 230
Subsurface or Underground Pests — Feed on Roots or Disrupt the Soil 230
     Asiatic Garden Beetle 230
     Ataenius 231
     European Chafer 231
     European Crane Fly or Leather Jacket 231
     Green June Beetle 232
     Ground Pearls 232
     Japanese Beetle 233
     May Beetles 233
     Masked Chafers 234
     Mole Crickets 234
     Oriental Beetle 235
     Shorttailed Cricket 235
References 235

0-87371-350-8/94/$0.00+$.50
© 1994 by CRC Press Inc.

Turfgrasses grown throughout the U.S. can be broken down into cool- or warm-season types. These two designations, cool- or warm-season turfgrasses, can be further broken down into cool, humid, and cool arid or semiarid as well as warm, humid and warm, arid-semiarid. There is considerable overlap in various regions of the U.S. as to the type of turfgrasses grown. In addition, it is somewhat difficult to give an accurate description for the boundaries of specific pests and virtually impossible to give comprehensive classifications for all turfgrass pests.

Certain insects prefer specific turf types, but others infest a great number of grasses and broadleaf plants. Pests such as ground pearls, mole crickets, and fire ants infest warm-season turfgrass whereas pests such as the European chafer infest cool-season areas. Sometimes this is a preference for a turf-type while other times it may be a reflection of the geographic range of the pest. Generally, insects are divided into two groups based upon the areas they inhabit. These groups are the surface or thatch dwellers and the subsurface or underground pests. The surface or thatch pests can be further divided into those that chew on leaves and stems, those that discolor the leaves and stems, and those that burrow into the stems. Some pests are not real pests of the grass at all but are considered to be a nuisance to the users of the turf.

Turfgrass managers must constantly be aware that even pests that occur sporadically may become quite abundant on a localized basis. The listing that follows provides background information for the major insect and mite pests attacking turfgrass in the U.S. collected from three main sources.[1-3]

NUISANCE PESTS — RARELY CAUSE ANY REAL TURF DAMAGE

Ants
Description
Ants have constricted "waistlines" and may be smooth or hairy; winged or wingless; red, brown, black, or yellow; and 1 to 10 mm long. Adult ants may be males (winged), females (winged), or workers (wingless females). Winged ants have two pairs of wings, the anterior pair being much larger than the second pair.

The white or pale-yellow eggs are almost microscopic in size and vary in shape according to the species. Larvae are translucent, soft, legless, and segmented. One of the most common forms is gourd- or squash-shaped with the tiny head located at the small end. Size varies with species. The larvae are small enough to be carried by the worker

ants. The soft, colorless pupae resemble adults in size and shape. They may or may not be enclosed in papery cocoons.

Biology
Although the distributions of particular species vary, ants in general are cosmopolitan insects. They are common from the Arctic to the tropics. Around buildings, ants may nest near sidewalks, foundations, and driveways as well as in turf.

The ants which damage turf do not actually feed on plants (although Texas leafcutting ants damage trees and shrubs obtaining leaves for fungus gardens to feed their larvae). Often, however, ants are attracted by sap from wounds or honeydew on the foliage of ornamental trees and shrubs. Ants feed primarily on seeds, small insects, and honeydew excreted by aphids associated with plant roots.

Ants build nests in the ground. They are particularly troublesome around the fringes of golf greens, on fairways, and in lawns. The anthills (ant-mounds) often smother the surrounding grass. If ants nest about the roots of grass, they may kill them. They also feed on grass seeds, thereby preventing good stands of newly reseeded grasses. Moisture stress is common in ant-infested lawns due to root damage and the many tunnels which promote water loss from the soil.

Ants live in colonies usually started by a fertile, winged female (queen) that makes a small nest and lays a few eggs. Such females eventually lose their wings. The queen cares for and feeds the first brood of larvae which develop into workers (wingless females). The workers construct and repair the nest, gather food and feed the immature and adult ants, care for the brood, and defend the nest. Workers may also lay eggs but egglaying is usually the queen's job. Male ants appear only in very large or old colonies and die soon after mating with the new queens.

Cultural practices can often reduce ant populations. If sap from trees is attracting ants, clean and paint the wounds.

Cicada Killer Wasp
Description
This large wasp has a rusty-red head and thorax, rust-colored wings, and a black- and yellow-striped abdomen. A length of 30 mm or more is not uncommon. The translucent, 3- to 6-mm-long, greenish-white egg is often described as "cigar-shaped." The larva may reach a maximum length of 28 to 32 mm. The mature, quiescent larva, however, is somewhat shrunken and leathery. The pupa remains undescribed. The woven, spindle-shaped cocoon

which surrounds the pupa averages 32 mm long and 11 mm wide and has a narrow band of pores along its center. The cocoon is often brown and stiff.

Biology

The cicada killer wasp occurs in all states east of the Rocky Mountains. It abounds in areas of full sun, scant vegetation, and light-textured, well-drained soils. Cicada killer larvae feed primarily upon cicadas. The adult wasp feeds on the nectar of flowers. In spite of its formidable size and burrowing habit, this wasp is unusually docile and harmless. Although capable of inflicting a painful sting, the cicada killer wasp is usually difficult to provoke. Mating males, however, are aggressive and more easily disturbed, although males cannot sting.

An unsightly mound of soil surrounds the burrow of each cicada killer. Since colonies of burrows are common, infested lawns usually have several mounds that can smother the grass. However, since cicada killers prefer to nest in areas of sparse vegetation, it is likely that an infested turf was already unthrifty before the wasps arrived. They rarely burrow in thick, vigorous turf.

The cicada killer wasp overwinters as a larva within a cocoon 15 to 25 cm deep in well-drained soil. Pupation occurs in the spring. The cicada killer's life history has not been closely studied in North Carolina, but the wasp appears as early as the first week of June in Arkansas and rarely before July 1 in Ohio. Emergence continues throughout the summer. The female adult feeds, mates, and digs burrows for several weeks before preying on cicadas. A vertical or slightly angled burrow 15 to 25 cm deep and 1.2 cm in diameter with broadly oval cells perpendicular to the main tunnel is excavated. The excess soil thrown out of the burrow forms a regular, U-shaped mound at the entrance.

Once cells have been constructed, the search for cicadas begins. Canvassing tree trunks and lower limbs, the wasp finds and stings its prey, turns the victim on its back, straddles it, and drags it or glides with it to the burrow. Each cell is furnished with at least one cicada (sometimes two or three) and a single egg before being sealed off. The egg hatches 2 to 3 days later. Depending on the number of cicadas in its cell, the larva feeds for 4 to 10 days until only the cicada's outer shell remains. During the fall, the larva spins a silken case, shrinks, and prepares to overwinter. Only one generation occurs each year.

Cultural practices can prevent or eliminate the establishment of cicada killer colonies. Adequate lime and fertilizer applications accompanied by frequently watering promote a thick growth of turf and can usually eliminate a cicada killer infestation in one or two seasons.

Fire Ants
Description

The various stages of development are similar to the general description for ants. All species of fire ants closely resemble the red imported fire ant. Identification of the species is difficult but can be made by a specialist.

The mound built by the red imported fire ant is somewhat larger than those built by other species. Mounds average 35 cm or more in diameter and 20 to 25 cm in height.

Biology

Since their introduction from South America in about 1918 through the port of Mobile, AL, imported fire ants have spread into Texas, Louisiana, Mississippi, Georgia, Florida, South Carolina, Oklahoma, Tennessee, North Carolina, and Arkansas. Although soil infesting, imported fire ants feed on a wide variety of insects, plant seeds, and parts. They are predaceous on some agricultural pests. Fire ant damage is most significant to agriculture in losses resulting from reduced efficiency of labor and machinery and perhaps reduced land value.

These ants prefer land exposed to the sun; consequently valuable farming and pasture land may be heavily infested. In urban areas, these ants invade lawns, parks, playgrounds, schoolyards, cemeteries, and golf courses. Fire ants have a severe bite and sting and will attack anything that disturbs their nest. Each ant is capable of stinging several times. The sting causes burning or itching followed by the formation of a small, white pustule. Scratching of the pustule may lead to secondary infection which can leave a scar. Some people who are allergic to wasp and bee stings may also be allergic to fire ant stings.

The red imported fire ant is the most troublesome fire ant species. It has taken over large areas in the South forcing out both the native and southern fire ant. Fire ant colonies are made up of three forms: the winged fertile females (queens) which lay the eggs, winged fertile males (kings), and three classes (sizes) of worker ants called minim, minor, and major workers. The queen lays and looks after the initial egg cluster. Later she only lays eggs as the workers take care of the other functions of the colony. The average colony may contain 100,000 to 500,000 workers and relatively few winged or reproductive forms. The short-lived males die shortly after mating with the queens in a large aerial mating swarm. The fertilized queens land and excavate a cell in which eggs are laid to start a new colony.

Frit Fly
Description
Several different species of frit flies attack grasses. These tiny flies are black or yellow in coloration and 1.5 to 2.5 mm long. The wings are short, broadly rounded at the tip and they lack closed cells except close to the base. Frit flies have small mouths and short antennae each with a stout, bare hair. The eggs are tiny and colored an off-white. The tiny larvae are off-white, legless maggots.

Biology
There are several species of frit flies that attack turf grasses throughout the U.S. All common lawn grasses are susceptible with bentgrass being most preferred. This pest may cause damage to bentgrass greens on golf courses and occasionally to turf grasses. The small larvae tunnel into and feed on the grass stems near the soil surface. Plants that are infested generally wilt and die. The presence of this pest may be detected if large numbers of small black flies are observed over the grass surface from mid- to late morning. The life history varies slightly between species but in general the female lays eggs on grass leaves or stems. The eggs hatch and the small larvae bore into the stems. After reaching the last larval instar, pupation takes place in the damaged stems or in the soil with adults soon emerging. There are generally several generations a year. Winter is passed in the larval stage at the base of the damaged plants.

Harvester Ants
Description
Harvester ants are large (5 to 6 mm long) and vary in color from reddish brown to yellow or black. The pedicel (or stalk) between the abdomen and thorax has two segments. They have elaborate fringes of hairs underneath the head. As is the case with other ants, adults may be winged males or females, or workers (wingless females). Winged ants have two pairs of wings, the anterior pair being much larger than the second pair.

The eggs are minute, less than 0.5 mm long, white and elliptical in shape. The larvae are white and legless. The larva is covered with short hairs and is shaped like a crookneck squash or gourd with a small distinct head. Found in a cocoon, pupae resemble the larvae except that the body is straight and rigid with legs and wings visible.

Biology
The harvester ants do not consume grass for food but denude areas around mounds and along trails. Their principal source of food is seeds of various grasses and the flesh of arthropods. The so-called harvester ants occur west of the Mississippi except for one species. The two most common species are the Texas harvester ant and Western harvester ant.

The harvester ants not only construct large mounds which cause loss of grass but also clear areas of grass around the nest and along the forage trails radiating from the central nest. Cleared areas around the nest may be 7 m or more in diameter. Generally, nests are constructed in open areas and are a real problem on golf courses, recreational areas, and occasionally in lawns. It is also thought that they may hinder reseeding of different grasses by collecting seed. In addition they can sting viciously. At least three small children have died from their stings in Oklahoma.

The harvester ants are social insects and live together in colonies. Winged males and females swarm from the parent colony, pair off, and mate. The males die soon after mating. The female loses her wings, finds a suitable nesting site in the soil, and begins laying a few eggs. After the eggs hatch, the queen feeds and cares for the young larvae. The larvae mature and change to pupae, then to adult workers. The adult workers then begin taking care of the queen as well as eggs and larvae of future generations.

Scoliid Wasps
Description
Scolia dubia is an attractive wasp 20 to 25 mm long, with black antennae and a shiny black head, thorax, and fore abdomen. A yellow spot (on each side of the abdomen) sometimes is faint or absent. The spots may appear as a band across the abdomen when the wasp is flying. Behind the yellow spots, the abdomen is brownish and the hairs on the body more noticeable. The wings are dark blue.

The larva of *Scolia dubia* is a legless, white grub with a brown head. It appears hairless and has no eyes. The antennae, maxillary, and labial palps are one-segmented. There is a slit-like silk gland on the labium.

Biology
Scoliid wasps (*Scolia dubia* in particular) range from New England to Florida and west to the Rocky Mountains. Other Scoliids are found throughout the tropics and temperate zones of the world. The wasps may be present from May to October. Green

June beetle and Japanese beetle grubs seem to be the primary host of *Scolia dubia*. No damage has been reported from these wasps. Homeowners, however, are often needlessly alarmed by the scoliid wasps which are often abundant in August and September. Although scoliid wasps are virtually harmless to humans and are beneficial in their control of beetle grubs, their presence in large numbers may indicate a large population of beetle grubs.

One of the most common scoliid wasps in North Carolina is *Scolia dubia*, often referred to as the blue-winged wasp. These beneficial wasps lay their eggs on soil-infesting white grubs such as the larvae of May and June beetles and green June beetles. The adults feed on the nectar and perhaps on the pollen of flowers. They do not sting people unless greatly aggravated or captured in the hands. They fly several centimeters (a few inches) above grub-infested soil in a more or less figure-eight pattern.

Female wasps work their way through the soil in search of grubs, burrowing their own tunnels or following those made by grubs. On locating a grub, she stings and paralyzes it. She may burrow 1.2 cm deeper to construct a cell around the host. Then she lays an egg on the outside of the grub. The parasitized larva provides a fresh food supply for the wasp larva. The stung grubs never recover. Many grubs are stung, but only a few eggs are laid. These wasps, therefore, are very important natural agents in the control of grubs in the soil.

Since these beneficial insects rarely, if ever, sting people, no control measures are needed. Their presence, however, means there is a green June beetle, Japanese beetle, or May beetle grub infestation that may be only partially controlled by these parasites.

Velvet Ants
Description
Velvet ants are wasps, not really ants. Adult females are 5 to 20 mm long, wingless, and ant-like. The head and body are covered with dense hairs. The color is generally a brilliant red, orange, or yellow with two black bands on the abdomen. The antennae and legs are black. Males are somewhat smaller and have two pairs of wings. The larva is a white legless grub.

Velvet ants are widely distributed throughout the South and West. They are usually found in open areas. Velvet ants are active throughout the warm periods of the year. They are most common during mid-summer. Usually, they are seen during the cooler parts of the day. Although usually found in relatively sandy areas, they occasionally may find their way indoors. Velvet ants are not phytopha-

gous. The adult is known to feed on bees and other wasps at times. The larva is an external parasite of various bee and wasp larvae and pupae. Some species are parasitic on various beetles and flies. No damage to turf by velvet ants has been reported. After bees or wasps have formed cocoons, adult female velvet ants enter the host nest by digging through the soil or breaking through nest wall. An individual chews a hole through the cocoon wall and deposits a single egg on the host larva. Usually the hole is sealed and the next cocoon is attacked. The egg hatches and the velvet ant larva feeds on the host. This host grub is ultimately killed. Females are capable of inflicting a very severe sting if handled or stepped on with bare feet, hence the common name, cow killer. The winged males are harmless. Since velvet ants do not cause any damage, no chemical control is recommended. Children should be cautioned not to handle velvet ants.

Wild Bees
Description
These small to medium-sized bees may be any of a wide range of colors: metallic red, black, blue, green, or copper. Usually no distinctive spots or bands are present. Length ranges from 8.5 to 17 mm. The tiny, white eggs may be up to 2.5 mm long and 0.5 mm wide. Each one is elongate, slightly curved, and attached at one or both ends to a pollen ball. Approximately egg-size when first hatched, the C-shaped larvae grow rapidly and, when mature, are about as long as adult bees. They have a distinct transverse arrangement of bumps across their backs. Pupae resemble pale, mummified adults and have no protective cocoon.

Biology
Since several bee species build nests in the soil, bee mounds in turf are nearly cosmopolitan in occurrence although only locally abundant. They are most common, however, in soils with sparse to moderate plant growth, little organic matter, and good drainage.

Essentially beneficial insects, wild bees feed on the nectar of many plants and gather pollen for the larvae to feed upon and are excellent pollinators of vegetable and fruit crops.

Bees do little significant injury to turf. Since they prefer to nest in soils with a sparse vegetative cover, only the unthriftiest of lawns become infested. As the bees tunnel in the soil, the excavated dirt forms mounds 1.5 to 6.0 cm wide and 0.25 to 1.5 cm high. The bees, however, are more often controlled out of fear of their stinging than to alleviate their detrimental effects upon turf.

Wild bees generally overwinter in their soil burrows as adults. They emerge by early April and begin digging new burrows. The burrow consists basically of a vertical shaft 8 to 15 cm deep. The number and size of side tunnels varies with the particular bee species. Unlike some bees, soil-nesting species are not social in that each female makes her own nest, provisions it with food, and lays eggs. There is no worker caste. The bees, however, are gregarious and often nest closely together. However, there is no "nest guarding" instinct. Wild bees seldom sting unless stepped upon or squeezed.

In central North Carolina, wild bees first begin to fly around mid-April. Mating takes place soon afterwards and females begin storing pollen in burrows. Furnishing each cell of their burrow with a pollen ball 3 to 5 mm in diameter, females then deposit a single egg on each pollen ball. Eggs hatch in early May. Throughout the summer, the larvae feed and develop within the burrows. Pupation occurs in later summer, usually in August. With some species, adult bees develop sometime in the fall but remain in their burrows to overwinter. Other species overwinter as larvae. A single generation is completed each year.

SURFACE AND/OR THATCH PESTS — FEED ON GRASS ABOVE THE GROUND

CHEW ON LEAVES AND STEMS
Armyworms
Description
The adult moth is pale brown to grayish brown in color, with a wingspread of about 40 mm. There is a characteristic white spot in the center of each forewing. Greenish white and spherical, the tiny eggs are laid in masses. The young larva is pale green. Full-grown larvae, approximately 35 to 50 mm long, are distinctly striped and yellow to brownish green in color. Like cutworms, armyworms may curl into a C-shaped ball when disturbed. Initially reddish brown, the pupa darkens until it is almost black.

Biology
Armyworms are found throughout the world; in the U.S. they are most common east of the Rocky Mountains. All common turfgrasses are susceptible. Varying in abundance from year to year, young armyworm caterpillars skeletonize the surface of leaf blades and the inner surface of the sheaths. Older caterpillars may eat circular bare areas in lawns. Armyworms overwinter as partially grown, inactive larvae that resume feeding in early spring. Their habit of moving *en masse* from one area to another accounts for their common name.

The first-generation moths appear in May or June and feed on nectar for several days, after which females begin laying eggs. Eggs are laid at night in clusters of as many as 130 between the sheath and the blade of growing grass or in other similar places. A female can lay up to 2000 eggs which hatch in 6 to 10 days. Second-generation larvae feed for a few weeks, then enter the ground to pupate in earthen cells 5 to 8 cm deep in the soil. The duration of the larval stage depends primarily upon food and temperature but generally lasts about 4 weeks. Outbreaks of the armyworm are characterized by their sudden appearance and disappearance.

Natural factors such as parasites and diseases usually check armyworm populations. Outbreaks of the armyworm usually occur after years of drought.

Burrowing Sod Webworm
Description
The moth is variable in size and coloration. The forewings are predominantly yellowish or grayish brown with a paler border along the folds and inner margins. Irregular dark brown spots are usually present but may be obscure. The hind wings are yellowish to bronzy brown. Wingspan ranges from 25 to 38 mm. The labial palps of the male are large, hairy, and recurved over the head and thorax. The grayish- or dirty-white larva has a brown head capsule and grows to 20- or 30-mm long when full-grown. The brown, chitinous pupa is 15 to 20 mm long.

Biology
Primarily tropical insects, burrowing sod webworms range from South America northward into the U.S. They occur in Arizona, New Mexico, and Texas, northward through Oklahoma, Kansas, and Nebraska, and eastward into Pennsylvania, New Jersey, North Carolina, and Florida. In addition to infesting the roots of most lawn grasses, these worms also feed on the roots of corn, bromeliads, orchids, and a large number of other hosts. Like sod webworms, burrowing sod webworms sever blades of grass near the thatch line, and feed upon these leaves within a burrow. Closely cut, irregularly shaped spots are indicative of burrowing sod webworm damage. The burrows of this webworm, however, tend to be deep and vertical whereas other webworms make their burrows nearer the soil surface. Little information concerning the life cycle of burrowing sod webworms is available. Occasionally emerging as early as May, the moths usually do not appear until June or July. Several days after the eggs are laid, larvae hatch. The worms construct burrows, approximately the diameter of a pencil, which extend 15 to 60 cm deep into the

ground. Fine, tubular webs mixed with soil particles and frass lead from the lower leaf blades into the burrow. When disturbed, larvae retreat into the burrow.

Cutworms
Description
When resting, cutworm moths hold their wings back in a triangular position. The moths are generally stocky and have a wingspan of about 40 mm. The forewings are dark brown and mottled or streaked; the hindwings are lightly colored and unmarked. The eggs are usually white (becoming darker prior to hatching), and round (0.5 to 0.75 mm in diameter). If disturbed, the larvae usually curl into a C-shaped ball. Cutworms are fat, smooth, dull-colored caterpillars that grow to about 45 mm long. Pupae are brown and 15 to 22 mm long.

Biology
Cutworms are caterpillars that feed on the stems and leaves of young plants and often cut them off near the soil line, hence their common name. Although there are many important species of cutworms, the black, granulate, and variegated cutworms are the ones most commonly encountered in the Southeast. Cutworms are found throughout the U.S. Cutworms attack most turfgrasses. Many cutworms prefer wilted plant material and sever the plants sometime prior to feeding. Stems are chewed off near the soil line. Some cutworms even climb shrubs and trees and feed on unopened buds.

Each cutworm differs slightly from the others in details of habits and appearance but their life histories are generally similar. Adults and larvae are nocturnal and hide during the day but may become active on cloudy days. Cutworms overwinter in the soil either as pupae or mature larvae. In the spring, the hibernating larvae pupate. Adults begin to appear in the middle of March. Female moths deposit eggs singly or in clusters, and each female can lay as many as 500 eggs. Under optimum conditions, the eggs hatch in 3 to 5 days, and larvae develop in 3 to 4 weeks, passing through six instars. Pupae mature in 2 weeks during the summer but take as many as 9 weeks in the fall. Some of the cutworms can produce as many as four generations each year in the southeastern U.S.

Some entomogenous nematodes, neem tree extract, azidiractin, and formulations of *B.t.* have proven effective for cutworm control.

Fall Armyworm
Description
The fall armyworm moth has a wingspan of about 38.5 mm. The hindwings are white; the front wings are dark gray, mottled with lighter and darker splotches. Each forewing has a noticeable whitish spot near the extreme tip.

Minute, light gray eggs are laid in clusters and are covered with grayish, fuzzy scales from the body of the female moth. The eggs become very dark just before hatching.

The mature green, brown, or black fall armyworm is 35 to 50 mm long, and has a dark head usually marked with a pale, but distinct, inverted "Y." Along each side of its body is a long black stripe. There are four black dots on the top of each abdominal segment.

The pupa, approximately 13 mm long, is reddish brown but darkens to almost black as it matures.

Biology
The fall armyworm is a continuous resident of the tropics of North, Central, and South America, and parts of the West Indies. With mild winters, it may persist year round along the Gulf Coast of the southern states. Each year it migrates as far northward as Montana, Michigan, and New Hampshire.

The fall armyworm has a wide host range but prefers plants in the grass family. Most grasses, including coastal bermudagrass, fescue, ryegrass, bluegrass, Johnsongrass, timothy, corn, sorghum, Sudangrass, and small grain crops, are subject to infestation.

Fall armyworms, often migrating in large armies, are potential turf and pasture pests in late summer and fall. Consuming all aboveground plant parts, they are capable of killing or severely retarding the growth of grasses.

Fall armyworms probably overwinter as pupae in the Gulf Coast region of this country. Egg-laying moths migrate northward throughout the spring and summer and arrive in North Carolina during mid-July. New moths may continue to appear into November. Each female lays about 1000 eggs in masses of 50 to several hundred. The small larvae emerge 2 to 10 days later, feed gregariously on the remains of the egg mass, and then scatter in search of food. Unlike the nocturnal true armyworms, fall armyworms feed any time of the day or night, but are most active early in the morning or late in the evening. When abundant, these caterpillars eat all plants at hand and then crawl in great armies to adjoining fields. After feeding for 2 to 3 weeks, the larvae dig about 2 cm into the ground to pupate. Within 2 weeks, a new swarm of moths emerges and usually flies several miles before laying eggs. Several generations occur each year in North Carolina.

During favorable seasons, a number of parasitic enemies keep fall armyworm larvae down to moderate numbers. Cold, wet springs seem to reduce the effectiveness of these parasites and allow large fall armyworm populations to develop.

Sod Webworm
Description

The moths, 13 to 19 mm long, have a wingspan of about 15 to 35 mm. They have a prominent forward projection (labial palps) on the head. The forewings are brown or dull ash gray, with a whitish streak from the base to the margin; the hindwings are brownish. When at rest, the moths fold their wings in a tentlike manner over their body.

The tiny, oblong eggs are white to pale yellow. Each egg is about 0.5 mm long and 0.3 mm wide. Most larvae vary from pinkish white to yellowish to light brown. They are 16 to 28 mm long when fully grown, with thick bodies, coarse hairs, and paired dorsal and lateral spots on each segment. The head is yellowish brown, brown, or black. Individual webworms often assume a C-shaped position. Tropical sod webworms are greenish and up to 19 mm long. The reddish-brown pupae are 11 to 13 mm long.

Biology

Many species of sod webworms occur in the U.S. The actual species present in any given area, however, is highly variable but may include bluegrass sod webworm, larger sod webworm (primarily on Kentucky bluegrass), and the tropical sod webworm. Whereas many sod webworms in the subfamily *Crambinae* are generally distributed, the tropical sod webworm is common primarily in Florida and the Gulf states and the larger sod webworm, *Pediasia trisecta*, abounds in the northern half of the U.S.

Sod webworms feed on lawns, golf course grasses, some clovers, corn, tobacco, bluegrass, timothy, as well as pasture and field grasses. They usually favor bluegrass and Tifdwarf hybrid bermudagrass, but will attack most grasses.

Larvae cut off grass blades just above the thatch line, pull them into their tunnels, and eat them. The injury appears as small brown patches of closely cropped grass. If many larvae are present, the patches run together to form large, irregular brown patches. Birds are commonly seen feeding in infested areas.

Webworms overwinter as young larvae a few centimeters below the soil line among the roots of weeds and grasses in silk-lined tubes. During early spring the larvae feed on the upper root systems, stems, and blades of grass. They build protective silken webs, usually on steep slopes and in sunny areas, where they feed and develop. In early May, they pupate in underground cocoons made of silk,

bits of plants, and soil. About 2 weeks later, adults emerge. Beginning in May, moth flights may occur until October. The moths, erratic and weak flyers, live only a few days and feed solely on dew. They are active at dusk, resting near the ground in the grass during the day.

The eggs, which are deposited indiscriminately over the grass, hatch in 7 to 10 days. Young larvae immediately begin to feed and construct their silken tunnels. The most severe damage occurs in July and August when the grass is not growing rapidly. During this hot weather, the larvae feed at night or on cloudy days. Most sod webworms complete two or three generations each year, with approximately 6 weeks elapsing between egg deposition and adult emergence. In Florida, tropical sod webworms may produce a new generation every 5 to 6 weeks. A sod webworm infestation in the lawn can be detected by applying 1 tablespoon of pyrethrin (insecticide) in 1 gallon of water per square meter (or square yard). The caterpillars will surface within a few minutes and can be found by separating the blades of grass, particularly at the interface between living and dead areas of turf. If three or four webworms are found in a 15-cm-square (6-in.-square) area, control is recommended.

Some resistant turfgrasses are available for sod webworms. In addition, parasites, nematodes, and neem tree extract, azidiractin, have been shown to be effective against these pests. Numerous natural enemies also attack sod webworms.

DISCOLOR LEAVES AND STEMS
Banks Grass Mite
Description

The adult females are broadly oval and about 0.4 mm long, while the males have a strongly tapered abdomen and are slightly smaller. During the spring to fall feeding period, the adults are a bright green with orange legs, but the green fades during the winter and the mites appear a bright orange color. The eggs are spherical and small (0.125 mm in diameter). They are pearly white at first, but change to a light straw-yellow color prior to hatching.

The larvae are oval in shape and when newly-hatched they have red eye spots and salmon-colored bodies. After feeding, the presence of plant juices in the mite causes its body color to become light green. The nymphs have four pairs of legs and are usually bright green.

Biology

The Banks grass mite is known to occur from Washington state to Florida and south, but not in the northeastern states. It commonly attacks Kentucky bluegrass in Oregon and bermudagrass in the

Southwest. Lightly infested plants have small yellow speckles along the grass blades. As damage increases, the leaves become more off-yellow colored and may wither and die. This often happens during hot, dry spells and is often mistaken for summer dormancy in cool-season grasses. This mite produces webbing at the base of the turf which is often seen in the morning dew.

Mature females overwinter at the base of grass plants and in the soil, although a few males and nymphs may be present. In the spring, the mites begin to feed on emerging grass, and gradually the mites turn green. The females begin to lay eggs in webbing or on the plants. The females lay 50 to 75 eggs. The eggs take anywhere from 4 to 25 days to hatch depending upon the weather. During hot weather, eggs can hatch in 4 days and development of the mite may be completed in 5 days. Six to nine generations a year may be completed. During hot, dry weather, immature mites and adults may migrate to the center of dormant grass clumps and rest until the grass returns to active growth.

Bermudagrass Mite
Description
Approximately 0.2 mm long, this mite is worm-like in shape and creamy white in color. The eggs are oval and about 0.07-mm long, and transparent to opaque white in color. The nymph stages are approximately two thirds the size of the adult and are whitish in color. They also appear worm-like or cigar-shaped and have four legs as do the adults.

Biology
Introduced probably from Africa, the bermudagrass mite may be found from Florida to Arizona in the U.S. It has also been collected from Las Vegas, NV. Initial damage is observed in the spring when lawns fail to begin normal growth. Some areas may be yellow or brown in color. Damage appears as a typical rosetting and tufting, caused by a shortening of the internodes. As the grass is mowed, infested clumps of grass become less vigorous. Individual grass plants will have large brown clumps of distorted stems that usually die. Grass in infested lawns usually thins, allowing weed development. The adult mites infest the grass in protected areas primarily under the leaf sheaths. The eggs are deposited in these areas. Complete development from egg to adult requires 5 to 7 days. This species is active primarily in the late spring and summer.

Bermudagrass Scale
Description
Bermudagrass scales are approximately 1.6 mm long and have a hard, white circular or clam-shaped covering. They are usually found on the stems, near nodes, and occasionally on leaves.

Biology
Females lay eggs under their own protective covering. The eggs hatch into "crawlers" that move to a new site on the plant and begin feeding. They soon lose their antennae and legs and become immobile. They quickly develop their waxy covering and feed for several months before laying eggs.

This insect can be found anywhere bermudagrass is grown, but is seldom a problem. Damage most frequently occurs when the grass is already under stress (environmental or other pest). Heavily thatched and/or shaded areas are most susceptible. Severe damage will result in slow growth and a dry appearance of the turfgrass.

Chinch Bug
Description
The adult chinch bug is about 4-mm long and black with opaque wings. The wings may be as long as the body or only $\frac{1}{3}$ to $\frac{1}{2}$ the length of the body. In any case, each wing bears a distinctive triangular, black mark.

Each egg, approximately 0.84 by 0.30 mm, is flattened at one end, that end bearing three to five minute projections. Its color gradually changes from pale yellow to red before hatching. The wingless nymph is smaller than but similar in shape to the adult. The head and thorax are brown; the eyes are dark red; and the abdomen is pale yellow or light red with a black top.

Biology
The southern chinch bug is common throughout the Gulf states and into Georgia and North and South Carolina. It is primarily a problem on thick mats of turf in sunny, open areas. The southern chinch bug is commonly a problem on St. Augustine grass but also infests angola, torpedo, centipede, and occasionally Bermuda grasses. The hairy chinch bug is present in the Northeast and Midwest. The hairy chinch bug prefers cool-season grasses.

Chinch bug populations are concentrated near the surface of the soil. The nymphs extract plant juices with their needlelike mouthparts and are primarily responsible for lawn damage. On St. Augustinegrass, feeding is primarily restricted to the tender basal area of the grass blades and to the nodes of runners. As the nymphs feed, yellowish spots appear in the grass and soon become brown dead areas. As the grass dies, the nymphs move to the periphery of the dead spots thereby causing them to enlarge. Chinch bug damage is greatest during hot, dry weather.

Except in southern Florida, where they remain active year round, chinch bugs overwinter as adults (hairy) or eggs and adults depending on species. Females insert eggs in crevices at grass nodes or between overlapping leaf blades. In spring, the eggs hatch releasing nymphs which subsequently infest lawns. Nymphs feed and develop for 2 to 6 weeks depending on weather conditions.

The new generation of adults causes little damage. Each female however, deposits an average of 300 eggs, which hatch 2 weeks later. With two to five chinch bug generations each year (depending on latitude), lawns may be infested from spring through late fall.

During cool, cloudy weather, chinch bugs are attacked by a fungal disease. If the weather is hot and dry, the operation of a sprinkler system helps prevent chinch bug damage.

Greenbug
Description
The greenbug is an aphid that is soft-bodied, pear-shaped, and usually 2 mm long. They are usually light green to greenish-yellow with the tips of the legs and the antennae usually black. They may be winged or wingless. Nymphs look similar to adults, but are smaller and never have wings.

Biology
Greenbugs are found throughout the U.S., but are probably the most serious problem on Kentucky bluegrass in the midwestern states. The greenbug also attacks most other cool-season grasses as well as small grains such as oats.

These insects have needlelike mouthparts to pierce into the leaf and remove plant juices. At the same time, they inject a toxin which causes yellow spots followed by death of the plant around the feeding site. Damage usually begins in shady spots and often occurs early or late in the season, as greenbugs prefer cool temperatures between 40 and 65°F. Warm weather often helps control a problem.

Leafhoppers
Description
Leafhoppers average 7 mm in length (rarely as long as 13 mm). They have a triangular, often elongated, head and may be yellow, green, or gray. Some species are mottled and speckled. The white, elongate eggs are 1 mm or less in length and are found inside grass stems and leaves. The pale, wingless nymphs are smaller than adults but similar in shape.

Biology
Leafhoppers are common throughout the eastern U.S. and occur in all areas of North Carolina. Leaf-hoppers infest several hundred kinds of cultivated and wild plants. Leafhoppers retard the growth of grass by piercing stems and leaves with their needle-like mouthparts and extracting sap. This type of feeding causes infested areas to have a whitened or bleached appearance. This symptom can be mistaken for drought or disease damage. Newly seeded lawns are sometimes killed by leafhoppers. Leafhoppers may overwinter as eggs or adults. Resuming activity in the spring, the adults begin to feed and mate. Females insert 75 to several hundred eggs, singly, into the leaf veins of tender new foliage. About 10 days later, nymphs emerge. Over a period of 12 to 30 days, these immature leafhoppers feed on new leaves and develop through five instars. The fully grown nymphs then molt to become adults and the life cycle is repeated. Leafhoppers produce one to four generations each year depending on the latitude and the particular leafhopper species.

Rhodesgrass Mealybug
Description
The Rhodesgrass mealybug is saclike and lacks appendages. The dark, purplish-brown body is oval to almost circular and averages 3 by 1.5 mm. The body is enclosed in a felted white waxy sac that yellows with age. These scales are born live; that is, no egg stage exists. The first instar or crawler is oblong-oval, cream colored, and very active. The second and third instars are saclike. They become sessile (nonmobile) and the body is enclosed in a felted waxy sac resembling that of the adult, except for size.

Biology
Rhodesgrass mealybug is a problem from Florida through the Gulf States to California. Rhodesgrass, Johnsongrass, bermudagrass, and St. Augustinegrass are preferred hosts of economic importance. There are more than 60 grass hosts, but other than those mentioned above, the other grasses are only occasionally or lightly infested. Heavy infestations can kill infested grasses. Infested grass plants gradually turn brown and die. St. Augustinegrass may become discolored and dead spots appear with severe infestations. Heavy infestations resemble an overdose of fertilizer that has caked around the grass nodes. The adults are parthenogenetic (no males) and reproduce ovoviviparously (live born) over a period of 50 days. The crawlers move about the plants actively, settling near the crown or lower nodes of the plants. The crawlers wedge themselves beneath a leaf sheath at the node, insert their mouthparts in the plant, and become sessile. Shortly afterwards, the felted white waxy sac is secreted.

The life cycle requires 60 to 70 days with five generations annually.

Cultural and biological controls have not proven practical. Control with granular or soil drench systemic is costly but effective.

Twolined Spittlebug
Description

The black, leafhopper-like adult is 6 to 10 mm long and has two red or orange lines across the wings. Its eyes are dark red. The bright yellow-orange, oblong egg is pointed at one end and is about 1 mm long. Red and black areas develop before the egg hatches. Enveloped in a white, frothy mass, the red-eyed immature form may be yellow, orange, or white with a brown head. The wingless nymph resembles the adult in shape but is slightly smaller.

Biology

The twolined spittlebug occurs from Maine to Florida and westward to Iowa, Kansas, and Oklahoma. This spittlebug feeds on many crops, ornamentals, and weeds in addition to turfgrasses. Ornamental and weedy grass hosts include Coastal Bermuda, St. Augustine, centipede, pangola, bahia, rye, crab, Johnson, and orchard grasses. Plants commonly infested by adults are holly, redbud, aster, gerbera, blackberry, pea, peach, honeysuckle, morning glory, and most small grain crops. Spittlebugs are seldom a problem on well-managed turf. Both nymphs and adults extract plant juices through their needlelike mouthparts. Such feeding by large numbers of spittlebugs may check, wither, or kill the growth of turf grasses. Twolined spittlebugs overwinter as eggs in hollow stems, behind leaf sheaths, or among plant debris. Emerging in the spring, the nymphs seek sheltered, humid, hiding places among plants and begin feeding. Soon they exude a white, frothy "spittle" mass which protects them from natural enemies and desiccation. After feeding for at least a month and developing through four instars, nymphs become adults. Most active in early morning, spittlebugs spend the warmer hours of the day hiding deep in the grass. Adults live about 23 days and females spend the last 2 weeks of this period depositing eggs. Hatching occurs about 2 weeks later. Two generations occur annually in the southeast.

Winter Grain Mite
Description

These tiny mites are 1 mm in length as an adult. They are dark brown with a rather distinct white dot on each side of the back. The body may have a tinge of green. The legs and mouthparts are usually reddish. This mite has a dorsal anus that is surrounded by a conspicuous reddish-orange spot. Despite their small size, this spot is a rather distinguishing characteristic.

The tiny eggs are kidney shaped and orange to reddish brown. They are found glued singly to the base of grass or thatch. There is one larval stage and it differs from other active stages by having six legs instead of eight. The larvae are initially reddish orange but become more brown. The nymphs have eight legs and are darker brown. The legs go from yellow-orange to reddish-orange as the nymph develops toward becoming an adult.

Biology

The winter grain mite is distributed widely throughout the U.S. They appear to prefer the cool-season grasses regardless of their location in the U.S. As the mites feed, they rasp the leaf surface and suck plant sap. This feeding gives the plant a scorched appearance that is the result of chlorophyll loss. The most severe damage occurs during the winter months when the mite is most active. Populations usually increase most rapidly during November and December and decline in April. The winter grain mite feeds during the winter and survives summers as eggs. Up to two generations per year are reported in southern states. Eggs hatch in the fall, and in areas where a second generation occurs, egg-laying begins by March. Females lay 30 to 65 eggs over approximately 40 days.

Many predatory insects and mites feed on winter grain mites. Avoid unnecessary insecticide treatments late in the season which may suppress predators and thus encourage winter grain mite populations.

BURROW INTO LEAVES AND STEMS
Billbugs
Description

Billbugs are brown to black, hard-shelled, snout beetles which are usually covered with clay or soil particles. They vary in length from 5 to 11 mm, depending on the particular species. The elongate, creamy white eggs are 2 to 3 mm long and turn yellow before hatching. The white, legless grubs have hard, yellowish- or reddish-brown heads. When mature, billbug grubs vary from 9 to 16 mm in length. The pale yellow or white pupae are similar to adults in size and shape.

Biology

The hunting billbug occurs along the eastern seaboard from Maryland to Florida. It has also been reported in California and Hawaii. Bluegrass billbugs are generally troublesome across the northern half of the U.S. Most species of grasses are subject to billbug infestation. However, hunting billbugs prefer zoysia and Bermuda grasses.

Both grubs and adults injure turf grasses. The larvae are root feeders; the beetles eat leaves and burrow in stems near the surface of the soil. Variously shaped, yellow or brown patches appear in infested lawns. The symptoms may resemble fertilizer burn. However, in the case of billbug-infested lawns, tufts of discolored grass can be easily pulled up (unlike grass burned by fertilizer).

In the southern states, billbugs may overwinter in any life stage, although adults are best able to withstand severe winters. Emerging in the spring, adults feed and mate. Females place eggs in cells cut into grass stems. Larvae appear 2 days to 2 weeks later and work their way down from the inner leaves to the root system. They feed for 3 to 5 weeks before pupating in cells in the soil. The pupal period lasts 3 to 7 days. Afterwards, the new adults may remain in the pupal cells to overwinter or may emerge and be active until the onset of cold weather. One generation is completed each year.

Billbug populations may be managed by using resistant turfgrass variations. Some perennial ryegrasses and fine fescues with fungal endophytes are resistant to billbugs.

Annual Bluegrass Weevil
Description
This weevil is about 4 to 5 mm long and is dark brown to black. A snout projects forward and downward from the front of the head like a small billbug. A conspicuous knobbed antenna is attached to each of the snout sides near the tip. Elongate, pale yellow to white eggs are found in grass stems. The eggs are less than 1 mm long.

The grub is creamy white with no legs and a rather pointed abdomen. The head is usually dark brown. Small larvae are about 1 mm long and mature larvae are between 4 and 5 mm long. Pupae look similar to adults, but are lighter in color at first and then gradually darken as they mature.

Biology
Annual bluegrass weevils are found in the New England area southward into northeastern Pennsylvania. Infestations are primarily confined to golf courses. This pest is capable of feeding on several hosts, but in reality is only a problem on close cut bluegrass such as golf course fairways, tees, and greens. The adults do feed on the leaves and stems, but the damage is usually of minimal importance.

The grub damage initially appears as small yellow-brown spots on the grass. Larvae sever stems from the plant and can kill many stems. The final instar grub is the most damaging to the grass. As adults spread and lay eggs in the spring, first-generation damage appears in late May and early June. Later-season damage from the second generation is usually less severe.

The annual bluegrass weevil overwinters as an adult under trees where there is adequate leaf litter. Migration from hibernating sites often coincides with the flowering of forsythia and all adults have dispersed by the time dogwoods bloom. Egg laying begins shortly after adults begin to feed, and development from egg to adult takes about 2 months. A second generation of larvae occurs usually in late July and early August.

Few natural enemies are known, although general predators such as carabid beetles and blackbirds commonly feed on the larvae.

SUBSURFACE OR UNDERGROUND PESTS — FEED ON ROOTS OR DISRUPT THE SOIL

Asiatic Garden Beetle
Description
The adults are 7 to 10 mm long and wedge shaped. They are a chestnut brown. The abdomen protrudes slightly under the wing covers, and the undersurface by the legs has a covering of short yellow hairs. The hind legs are distinctly larger and broader than the others.

The eggs are laid in small clusters of up to 15. These are loosely held together in a gelatinous mass. The individual eggs are oval and about 1 mm long. The larvae are about 1.4 mm long and have light-brown heads when they first hatch. Full-grown larvae are about 15 to 18 mm when stretched out. These grubs are identified by a curved row of brown spines across the anal area. The pupae are in the last larval skin and are about 8 to 10 mm long. They are initially white but gradually turn tan.

Biology
This insect is most common in the northeastern U.S. from New England across to Ohio and down into South Carolina. The grubs may be grouped around weedy areas and next to flower beds or gardens since the adults prefer these areas. The adults strip the foliage off plants, leaving them looking ragged and much different from the skeletonized appearance of the Japanese beetle feeding.

Most of the adults are found from mid-July to mid-August. The females tend to seek out turf and pastures to lay eggs and deposit them 1 to 2 in. deep in the soil. The females lay about 60 eggs over a several-week period. The eggs hatch in about 10 days during the summer. The young larvae dig to the surface where they feed on roots and decomposing organic matter. Most first instar larvae are found in August and early September. The second instars are found in September, and many do not reach third instar until the following spring. As cool temperatures arrive in the fall the larvae burrow deep in the soil to pass the winter. They return to the soil surface in the spring and are mature by mid-June at which time they pupate in the soil. The pupal stage only lasts 8 to 15 days.

Ataenius
Description

The genus *Ataenius* includes 40 species of small black scarab beetles about 5 mm long. Although small, their form is blocky and robust like that of most other scarabs such as Japanese beetles, May beetles, and June beetles. *Ataenius* larvae are white grubs resembling typical white grubs but are only about 5 mm long when full-sized.

Biology

Species of *Ataenius* occur throughout the U.S., but most species are eastern and southern. *Ataenius spretulus*, the black turfgrass ataenius, is most often associated with turf damage, and has been reported from all but seven states. So far it has been regarded as a turf pest only from about 20 states, mostly north of and including Missouri, Kentucky, and Virginia. As more attention is directed toward these pests, the distribution of the black turfgrass ataenius and related species as turf pests may be found to extend further south.

Many species of *Ataenius* typically feed on cow dung and on fungi. In recent years the larvae of *Ataenius* have been found to feed on grass roots, particularly those of bluegrass and bentgrass. Presumably turf with thick root thatch resembling the cow dung habitat would be more attractive to the pests.

Infested sod wilts and dies particularly during dry weather due to damage to the roots just below the thatch line. The sod is loosened and cut off from deeper water sources. As many as 500 larvae per square foot may be found.

In Ohio the insect overwinters as an adult 2.5 to 3.5 cm deep in well-drained soil near turf areas. The beetles emerge from late March through April, and lay eggs in turf throughout May. The larvae of this generation are most numerous from early June to early July. The larvae pupate, and then adults emerge from late July through mid-August. During this time the eggs for the second generation are laid. Second generation beetles emerge, mate, and fly to overwintering sites during September and October. When *Ataenius* infestations are studied farther south, slight variations in the cycle may be expected due to differences in climate.

The only natural control observed has been the high incidence of milky disease in some populations of black turfgrass ataenius.

European Chafer
Description

Adult European chafer beetles look like small versions of what most people refer to as June beetles. The reddish-brown beetles are about 13 mm long with a more yellow head. The pearly white, slightly oblong eggs are less than 1 mm long when first laid and expand to almost 3 mm long just before hatching.

The grub have the typical C-shaped appearance and grow to be about 23 mm long. Bristles on the tail end of the grub allow separation from other species. Most prominent is a Y-shaped end slit unlike any other grub.

Biology

The European chafer was introduced from Europe sometime around 1940. Its range is restricted to the Northeast with some infestations present in northern Ohio, Pennsylvania, and New Jersey. The grubs are considered very damaging and feed on all turfgrasses and nursery stock. They appear earlier and feed longer than the Japanese beetle. The grubs feed throughout the root zone and are considered more destructive than other white grubs. Adults may congregate by the thousands on trees and shrubs in the summer but are not damaging. Turfgrass fed upon by European chafer grubs exhibits typical grub symptoms of spongy sod, wilting grass, and irregular dead patches. This may be more common in acidic, wet soils and near trees.

Adults are seen in June and July and may be seen swarming around trees and shrubs in the early evening. The females then soon lay about 20 eggs in the soil. First instar grubs may be found as early as late July. By fall the grubs begin to migrate down below the freeze line and in May move upward again.

Some strains of milky spore infect European chafer grubs. In the field, such levels of infection have always been low, so milky spore is probably not a major factor in providing biological control. Some parasitic insects from Europe have been released in New York, but there is little evidence of any effect. Chemical controls must be applied in the late summer or early fall.

European Crane Fly or Leather Jacket
Description

These large crane flies have slender bodies almost 20 mm long with very long legs. The adults are usually tan to brown with smokey-brown wings. The eggs are black, 1 mm long and oval with one side flattened and the other pointed. They are laid in the turf. The eggs hatch into maggots which are white and worm-like.

As these larvae grow they turn grayish brown and develop a tough skin and thus take on the name "leather jackets." The larvae may exceed 1 in. in length. They have two breathing holes at the end of the anal segment and are surrounded by six fleshy lobes. The inch-long brownish pupae have the legs, wing pads, and antennae glued down and are found just below the soil surface.

Biology

This insect is primarily a pest in the Northwest. Crane flies look like giant mosquitoes when they emerge from lawns, pastures, and ditches anytime from late August to mid-September. The harmless adults mate, and females begin laying eggs immediately. In about 2 weeks the eggs hatch into small brown maggots which begin feeding on plant roots and foliage. Larvae feed slowly during the winter. Turf damage is most evident in the spring. The larvae stay underground during the day but come to the surface to feed on damp, warm nights. Feeding damage usually stops by late May into June. The larvae then become pupae. These remain just below the soil surface for almost 2 weeks during August. At emergence, the pupae wriggle to the surface of the turf and the adults appear in late August through September.

Green June Beetle
Description
The beetle is 15 to 22 mm long with dull, velvety green wings. Its head, legs, and underside are shiny green, and its sides are brownish yellow. When the egg is first laid, it is pearly white and elliptical (1.5 by 2.1 mm). It gradually becomes more spherical as the larva inside develops. The newly hatched larva is 8 mm long and grows to a length of about 40 mm. Whitish with a brownish-black head, the grub has conspicuous brown spiracles along the sides of its body. The brown pupa, approximately the same shape as the adult, becomes metallic green just before the adult emerges. It is about 15 mm long and 15 mm wide.

Biology

The green June beetle occurs in the eastern U.S. westward to Kansas and Texas. Green June beetles prefer ripening fruits of many plants. The grubs feed on decaying organic matter in the thatch and root zone of many grasses, as well as on the underground portions of other plants such as sweet potatoes and carrots.

Adult and larval feeding on economic crops causes some financial loss; however, the grubs tunneling for feed and the adults burrowing into the soil each night cause more serious destruction. The tunneling uproots young plants. The many exit holes of the adults and larvae resemble ant hills and mar lawns and golf course greens.

Unlike the May beetles, only 1 year is required for these beetles to complete their life cycle. They overwinter as grubs that may become active on warm winter days. They increase their activity in the spring, and in June pupate in earthen cells several centimeters underground. The pupal stage lasts about 18 days; adults appear in July and August. In midsummer, adults lay eggs underground in earthen balls. Each female lays 60 to 75 eggs over a span of about 2 weeks. About 18 days after the eggs are laid, they hatch into small white grubs. The larvae molt twice before winter. The third larval stage lasts nearly 9 months, after which pupation occurs. At night, the larvae may be found on the ground crawling on their backs. This curious form of locomotion is peculiar to the green June beetle.

Sections of turf 929 cm^2 (about 1 ft^2) and 5 to 10 cm (about 2 to 4 in.) deep should be examined for green June beetle grubs. On golf course fairways 10 to 20 samples of this size should be taken. If examination reveals an average of six to eight larvae per 929 cm^2, treatment is usually necessary. Do not apply manures or organic fertilizers during the summer months since adults are attracted to these materials.

Ground Pearls
Description
About 1.6 mm long, the adult female is a pinkish scale insect with well-developed forelegs and claws. The male is a gnatlike insect varying from 1 to 8 mm in length. Clusters of pinkish-white eggs are laid in a white waxy sac. Commonly referred to as a ground pearl, the slender nymph is covered with a hard, globular, yellowish-purple shell 0.5 to 2.0 mm in diameter.

Biology

Ground pearls are potentially serious problems in both southeastern and southwestern states. The roots of bermuda, St. Augustine, zoysia, and centipede grasses are most commonly infested with ground pearls. The ground pearl nymphs extract juices from underground plant parts. The damage is most apparent during dry spells when irregularly shaped patches of grass turn yellow. The grass in these spots eventually turns brown and usually dies by fall.

Overwintering takes place in the ground pearl stage. Females usually reach maturity in late spring and emerge from their cysts. After a brief period of mobility, the wingless females settle 5 to 7.5 cm deep in the soil and secrete a waxy coat. Within this protective covering, females develop eggs (without mating) and deposit them throughout early summer. Approximately 100 eggs are laid by each female. The slender nymphs emerge in mid-summer and infest grass rootlets. Once they initiate feeding, nymphs soon develop the familiar globular appearance. There is usually one generation each year. However, if conditions are not favorable for emergence, female nymphs may remain in the ground pearl stage for several years.

Insecticides have not been effective against ground pearls. Good cultural practices, such as watering and fertilization may help lawn grasses recover from injury, but such beneficial effects may be only temporary.

Japanese Beetle
Description
About 13 mm long, this shiny, metallic green beetle has coppery brown wing covers which extend almost to the tip of the abdomen. Two small tufts of white hairs occur at the tip of the abdomen just behind the wing covers. Five more white patches are located on each side of the abdomen.

The translucent white to cream-colored egg is elliptical and about 1.5 mm in diameter when first laid. By the time it is ready to hatch, the egg is more spherical in shape and has doubled in original size.

The grayish-white, slightly curled grub has a yellowish-brown head and measures about 26 mm long when mature. It can be distinguished from other white grubs by two rows of spines which form a "V" on the underside of its last abdominal segment.

The cream-colored pupa, approximately 13 mm long and 6 mm wide, gradually turns light brown and finally develops a metallic green cast.

Biology
First reported in North America in 1916, the Japanese beetle now occurs in over 20 states from southern Maine southward into Georgia westward into parts of Kentucky, Tennessee, Illinois, Michigan, and Missouri. The grubs are serious pests of lawns, other grasses, and nursery stock. Tender grasses are preferred to tougher varieties. The adult beetles infest over 275 different plants, including many shade and fruit trees, ornamental shrubs, small fruits, garden crops, and weeds.

Japanese beetle grubs may be abundant in well-kept lawns, pastures, and golf courses. They burrow through the soil, severing and consuming roots. Large areas of dead brown grass often appear in infested lawns when large numbers of grubs (ten or more per square foot) are present or during dry spells. Such dead areas are noticeable by September or early October. Unlike the grubs, Japanese beetle adults eat foliage, leaving only a lacy network of leaf veins.

The grubs overwinter in cells within 13 cm of the soil surface. In spring, they move upward, almost to ground level, where they complete feeding and pupate. About 140 days elapse from egg hatch to larval maturity. Adults emerge as early as mid-May in eastern North Carolina and as late as July in New England. However, even in the beetles' southernmost range, peak emergence occurs in July. Throughout summer, the beetles attack the fruit and foliage of many plants. Soon after emerging, females deposit 40 to 60 eggs in small batches 5 to 8 cm deep in the ground. Under extremely dry conditions, many eggs and young grubs perish. However, during warm, wet summers populations thrive and eggs hatch about 2 weeks after deposition. The newly emerged larvae feed until cold weather forces them into hibernation. One generation occurs each year.

Commercial preparations of the milky spore disease may offer a long-lasting method of soil treatment. These bacterial spores infect and kill Japanese beetle grubs. Multiplying within the grubs, the bacteria provide semipermanent protection against these pests. Milky spore preparations can be applied from July until the first hard freeze to areas of turfgrass which are mowed 5 to 8 cm (2 to 3 in.) high at regular intervals. These areas are preferred egg-laying sites for adults, and during the warm months, grubs feed close to the soil surface. Though long-lasting, the milky spore application is expensive.

May Beetles
Description
Many species of May beetles (also known as June beetles) occur in any given area. They are shiny, robust insects reddish brown to black in color. Oblong in shape, they reach a length of 20 to 25 mm. The eggs are pearly white and oblong. Each one, initially about 2.5 mm long and 1.5 mm wide, becomes slightly larger as the larva inside grows. Commonly called white grubs, the larvae are white and C-shaped, with a distinct, brown head. Young larvae are about 5 mm long, but attain a length of about 25 mm. Two rows of hairs on the underside of the last abdominal segment distinguish true white grubs from similar grubs. The oval, brownish pupae occur within earthen cases.

Biology
More than 200 species of May beetles occur throughout North America. Therefore, a single species population is seldom found. In North Carolina they are most numerous from the piedmont to the coast. Although oaks are the favorite food source, adult May beetles also feed on the foliage of many

other trees. Larvae prefer lespedeza, sod, and corn, but they too have additional host foods which include lawn grasses and nursery plantings. Both larvae and adults are destructive. The adults are defoliators, chewing the leaves of various hardwood trees. The grubs feed on and injure the root systems of grasses and other plants. Heavily infested turf can often be rolled up like a carpet, exposing the white grubs.

May beetles have a 2- or 3-year life cycle, depending upon the species. They overwinter in the soil as larvae in two distinct sizes and as adults that have never flown. In the spring, the adults emerge from the ground in the evening, feed on tree leaves, and mate during the night. They return to some sheltered site in the morning. Females then enter the ground to deposit about 50 eggs in earthen balls. The egg-laying period lasts a couple of weeks. In 3 to 4 weeks, grubs hatch from the eggs and feed on dead organic matter, later moving to the roots of plants. The larvae molt twice, the second and third instars being the overwintering forms. In late August, the second and third instars burrow over 1 m deep into the ground to hibernate. The larvae do the most damage during the second year. In early spring, third instar larvae construct earthen cells in which they pupate. Adult beetles emerge from pupal cases in late summer, but do not leave the ground; instead, they overwinter there and emerge the next spring.

Sections of turf approximately 929 cm² (1 ft²) and 5 to 10 cm (2 to 4 in.) deep should be examined for May beetle grubs. On golf fairways, 10 to 12 samples of this size should be taken. If examination reveals an average of three grubs per 929 cm² (roughly 1 ft²), treatment is probably necessary.

Masked Chafers
Description
Two separate species, the northern masked chafer and the southern masked chafer, are important turfgrass pests. The life stages of both species are very similar. The adults are about 12 mm long and are yellowish brown. They both have dark brown heads. Pearly-white oval eggs less than 2 mm long are laid in the soil.

The grubs are similar in appearance to typical Japanese beetle grubs with the C-shaped appearance. Grubs of the two species are virtually impossible to tell apart. A specialist can readily separate these grubs from the species based on the bristle pattern on the rear of the abdomen. With training, turfgrass managers can accurately determine these differences.

Biology
Both masked chafers are native to the U.S. and found throughout the eastern two thirds of the coun-

try. The northern masked chafer is commonly found in the northern states but extends south and west to Alabama and Missouri. The southern masked chafer is most common in Kentucky, Indiana, Illinois, Missouri, and to some degree further west and south.

Masked chafer grubs attack all turfgrasses like other white grubs, but are especially damaging to Kentucky bluegrass. However, the grubs of masked chafers feed more extensively on organic matter in the soil than most other grubs. Consequently, higher populations of masked chafers can be tolerated before treatment is justified. Damage turf exhibits typical grub damage symptoms of wilting grass, spongy sod, and dead patches. The grubs cause most of their damage in the late summer and early fall and little damage in the spring. The adults are not damaging.

Adults are usually seen in July when they are flying over the turf and laying eggs. Flights may begin in mid-June and continue through mid-August. The small grubs are present in the soil as early as late June and are fully developed grubs by September. They overwinter in the soil and return near the surface to feed again in March and April. In May, they pupate to form the adult stage.

Commercial milky spore has never proven to be effective against masked chafers despite the fact the disease is found naturally occurring in chafer grub populations. At least one parasitic wasp has demonstrated a preference for attacking chafer grubs.

Mole Crickets
Description
Mole crickets are covered with fine hairs which give them a velvety appearance. These 3.8-cm long, brown mole crickets have beady eyes and stout, shovel-like front legs for digging. The greenish, oval eggs are about 3.2-mm long. Although wingless and slightly smaller, the nymphs resemble adults in shape.

Biology
Thriving best in moist light soils, mole crickets are common in Coastal Plain areas from North Carolina to Texas. They are particularly attracted to soil containing manure or rotting crop debris. Their vertical distribution in the soil varies with temperature and moisture. Mole crickets have a wide host range and are particularly damaging to the underground plant parts of vegetable crops, tobacco, peanuts, strawberries, and all grasses. Burrowing in the soil, mole crickets feed at night on roots, stems, and tubers. The burrows cause the soil to dry out, thereby affecting plants that are not actually fed upon. Some plants may be up-

rooted, but turf usually dies from root damage and drought. Small numbers of mole crickets are capable of extensive damage especially on newly seeded turf. One mole cricket may cover several meters (or several yards) each night. On golf greens, the raised tunnels made by mole crickets are skimmed off by the mower.

Mole crickets generally overwinter as nymphs 7.5 to 25.0 cm (3 to 10 in.) deep in the soil. These nymphs become active in March and feed until they mature in late spring. In May or June, the new adults emerge from the soil and are attracted to lights as they engage in mating flights. Eggs are laid in the soil in cells constructed by the females. Approximately 35 to 50 eggs are placed in each cell. Hatching occurs in 10 to 40 days depending on temperature. Nymphs develop through eight instars and may become adults by winter or many overwinter as immatures. One generation occurs each year.

Oriental Beetle
Description
The adult oriental beetles are typical oval scarabs about 13 mm long. Adults are marked in a variety of patterns, but most are straw-colored with varying black markings on the back. Eggs are small (1.5 mm), oval, and smooth.

First instar grubs are only about 4 mm long, but grow to 8 mm before becoming a second instar grub. The fully grown third-instar grubs may be 25 mm in length. These grubs look very similar to the Japanese beetle grub (identification should be done by a person familiar with the characteristics).

Biology
Oriental beetles apparently came to the U.S. from Japan in 1920. They are probably native to the Philippines. Oriental beetles are currently found in six northeastern states plus North Carolina. Adults sometimes feed on numerous flowering plants, but are rarely a real pest. Grubs, however can be quite damaging to the roots of turfgrass, primarily cool-season grasses. The feeding results in root pruning very near to the soil surface. Damage symptoms are similar to the Japanese beetle. The life cycle is also very similar to the Japanese beetle. Third instar grubs overwinter and pupate in May or June. The adults are present in June through August with most eggs laid in July and early August. The grubs then appear in August and feed before overwintering. Since this is an introduced pest, few natural enemies are present. Oriental beetles are commonly confused with Japanese beetles and proper identification is important since they can react differently to chemical and/or biological controls. Oriental beetle grubs are frequently found in turfgrass in close proximity to flowers and shrubs.

Shorttailed Cricket
Description
Shorttailed crickets are similar to field crickets except for the short ovipositor. They are light brown with fully developed black wings. The body length ranges from 14 to 17 mm. The off-white, oblong egg is 1.8 to 2.8 mm long. The color of the nymphs is generally light brown in all stages.

Biology
In the U.S., shorttailed crickets occur along the Atlantic coast from New Jersey to Florida and west into Louisiana, southeastern Texas, and the eastern half of Oklahoma. This insect feeds on grasses, weeds, pine seeds, and pine seedlings. In turf, burrows constructed by the nymphs and adults result in unsightly mounds of small soil pellets which may smother the surrounding grass. Burrows are most abundant when soil is moist and just after a rain during the warm part of the season. Nymphs and adults may feed upon blades of grass at night but the extent of damage is apparently negligible. Shorttailed crickets overwinter as nymphs in the next-to-last stage. After several molts in early spring the insect reaches the adulthood stage. Mating occurs and oviposition begins in late spring or early summer. Hatching takes place in a multichambered burrow constructed by the adult. For a short period of time both eggs and nymphs may be found in the burrow. Between the fourth and sixth instars, nymphs begin to leave the parent burrows and construct burrows of their own. At first burrows are small, 2 to 3 mm in diameter and 5 to 10 cm deep. As the shorttailed cricket matures, its burrow is enlarged and may reach a depth of 30 to 51 cm (12 to 20 in.). Only one shorttailed cricket is found per burrow except for a short time when certain burrows contain eggs and nymphs.

Certain species of spiders are suspected predators of shorttailed crickets, and several tachinid flies are parasites of the cricket. At present there is no chemical registered specifically for shorttailed cricket control.

REFERENCES

1. **Baker, J. R.,** *Insects and other Pests Associated with Turf,* AG-268, N.C. Agricultural Extension Service, Raleigh, NC, 1986, 188 pp.
2. **Shetlar, D. J., Heller, P. R., and Irish, P. D.,** *Turfgrass Insect and Mite Manual,* The Pennsylvania Turfgrass Council, 1983, 67 pp.
3. **Tashiro, H.,** *Turfgrass Insects of the United States and Canada,* Cornell University Press, Ithaca, NY, 1987, 391 pp.

Chapter 22

Major Insect Pests of Ornamental Trees and Shrubs

Peter B. Schultz, *VPI & SU, Hampton Roads Agricultural Experiment Station, Virginia Beach, VA*

David J. Shetlar, *Department of Entomology, The Ohio State University, Columbus, OH*

CONTENTS

Introduction 237
Chewing Insects — Foliage Feeders 238
    Defoliators 238
        Cankerworms 238
        Gypsy Moth 239
    Specialized Feeders 239
        Elm Leaf Beetle 239
    Case Makers and Nest Formers 240
        Bagworm 240
        Fall Webworm 240
        Tent Caterpillars 241
    Leafrollers and Leaftiers 241
Chewing Insects — Borers 241
    Root Weevils 242
    Moth Borers 242
        Dogwood Borer 243
        Lilac (Ash) Borer 243
        Pine Tip Moths 243
Chewing Insects — Leafminers 244
        Boxwood Leafminer 244
        Holly Leafminer 244
Sucking Insects 245
    Aphids 245
    Lace Bugs 245
        Azalea Lace Bug 245
    Scales 246
        Euonymous Scale 246
    Plant Bugs 246
    Mites 246
        Two-Spotted Spider Mite 247
Gall Insects 247
References 247

INTRODUCTION

Insect and mite pest control on ornamental trees and shrubs presents special problems for the plant care manager. The vast number of plant species makes it difficult to identify the multitude of insects and mites which may attack each plant. Unlike agronomic or greenhouse crops, ornamental trees and shrubs are established in an area for a long period of time, and pest infestations may be sporadic from year to year.

There are only a few species of insect and mite pests which attack a large number of different ornamental plants. Most pests are host specific, or at least, attack plants which are closely related and cause similar types of damage on respective host

0-87371-350-8/94/$0.00+$.50
© 1994 by CRC Press Inc.

plants. Thus, the yellow specks found on a leaf caused by spider mites may look the same on oak, boxwood, roses, or spruces. However, the oak spider mite, the boxwood spider mite, European red mite, and spruce spider mite are the respective pests. Fortunately, the similar types of damage coupled with host specificity means that the majority of pests can be identified by knowing the type of damage and the host plant.

A good knowledge of many kinds of insects would be helpful in identifying pests of ornamentals. However, by looking for damage and associated insects, a correct determination can be made. Insects and mites attacking ornamental plants are generally classed into the following categories according to the type of damage caused:

> **Chewing pests — foliage feeders** have mouthparts which chew holes in foliage (skeletonizers, shot holes, notching) or eat all of the plant foliage (defoliators). Many pests in this group use plant parts in which to hide (leafrollers and leaftiers) or make nests with silk (bagworms, tent caterpillars, and webworms).
> **Chewing pests — borers** tunnel into stems and trunks. The shape, size, and direction of the tunnels are usually distinct for each pest. Larvae of some weevil species drop to the ground, enter the soil and feed on roots.
> **Chewing pests — leafminers** are small insects with larvae which feed between the upper and lower leaf layers. This mining produces characteristic trails and marks visible on the surface of the leaf.
> **Sucking pests** are a bit harder to diagnose but they generally cause leaf curl, spotting, and speckling of the foliage, deposits of excrement (honeydew or tar spots), or growth of sooty mold on the honeydew. There is a vast array of insects and mites in this category and they are better identified by taxonomic categories rather than by damage symptoms.
> **Gall pests** form species-specific abnormal growths on leaves, stems, or roots. These pests feed on the insides of the abnormal growths.

The following sections will discuss these types of pests and their signs of damage. Early detection of pest damage is important so that population levels can be assessed. Also, a vigorously healthy plant will withstand a considerable pest load while a damaged, unhealthy plant may readily succumb. And finally, many insects (or damage) will remain at low levels and cause relatively little damage in most years, especially if pesticides are only used as spot or target treatments and the parasite predator complex is diverse.

CHEWING INSECTS — FOLIAGE FEEDERS

DEFOLIATORS
Insects which feed on the entire leaf or most of the leaf are often difficult to identify unless an actual specimen can be found. Their damage is *defoliation* and they are commonly called *defoliators*. They usually hide under leaves or on stems and trunks of the host plant during the daytime. Most of these pests are the large caterpillars and sawfly larvae. Examples of these pests include: redhumped caterpillar (*Schizura concinna*), mourningcloak butterfly (*Nymphalis antiopa*), yellownecked caterpillar (*Datana mimistra*), Douglas-fir tussock moth (*Orgyia pseudotsugata*), and the sawflies — redheaded pine sawfly (*Neodiprion lecontei*) and willow sawfly (*Nematus ventralis*).

Cankerworms
Cankerworms are commonly known as loopers, spanworms, inch worms, and measuring worms. The worms drop from trees on their threads creating a nuisance in urban areas. Both the fall and spring cankerworms are distributed from Maine to North Carolina and west to Texas and north to Montana.

Description and Life Cycle
Moths of the fall cankerworm emerge in December (later in cool climates) from pupae in the ground. The wingless female moth crawls up the tree and deposits a single layer of gray eggs on the trunk, branches, or twigs. The spring cankerworm moth appears in early spring. The female is also wingless and resembles the fall cankerworm. The eggs are laid in a cluster under bark scales. The eggs of both species hatch in the spring when leaves first appear. Larvae are brown and green, with three narrow stripes along their body and a yellow stripe beneath. Mature larvae drop to the ground and pupate in June, 1 to 4 in. under the soil surface. There is only one generation per year. Cankerworm numbers fluctuate greatly from year to year.

Damage
The fall cankerworm prefers to feed on apple, oak, and elm. However, it will also eat many other tree species. The spring cankerworm also prefers to feed on apple and elm trees. Both species may completely defoliate trees by skeletonizing the

leaves. Trees that are heavily infested have a scorched appearance, and are weakened by repeated defoliation.

Control

Sticky bands can be placed around those trees preferred by the fall cankerworm in mid-December, and in mid-March for spring cankerworms. These bands will trap female moths as they attempt to migrate to the tops of trees to lay eggs. In heavy infestations, the microbial insecticide *Bacillus thuringiensis* can be sprayed.

Gypsy Moth

The gypsy moth, *Lymantria dispar* (L.) is extremely destructive to most trees and shrubs. Gypsy moth larvae eat the foliage of nearly all forest, shade, and fruit trees. Damage can be extensive if populations are not controlled. Introduced into Massachusetts in 1869, the gypsy moth has spread south and west.

Description and Life Cycle

The adult male gypsy moths are brownish gray with irregular, dark markings. The female adult moths are larger, with a wingspan of 2 in., and are creamy white with contrasting dark markings. Moths emerge in the middle of July. The female does not fly and attracts the male by emitting a sex attractant. Eggs are laid in summer on trees or nearby objects, and are covered with tan, hair-covered masses. Caterpillars hatch in April or May depending on the latitude and grow to 3 in. long by early summer. The larvae are brown-black and very hairy. The older larvae have five pairs of blue spots and six pairs of red spots on their backs. The larvae feed at night and descend to a cooler, shadier place during the day. Gypsy moth pupae are reddish brown and about 1 in. long. Pupae attach to all types of objects and are easily transported to new regions.

Damage

The gypsy moth larvae feed on the foliage of many different trees and shrubs and can defoliate a plant. Several years of defoliation may kill the tree or shrub. Evergreens may be killed by 1 year of defoliation.

Control

Egg masses may be collected and destroyed. Larvae may be collected and destroyed by banding trees with a coarse cloth. Individual trees can be protected by banding them with a sticky barrier band. This band traps and prevents larvae from migrating to the canopy of the tree. Parasitic and predatory insects, e.g., *Calosoma sycophanta* (L.), that feed on the pest in Europe have been intro-duced into the U.S. These insects and a viral disease control the moth under some conditions. Recently, a fungal disease has greatly reduced populations in the northeastern U.S. Applications of *Bacillus thuringiensis* may be used to control the gypsy moth. Spraying should be done when leaves have expanded but caterpillars are small, usually in early or mid-May. Pheromone traps can also be used to monitor adult populations, and are also used to confuse the male's ability to locate females for mating.

SPECIALIZED FEEDERS

Many chewing insects do not feed on the entire leaf of a plant but feed on certain parts. Many beetles and caterpillars eat the soft tissues between the veins of a leaf. This type of feeding which leaves behind the lacy skeleton of a leaf is called **skeletonizing**. Common skeletonizers are the leaf beetles in the family Chrysomelidae, such as the elm leaf beetle (*Pyrrhalta luteola*) and the gray willow leaf beetle (*Tricholochmaea decora*); small caterpillars such as the oak skeletonizer (*Bucculatrix ainsliella*) and birch skeletonizer (*B. canadensisella*); and the slug sawflies such as the pear sawfly (*Caliroa cerasi*). The Japanese beetle (*Popillia japonica*) is the most common skeletonizer in the eastern states.

Some chewing insects feed from the side of the leaf causing a **notching**-type of damage. Most of the root weevils, a few caterpillars, and leaf cutting bees do this type of damage. A common example is the black vine weevil (*Otiorhynchus sulcatus*).

Shot hole damage is caused by small caterpillars and some beetles — especially flea beetles and May/June beetles. This damage consists of small round holes chewed in the leaf. Often this damage occurs while leaves are still growing. After the leaf is fully extended the holes become much larger and irregular in shape.

Elm Leaf Beetle

The elm leaf beetle was introduced into the U.S. from Europe and has now spread throughout the country and into Canada. It occurs on almost every species of elm.

Description and Life Cycle

The elm leaf beetle adult is $1/4$ in. long. They are yellow to olive green with black stripes along the outer margin of each wing, and have yellow legs. In spring, adults feed for some time before laying eggs. This causes irregular holes in leaves. The adult female lays eggs in the spring (later in northern latitudes) on the underside of the elm leaf. The eggs are yellow-orange, and occur in groups of 5 to 25. Young larvae are black, later becoming dull yellow and $1/2$ in. long. They feed on the lower leaf surface. The larvae migrate to the base of the tree to pupate on the ground or in bark crevices and protected areas. Pupae are bright orange to yellow, and covered with black bristles. After emergence, the beetle moves to the leaves to feed and lays eggs for a second generation. The beetle overwinters in protected areas and may become a problem inside homes.

Damage

Larvae injure the foliage by skeletonizing the lower leaf surface. Damage occurs primarily to the foliage of the various elm species. Leaves may turn brown and drop. Trees which have been defoliated several consecutive years may be weakened or die.

Control

Egg parasites for control of elm leaf beetle are currently being evaluated and may eventually provide biological control. Resistant elm cultivars are available for new plantings. A strain of *Bacillus thuringiensis* is also available for controlling larvae.

CASE MAKERS AND NEST FORMERS

Some beetle and moth larvae construct cases around themselves by gluing together pieces of plants, or spin cases out of silk and attach plant material to the outside of the cases. Examples of these chewing insect pests include the bagworm (*Thyridopteryx ephemeraeformis*), the pistol casebearer (*Coleophora malivorella*), and the elm casebearer (*C. ulmifoliella*). Several common caterpillars live together in nests constructed of silk on their host plants. Examples of this group include the eastern tent caterpillar (*Malacosoma americanum*) and Pacific tent caterpillar (*M. constrictum*), the fall webworm (*Hyphantria cunea*), the mimosa webworm (*Homadaula anisocentra*), and the uglynest caterpillar (*Archips cerasivoranus*). Because this group is often protected by their cases and webs, they can be difficult to control with contact poisons. The following species have typical habits for this group.

Bagworm

The bagworm, *Thyridopteryx ephemeraeformis* (Haworth), is found throughout the eastern two-thirds of the U.S. and has a wide host range. It is particularly damaging to evergreens such as arborvitae and juniper. It is often overlooked until after serious damage has been done and it is too late for control measures.

Description and Life Cycle

Bagworms are easily identified by the spindle-shaped bag of silk covered with bits of needles or leaves that the larvae construct around themselves. The fragments of needles and leaves on the bags provide such an excellent camouflage that these pests are often not seen until considerable defoliation has occurred. Eggs are laid and overwinter inside the bag of the female; larval hatching begins in May and extends into June in the mid-Atlantic region. Newly hatched larvae immediately spin a bag of silk around themselves, attaching bits of foliage to the outside as they feed and enlarging the bags as they grow. The full-grown larvae are $3/4$ to 1 in. in length and are enclosed in bags that are $1^1/2$ to 2 in. long. At maturity the larvae attach their bags to twigs with a strong band of silk. The larvae transform into the pupal stage inside the bags in late summer, and become adults in the fall. The male moth flies to mate with the wingless and legless female which remains within the bag. Each female lays 500 to 1500 eggs in the bag, then dies. There is only one generation per year.

Damage

Bagworms feed on many kinds of evergreen and deciduous ornamentals. If there is a heavy infestation, bagworms will strip evergreens of their foliage and cause mortality. Deciduous trees are rarely damaged to that extent.

Control

Infestations on small trees and shrubs can be controlled by hand-picking the bags. For best results this should be done in the winter and spring before the eggs hatch. Bags should be collected and destroyed. Native parasites provide a degree of natural control. Chemical control is only effective in the spring and early summer, when the bags are $1/4$ to $1/2$ in. long, or as soon as there is noticeable damage to plants. The microbial pesticide *Bacillus thuringiensis* is a biorational product labeled for control.

Fall Webworm

The fall webworm, *Hyphantria cunea* (Drury), appears in summer and early fall. Its unsightly webs are more cosmetic than detrimental to the tree or shrub. The fall webworm can be found throughout the U.S. There is one generation annually in northern areas, two generations in the south.

Description and Life Cycle

Moths emerge in the spring. Their wings are satiny white, with some brown or black spots, and 2 to $2^1/2$ in. across. The abdomen is yellow to orange with some black spots. The adult lays 200 to 500 pale greenish-white eggs, in clusters on the leaves of the host plant. Larvae hatch in a few days and begin to spin silken webs over the foliage. The larvae are pale green or yellow with a dark stripe down the back and a yellow stripe along each side. The body is covered with long silky hairs. The larvae grow to a length of about 1 in., and feed in colonies within the web. They are fully grown in 4 to 6 weeks, and crawl down the tree or shrub to spin cocoons on tree trunks or in ground litter. Pupation occurs in July and September where two generations occur.

Damage
The fall webworm feeds on at least 120 varieties of deciduous trees and shrubs. Unlike the tent caterpillar which makes its web in the crotch of trees, the fall webworm makes its webs on the ends of branches. They feed in colonies within the webs and are most active at night. As the food within the web is consumed, the caterpillars envelop additional branches.

Control
One control technique is to remove web-enclosed branches as soon as the web appears and destroy them. There are a number of natural enemies, e.g., *Itoplectis conquisitor* (Say), that keep the fall webworm in balance. If the site requires chemical treatment, *Bacillus thuringiensis* should be applied when the webs appear, and the larvae are small.

Tent Caterpillars
Tent caterpillars are considered by some to be the most widely distributed defoliator of deciduous trees in the U.S. They include the eastern tent caterpillar, *Malacosoma americanum* (F.), which feeds on wild cherry, apple, and crabapple trees; the forest tent caterpillar, *Malacosoma disstria* (Hubner), which feeds on maple, oak, poplar, ash, birch, and gum, and several species of *Malacosoma* in the western states that feed on oak and redbud.

Description and Life Cycle
Tent caterpillars overwinter as eggs which are laid in clusters around smaller branches or on bark at branch crotches. The eggs hatch in early spring. The larvae are blue to brown. Eastern tent caterpillars have a white stripe down their back while forest tent caterpillars have keyhole-shaped white spots in the same location. Eastern tent caterpillars construct a small web on the tip of the branch near where the eggs were laid. Later, as they grow, they construct a tent in a fork in the tree. By contrast, the forest tent caterpillars create a layer of webbing on the trunk of the tree. They complete their larval stage in about 6 weeks. The pupae are red-brown

and within 10 to 12 days emerge as moths. Eggs are laid at the end of the summer. There is only one generation per year.

Damage
Damage caused by the eastern tent caterpillar is most often of aesthetic value. The defoliation produced from the feeding and the tents built by the larvae are unsightly nuisances in parks and other recreation areas. During population explosions of forest tent caterpillar, which occurs at intervals of 10 to 15 years, millions of acres of forest may be defoliated. Urban trees can be similarly defoliated. The hosts are seldom killed even by successive annual defoliation.

Control
Large outbreaks of the forest tent caterpillar have been controlled by unfavorable weather conditions, food depletion, and parasites. Egg masses can be crushed or pruned from branches during the fall and winter. Tents can be physically removed by hand or by pruning when small. If population levels of either species warrant treatment, it should be accomplished when foliage is about one-third developed and larvae are young. The microbial insecticide *Bacillus thuringiensis* is used to control the young larvae. Small numbers of tents can be removed by pruning.

LEAFROLLERS AND LEAFTIERS
Many small caterpillars, usually in the family Tortricidae, fold over the edge of a leaf and attach it with silk. Others actually roll up an entire leaf, fixing it with silk. These larvae feed on the inside leaf surfaces, and in some cases, eat the entire leaf by the time they are grown. Common examples of this group are the fruit tree leafroller (*Archips argyrospilia*), the oak leafroller (*A. semiferanus*), and the oak leaftier (*Croesia semipurpurana*). The fruit tree leafroller is a good example of this group.

CHEWING INSECTS — BORERS
Borers are usually in the orders Coleoptera (beetles) and Lepidoptera (butterflies and moths). The larvae of borers feed on the fibrous stems, branches, roots, and trunks of trees and shrubs. They can usually be detected by the sap or sawdust which comes from their entry and exit holes. They often do minor structural damage if in low numbers. However, major damage can result if extensive feeding takes place or if the borers girdle the cambium layer and kill the terminal plant growth. Most borers do not attack plants unless the plants are under stress or have been physically

damaged, such as by lawn mower nicks or broken limbs from storm damage.

Roundheaded borers usually attack dead or dying trees or tree parts. Those that attack healthy plants usually bore into the heartwood and may cause structural damage. Sudden death of the plant is rarely caused by insects in this group, but the plants may be broken by wind or ice storms. Common examples are the locust borer (*Megacyllene robiniae*), dogwood twig borer (*Oberea tripunctata*), roundheaded apple tree borer (*Saperda candida*), poplar borer (*S. calcarata*), and sugar maple borer (*Glycobius speciosus*).

Flatheaded borers often attack healthy trees, and the larvae feed in the cambium and sapwood areas. If the larvae completely tunnel around a branch or trunk, girdling and sudden death will occur. Common examples are the bronze birch borer (*Agrilus anxius*), the flatheaded appletree borer (*Chrysobothris femorata*), and twolined chestnut borer (*A. bilineatus*).

Bark beetles usually attack weakened trees and may do so in mass. When a female beetle finds a weakened tree, she may release a congregation pheromone which attracts many others to the site. This often results in numerous "gum drops" or pitch masses on the bark. The larval tunneling from such a mass attack usually girdles and kills the tree. Common examples are the European elm bark beetle (*Scolytus multistriatus*), the shothole borer (*S. rugulosus*), mountain pine beetle (*Dendroctonus ponderosae*), and the red turpentine beetle (*D. valens*).

Weevils which bore into trees usually attack conifers. These are called "terminal weevils" because they usually attack the top branches of pines and spruces. The white pine weevil (=Sitka spruce weevil, *Pissodes strobi*) and the northern pine weevil (*P. approximatus*) are the most commonly seen pests in this group.

ROOT WEEVILS

Several species of root weevils attack ornamentals. These include the two-banded Japanese weevil, *Callirhopalus bifasciatus* (Roelofs), which was first reported in the U.S. in the early 1900s. It is now found in New England, the mid-Atlantic states, and west to Kentucky and Indiana. The black vine weevil *Otiorhynchus sulcatus* (F.) occurs in the northern half of the U.S. Both weevil species feed on numerous ornamentals.

Description and Life Cycle

The adult two-banded Japanese weevil is $1/4$ in. long, tan to brown with darker markings. It is unable to fly because its wings are fused. The black vine weevil is black and nearly $1/2$ in. long. Both species of weevil are slow moving and, when disturbed, will drop to the ground and lay motionless. Japanese weevil eggs are deposited in the folds of dead leaves which are then fused by the female to form a pod. Hatching larvae enter the soil and feed on the roots of the host. The adults appear in July. There is only one generation per year. Black vine weevil adults emerge as early as March in California. Eggs are laid in the soil near the base of a host plant. Larvae feed and injure the roots.

Damage

Feeding by the weevils produces notches in the leaves, and eventually only the petiole remains. Heavily infested plants are usually badly defoliated by the fall. New leaves and shoots are commonly attacked. Larvae consume the roots of plants, and may girdle plants at the soil line. This may result in plant death especially in dry, landscape situations. Black vine weevil is most destructive on taxus and rhododendron, while the two-banded Japanese weevil can be a serious pest of azaleas.

Control

Recent research efforts have evaluated entomopathogenic nematodes as biological control agents. Several synthetic insecticides are registered for control. Controls should be applied when notching of leaf margins is initially observed. Rhododendrons resistant to adult root weevils have been identified, and their use would reduce the need for pesticide applications.

MOTH BORERS

The common Lepidoptera which bore into stems and wood of ornamentals usually belong to three families: the clearwing moths (Family Sesiidae), the carpenterworms and leopard moths (Family Cossidae), and twig, tip or shoot moths (Family Torticidae).

The **clearwinged moths** are active daytime fliers which look like bees or wasps. The white larvae bore into stems, roots, and trunks of plants. Common examples are the ash/lilac borer (*Podosesia syringae*), dogwood borer (*Synanthedon scitula*), rhododendron borer (*S. rhododendri*), and peachtree borer (*S. exitiosa*).

The **carpenterworm** (*Prionoxystus robiniae*) and related leopard moths are heavy bodied moths with small wings. The larvae are large white caterpillars with black spots. They excavate large galleries in wood of a variety of trees, often causing considerable structural damage.

The **twig**, **tip**, and **shoot moths** include a large number of small moths whose larvae bore into the buds and developing shoots of plants. The larvae are small, pinkish or tan bodied, and have dark heads. Common examples are the European pine shoot moth (*Rhyacionia buoliana*), Nantucket pine tip moth (*R. frustrana*), spruce budworm (*Choristoneura fumiferana*), and maple twig borers (*Proteoteras* spp.).

Dogwood Borer

The dogwood borer, *Synanthedon scitula* (Harris), also referred to as the pecan borer, is a native pest of flowering dogwood in the eastern half of the U.S. It is a greater problem in cultivated plantings of homes and parks than of woodlands.

Description and Life Cycle

The adult dogwood borers are small, blue-black moths with yellow banded legs and yellow stripes on segments two and four of their abdomen. Their wings are transparent with blue-black margins, and have a span of $^5/_8$ to $^7/_8$ in. Adults emerge from May to October and lay eggs on bark, usually in roughened areas or in wounds. The eggs soon hatch into white larvae with pale brown heads. After hatching, the larvae enter the bark through openings and feed in the cambial area. These borers remain under the bark throughout their developmental period.

Damage

An early symptom of borer attack is the sloughing of loose bark; dieback and adventitious tree growth are more advanced symptoms. Since dogwood borers live beneath the bark, their presence may be indicated by small, wet areas on the bark. Later in the summer, fine, white dust-like borings that have been pushed from burrows may be seen. Dogwood borers can kill young or unhealthy trees; however, old trees, even when infested annually, are rarely killed.

Control

Control should be concentrated on preventing the entrance of dogwood borers into injured bark surfaces. Mechanical injury from lawn mowers and landscape maintenance equipment can be avoided by planting groundcovers or installing mulch or landscape barriers around the tree base. Pheromone traps are often used to provide optimum timing for emergence of male moths. If residual insecticide applications are required they should begin 7 to 10 days after the first males are captured, approximately when eggs hatch. Recently, entomopathogenic nematodes have been used to control dogwood borer larvae beneath the bark.

Lilac (Ash) Borer

The lilac or ash borer, *Podosesia syringae* (Harris), is active throughout the eastern U.S. and from Texas north through the midwestern states. It attacks green, red, white, and European ash trees.

Description and Life Cycle

The borer larva is white, $1^1/_2$ in. in length with a brown head. The adult is a clearwing moth that resembles a wasp. The front wings are opaque blackish brown with streaks between the veins to the wing base and the fringe is dark brown. The hind wings are transparent with a narrow black border and yellowish brown veins. Its thorax is brownish black with chestnut red markings, and the abdomen is black or brown, but not banded. The females lay eggs from mid-April to late June, depending on the region. The larvae enter the tree through cracks in the bark and bore into the living wood, excavating galleries several inches long. After overwintering in the tunnel, the larvae pupate just beneath the bark, and emerge the following year.

Damage

The borer tunnels into the tree trunk just below ground level or near the base, making so many tunnels that the tree easily breaks over in a wind. The borer's damage also renders the tree vulnerable to infection.

Control

Control strategies include the removal of infested trees to reduce further infestation. Chemical measures are preventive treatments aimed at egg-laying adults and/or newly hatched larvae prior to entering the tree. Pheromone traps can be utilized to optimize timing of treatments.

Pine Tip Moths

Pine tip moths, *Rhyacionia* spp., attack all pine species except slash and longleaf. Damage is most severe on trees from seedling stage to about 12-ft tall.

Description and Life Cycle

Adult tip moths are small and not easily seen, flying mostly at dusk or at night. The moths can be seen during the day if disturbed by shaking young pines. They will fly for a short distance and return to the tree. The moths lay their eggs singly on needles, buds, or shoots. Upon hatching, the larvae feed at the base of shoots or buds; feeding sites usually are marked by exuding pitch and frass. The growing larvae tunnel into buds or shoot tips and pupate. Except for the overwintering generation, the moths emerge from the pupae in 1 to 2 weeks. As many as four generations per year may occur, depending on the species and climatic conditions.

Damage

Tip moth larvae injure pines by boring into the tender, growing shoots of the branches. Several inches of the shoot may be killed; and repeated, heavy attacks cause pines to be bushy, crooked, or distorted.

Control

Damage can be minimized by planting resistant varieties. Natural enemies can reduce populations. Infestations of established trees can be managed with chemical treatments. Timing of insecticide applications is very important. Pheromone traps and degree day models can assist in optimizing management.

CHEWING INSECTS — LEAFMINERS

Leafminers are leaf-feeding insects that feed on the succulent, interior leaf tissue while tunneling between the upper and lower epidermal layers. They attack both coniferous and deciduous trees leaving characteristic damage. This damage generally appears as follows: serpentine leaf mines, blotch leaf mines, comma leaf mines, and needlemines.

Serpentine leafminers are usually caterpillars, such as the aspen serpentine leafminer (*Phyllocnistis populiella*), which leave a long winding trail in the leaf. Most of these are not damaging to the plants and rarely occur in enough numbers to warrant control.

The **blotch leafminers** are usually beetle, sawfly, midge, or small moth larvae that tunnel out a large area in the leaf. Common examples are the birch leafminer (*Fenusa pusilla*), the elm leafminer (*F. ulmi*), the yellow poplar weevil (*Odontopus calceatus*), the boxwood leafminer (*Monarthropalpus buxi*), and the solitary oak leafminer (*Cameraria hamadryadella*). Extensive blotch mines turn brown and cause the trees to look like they are under severe drought stress or have extensive leafspot diseases. Most blotch leafminers are damaging to the plants because they cause premature leaf drop.

Comma leafminers are small flies with larvae that mine by starting out in the serpentine pattern and later become blotch-like. The most common example is the native holly leafminer (*Phytomyza ilicicola*). The female flies use their ovipositor to punch holes (**pinholes**) into the leaves. The mines and pinholes make the foliage look unsightly and may cause premature leaf drop.

Needleminers are usually small caterpillars which feed within the needles of pines and spruces. Common examples are the pine needleminer (*Exoteleia pinifoliella*) and the spruce needleminer

(*Endothenia albolineana*). The spruce needleminer group can be especially damaging by turning most of the 1-year and older needles brown.

Boxwood Leafminer

The boxwood leafminer, *Monarthropalpus buxi* (Laboulbene), was first reported in the U.S. in 1910. It is now found on both coasts, wherever boxwood is grown, and is considered to be a major insect problem of boxwood.

Description and Life Cycle

The adult boxwood leafminer is a small, delicate fly $1/16$-in. long. The female lays her eggs in the leaf tissue from the underside in May and dies soon after. The eggs hatch in 3 weeks. The larvae spend the summer, fall, and winter in the leaf tissue. When spring temperature rises larvae pupate. The pupae are yellow with dark heads. They emerge over a 2-week period from the end of April to mid-May during the period of shoot elongation. Pupal cases are left clinging to the leaves.

Damage

The boxwood leafminer feeds on the soft tissues of the leaves causing the leaves to yellow and drop prematurely. The larvae also produce small blisters on the underside of the leaves. Their damage is evident from mid-summer to the following spring. Continuous heavy infestations result in loss of vigor and plants become more susceptible to diseases.

Control

Foliar systemic insecticides applied when young larvae are present provides good control.

Holly Leafminer

The native holly leafminer, *P. ilicicola* Loew, is one of several fly species which feed on hollies. This insect prefers *Ilex opaca*.

Description and Life Cycle

The holly leafminer adult emerges in the spring when budbreak occurs and may continue for several weeks. Eggs are deposited in the underside of newly developing leaves. The larvae mine just under the upper leaf surface. Initially, the mines are thread-like and inconspicuous, but by late autumn they widen, into blotches or blisters. The larvae overwinter in the mines and pupate in March. Only one generation occurs per year.

Damage

Many year-old holly leaves which are heavily infested may drop off in the spring. Damage is confined to the upper side of the leaves. Frequently several mines will coalesce producing one very large mine which may contain several larvae. These mines frequently destroy the appearance of an entire leaf.

Control

Parasites partially control the leafminer population. For infestations resulting in aesthetic injury, chemical treatments can be applied when the fly emerges in the spring but before the eggs are laid. The presence of adult flies can be determined by the presence of small pits in the new leaves, the result of adult feeding, or with yellow cards. Systemic insecticides are also effective in controlling larvae inside the leaves.

SUCKING INSECTS

Insects in the orders Thysanoptera (thrips), Hemiptera (true bugs), and Homoptera (hoppers, aphids, scales, and mealybugs), as well as mites, generally suck plant juices. Injury may be caused by direct injury to cells and tissues from the piercing stylets. Often extra injury occurs when these pests inject salivary toxins which block the vascular system. Other sucking pests such as aphids, scales, and mealybugs excrete a sugary solution of excess sap, called honeydew. This coats the foliage and promotes the growth of sooty mold and visitations by ants and bees. Many sucking pests have been shown to transmit plant diseases by feeding on multiple hosts. Typical signs of damage are blotches of yellow or brown on older foliage, leaf curl or distortion, and wilting.

APHIDS

This is one of the largest groups of insects which inflict damage through sucking mouthparts. Almost no plant is unaffected by aphids or plant lice. Most aphids belong to the family Aphidae of the order Homoptera. Many aphids have very complex life cycles which include generations of different forms and different host plants. Aphids are copious producers of honeydew and often transmit diseases.

Description and Life Cycle

Aphids have pear-shaped, soft bodies with long slender legs. They are about $1/8$ in. long and may be green, black, red, yellow, blue, or white. Two secreting tubes project from the rear of their body. Most aphids are wingless. Wooly aphids are covered with white waxy threads. Some species of aphids congregate on buds, stems, and the underside of leaves, but may attack any part of the plant. Males are produced in the fall and mating results in egg production. After overwintering as eggs, spring egg hatch results in females. These females give live birth to females, and this cycle repeats several times during the season. The first few generations are wingless, but as the population increases and becomes crowded, or fall migration nears, the aphids become winged. Many aphid species have alternate plant hosts during the season.

Damage

Aphids can cause extensive damage to plants. They pierce the leaves to feed and withdraw plant fluids. This weakens the plant and may also distort the leaves. Honeydew, a sugary substance excreted by the aphids, attracts ants and encourages the growth of sooty mold. Several aphids feed on roots of plants such as elm and apple. Some aphids are also capable of transmitting plant viruses. Others species produce galls on plant tissue.

Control

Damaging infestations of aphids may be controlled in some plants by using a stream of water to forcibly remove the aphids. Lady beetles and lacewings are natural predators of aphids, and may be purchased from biological control firms. Small wasps also parasitize aphids. Possible insecticides include biorational materials such as soaps, oils, and pyrethrins.

LACE BUGS

These bugs have sculptured lacelike patterns on the body and wings. The nymphs usually have spiny projections sticking out from the sides of the body. They belong to the family Tingidae. Common lace bugs include the sycamore lace bug (*Corythucha ciliata*), the azalea lace bug (*Stephanitis pyrioides*), and the rhododendron lace bug (*S. rhododendri*). All of these species occur wherever their host plants are found, and most prefer sunny habitats.

Azalea Lace Bug

The azalea lace bug, *S. pyrioides* (Scott), is a major pest of azaleas in the eastern region of the country. They do the most damage to azaleas growing in full sunlight. The azalea lace bug is similar in appearance, habits, and life history to the rhododendron lace bug, but each attacks only its respective host.

Description and Life Cycle

Adult lace bugs are $1/8$ in. long, flat, and brown to black. They are oval or rectangular and have netted wings. They have a hood-like covering on their heads and fold their wing flat over their abdomen. The azalea lace bug overwinters in the egg stage inside the leaf tissue along the leaf midrib. The nymphs hatch in the early spring. They are colorless but later become black and spiny. The nymphs develop into adults within 30 days. The azalea lace bug may have up to four generations per year depending on the location.

Damage
The nymphs and adults of the azalea lace bug withdraw the sap from the underside of the leaf. The upper surface of the damaged leaves has a stippled appearance. The underside of the leaves becomes covered with dark spots of tarlike excrement. These spots provide a good method for diagnosing infestations of azalea lace bug. Caste skins of molting nymphs can also often be visible. Plant vigor is greatly reduced in severely infested plants.

Control
Nymphs that hatch from overwintering eggs should be controlled before they become adults and begin to lay eggs. During the summer, reinfestations from nearby plantings may occur. Applications of insecticidal soap in mid-May and a second application 10 days later will reduce populations. Additional treatments may be needed throughout the summer. Sprays should be directed to the underside of the leaves, and are most effective when used at high pressure. Systemic insecticides also provide good control of aphids.

SCALES
The scale insects comprise a large group of highly specialized insects. In most species, the females are wingless and sessile, while the males have one pair of wings and nonfunctioning mouthparts. Scales often produce honeydew. The normal life cycle includes a mobile first nymph called a crawler, several sessile nymph stages, and formation of a nymphlike adult female or formation of resting instar, often called a pupa, which gives rise to the winged male. Scales are generally categorized as either soft scales, Coccidae, in which the covering is a smooth leathery exoskeleton, or armored scales, Diaspididae, in which the body is small and covered with wax. Soft scales usually overwinter as nymphs or as mated but underdeveloped females. These females complete their growth in the spring before laying eggs. Armored scales overwinter as eggs or as mated females.

Euonymus Scale
The euonymus scale, *Unaspis euonymi* (Comstock) is one of the most destructive scales and is a major pest of euonymus. It is more prevalent in cool climates.

Description and Life Cycle
The adult female euonymus scale is dark brown or gray, $^1/_{16}$ in. long, and shaped like an oyster shell.

The male is narrow and white and may entirely cover the stems and leaves of infested shrubs. The females are fewer in number on leaves and may be scattered along the stems and leaf veins. Mature fertilized females overwinter under the scale covering. Orange-yellow eggs are laid under the female shell in early spring and hatch in the late spring. Yellow crawlers are visible in late May and June in the mid-Atlantic region. A second generation appears in August and September in the mid-Atlantic region. Additional generations have been reported in lower latitudes.

Damage
Scale insects inflict their damage to the stems and leaves due to their feeding. Yellow and white spots on the leaves are an early sign of the scale. Heavy infestations will cause the leaves to yellow and drop. Branches or entire shrubs can be killed by heavy infestations.

Control
Both naturally occurring parasites and introduced predators suppress this scale. Oil sprays are effective during the dormant period, and would not impact beneficial insects. Control with insecticides is also effective during the crawler periods of each generation.

PLANT BUGS
These true bugs generally belong to the family Miridae and have long antennae, oval bodies, and the wings cross to form a diamond pattern on the back. Common plant bugs include the honeylocust plant bug (*Diaphnocoris chlorionis*), the ash plant bug (*Tropidosteptes amoenus*), the fourlined plant bug (*Poecilocapsus lineatus*), and the tarnished plant bug (*Lygus lineolaris*). Other hemipterans which occur on plants and cause similar damage are the seed bugs (Lygaeidae), coreid bugs (Coreidae) such as the leaffooted bugs and boxelder bug, and the stink bugs (Pentatomidae).

MITES
There are two major groups of mites which attack plants. The spider mites, family Tetranychidae, cause chlorosis and bronzing of foliage. Rust mites, family Eriophyidae, sometimes feed on leaf surfaces, but most cause galls or other unusual plant growths and will be discussed under galls.

Spider mites tend to be warm-season pests. The twospotted spider mite (*Tetranychus urticae*) occurs on a wide variety of plants while most other species are fairly host specific. Other warm-season spider mites include the oak mite (*Oligonychus bicolor*), the honeylocust spider mite (*Platyetranychus multidigitali*), and the European red mite (*Panonychus ulmi*). Other spider mites seem to prefer the cool spring and fall temperatures and are considered cool-season mites. The spruce spider mite (*O. ununguis*) and southern red mite (*O. ilicis*) are the most common species in this group.

Two-Spotted Spider Mite

The two-spotted spider mite, *Tetranychus urticae* (Koch), is one of the most serious pests in floriculture, damaging many different plants. Fuchsia and impatiens are common hosts but the most severely injured are carnations, chrysanthemums, sweet peas, snapdragons, violets, diffenbachia, and roses. Cedar, evergreens, hydrangeas, clematis, and many other species are also damaged by the mite.

Description and Life Cycle

Mites are not insects in that they possess eight legs and only two body regions. The body of the two-spotted mite is very small, oval, and covered with spines. The female may range in color from yellow to green to brown. The male is slightly smaller and narrower with a pointed abdomen. The mite uses its sucking mouthparts to pierce the leaf. The female mite lays several eggs a day for about 2 months. The eggs are round and shiny and are located on the underside of the leaf. The eggs may be covered with a fine web produced by the mite. Eggs hatch after 3 to 19 days, depending on the temperature.

Damage

Plants with low populations of mites will have pale spotted or flecked leaves. The webbing produced by the mites may entirely cover a plant. Plants with heavy infestations weaken and eventually die.

Control

Mite infestations may be controlled with biorational applications of insecticidal soaps and horticultural oils, though synthetic miticides are most effective. Broadleaf and needled evergreens should be sprayed in May and in the fall. Deciduous trees and shrubs should be monitored closely in June and July, and action should be taken if populations reach damaging levels. Applications of several insecticides may result in secondary outbreaks.

GALL INSECTS

Certain insects and mites have developed special relationships with their host plants. These pests inject or secrete chemicals into plant tissues that cause the plant to undergo abnormal growth. The growth usually surrounds the insect or mite, enclosing it in a protective chamber. Inside this protective chamber, the pest may suck plant juices, eat succulent growth, or absorb plant secretions through their cuticle.

Galls can usually be placed into two general categories, open and closed, according to the way the pest exists within the gall. Pests with sucking mouthparts produce open galls in which there is a slit or hole formed by the plant. This exit slit is usually tightly closed during the development of the insect or mite, but the slit opens after the tissues harden or dry out.

Those with chewing mouthparts produce closed galls and they generally chew an exit hole through the gall in order to emerge. The adults insert their eggs into plant tissues. The larvae or egg-laying procedure causes the plant to grow around the developing pest and no outside opening is present until the maturing larva or adult chews an exit hole through the gall's wall. Galls are usually named according to their location on the plant (leaf, petiole, or stem), their host plant, and shape (bladder, cone, shot etc.). Galls are produced by a very diverse group of pests and have representatives from the mites (usually eriophyids), Homoptera (aphids and adelgids), Coleoptera (beetles), Lepidoptera (moths), Diptera (gall midges), and Hymenoptera (gall wasps).

REFERENCES

1. **Baker, J. R.,** *Insects and Related Pests of Shrubs,* North Carolina Agricultural Extention Service, Raleigh, NC, 1980.

2. **Davidson, R. H. and Lyon, W. F.,** *Insect Pests of Farm, Garden and Orchard,* John Wiley & Sons, New York, 1987.

3. **Johnson, W. T. and Lyon, H. H.,** *Insects that Feed on Trees and Shrubs,* Cornell University Press, Ithaca, NY, 1988.

4. **Koehler, C. S.,** *Insect Pest Management Guidelines for California Landscape Ornamentals,* Cooperative Extension, University of California Publishers, Oakland, CA, 1987.

5. **Olkowski, W., Daar, S., and Olkowski, H.,** *Common-Sense Pest Control,* Taunton Press, Newtown, CT, 1992.

Chapter 23

Symptomology and Management of Common Turfgrass Diseases in the Transition Zone and Northern Regions

Peter H. Dernoeden, *Department of Agronomy, University of Maryland, College Park, MD*

CONTENTS

Introduction 249
Spring and Fall Diseases 250
    Dollar Spot 250
    Helminthosporium Leaf Spot and Melting-Out 250
    Spring Dead Spot 252
    Necrotic Ring Spot 252
    Stripe Smut and Flag Smut 253
    Red Thread and Pink Patch 253
    Powdery Mildew 254
    Rusts 254
Summer Diseases 254
    Brown Patch/Rhizoctonia Blight 254
    Pythium Blight 255
    Summer Patch 256
    Fusarium Blight 256
    Melting-Out 256
    Fairy Rings 256
Winter Diseases 257
    Snow Mold Diseases 257
    Yellow Patch 258
Factors Associated with Fungicide Use 258
The Fungicide Dilemma in Lawn Care 259
The Fungicide Dilemma on Golf Courses 260
Nematodes 261
References 263

INTRODUCTION

Most turfgrass diseases are caused by pathogenic fungi that invade leaves, stems, and roots of plants. As a result of the injurious effects of a disease, the plant will exhibit various symptoms such as leaf spots, root rots, or death of leaves, tillers, or entire plants. Sometimes these fungi produce visible signs such as mushrooms; white powdery mildew; white, fluffy mycelial growth; pink gelatinous mycelial growth; red or black pustules on leaves, etc. Mycelium is the vegetative body of a fungus that is composed of a network of fine tubes that often appear cottony. It is through the use of these symptoms and signs that disease problems are diagnosed. Time of year and turfgrass species also provide important clues in diagnosing diseases. For example, brown patch and Pythium blight are seldom a problem when night temperatures fall below 65°F, and Pythium blight rarely causes severe injury to mature stands of Kentucky bluegrass. Conversely, dollar spot tends to be more damaging under conditions of cooler night temperatures in late spring and early fall than during the hottest weeks of summer. Summer patch is strictly a high temperature, summer disease of Kentucky bluegrass, fine leaf fescues, and annual bluegrass. Species such as perennial ryegrass, creeping bentgrass, and tall fescue are resistant to summer patch.

Most turfgrass diseases are caused by fungi rather than bacteria or viruses. In the pages to follow, the most common fungal diseases of turfgrasses, as well as plant parasitic nematodes, are described and cultural and chemical approaches to their management are outlined. A more complete list of turfgrass diseases, hosts, and pathogens is provided in Table 1.

SPRING AND FALL DISEASES

DOLLAR SPOT

The dollar spot pathogen (*Sclerotinia homoeocarpa*) is widespread and extremely destructive to turfgrasses. The taxonomy of *S. homoeocarpa* is unclear, and this fungus may be referred to as an unknown species of either *Moellerodiscus* or *Lanzia*. Dollar spot is known to attack most turfgrass species including annual bluegrass, bentgrasses, fescues, Kentucky bluegrass, perennial ryegrass, bermudagrass, zoysiagrass, centipedegrass, and St. Augustinegrass. The symptomatic pattern of dollar spot varies with turfgrass species and management practices. Under close mowing conditions, as with intensively maintained bentgrass or zoysiagrass, the disease first appears as small, circular, straw-colored spots of blighted turfgrass about the size of a silver dollar. With coarser textured grasses such as Kentucky bluegrass or perennial ryegrass suited to higher mowing practices, the blighted areas are considerably larger, irregularly shaped, straw-colored patches 3 to 6 in. in diameter. Affected patches frequently coalesce and involve large areas of turf. Grass blades often dieback from the tip, and have straw-colored or bleached-white lesions shaped like an hourglass. The hourglass banding on leaves is often made more obvious by a definite narrow brown, purple, or black band which borders the bleached sections of the lesion from the remaining green portions. When the fungus is active and moisture is present, a fine, cobwebby mycelium may cover the diseased patches during early morning hours. Disease severity usually peaks in late spring-early summer and again late summer-early fall. In the upper Midwest, however, the disease tends to be most damaging during the autumn. Dollar spot can

remain active during mild periods throughout fall and into early winter in some regions.

Dollar spot tends to be most damaging in poorly nourished turfs, particularly if soils are dry and humidity is high or a heavy dew is present. Avoiding drought stress, irrigating deeply during early morning or daytime hours, maintaining a balanced N-P-K fertility program, and controlling thatch and compaction are cultural approaches that minimize dollar spot injury. Irrigating between 5 a.m. and 8 a.m., when dew is present on leaves, does not extend the fungal infection period. Hence, where water use is restricted to between sundown and sunup, pre-dawn irrigation will not promote disease while at the same time local watering laws are not violated. Most fungicides that are labeled for dollar spot provide effective disease control.

HELMINTHOSPORIUM LEAF SPOT AND MELTING-OUT

Many of the fungi that cause leaf spotting and melting-out diseases of turfgrasses were once assigned to the genus *Helminthosporium*. Today, these fungi are more appropriately referred to as species of *Drechslera* or *Bipolaris,* but the common names of the diseases they cause remain Helminthosporium leaf spot, melting-out, or net-blotch.

Among the most important spring and autumn diseases of Kentucky bluegrass is leaf spot, which is caused by *Drechslera poae*. This disease is not as devastating as it once was because of the development and widespread use of resistant bluegrass cultivars. South Dakota, Kenblue, Park, and other "common" types of Kentucky bluegrass are very susceptible to leaf spot. The common types, which generally survive extreme environmental stresses, are still used today as components of bluegrass blends in some regions of the U.S. because they lend genetic diversity to the stand.

Drechslera poae is a cool-weather pathogen that is most active during the spring (especially April and May), autumn (especially September and October), and throughout mild winter periods. *D. poae* causes disease that may occur in two phases: the leaf spot phase and the melting-out phase. In the leaf spot phase, distinct purplish-brown leaf spot lesions with a central tan spot are produced on the leaves of affected plants. In a heavily infected stand, the turf appears yellow or red-brown in color when observed from a standing position. During favorable disease conditions, lesions may increase in size to encompass the entire width of the blade causing a die-back from the tip. Leaf spot lesions are initially associated with older leaves, which die prematurely as a result of the invasion. If favorable

Table 1 **The most frequently encountered diseases of lawns, golf courses, athletic fields, and other turf areas in transition zone and northern regions of the U.S.**

Disease	Primary Hosts[a]	Pathogen(s)
I. Patch Diseases Caused by Root Invading Pathogens		
Summer patch	ABG, FLF, KBG	*Magnaporthe poae*
Necrotic ring spot	KBG, FLF	*Leptosphaeria korral*
Take-all patch	CBG	*Gaeumannomyces graminis* var. *avenae*
Spring dead spot	Bermudagrass	*Leptosphaeria korrae* or *Ophiospharella herpotricha*
II. Foliar Diseases		
Dollar spot	All species	*Sclerotinia homoeocarpa*
Red thread	Most species	*Laetisaria fuciformis*
Pink patch	CBG, PRG, FLF	*Limonomyces roseipellis*
Brown patch/Rhizoctonia blight	PRG, TF, CBG, Bermuda	*Rhizoctonia solani* and *R. zeae*
Leaf spot (Helminthosporium)	All species	*Drechslera* or *Bipolaris* spp.
Gray snow mold	All species	*Typhula incarnata* and *T. ishikariensis*
Yellow patch (cool temperature brown patch)	ABG, KBG, CBG, PRG	*Rhizoctonia cerealis*
Gray leaf spot	St. Augustine, PRG	*Pyricularia grisea*
Large patch	Zoysiagrass	*Rhizoctonia solani*
White blight	TF	*Melanotus phillipsii*
Nigrospora blight	KBG, PRG, FLF	*Nigrospora sphaerica*
Leptosphaerulina blight	ABG, KBG, PRG, CBG	*Leptosphaerulina trifolii*
III. Foliar/Stem or Root Diseases		
Pythium blight	ABG, CBG, PRG	*Pythium aphanidermatum, P. myriotylum P. ultimum*
Pink snow mold	All species	*Microdochium nivale*
Anthracnose	ABG, CBG, FLF	*Colletotrichum graminicola*
Pythium-induced root dysfunction	ABG, CBG	*Pythium* spp.
Nematodes	Most species	Many species
Melting-out	KBG, FLF, PRG	*Drechslera* or *Bipolaris* spp.
Fusarium blight	KBG, FLF	*Fusarium culmorum* and *F. poae*
IV. Diseases Caused by Obligate Parasites		
Stripe smut	KBG, CBG	*Ustilago striiformis*
Flag smut	KBG	*Urocystis agropyri*
Powdery mildew	KBG in shade	*Erysiphe graminis*
Rust	PBG, KBG, Zoysia	*Puccinia* spp.
Yellow tuft	All species	*Sclerophthora macrospora*
V. Diseases Causing Hydrophobic Thatch or Soil, or Thatch Shrinkage		
Fairy ring	All species	Many basidiomycetes
Yellow ring	KBG	*Trechispora alnicola*
Superficial fairy ring	Putting greens	Basidiomycetes
Localized dry spot	Putting greens	Unknown basidiomycetes

Table 1 (Cont.) **The most frequently encountered diseases of lawns, golf courses, athletic fields, and other turf areas in transition zone and northern regions of the U.S.**

Disease	Primary Hosts	Pathogen
	VI. Seedling Disease	
Damping off	All species	*Pythium* spp.
		Fusarium spp.
		Rhizoctonia solani
		Bipolaris spp.
		Drechslera spp.
		Curvularia spp.
		Others

[a] Hosts: ABG = annual bluegrass; KBG = Kentucky bluegrass; CBG = creeping bentgrass; PRG = perennial ryegrass; FLF = fine leaf fescue; and TF = tall fescue.

environmental conditions for disease continue, particularly overcast, cool, and drizzling weather, successive layers of leaf sheaths are penetrated, and the crown is invaded. Once the crown is invaded the disease enters the melting-out phase. During this phase, entire tillers are lost, and the turf loses density. Hence, it is the melting-out phase that is most damaging to the sward.

Net-blotch disease of tall fescue and perennial ryegrass is caused by another of the helminthosporia, *Drechslera dictyoides*. *D. dictyoides* is also a cool, wet weather pathogen that attacks turf primarily during cool and moist periods of spring and fall. Initially, symptoms appear as minute, purple-brown specks on leaves. As the disease progresses, a dark brown, netlike pattern of necrotic lesions develops on tall fescue. These net blotches may coalesce, leaves turn brown or yellow, and die-back from the tip. On leaves of perennial ryegrass, numerous oblong, brown lesions are produced. Under ideal environmental conditions the fungus may invade crowns, causing a melting-out of the stand. Both diseases can be active during relatively warm, rainy periods of winter.

Cultural practices that minimize injury from leaf spot or melting-out diseases are as follows: raise the mowing height; avoid spring and summer applications of water soluble nitrogen fertilizers; avoid light, frequent irrigations; control thatch; overseed with resistant cultivars, and avoid the use of broadleaf phenoxy herbicides during periods when these diseases are active. Fungicides that effectively control leaf spot diseases include chlorothalonil, iprodione, mancozeb, and vinclozolin.

SPRING DEAD SPOT
Spring dead spot (SDS) is perhaps the most damaging disease of bermudagrass turf. In the U.S. it is caused by *Leptosphaeria korrae, Ophiospharella*

herpotricha, and possibly other fungi. As the name implies, SDS injury becomes apparent in the spring. The actual infection, however, may begin as early as autumn, but root injury by the pathogen becomes rapid prior to spring green-up, during late winter, or early spring. As bermudagrass breaks dormancy, circular patches of brown, sunken turf 2 in. to 3 ft in diameter become conspicuous. Rhizomes and stolons from nearby, healthy plants eventually spread into and cover the dead patches. This filling-in process is slow, a period which may last 4 to 8 weeks following spring green-up. The slowness of the filling-in process is believed to be due to toxic substances generated in the soil below the dead patches. Weeds commonly invade the dead patches. These weeds should be controlled to reduce competition with the bermudagrass and thereby speed up the recovery process.

SDS is most commonly associated with mature bermudagrass turfs older than 3 years. The disease, however, may appear the spring following sprigging with stolons from sites that previously had SDS. SDS injury is most likely to occur where thick thatch layers exist and where nitrogen fertilizers have been heavily applied during late summer or fall. Fenarimol applied in mid- to late September helps to alleviate SDS. Ammonium sulfate or ammonium chloride (applied at 1.0 lb N/1000 ft^2) and potassium chloride (applied at 1.0 lb K/1000 ft^2) applied on monthly intervals from mid-May to mid-August speed recovery of turf injured by SDS, and help to alleviate disease severity over time.

NECROTIC RING SPOT
Necrotic ring spot (NRS) is a disease of Kentucky bluegrass and creeping red fescue, and is caused by *Leptosphaeria korrae*. Although first described in 1986, NRS probably has been around for some time, but has been confused with Fusarium blight.

NRS was first suspected as being a new disease in Wisconsin, where the fungicide triadimefon had been applied unsuccessfully to control what was believed to be Fusarium blight. Unlike Fusarium blight, NRS primarily begins during cool, wet weather of spring and fall. Fusarium blight and another recently described patch disease that mimics Fusarium blight, summer patch, are both high-temperature diseases of summer. Frequently, symptoms of NRS do not become evident until summer, when environmental stresses kill plants whose root systems were damaged by *L. korrae* in the spring. The confusion over NRS and summer patch is further compounded because both diseases can appear as rings of dead grass with living turf in the center. In chronically affected areas, NRS patches tend to be large, often greater than 1 ft in diameter, and distinct frog eyes are a common symptom. Initially, leaves of affected plants may display a purple color and plants may appear stunted in a circular pattern. Sodded lawns appear more likely to be affected, but NRS also occurs where lawns are seeded. To date, NRS has been observed mainly in Kentucky bluegrass turf grown in the Northeast, upper Midwest, and Pacific Northwest regions. Its occurrence is relatively uncommon in transition zone regions of mid-Atlantic states.

Because NRS is a root disease, judicious irrigation is required during summer months when heat and drought severely stress injured plants. Use of slow release and bio-organic fertilizers helps to reduce disease severity. The Kentucky bluegrass cultivars Wabash, Vantage, I-13, and Adelphi were reported to have some NRS resistance. Early spring applications of fenarimol or thiophanate reduce the severity of NRS.

STRIPE SMUT AND FLAG SMUT

Stripe (*Ustilago striiformis*) and flag smuts (*Urocystis agropyri)* are diseases that occur primarily in mature Kentucky bluegrass stands and occasionally in bentgrass and perennial ryegrass turf. Symptoms are most conspicuous during the cool, moist seasons of spring and fall. Infected plants are often stunted and pale green or yellow in color. Narrow, silvery or gray-black streaks will appear on the leaves. These streaks are fruiting structures (sori) in which large masses of spores (teliospores) are produced. When sori mature, the cuticle and epidermis rupture, and the leaves shred and curl releasing the teliospores. Although the disease is systemic and persistent in surviving plants, during summer months infected plants may appear amazingly healthy if properly maintained. In spring or autumn, however, badly infected stands may appear chlorotic and in need of nitrogen fertilizer.

During winter, leaves that had been shredded by matured fruiting bodies develop a gray-brown, desiccated appearance.

Stripe and flag smuts can be very damaging to infected plants during periods of heat and drought stress. If properly irrigated and fertilized, however, badly smutted stands often survive, exhibiting only a decrease in turf quality and some thinning during stressful summer months. These smut diseases most commonly occur in mature (2 to 4 years and older) stands that have been managed with high levels of nitrogen fertilizer. Merion, Windsor, and Fylking Kentucky bluegrasses are among the most susceptible cultivars. Recently introduced cultivars are less susceptible to these diseases, which has led to a reduction in the occurrence of stripe and flag smut. Using a balanced N-P-K fall fertilizer program, increasing the mowing height in summer, and deep irrigation at the first sign of drought stress are effective management practices that greatly minimize smut injury in Kentucky bluegrass. Symptoms of smut diseases are effectively suppressed with a spring or fall application of a systemic fungicide such as propiconazole or triadimefon.

RED THREAD AND PINK PATCH

Red thread (*Laetisaria fuciformis*) and pink patch (*Limonomyces roseipellis*) are common diseases of turfgrasses, and development is favored by cool (65 to 75°F), wet and extended overcast weather of spring and fall. The diseases may also occur during warm to very cold weather in the presence of plenty of surface moisture or at snow melt in winter. This disease may become widespread among turfgrass species during mild winters.

Red thread and pink patch often occur together, and are most damaging to perennial ryegrass, common-type Kentucky bluegrasses, and the fine leaf fescues. Improved cultivars of Kentucky bluegrass, tall fescue, bentgrass, and bermudagrass may also be affected, but these grasses do not usually sustain a significant level of injury if they are sufficiently fertilized with nitrogen.

Symptoms of red thread and pink patch are concentrated in circular or irregularly shaped patches 2 in. to 3 ft in diameter that frequently coalesce to involve large areas. From a distance, affected turf has a straw-brown, tan, or pinkish color. The signs of red thread are distinctive and unmistakable. In the presence of morning dew or rain, a coral pink or reddish layer of gelatinous fungal growth (mycelium) can easily be seen on leaves and sheaths. The infested green leaves of these plants soon become water-soaked in appearance. When leaves dry, the fungal mycelium becomes pink in color and is easily seen on the straw-brown or tan tissues of dead leaves and sheaths. During the final phase of disease activity, bright red, hard, and brittle strands of fungal mycelium called "red threads" or sclerotia may be seen extending from leaf surfaces, particularly cut leaf tips. These red threads fall into the thatch and serve as resting structures for the fungus by surviving long periods that are unfavorable for growth of the pathogen. The visible signs of pink patch appear as a gelatinous mass of pinkish mycelium associated with a water-soaked appearance of leaves. The pink patch fungus does not produce red threads.

Red thread and pink patch are generally most injurious to poorly nourished turfs. Frequently, the disease is best controlled by an application of 0.5 to 1.0 lb N/1000 ft^2. Application of nitrogen during periods too cool for turf growth will not aid in reducing disease severity. This is because nitrogen alleviates red thread or pink patch disease symptoms by stimulating plant growth and vigor. The nitrogen-stimulated plants are able to replace damaged tissues more rapidly than these fungi can inflict injury. Fungicides that control red thread and pink patch are chlorothalonil, iprodione, flutalonil, and triadimefon.

POWDERY MILDEW

Powdery mildew (*Erysiphe graminis*) is a disease confined to shaded environments. The presence of grayish-white mycelium and spores on the upper surfaces of leaves is a conspicuous, diagnostic sign of the disease. The lower, older leaves of the plant are generally more heavily infected than upper, younger leaves. In heavy infestations, leaves appear to have been dusted with ground limestone or flour. The abundant surface mycelium absorbs nutrients from the epidermal cells and the leaves turn yellow. Eventually leaves and tillers may die and the turf will exhibit poor density. The fungus seldom kills plants; however, its presence weakens plants and therefore can predispose them to injury from environmental stresses or other diseases.

Powdery mildew can be found at almost any time of year, but peak activity normally occurs in the fall. Spores are produced in abundance on leaf surfaces and they are easily moved to adjacent, healthy leaves. The spores germinate rapidly, even in the absence of dew or water. Disease activity is most prevalent during cool, humid, and cloudy periods of spring and fall. Because shade is the primary predisposing factor for powdery mildew, reducing shade and improving air circulation is a cultural, but often impractical approach to reduce damage. Planting or overseeding with shade-tolerant cultivars, increasing mowing height, avoiding drought stress, and using a balanced fertilizer pro-

gram will promote turfgrass growth and help to minimize injury from powdery mildew. Fungicides may be applied in situations where the disease is yellowing plants and thinning the stand. Some effective fungicides are propiconazole and triadimefon.

RUSTS

Rust diseases in turf are caused by several *Puccinia* spp. and they are most damaging to poorly nourished turfs and turfs grown under a low mowing height. Stem rust (*P. graminis*) of Kentucky bluegrass, crown rust (*P. coronata*) of perennial ryegrass, and zoysiagrass rust (*P. zoysiae*) are most commonly observed during cool, moist periods of fall. Rust-affected turfs exhibit a yellowish or reddish-brown appearance from a distance. Close inspection of diseased leaves reveals the presence of conspicuous red, black, orange, or yellow pustules. Sterol-inhibiting fungicides (such as propiconazole and triadimefon) effectively control rust. Contact fungicides (such as chlorothalonil and mancozeb) will help somewhat to alleviate injury. A balanced fertility program may be preferred to fungicides in situations where rust is damaging poorly nourished turfs. Also, irrigate early in the day to ensure leaf dryness prior to nightfall, irrigate deeply but infrequently, increase mowing height, and increase mowing frequency. By increasing mowing frequency, leaves bearing immature spores are removed and this reduces the potential for more leaf infections.

SUMMER DISEASES

BROWN PATCH/RHIZOCTONIA BLIGHT

Brown patch, also known as Rhizoctonia blight, is caused by *Rhizoctonia solani* and it is a common, summertime disease of turfgrasses. In southeastern states, *R. zeae* also may produce symptoms that are typical of brown patch. *R. solani* attacks nearly all grasses used as turf, but is most damaging to tall fescue, perennial ryegrass, creeping bentgrass, and annual bluegrass. *R. solani* attacks sheaths of zoysiagrass during autumn or spring causing a disease known as large patch.

The symptoms of Rhizoctonia blight vary according to host species. On closely mown turf, affected patches are roughly circular and range from 3 in. to 3 ft or greater in diameter. The outer edge of the patch may develop a 1- to 2-in.-wide smoke ring. The smoke ring is blue-gray in color and is caused by mycelium in the active process of infecting leaves. On high-cut turfs, smoke rings are not usually present and patches may have an irregular rather than circular shape. Close inspection of leaf blades reveals that the fungus primarily causes a blight or dieback from the tip down, to give affected turf its brown color. *R. solani* produces distinctive and often greatly elongated lesions on tall fescue leaves. The lesions are a light, chocolate brown color, and are bordered by narrow, dark-brown bands of tissues. On perennial ryegrass, smaller leaf lesions are produced and tip dieback commonly occurs. During early morning hours, when the disease is active, the cobweb-like mycelium may be observed on leaves in the presence of water or heavy dew.

Environmental conditions that favor disease development are day temperatures above 85°F and high relative humidity. A night temperature above 68°F is perhaps the most critical environmental requirement for disease development. Long leaf surface wetness periods and very high night humidity are required for severe brown patch blighting to occur. Summer application of fertilizers, in particular water-soluble N fertilizers, may enhance disease injury from brown patch. Avoiding nitrogen when the disease is active and irrigating early in the day (i.e., between dawn and 8 a.m.) are the only cultural practices that may help alleviate brown patch. Chlorothalonil, iprodione, flutalonil, mancozeb, vinclozolin, and the thiophanates effectively control brown patch.

PYTHIUM BLIGHT

Pythium blight is among the most destructive turfgrass diseases. During periods of high relative humidity, night temperatures above 70°F, and abundant surface moisture, the disease may progress rapidly, destroying large turf areas within 24 hours. The disease often is first observed in areas that are shaded, low lying, and adjacent to water where air circulation is poor. While there are several species capable of causing the disease, *P. aphanidermatum, P. ultimum,* and *P. myriotylum* are the most common.

It is a general misconception that Pythium blight is a common, wide-spread disease. Although *Pythium* spp. can cause damping-off of any seedling species, it rarely attacks mature lawns comprised of Kentucky bluegrass, tall fescue, fine fescue, or zoysiagrass. Pythium blight is most likely to attack creeping bentgrass or perennial ryegrass grown under the high management (i.e., frequent night irrigation, low mowing, and high nitrogen fertility) conditions commonly found on golf courses.

On closely mown bentgrass putting greens, the disease kills turf in circular patches, rings, or streaks that follow the water drainage pattern. During morning hours when the disease is active, bentgrass turf displays an orange-bronze color and there may be a gray smoke ring on the periphery of affected patches. In low-lying areas where water collects, the patches are brown and all plants are often killed. In perennial ryegrass, affected foliage develops an oily or dark-gray color, and leaf blades have a water-soaked appearance. Blades then collapse, mat together, and turn brown.

A cottony web of mycelium covers the grass leaves and is visible during early morning hours when leaves are wet. *Pythium* spp. are capable of producing an abundance of mycelium in a few hours, which bridges leaf blades, giving turf the cottony appearance. The fungus primarily spreads over a turf by rapid mycelial growth or by movement of mycelial fragments and motile spores in rain or irrigation water.

Water management may greatly influence disease severity. It is therefore helpful to irrigate early in the day to avoid moist foliage prior to nightfall. Improving water drainage and air circulation will help reduce disease development, but these cultural measures are often expensive and difficult to achieve. A fall fertilization program using a balanced N-P-K fertilizer, avoiding the use of lime in alkaline soils, and avoiding nitrogen fertilizers during summer stress periods may help to reduce disease incidence and severity.

While fungicides are not generally used in lawn care for Pythium blight control, they are considered a necessity in golf course management in many regions of the U.S. Before the advent of systemic fungicides targeted for Pythium blight in the early 1980s, the disease was combatted with short residual chemicals such as chloroneb and etridiazol. Metalaxyl was registered for use on turf in 1981; it provides over 20 days of control and can be used either preventatively or curatively. The widespread reliance and continuous usage of metalaxyl on golf courses has led to reduced effectiveness and in some cases the selection of *Pythium* spp. biotypes resistant to metalaxyl. Reduced residual effectiveness may also be attributable to a build up of microorganisms that degrade the active ingredient of the fungicide. Propamocarb and fosetyl-Al are other fungicides that provide long, residual Pythium blight control, but they should be applied preventatively and they are expensive to use.

To avoid the build-up of fungicide resistant biotypes and to avoid reduced residual effectiveness of compounds due to microbial build-up, fungicides should always be rotated and they should be applied in tank-mix combinations whenever economically feasible. Recent research indicates that the aforementioned problems can be avoided by tank mixing low label rates of metalaxyl and propamocarb or reduced rates of metalaxyl + propamocarb + fosetyl-Al. Tank mixing metalaxyl with mancozeb also helps reduce the probability of resistant *Pythium* biotypes from dominating. Alternating systemics with contact sprays, although the latter may only provide 3 to 7 days of control, also will help to reduce these potential problems from occurring.

SUMMER PATCH

Summer patch is a destructive disease of Kentucky bluegrass, creeping red fescue, and annual bluegrass turf and is incited by *Magnaporthe poae*. Symptoms of summer patch initially appear as wilted, dark-green areas of turf. Initially the straw-brown, dead patches resemble the symptoms of dollar spot disease, but these patches soon increase in size and may take on crescent shapes, elongated streaks, or circular patches. Healthy turf may persist in the center of blighted patches producing rings or "frog-eye" symptoms. In some regions, the frog-eye symptom is only occasionally observed and the dead patch symptom is more commonplace. Affected regions may coalesce, and large areas of turf are destroyed within a 7- to 10-day period. There are no distinctive leaf lesions associated with the disease, but leaves generally die back from the tip and plants at the periphery of affected patches display a bronze or copper color.

Summer patch most commonly occurs in Kentucky bluegrass and creeping red fescue turfs that are 2 years of age or older. To date, the disease has principally been a problem in Kentucky bluegrass, annual bluegrass, and fine leaf fescues. Environmental conditions play a significant role in the predisposition of turf to the disease. The disease generally appears in late June or early July when daytime temperatures above 90°F prevail. The disease is most severe on sunny, exposed slopes or other heat-stressed areas of lawns such as those adjacent to paved walks and driveways. The disease most frequently occurs during periods of drought stress that were preceded by a wet spring or early summer. Turf allowed to enter drought-induced dormancy is often severely damaged. Mysteriously, the disease may flair up following rainy periods in late summer and September. Other predisposing factors include: spring applications of high levels of nitrogen fertilizer, accumulation of thatch, frequent light irrigations or rain storms, and soil compaction. The most important environmental factors required for disease development are for soil to be moist and for root zone temperatures to exceed 78°F. Low mowing (especially <2.0 in.) is the major cultural practice that exacerbates summer patch. For home lawns, increasing mowing height to 3.0 in. in late spring, applying water deeply and only at the onset of wilt, and use of a slow-release nitrogen fertilizer, such as sulfur coated urea, are the best approaches to minimizing summer patch. Preventative applications of fenarimol, propiconazole, triadimefon, or curative applications of thiophanate drenches may provide a satisfactory level of control on close-cut Kentucky bluegrass fairways. To control the disease in annual bluegrass on putting greens, propiconazole or triadimefon should be applied at 28-day intervals from mid-May through August. Fungicides are ineffective if turf is allowed to enter drought-induced dormancy.

FUSARIUM BLIGHT

The name Fusarium blight is presently used to describe foliar blight, crown rot, and root rot symptoms in situations where signs of *Fusarium* spp., such as spores and pink mycelial growth, are abundant. Summer patch and necrotic ring spot, described previously, are recognized as diseases with symptoms similar to Fusarium blight, but where *Fusarium* spp. are not involved and roots bear dark-brown, runner hyphae. The control measures for Fusarium blight are the same as those described for summer patch.

MELTING-OUT

During the summer, Kentucky bluegrass, perennial ryegrass, fine leaf fescues, and other grasses may decline due to invasion by *B. sorokiniana*. This fungus also may cause a leaf spot and melting-out phase. *B. sorokiniana* is normally most severe during wet summers when temperatures exceed 85°F and humidity is high. This disease is generally aggravated when infected stands are subsequently subjected to drought stress. Also, during warm and wet periods *B. cynodontis* may become a severe crown, stolon, and root rot pathogen of bermudagrass. See the "Helminthosporium Leaf Spot and Melting-Out" section for management of melting-out.

FAIRY RINGS

Fairy rings may be caused by any one of 60 or more species of fungi. The activity of these fungi in thatch and soil results in rings or arcs of dead or unthrifty turf, or rings of dark green, luxuriantly growing grass. The most destructive rings are classified as Type 1 rings. Type 1 rings are very common, especially in old pasture turfs or any mature turf with a lot of thatch. Type 1 rings normally appear as circles or arcs of dark green, fast-growing grass. The most common fungus known to cause Type 1 fairy ring is *Marasmius oreades*. These rings are distinguished by three distinct zones: an inner lush zone where the grass is stimulated and grows luxuriantly; a middle zone where the grass may be dead; and an outer zone in which the grass is stimulated. The distance from the inside of the inner zone to the outside of the outer zone may range from a few inches to one or more feet wide. The dead zone is due to a massive buildup in the thatch and soil of fungal mycelium. It accumulates in sufficiently large amounts to form a hydrophobic barrier that prevents entry of rain or irrigation water, thus killing the plants by drought. The formation of the three zones is noticeable from early spring to winter.

Control of fairy rings is made extremely difficult due to the impermeable nature of the infested soil. Chemical control has been ineffective because the fungus grows so deeply into the soil that lethal concentrations of fungicide do not come into contact with the entire fungal body. Suppression is the most practical approach to combating fairy rings in most situations. The suppression of symptoms approach is based upon the premise that fairy rings are less conspicuous and less numerous where turf is well watered and fertilized. This method of control involves a combination of acration, deep watering, and proper fertilization. Aeration is beneficial, as it aids in the penetration of air and water. The entire area occupied by the ring, to include a 2-ft periphery beyond the ring, should be core aerified on 2 to 4 in. centers. The soil in the area should then be irrigated to a depth of 4 to 6 in. Use of a wetting agent should help improve water infiltration. The ring area should be re-treated in a similar fashion at the earliest indication of drought stress; that is, repeat the process whenever the dark green grass turns blue-gray and begins to wilt. When an aerator is not available, a deep root feeder with a garden hose attachment may be useful to force water into the dry soil. About 3.0 to 4.0 lb N/1000 ft^2 should be applied to Kentucky bluegrass or perennial ryegrass in three to four applications during the fall.

There are two methods of eradication: fumigation and excavation. Both methods are laborious, costly, and not always successful. To fumigate, the sod must be removed from an area 2 ft to the inside and at least 2 ft to the outside of the rings. It is essential not to spill any soil or sod onto the healthy grass. The most commonly used fumigants are methyl bromide and metam. Fumigation with methyl bromide must be carried out by a licensed pesticide applicator. Special precautions must be taken to ensure that children or pets do not come into contact with these fumigants. The second alternative for fairy ring eradication is to carefully dig out and discard all infested soil in the ring. This would involve removal of soil to a 12-in. depth, and the excavation should be wide enough to extend at least 2 ft beyond the outermost evidence of the ring. The excavation must then be filled with fresh, uncontaminated soil and the area reseeded or sodded. Hence, the eradication approach for fairy ring control is impractical and seldom performed.

WINTER DISEASES

SNOW MOLD DISEASES

Snow protects dormant turfgrass plants from desiccation and frost, but it also provides a microenvironment conducive to development of some low-temperature, pathogenic fungi. Like most other disease problems, there is no shortage of fungal species capable of damaging turf during cold periods between late fall and early spring. The most common low-temperature fungal diseases are pink snow mold (*Microdochium nivale*) and gray snow mold (*Typhula incarnata* and *T. ishikariensis*). Other diseases known to be active under snow cover or during winter months include red thread (*Laetisaria fuciformis*) and leaf spot (*Drechslera* spp.).

Snow mold fungi are remarkable in being active at temperatures slightly above freezing. Snow molds are damaging when turf is dormant or when growth of turf has been retarded by low temperatures. Under these conditions, turfgrasses cannot actively resist fungal invasion. Although known as snow molds, these fungi can attack turf with or without snow cover. In general, these diseases develop whenever temperatures are cool (32 to 55°F) and there is an abundance of surface moisture.

Conditions favoring pink snow mold include low to moderate temperatures; plenty of moisture; prolonged deep snow; snow fallen on unfrozen ground; lush turf stimulated by late season application of high amounts of nitrogen fertilizer, and alkaline soil conditions. Prolonged periods of cold, rainy weather are particularly conducive to disease development on putting greens. Symptoms of this disease appear as small, water-soaked patches 2 to 3 in. in diameter that may increase in size to 1 to 2 ft in diameter and coalesce. The pink coloration of affected turf at the edge of the patches is produced by the pinkish color of the mycelium. The mycelium mats the leaves, and plants eventually collapse and die. In the absence of snow on putting greens the disease often appears as 1- to 3-in. diameter patches that have a reddish-brown color. Mycelium on the leaf blades produce fruiting bodies (sporodochia) upon which spores are borne in prodigious numbers. These spores are easily spread by machinery and foot traffic. When damage occurs under snow the extent of injury is usually more severe than without snow cover. The pathogen, once known as *Fusarium nivale*, is able to survive unfavorable environmental conditions as spores and as resting mycelium that remain viable in plant debris.

Pink snow mold attacks a wide range of turfgrass species under snow including perennial ryegrass, Kentucky bluegrass, bentgrass, and the fescues. This disease is generally most destructive to annual bluegrass and bentgrass. Pentachloronitrobenzene (PCNB), iprodione, and triadimefon provide good control. Fungicidal control is best achieved with a preventative application prior to the first major snowstorm of the year. Subsequent applications should be made during mid-winter thaws and early spring snowmelt in areas where the disease is a chronic problem.

Gray snow mold, or Typhula blight, is also a serious disease of turfgrasses. Initially, symptoms appear as light brown or gray patches 2 to 4 in. in diameter enlarging to 2 ft in diameter and coalescing. Gray snow mold also occurs with and without snow cover; however, damage is usually minimal in the absence of snow. Like pink snow mold, Typhula blight is more damaging under prolonged deep snow, particularly when heavy snow accumulates on unfrozen ground.

Gray snow mold fungi initially begin the disease cycle as saprophytes colonizing dead organic matter. Under snow, however, the fungus moves onto living leaves, sheaths, and may ultimately invade the crown. Normally, *Typhula* spp. do not completely kill crowns, so plants generally recover during the spring. Conversely, pink snow mold more frequently invades crown tissues and kills turf. *Typhula* spp. survive unfavorable environmental conditions as sclerotia. Sclerotia are compact masses of fungal mycelium covered with a dark-colored protective rind. Sclerotia are chestnut brown or black in color and are often less than $1/8$ in. in diameter. When cool, moist weather conditions return in late fall, these sclerotia germinate to produce fungal mycelium or a specialized fruiting body upon which spores are borne. All species of *Typhula* that attack turf produce similar symptoms. Sclerotial color is one of the primary characteristics pathologists use to differentiate between the two species of *Typhula* known to cause gray snow mold.

Chloroneb, iprodione, triadimefon, and PCNB provide effective control of gray snow mold. Snow mold prevention with fungicides **is generally only warranted for golf course turf** in most regions of the U.S.

Researchers have shown that the fungus *Typhula phacorrhiza* applied on grain inoculum suppressed gray snow mold on creeping bentgrass greens by 44 to 70%. Although not yet commercially available, these researchers believe that this biological control agent can be formulated into pellets and be applied by standard fertilizer spreaders.

Snow mold injury can be reduced by applying a balanced N-P-K fertilizer in fall. Ammonium sulfate use was associated with reduced pink snow mold injury in Washington. Continue to mow late in the fall to ensure that snow does not mat a tall canopy. On golf courses, snow fences and windbreaks should be used to prevent snow from drifting on chronically damaged greens. Avoid compaction of snow by skiers and snow mobiles on greens. Also, avoid the use of limestone where soil pH is above 7.0, as soil alkalinity may encourage pink snow mold.

YELLOW PATCH

During cool, prolonged overcast and moist periods of late fall or early spring, yellow patch (sometimes called cool temperature brown patch), caused by *Rhizoctonia cerealis* appears. Yellow patch is a disease of bentgrass, annual bluegrass, and sometimes perennial ryegrass turf. It is most frequently observed on putting greens producing rusty-brown and yellow rings or yellow patches a few inches to one or more feet in diameter. Damage is generally superficial, but significant thinning of turf may occur during prolonged wet and overcast weather of late winter and early spring. A broad-spectrum fungicide such as chlorothalonil or iprodione will prevent severe thinning, but no fungicides or cultural practices are known to prevent the formation of rings and patches.

FACTORS ASSOCIATED WITH FUNGICIDE USE

Arriving at the decision of whether to apply a fungicide to any turf area is often difficult and generally based upon economic considerations. Aside from cost, the primary determinants in using a fungicide are the prevailing environmental conditions, host species and cultivars present, and the pathogen. The environmental factor has unique implications in turfgrass pathology because the intensity and nature of turfgrass management greatly influences plant vigor and therefore the severity of diseases.

Promoting vigorous growth through sound cultural practices is the first step in minimizing disease injury. Frequently, however, environmental stresses, traffic, and poor management weaken plants, predisposing them to invasion by fungal pathogens. When disease symptoms appear, it is imperative that a rapid and accurate diagnosis of the disorder be made. The prudent manager also attempts to determine those factors that have led to the development of the disease. The most common cause for extensive disease injury in lawn turf can frequently be related to poor management practices by the homeowner. Such practices include frequent and close mowing, light and frequent irrigations, and inadequate or excessive nitrogen fertility. The development of excessive thatch layers, shade, poor air or water drainage, and traffic may also contribute significantly to disease problems. A good case in point is Helminthosporium diseases, which are particularly damaging when turf is mown too closely, given light and frequent irrigations, and when turf is excessively fertilized. Despite hard work and adherence to sound management practices, diseases may become serious problems. This normally occurs when environmental conditions favor disease development, but not plant growth and vigor. For example, summer patch and brown patch are most damaging when high summer temperatures stress plants and impair their growth and recuperative capacity. In this situation, fungicides may be recommended in conjunction with cultural practices that promote turf vigor.

Fungicides may be applied preventatively (i.e., before anticipated disease symptoms appear) or curatively (i.e., when disease symptoms first become evident). Applying a fungicide after the turf has been damaged significantly is generally a waste of time, money, and effort. Curative applications are more economical and environmentally wise, but only if the disease can be treated before damage is excessive. In general, a single, or possibly two, properly timed applications will provide effective control of most disease problems encountered on home lawns. Contact fungicides are generally less expensive and provide good control. Contact fungicides, however, may only provide 7 to 14 days of control under high disease pressure conditions. Where sudden and severe, or chronic disease problems occur, a systemic alone, or a systemic plus a contact may be needed. Systemic or local systemic fungicides will provide 14 to 21 days protection during high-pressure disease periods. Tank mixing a systemic plus a contact fungicide provides a slightly longer residual effect and a wider spectrum of control. Frequently, a fungicide may only be needed to help the turf better survive a high pressure disease period. Favorable changes in weather such as alternating hot-humid and cooler periods,

however, provide the most effective means of reducing or eliminating disease problems in the summer.

THE FUNGICIDE DILEMMA IN LAWN CARE

Proper utilization and selection of fungicides is too difficult and complicated for the vast majority of homeowners. Because of this, only lawn care companies can provide the most reliable lawn disease service. Fungicides, however, should not become a part of a normal application schedule. As a general rule, use of fungicides is not recommended in most home lawn situations because: (1) proper diagnosis and proper fungicide selection are difficult, (2) it is generally too late to achieve the economic and aesthetic benefits of a fungicide once extensive injury has occurred, (3) lawn care companies capable of only dry or granular applications do not have the proper spray equipment or they cannot obtain the desired fungicide(s) in granular form, and (4) it may be less expensive, and better in the long run, to overseed a damaged turf area in the autumn with disease-resistant cultivars.

There are several disease situations of lawn turf that are best controlled through a preventative fungicide application. These disease situations are: (1) Kentucky bluegrass lawns injured in previous years by summer patch, necrotic ring spot, stripe smut, and perhaps dollar spot, (2) perennial ryegrass lawns injured in previous years by Pythium blight, brown patch, or dollar spot, and (3) tall fescue lawns chronically damaged by brown patch. Many diseases, however, are effectively controlled with curative fungicide applications when disease symptoms first appear. For example, leaf spot in lawns of common-type Kentucky bluegrasses (e.g., Kenblue, Newport, Park, South Dakota, etc.) and fine leaf fescue (e.g., Pennlawn and Jamestown), and brown patch in tall fescue or perennial ryegrass can be controlled effectively with a curative fungicide application. Dollar spot disease is extremely common and if allowed to go unchecked, may cause extensive injury to Kentucky bluegrass, perennial ryegrass, red fescue, and zoysiagrass lawns. When diagnosed in its early stages, however, dollar spot is also controlled effectively by fungicides. Given these situations, it becomes obvious that effective fungicide programs hinge upon: (1) a knowledge of past disease problems in a particular lawn or neighborhood, (2) an ability to distinguish between turfgrass species and sometimes cultivars within a species, and (3) an ability to diagnose turfgrass diseases. Hence, in addition to the expense of fungicides and logistical problems associated with sending trucks to specialized fungicide accounts, the lawn care company must also educate its employees to diagnose diseases. This educational process is best achieved by in-house training programs. This knowledge must be reinforced by encouraging employees to attend turf workshops and conferences held by state universities and other organizations, as well as maintaining subscriptions and reading articles in trade magazines. Expecting the employee to be proficient by reading alone is not realistic, even for highly motivated individuals.

In Table 2, the most common lawn diseases are listed, as well as their primary season(s) of occurrence, and those species that are most commonly damaged. Remember, once a disease has severely reduced stand density, fall overseeding with resistant cultivars is normally suggested. Fact sheets describing the disease and a list of cultural practices that will help minimize disease injury should be provided by the lawn care company to homeowners.

THE FUNGICIDE DILEMMA ON GOLF COURSES

Where extremely high-quality turf is desired, fungicides will be needed in most years, and in nearly all areas of the U.S. The indiscriminate use of fungicides or employment of numerous, preventive applications of fungicides for many diseases should be discouraged. Other than economic restraints, reasons why repeated fungicide applications may not be desirable include:

1. Development of fungicide-resistant pathogens.
2. Fungicides may disturb a delicate balance among microorganisms that compete with and antagonize disease-causing fungi. This may explain why some diseases recur more rapidly and cause more injury in turfs previously treated with fungicides.
3. A fungicide may control one disease, but encourage other diseases.
4. Fungicides may reduce the population of beneficial microorganisms in the soil, which could lead to a thatch buildup.

The development of fungal strains resistant to fungicides has been well documented. Resistant strains of the dollar spot fungus first developed as a result of repeated usage of cadmium-based fungicides and benomyl on golf courses. In addition to cadmium and benzimidazole resistance, strains of the dollar spot fungus resistant to anilazine, dicarboximide, and sterol-inhibiting fungicides also have been reported. The buildup of resistant strains of fungi occurs in response to a selection process that eventually enables a small, but naturally occurring subpopulation of resistant biotypes to predominate in the fungicide-treated turfgrass microenvironment.

Turf managers have observed that some diseases may recur more rapidly and severely in turfs

Table 2 **Common diseases of transition zone and northern lawn grasses, the primary season of occurrence, and the primary grasses most likely to be damaged**

Disease	Primary Season(s) of Occurrence	Primary Lawn Grasses Damaged
Helminthosporium leaf spot, melting-out, and net-blotch	Spring, fall, summer	Common-type Kentucky bluegrass, fine leaf fescue, perennial ryegrass, tall fescue
Red thread	Spring and fall	Perennial ryegrass, fine leaf fescue, Kentucky bluegrass
Necrotic ring spot	Early spring to early winter	Kentucky bluegrass
Dollar spot	Early spring to early winter	Perennial ryegrass, Kentucky bluegrass, fine leaf fescue
Brown patch	June, July, August	Perennial ryegrass, tall fescue
Summer patch	June, July, August	Kentucky bluegrass, fine leaf fescue
Pythium blight	June, July, August	Perennial ryegrass and seedlings
Stripe smut	Spring and fall	Kentucky bluegrass
Rust	Spring and fall	Perennial ryegrass, Kentucky bluegrass, zoysiagrass
Powdery mildew	Spring and fall	Kentucky bluegrass, fine leaf fescue
Fairy rings	Spring to early winter	All species

previously treated with fungicides, as compared to adjacent untreated areas. Dollar spot and brown patch are probably the most common diseases to exhibit this phenomenon. Disease resurgence is attributed to fungicides being toxic to beneficial microorganisms, which naturally antagonize and keep disease-causing fungi in abeyance.

Fungicides applied to control one disease may encourage other diseases. Benomyl has been reported to enhance red thread, Helminthosporium leaf spot, and Pythium blight. Thiophanate-methyl may increase crown rust in perennial ryegrass, iprodione can increase yellow tuft, maneb may enhance dollar spot, and chlorothalonil can increase summer patch in Kentucky bluegrass. Encouragement of disease in these situations may again be attributed to offsetting the delicate balance between antagonistic and pathogenic microorganisms in the ecosystem.

When used repeatedly, certain fungicides have been shown to enhance thatch accumulation. Benzimidazole fungicides, such as benomyl and the thiophanates, and sulfur-containing fungicides such as mancozeb, maneb, and thiram can cause thatch to accumulate by acidifying soil. The effect of these fungicides is indirect, that is they inhibit the thatch decomposition capacity of beneficial microorganisms by lowering soil pH. Some fungicides can enhance stem tissue production and also indirectly cause thatch to accumulate. Fungicides also may contribute to thatch buildup by being toxic to earthworms. Earthworms help reduce thatch by mixing soil with organic matter. Benomyl, mancozeb, anilazine, chlorothalonil, and various insecticides and nematicides have been shown to be toxic to earthworms.

The phytotoxicity that accompanies the usage of some fungicides is generally not severe. Most phytotoxicity problems occur when fungicides are applied to bentgrasses, particularly during periods of high-temperature stress. Repeated applications of sterol-inhibiting fungicides such as fenarimol, propiconazole or triadimefon may elicit a blue-green color in foliage of creeping bentgrass and other turfgrass species. Tank-mixing sterol-inhibiting fungicides with some plant growth regulators may also elicit an objectional level of discoloration.

It should be noted that many of the harmful side effects just described were either isolated events or occurred only after repeated use of one fungicide over the course of several years. Experienced turfgrass managers have long recognized that tank mixing known synergists and rotating fungicides of different modes of action greatly minimizes these potential problems. The importance of rapid and accurate disease diagnosis, and the judicious use of fungicides are integral in management programs where fungicides are commonly employed.

NEMATODES

Nematodes are known to cause extensive injury to turfgrasses in warm temperate and subtropical areas in the U.S. In northern climates, the loss of turfgrass that can be attributed to nematodes is unknown. However, nematodes may be more troublesome in the transition zone and Mid-Atlantic area than is commonly believed. This would be particularly true following a mild winter. In general, nematodes most actively feed on turfgrass during environmental periods favorable for growth of the grass. Hence, feeding would be more active on cool season species in spring and fall, while on warm season grasses heaviest feeding would occur during summer. The injurious effects of this feeding, however, may not become noticeable until turf is subjected to environmental stresses in mid- to late summer.

Nematodes are very small eel-like worms, ranging from $\frac{1}{50}$ to $\frac{1}{8}$ in. in length. Nematodes reproduce by eggs, which hatch to liberate larvae. Larvae molt four times before reaching adult size. Each female is capable of producing hundreds of eggs, and the entire life cycle is completed in 5 to 6 weeks under suitable conditions for most species. Because plant pathogenic nematodes are obligate parasites, they must feed on living tissues in order to grow and reproduce. Most nematodes are capable of attacking a wide range of plant species, and can survive on weeds in the absence of turfgrasses. Most nematodes store large quantities of food, which enables them to survive long periods in soil in the absence of suitable plants. Many also survive in frozen soils and may overwinter in living roots or in dead plant tissues.

Literally millions of nematodes can inhabit a few square feet of soil, but most nematodes are nonpathogenic. Some nonpathogens actually perform a beneficial service in soil by helping to degrade organic matter. Although there are thousands of species, only about 50 species are known to parasitize turfgrasses. All plant parasitic nematodes bear a hollow, spear-like structure called a stylet. The stylet is similar to a hypodermic syringe, and is used to inject enzymes into plant cells. Simultaneously, partially digested food is withdrawn. Plant pathogenic nematodes are grouped according to feeding habit. Endoparasitic nematodes partially or totally burrow into plant tissues and feed primarily within; whereas, ectoparasitic nematodes feed from the plant surface, although a small portion of the body may be embedded. The ectoparasitic are more commonly injurious to turfgrasses than endoparasitic nematodes. Nematode activity is favored by warm and moist soil conditions. Nematode populations generally peak in June or July and again in late August or early September. Their activity also is enhanced in light textured soils and reduced in compacted or heavy soils where aeration becomes restricted. Nematodes are unable to move more than a few millimeters in soil, but they may be transported farther by moving water and soil.

The symptoms of nematode injury include yellowing, stunting, wilting or early signs of drought stress, and thinning of the stand. These symptoms are related to the injury nematodes inflict upon the root system. Therefore, symptoms of injury may not become noticeable until water becomes limiting.

Due to the similarities between environmental stress symptoms and nematode injury, the source of the problem is difficult to diagnose. Like many so-called weak or secondary fungal pathogens, nematodes may not cause much of a problem until environmental extremes reduce the vigor of a turf. There is often no pattern to nematode injury, but generally affected areas may appear in streaks or oval-shaped areas. Severe infestations may result in a nearly total loss of grass plants, which are soon replaced by weed species. Inspection of roots may or may not reveal some indication of nematode feeding. Roots may exhibit one or more of the following symptoms: swellings, red or brown lesions, excessive root branching, necrotic root tips, and root rot. Some of the more common plant parasitic nematodes, which are known to injure turfgrasses, are listed in Table 3.

Table 3 **Common plant pathogenic nematodes known to be injurious to turfgrasses**

Common Name/Genus	Feeding Group	Turfgrasses Injured/Root Symptoms
Lesion (*Pratylenchus* spp.)	Endoparasitic	Cool- and warm-season grasses are injured. Root lesions initially minute and brown, but enlarge and may prune the root system.
Lance (*Hoplolaimus* spp.)	Ectoparasitic	Warm-season grasses, creeping bentgrass, and annual bluegrass are injured. Causes swelling of roots followed by necrosis and sloughing of cortical tissues. Can be endoparasitic and migratory.
Ring (*Macroposthonia* spp.)	Ectoparasitic	Especially important in centipedegrass, also injured are Kentucky bluegrass, bentgrass, bermudagrass, and zoysiagrass. Roots are stunted; produces brown lesions on roots.
Spiral (*Helicotylenchus* spp.)	Ectoparasitic	Cool- and warm-season grasses injured. Roots are poorly developed with premature sloughing of cortical tissues.
Stunt or stylet (*Tylenchorhynchus* spp.)	Ectoparasitic	Warm- and cool-season grasses injured. Roots shortened, shriveled, brown; no lesions evident on roots.
Root-knot (*Meloidogyne* spp.)	Endoparasitic	Warm-season (especially zoysiagrass) and cool-season grasses (especially bentgrass) injured. Galls (i.e., swellings or knots) on roots. Galls may be small and difficult to see.
Stubby root (*Paratrichodorus* spp.)	Ectoparasitic	Warm-season grasses, Kentucky bluegrass, and fescue injured. Large brown lesions on roots; swelling on root tips.
Sting (*Belonolaimus* spp.)	Ectoparasitic	Bermudagrass, zoysiagrass, and St. Augustine grass are injured. Lesions evident, especially root tips.
Dagger (*Xiphinema* spp.)	Ectoparasitic	Warm-season grasses (especially zoysiagrass) and perennial ryegrass are injured. Root lesions are reddish-brown to black and sunken.
Pin (*Paratylenchus* spp.)	Ectoparasitic	Kentucky bluegrass, fine and tall fescues are injured. Tillering and rooting increased, but fewer lateral roots. Distinct lesions on roots.

Turfgrass areas damaged by nematodes are especially prone to wilt and do not respond readily to an application of fertilizers or fungicides. This lack of response may be a good indicator of a nematode problem. In this situation a soil sample should be sent to a nematologist for analysis. Soil should be collected from a dozen or more areas at the edge or interface between healthy and injured turf. Sampling from severely thinned areas may yield unreliable results, because these obligate parasites will not survive in large populations in the absence of living plants. Soil samples should be collected in the root zone region, normally the upper 3 to 6 in. in heavy soils, but 6 to 12 in. or deeper in sandy soils. The samples should be combined, as in a routine soil-fertility sample, and at least a pint of soil is needed. Samples should also be taken from nearby, healthy turf so that the nematologist can compare numbers and species of nematodes between the two areas. The soil must be kept moist and given to the nematologist as soon as possible. Refrigerate the sample, but do not freeze it, if there will be a delay in transport to the lab. In the laboratory, various methods for extraction are used. Unfortunately, there are no reliable data correlating nematode number per sample and expected degree of turf injury in the field. The nematologlst, however, will generally be able to make a relatively good management recommendation based on species and number of nematodes present in a sample.

An absolute determination of a nematode problem from visual symptoms and even soil analysis is difficult. Frequently, the best indication of a nematode problem is a positive response from a nematicide. It should be pointed out, however, that turf green-up frequently occurs following a nematicide application, presumably because of the death of nematodes as well as other invertebrates which liberate nitrogen upon decay. Nematicides are highly toxic, organophosphate derivatives. They must be handled with extreme caution and used only according to the procedure and at the rates given on the label. Because the target of a nematicide is in soil, it is essential the chemical be thoroughly watered-in, and aerification prior to application will facilitate the downward movement of the chemical. Ethoprop and fenamiphos are the only nematicides registered for turf, and they are primarily used on golf courses. Both are available in granular form. Fumigants are of little practical value because they kill turf as well as all other living organisms with which they come into contact. Obviously, fumigants would only be used prior to establishing a turf or if total renovation is desired.

Because nematodes only may be injurious in northern regions during summer stress periods, cultural practices that alleviate stress may help minimize injury. Such practices would include judicious irrigation, increasing the mowing height, and use of a balanced fertilizer. Application of soluble nitrogen fertilizer during environmental stress periods of summer will place an additional (and perhaps lethal) stress on an already dysfunctioning root system under attack from nematodes. There are some research data indicating that use of organic forms of nitrogen fertilizer (e.g., sewage sludges) can discourage development of high populations of some parasitic nematodes, when compared to the use of inorganic forms. Granular chitin-protein materials, such as Clandosan, are reported to control nematodes in some crops. These materials, however, do not appear to have any activity on plant parasitic nematodes found in turfgrasses.

REFERENCES

Smiley, R. W., Dernoeden, P. H., and Clarke, B. B., **Eds.,** *Compendium of Turfgrass Diseases*, 2nd ed., American Phytopathological Society, St. Paul, MN, 1992.

Smith, J. D., Jackson, N., and Woolhouse, A. R., *Fungal Diseases of Amenity Turf Grasses*, Routledge Chapman & Hall, New York, 1989.

Vargas, J. M., Jr., *Management of Turfgrass Diseases*, Lewis Publishers, Boca Raton, FL, 1993.

Integrated Management of Weeds: Evaluation, Modification of Management, Control Decisions

Weeds invading a bare soil area in grass

Application of mulch to an ornamental border to control
weed infestation (photo courtesy of Tim Rhay)

Cemetery at West Point, NY — typical site where growth regulators
are used to minimize mowing

Cosmos — part of a wildflower landscape as an
alternative to grass (photo courtesy of John Krause)

Chapter 24

Understanding Turfgrass Growth Regulation

John E. Kaufmann, *Monsanto Company, St. Louis, MO*

CONTENTS

Introduction 267
Review of Turfgrass Responses to PGRs 268
PGR and Turfgrass Growth Dynamics 268
Characterizing the Growth Regulators 268
    Type I Growth Regulators 268
    Type II Growth Regulators 269
Timing Turfgrass PGR Applications 269
    Application Timing to Cool-Season Grasses 269
    Application Timing to Warm-Season Grasses 269
Specific Cool-Season Grass Responses to PGR Types 269
    Type I PGR Timing 269
    Type II PGR Timing 270
    Consequences of Stage IV Application 270
Biosphere Targeting of PGRs 270
    Soil-Targeted PGRs 271
    Foliage-Targeted PGRs 271
    Herbicide/Fungicide Targeting 271
Equipment for Proper Targeting 271
    Soil, Thatch, and Root Targeting 271
    Foliage Targeting 271
Classifying Turf Areas Relative to PGR Types 272
    Class A Turfs 272
    Class B Turfs 272
    Class C Turfs 272
    Class D Turfs 272
Summary 272
Trademark Identification 273
References 273

INTRODUCTION

Plant growth regulators (PGRs) have been applied to turfgrass since the late 1940s. While the earliest use was to reduce roadside mowing requirements, many other uses have now been proposed including reduced clipping disposal, enhanced turf quality, minimizing fire hazard, reduced trimming, and reduced chilling injury. However, the most widespread uses of PGRs remain (1) to suppress normal growth of turfgrass to reduce mowing requirements or volume of clippings, (2) to suppress undesirable turfgrass development such as seedhead formation, and (3) to suppress competition from undesirable grass species such as annual bluegrass (*Poa annua* L.).

Thus, turf PGRs are positioned to alter the appearance of or competitiveness within the turfgrass ecosystem and not to eliminate a species as is the case with herbicides. Some people have suggested that PGR use better fits the IPM concept than herbicide use. However, to argue distinctions between the ecological effects of these two types of chemicals is very difficult.

0-87371-350-8/94/$0.00+$.50
© 1994 by CRC Press Inc.

REVIEW OF TURFGRASS RESPONSES TO PGRs

Several hundred papers have been written on PGR effects on turfgrass and these have been reviewed in detail by DiPaola,[1] Elkins,[2] and Watschke.[4] If a single conclusion can be drawn from the PGR literature, it is that turfgrass responses to growth regulators have been inconsistent and many of the responses have been aesthetically undesirable. While applicator knowledge of the proper use and application timing has improved greatly in the past few years, uncertainty of turfgrass response continues to limit widespread use of PGRs compared to herbicides.

In these reviews, PGRs are reported to suppress grass growth anywhere from 4 to 10 weeks after application, occasionally with a period of post-retardation shoot growth and color enhancement. Seedhead suppression has ranged from none to complete season-long control. Suppression of new leaves, tillers, stolons, rhizomes, and roots has been observed in some cases while enhancements of nearly all of the above have also been observed.

The various types of turfgrass injury reported with PGR use include chemical phytotoxicity, discoloration, droughty appearance, off color, and tip die back. Sometimes the injury is latent, occurring 4 to 5 weeks after treatment. But many times there is no injury and enhanced dark green color occurs throughout the entire period of growth regulation.

Turfgrass ontogeny and physiology has been identified as very important to characterize turfgrass response to PGRs.[3] For years the turfgrass research community has suggested that the process of mowing turfgrasses results in synthesis of growth hormones such as ethylene induction. Further, change in hormone balance due to mowing results in developmental responses such as increased tillering. Therefore it is proposed in this paper that physiological and developmental processes associated with mowing are again altered when mowing is reduced or eliminated. It is further suggested that alteration of these processes is important in understanding a large part of the inconsistent responses to PGRs, both in turfgrass quality and in shoot growth suppression. (A good review of Chapter 10 will enhance reader understanding of the material presented in this chapter.)

This chapter discusses (1) the relationship of physiological processes of turfgrass growth and development with the observed turfgrass responses to PGR, (2) how to time PGR application for maximum benefit and consistent performance, (3) how to target specific PGRs in the turfgrass ecosystem to obtain maximum results with minimum chemical, and (4) how to select the best PGR for the different cultural levels at which turfgrass is maintained.

PGR AND TURFGRASS GROWTH DYNAMICS

Given the importance of the regeneration of new, young plants for maintenance of the turfgrass ecosystem, it is easy to see that prolonged use of a PGR that stops new growth will result in a natural loss of turfgrass color as the turf ages. Based on the estimates of leaf duration, a PGR that stops growth completely will cause the turf to begin a natural leaf aging approximately 4 to 6 weeks after treatment. Natural leaf aging begins with tip die back and proceeds to the base of the leaf. As a result, where growth control is the desired PGR benefit, it is preferable to suppress growth rather than inhibit growth completely. PGRs that allow for some new leaf appearance provide for continuity of green color.

CHARACTERIZING THE GROWTH REGULATORS

Growth can be defined as irreversible enlargement in size, while development is the transformation of apparently identical cells into diversified cells and plant organs. Based on these definitions, turf PGRs can be divided into two basic types.

TYPE I GROWTH REGULATORS

Type I PGRs suppress both growth and development (Table 1). The chemicals in this group act as inhibitors of cell division or cell cycle. Examples include mefluidide and maleic hydrazide. These regulators suppress leaf growth and initiation of new turfgrass leaves for about 6 weeks. These PGRs have good overlap safety to meristems of the plant, but may result in immediate but transient injury to the leaves.

Other chemicals known to inhibit growth and development of cool-season grasses are labeled

Table 1 **Type I growth regulators**

Common Chemical Name	Trade Name
Mefluidide	Embark® Lite
	Embark® Trim-Cut
Maleic hydrazide	Royal Slo-Gro®

herbicide Type I regulators, which inhibit enzymes of amino acid or organic acid biosynthetic pathways critical to growth (Table 2). Thus, low rates can act as PGRs. Examples of this type of chemical include glyphosate, sulfonyl ureas, imidazolinones, and sethoxydim. These compounds have a very narrow margin of safety on cool-season grasses, and accidental overdoses severely injure or kill turf. Within herbicide Type I PGRs, sulfonyl ureas and imidazolinones have residual soil action while glyphosate and sethoxydim do not.

TYPE II GROWTH REGULATORS

Type II PGRs suppress growth only (Table 3). The developmental sequence of the plant continues, however, new plant organs develop in miniature size. Examples of this type include flurprimidol, paclobutrazol, trinexapac-ethyl, and uniconazole. These compounds have primary action as gibberellin biosynthesis inhibitors and are effective in suppressing elongation of cells and internodes.

Fungicide Type II growth regulators are those that are primarily used as a fungicide but have been observed to modify the host (Table 4). Darker green color and growth reduction has been observed with triadimefon and fenarimol. The latter has been recommended for use on putting greens for selective suppression of annual bluegrass.

TIMING TURFGRASS PGR APPLICATIONS

It is important to understand seasonal growth patterns of turfgrasses before one can understand the proper application timing of growth regulators. Important details of the stages of turfgrass growth and development have been outlined in Chapter 10, and reference will be made to that information.

APPLICATION TIMING TO COOL-SEASON GRASSES

Since up to 50% of lawn clippings are cut during a 6-week period in the spring, the ideal time to apply a PGR to cool-season grasses is early spring to prevent the greatest amount of vertical growth. Often there is a period of growth enhancement after chemical suppression. However, postretardation growth enhancement is usually of little consequence since the hot summer environment is already reducing growth. In fact, improved green color at this time is desirable. Because most cool-season grasses only grow seedheads once per year, properly timed application of PGRs that control seedheads can create the effect of season-long control without residual chemical action.

Table 2 Herbicide Type I growth regulators

Nonselective herbicides
   Example: Glyphosate — Roundup®
                  Imidazolinones — Event®
Selective broadleaf herbicides
   Examples: Sulfonyl ureas — Telar®, Oust®
Selective narrowleaf herbicides
   Example: Sethoxydim — Poast®

Table 3 Type II growth regulators

Common Chemical Name	Trade Name
Flurprimidol	Cutless
Paclobutrazol	Scotts TGR®
	TGR Turf Enhancer®
Trinexepac-ethyl	Primo®
Uniconazole	Sumagic®

Table 4 Fungicide Type II growth regulators

Common Chemical Name	Trade Name
Triadimefon	Bayleton®
Fenarimol	Rubigan®

APPLICATION TIMING TO WARM-SEASON GRASSES

The ideal time to apply a PGR to warm-season grasses is in early summer as growth begins. For most of the warm-season region, a single application is often not sufficient to regulate growth for the entire period of high vertical shoot growth, and reapplication is necessary for season-long control. Warm-season grasses grow seedheads throughout the year. Thus, unless the PGR has a residual chemical effect, reapplication will also be necessary for season-long seedhead control.

SPECIFIC COOL-SEASON GRASS RESPONSES TO PGR TYPES

Since cool-season grasses exhibit a wide variation of growth responses to environmental stimuli such as daylength and cold temperature, PGR application designed to slow growth has resulted in a wide range of responses. Thus specific growth stage responses to types of PGRs are detailed below.

TYPE I PGR TIMING

If a Type I PGR is applied at Stage I (see Figure 3 in Chapter 10), the most noticeable effect is a delay of spring greenup. Since development is slowed as well as growth, the rate of appearance of new green leaves is slowed and leaf size is diminished. Root active PGRs are effective in reducing growth when applied at this stage while foliarly active PGRs require green leaves to absorb the chemical.

Application at Stage II results in delay of further greenup and subsequent growth suppression. Application at this stage is desirable since the turf often has greened sufficiently and rapid spring growth has not yet begun.

Stage III is considered the optimum time for application of Type I PGRs to provide good turfgrass quality and the normal 5- to 6-week duration of vegetative suppression or inhibition. Often there is a slight loss of turf quality during the 3rd and 4th weeks from leaf aging, and enhanced dark green color from the 7th to 10th weeks or longer. Seedhead control is usually greater than 90% for applications made during this stage. The end of this stage can be identified by seedhead formation at the base of the plant. While it is too late to control those seedheads, a high number of later-forming seedheads can still be controlled with a Type I PGR.

Applications of any Type I PGR at Stage IV is often detrimental to the appearance of the turfgrass area, especially if the grass is a stemmy type. PGRs do not reverse the negative effects of reproductive physiology, but exacerbate the situation to completely inhibit growth of existing leaves. Likewise, they greatly slow tiller development, at least for a time. Eventually one or more lateral buds, deep in the thatch, begin to grow very rapidly and develop into elongated tillers. Thus, application of Type I PGRs at Stage IV results in undesirable turfgrass responses: (l) excessive growth inhibition for a short period, (2) severe loss of turfgrass quality as leaves senesce and die, and (3) early termination of activity due to rapid growth of escaped tillers not affected by the chemical.

TYPE II PGR TIMING

Because Type II PGRs do not suppress plant development, applications at any of the stages from I through IV can result in (l) diminutive seedhead expression below the mowing height, (2) natural senescence and death of the main tiller, and (3) suppression of the size of new tillers that normally grow large enough to mask the dying leaves. Therefore, no stage of application on stemmy, prolific seeding turfgrass varieties in the spring is acceptable for Type II PGRs.

It is important to state that the Type II PGRs do show acceptable results on non-stemmy, highly vegetative species and varieties. For instance, tall fescue (*Festuca arundinaceae* Schreb.) seedheads apparently can quite easily be mowed off prior to the reproductive physiology, even when stunted by a Type II PGR, and good results have been achieved. Type II PGR use on Kentucky bluegrass (*Poa pratensis* L.) cv. "Baron," however, has not been as successful. Apparently when the seedhead height is stunted, the mower does not remove the seedhead soon enough to prevent natural reproductive physiology, and leaves usually senesce and brown off rapidly during Stage IV. Finally, it should also be noted that Type II PGRs have shown excellent turfgrass safety in Stage VII (during fall) when perennial cool-season species do not exhibit a reproductive growth stage.

CONSEQUENCES OF STAGE IV APPLICATION

The problems encountered in Stage IV reinforce the fact that Stage II or III is the preferred time for application. Since developmental inhibitors applied at these stages prevent seedheads from developing, they also prevent reproductive physiology and associated negative turfgrass quality consequences. As a result there is improved turfgrass quality compared to a nontreated area undergoing the "stemmy" reproductive physiology phase. Further, the effect of preserving leaves seems to be accompanied by a preservation of existing roots. As a result, improved summer growth, color, rooting, and tolerance to summer stresses (heat, drought, and diseases) have been observed when using Type I growth regulators.

BIOSPHERE TARGETING OF PGRs

Just as it is important to know where a pest lives in the turfgrass biosphere for targeting a pesticide application, the site of uptake of each PGR needs to be identified and the PGR targeted to that site for consistent turfgrass response. Table 5 shows the preferred site for optimum uptake of five key PGRs and for herbicides and fungicides.

Table 5 **Biosphere targeting of chemicals**

| Chemical | Biosphere | |
	Soil	Foliage
Mefluidide		X
Maleic Hydrazide		X
Paclobutrazol	X	
Flurprimidol	X	
Trinexapac-ethyl		X
Herbicides	PRE	POST
Fungicides	Systemic	Contact

SOIL-TARGETED PGRs

Paclobutrazol, flurprimidol, and uniconazole must be targeted to the crown or root system of the plant to achieve results. This does not mean these compounds are not absorbed by the leaves. In fact, leaf absorption may reduce product performance, since once inside the leaves downward movement is minimal. Thus the only route to the active site is through root absorption and upward translocation. Irrigation or rainfall soon after application is actually desirable.

The root-absorbed chemicals are "weather proof" in that immediate rainfall actually hastens and enhances activity. Given the frequency of rain in April, when cool-season turfgrass application should occur over much of the midwestern and northeastern U.S., the probability of rainfall soon after application is high. Many turf managers are applying root-active products during the rain and thereby taking advantage of an otherwise "down" day.

Further, research has shown that root-absorbed growth regulators do not exhibit direct chemical burn to the leaves and may be overlapped without noticeable injury. However, overlaps may result in double the amount and duration of PGR activity, that may result in uneven regulation of shoot growth.

FOLIAGE-TARGETED PGRs

Mefluidide, maleic hydrazide, and trinexapac-ethyl need to be targeted to the foliage. Uniform and complete coverage is important to achieve a uniform response from these products. All of these products require a period of time on the leaves for absorption before rainfall occurs, or they can be washed off and thus inactivated. This is especially true for maleic hydrazide which is slowly absorbed over a period of 24 hours. In addition, high humidities during the 24 hours are preferred to enhance absorption. Trinexapac-ethyl is unique in that it is the only foliar applied Type II PGR.

HERBICIDE/FUNGICIDE TARGETING

The herbicide and fungicide information in Table 5 is given to show proper biosphere targeting for these types of chemicals. Preemergence herbicides, systemic fungicides, and most insecticides perform best when they reach the soil or thatch soon after application, while postemergence herbicides and contact fungicides generally need to be on the foliage for a period of time. Therefore, after checking for chemical compatibility, it is also important to consider timing compatibility and biosphere targeting compatibility before tank mixing PGRs with fungicides or herbicides.

EQUIPMENT FOR PROPER TARGETING

Responsible use of all chemicals demands that they are targeted to the zone of maximum effectiveness. Therefore the first rule for equipment is to make sure it is in good working order and calibrated to deliver a uniform and proper dose. However, certain types of equipment are more suited for targeting chemicals to the foliage and others to the soil (Table 6).

Table 6 **Biosphere targeting equipment**

Soil	Foliage
High Gallonage	Low Gallonage
– Flood jet nozzle	– Flat fan nozzle
– Rain drop nozzle	– Micro-drop ULV
– Lesco Gun®	– Expedite® System
– Hose end sprayer	– Mist blower

SOIL, THATCH, AND ROOT TARGETING

Equipment such as single- or multiple-flood jet nozzles, raindrop nozzles, the Lesco Gun®, and hose-end sprayers all work best for soil-targeted chemicals. The application is characterized by large droplet sizes, higher carrier (water) volume, and low operating pressures. The objective is to provide uniform soil coverage while minimizing the amount of product remaining on the leaves. For turf professionals who desire to apply liquid forms of nitrogen this targeting objective is critical to minimize leaf burn. Increasingly, turf managers are also improving performance of soil-active products by scheduling irrigation immediately after application or making the application during rainfall.

FOLIAGE TARGETING

Equipment types available for foliar targeted products include flat fan nozzles or the ultra-low-volume applicators such as the Expedite® System. Mist blowers also fall into this category, but are usually not used in turf culture due to the chance of drift to nontarget plants. The spray operation is generally characterized by finer droplet sizes, higher operating pressures, and lower carrier (water) volumes. The objective of foliar application is to uniformly apply product to each leaf while minimizing runoff. Foliage-targeting equipment can work well for soil-targeted or root-absorbed chemicals providing irrigation or rainfall washes the chemical down to the soil within a reasonable time.

CLASSIFYING TURF AREAS RELATIVE TO PGR TYPES

A most important step in understanding where to use growth regulators is to classify the areas according to level of management. The following classification scheme proposes relative management intensities and suggests the types of PGRs most commonly used at each level.

CLASS A TURFS

Class A turf is that receiving high levels of input. Mowing is done on a frequent basis to maintain a groomed appearance at all times. Fertilizers are usually applied two to four times per year. Pests are generally controlled on a curative program and the areas are often, but not necessarily, irrigated. Examples of Class A turf include golf greens, tees and fairways, sportsfields, high-quality home lawns, and improved sections of industrial grounds, parks, and cemeteries. Widespread use of PGRs on Class A turfs has been limited due to off-color associated with leaf aging. However, as applicators gain knowledge of the role of turfgrass ontogeny and learn the proper timing and targeting of PGRs, use on Class A turfs will grow, especially on hard-to-mow areas such as on steep slopes and around obstacles. Specialty uses of Type II PGRs, such as suppression of annual bluegrass on golf course fairways, will continue to grow, and annual bluegrass seedhead suppression with mefluidide will also grow. Undoubtedly, trinexapac-ethyl will open new opportunities given its unique position of foliar absorption and Type II activity.

CLASS B TURFS

Class B turfs are those that for reasons of aesthetics need to be mowed on a frequent basis but other management inputs are somewhat limited. The key objective remains vegetation height control similar to that required for Class A turfs. Perhaps once a year or every 2 years, these areas are fertilized and broadleaf weeds are controlled. Mowing frequency is equal to that for Class A turf when based on turfgrass growth rate, but may be somewhat less based on calendar days. Examples of these areas include the major portion of industrial grounds, parks, cemeteries, golf course roughs, and home lawns. Both Type I and II PGRs can be used successfully on Class B turfs. However, the key value of a PGR on these turfs is mowing replacement, and for large open lawns, three to four mowings need to be saved in order to be cost effective. Where obstacles and slopes reduce mowing efficiency, a savings of one or two mowings may be cost effective. PGRs only need to save one string-

trimming operation. It is important to select PGRs with good overlap safety and uniform overlap activity where large areas require numerous side-by-side passes with the equipment.

CLASS C TURFS

Class C turf is mowed two to three times per year, usually never fertilized, but control of certain broadleaf weeds may occasionally occur when infestations become severe. The key objective with this mowing frequency is to cut down seedheads (which result in brown color) and excessive vegetation which may harbor unwanted animals. An example of this type of turf area would be highway roadsides and remote industrial sites. Generally, cost effective PGR application can only be achieved with Type I and herbicide Type I PGRs. One pass of the spray boom along each side of the highway provides sufficient vegetation control without the possibility of overlap injury. In other cases, overlap injury may not be a concern. PGR use on Class C turfs is well established.

CLASS D TURFS

Vegetation control along these areas is usually done with a "brush-hog" or chemicals known as "total veg" control materials. Examples of this class include railroad and power line right-of-ways as well as the more obscure parts of highway right-of-ways. Herbicide Type I PGRs are widely used on Class D vegetation, including those with residual action. Mobility of residual chemicals to surrounding desired vegetation has resulted in the use of mixtures of residual and nonresidual chemicals to reduce the chance of offsite injury.

SUMMARY

The reasons to consider growth regulators for lawns and turfgrass continue to grow. With increasing concern about the amount of landscape debris in landfills, eliminating clipping removal from lawns seems a good first start. Yet, many lawn areas are not mowed frequently enough during periods of high vertical growth, and residual clippings smother the lawn. The cost of both hand and mechanical labor for vegetation control continues to rise. New ideas for PGR use on Class A turfs continue to emerge. For these, and many other reasons, PGR use on turfgrasses will continue to grow. In order to make use of PGRs to enhance our environment, it is critical to understand the turfgrass ecosystem, the seasonal aspects of turfgrass growth and development, and proper use of the various PGR chemicals.

TRADEMARK IDENTIFICATION

Embark® Lite and Embark® Trim-Cut are registered trademarks of the PBI Gordon Corporation.

Royal Slo-Gro® is a registered trademark of the Uniroyal Chemical Company, Inc.

Roundup® is a registered trademark of the Monsanto Company.

Telar® and Oust® are registered trademarks of the E.I. Du Pont De Nemours & Company.

Event® is a registered trademark of the American Cyanamid Company.

Poast® is a registered trademark of the BASF Corporation.

Cutless® and Rubigan® are registered trademarks of DowElanco.

Scotts TGR® and TGR Turf Enhancer® are registered trademarks of the Scotts Company.

Primo® is a registered trademark of the CIBA-GEIGY Corporation.

Sumagic® is a registered trademark of the Valent USA Corporation.

Bayleton® is a registered trademark of Miles Inc.

REFERENCES

1. **DiPaola, J. M.,** Turfgrass growth regulation, in *Chemical Vegetation Management*, Kaufmann, J. E. and Westerdahl, H. E., Eds., Plant Growth Regulator Society America, Athens, GA, 1988, 155.
2. **Elkins, D. M.,** Growth regulating chemicals for turf and other grasses, in *Plant Growth Regulating Chemicals*, Vol. 2, Nickell, L. G., Ed., CRC Press, Boca Raton, FL, 1983, 113.
3. **Kaufmann, J. E.,** Biological responses of amenity grasses to growth regulators, in *Plant Growth Regulators for Agricultural and Amenity Uses*, Hawkins, A. F., Stead, A. D., and Pinfield, N. J., Eds., Symposium Proceedings, British Crop Protection Council and the British Plant Growth Regulator Group. Monograph No. 36, University of Reading, UK, 1987, 99.
4. **Watschke, T. L.,** Turfgrass weed control and growth regulation, in Proceedings Fifth International Turfgrass Research Conference, Lemaire, F., Ed., Avignon, France, 1985, 63.

Chapter 25

Turfgrass Weed Management — An IPM Approach

Joseph C. Neal, Department of Floriculture and Ornamental Horticulture, Cornell University, Ithaca, NY

(This chapter is a revised reprint of *WeedFacts No. 8,* a Cornell Cooperative Extension publication, Cornell University, Ithaca, NY, 1992. With permission.)

CONTENTS

Weed Scouting: What, When, and How 276
    What to Look For 276
    When to Scout 276
    How to Scout for Weeds 276
Using the Scouting Report for Decisions and Action 278
    Are Your Weeds Trying to Tell You Something? 278
    Weed Control: Do I Need a Herbicide and, If So, Which One? 279

Weed management is a component of any sound turfgrass management program and is dependent upon the maintenance of a dense and healthy turf. A thin, weak turf is an invitation to weeds! Weed invasions can be minimized by selecting the most appropriate turfgrasses for the site, preparing the seedbed properly, optimizing turfgrass maintenance (fertilization, irrigation, mowing, etc.), and controlling insect pests and diseases. However, regardless of the maintenance regime, some weeds will encroach. An effective weed management program will optimize the competitiveness of the turf to minimize the number of weeds encroaching, then control guidelines may be developed for the remaining species based upon knowledge — of the weeds, the turf management system, and the available weed management options.

To develop a pest management program you must first understand the pest problem. This is accomplished by first (if possible) developing a historical account of the turfgrass and weed management program and problems, and second, by scouting the property. Record the turfgrass species, management practices (particularly fertilization, irrigation, and mowing practices), and previous pests and pest management practices. Develop a map of the property identifying historical "hot spots" for weeds, as well as other pests and thin turf (each of which will lead to weed encroachment). This historical account can help make your weed scouting efforts more efficient and effective, and provide a basis for evaluating the success of future weed management decisions and actions. Weed scouting is the "tool" with which we will accumulate the remaining information needed to develop a customized and effective weed management program. Once you have identified the weed species present (see suggested reference guidebooks listed below), and documented their distribution, abundance, **and** possible reasons for their occurrence, you will be better able to make the most effective weed management decisions.

Selected Weed Identification References:

Common Weeds of the U.S.
    U.S. Department of Agriculture
Guide to the Identification of Dicot Turf Weeds
    O.M. Scott and Sons, Marysville, OH
Guide to the Identification of Grasses
    O.M. Scott and Sons, Marysville, OH
Identifying Seedling and Mature Weeds in the S.E.
    Stucky et al.; North Carolina State Agricultural Research Service, AG-208
Weeds of the North Central States
    University of Illinois, Ag. Exp. Sta. Bull. 772
Weeds of Southern Turfgrass
    Murphy et al.; University of Georgia, Cooperative Extension Service

(For illustrative drawings of many of the weeds mentioned in this chapter, see Appendix A.)

0-87371-350-8/94/$0.00+$.50
© 1994 by CRC Press Inc.

WEED SCOUTING: WHAT, WHEN, AND HOW

WHAT TO LOOK FOR

Identify and map the weeds and thin turf areas (which will sooner or later harbor weeds). Identify the weeds present, paying particular attention to those which have escaped previous control procedures, those which occur in patterns, species known to be troublesome in your area, and species new to the site. If an unknown species is encountered, take a sample for identification. Also learn to recognize "diagnostic weeds," or those which by their presence may indicate correctable soil, site, or management problems (this is discussed below in the "Using the Scouting Report for Decisions and Action" section). It is often helpful to refer to the history for guidance on what to expect and where to look for the first and most severe weed outbreaks.

On the scouting map, record the following:

- the weed species, location, distribution, relative abundance, and patterns (if present)
- areas of thin or damaged turf
- drainage patterns, dense shade from trees, or other factors which might directly or indirectly affect turfgrass vigor

It is not necessary to record every species encountered. Some knowledge of control procedures can help streamline the scouting process. For example, if a property is infested with a complex of easily controlled broadleaf weeds, it is sufficient to note the species present but to identify them on the map as "broadleaves." On the other hand, if a single weed species dominates the population, it certainly should be identified and mapped separately. Also, species requiring different management procedures should be individually identified. For example: A mixed population of crabgrass, foxtail, and panicum could be recorded as "summer annual grasses;" however, goosegrass generally requires a different management strategy and should be recorded separately. A scouting form can be designed to facilitate these groupings. An example of one such form is shown in Figure 1. Note the weeds are grouped or separated by type, growth habit, and control recommendation (based on New York State guidelines).

WHEN TO SCOUT

Weed scouting (or mapping) is a continuous process; each time you (or your employees) are on a property, look for pest and turf problems. However,

a once- or twice-a-year comprehensive weed scouting is recommended to provide the information you need to reassess the effectiveness of your current weed management program and to identify serious weed problems while they are small and more easily managed. Timing of the comprehensive scouting is critical. In the Northeastern U.S., I recommend once-a-year comprehensive scouting in the late summer or early fall. Advantages of this timing are that:

1. summer annual, winter annual, biennial, and perennial weeds are all present and easily identified
2. winter annuals are young and more easily controlled in the fall than in the spring
3. perennial broadleaf weeds are also more easily controlled in the fall
4. site modification or repair work can be accomplished and turf reseeded in the fall
5. ample time is available during the winter to reassess your current weed management program.

Less comprehensive scouting in the spring and summer months should be conducted to assess the effectiveness of fall treatments, identify seedling weeds when they are young and more easily controlled (particularly important for summer annuals such as crabgrass, goosegrass, spurge, etc. ...), and to gauge the density and quality of the turf. Owing to the longer growing season and different weed species, more frequent scouting may be necessary to accurately document the weed infestations in warmer climates.

HOW TO SCOUT FOR WEEDS

The actual scouting process is simple. It is useful to divide the property into management units. In a home lawn, these units might be as simple as the front, back, and side yards. On a golf course, tees, greens, fairways, and roughs are easily distinguished management units each with its own unique management inputs and options. On commercial properties, campuses, and parks the management units should reflect the priority status of the area. High priority areas will have a more intensive maintenance regime and a lower threshold for weed infestations; whereas, low priority/lower maintenance areas may have a higher threshold for weeds.

Walk or ride the area in a zig-zag pattern, stopping at key indicator areas for a closer look when early detection is essential. If you encounter an unknown weed, take a sample for identification. High priority areas require a more comprehensive

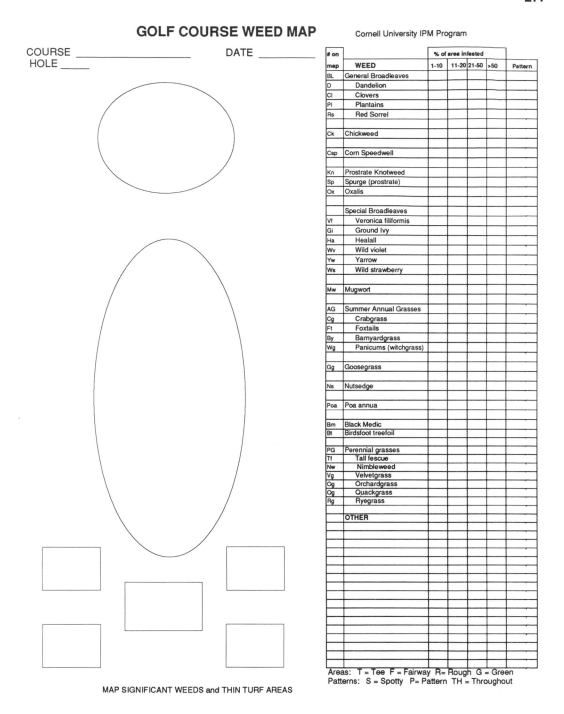

Figure 1 Example of a generic golf course weed map and scouting form used in the Cornell University golf course IPM scouting program.

procedure while lower priority areas will require a concomitantly less comprehensive scouting. On the property map, record the information previously described. By mapping the species and locations, you build a "picture" of the weed population and distribution, as well as identify patterns which might suggest an underlying reason for the weed's presence. An example of a golf course scouting form used in New York is provided. Note the generic nature of this map allows for flexibility, but a more accurate map of the individual hole would provide greater precision.

The accompanying table lists the most common turf weeds in the area grouped by type and control procedures. This should simplify the next step — which is using this information to make management decisions.

USING THE SCOUTING REPORT FOR DECISIONS AND ACTION
(or now that I have all this information — what do I do with it?)

ARE YOUR WEEDS TRYING TO TELL YOU SOMETHING?

Weeds occur in turf for several reasons. The weed may be (a) well adapted to persist in the closely-mowed plant community (like annual bluegrass), (b) exploiting a unique niche created by management procedures (like pearlwort in a heavily irrigated green), or (c) the turfgrass has been weakened by some site, environmental, or management factor(s) consequently favoring weed growth over that of the desirable turfgrass. Many weed infestations can be minimized by altering the site or management practices to tip the competitive balance in favor of the turfgrass. How do we identify such instances? One way is to take your cues from the "diagnostic weeds," the presence of which may indicate soil, site, or management problems. For instance: the following species are well adapted to extremes in soil moisture conditions, drought-prone or excessively moist sites.

Drought-prone sites	Wet sites
Prostrate spurge	Moneywort
Black medic	Annual sedge
Yellow woodsorrel	Annual bluegrass
Goosegrass	Alligatorweed
Annual lespedeza	Pearlwort
Birdsfoot trefoil	Moss
Prostrate knotweed	Liverwort
Bracted plantain	Rushes (*Juncus* spp.)

Encountering one or more of these species **as the predominant weed(s) suggests** that the site conditions may be too dry or too moist for optimum turfgrass growth. However, you must remember that the presence of one or more of these weeds is not proof of a moisture problem as these weeds will grow quite well under optimum water availability. Also, there may be other reasons for a weed's presence; for example: just because you find annual bluegrass you cannot automatically assume that the site is too wet for there are other conditions which can contribute to this species being more competitive than the desirable turf, such as compaction.

The following weeds are adapted to the shallow rooting conditions associated with compacted soils.

Annual bluegrass	Corn speedwell
Annual sedge	Goosegrass
Annual lespedeza	Prostrate knotweed
Broadleaf plantain	Prostrate spurge

Several of these species are better adapted to compacted, dry soils, like annual lespedeza, goosegrass, and spurge, while annual bluegrass and annual sedge are better adapted to compacted, moist soils. Either extreme results in shallow rooting and poor turfgrass vigor, leading to a competitive advantage for the weed(s).

Core cultivation is used to alleviate compaction. I am often asked if core cultivation following preemergent herbicide applications will reduce weed control. Under most conditions, herbicide efficacy is not reduced by this procedure. Another concern often raised is whether core cultivation will increase weed germination by opening the turf. When done at a time of year when the turfgrass is growing vigorously, no increase in weed germination should occur. However, if coring is done when turf vigor is low and weed germination is high, some increase in weed germination can be observed. For example, core cultivating Kentucky bluegrass in early summer will often lead to increased crabgrass emergence. When core cultivation must be done at nonoptimal times, use a narrow tine cultivator; the turf will fill the smaller holes more quickly thus minimizing the opportunity for weed emergence. In general, core cultivation will improve turf vigor and competition, effectively offsetting any increased weed emergence which might occur.

Soil pH and fertility can affect turfgrass vigor and the weed species present. The presence of red sorrel often indicates acid soil conditions (however, I have observed this species growing in pH 7 soils). When we examine the effect of soil fertility on turfgrass vigor, nitrogen emerges as the most important element to consider. The following species are well adapted to extremes of soil fertility, very low to very high nitrogen.

Low N	High N
Birdsfoot trefoil	Annual bluegrass
Black medic	Chickweed
Broomsedge	Moss
Chicory	Ryegrass
Common Speedwell	

Phosporous and potassium have also been shown to influence weed populations; however, in each case the impact of nutrient treatments on "weedyness" could be linked to the overall health and vigor of the turfgrass. In other words — fertility regimes should be designed for optimum turfgrass health and vigor, based on soil test results and the optimum fertility recommendations for the turf type and use.

Altering the fertilization program to favor the desirable turfgrass is the first step in managing these weeds. Weeds of low fertility sites can eventually be controlled strictly through improved turfgrass management. This may also involve overseeding with a turfgrass better adapted to the site. In contrast, those species adapted to high fertility sites can be reduced but generally not eliminated by cultural means alone. Some other control procedure will be necessary to eliminate these species.

Mowing height and frequency can also alter the weed population. The following species are adapted to different mowing regimes.

High/Infrequent	Close/Frequent
Bull thistle	Annual bluegrass
Burdock	Chickweeds
Chicory	Pearlwort
Sweet clover	Thymeleaf speedwell
Teasel	

The species adapted well to high/infrequent mowing are common roadside weeds. If these species are present in abundance in a lawn situation, it would be wise to evaluate the current mowing regime. Weed species adapted to close/frequent mowing are commonly found on golf greens. These latter species will also grow quite well at higher cutting heights but tend to be less competitive with turf as mowing heights are increased. As with fertility, mowing height should favor that which is optimum for the turfgrass species, management, and site conditions to ensure the maximum competitiveness and quality.

Close examination of these lists reveals that several weeds are adapted to a range of conditions. This is often associated with the broad adaptation of a species to many ecological niches. Examples include annual bluegrass tolerance of shallow rooting caused by excessive moisture or compaction, and black medic tolerance of drought-prone and low fertility sites. The presence of one or more of these weeds is not proof-positive of the stated conditions, but rather suggests potential reasons for their presence. Evaluate these conditions and correct them if possible.

In summary, turfgrass management procedures and site conditions can increase or minimize weed problems, depending upon how well the conditions and management are optimized for the turfgrass.

Turf management procedures or conditions that promote weed invasions include:

- Wear (foot, tire, use, etc.)
- Environmental stresses (drought, waterlogging, etc.)
- Introducing weed seed or other propagules (on equipment, in top soil, etc.)
- Aerating or dethatching during peak weed germination and low turfgrass vigor
- Any management input not matched to soil tests, the turfgrass species, and management of the site.

Good cultural procedures for preventing and minimizing weed invasions include:

- Proper species and variety selection for the site and use
- Optimum fertility, pH, and irrigation management for the turfgrass and soil properties
- Correct mowing height, equipment, and regime
- Traffic, thatch, and compaction control and management
- Effective management of other turfgrass pests.

Scouting can greatly assist in identifying the underlying reasons for weed infestations. Correcting the site or management problems contributing to weed encroachment should be the first priority in your weed management program. While correcting the problem and increasing turfgrass vigor and competitiveness will generally not eliminate the weed, failure to do so will almost ensure that the weed will return after herbicide applications. Therefore, control procedures will need to be coupled with site or management modification for effective long-term weed management.

WEED CONTROL: DO I NEED A HERBICIDE AND, IF SO, WHICH ONE?

If the scouting report contains both the identification and some estimates of distribution, relative abundance, and patterns, the decision-making procedure should be simple. Based on the priority status of the pest management unit and the relative abundance of the weeds, ask yourself — **do I need to treat for weeds?** If the answer is yes, the distribution estimates should tell you whether broadcast (distributed throughout) or limited area or spot applications (spotty distribution) are warranted. If spot or limited area treatments are to be conducted, the scouting map should help you estimate the area to be treated and the amount of product needed, and guide the applicator to the infestations.

At the present time, when weeds must be controlled we currently have only two options: physical removal and herbicides. As strange as it may sound, there are situations where hand weeding is still practical. If only a few weeds are present in a high priority area, hand weeding can be the fastest and most economical alternative. For example, a few goosegrass plants on a golf green or a half dozen dandelions in a home lawn can be hand weeded faster than you could get the sprayer ready (and time is money). Where this is not practical, proper selection and use of appropriate herbicides is essential.

Selecting the most appropriate herbicide from among the many that are available can be a challenge. The following are some suggested guidelines for choosing the best product for the job.

- **Efficacy on the target weed species** — Will it control the target weed?
- **Longevity of residual control** (if preemergent) — How long will the control last? Will multiple treatments be necessary? Will the residual interfere with my overseeding program?
- **Turfgrass species and management** — Will it injure the turfgrass species on my site? Is it registered for close cut turf such as tees and greens?
- **Weed status or growth stage** — Do I need a preemergent or postemergent product? For a postemergent treatment, does weed age or size (tiller number) affect product or rate selection?
- **Weed control spectrum** — Will other (incidental but important) weeds also be controlled? For example, crabgrass will be controlled but what about the spurge or goosegrass which occurs sporadically?
- **Available equipment** — Do I want to spray or spread granules? Do I need to alter my equipment (such as changing spray pressure and nozzles) to achieve adequate control?
- **Proximity of susceptible species** — Are there susceptible landscape plants nearby? Would another product reduce the chances for nontarget effects?

- **Environmental impact and mammalian toxicity** — Is this product the least toxic or most environmentally friendly option?
- **Economics** — How much does it cost for an "acre treatment" and how many "acre treatments" will be necessary for season-long control with each option? What are the labor and equipment costs associated with repeat or sequential applications?

Consult your state or local Cooperative Extension Service for the products, rates, and application methods recommended in your area.

Once a herbicide has been chosen, it must be applied correctly. Misapplication increases the chances of turfgrass injury, lack of control, skips in control and injury to off-target species; wastes herbicides, applicator time and money; produces unsatisfied customers; and invites criticism and regulation. Information on the subject of sprayer or spreader maintenance, calibration, and use is available from equipment manufacturers and from your local Cooperative Extension office.

Weed mapping can be used to identify not only weed populations but also turfgrass management or site conditions which may be contributing to the weeds' success; in other words, **your weeds may be trying to tell you something** about the site, soil, or management which are allowing weeds to be successful. If we "listen to our weeds," follow soil test and best-management recommendations for our turfgrasses and regions, we can effectively minimize the number and severity of weed infestations. After all — **a dense and healthy turf is our first line of defense against weeds.** However, it is nearly impossible that improved turfgrass management alone will completely eliminate weeds. If a reasonably weed-free turf is desired, an integration of cultural procedures and herbicides will be necessary and will produce the best long-term solution to our turfgrass weed management problems.

Appendix A

Weed Illustrations

(Excerpted from *Selected Weeds of the United States,* Agriculture Handbook No. 366, by the Agricultural Research Service, U.S. Department of Agriculture, Washington, DC: March 1970.)

CONTENTS

Alligatorweed 283

Barnyardgrass 283

Black Medic 283

Broomsedge 283

Buckhorn Plantain 284

Bull Thistle 284

Burdock 284

Carpetweed 284

Chickweed 285

Chicory 285

Corn Speedwell 285

Dandelion 285

Fall Panicum 286

Field Chickweed 286

German Velvetgrass 286

Giant Foxtail 286

Ground Ivy 287

Goosegrass 287

Healall 287

Henbit 288

Lambsquarters 288

Large Crabgrass 288

Mallow 288

Mouseear Chickweed 289

Mugwort 289

Nimblewill 289

Prostrate Knotweed 289

Prostrate Spurge 290

Quackgrass 290

Red Sorrel 290

Shepherdspurse 290

Speedwell 291

Teasel 291

Wild Garlic 291

Yarrow 292

Yellow Nutsedge 292

Yellow Woodsorrel 292

0-87371-350-8/94/$0.00+$.50
© 1994 by CRC Press Inc.

Alternanthera philoxeroides (Mart.) Griseb. Alligatorweed. *A,* Habit— × 0.5; *B,* roots and young plant— × 0.5; *C,* part of aquatic growth, new shoot from rooting node— × 0.5; *D,* flower— × 2.5; *E,* persistent chaffy flower with the single mature achene— × 2.5; *F,* achenes— × 2.5; *G,* seeds— × 5.

Alligatorweed

Echinochloa crus-galli (L.) Beauv. Barnyardgrass. *A,* Habit, forma *longiseta* (Trin.) Farw.— × 0.5; *B,* spikelet— × 2.5; *C,* ligule— × 2.5; *D,* florets— × 4; *E,* caryopses— × 4; *F,* spike, var. *mitis* (Pursh) Peterm.— × 0.5; *G,* floret of awnless variety— × 4.

Barnyardgrass

Andropogon virginicus L. Broomsedge. *A,* Habit— × 0.5; *B,* ligule (left, opened; right, compressed as in nature)— × 2.5; *C,* inflorescence— × 2.5; *D,* florets— × 7.5.

Broomsedge

Medicago lupulina L. Black medic. *A,* Habit— × 0.5; *B,* flower raceme— × 6; *C,* fruiting raceme — × 3; *D,* flower— × 10; *E,* legume— × 5; *F,* seeds— × 5.

Black Medic

Habit for all— × 0.5. *A, Plantago lanceolata* L. Buckhorn plantain. *a,* Flower—
× 2.5; *b,* capsule— × 3; *c,* seed— × 5. *B, Plantago major* L. Broadleaf plantain.
a, Flower— × 2.5; *b,* capsule— × 3; *c,* seeds — × 5. *C, Plantago rugelii* Decne.
Blackseed plantain. *a,* Flower— × 2.5; *b,* capsule— × 2.5; *c,* seeds— × 3.

Buckhorn Plantain

Cirsium vulgare (Savi) Tenore. Bull thistle. *A,* Habit— × 0.5; *B,* flower— × 5;
C, immature fruit— × 1.75; *D,* achenes— × 3.5.

Bull Thistle

Arctium minus (Hill) Bernh. Common burdock. *A,* Habit: root, leaf, upper raceme of heads—
× 0.5; *B,* flower and phyllaries— × 4; *C,* achene— × 3.

Burdock

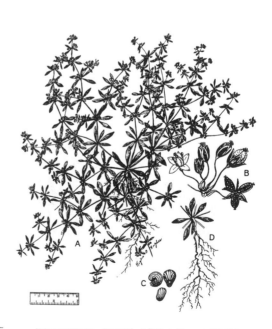

Mollugo verticillata L. Carpetweed. *A,* Habit— × 0.5, as seen from above;
B, flowers and fruits— × 5; *C,* seeds— × 6; *D,* taproot— × 0.5.

Carpetweed

Stellaria media (L.) Cyrille. Common chickweed. *A,* Habit—× 0.5; *B,* flower—× 3; *C,* capsule—× 3.5; *D,* seeds—× 7.5.

Chickweed

Cichorium intubus L. Chicory. *A,* Habit—× 0.5; *B,* terminal portion of inflorescence; *C,* involucre—× 2.5; *D,* flower—× 2.5; *E,* achenes—× 7.5.

Chicory

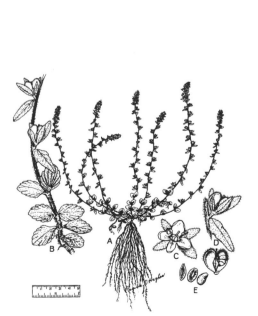

Veronica arvensis L. Corn speedwell. *A,* Habit—× 0.5; *B,* enlarged branch—× 2.5; *C,* flower× 7.5; *D,* capsules—× 3; *E,* seeds—× 5.

Corn Speedwell

Taraxacum officinale Weber. Dandelion. *A,* Habit—× 0.5; *B,* flower—× 3; *C,* achenes *D,* achenes with pappus—× 1.

Dandelion

Panicum dichotomiflorum Michx. Fall panicum. *A*, Habit— × 0.5; *B*, spikelet, showing florets— × 7.5; *C*, ligule— × 4; *D*, caryopses— × 7.5.

Fall Panicum

Cerastium arvense L. Field chickweed. *A*, Habit— × 0.5; *B*, enlarged leaves— × 1; *C*, flower— × 1.5; *D*, capsule— × 3; *E*, seeds— × 10.

Field Chickweed

Holcus mollis L. German velvetgrass. *A*, Habit— × 0.5; *B*, inflorescence— × 2.5; *C*, spikelet (lower floret perfect, awnless; upper floret staminate, awned)— × 5; *D*, glumes— × 5; *E*, ligule— × 2.5; *F*, leaf sheath (retrorse hairs of sheath at joint)— × 2.5.

German Velvetgrass

A, *Setaria faberi* Herrm. Giant foxtail. *a*, Habit— × 0.5; *b*, spikelet, showing subtending bristles — × 5; *c*, ligule— × 1.5; *d*, caryopses— × 5. *B*, *Setaria viridis* (L.) Beauv. Green foxtail. *a*, Habit— × 0.5; *b*, spikelet— × 5; *c*, ligule— × 1.5; *d*, caryopses— × 5. *C*, *Setari glauca* (L.) Beauv. Yellow foxtail. *a*, Habit— × 1; *b*, spikelet— × 5; *c*, ligule— × 1.5; *d*, caryopses— × 5.

Giant Foxtail

Glechoma hederacea L. Ground ivy. *A*, Habit—× 0.5; *B*, flower cluster—× 2.5; *C*, flower diagram, showing the four ascending stamens—× 2.5; *D*, nutlets—× 6.

Ground Ivy

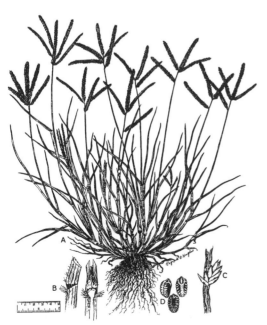

Eleusine indica (L.) Gaertn. Goosegrass. *A*, Habit—× 0.5; *B*, ligules—× 2.5; *C*, spikelet—× 3 *D*, caryopses—× 12.5.

Goosegrass

Prunella vulgaris L. Healall. *A*, Habit—× 0.5; *B*, flower—× 2; *C*, persistent calyx—× 2.25; *D*, nutlets—× 4.

Healall

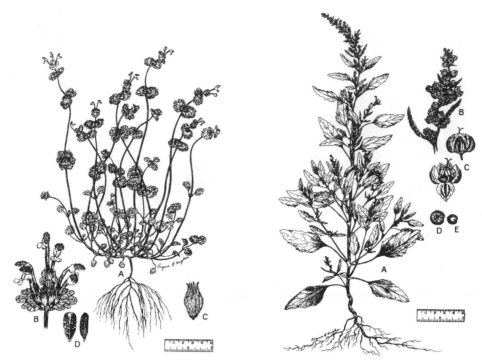

Lamium amplexicaule L. Henbit. *A*, Habit— × 0.5; *B*, flower clusters, showing very short upper internodes— × 1.5; *C*, calyx surrounding nutlets— × 4; *D*, nutlets— × 7.5.

Henbit

Chenopodium album L. Common lambsquarters. *A*, Habit, small plant; *B*, floral spike— × 2.5; *C*, flowers— × 7.5; *D*, utricle— × 4; *E*, seed— × 4.

Lambsquarters

Digitaria sanguinalis (L.) Scop. Large crabgrass. *A*, Habit— × 0.5; *B*, florets, front and back views— × 5; *C*, caryopsis— × 6.

Large Crabgrass

Malva neglecta Wallr. Common mallow. *A*, Habit— × 0.5; *B*, enlarged branchlet— × 2; *C*, flower diagram— × 5; *D*, carpel— × 5; *E*, seeds— × 5.

Mallow

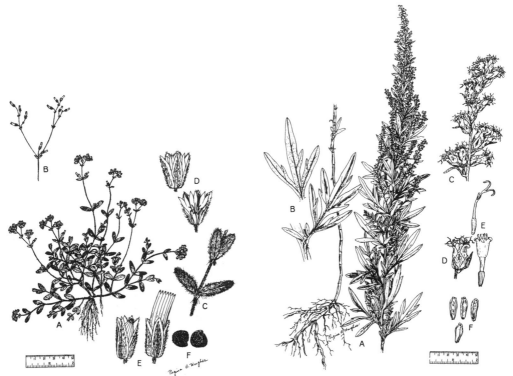

—*Cerastium vulgatum* L. Mouseear chickweed. *A*, Habit—× 0.5; *B*, mature dichotomous—
× 0.5; *C*, enlarged leaves—× 1.25; *D*, flowers—× 3.5; *E*, capsules—× 3.5; *F*, seeds—× 15.

Mouseear Chickweed

1.—*Artemisia vulgaris* L. Mugwort. *A*, Habit—× 0.5; *B*, enlarged leaves—× 1; *C*, panicle—× **3**
D, flower head—× 4; *E*, flowers—× 7.5; *F*, achenes—× 5.

Mugwort

Muhlenbergia schreberi J. F. Gmelin. Nimblewill. *A*, Habit—× 0.5; *B*, ligules—× 2.5; *C*,
spikelet to show glumes—× 17.5; *D*, florets—× 17.5.

Nimblewill

Polygonum aviculare L. Prostrate knotweed. *A*, habit—× 0.5; *B*, flowering branch, enlarged—×
2.5; *C*, flower—× 7.5; *D*, fruiting calyx—× 7.5; *E*, achenes—× 7.5.

Prostrate Knotweed

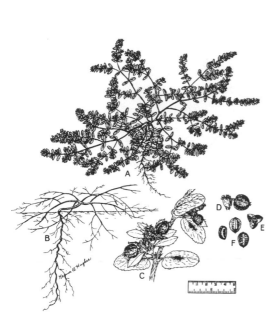

Euphorbia supina Raf. ex Boiss. Prostrate spurge. *A*, Habit—× 0.5; *B*, root—
× 0.5; *C*, branch with cyathia—× 3.5; *D*, gynophore—× 6; *E*, capsule from top,
showing subacute parted cleft styles—× 6; *F*, seeds—× 7.5.

Prostrate Spurge

Agropyron repens (L.) Beauv. Quackgrass. *A*, Habit—× 0.5; *B*, spikelet—× 3; *C*, ligule—
D, florets—× 3.25.

Quackgrass

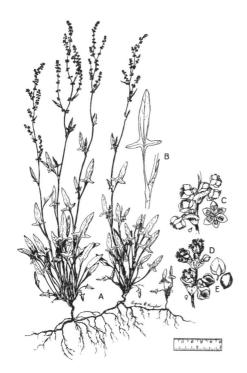

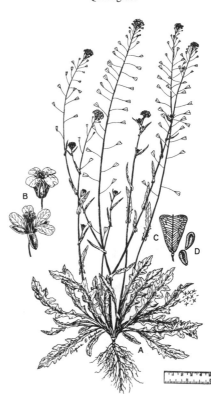

Rumex acetosella L. Red sorrel. *A*, Habit—× 0.5; *B*, leaf detail—× 1.5; *C*, staminate
flowers— × 7.5; *D*, pistillate flowers— × 7.5; *E*, achenes, in and out of calyx—× 10.

Red Sorrel

Capsella bursa-pastoris (L.) Medic. Shepherdspurse. *A*, Habit—× 0.5; *B*,
flowers— × 5; *C*, silicle — × 4; *D*, seeds— × 10.

Shepherdspurse

Veronica officinalis L. Common speedwell. *A*, Habit—× 0.5; *B*, flowers—× 2.5; *C*, capsules—× 2.5; *D*, seeds—× 12.5.

Speedwell

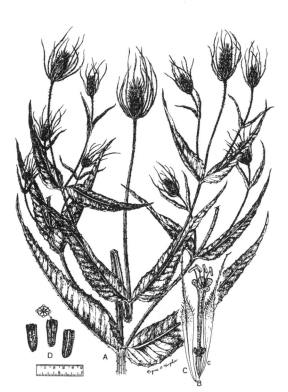

Dipsacus sylvestris Huds. Teasel. *A*, Habit—× 0.5; *B*, flower—× 5; *C*, and *c*, bracts (chaff), × 5; *D*, achenes—× 4.

Teasel

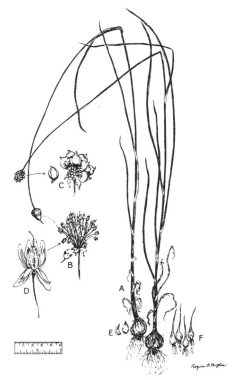

Allium vineale L. Wild garlic. *A*, Habit—× 0.5; *B*, flower cluster—× 0.5; *C*, bulblets—× 2.5; *D*, flower—× 3.5; *E*, bulbs, hard-shell—× 0.5; *F*, bulbs, soft-shell—× 0.5.

Wild Garlic

Achillea millefolium L. Common yarrow. *A*, Habit— × 0.5; *B*, enlarged leaves and stem— *C*, flower head— × 4; *D*, female and male flowers— × 5; *E*, seeds— × 6.

Yarrow

Cyperus esculentus L. Yellow nutsedge. *A*, Habit— × 0.5; *B*, spikelet— × 5; *C*, achene

Yellow Nutsedge

Oxalis stricta L. Common yellow woodsorrel. *A*, Habit— × 0.5; *B*, leaves enlarged— *C*, flower diagrams— × 2.5; *D*, capsule— × 1.5; *E*, seeds— × 10.

Yellow Woodsorrel

Chapter 26

Plan Before You Plant: A Five-Step Process for Developing a Landscape Weed Management Plan

Joseph C. Neal, *Department of Floriculture and Ornamental Horticulture, Cornell University, Ithaca, NY*

(This chapter is a revised reprint of *WeedFacts No. 5,* a Cornell Cooperative Extension publication, Cornell University, Ithaca, NY, 1992. With permission.)

CONTENTS

Step 1: Site Assessment 293
Step 2: Define the Type of Planting 294
    Woody Tree and Shrub Beds 294
    Woody Groundcovers 294
    Annual Flower Beds 294
    Herbaceous Perennial Beds 294
    Mixed Plantings 295
Step 3: Selecting the Ornamental Species and Weed Management Options 296
    Mulches 296
    Geotextiles 297
    Herbicides 297
Step 4: Site Preparation — Weed Management Concerns 299
Step 5: Installation and Implementation 300
Suggested References 300

Supplemental hand weeding accounts for the majority of landscape bed maintenance costs. When used exclusively, it can cost 10 to 100 times as much as an effective herbicide or mulching program. However, many of the costly and unsightly weed problems can be avoided or at least minimized with a little planning. Developing a landscape weed management plan involves five basic steps.

1. **Site assessment.** Survey the site for cultural aspects as well as weed species.
2. **Define the type of planting.** The type of planting, woody shrubs vs. bedding plants, etc. will define the post-plant weed management options available.
3. **Selection of ornamental species and compatible weed management options** based upon design, cultural, and weed management criteria.
4. **Site preparation.** Control weeds that cannot be controlled after planting.
5. **Installation and implementation** of the plants and the plan.

These steps will be discussed separately, but keep in mind that each step relates to and is dependent upon the decisions made in the other steps. The goal is to provide a process by which an effective and economical weed management plan may be developed.

STEP 1: SITE ASSESSMENT

Adequate site assessment will allow proper species selection based upon design criteria, cultural

293

suitability, and management regime, including weed management. Take soil samples for pH and nutrient analysis. Note soil type and physical condition, drainage patterns, exposure and edaphic aspects, and potential maintenance problems such as gutter down-spouts, chlorine from the pool, or traffic patterns. Identify the weeds in the area, with particular emphasis on perennial weeds. Ask yourself: "Can these weeds be controlled after planting?" Some species which are difficult or impossible to control after planting include bindweed, nutsedge, mugwort, Canada thistle, goldenrod, bamboo, Japanese knotweed, wild violet, and field horsetail. Also, inspect the surrounding areas for weeds which may encroach, such as ground ivy, wild strawberry, yarrow, buffalograss, bermudagrass, creeping speedwell, quackgrass, or other creeping perennials.

The best time to scout for weeds is in mid- to late summer, when annual and perennial weeds can be identified. Scouting in late fall or early spring is likely to miss many of the important species. Also, scouting in the summer will allow adequate time for decision-making and site preparation before planting.

STEP 2: DEFINE THE TYPE OF PLANTING

The species to be planted will define the intensity of management required and, to a large degree, govern your future weed management options. A planting of woody trees and shrubs will allow the most post-plant weed management options. In contrast, a mixed planting of woody and herbaceous plants will have fewer post-plant options. Where herbaceous perennials or annuals are included in a permanent landscape bed, geotextiles often cannot be used and herbicide choices are limited. Table 1 provides an overview of the weed management limitations and options for the different types of plantings. Tables 2 and 3 list the herbicides registered for use in landscape ornamentals and the suggested types of plantings where they can be used. Understanding these limitations and options will help guide you in the following steps toward developing an effective weed management plan.

WOODY TREE AND SHRUB BEDS
Two options set woody tree and shrub beds apart from the rest; one is the possibility of post-plant perennial weed control, the other is the use of geotextile fabrics for annual weed control. Perennial weeds may be controlled by manual removal, spot applications of glyphosate (Roundup), or, in some instances, dormant-season applications of diclobenil (Casoron). Care should be taken not to contact desirable foliage with Roundup; also, use diclobenil only on labeled species as injury is likely to result if applied to other plants. Annual weeds may be controlled using geotextile fabrics, organic or inorganic mulches, and/or herbicides. It is often necessary to combine these treatments for complete weed control. Geotextile fabrics are somewhat expensive to install but become cost-effective if the landscape bed is to remain in place for more than 4 years. Preemergent herbicides are less expensive and equally effective, but must be reapplied annually (usually two applications per year). Escaped weeds may be controlled manually or with spot applications of postemergent herbicides.

WOODY GROUNDCOVERS
Ultimately, woody groundcovers should exclude most weeds; however, weed encroachment during establishment is likely. After planting, it is difficult to make spot applications of Roundup or other nonselective herbicides without injuring desirable plants; therefore, perennial weeds should be eliminated before planting. Exceptions are perennial grasses, which can be selectively controlled after planting with sethoxydim (Vantage) or fluazifop-p (Ornamec). Annual weeds may be controlled with mulch plus a preemergent herbicide, supplemented with some hand weeding.

ANNUAL FLOWER BEDS
Weed control in annual flower beds can be simple if perennial weeds are eliminated before planting. Perennial grasses can be selectively controlled with sethoxydim or fluazifop-p, but other perennial weeds cannot be selectively controlled after planting. Nonselective herbicides will kill or severely injure annual bedding plants and should be avoided after planting. Geotextiles are generally not used in annual flower beds but can be useful if no other options are available. Annual weeds may be controlled with mulches, preemergent herbicides, and/or hand weeding.

HERBACEOUS PERENNIAL BEDS
Weed management options in herbaceous perennial beds are similar to those for annual flowers, except (1) it is more important to eradicate perennial weeds, as there will be no opportunity to cultivate or renovate the bed for several years; and (2) fewer species are included on herbicide labels. Geotextile mulches may be used around clump-type plants, but not around spreading species.

Table 1 **Weed management options and limitations for the five types of landscape plantings**

Tree and Shrub Beds: Densely shaded plantings exclude weeds.
–   Geotextiles and mulches are useful.
–   Many broad-spectrum herbicides are available for pre- and postemergent control.
–   Spot or directed applications of nonselective herbicides, like Roundup, are possible.
–   Species selection is flexible and pre-plant weed control is not as critical as in other types of plantings.
Recommendations: Control perennial weeds before planting (although control may be possible after planting), use geotextiles with a shallow layer of mulch, use a preemergent herbicide if needed, and supplement with spot applications of postemergent herbicides and/or hand weeding.

Woody Ground Cover Beds: The ground cover should ultimately exclude most weeds.
–   Limited uses for nonselective herbicides; therefore, control perennial weeds before planting.
–   Do not use geotextiles where ground covers are expected to root and spread.
–   Control annual weeds with mulching, hand weeding, and/or herbicides.
–   Several preemergent herbicides are available.
–   Few uses for postemergent herbicides.
–   Postemergent control of annual and perennial grasses is possible.
Recommendations: Control perennial weeds before planting, use geotextiles where possible or else use mulches with a preemergent herbicide and supplement with hand weeding.

Annual Flower Beds: A closed canopy will shade out many weeds.
–   Periodic cultivations (annually or between display rotations) will suppress many weeds.
–   Very limited use of nonselective herbicides; control perennial weeds before planting.
–   Geotextiles generally are not useful (due to the short-term nature of the planting).
–   Few preemergent herbicides are safe; careful species and product selection are required.
–   Mulches will suppress many annual weeds.
Recommendations: Control perennial weeds before planting, carefully select species for weed management compatibility, use mulches and a preemergent herbicide, and supplement with hand weeding.

Herbaceous Perennial Beds: Similar to annual flower beds except:
–   Lack of periodic cultivations will encourage perennial weed encroachment.
–   Fewer herbicides are labeled; check the labels carefully.
–   Geotextiles may be useful in clump-type plantings or to restrict growth of spreading-types.
–   Very limited use of nonselective or postemergent herbicides.
Recommendations: Control perennial weeds before planting, use geotextiles where possible, use mulches with a preemergent herbicide, and supplement with hand weeding.

Mixed Plantings (of woody and herbaceous plants):
–   More complex due to the diversity of species.
–   Different areas of the bed could receive different treatments.
–   Site preparation is usually critical.
–   Few herbicides are registered for a wide spectrum of ornamental plant types.
–   Geotextiles may or may not be useful.
Recommendations: Maximize the number of weed control options by compatible species selection. Control perennial weeds before planting, use geotextiles where possible, use mulches with a preemergent herbicide where possible, and supplement with hand weeding.

MIXED PLANTINGS

In mixed plantings of woody and herbaceous ornamentals, site preparation is as critical as for herbaceous perennials because post-plant choices are few. One option for such areas is to plant the woody species first, control the perennial weeds in the first two growing seasons, then introduce the herbaceous species. Another option may be to define use areas within the bed which will receive different weed management programs — such as a section

Table 2 Preemergent herbicides registered for use in landscape plantings

Common Name	Trade Name(s)	Suggested Use Sites			
		Trees & Shrubs	Ground-covers	Annual Flowers	Herbaceous Perennials
Bensulide	Betasan, Lescosan, Pre-San, Others	✓	✓	✓	✓
DCPA (chlorthal dimethyl)	Dacthal	✓	✓	✓	✓
Diclobenil	Casoron, Dyclomec, Norosac	✓	Few	No	No
EPTC	Eptam	✓	✓	No	No
Isoxaben	Gallery	✓	✓	No	No
Napropamide	Devrinol	✓	✓	✓	✓
Metolachlor	Pennant	✓	✓	No	Few
Oryzalin	Surflan	✓	✓	✓	✓
Oryzalin + isoxaben	Snapshot DF	✓	✓	No	No
Oryzalin + benefin	XL	✓	✓	✓	✓
Oxadiazon	Ronstar	✓	✓	No	No
Oxyfluorfen	Goal	✓	Few	No	No
Oxyfluorfen + pendimethalin	Scotts ornamental herbicide 2 (OH2)	✓	✓	No	No
Oxyfluorfen + oryzalin	Rout	✓	✓	No	No
Pendimethalin	Pendulum, Southern Weedgrass Control, others	✓	✓	No	No
Prodiamine	Barricade	✓	✓	No	Few
Pronamide	Kerb	✓	No	No	No
Simazine	Princep, Caliber 90, Simazine, others	✓	Few	No	No
Trifluralin	Treflan, Preen, others	✓	✓	✓	✓
Trifluralin + isoxaben	Snapshot TG	✓	✓	No	No

Note: Always check the herbicide label for list of registered species, directions for use, and precautions.
Key to symbols: ✓ = Registered for numerous species; Few = registered on a few species; No = not recommended as most species in this category would be injured.

devoted to annual flowers in an otherwise woody tree and shrub bed.

STEP 3: SELECTING THE ORNAMENTAL SPECIES AND WEED MANAGEMENT OPTIONS

Based upon the type of planting desired and the site assessment, we can now select the species for planting. The criteria for species selection should include design and site suitability, and maintenance aspects including disease and insect resistant species/varieties, as well as weed management option compatibility. Selecting the proper weed management option(s) will depend upon weed species present, your flexibility in planting design and species selection, economics, as well as some personal choice. Generally **it is best to**

control perennial weeds before planting (see "Site Preparation"). If perennial weeds cannot be controlled before planting, carefully evaluate the type of planting and species selection to ensure that weeds can be selectively controlled after planting. Geotextile fabric mulches are generally not used in annual flower beds, but may be useful if no other options are available. Annual weeds may be controlled after planting using mulch, preemergent herbicides, by hand weeding, or a combination of these options.

MULCHES

Many types of **mulches** are available including barks, various hulls (pecan, cocoa, buckwheat, etc.), municipal composts, crushed rocks, and others. All types suppress annual weeds by excluding light, which is required for seed germination. When

Table 3 Postemergent herbicides registered for use in landscape plantings

Common Name	Trade Name(s)	Trees & Shrubs	Ground-covers	Annual	Herbaceous Perennials
			Suggested Use Sites		
Bentazon	Basagran T/O	Directed	Few	No	No
Clethodim	Prism	✓	✓	✓	✓
Diclobenil	Casoron, Dyclomec, Norosac	✓	Some	No	No
Diquat	Reward	Directed	No	No	No
Fenoxaprop	Acclaim	✓	✓	✓	✓
Fluazifop-p	Fusilade, Ornamec, Grass-B-Gon	✓	✓	✓	✓
Glufosinate	Finale	Directed	No	No	No
Glyphosate	Roundup, Kleenup, others	Directed	No	No	No
Oxyfluorfen	Goal	Few	Few	No	No
Pronamide	Kerb	✓	No	No	No
Salts of fatty acids	Sharpshooter	Directed	No	No	No
Sethoxydim	Vantage (formerly Poast)	✓	✓	✓	✓

Note: Always check the herbicide label for the list of registered species, directions for use, and precautions.
Key to symbols: ✓ = Registered for over-the-top or directed applications on many species within the category; directed = do not contact desirable foliage; Few = registered for use over only a few species in the category; No = not recommended as most species in this category would be injured.

mulches are too fine, too thick, or begin to decompose, they stay wet between rains, allowing weeds to germinate and grow directly in the mulch. Therefore, for weed control, a mulch which is fairly coarse-textured with a low water-holding capacity would be preferred. To effectively suppress weeds, organic mulches should be about 4 in. thick. Inorganic mulches should be 3 to 4 in. thick, as they do not decompose or settle as quickly. Plan for periodic replenishment. When used alone, mulches rarely provide 100% weed control. Supplemental hand weeding or spot spraying are generally necessary. In many situations, the amount of supplemental weeding required can be burdensome and expensive; therefore, most landscapers will choose to use geotextiles and/or preemergent herbicides with a thin layer of a decorative mulch.

GEOTEXTILES
Geotextiles are synthetic fabrics which allow water and air to pass, but prevent weed seedling emergence. Although these materials are relatively expensive and time-consuming to install, they become cost-effective if the planting is to remain in place for 4 or more years. These fabrics are as effective as a good preemergent herbicide, but, in contrast to herbicides, without the need to reapply or the worry of potential herbicide injury to nonlabeled species. Geotextiles are not suggested where the area is to be replanted periodically, as in annual flower beds,

or where the fabric would inhibit rooting and spread of ground covers. Geotextiles must be covered by a mulch to prevent photodegradation. Use a shallow layer of mulch, as roots of weeds germinating on top can penetrate the fabric. Many perennial weeds can grow through plastic or geotextile mulches; therefore, those species must be controlled during site preparation. If weeds do grow into or through the geotextile, remove them when they are small to prevent holes in the fabric.

HERBICIDES
Herbicides are relatively inexpensive and effective, and, when properly chosen and applied, can be used in nearly any type of planting. If you decide to use a herbicide, consider the following selection criteria:

1. select one that controls most of the weeds present (no herbicide will control all weeds)
2. be sure the ornamental species are on the herbicide label
3. proximity of susceptible species, and the likelihood of exposure
4. potential residual effects on subsequent plantings (such as an annual flower bed)
5. type of application equipment (granular or spray)
6. economics (don't forget the cost of supplemental hand weeding)

Table 4 **Effectiveness of preemergent herbicides on common landscape weeds**

Herbicides	Annual Grasses	Chickweed	Galinsoga	Groundsel	Morningglory	Oxalis	Purslane	Spurge
Barricade	G	G	N	P	?	F	P-F	F
Betasan	G	F	P	?	N	N	F	P
Casoron	F-P	G	P	F	P	P	F	P
Dacthal	F	G	N	F-P	N	F	F	G
Devrinol	G	G	F	F	P	P	G	?
Eptam	G	G	N	?	P	?	G	?
Gallery	F	G	G	G	F	G	G	G
Goal	G	G	G	G	G	G	G	G
Kerb	G	?	G	G	?	?	G	?
Scotts Ornamental Herbicide 2	G	G	G	G	F	G	G	F
Pennant	G	F	G	P	N	P	F	P
Princep	F	F	F	F	G	G-F	G	G
Ronstar	G	P	F	F	F-G	G	G	F
Rout	G	G	G	G	F	G	G	F
Snapshot	G	G	G	G	?	G	G	G
Surflan	G	G	F	F	N	G	G	F
Pendulum, Southern Weedgrass Control	G	G	F	P	P	F	F	F
Treflan, Preen	G	G	F	P	P-N	F	F	F
XL	G	F	P	P	N	F-P	F	F-P

Note: Weed control rankings based on label information and author's experiences: G = good, F = fair, P = poor, N = no control expected, ? = unknown.

Consider using the ornamental species listed on the label as a selection criteria, for the herbicide as well as for the ornamental species. In this way you can obtain maximum weed control and avoid injury to desirable plants.

Preemergent herbicides are applied after planting but before weeds germinate. Some of the more popular preemergent herbicides used in the landscape plantings include trifluralin (Treflan), oryzalin (Surflan, XL), and metolachlor (Pennant). Treflan and Surflan control annual grasses and many broadleaf weeds, but can be used safely around many woody and herbaceous ornamentals. Metolachlor has become popular because it controls yellow nutsedge as well as most annual grasses. Table 4 provides information on the effectiveness of preemergent herbicides on several common landscape weeds.

When weeds escape the preemergent herbicides or geotextile fabrics, **postemergent herbicides** are often used. There are selective and nonselective types. Roundup, Diquat, and Sharpshooter are nonselective, injuring any vegetation contacted. Roundup is systemic, translocating to the roots,

thereby killing the entire plant. It is effective on annual and perennial weeds. Diquat and Sharpshooter are contact-type herbicides, controlling small annual weeds but only "burning-back" perennial or large annual weeds. The other postemergent herbicides listed in Table 3 are selective, that is, they kill or injure some species but not others. Before using these products, carefully check the lists of weeds controlled and the ornamental species over which the herbicide may be safely used. For example, Casoron will control many perennial weeds but will injure most herbaceous and some woody ornamentals. Oxyfluorfen (Goal) can be used over many conifers but will injure many other ornamentals. Ornamec and Vantage selectively control annual and perennial grasses, but check the labels for species and varietal differences in plant safety.

For more detailed and extensive information on herbicides registered for landscape uses, see your local Cooperative Extension office. Also, two good references on herbicides used in landscapes are the *Weed Management Guide for Herbaceous Ornamentals* (see Chapter 27) and the *Weed Control*

Table 5 **Effectiveness of pre-plant weed control measures on certain hard-to-kill perennial weeds**

Species	Roundup-Fall	Roundup-Spring	Fumigation	Cultivation
Bindweed	Fair	Fair	Good	Poor
Japanese knotweed	Very good	Poor	Good	Good
Mugwort	Good	Poor	Good	Poor
Canada thistle	Good	Fair	Good	Fair
Wild Violet	Fair	Fair	Excellent	Fair
Goldenrod	Very good	Fair	Excellent	Good
Nutsedge	Poor	Poor	Very good	Poor
Bamboo	Poor	Poor	Fair	Fair
Quackgrass	Good	Good	Very good	Poor
Bermudagrass	Good	Fair	Very good	Poor
Buffalograss	Poor	Poor	Very good	Poor
Equisetum	Poor	Poor	Good	Poor

Suggestions for Christmas Trees, Woody Ornamentals, and Flowers (see Suggested References).

When selecting planting types and species to be included, use the weed control options as one of the selection criteria. The following two examples demonstrate how this can result in improved weed management.

Example 1. In a potential planting bed you identified yellow nutsedge as a major weed. Since yellow nutsedge is not controlled by mulches, geotextiles, or most preemergent herbicides, and postemergent herbicides applied in late summer or fall would have little or no effect on tuber emergence in the spring, your options are limited. Preemergent applications of metolachlor (Pennant) can provide adequate control, but this herbicide will injure many annual flowers and ornamental grasses. For example, Pennant should not be applied to begonia or impatiens, but it is safe on many woody species and herbaceous perennials. Therefore, an important species selection criteria would be to choose from those listed on the Pennant label. If you must plant begonias, then you would have to fumigate the site to eliminate nutsedge tubers.

Example 2. In a new landscape, a *Pachysandra* bed interplanted with daffodils for spring color was specified. Site analysis shows *Oxalis* (woodsorrel) and common groundsel to be the predominant weeds in the area. Since geotextiles would prevent spread of the *Pachysandra*, they should not be used. Additionally, both of these weeds grow quite well in organic mulches and are difficult to hand weed (particularly the woodsorrel). The only herbicides labeled for both *Pachysandra* and daffodils do not control these weeds. Switching to *Vinca* as the ground cover species would allow the use of Surflan (oryzalin), which controls both weeds and is labeled for both ornamental species.

Once the species selection and weed management options have been chosen, you are ready to prepare the site for planting.

STEP 4: SITE PREPARATION — WEED MANAGEMENT CONCERNS

The best time to control perennial weeds is before planting. There are basically three options: repeated cultivation, fumigation, or Roundup. Table 5 provides guidelines for controlling several perennial weeds that are difficult or impossible to control after planting. Note that spring applications of Roundup are less effective than fall applications on several species. Therefore, where possible, plan ahead and do your site preparation in the fall. Note also that even Roundup does not control all weeds. For species like nutsedge, field horsetail (*Equisetum*), and wild violet, other measures may be necessary, such as fumigation. Fumigation is also useful if annual weeds are present which cannot be controlled after planting. (Of course, here is an opportunity to utilize the planning process to select ornamental species or types in which you can control these weeds.)

If the site is to be amended with top soil or organic matter, inspect the sources of these materials for noxious weeds. Top soil from farm land or stream banks is notorious as a source of nutsedge tubers and seeds of many annual weeds. Inspect piles of compost or mulch for signs of weeds. Some species frequently found in mulch piles include mugwort, thistle, bindweed, and field horsetail. If these weeds are present, find an alternate source!

STEP 5: INSTALLATION AND IMPLEMENTATION

Once you have gone to so much trouble to prepare a weed-free planting area, **don't introduce weeds**. Many perennial weeds are introduced in soil balls of field-grown nursery stock. Look for signs of mugwort, bindweed, field horsetail, or nutsedge. Again, inspect your source of mulch for weeds. If geotextiles are to be used, install them properly. If herbicides are to be used, apply them carefully. In short, implement the plan that you have developed.

No single weed management strategy will control all weeds. An integrated approach, utilizing all options at your disposal is the most economical and effective means of controlling weeds. To achieve this, remember to **PLAN BEFORE YOU PLANT.**

SUGGESTED REFERENCES

Cornell Pest Management Recommendations for Commercial Production of Trees and Shrubs. Available from Cornell Cooperative Extension offices.

Weed Control Suggestions for Christmas Trees, Woody Ornamentals, and Flowers, by Walter A. Skroch, available from the Department of Horticultural Science, Box 7609, North Carolina State University, Raleigh, NC 27659-7609.

Nursery and Landscape Weed Control Manual, by Robert P. Rice, Jr., Thomson Publications, P.O. Box 9335, Fresno, CA 93791.

Chapter 27

Weed Management Guide for Herbaceous Ornamentals

Andrew Senesac, *Cornell Cooperative Extension of Suffolk County, Long Island Horticultural Research Lab, Riverhead, NY*

Joseph Neal, *Department of Floriculture and Ornamental Horticulture, Cornell University, Ithaca, NY*

(This chapter is a reprint of *WeedFacts No. 1,* a Cornell Cooperative Extension publication, Cornell University, Ithaca, NY, 1992. With permission.)

CONTENTS

Weed Management Options 301
Types of Herbaceous Ornamentals 311
Growing Situations 312
Herbicide Application 312
     Injury Symptoms 312
     Formulations and Application 312
Selected References 313

WEED MANAGEMENT OPTIONS

This guide is intended to help the commercial grower and landscaper choose a safe and effective weed management program for herbaceous ornamentals. Every attempt has been made to provide updated information on the current registered uses. However, it is the applicator's responsibility to read and follow the label.

Table 1 has information about application and use characteristics of each herbicide. Tables 2 and 3 contain information on herbicides that are currently registered for use in the U.S.

Weed management is an integral and important part of all commercial production of herbaceous ornamentals. Weeds compete and interfere with plant growth and devalue the yield and quality of landscape, container and field grown ornamentals. It is important to develop a weed control strategy that utilizes all the available options at your disposal. These include **preventive** measures such as organic and inorganic **mulches**, preemergence **herbicides** and **sanitary practices** that prevent weed seeds and vegetative parts from spreading. This is especially important in container operations where the potting media is often soilless and relatively weed-free.

Several pictorial guides and botanical identification keys are available to identify the most common weeds. It is essential to know the correct names to understand the herbicide labels and control recommendations. Most weeds that infest ornamentals have one of four life cycles: **summer annuals** which emerge in the spring, flower and set seed before the first frost; **winter annuals**, which germinate at the end of the summer and over winter as small dormant, but green plants; **biennials** are similar to winter annuals, but germinate earlier in the summer; or **perennials** which, as the name suggests, survive more than two seasons, and can propagate by seed or vegetative reproduction. Knowing the weed life cycle is key to determining the optimal timing of a herbicide application or cultural practice. It is important to scout the weed population during and after the growing season in order to assess the success of the weed control program. For instance, at the end of the season in the fall, escaped summer annuals and some perennials will be dead but can be identified by their characteristic "skeletons." Escaped winter annuals, biennials and most perennial weeds will survive the winter as dormant rosettes, crowns or underground rhizomes.

Table 1 Description of herbicides registered for herbaceous ornamentals

Trade name	Acclaim
Common name	*fenoxaprop*
Formulations	1-EC
Rate range (a.i. lb/a)	0.12 - 0.35 lb/a
Product per ACRE	0.96 - 2.8 pints
Product per 1,000 sq. ft	0.365 - 1.02 fl oz
Pre vs. post activity	Postemergence only.
Mechanism of action	Inhibits fatty acid synthesis. Growth of susceptible grasses is arrested soon after application. Symptoms take several days to appear.
Weed control spectrum	Active on grasses only. Broadleaves and sedges aren't susceptible.
Volatility potential	Loss from volatility is minimal.
Leaching potential	Leaching is negligible.
Half life/mechanism of loss	Soil half life of is short: 1-14 days. Chemical decomposition accounts for much of the loss.
Landscape vs. turf use	Can be used over the top of many ornamentals and established turf species.
Application restrictions	Control is optimal when grass weeds are young and not drought stressed.
Labeled combinations	Several tank mixes are labeled - bensulide, pendimethalin and DCPA.
Weaknesses	No broadleaves or sedges are controlled.

Trade name	Barricade
Common name	*prodiamine*
Formulations	65 WDG(water-dispersible granules)
Rate range (a.i. lb/a)	0.325 - 1.5 lb/a
Product per ACRE	0.5 - 2.3 lb/a
Product per 1,000 sq. ft	0.18 - 0.40 oz
Pre vs. post activity	Preemergence only.
Mechanism of action	Inhibits cell division and root growth - prevents weed seedling establishment.
Weed control spectrum	Annual grasses and some important annual broadleaves (prostrate spurge, knotweed, chickweed).
Volatility potential	Slight loss from soil can result from photo degradation and volatility.
Leaching potential	Strongly adsorbed to clay and organic matter and not leached.
Half life/mechanism of loss	3-6months of weed control can be expected at usage rates. Metabolized by soil microbes.
Landscape vs. turf use	Can be used on cool and warm season turf and ornamentals.
Application restrictions	Can be used near many established woody and herbaceous landscape ornamentals.
Labeled combinations	No combinations specifically prohibited or labeled for ornamentals.
Weaknesses	No label rates for use on ORNAMENTALS. A number of tolerant established ornamentals are listed .

Trade name	Basagran T/O
Common name	*bentazon*
Formulations	4-WS(water soluble)
Rate range (a.i. lb/a)	0.75 - 1.0 lb/a
Product per ACRE	1.5 - 2 pints
Product per 1,000 sq. ft	0.55 - 0.73 fl oz
Pre vs. post activity	Postemergence only.
Mechanism of action	Inhibits photosynthesis. Foliar applications translocate to roots.
Weed control spectrum	Controls broadleaves and sedges only.
Volatility potential	No loss from volatility or photo decomposition.
Leaching potential	Not adsorbed to soil, but rapidly incorporated into organic matter.
Half life/mechanism of loss	Rapidly broken down in soil by microbes.
Landscape vs. turf use	Can be used in both landscape and cool season turf.
Application restrictions	With a few exceptions, applications to ornamentals should be directed and *not* over the top.
Labeled combinations	A crop oil concentrate should be added for some weed species.
Weaknesses	Two applications are needed for some difficult weed species such as yellow nutsedge.

Trade name	Betasan, Betamec, Lescosan, Pre-San, Pro-Turf, many others
Common name	*bensulide*
Formulations	2.9-EC, 4-EC, 3.6-G, 7-G, 12.5-G
Rate range (a.i. lb/a)	7.5 -12.6 lb/a
Product per ACRE	4-EC = 15 - 25 pints, 3.6-G = 209 - 350 pounds
Product per 1,000 sq. ft	4-EC = 5.5 - 9.2 fl oz, 3.6-G = 4.8 - 8 pounds
Pre vs. post activity	Preemergence only.
Mechanism of action	Inhibits root growth, not translocated.
Weed control spectrum	Controls annual grasses and a few annual broadleaves, but no perennials.
Volatility potential	Low volatility but must irrigate for activation.
Leaching potential	Low leaching potential, inactivated in soils high in organic matter.

Table 1 (Cont.) Description of herbicides registered for herbaceous ornamentals

bensulide continued

Half life/mechanism of loss	Half life 4–6 months, degraded slowly by soil microorganisms.
Landscape vs. turf use	Several herbaceous ornamentals are labeled including some bulbs. Bensulide can be used in turf and landscape situations.
Application restrictions	Apply to well established ornamentals.
Labeled combinations	No combinations specified on label.
Weaknesses	Controls only a few annual broadleaf species and no perennials.
Trade name	Dacthal, many others
Common name	*DCPA or chlorthal-dimethyl*
Formulations	75-WP, 5-G, 6-Fl
Rate range (a.i. lb/a)	10.5 - 12 lb/a
Product per ACRE	75-WP = 14 – 16 pounds, 5-G = 210 – 240 pounds
Product per 1,000 sq. ft	75-WP = 5.14 – 5.9 oz, 5-G = 4.8 – 5.5 pounds
Pre vs. post activity	Preemergence only with no foliar absorption.
Mechanism of action	Inhibits cell division in the roots of germinating seedlings.
Weed control spectrum	Annual grasses, spurge, purslane, chickweed, carpetweed. (*Veronica filiformis* is controlled with post applications of WP or FL formulations).
Volatility potential	Nonvolatile.
Leaching potential	Very low leaching potential, very insoluble.
Half life/mechanism of loss	100 day half life, effective control approximately 9 weeks at usage rates. Microbial breakdown is the primary means of degradation.
Landscape vs. turf use	Can be used on turf or ornamentals, newly planted or established plants.
Labeled combinations	No combinations labeled and no incompatibilities.
Application restrictions	No longer registered in Suffolk County, NY.
Weaknesses	The high usage rates mean that a large amount of product must be handled.
Trade name	Devrinol
Common name	*napropamide*
Formulations	50-WP, 2-G, 5-G, 10-G, 2-E
Rate range (a.i. lb/a)	4 - 6 lb/a
Product per ACRE	50-WP=6 – 8 pounds, 5-G 80 –120 pounds
Product per 1,000 sq. ft	50-WP=3.0 – 4.4 oz, 5-G=20 – 44 oz
Pre vs. post activity	Preemergence activity only.
Mechanism of action	Inhibits root growth. It is translocated but is not an effective postemergence herbicide.
Weed control spectrum	Grasses and some important broadleaves including common chickweed and groundsel.
Volatility potential	Little loss from soil by volatility, but photodecomposition can occur if not incorporated within 4 days
Leaching potential	Resists leaching in mineral soils.
Half life/mechanism of loss	Half life of 8-12 weeks under most conditions, slowly broken down by soil microorganisms.
Landscape vs. turf use	Ornamentals and warm season turf only. Will injure cool season turf.
Application restrictions	Can be applied to newly planted container-or field-grown herbaceous ornamentals. Needs 1 to 2 inches of water incorporation for good contact with germinating seeds.
Labeled combinations	Can be tank mixed with Betasan. No known incompatibilities.
Weaknesses	Higher rates are needed for full season control of the labeled annual broadleaves.
Trade name	Eptam
Common name	*EPTC*
Formulations	2.3-G, 5-G, 10-G
Rate range (a.i. lb/a)	5 lb/a - up to 15 lb/a for some perennial weed control.
Product per ACRE	2.3-G=200 pounds, 5-G=100 pounds
Product per 1,000 sq. ft	2.3-G=5 pounds, 5-G=2.5 pounds
Pre vs. post activity	Preemergence only.
Mechanism of action	Absorbed by roots and translocated-inhibits underground shoot growth of annuals and some perennials.
Weed control spectrum	Controls annual grasses and some important annual broadleaves and perennials: common chickweed, nightshade, lambsquarters, purslane, yellow nutsedge, mugwort and quackgrass.
Volatility potential	Highly volatile when applied to moist soils, must be incorporated mechanically or with water, soon after application.
Leaching potential	Leaching may occur in sandy soils with excessive rainfall.
Half life/mechanism of loss	1 week half life, average soil life is 2 to 6 weeks. Rapid soil microbial breakdown is common.
Landscape vs. turf use	Several herbaceous ornamentals are labeled. Will injure most turf species.
Application restriction	Wait 2 weeks after transplanting herbaceous plants. For perennial weeds, incorporate deeply (3–6 in.). Can be applied preplant and incorporated 4 weeks before planting some ornamentals(see label).
Labeled combinations	No combinations or incompatibilities are noted on label.
Weaknesses	Injures some ornamentals (especially bulbs). The need to incorporate limits its practicality.

Table 1 (Cont.)   **Description of herbicides registered for herbaceous ornamentals**

Trade name	(*formerly FUSILADE*) Ornamec, Grass-B-Gon, Takeaway
Common name	*fluazifop-p*
Formulations	1-EC
Rate range (a.i. lb/a)	0.25 - 0.375 lb/a - use with a nonionic surfactant
Product per ACRE	2 - 3 pints
Product per 1,000 sq. ft	2.2 - 3.3 teaspoons
Pre vs. post activity	Postemergence. There may be a preemergence residual effect on susceptible grass cover crops.
Mechanism of action	Enters foliage rapidly and stops growth of susceptible grasses within hours of application. Inhibits fatty acid synthesis.
Weed control spectrum	Active on grasses only. Broadleaves and sedges are not susceptible.
Volatility potential	Negligible. ·
Leaching potential	Low to moderate potential for leaching in soil. Rain-fast on plants within one hour.
Half life/mechanism of loss	Soil half life of 1-3 weeks. Rapid microbial breakdown occurs in moist soils.
Landscape vs. turf use	Many ornamentals are labeled for directed or over-the-top applications. Will injure most turf.
Application restrictions	Apply to actively growing grass weeds.
Labeled combinations	If possible, use in a program which controls broadleaf weeds.
Weaknesses	No broadleaf or sedge control. Some cultivars of some ornamentals may be injured.

Trade name	Gallery
Common name	*isoxaben*
Formulations	75-WDG (water dispersible granule)
Rate range (a.i. lb/a)	0.5 - 1.0 lb/a
Product per ACRE	0.66 - 1.33 pounds
Product per 1,000 sq. ft	0.25 - 0.5 oz
Pre vs. post activity	Preemergence primarily. Gallery can injure some species with postemergence applications.
Mechanism of action	Inhibits normal root cell development.
Weed control spectrum	Primarily annual BL, some grass suppression. some perennial broadleaves from seed.
Volatility potential	Negligible volatility under field conditions.
Leaching potential	Low water solubility and low Leaching potential.
Half life/mechanism of loss	Soil half life 5 - 6 months. Microbial breakdown is primary means of loss from soil.
Landscape vs. turf use	Can be used on many herbaceous ornamentals and cool season turf. Do NOT use on bedding plants.
Application restrictions	Wait for soil to settle after planting - ground covers should be well rooted. Not labeled for any food crops in US.
Labeled combinations	Compatible with other herbicides registered for use on ornamentals.
Weaknesses	May be too long lasting in some situations, however annual grass control is poor .

Trade name	Pennant
Common name	*metolachlor*
Formulations	7.8-EC, 5-G
Rate range (a.i. lb/a)	2.0 - 4.0 lb/A
Product per ACRE	7.8-EC=2 - 4 pints, 5-G=40 - 80 pounds
Product per 1,000 sq. ft	7.8-EC=2.25 - 4.5 fl oz, 5-G= 1 - 2 pounds
Pre vs. post activity	Preemergence only, although post applications may cause temporary leaf cupping.
Mechanism of action	General growth inhibition, especially of root elongation.
Weed control spectrum	Annual grasses, a few important broadleaves (especially galinsoga) and yellow nutsedge.
Volatility potential	Relatively nonvolatile.
Leaching potential	Moderately water soluble, but highly adsorbed to soil organic matter.
Half life/mechanism of loss	If organic matter approaches 2% in the soil, leaching is not expected to be significant. Soil half life is from 15-50 days, depending on climatic conditions and soil type.
Landscape vs. turf use	Cannot be used on cool season turf species.
Application restrictions	Apply after soil has settled around roots of newly transplanted ornamentals.
Labeled combinations	Compatible with other herbicides registered for use on ornamentals.
Weaknesses	Full season control of heavy yellow nutsedge infestations may require two applications.

Trade name	Ronstar
Common name	*oxadiazon*
Formulations	2-G, (50-WP is not labeled for herbaceous species)
Rate range (a.i. lb/a)	2.0 - 4.0 lb/A
Product per ACRE	100 - 200 pounds
Product per 1,000 sq. ft	2.75 - 4.5 pounds
Pre vs. post activity	Primarily preemergence, some very early postemergence activity on seedling weeds. foliar burn of some ornamentals is possible if granule settles in plant whorl.
Mechanism of action	Controls seedlings as they emerge at the soil surface.
Weed control spectrum	Annual grasses and some important broadleaves including: bittercress, oxalis and liverwort.

Table 1 (Cont.) **Description of herbicides registered for herbaceous ornamentals**

oxadiazon continued	
Volatility potential	Loss by volatility not a problem.
Leaching potential	Very low water solubility and strong adsorption to soil colloids prevent leaching.
Half life/mechanism of loss	Half life of 3-6 months depending on soil type.
Landscape vs. turf use	Can be used on some ornamentals and cool season turf, except red fescue.
Application restrictions	Can be applied to some newly transplanted species.
Labeled combinations	No tank mixes are labeled.
Weaknesses	Wettable formulation is not labeled for herbaceous plant species.
	Will stand alone for many ornamental weed control problems *except* common chickweed.

Trade name	Rout
Common name	*oxyfluorfen (Goal) and oryzalin (Surflan)*
Formulations	3-G. (oxyfluorfen 2% and oryzalin 1%)
Rate range (a.i. lb/a)	3.0 lb/A
Product per ACRE	100 pounds
Product per 1,000 sq. ft	2.2 pounds
Pre vs. post activity	Preemergence, but foliar burn is possible if granule settles in plant whorl.
Mechanism of action	Contact action on seedlings as they emerge from the soil surface.
Weed control spectrum	Annual grasses and some important broadleaves including: bittercress, oxalis and chickweed
Volatility potential	Loss by volatility not a problem under most conditions.
Leaching potential	Very low water solubility and strong adsorption to soil colloids prevent leaching.
Half life/mechanism of loss	Half life of oxyfluorfen is 30 -40 days.
Landscape vs. turf use	Can be used on a few herbaceous and many woody ornamentals. Will injure turf.
Application restrictions	Do not apply to whorled plants or wet foliage. Do not mechanically incorporate.
Labeled combinations	No combinations with other pesticides are labeled.
Weaknesses	Only a few herbaceous species are labeled or tolerant.

Trade name	Surflan
Common name	*oryzalin*
Formulations	4-AS, 0.25-AS (aqueous solution), 75-WP.
Rate range (a.i. lb/a)	2-4 lb/a
Product per ACRE	4-AS=2 – 4 qts
Product per 1,000 sq. ft	4-AS=1.5 – 3 fl oz, 0.25-AS=1.5 – 3.0 pints
Pre vs. post activity	Preemergence only, needs one half inch of water to incorporate for good weed seed contact.
Mechanism of action	Inhibits cell division and root growth; prevents weed seedling establishment
Weed control spectrum	Annual grasses and several important broadleaves including: bittercress, common chickweed, groundsel and oxalis.
Volatility potential	Little volatility or photo decomposition — remains stable on soil surface for 21 days.
Leaching potential	A limited amount of leaching can occur under normal rainfall on coarse soils with low organic matter.
Half life/mechanism of loss	Soil half life is 2-6 months. Biodegraded by soil microbes.
Landscape vs. turf use	Landscape, field or container nurseries – oryzalin will injure cool season turf.
Application restrictions	Rooted cuttings should be established 2 weeks or more.
Labeled combinations	Can be tank mixed with several herbicides – Roundup, Paraquat, Fusilade, and Poast.
Weaknesses	No granular formulation of oryzalin is currently available, although the combination product 'XL' (Surflan/Balan) is registered for ornamentals.

Trade name	S.W.C. (Southern Weedgrass Control), Pre-M, Stomp
Common name	*pendimethalin*
Formulations	2.7-G, 60-WDG, 4-EC
Rate range (a.i. lb/a)	2-3 lb/a
Product per ACRE	2.7-G=76 – 113 pounds, 60-WDG=3.3 – 5 pounds, 4-EC=2 – 3 pints
Product per 1,000 sq. ft	2.7-G=1.7 – 2.6 pounds, 60-WDG=1.2 – 1.8 oz, 4-EC=4.4 – 6.6 teaspoons
Pre vs. post activity	Primarily preemergence, but there is some very early postemergence activity.
Mechanism of action	Inhibits cell division and root growth - prevents weed seedling establishment.
Weed control spectrum	Annual grasses and some important annual broadleaves (prostrate spurge, bittercress, oxalis, pearlwort, chickweed).
Volatility potential	Slight loss from soil can result from photo degradation and volatility.
Leaching potential	Strongly adsorbed to clay and organic matter and not leached.
Half life/mechanism of loss	3-5 months of weed control can be expected at usage rates.
Landscape vs. turf use	Can be used on cool season turf and ornamentals.
Application restrictions	Field grown ornamentals may be treated right after planting. Allow bare root containerized liners 2 - 4 weeks to establish, and remove weeds prior to application.
Labeled combinations	No combinations specifically prohibited or labeled for ornamentals.
Weaknesses	Persistent orange color can stain landscape areas.

Table 1 (Cont.) **Description of herbicides registered for herbaceous ornamentals**

Trade name	Treflan, Preen, and others
Common name	*trifluralin*
Formulations	4-EC, 5G
Rate range (a.i. lb/a)	4 lb/a
Product per ACRE	5-G=80 pounds
Product per 1,000 sq. ft	5-G=1.8 pounds
Pre vs. post activity	Preemergence only.
Mechanism of action	Inhibits cell division of germinating seedlings.
Weed control spectrum	Annual grasses and a few important annual broadleaf weeds including: common chickweed, purslane, prostrate knotweed, lambsquarters, pigweeds.
Volatility potential	Moderately volatile and subject to photo decomposition. Needs to be incorporated into soil mechanically or with water.
Leaching potential	Insoluble, leaching potential is negligible.
Half life/mechanism of loss	Average soil persistence and acceptable weed control ranges from 6 -8 weeks.
Landscape vs. turf use	Landscape usage rate is 4 lb/a. It is combined at a lower rate with Balan as a combination granular(Team) for turf.
Application restrictions	Can be applied prior to planting or on established plants, many herbaceous species are labeled.
Labeled combinations	Combined formulations are available with Balan, Gallery and fertilizers.
Weaknesses	Some important broadleaf weed species are tolerant. Can cause severe injury through volatility if applied in closed greenhouses.

Trade name	Vantage (POAST)
Common name	*sethoxydim*
Formulations	1-EC, 1.5-EC
Rate range (a.i. lb/a)	0.3 - 0.5 lb/A. wit h Poast, use with crop oil concentrate or nonionic surfactant .
Product per ACRE	Vantage (1-EC) 2.25 - 3.75 pints, Poast (1.5-EC) 1.5 - 2.5 pints
Product per 1,000 sq ft	Vantage (1-EC) 1.6 - 2.8 Tablespoons, Poast 1 - 1.6 Tablespoons
Pre vs post activity	Postemergence herbicide, but can have a short residual effect on grass germination.
Mechanism of action	Translocates rapidly in plant and kills the growing points of susceptible grasses soon after application. Inhibits fatty acid synthesis.
Weed control spectrum	Active on grasses only. Broadleaves and sedges aren't susceptible.
Volatility potential	Loss from volatility is minimal.
Leaching potential	Little potential for leaching because of rapid breakdown in soil.
Half life/mechanism of loss	1-2 week half life. Broken down rapidly in soils.
Landscape vs turf use	Many ornamentals are labeled for over the top applications. Most turf is injured, but *Poa annua* and some fescues are tolerant.
Application restrictions	Can be used right after transplanting but application is timed to control actively growing grass weeds.
Labeled combinations	Can be tank mixed with Surflan.
Weaknesses	Controls grasses only.

Trade name	XL (Balan plus Surflan)
Common name	*benefin plus oryzalin*
Formulations	2-G (1:1 benefin:oryzalin)
Rate range (a.i. lb/a)	2-3 lb/a
Product per ACRE	100 - 150 pounds
Product per 1,000 sq ft	37 - 55 oz
Pre vs post activity	Preemergence only.
Mechanism of action	Both benefin and oryzalin are dinitroanilines and inhibit cell division.
Weed control spectrum	Annual grasses and a few broadleaves including: common chickweed and purslane.
Volatility potential	Slight volatility and photo decomposition may occur.
Leaching potential	A limited amount of leaching can occur under normal rainfall on coarse soils with low organic matter.
Half life/mechanism of loss	Soil half life is 1-3 months. Biodegraded by soil microbes.
Landscape vs turf use	Landscape, field or container nurseries – oryzalin will injure cool season turf.
Application restrictions	Can be applied to newly planted or established plants. Do not apply to emerged tulips in the spring.
Labeled combinations	No known incompatibilities.
Weaknesses	Only a few broadleaf species are controlled.

Table 2 **Herbicides registered for use on herbaceous ornamentals (bedding plants, bulbs, cutflowers, perennials, and groundcovers)**

Common Name	BASAGRAN / BETASAN	DACTHAL / DEVRINOL	EPTAM	PENNANT / GALLERY	ROUT / RONSTAR	SNAPSHOT	SURFLAN / SWC?	TREFLAN / XL	ACCLAM / fusilade*	VANTAGE	Genus
African Daisy		✔	✔	✔			✔	✔	✔	✔	Arctotis
African Lily			✔	✔✔ ✔			✔ ✔	✔ ✔		✔	Agapanthus
African Violet	✔										Saintpaulia
Ageratum		✔✔	✔	✔				✔		✔	Ageratum
Allium			✘	✔							Allium
Alyssum (Golddust)	✔	✔	✔	✔						✔	Aurinia
Amaranthus											Amaranthus
Asparagus Fern			✔				✔	✔		✔	Asparagus
Aster	✔	✔	✔	✔	✔			✔			Callistephus
Astilbe						✔			✔	✔	Astilbe
Baby's Breath		✔			✔	✔			✔		Gypsophila
Baby-Blue-Eyes									✔		Nemophila
Bachelor's Button	✔						✔				Centaurea
Balsam			✔			✔		✔			Impatiens
Beardtongue (Penstemon)						✔ ✔			✔		Penstemon
Begonia, fibrous			✔						✔	✔	Begonia
Bellflower	✔	✔			✔			✔	✔	✔	Campanula
Bird of Paradise			✔				✔		✔	✔	Strelitzia
Black-eyed Susan								✔	✔		Rudbeckia
Bleeding Heart		✔					✔		✔ ↳	✔	Dicentra
Bluebells									✔		Mertensia
Bugleweed (Ajuga)	✔	✘	✔	✔ ✘	✔ ✔	✘	✔ ✔		✔ ↳	✔	Ajuga
Burnet									✔		Sanguisorba
Butterflyweed (milkweed)				✔							Asclepias
Cactus										✔	Cactus
Calendula	✔						✔	✔		✔	Calendula
Calliopsis (Coreopsis)		✔		✔			✔	✔	✔	✔	Coreopsis
Candytuft	✔	✔		✘	✔✔ ✘				✔	✔	Iberis
Canna				✔					✔		Canna
Cape Marigold				✔		✔ ✔	✔ ✔	✔			Dimorphotheca/Osteospermum
Carex				✔							Carex
Carnation	✔				✔		✔				Dianthus
Catchfly									✘ ✔		Silene
Chamomile									✔		Matricaria
Chives, ornamental										✔	Allium
Chrysanthemum		✔✔	✔	✔	✔		✔	✔ ✔	✔		Chrysanthemum
Cockscomb										✔	Celosia
Coleus									✔	✔	Coleus
Columbine		✔		✔					✔	✔	Aquilegia
Coneflower		✔					✔		✔		Echinacea
Coral Bell	✔	✔							✔		Heuchera
Cornflower								✔	✔		Centaurea
Cosmos		✔						✔	✔		Cosmos
Cranesbill										✔	Geranium
Crocus			✘	✔						✔	Crocus
Crown vetch									✔	✔	Vicia
Daffodil	✔		✘	✔			✔	✔	✔		Narcissus
Dahlia	✔	✔✔	✔	✔			✔	✔✔			Dahlia
Daisy	✔		✔	✔			✔	✔✔	✔		Chrysanthemum
Dames rocket									✔		Hesperis
Daylily			✔	✔			✔ ✔		✔	✔	Hemerocallis
Dusty Miller				✔				✔		✔	Centaurea
Evening Primrose									✔	↳	Oenothera
False Dragonhead											Physostegia
Fescue, Blue								✔			Festuca
Feverfew		✔									Matricaria
Forget-Me-Not									✔		Myosotis
Fortnight Lily				✔			✔				Morea

† Southern Weedgrass Control (pendimethalin). See text for other trade names.

*Takeaway or Ornamec (formerly Fusilade)

✔	Registered for this species. Can be applied over the top.
↳	Registered for this species. Directed application only.
?	Registered, but research indicates that injury can occur.
✘	Label PROHIBITS use on this species.

Table 2 (Cont.)   **Herbicides registered for use on herbaceous ornamentals (bedding plants, bulbs, cutflowers, perennials, and groundcovers)**

Common Name	BASAGRAN	DACTHAL	BETASAN	DEVRINOL	BPTAM	GALLERY	PENNANT	XONSTAR	ROUT	SNAPSHOT	SURPLAN	SWCt	TREFLAN	XL	ACCLAIM	Densiflog	VANTAGE	Genus
Fountain Grass, green		X															↓	*Pennisetum*
Fountain Grass, red																	✔	*Pennisetum*
Four O'Clock		✔										✔						*Mirabilis*
Foxglove		✔																*Digitalis*
Freesia		✔																*Freesia*
Gaillardia		✔				✔					✔							*Gaillardia*
Gayfeather (Liatris)											✔							*Liatris*
Gazania		✔		✔	✔	✔	✔		✔	✔	✔	✔	✔	✔	✔		✔	*Gazania*
Geranium		✔		✔		✔					✔			✔	✔		✔	*Pelargonium*
Gerbera Daisy		✔															✔	*Gerbera*
Germander		X										✔						*Teucrium*
Geum (Avens)		X				✔												*Geum*
Gilia															✔			*Gilia*
Gladiolus		✔		✔			?				✔		✔	✔			✔	*Gladiolus*
Godetia														✔				*Clarkia*
Heath		✔					✔											*Erica*
Heather				✔			✔										✔	*Calluna*
Heather, False																	✔	*Cuphea*
Hens and Chickens																	✔	*Sempervivum*
Hollyhock																	✔	*Alcea*
Hosta (Plantain-Lily)				✔		✔			✔	✔	✔			✔			✔	*Hosta*
Hyacinth				X		✔				✔								*Hyacinthus*
Hyacinth, Grape				X		✔												*Muscari*
Hyacinth, Wood				X		✔												*Endymion*
Iceplant		✔	X	✔	✔	✔			✔	✔	✔		✔	✔			✔	(several genera)
Impatiens						?					✔		✔				✔	*Impatiens*
Iris, bulbous			✔	X							✔			✔				*Iris*
Iris, rhizomatous		✔													✔	✔	✔	*Iris*
Ivy		✔		✔	✔	✔		✔		✔	✔		✔	✔			✔	*Hedera* spp
Jack-in-the-Pulpit																	✔	*Arisaema*
Joseph's Coat		X																*Alternanthera*
Lamb's ear (Stachys)						✔												*Stachys*
Lantana		✔	✔						✔									*Lantana*
Larkspur (Delphinium)		✔				✔												*Delphinium*
Lavender																	✔	*Lavendula*
Lavendercotton		✔														✔	✔	*Santolina*
Leatherleaf Fern						✔										✔		*Rumohra*
Leopards-Bane		✔				✔									✔			*Doronicum*
Lily		✔		X		✔												*Lilium*
Lily-of-the-Valley																	✔	*Convallaria*
Liriope	✔				✔	✔			✔	✔	✔	✔	✔	✔	✔		✔	*Liriope*
Lobelia											✔		✔	✔			✔	*Lobelia*
Lupine		✔				✔					✔							*Lupinus*
Marguerite, Golden		✔																*Anthemis*
Marigold		✔		✔		?					✔			✔		✔	✔	*Tagetes*
Miscanthus					✔	✔				✔		✔						*Miscanthus*
Mondo Grass					✔	✔			✔		✔					↓	✔	*Ophiopogon*
Moneywort																	✔	*Lysimachia*
Morning Glory		✔									✔							*Convolvulus*
Moss-Rose		✔											✔			✔	✔	*Portulaca*
Mother of Thyme		✔																*Thymus*
Mourning-bride		✔											✔					*Scabiosa*
Narcissus		✔			✔	X				✔					✔			*Narcissus*
Nasturtium		✔										✔						*Nasturtium*
Nicotiana												✔					✔	*Nicotiana*
Nightshade											✔		✔					*Solanum*
Pachysandra	✔	✔		✔	✔	✔	✔		✔	✔	✔		✔	✔		✔	✔	*Pachysandra*
Pampas Grass						✔	?				✔		✔				↓	*Cortaderia*
Pansy		✔	X			✔					✔			✔			✔	*Viola*

† Southern Weedgrass Control (pendimethalin). See text for other trade names.

* Takeaway or Ornamec (formerly Fusilade)

✔	Registered for this species. Can be applied over the top.
↓	Registered for this species. Directed application only.
?	Registered, but research indicates that injury can occur.
X	Label PROHIBITS use on this species.

Table 2 (Cont.) Herbicides registered for use on herbaceous ornamentals (bedding plants, bulbs, cutflowers, perennials, and groundcovers)

Common Name	BASAGRAN	BETASAN	DACTHAL	DEVRINOL	EPTAM	GALLERY	PENNANT	RONSTAR	ROUT	SNAPSHOT	SURFLAN	SWC†	TREFLAN	XL	ACCLAIM	fusillade*	VANTAGE	Genus
Peony			✔												✔			Paeonia
Pepper, ornamental					X												✔	Capsicum
Periwinkle (bedding plt)		✔											✔	✔	✔			Catharanthus
Periwinkle (ground cvr)	✔	✔			✔		✔	✔			✔		✔	✔	✔	✔	✔	Vinca minor
Petunia			✔	✔	✔		✔	✔			✔		✔	✔	✔	✔	✔	Petunia
Phlox			X		X		✔	✔					✔					Phlox
Pimpernel															✔			Anagallis
Pink			X		✔								✔		✔		✔	Dianthus
Pink clover																✔		Polygonum
Poker-plant		✔																Kniphofia
Poppy															✔			Papaver
Poppy, California													✔		✔			Eschscholzia
Prickly Pear																↳		Opuntia
Primrose	✔						✔											Primula
Queen Anne's Lace							✔											Daucus
Ranunculus	✔										✔							Ranunculus
Rose		✔	✔	✔	✔	✔	✔	✔	✔	✔	✔		✔	✔	✔	✔		Rosa
Rosemary									✔				✔			✔		Rosmarinus
Rupturewort																	✔	Herniaria
Sage, Sweet or Texas											✔					✔	✔	Salvia
Salvia				X			✔						✔		X	✔	✔	Salvia
Sandwort																✔		Arenaria
Scarlet Flax													✔					Linum
Sea Pink													✔				✔	Armeria
Sedum (Stonecrop)	✔		✔	✔	✔	X	✔	✔		X	✔		✔	✔	✔		✔	Sedum
Shasta Daisy											✔		✔		✔		✔	Chrysanthemum
Snapdragon			✔		X		✔				✔		✔		✔		✔	Antirrhinum
Snow-in-Summer											✔		✔		✔		X	Cerastium
Snow-on-Mountain					X				X				✔					Euphorbia
Soapwort																✔		Saponaria
Speedwell (Veronica)								✔					✔					Veronica
Spiderwort		✔															✔	Tradescantia
Squill (Scilla)				X			✔											Scilla
St. John's-wort		✔		✔		✔	✔		✔	✔			✔		✔			Hypericum
Star-of-Bethlehem							✔											Ornithogalum
Starflower															✔			Trientalis
Statice							?								✔	✔	✔	Limonium
Stock		✔															✔	Matthiola
Stokes Aster											✔	✔						Stokesia
Strawberry, ornamental		✔			✔								✔			✔		Fragaria
Strawflower			✔															Helichrysum
Sundrops			✔										✔					Oenothera
Sunflower			✔										✔					Helianthus
Sweet Alyssum		✔					✔									✔		Lobularia
Sweet Flag												✔					✔	Acorus
Sweet Pea		✔	✔										✔					Lathyrus
Sweet William			X				✔				✔		✔		✔		✔	Dianthus
Sword fern															✔	✔		Nephrolepsis
Tidy tips															✔			Layia
Tulip		✔			X		✔				✔		✔	✔				Tulipa
Verbena													✔				✔	Verbena
Vinca (ground cover)							✔	✔			✔		✔	✔	✔		✔	Vinca major
Wallflower		✔													✔			Cheiranthus
Wild Thyme																		Thymus
Wormwood			✔															Artemisia
Yarrow			✔								✔	✔	✔			✔		Achillea
Yucca												✔				✔		Yucca
Zinnia		✔	✔	✔	✔		?				✔		✔		✔	✔	✔	Zinnia
Zinnia, creeping															✔			Sanvitalia

† Southern Weedgrass Control (pendimethalin). See text for other trade names.

* Takeaway or Ornamec (formerly Fusilade)

✔ Registered for this species. Can be applied over the top.
↳ Registered for this species. Directed application only.
? Registered, but research indicates that injury can occur.
X Label PROHIBITS use on this species.

Table 3 Weed susceptibilities to herbicides registered for herbaceous ornamentals

Weed Species Broadleaves	BASAGRAN BETASAN	DACTHAL DEVRINOL	EPTAM	PENNANT GALLERY	ROUT RONSTAR	SURFLAN SNAPSHOT	TREFLAN SWCT	XL	ACCLAIM Fusilade*	VANTAGE
Bittercress, Hairy				●	❑	● ●	●		O	O
Carpetweed		●	●	● ●	●	● ●	●	●	O	O
Carrot, Wild			❑			❑	●	●	O	O
Chickweed, Common	●	●	●	● ●	● O	●	●	●	O	O
Chickweed, Mouseear	●		❑	●		❑	●		O	O
Clover, White			❑	●		● ●	●		O	O
Cocklebur, Common	●	❑	❑	❑			❑	O	O	O
Cudweed				●		● ❑	●		O	O
Dandelion (seedling)	●		❑	) ❑	●	●	●		O	O
Dodder		●	❑			❑			O	O
Dogfennel				●		❑			O	O
Evening Primrose			❑		● ●	●	●		O	O
Filaree, Redstem			●	●		●	●		O	O
Fleabane				●				●	O	O
Galinsoga, Hairy	●	O	❑	● ●	)	●	❑ ❑		O	O
Geranium, Carolina									O	O
Groundsel, Common				●)	● ●	●	❑		O	O
Henbit	●		●	●		● ●	●	●	O	O
Horseweed (Marestail)				●	●	●)			O	O
Jimsonweed	●	O	❑	❑	❑	● ❑	❑ O		O	O
Knotweed, Prostrate		●		)	●	● ●	●		O	O
Lambsquarters	)	) ●	●	● ●	● ●	● ●	● ●		O	O
Lettuce, Prickly			●	●		●)			O	O
Liverwort		❑			)				O	O
Mallow	●	)		)		●)	O		O	O
Morningglory, Annual	) ❑	❑	❑	❑		))			O	O
Moss		❑			●				O	O
Mugwort		❑	●			❑			O	O
Mustard, Wild	●	O		●	❑	●)			O	O
Nightshade, Black	❑	❑ ❑	●	●		)			O	O
Pearlwort		❑			●		●		O	O
Pepperweed									O	O
Pigweed spp.	) ●	●	● ●	● ●	● ●	●	● ●	❑	O	O
Pineappleweed			●	●		●			O	O
Plantain	●			●		●			O	O
Purslane, Common	●	●	●	●	)	● ●	● ●	●	O	O
Pusley, Florida	●	●	❑	●	●)	●			O	O
Ragweed, Common	●	❑	O)	)	●	)	● O		O	O
Shepherdspurse	●	●		●	●	●			O	O
Sida, Prickly	●		❑			)			O	O
Smartweed	●	❑	O	●		●)	●		O	O
Sorrel, Red		❑							O	O
Sowthistle, Annual		●	●	)	● ●	●)			O	O
Speedwell spp.	❑	●	❑	●		●			O	O
Spurge, Prostrate	●	)	❑	●	● ●	● ●	●	●	O	O
Spurge, Spotted	●	)	❑	●	● ●	) ●	●		O	O
Sourry, Corn			❑	●					O	O
Thistle, Canada	●		❑						O	O
Velvetleaf	●	❑	O	❑	●	)	O	❑	O	O
Y. Woodsorrel (Oxalis)	●		❑	●	❑ ●	●	●	●	O	O

† Southern Weedgrass Control (pendimethalin). See text for other trade names.

*Takeaway or Ornamec (formerly Fusilade)

Based on labels:
● Full control (80 to 100%)
) Partial control (30 to 80%)
O No control (0 to 30%)
❑ No control (based on research)

Table 3 (Cont.) **Weed susceptibilities to herbicides registered for herbaceous ornamentals**

Weed Species / Grasses or grasslike	BASAGRAN / BETASAN	DACTHAL / DEVRINOL	EPTAM	PENNANT / GALLERY	RONSTAR	ROUT	SURFLAN / SNAPSHOT	TREFLAN / SWC†	ACCLAIM / XL	Quasitop* / VANTAGE
Barnyardgrass		●	●	▶	●	●	●	●	●	●
Bermudagrass	❑	❑	❑	❑	❑	❑	❑	❑		◆
Bluegrass, Annual		●	●	▶	●	●	●	●	❑	
Brome, Downy			●							
Crabgrass, Large		●	●	▶	●	●	●	●	●	●
Crabgrass, Smooth	❑	●	●	▶	●	●	●	●	●	●
Fescue, Tall	❑	❑	❑		❑		❑	❑		▶
Foxtail (Yellow, Green)		●	●	▶	●	●	●	●	●	●
Foxtail, Giant		●	●	▶		●	●	●	●	●
Goosegrass		▶	●	▶	●	●	●	●	●	●
Horsetail (Equisetum)		❑	❑		❑	❑		❑	❑	❑
Johnsongrass (rhizome)		○				❑				
Johnsongrass (sdlg)		▶	▶		▶		●		●	●
Nutsedge, Yellow	●	○		❑	❑	❑	❑	○	❑	❑
Orchardgrass						●				▶
Panicum, Fall	●		●		●	●	●	●	●	●
Quackgrass	❑	❑	❑	❑	❑		❑	❑	❑	◆
Ryegrass, Annual	❑		●	▶		●	●		●	
Sandbur		▶	●	●			●	●	●	
Signalgrass, Broadleaf			●		◆		●		●	●

† Southern Weedgrass Control (pendimethalin). See text for other trade names.

*Takeaway or Ornamec (formerly Fusilade)

Based on labels:
● Full control (80 to 100%)
▶ Partial control (30 to 80%)
○ No control (0 to 30%)
❑ No control (based on research)

There are several herbicides available that can be used safely and legally to control weeds in herbaceous ornamentals. Herbicides are commonly classified by their mechanism of action and use pattern. **Preemergence** herbicides are applied before weeds emerge and generally provide residual control of weed seedlings for several weeks.

Postemergence herbicides, applied after the weeds have emerged, are of two types. **Contact** herbicides kill only the portion of the plant that is actually contacted by the herbicide. Good spray coverage is important when using contact herbicides. **Systemic** herbicides are absorbed and move through the plant. These are useful for controlling the creeping roots and rhizomes of perennial weeds. With systemic herbicides, the weeds must be actively growing so that the herbicide can be fully translocated. The postemergence herbicides that are labeled for herbaceous ornamentals are non-residual and have little or no soil activity.

In many situations, herbicides cannot be used or are not effective in controlling all the weeds. In these situations, **cultivation** and **hand pulling** are often the only available options. There are two important facts to remember about mechanical cultivation. Hoeing and tilling will control small annual weeds fairly well. However, successive flushes of germinating weeds, stimulated by the cultivation itself, need to be controlled on a two to three week cycle. Once residual herbicides are applied and activated with water, they need to be in intimate contact with the germinating weed seedlings to work well. Mechanical cultivation will destroy this contact in many cases.

Hand pulling is often an important, if back-breaking component of a weed management program. It should be considered when there are no other cultural or herbicide options and when there are weeds that will disperse their seed by wind or other means to weed-free areas.

TYPES OF HERBACEOUS ORNAMENTALS

Plant species which are listed on herbicide **labels** have been tested by independent researchers and approved or **registered** by state and federal agencies. Because of the great number of herbaceous species, it is possible to test only a small fraction of all that are commercially grown. If a species does not appear on the herbicide label, it is not legal to use on that species even though the applicator assumes all risks and liabilities. See Table 1 for application restrictions.

Spring-flowering **bulbs** that are planted in the fall can be treated with preemergence herbicides shortly after planting and again in the spring. If the planting is late, herbicides can be applied in early spring before summer annual weeds germinate.

Annual bedding plants are generally seeded in the greenhouse and transplanted in the landscape bed in mid-spring. In most cases, preemergence herbicides should be applied after transplanting to weed-free soil and then irrigated in. Some research has shown that cultivars of a species can respond quite differently to the same herbicides. If possible, always test any herbicide on a small area first.

Perennials are propagated in several ways — seed, transplants, vegetative division, etc. and are grown in the landscape as well as containers and the field. Most preemergence herbicides should be applied soon after transplanting.

Cutflowers are usually started from transplants, divisions or tubers, but sometimes are grown in the field from seed. For the most part, preemergence herbicides should be applied after transplanting. Research has shown that most field-seeded flowers are *not* as tolerant of the same herbicides that are safe on transplants. To achieve the same level of safety, the herbicide usually should not be applied until *after* plants emerge and are established.

Herbaceous and semi-woody **groundcovers** are generally fairly tolerant of preemergence herbicides. In the landscape, it is crucial that weeds are controlled for the first two years of establishment. Using organic mulches in combination with pre- and postemergence herbicides is usually the most successful strategy. Whatever the choice of mulch material, it should not inhibit rooting or spread of the groundcover.

GROWING SITUATIONS

A chemical weed control program in a commercial or home **landscape** is complicated by the diversity of plants being grown. Bulbs, annual bedding plants, perennials and groundcovers are often planted within a single bed. Good record-keeping — of weeds, herbicides and ornamentals — is important for site preparation, planning the planting and avoiding injury to sensitive species.

A herbicide program should be devised for multi-species **container** operations so that ornamentals tolerant of similar herbicides are grown in the same area. Since preemergence herbicides will not control emerged weeds, containers should be treated shortly after the plants are potted up or the weeds are removed from established plants. For winter annual weed control in the fall, preemergence herbicides should be applied at least two weeks before container houses are covered. *Never apply preemergence herbicides in heated or unheated covered houses or greenhouses. Several herbicides that are otherwise safe, can volatilize under these conditions and cause injury.*

When growing any herbaceous ornamentals in the **field**, the most important weed management jobs are done *before* planting. Good site preparation includes scouting for perennial weeds and controlling them with cultivation and herbicides the season before planting. If the field is heavily infested, soil fumigation should be considered — especially if there are no effective herbicides registered for the crop.

HERBICIDE APPLICATION

INJURY SYMPTOMS

Most of the preemergence herbicides that are registered for herbaceous ornamentals act by inhibiting the normal root development of small weeds before they emerge. In some cases, the ornamental species are inherently tolerant of the chemical, but more often, selectivity and safety is attained by placement. Because most weed seeds germinate within the upper half inch of the soil, surface herbicide applications control them without injury to the ornamental, which has roots normally growing well below the treated zone. Therefore, when injury does occur, it is often manifested by stunted and malformed roots and general failure to thrive. This may be difficult to detect sometimes if, for instance, all the plants in a bed are uniformly injured. Any stress to the plants will exaggerate the symptoms and worsen the injury.

FORMULATIONS AND APPLICATION

For several preemergence herbicides, there is a choice of formulation available. Usually, the **sprayable formulations** (emulsifiable concentrates, wettable powders, dry flowables, water-dispersible granules) are less expensive than granular formulations. These can be applied through a tractor mounted sprayer in field and container operations or by hand-held backpack sprayers equipped with a spray boom. It is important with backpack sprayers to get as uniform an application as possible by maintaining a constant foot pace, even spray pressure and uniform nozzle orifices.

Granular formulations are often used in landscapes and containers where spraying is not possible. Rates for granules should be calculated on an area basis and applied uniformly over the entire area. Granular herbicides should not be applied with a tablespoon to individual containers. This will concentrate the herbicide and increases the chance of plant injury at the same time that it decreases weed control.

SELECTED REFERENCES

Title	Author(s)	Publisher	Date
Weed Control Guides			
Cornell Recommendations — Trees and Shrubs	Cornell Coop Ext	Cornell Coop Ext	1992
Nursery and Landscape Weed Control Manual	Rice	Thomson	1986
Turf and Ornamentals Chemical Reference	C & P Press	John Wiley & Sons	1991/92
Weed Control Suggestions for Christmas Trees, Woody Ornamentals and Flowers	Skroch et al.	N.C. State Ag Ext. Serv. AG-427	1992
Weed Identification Guides			
Guide to the Identification of Dicot Turf Weeds	Scotts	O.M. Scott & Sons	1985
Guide to the Identification of Grasses	Scotts	O.M. Scott & Sons	1985
Identifying Seedling & Mature Weeds in S.E.	Stucky et al.	N.C. State Ag. Res. ServAG-208	1980
The Weeds	Wilkinson, Jaques	Wm. C. Brown Co.	1978
A Field Guide to Wildflowers	Peterson, McKenny	Houghton Mifflin Co	1968
Weeds of the West	Whitson et al.	WSWS/U. Wyoming	1991
In the WeedFact Series ...			
Greenhouse Weed Control	Neal and Senesac	Cornell Cooperative Extension*	
Conducting a Bioassay for Herbicide Residues	Neal and Senesac	Cornell Cooperative Extension*	

* For copies, send stamped self-addressed envelope to: Cornell University Department of Floriculture and Ornamental Horticulture, Plant Science Bldg., Ithaca, NY 14853.

Chapter 28

Integrated Pest Management of Wildflower-Grass Mixes in the Eastern U.S.

John M. Krouse, *Department of Agronomy, University of Maryland, College Park, MD*

CONTENTS

Introduction 315
Goals Selection and Species Selection in Wildflower-Grass Mixes 316
Natural Grasslands, Native Species, and Wildflower-Grass Mixes 317
Precipitation in Natural Grasslands Considered in Wildflower-Grass Mix IPM 317
Fire in Natural Grasslands Considered in Wildflower-Grass Mix IPM 319
Mowing in Natural Grasslands Considered in Wildflower-Grass Mix IPM 319
Soil Disturbance in Natural Grasslands Considered in Wildflower-Grass Mix IPM 319
Grasses for Use in Wildflower-Grass Mixes 320
Pests of Wildflower-Grass Mixes 321
Animal Pests of Wildflower-Grass Mixes 321
Weed Pests of Wildflower-Grass Mixes 321
Weed Control with Tillage 321
Weed Control with Manual Methods 322
Weed Control with Herbicides 322
Weed Control with Fumigant Herbicides 323
Weed Control with Preemergent and Preplant Incorporated Herbicides: Options and Uses 323
Weed Control with Preemergent and Preplant Incorporated Herbicides: IPM Practice 323
Weed Control with Postemergent Herbicides 324
Role of Pesticides in Wildflower-Mix IPM, with Some Concluding Comments 324
References 325

INTRODUCTION

Wildflower-grass mixes are used as ground cover in the eastern U.S. for many reasons. One of the most commonly suggested involves some reduction of the water, pesticide, and mowing inputs associated with turfgrass culture. While wildflower-grass mixes may require fewer of these inputs than conventionally maintained turfgrass, these mixes also present many unique and unconventional problems which complicate their management. Difficulties have often arisen because of inadequately formulated or conflicting goals. As is often the case with "new ideas," the popular acceptance of wildflower-grass mixes has far outdistanced the development of techniques for their effective management.

Many problems associated with the management of wildflower-grass mixes have yet to receive formal study. Technical developments have often been poorly reported, which has slowed the development of management strategies. It is the intention of this discussion to address pest control in wildflower-grass mix plantings from an integrated pest management (IPM) perspective.

An examination of the goals that wildflower-grass mixes are expected to achieve is the first consideration. Well-formulated goals enable the selection of species compatible to the use of the planting. Once the goals of the planting and its component species are established, the next need is to examine the ecology of natural grasslands, as natural factors greatly influence management practices. Since pests are judged as problems to the extent that they interfere with the goals of the planting, they suggest modifications to the management scheme and strategies for their avoidance or control.

GOALS SELECTION AND SPECIES SELECTION IN WILDFLOWER-GRASS MIXES

Since wildflower-grass ground cover is chosen for many reasons, the effective management of these plantings depends on clearly identified goals. Wildflower-grass plantings in the East are typically undertaken in order to accomplish one or more of the following goals:

1. To create an attractive ground cover.
2. To create a ground cover that needs less management input.
3. To protect soil and water from pollution or erosion.
4. To improve or restore plant and animal habitats.
5. To educate and demonstrate prairie-type ecosystems.

It seems likely that the greatest number of people now involved in planting wildflower-grass mixes are those primarily interested in goal #1 (above). Those interested primarily in goals #2 and #3 probably plant the most acreage, while those most interested in goals #4 and #5 are usually engaged in specialized ecological projects which rarely emphasize ornamental landscaping.

Only goal #1, the creation of an attractive ground cover, will be considered in detail here. Though wildflower-grass mixes are commonly planted to achieve a reduction of management inputs (goal #2), the success of this depends heavily on expectations placed towards the achievement of goal #1. Goal #3 will be presumed to be satisfied by typical wildflower-grass ground covers, and not considered further. Goals #4 and #5, though certainly worthy, will be left for experts in natural areas restoration.

The concerns of those interested in creating attractive ground cover (goal #1) typically involve the flowers, the plants, and the whole display. These areas of concern establish the framework for the consideration of IPM methodology since they define the aesthetic and functional expectations of ornamental wildflower-grass plantings. At the University of Maryland, the species in wildflower-grass mixes are evaluated in these three concerns, each with several characteristics required of successful ornamental wildflower mixes:

The Flowers

1. Flowers of bright, showy colors.
2. Flowers of large size (either singly or in groups).
3. Flowers concentrated at the tops of plants.
4. Flowers in bloom over extended periods of time.

The Plants

1. Long-lived perennials, or annuals which will reseed.
2. Bunch-type or weakly spreading vegetative growth pattern.
3. Erect stems of low woodiness and rapid decomposition.
4. Leaf canopy of moderate density.
5. Vigorous growth, but not ahead of species in bloom.
6. Maximum height less than 4 ft.

The Display

1. Species are diverse, each with some unique contribution.
2. Flowers with contrasting colors bloom simultaneously.
3. Flower and foliage colors change through seasons.
4. Floral display continues through entire growing season.
5. Stand density remains high, with uniform component heights.
6. Total live ground cover is high, preferably exceeding 70%.

These characteristics of successful wildflower-grass mixes guide the selection of species, management strategies, and the evaluation of the success of the planting. Wildflower-grass mix IPM must begin with an assessment of these characteristics, so that management plans and thresholds can be determined. To the extent that weeds or other factors interfere with floral displays, they are considered pests. The relative importance of goals #1 and #2 are important in the decision-making process, and must be balanced in the goal assessment of the site.

Those only interested in creating a ground cover that requires less management input (goal #2) or those only interested in protecting soil and water from pollution or erosion (goal #3) may well exclude wildflower-grass mixes from consideration, since wildflowers may not be optimally useful for either purpose. Attractive wildflower-grass plantings often require a significant increase of management skills over that needed to maintain low-maintenance grasses,[60] and may not always be less expensive.[37,44]

However, wildflower-grass mixes are functionally equivalent to low maintenance turfgrass in sites where woody plants must be excluded and where vegetation up to 4 ft tall can be tolerated. When seasonal floral displays are valued, then wildflower-grass mixes may be preferable to turfgrass. The balance between goal #1 and #2 defines the role of the wildflower-grass mix manager. This is the context of the IPM challenge.

Since many factors influence the management of these plantings, an understanding of what is possible to modify and what is necessary to accept is vital to the management of wildflower-grass mixes. After the formulation of goals and the concerns that define them, the next step is to evaluate the environmental conditions that influence wildflower-grass mixes, in order to anticipate problems and to maximize successes. For this, some consideration of the environmental factors at work in the natural grasslands of North America will be found invaluable.

NATURAL GRASSLANDS, NATIVE SPECIES, AND WILDFLOWER-GRASS MIXES

For many wildflower enthusiasts, the re-creation of grassland communities that are now nearly extinct in their natural ranges has been regarded as one of the most important functions of wildflower-grass plantings. The selection of native rather than non-native plants has therefore become a hotly debated topic between those concerned primarily with ecosystem restoration, and those interested in establishing attractive, functional ground covers.[1,2,23,39,59]

Demands for the increased use of native wildflowers in the East are often difficult to reconcile. Many wildflower species have been advocated for use in the East because they are native to North America. Whether the species were originally found in the areas proposed for their establishment is often given little consideration. Many people seem unaware that most of the area east of the Mississippi River Valley was heavily forested, and that the native species were typically adapted to woodland habitats and rarely found elsewhere.[10,57]

In Map 1, the approximate eastern boundary of the natural grasslands is shown. To the north, coniferous forest borders the grassland, and between these regions and the Atlantic Ocean is deciduous forest. The boundaries between forest and grassland are never sharp, especially at the deciduous forest border.[7-10,55,57]

Grassland is not native in the East. The climatic conditions needed to maintain grassland vegetation are not present and many other environmental factors are absent. Thus grasslands and wildflower-grass mixes are out of place over much of the area proposed for their establishment. As a consequence, most of the species now commonly used in eastern wildflower-grass mixes are not native to the areas where they are being planted.

Many of the native eastern species are difficult to establish in the often stressful environments chosen for wildflower-grass plantings. Many of the natives are also less showy than some of the better

adapted introduced species, and so are rarely selected for use. Furthermore, seeds of many native species are simply not available.[2,27]

The use of native species cannot be criticized. Many native species are interesting and showy when reestablished in their natural environments, and are often in danger of being lost otherwise. Roadside banks, golf course roughs, and vacant urban lots, however, are not natural environments, and will not be used for habitat restoration. Rather, what will be planted on these sites will be chosen for its functional and ornamental value. The species best adapted to the conditions and goals of the site will be selected, and the best management practices will be used to maximize their success.

PRECIPITATION IN NATURAL GRASSLANDS CONSIDERED IN WILDFLOWER-GRASS MIX IPM

In the grasslands of the central U.S., several associations of grass and broadleaf plants intergrade with each other and the woodland communities that border this region.[9,10,14,57] The differences between these associations reflect many differences in the natural environment, of which precipitation may be the most important.[10,57] Natural grasslands are rarely observed where precipitation exceeds 40 in. per year. As annual precipitation decreases below 40 in., woody plant competition declines markedly, so that by 30 in. of precipitation grasslands are the predominant vegetation type.[10,56] The transition from 40 to 30 in. of precipitation is therefore critical to the existence and maintenance of natural grasslands, and may be viewed as an indicator of the fitness of a site to support a permanent wildflower-grass community.

The relative drought tolerance of naturally-occurring wildflower-grass mixtures has been a significant factor in their development as low-maintenance ground cover. Early efforts to intentionally establish similar mixes were often made as a response to the inability of turfgrasses to succeed on the same sites.[3,13] In consequence, much of the popularity of wildflower-grass mixes in the East may be traced to their successful "imitative" use on dry or infertile sites with species native to the western U.S.

Though many western wildflower species are intolerant of the relatively wet conditions of the East, many introduced and some native species are well adapted. The selection of sites with well-drained conditions in the full sun enables the use of a wide variety of wildflower species, while reducing top growth and keeping woody plant and other weed competition to a minimum. The exploitation of sunny and dry sites in the eastern U.S. for wildflower-grass

Grassland

Map 1 Approximate eastern limits of U.S. grassland.

mixes has therefore been a successful IPM strategy, and should remain a priority technique.

FIRE IN NATURAL GRASSLANDS CONSIDERED IN WILDFLOWER-GRASS MIX IPM

After rainfall, the frequency of fire is the most significant factor in the ecology of grasslands. Fire becomes much more frequent as precipitation decreases in the natural grasslands. Where the amount of precipitation is sufficient to support trees, fire may be of overriding importance to the maintenance of grassland vegetation.[5,10] A large extension of the grasslands called the Prairie Peninsula, which extended from Illinois to Ohio, may have existed entirely due to the effect of fire on woody species otherwise adapted to the region.[10,55]

When fire is suppressed in natural grasslands, the plant communities experience subtle changes which result in the loss of species dependent on fire for their propagation or establishment. A management technique known as prescribed burning has been developed for the intentional use of fire in natural grasslands.[5,8] This technique has enabled the restoration of mismanaged grasslands and the establishment of new grassland preserves. As a method of rapid grassland regeneration and woody plant control, prescribed burning has no effective equal.

Periodic fires remove vegetation cover and debris at the soil surface, creating the space needed for the germination and establishment of annual broadleaf species. In the natural grasslands, the spectacular displays of color that occur after fires are largely due to the growth of these annual plants.[14,57] In the absence of fire, grasses compete vigorously with broadleaf species, and color displays are substantially reduced.

In eastern urban or suburban settings, however, burning may not be feasible, particularly because of problems with smoke. The smoke produced by intentional grass burns has caused serious problems even in rural areas.[32] Nevertheless, the prescribed burning of small areas under conditions favoring less smoke production may enable more widespread use of this otherwise natural and useful technique.

MOWING IN NATURAL GRASSLANDS CONSIDERED IN WILDFLOWER-GRASS MIX IPM

As a simulation of the effects of grazing and trampling by buffalo (*Bison bison*),[14] mechanical mowing techniques may have some value. Nevertheless, mowing must be regarded as a poor substitute for burning, and one that is more accommodating to the constraints of Eastern wildflower-grass culture than as a management technique of choice in its own right. Eastern wildflower-grass plantings that involve only mowing for vegetation and weed control may have a maximum lifespan of only 8 years.[26] As little as 1 or 2 years is not uncommon when weed pressure is high.

Nevertheless, mowing techniques for wildflower-grass mix management will probably remain much more important than fire in the East. A single mowing at the end of the growing season is now common practice for both vegetation management and weed control. Mowing is usually performed in the late fall after the wildflower seeds have matured and the stems have dried down. This technique suppresses many woody weed species, and usually improves winter appearance, but does little to clear ground for the germination of wildflower seedlings. The cuttings generated by once-per-year mowing may occasionally be very heavy, and may not only inhibit new seedlings, but also harm established plants.

Techniques that involve mowing once at some point in the dry midsummer season,[15,16] or even at several times through the growing season have enabled the use of lower-growing and more consistently blooming wildflower-grass mixes[25,46] and have allowed the use of many less common species.[6] The long-term effects of frequent mowing on the development of short but aggressive weeds such as bermudagrass [*Cynodon dactylon* (L.) Pers.] has yet to be evaluated. In the future, the adoption of a standard twice-per-year mowing schedule seems likely, and may represent the most feasible way to reduce top growth, weeds, and cuttings, while encouraging more constant flowering and diverse mixes.

The choice of mowing machinery may be an important factor in wildflower-grass mix management. Both rotary and flail mowers tend to produce cuttings that dissipate more quickly than those produced by sickle-bar type mowers. Finer cuttings effectively reduce the time plants are covered by cuttings, which enhances their vigor and improves the appearance of the mowed area. In plantings on very fertile or moist soil, or in other sites where plants have grown over 4 ft tall (not uncommon in the East), the use of flail mowers may be superior to other mowing methods.

SOIL DISTURBANCE IN NATURAL GRASSLANDS CONSIDERED IN WILDFLOWER-GRASS MIX IPM

Light plowing for the purpose of disturbing limited areas of an established wildflower-grass mix is a way to simulate some of the agents of soil disturbance in natural grasslands. The buffalo and burrowing rodents such as the prairie dog (*Cynomys ludovicianus*) and woodchuck (*Marmota monax*) were responsible for creating bare-soil germination sites over thousands of acres of natural grassland. Soil disturbance as a management technique may rely on natural reseeding of wildflower-grass species, or seed may be added to the cleared areas. By careful selection of planting dates and the species included in reseeding mixes, precise control of floral displays may be achieved, and management problems reduced.

Skip plowing and other selective ground disturbance techniques will allow the establishment of colorful young plants in sites low in flowering species. Thus, the regeneration of small areas and the removal of weedy spots may be achieved within larger plantings without extensive renovations. Management techniques that combine mowing and ground disturbance operations with replanting in the spring or fall may overcome the hostile seedling environment of many established eastern wildflower sites. Such a series of simple, properly timed, operationally related steps may be necessary to achieve sustainable and colorful eastern wildflower-grass mixes in the absence of fire.

As more specialized ground disturbance and planting techniques are incorporated into the management scheme, large homogenous plantings will inevitably be replaced by smaller units with different wildflower-grass mixes. As a consequence, management intrusions will become less conspicuous, and the diversity and visual appeal of the plantings will increase.

GRASSES FOR USE IN WILDFLOWER-GRASS MIXES

In the grasslands of North America, a few perennial grasses dominate the plant communities. Warm-season grasses are most abundant, and include big bluestem (*Andropogon gerardi* Vitman), little bluestem (*Andropogon scoparius* Michx.), indian grass (*Sorghastrum nutans* Nash), and switchgrass (*Panicum virgatum* L.). Cool-season grasses are less abundant, and include junegrass [*Koeleria cristata* (L.) Pers.] and prairie dropseed (*Sporobolus heterolepis* Michx.).[10,14] Broadleaf plants typically outnumber grass species by three or four to one, but contribute only 10 to 20% of total biomass.[11] Clearly, natural grasslands are aptly named.

Wildflower-grass mixes are not grasslands. As plantings designed primarily for aesthetic appeal they may include grasses for their unique contributions to the mix, but grasses are typically neither as varied nor as colorful as flowering broadleaf plants. They are therefore somewhat less desirable, and tend to be used considerably less than flowering broadleaf species. It is also well to remember that the reason that grasses so dominate the prairies is because they are very hardy, and some are very aggressive. Many are also quite tall: big bluestem easily exceeds 6 ft in height, and other prairie species are even taller.[14] For use in wildflower-grass mixes, such grasses are not only out of place, they are unmanageable.

For wildflower-grass mixes, grasses are needed that are compatible with the growth and display of the principal plants in the mix: the wildflowers. Their use is conditional on their conformance to the characteristics of successful wildflowers, previously discussed. Many grasses possess significant ornamental value, including showy inflorescences and bright autumnal colors. Some must be suited for use in wildflower-grass mixes without being too tall, too wide, or too aggressive.

Many early recommendations for wildflower-grass plantings included some of the prairie species mentioned, while others included such common turf and pasture species as tall fescue [*Fescue elatior var. arundinacea* (Schreb.) Wimm.], orchardgrass (*Dactylis glomerata* L.), and Kentucky bluegrass (*Poa pratensis* L.).[1,13,59] These species are now typically considered pests, and only a few grasses are currently recommended, including sheep fescue (*Festuca ovina* L.), hard fescue [*Festuca ovina var. duriuscula* (L.) Koch.], and chewings fescue (*Festuca rubra var. commutata* L.).[6,12,33,37,52,54] Nevertheless, some authorities still recommend that Eastern wildflower-grass plantings be composed of 50 to 80% midwestern "native grasses."[8,9,39] Such an emphasis on grasses seems counterproductive and unnecessary.

In a study conducted at the University of Maryland from 1989 to 1991, sheep fescue was used in spring- and fall-planted wildflower mixes at the rate of 10 lb/acre (vs. the common turf rate of 160 lb). By the end of the first growing season, the fall-planted mixes had an average of 47% sheep fescue ground cover, and the spring-planted mixes had an average of 20% ground cover. By the end of the second growing season, competition from sheep fescue had become so intense that there was virtually no bare ground, and the wildflower species appeared thinned and stunted. Current studies at the University of Maryland now use only 1 lb of sheep fescue per acre. Nevertheless, sheep fescue seems a marginally useful species for wildflower-grass mixes, because it is usually invisible until after the annual mowing.

There is currently a significant lack of information regarding appropriate grasses for inclusion in wildflower-grass mixes. Of the many grasses found in the U.S., only a relatively small number are grown to supply the seed required for direct seeded plantings, and most of these have been selected for success in turf or pasture uses, not in wildflower-grass mixes.

Many eastern native or naturalized grasses may be valuable for this use. For example, broomsedge (*Andropogon virginicus* L.) is a currently unutilized species which has many of the characteristics needed for successful use in wildflower-grass mixes. A brief description follows, which is included in order to highlight this species, and to suggest some of the qualities needed in grasses chosen for use in wildflower-grass mixes.

Broomsedge is a perennial warm-season grass that makes little growth until the end of the summer, when slender bluish stems bearing inconspicuous flowers are sent straight up to a height of about 3 ft. The plants are compact and do not spread, and each rarely occupies more than 1 ft² of space. When the stems and leaves are killed by frost in the autumn, they turn a bright orange-brown color, which softens through the winter. The color is distinctive and ornamental, and the plants remain attractive as they wave gracefully in the wind. During this time, small fluffy seeds are released, and may often be seen at the tops of the plants. The stems typically remain erect long after new shoots emerge the following spring. Broomsedge is widespread, and will grow nearly anywhere, though it is usually found in drier sites where it can escape more aggressive, taller plants. It is very persistent, and never weedy or invasive.

If grasses such as broomsedge are included, the wildflower mixes will be sufficiently enhanced to justify their inclusion. Most grasses now used in eastern wildflower-grass mixes tend to be rather low in ornamental value, and tend to be aggressive and dominating, and are ultimately about as damaging to the mix as the worst weed species.

PESTS OF
WILDFLOWER-GRASS MIXES

Of all the pests harmful to wildflower-grass mixes, only the problem of weeds has received much consideration. Perhaps weed infestation is so widespread and so frequently overwhelming, that other pest problems have seemed insignificant by comparison. Nevertheless, other pests do affect wildflower-grass plantings.

ANIMAL PESTS OF
WILDFLOWER-GRASS MIXES

Perhaps the most important nonweed pests of wildflower-grass plantings are mammals. Ordinarily, the addition of mammals might be regarded as evidence of the habitat-enhancing role of wildflower-grass plantings. The habitat provided by wildflower-grass plantings has been observed to be heavily utilized by many species of songbirds, including the American goldfinch (*Carduelis tristis*), the Eastern bluebird (*Sialia sialis*), and the mourning dove (*Zenaida macroura*) without any detrimental effect.

One of the most persistent and abundant mammal species, the woodchuck (*Marmota monax*), has been mentioned as a factor in the ecology of native grasslands. In plantings maintained for their ornamental value, however, the large burrows and tremendous appetite of the woodchuck can be very damaging. Since its preferred habitat along roadsides and other areas near human habitation makes it safe from most of its natural predators, the woodchuck can develop large and destructive colonies in wildflower-grass plantings. The woodchuck is probably the most significant nonweed pest in the East.

Compared to woodchucks, the damage done by whitetail deer (*Odocoileus virginianus*) and the cottontail rabbit (*Sylvilagus floridanus*) is less significant, but may still be troublesome. High populations of either may cause considerable damage to wildflower-grass mixes. The trampling and selective browsing of deer is often highly visible and long-lasting. Rabbit-feeding damage is less noticeable, but may be very destructive to seedlings. Fencing is an effective control method for deer, but rabbits can rarely be fenced out. The natural predators of rabbits rarely allow high populations to persist.

WEED PESTS OF
WILDFLOWER-GRASS MIXES

Weed control is of overwhelming importance to the establishment and persistence of wildflower-grass mixes. Though selective postemergent weed control strategies are under development, the simplest and most effective weed control techniques still rely on the reduction of weed populations at the time of establishment. Weed control initiated after establishment is usually more difficult, and complicated by problems of identification and selection. Substantial competitive damage is often done before weed problems are recognized and controlled.

WEED CONTROL WITH TILLAGE

The preparation of a seedbed that has no standing weeds and a low weed seed content, particularly of troublesome species such as Canada thistle [*Cirsium arvense* (L.) Scop.], Johnson grass [*Sorghum halepense* (L.) Pers.], quackgrass (*Agropyron repens* L.), and clovers (*Trifolium repens* L. and others) is essential to both the short- and long-term success of a wildflower-grass planting. All perennial plants should be removed, and especially widespreading colonial weeds and vines such as grape (*Vitis* spp.), Virginia creeper (*Parthenocissus quinquefolia*), poison ivy (*Toxicodendron* spp.), kudzu [*Pueraria lobata* (Willd.) Ohwi.], Japanese honeysuckle (*Lonicera japonica*), and brambles (*Rubus* spp.).

Repeated deep tillage at intervals timed to maximize weed seed germination and prompt destruction is a time consuming but effective technique.[24,33,53] As an alternative to deep tillage, however, it has frequently been observed that shallow or even minimum tillage techniques often result in less weed competition, and that such techniques may enable more rapid and more successful establishment of wildflower-grass mixes.[17,24,35] A single deep tilling without other follow-up cultivation or other weed control methods is rarely successful in the East.[16,24,41] Total preparation time with intensive management techniques may require a minimum of 4 to 6 weeks,[24,53] though many months may not be unreasonable if weed pressure is high.[9,67]

WEED CONTROL WITH MANUAL METHODS

Manual removal has often been suggested for the selective control of weeds in established wildflower-grass mixes. Such methods are often unpleasant and difficult, and the proper identification and removal of weeds from a stand of wildflower plants requires training and a good sense of discrimination. If the weeds are thickly established, considerable damage may be done to the stand by trampling, with little net improvement.

Pulling weeds up by the roots often results in significant damage to neighboring plants, which are frequently uprooted along with the weeds. Basal cutting may be difficult to accomplish due to the interference of surrounding plants, and depending on the species of weed and the timing of the cutting, may allow more or less weed regrowth. The disposal of cuttings, especially when they are abundant, may also hinder efforts at hand weeding.

Though many annual broadleaf weeds may be effectively removed from wildflower-grass plantings with basal cutting, most perennial broadleaf species will regrow, and vegetative expansion will continue. Grass weeds are rarely restrained for long by this method, especially perennial species such as quackgrass and Johnsongrass. Low-growing broadleaf weeds and turf-type grasses are practically impossible to remove by manual methods.

Manual removal of certain weeds can be very effective if they are not too thickly established in the planting, and if they possess stout, solitary stems. Horseweed (*Erigeron canadensis* L.), daisy fleabane [*Erigeron annuus* (L.) Pers.], and a few aster species (*Aster* spp.) can be quickly removed from wildflower-grass mixes with a single one-handed pruner cut made low on the stem. Since these species are often damaging because of their height (up to 6 ft or more), the identification and determination of location and density of these species is relatively simple. The majority of overtopping horseweed and daisy fleabane plants may be rapidly removed from large areas with little effort. A moderately large population of these weeds can be removed from an acre-sized planting in about 1 hour. When undertaken in July or early August the entire plant and its mass of flowers and developing seeds may be removed in a single cutting. Regrowth will occur after a short delay, but the

plants will be largely unable to reach pest size again, and seed production will be greatly reduced. The appearance of the wildflower-grass mix may be dramatically improved. By removing just these species, the maximum height of the stand can typically be reduced by as much as 50%, thereby restoring the underlying height uniformity of the wildflower-grass mix and revealing many more flowers to view. Other competitive broadleaf weeds, including common ragweed (*Artemisia artemisiifolia* L.), pigweed (*Amaranthus hybridus* L.), lambsquarters (*Chenopodium album* L.), and several thistles (*Sonchus* spp., *Cirsium* spp.) that are tall or bushy may also be manually removed, but with considerably more effort due to the lack of high solitary stems.

Annual weeds are most abundant in the first season after planting, and manual weed control techniques are most effective and most justifiable during that time. As the planting matures, perennial weeds typically become more troublesome, and hand methods are much less effective and often unjustifiably expensive in time and effort. When the planting has become thickly invaded by these weeds, a selective herbicidal treatment may provide the only effective control.

However, if the site is also low in flowering species, the re-establishment of a new wildflower-grass mix may be the best management strategy.

WEED CONTROL WITH HERBICIDES

The use of herbicides to control unwanted plants in wildflower-grass mixes is a subject that has only recently received much attention. Though disinclination to use pesticides in naturalized areas has perhaps been significant, even more important has been the fact that few pesticides of any class have been evaluated for safety to the species in wildflower-grass mixes. This, and the fact that few herbicides have been labeled for use in wildflower-grass plantings has made their use virtually impossible. Yet, selective chemical weed control in wildflower-grass mixes offers potentially powerful tools to the wildflower manager. Continued progress and refinement of technique in this area seems certain.

WEED CONTROL WITH FUMIGANT HERBICIDES

Soil fumigation with methyl bromide is an older technique useful for the reduction of weed competition from sites prior to the establishment of wildflower-grass mixes. Metham-sodium and dazomet are somewhat simpler and less hazardous to use, but do not kill seeds as effectively.[19,24,49] Where the increased costs of application can be justified, any of these compounds may be valuable in areas known to be infested with difficult-to-control weed plants or their seeds. Nonchemical soil sterilization with black plastic tarping is a well-developed alternative technique, though it requires careful attention and warm soil conditions to achieve comparable results.[16,33,53,67]

WEED CONTROL WITH PREEMERGENT AND PREPLANT INCORPORATED HERBICIDES: OPTIONS AND USES

The use of selective herbicides applied prior to the germination of weeds is under development as a control strategy. The control of troublesome weed populations before they become damaging is an efficient approach, and may greatly improve establishment and long-term success of wildflower-grass plantings.

As in other crops, the herbicides may be applied prior to the germination of the weeds (preemergently) or prior to both the germination of weeds and wildflower species (preplant incorporated). Either technique may be successful depending on the tolerances of the wildflower species, and of the targeted weeds.[3,16,19,21,26] Since variable damage to desirable species may result, consideration of the risks posed towards each of the component species is needed before these compounds may be used. Some change in the composition of the wildflower-grass mix may be necessary in order to avoid intolerable damage or death of desirable species, particularly when preplant-incorporated herbicides are involved.[19]

Tolerance to a mixture of EPTC and trifluralin (for the preemergent control of grasses) has been evaluated for 50 commonly planted wildflower species, with about 33 species showing minimal or no adverse effects when applied to established plants.[21] Sulfometuron methyl has been reported to give preemergent Johnsongrass [*Sorghum halepense* (L.) Pers.] control for up to 1 year after application to established and newly germinated wildflowers,[3] and will control other weed grass species as well,[26] but is not recommended for preplant incorporation.[15,16] Chlorsulfuron is likewise nonphytotoxic only when applied postemergently to established plantings.[19] Metolachlor,[16,19] pronamide,[16,19] alachlor,[16,19] benefin,[16,19] trifluralin,[16] DCPA,[17] metribuzin,[15] and simazine[49] have been reported effective and without excessive phytotoxicity for the control of grassy weeds when used in preemergent or preplant-incorporated applications in certain mixtures of wildflowers, though several commonly used annual species are intolerant, including cornflower (*Centaurea cyanus* L.), corn poppy (*Papaver rhoeas* L.), and California poppy (*Eschscholzia californica* Cham.).[19] Oryzalin[16,17] and metsulfuron methyl[16] may be unacceptably phytotoxic to wildflower species for any preplant-incorporated or preemergent use.

Since grass species are physiologically quite different from most broadleaf (wildflower) species, the use of preemergent herbicides for grass weed control has shown great promise.[16,24,26] Studies of the ability of herbicides to selectively control unwanted broadleaf weed seedlings in wildflower-grass-mixes by preplant-incorporated or preemergent applications have not been as successful, though unwanted selective effects have been deliberately minimized by excluding sensitive species from the wildflower-grass mix.[26]

WEED CONTROL WITH PREEMERGENT AND PREPLANT INCORPORATED HERBICIDES: IPM PRACTICE

In wildflower-grass mix trials at the University of Maryland, the value of preemergent control for giant foxtail (*Setaria faberii* Herrm.) became apparent after this grass became an overwhelming pest in the first season following early spring planting. Giant foxtail was not a problem after late fall planting, nor in any following season regardless of original planting date. Since sheep fescue (*Festuca ovina* L.) was a constituent of the mixes, the use of herbicides active against cool-season grasses (e.g., metolachlor, EPTC, pronamide) would have been contraindicated. In this situation, knowledge of the mix and the presence and development of weeds at the site would have enabled the avoidance of an unwise planting date, or guided the choice of a selective preemergent herbicide for use in the season after a spring planting. The use of herbicides would have been unnecessary with fall planting or in established wildflower-grass mixes.

Several broadleaf species were troublesome from fall plantings, including spring vetch (*Vicia sativa* L.) and white heath aster (*Aster pilosus* L.). White heath aster, a common and very aggressive late-blooming perennial daisy, may have been the most damaging weed in our trials, and was the only species to substantially increase after the first growing season. It was also the only broadleaf species to become so thickly established that manual removal or spot treatment was impractical. In this case, the use of a fall-applied preemergent herbicide for the selective control of white heath aster would have improved establishment of the planted species, but also significantly enhanced the long-term success of the mix. As in the case of giant foxtail, a change of planting season may have reduced the severity of the weed problem, and herbicide application would likely have been warranted only at the time of planting.

It has been observed that relatively few weeds or desirable wildflower plants germinate and emerge through the dense ground cover of established and undisturbed wildflower-grass mixes. As a consequence, it seems likely that preemergent and pre-plant-incorporated herbicides will be utilized most in new plantings, and not in the regular management of established plantings.

WEED CONTROL WITH POSTEMERGENT HERBICIDES

Postemergent herbicides have long been used in wildflower-grass mixes, particularly for spot weed control. Glyphosate has probably been the most used and most recommended nonselective herbicide, particularly for the control of herbaceous perennial weeds. Many wildflower seed companies and other authorities encourage the use of glyphosate for spot weed control, and to remove unwanted vegetation prior to planting.[6,20,50,58,65]

The postemergent removal of grass weeds from broadleaf wildflower species is well documented and may represent some of the most reliable selective herbicide techniques. There are many currently available compounds with potential for this use. Though sethoxydim and fluazifop have been most recommended for the control of grass weeds,[12,15,16,19,24,41,49] other herbicides may allow even more precise selection. For instance, MSMA has been used to control warm-season grasses such as giant foxtail,[15,16,66] though fenoxaprop may be even more effective and less phytotoxic to both broadleaf wildflower species and cool-season grasses.

The control of broadleaf weeds in wildflower-grass mixtures remains difficult to achieve without adverse effects on nontarget plants' safety, due to the wide spectrum of control of most broadleaf herbicides. The presence of many unrelated plant species in the typical wildflower-grass mix complicates precise selection. Species within the same genus, and often genera within the same plant family are typically affected similarly by most herbicides.[24,26] Ongoing research indicates that the development of herbicide combinations capable of removing selected broadleaf weeds from a stand of other broadleaf plants may be feasible with available compounds.[16,24,26,61] The likelihood of significant development in this area seems particularly promising.

ROLE OF PESTICIDES IN WILDFLOWER-GRASS MIX IPM, WITH SOME CONCLUDING COMMENTS

While questions regarding the proper role of pesticides in naturalized plantings will not be resolved here, it seems appropriate to note again that wildflower-grass mixes are human-engineered environments. Decision making is essential: choices must be made in the selection of plant species and in every aspect of establishment and maintenance. The alternative, at least in the eastern U.S., is a more or less rapid succession of less desirable species, leading ultimately to trees.

The goals of the wildflower-grass planting must be determined at the outset so that appropriate management actions can be undertaken later. In this context, the careful selection of wildflower species, seeding rates, planting dates, and non-chemical vegetation management methods may be all that is needed for the establishment and maintenance of many attractive and functional wildflower-grass plantings. Under these circumstances, the use of any pesticides will be largely unnecessary, and this reduced need must be considered a special virtue of these plantings. Nevertheless, in the environment of the eastern wildflower-grass mix, where human-defined objectives dominate the ecosystem, herbicides will likely remain the most used and useful class of pesticides in the management scheme, however infrequently they may be employed.

The IPM of eastern wildflower-grass plantings depends on an understanding of environmental processes affecting natural grasslands, as well as of management strategies to modify plant population dynamics within wildflower-grass mixes. Recognition of the need for broad interdisciplinary knowledge, careful observation, and flexible management schemes effectively brings to a close the days of the casual "wildflower meadow," and opens instead the new science of wildflower-grass mix cultivation — a science which is quickly claiming a place in the field of turf and ornamental horticulture as unique and fully complex as any other. The developments in this new science, as in every new field, are both rapid and revolutionary, and are of vital importance to the wildflower-grass manager.

In this science, an integrated approach that recognizes both the realities and potentialities of wildflower-grass mix cultivation in the east will not be optional, it will be a defined part of the concept of the wildflower-grass mix.

REFERENCES

1. **Anonymous.** 1980. For the East Coast: meadow lawns. *Rodale's New Shelter,* May/June 1980 pp. 26-29.
2. **Anonymous.** 1990. Wildflowers: is it best to plant native or non-native species? *Landscape Mgt.,* Oct. 1990 pp. 11, 14.
3. **Anonymous.** 1991. Flower Beds, Texas-style. *Vegetation Management,* DuPont Magazine.
4. **Anonymous.** 1992. Wildflowers: gaining popularity in ornamental landscapes. *Turf North,* Aug. 1992 p. 13.
5. **Anderson, B.** 1990. Using prescribed burns as a prairie management tool. *Wildflower,* vol. 3 num. 2 pp. 27-33.
6. **Applewood Seed Company.** 1992. Wildflower Seeds for Landscaping, Wholesale Catalog. Arvada, CO. 23 pp.
7. **Art, H. W.** 1986. *A Garden of Wildflowers.* Storey Communications, Pownal, VT. 290 pp.
8. **Art, H. W.** 1987. *The Wildflower Gardener's Guide,* Northeast, Mid-Atlantic, Great Lakes, and Eastern Canada Edition. Storey Communications, Pownal, VT. 180 pp.
9. **Art, H. W.** 1991. *The Wildflower Gardener's Guide,* Midwest, Great Plains, and Canadian Prairies Edition. 1991. Storey Communications, Pownal, VT. 192 pp.
10. **Barbour. M. G., J. H. Burk, and W. D. Pitts**. 1980. *Terrestrial Plant Ecology.* Benjamin/Cummings, Menlo Park, CA. 604 pp.
11. **Barnes, C. P.** 1948. Soil-grass-conservation: environment of natural grassland. pp. 45-49 in: A. Stefferud, ed. *Grass, The Yearbook of Agriculture.* USDA.
12. **Borden, P.** 1990. Low Growing Wildflowers. *Grounds Maint.,* Feb. 1990 pp. 78, 80, 84, 86, 90, 114, 115.
13. **Brant, F. H., and M. H. Ferguson.** 1948. Safety and beauty for highways. pp 315-319 in: A Stefferud, ed. *Grass, The Yearbook of Agriculture.* USDA.
14. **Brown, L.** 1985. *The Audubon Society Nature Guides: Grasslands.* Alfred Knopf, New York, NY. 606 pp.
15. **Corley, W. L., and A. E. Smith, Jr.** 1990. Evaluation of Wildflower Plant Species and Establishment Procedures for Georgia Road Sites. GDOT Research Project No. 8604 Final Report. University of Georgia. 22 pp.
16. **Corley, W. L., and A. E. Smith, Jr.** 1991. Establishment and Management Methodology for Roadside Wildflowers. GDOT Research Project No. 8909 Final Report. University of Georgia. 21 pp.
17. **Corley, W. A.** 1992. Wildflower Establishment and Culture. Fact Sheet H-92-010 Extension Horticulture, University of Georgia. 4 pp.
18. **Dayton, W. A.** 1948. Cousins and companions: weeds are plants out of place. p. 727-729 in: *Grass: The Yearbook of Agriculture.* USDA.
19. **Dickens, R.** 1992. Wildflower weed control. *Grounds Maint.,* Apr. 1992 pp. 66, 68, 72.
20. **Environmental Seed Producers.** 1991. Landscaping with Wildflowers. Unbound, with Catalog 1991. Lompoc, CA. 58 pp.
21. **Erusha, K. S, C. Fricker, R. C. Shearman, and D.H.Steinegger.** 1991. Effect of preemergence herbicide on wildflower establishment. *HortScience* 26: 209.
22. **Evans, J. A.** 1991. Eye-catching wildflowers offer an attractive landscape alternative. *Lawn & Landscape Maint.,* June 1991 pp. 40-42.
23. **Field, S. S.** 1991. Putney Nursery, Inc.: Native wildflowers for more than 60 years. *Turf North,* Sept. 1991 pp. 38-41.
24. **Gallitano, L., W. A. Skroch, and D. A. Bailey.** 1992. Weed Management for Wildflowers. North Carolina Cooperative Extension Service Leaflet No. 645. 6 pp.
25. **Gillespie, A. R.** 1990. Home meadows confronting weed ordinances. *Wildflower,* vol. 3 num. 2 pp. 12-19.
26. **Harlow, S.** 1991. Stopping weeds in wildflowers. *Turf North,* Aug. 1991 pp. 18-21.
27. **Harlow, S.** 1992. Wildflower seed in short supply. *Turf North,* Mar. 1992 pp. 12-14, 16, 17.
28. **Holcomb, G. B., and H. B. Kerr.** 1987. A small scale agriculture alternative: Wildflowers. USDA Office for Small Scale Agriculture, CSRS. 2 pp.
29. **Hottenstein, W. L.** 1969. Highway Roadsides. pp. 603-637 in: A. A. Hanson and F. V. Juska, eds. Am. Soc. Agron. Series #14.
30. **Leslie, A. R.** 1989. Development of an IPM program for turfgrass. pp. 315-318 in: A. R. Leslie and R. L. Metcalf, eds. *Integrated Pest Management for Turfgrass and Ornamentals.* USEPA, Washington, DC.

31. **Livingston Landscape Architects.** 1992. Why wildflowers... why not? *Landscape Mgt.*, Mar. 1992.p. 42.

32. **Mackin, A.** 1991. Field burning: is it a smoke screen or firestorm? *Lawn & Landscape Maint.*, Nov. 1991. pp. 82-88.

33. **Martin, L.** 1990. *The Wildflower Meadow Book, 2nd Ed.* Globe Pequot Press, Chester, CT. 320 pp.

34. **Martin, L.** 1991. Wildflowers in residential areas adding a natural touch. *Turf North*, Sept. 1991 pp. 10-11.

35. **Martin, L.** 1992. The latest in wildflower seeding techniques. The Wildflower Group of the American Seed Trade Association, Washington, DC. 5 pp.

36. **McIver, T.** 1990. Like a good neighbor. *Landscape Mgt.*, July 1990 pp. 22-25.

37. **Merrill, L. S.** 1990. Wildflowers... gaining ground: Management is a must for success. *Turf North*, Mar. 1990 pp. 4-7.

38. **Merrill, L.S.** 1991. Flowering Meadows. 45 Acres of wildflowers. *Turf North*, March 1991 pp. 4, 5, 7.

39. **National Wildflower Research Center.** 1992. *Wildflower Handbook, 2nd Ed.* Voyageur Press, Stillwater, MN. 304 pp.

40. **North Carolina Department of Transportation Landscape Unit.** 1989. Wildflowers on North Carolina Roadsides. Raleigh, NC. 28 pp.

41. **Ohio Department of Transportation Bureau of Maintenance.** 1988. Wildflower Research on Ohio Roadsides. 31 pp.

42. **Prairie Ridge Nursery.** 1992. Catalog. Mt. Horeb, WI. 23 pp.

43. **Rees, Y.** 1991. *Wildflower Gardening.* Crowood Press, Wiltshire, England. 128 pp.

44. **Roche, J.** 1989. The best of the roadsides: innovation in North Carolina. *Landscape Mgt.*, July 1989 pp. 34-35.

45. **Rodale, R.** 1980. Starting thought: getting value from our lawns. *Rodale's New Shelter*, May/June 1980 pp. 5-8.

46. **Rodgers, P.** 1990. From weedy lawn to flowering glade. *Niche Notes*, vol. 2 Sp. 1990. Niche Gardens Chapel Hill, NC pp. 1-2.

47. **Rose-Fricker, C.** 1991. A scientific approach: national wildflower trial. *Grounds Maint.*, July 1991 pp. 20, 22 ,26, 28.

48. **Sanford, D. L.** 1991. Establishment and evaluation of selected Northeastern wildflower seed mixes. *Wildflower*, vol. 4 num. 1 pp 25-30.

49. **Skroch, W. A., and L. Galitano.** 1991. *Weed Control Plan for Wildflower Plant Beds.* North Carolina Department of Transportation. Raleigh, NC. 128 pp.

50. **Smith, S.W., Sen. Ed.** 1984. *Landscaping with Wildflowers & Native Plants.* Ortho Books, San Francisco, CA. 120 pp.

51. **Sperka, M.** 1973. *Growing Wildflowers.* Charles Scribner's Sons, New York, NY. 277 pp.

52. **Stevenson, V.** 1985. *The Wild Garden.* Viking Penguin, New York, NY. 168 pp.

53. **Stokes, D., and L. Stokes.** 1992. *The Wildflower Book, East of the Rockies.* Little, Brown, and Co., Boston, MA. 95 pp.

54. **Stroud, T.** 1989. Rough and wild: using wildflowers on golf courses. *Grounds Maint.*, Jan. 1989 pp. 28, 30, 32.

55. **Transeau, E. W.** 1935. The prairie peninsula. *Ecology* 16: 423-437.

56. **U.S. Department of Commerce.** 1968. Climatic Atlas of the United States. ESSA, Environmental Data Service. 80 pp.

57. **Vankat, J. L.** 1979. *The Natural Vegetation Of North America.* John Wiley & Sons, New York, NY. 261 pp.

58. **Vermont Wildflower Farm.** 1992. Wildflower Seeds and Gifts Catalog. Carlotte, VT. 11 pp.

59. **Vining, D. M.** 1980. A patch of prairie, creating a lawn of wildflowers and grasses. *Rodale's New Shelter,* May/June 1980 pp. 20-25.

60. **Watschke, T. L.** 1990. Low-maintenance grasses for highway roadsides. *Grounds Maint.*, Aug. 1990 pp. 40, 42.

61. **Watschke, T. L., L. J. Kuhns, N. L. Hartwig, G. T. Lyman, A. E. Gover.** 1989. Herbaceous weed control study. pp. 11-18 in: Roadside Vegetation Management, 3rd year report Penn-DOT.

62. **Watschke, T. L., L. J. Kuhns, N. L. Hartwig, G. L. Lyman, A. E. Gover.** 1989. Wildflower Evaluation Research. pp. 44-56 in: Roadside Vegetation Management, 3rd Year Report Penn-DOT.

63. **Watschke, T. L., L. J. Kuhns, N. L. Hartwig, G. L. Lyman, A. E. Gover.** 1990. Wildflower Evaluation Research. pp. 37-50 in: Roadside Vegetation Management, 4th Year Report Penn-DOT.

64. **Westrick, D.** 1990. Wildflowers bring color, variety to the landscape. *Lawn & Landscape Maint.*, Mar. 1990 pp. 30-32.

65. **Wildseed Company.** 1992. Wildseed Catalog. Eagle Lake, TX. 45 pp.

66. **Wilson, D.** 1990. Wildflower benefits always in bloom. *Landscape Mgt.*, Aug. 1990 p. 54.

67. **Wilson, J.** 1992. *Landscaping with Wildflowers*. Houghton Mifflin Co., Boston, MA. 244 pp.

Integrated Management of Insects: Lawn and Garden

Pruning to control boxwood psyllid, effective
alternative when timing for dormant
oil spraying is missed

Dormant pruning of rose bushes
(photo courtesy of Tim Rhay)

Soap flush to determine webworm populations (photo courtesy of
Grounds Maintenance magazine and Lee Hellman)

Chapter 29

Turfgrass Insect Detection and Sampling Techniques

Lee Hellman, *Department of Entomology, University of Maryland, College Park, MD*

CONTENTS

Introduction 331
Key Locations Monitoring 331
Sampling Techniques 332
    Visual Inspection Techniques 332
        Spot Sampling 332
        Irritant Sampling 332
        Flotation Sampling 333
        Soil Pest Sampling 333
    Passive Sampling Techniques 333
        Black Light Traps...333
        Pheromone or Floral Lure Trap 334
        Pitfall Traps 334
Additional Benefits of Monitoring..335
    Prediction of "Pest Outbreaks" by Adult Activity 335
    "Outbreak" Detections for Adjacent Sites 335
    Habitual Recurrences 335
References 336

INTRODUCTION

Most detection and sampling methods can be classified as either active or passive techniques. Both have the capability to help in *predicting* pest problems or *quantifying* existing damage and pest infestations. The most popular and efficient active sampling method involves visual inspections (scouting). Active sampling systems may be directed at the damaging larval stage. In turfgrass IPM programs, passive systems can use black light or pheromone or mechanical traps to supplement scouting data. Traps best indicate if the pest is present at a specific location or active during the sampling period. Answers to questions such as "What other life stages are present?," "Is the pest causing damage?," or "How extensive is the infestation?" ultimately require a visual inspection. The best monitoring programs provide the most complete information. Such programs utilize both visual inspections and a variety of passive methods.

Regardless of the sampling methods used in commercial turf IPM programs, the classic trade-off becomes time against accuracy or costs. Scouting costs and time can be reduced, and accuracy not sacrificed, if monitoring is concentrated on key pests in key locations. Key pests are those responsible for major turf losses at one particular site. Key locations reflect the behavior of the key pest to habitually select and damage the same landscape sites over time. For example, we often hear the saying, "If you want grubs, I've got the spot, they are always killing the turf at this site." These key sites are considered at high risk for insect damage because they quickly attract and maintain pests due to the combination of turfgrass species, soil characteristics, site topography, and habitat microclimate. In addition, because of the landscape structure, rarity of plant material or sentimental value to the owner, these key areas may also require increased monitoring and lower damage thresholds in order to meet the aesthetic expectations of clients.

KEY LOCATIONS MONITORING

We recommend an intensive examination of these predetermined *key locations* to detect the presence

0-87371-350-8/94/$0.00+$.50

of pests or first signs of damage. Many turf insect pests require warm, moderately dry turf conditions for optimal development. For example, chinch bugs prefer the full sunlight areas on southern and eastern turf exposures. Another prime example of a key location is an area receiving intense reflected sunlight from adjacent glass-walled commercial buildings. These situations frequently have both increased insect and disease problems because of the added heat, extended growing season, and plant stress. For example Japanese beetle adults lay more eggs, and young grubs survive better, in well-watered turf areas in sunlight as compared to dry, droughty areas. However, damage is exacerbated on sandy hillsides receiving intense summer sunlight. During the spring, billbug species will lay more eggs in turfgrass growing adjacent to driveways, sidewalks, or stonewalls rather than in the more open turf. These adjacent areas are generally warmer in the early spring, so the newly emerged, overwintered females concentrate their first egg-laying activities there. First signs of billbug larval damage occur 2 to 3 weeks later at such sites. In southern regions, areas first encountering migrant adult mole cricket damage in the spring will become the most severely damaged in late summer. These examples should demonstrate the need for understanding the biology and behavior in relation to the site.

Another key location that requires special monitoring is an historically infestated area. Every year these areas have higher populations or more frequent occurrences of pest problems. Historical computer records or an experienced manager's memory are important in identifying these situations. If historical documentation is unavailable, you must start correlating data on pest damage to site, weather, and turf characteristics. These sites will, over time, become very evident. In addition to identifying the key locations, this type of detailed factual record keeping will better justify your selection of control tactics. Once a key site is identified, the contributing habitat structure or plant material can be modified to reduce these reinfestation problems and thus in the long term help reduce pesticide usage. In many circumstances the complete elimination of turfgrass or use of a better adapted turfgrass species will solve the problem.

SAMPLING TECHNIQUES

VISUAL INSPECTION TECHNIQUES

To date, visual inspection is the quickest, most accurate, and most frequently used technique for the diagnosis of turf insect problems. The observer, however, must have adequate training in turf management, turfgrass pathology, and IPM in order to make the correct diagnosis. Detection of insect damage or the prediction of pest outbreaks also depends on repeated observation of insect adult activity as well as recognizing "off" colors and other diagnostic symptoms caused by pests in the turfgrass. The frequency of scouting visits depends on the pest complex, client expectations, thresholds, and costs. Our experiences in Maryland indicate a weekly or bimonthly schedule in spring and summer is needed to provide adequate coverage. A monthly schedule is acceptable after mid-August into the fall. The following examples are situations and conditions where direct observations are essential to a successful IPM program.

Spot Sampling

Trained individuals can quickly make accurate observations and pest counts. A 30-second spot count per square foot sample site in 20 or so lawn locations should provide information on the scope of damage, stages of the pest present, and population estimates. Spot counts take relatively more time and require searching the thatch and root zone thoroughly in several locations. All pest species can be detected with this method. Accurate visual inspections only reflect a population response to the environmental and site conditions on the day you scout. For example, at the time of sampling, the weather may be cool or excessively wet. These conditions tend to slow down insect development, daily movements, and response to flushing agents. Generally, samples taken under extremes in temperature and moisture tend to underestimate population levels. This effect can be reduced by increasing the number of sample dates each month, or avoiding sampling in weather extremes. Supplemental sampling with traps should also help provide a better population estimate because the trap sample counts reflect an average over longer periods of time. Traps sample day and night regardless of environmental conditions, personal discomfort, and variability in human sampling techniques. Scouting is usually accomplished during daylight hours, but in hot weather, insects may not be active until late evening and traps will alert you to problems. If a problem is identified via a spot count, one of the following methods is recommended to better estimate threshold populations.

Irritant Sampling

This is usually employed after spot sampling detects problems. This method is more accurate than the 30-sec spot count, particularly when insects are hidden in thatch or cracks in the soil. The irritants are only recommended for sampling highly mobile insect pests.[1,2] Soil pests such as white grubs and billbug larvae are not detected with irritants. Thus we recommend other special soil sampling methods.

Species living in the thatch or soil surface such as sod webworms, cutworms, chinch bugs, and billbug adults respond to both soap and pyrethrin irritant flushing agents. The use of soap to flush mole crickets from the soil is a common practice throughout the South. However, accuracy may be diminished because of variations in thatch thickness, soil temperature, soil moisture, and depth of feeding activity. The following flushing agents are recommended in several turfgrass IPM scouting programs.

> **Soap flush**: Joy®, Ivory®, and Wisk® brands of liquid detergent at the same concentration of 0.25% in 1 gallon of water per square yard sample is highly effective in nonthatch turf situations.
> **Pyrethrin (Pyrenone 6% EC)**: About 0.002% water solution of 6% Pyrethrins will provide comparable results.
> **Synthetic Pyrethrins**: 1 to 2 drops of permethrin or other pyrethroid insecticide in 1 gallon of flush water is also effective. Insecticidal activity will vary among compunds, so test each new compound and adjust the concentration.

Use of the irritant method requires a thorough soaking of both the thatch layer and soil surface. Close observation is needed in order to detect the excited insects. Insects normally exit the thatch within 5 to 6 min, but may also move out of the sample area quickly. Be careful in mixing because the higher concentrations of synthetic pyrethrins will kill the younger insects, and the population density may be underestimated.

Regardless of irritant used, the turf should be mowed or clipped before making irritant treatments.

Flotation Sampling

This is used primarily for estimating chinch bug populations. A 1- or 2-lb size open-ended coffee can, or special 1-ft^2 area cylinder with handles can be used. The cylinder is forced into the soil through the thatch layer, and 3 to 4 in. of water is added inside. If the water recedes, add more water to maintain a high level. All stages of the chinch bug and the principal predators such as the big-eyed bug usually float to the surface within 5 to 10 min.

Although this method is very accurate, it is also time consuming. This method is best suited to confirm chinch bug infestations and determine the extent and precise population density of infestations.

Soil Pest Sampling

This is the most difficult and time consuming of all the above-mentioned methods. However, it remains to date the most accurate for determining white grub and billbug larval population densities.

The quickest soil sample is made with a sod cutter. During the summer and early fall, population estimates are made by removing a series of 1- to 2-in.-thick sod slabs and counting the unearthed grubs. Unfortunately some grubs will be missed if they feed at the thatch level, or well below the 2 to 3 in. level. This becomes a particular problem with species that move rapidly up and down in the soil profile in response to varying moisture levels. European chafer, green June beetle, and oriental beetles require more extensive digging. However, Japanese beetles and masked chafers, the two most important species in the mid-Atlantic area, tend to concentrate at the 1- to 2-in. soil level and are easily sampled. This method becomes less accurate in the late fall and early spring because grubs are moving up and down within soil profiles.

The 1-ft^2 sod spade method is less destructive. A square-foot sample is cut on three sides to a depth of 3 to 4 in. and the sod square is folded back to expose the soil. The soil is then broken apart, the grubs counted, and the soil returned to the hole. The grub counts are recorded, and the sod flap is tamped down on top of the loose soil. This method is more accurate than the sod cutter, but many samples are required for a good population estimate. Adjustments in the depth of cut can be made for fall and spring population estimates, when grubs are more widely distributed throughout the soil profile.

The quickest and least destructive method utilizes a cup cutter. A standard golf course cup cutter will allow more samples to be taken and thus increase the sampling accuracy. The standard cup cutter is 4.25 in. in diameter. To convert grub counts to a square foot basis, multiply each sample by 10.15. Soil cores to a depth of 3 to 4 in. are first taken at key locations. If grub activity is detected, more extensive sampling will be required.

PASSIVE SAMPLING TECHNIQUES

Light traps, mechanical traps, and pheromone traps are best used in areawide (e.g., golf course, community, or public park) IPM programs. Although less accurate than the visual method, they are still useful in monitoring a pest's presence or yearly and seasonal population fluctuations. They also aid in scheduling peak scouting activities.

Black Light Traps

These systems can collect large numbers of sod webworm, masked chafer, and other white grub species. However, to date we have no way of estimating the damage potential, or the resulting larval population, from these adult counts. A 15-W black light size and trap design standard does exist for monitoring agricultural pests, but the high cost of $200 to $300 per trap makes this option prohibitive for most individual lawns and small landscape situations. Cheaper commercial or homemade traps have also worked well. Regardless of design, the light trap can be very useful for monitoring first occurrences of pests, or delineating the species distribution over a large geographic area, e.g., neighborhood, golf course, national park, community, county, or state. This survey tool also helps in determining the relative abundance of species and the risk of damage from one year to another. Regardless of which trap is used, homemade or the USDA standard, the light trap must be uniform in design, light wattage, height above the turf, and site location in order to make these comparisons. We highly recommend the 15-W black light fluorescent tube regardless of trap design.

Pheromone or Floral Lure Trap

The most widely used trap in turf IPM programs is the Japanese beetle trap. This trap utilizes a combination of floral lure and a female sex pheromone. As a control tactic, its use is limited by the high price per trap and the excessive number required per acre to significantly reduce the adult beetle populations. But used as a monitoring tool similar to the blacklight traps, the floral lure baited traps will be invaluable in new neighborhoods to detect the buildup of Japanese beetle populations, and to monitor population variations between geographic locations and between years or from one year to another. The sex lure actually increases (10 to 40%) the number of male beetles collected. When using the traps for population monitoring we recommend not using the sex lure component. Without the sex lure, trap catches will more accurately reflect the normal 50:50 sex ratio. Monitoring the females is more important because they lay the eggs that give rise to the damaging grub population. In addition to adult beetle monitoring, the Winsome fly, *Istocheta aldrichi*, an important adult parasite, can also be monitored and collected alive inside beetles for redistribution into new communities or geographic locations.

Daily beetle trap catches will be strongly influenced by temperature, rain, location to host plants, soil type, groundwater levels, and natural enemies. Using several traps at each sampling location will help reduce this variability due to unique trap site locations.

Pheromones for several other species of annual white grubs are now being tested and may be on the market within a few years. Otherwise, lures for species such as black cutworm, true armyworm, and fall armyworm are now available for use. Unfortunately, little progress is being made on the sod webworm species other than the cranberry girdler, *Chrysoteuchia topiaria* (Zell).

This sampling tool will become more important in the near future when additional pheromone systems are marketed for turf insects.

Pitfall Traps

Pitfall traps are primarily used to monitor billbugs, mole crickets, chinch bugs, and other highly mobile arthropods. The traditional design is a small hole or pit lined with a smooth sided container. The bottom is filled with a killing liquid such as soapy water or alcohol. The killing agent is not required for our standard turf pitfall trap. We suggest that small pin holes be punched into the bottom of all the cups to allow rain water drainage.

The trap line set-up is time consuming but the daily or weekly season-long inspection is quick. This trap style is best utilized for monitoring first occurrence or the length of the adult billbug activity period during the spring. The basic design is simple: insects drop inside the trap and cannot crawl out. Two variations are figured, however the 16-oz. plastic cup trap is the most commonly used of the two (Figures 1 and 2).

With the specialized mole cricket drop, the mole crickets work their way through the thatch layer, fall into the 1 in. slots and eventually crawl down the pipe into the plastic jug. Traps should be emptied once or twice a week and the females dissected to check for egg development. Maturing eggs indicate that irritant flushes should be applied within 1 to 2 weeks beyond the peak egg-laying period to verify threshold population levels.

A unique modification of the pitfall trap is used in the southern states to monitor mole crickets.[3] A wading pool 1.5 m in diameter, partially filled with water, and an electric calling device

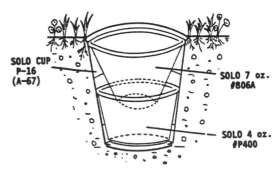

Figure 1 Standard Solo® cup pitfall trap.

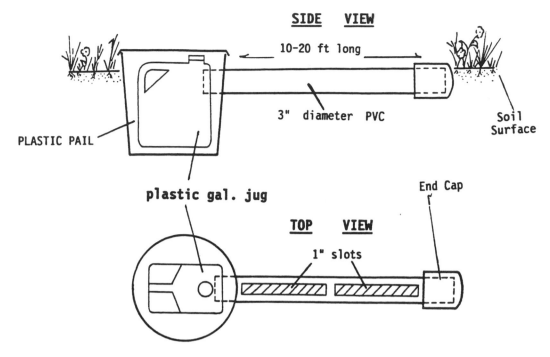

Figure 2 The Lawrence soil arthropod linear pitfall trap.[4] *(From Lawrence, K. O.,* Fla. Entomol., *65, 376, 1982. With permission.)*

suspended over the pool are used to attract adult female crickets. Each mole cricket species responds to a species-specific sound. Attracted females fly to the synthetic call and collect on the water surface. Daily counts during the flight period help determine first occurrence as well as seasonal and site abundance in comparison to pervious seasons' trapping data. At present, accurate damage predictions based on this trap type are not reliable.

ADDITIONAL BENEFITS OF MONITORING

Trap methods become more valuable if used in conjunction with a degree-day heat unit life history model. Degree day models are being developed for the bluegrass billbug, hairy chinchbug, and sod webworm. Publication is expected in 1993–94.

PREDICTION OF "PEST OUTBREAKS" BY ADULT ACTIVITY

Regardless of the method, traps or visual observations of adult activity of mole crickets, Japanese beetle, sod webworm, or billbugs can help in predicting problems. High counts may or may not imply high risk of damage later in the season. Generally, adult females are active and lay eggs 2 to 4 weeks before the immature stages start to cause noticeable damage. For example, if one billbug adult is observed per minute of observation, sufficient egg laying and subsequent larval damage will occur and preventive treatments may be required. Dr. David Shetlar[5] observed in Ohio that if billbug pitfall trap counts range from 2 to 5 adults per day over the peak egg-laying period, this could equate to moderate or spotty turf injury later in July. More severe losses can occur if counts exceed 7 to 10 per day over several days. Our observations with billbugs in Maryland sod farms tend to substantiate these threshold levels.

"OUTBREAK" DETECTIONS FOR ADJACENT SITES

Detecting injury or pest activity in adjacent sites may indicate future problems. Damage to adjacent lawns may in time spread to undamaged turf by migrating pest populations and subsequent generations. This frequently occurs with chinch bugs and billbugs because they rarely fly (they slowly walk from one area to another). Thus, the lawn areas immediately adjacent to severely damaged turf will be at risk and should be monitored more frequently.

HABITUAL RECURRENCES

Many turf sites may be at high risk for reinfestation. Often moderate to severe outbreaks of a pest occur repeatedly in the same site over several years. These key locations should receive more intensive special monitoring considerations. Pests are cyclic and outbreaks may run for 2 to 4 years in succession and then disappear. Predictions and management decisions based solely on historic occurrences are not recommended because changing biological and environmental conditions can rapidly eliminate insect populations.

REFERENCES

1. **Tashiro, H., Murdoch, C. L., and Mitchell, W. C.,** Development of a survey technique for larvae of the grass webworm and other lepidopterous species in turfgrass, *Environ. Entomol.*, 12, 1428, 1983.

2. **Niemczyk, H. D.,** *Destructive Turf Insects,* H.D.N. Booker, Wooster, OH, 1981, 48 pp.

3. **Walker, T. J.,** Sound traps for sampling mole cricket flights (Orthoptera: Gryllotolpidae: *Scapteriscus*), *Fla. Entomol.*, 65, 105, 1982.

4. **Lawrence, K. O.,** A linear pitfall trap for mole crickets and other soil anthropods, *Fla. Entomol.*, 65, 376, 1982.

5. **Shetlar, D.,** personal communication.

Chapter 30

Use of Insect Attractants in Protection of Ornamental Plants

Whitney S. Cranshaw, Department of Entomology, Colorado State University, Fort Collins, CO

CONTENTS

Introduction 337
Use of Pheromones in Management of Insect Pests of Ornamentals 338
    Detection of Exotic Species 338
    Monitoring Endemic Species 339
Use of Pheromones and Attractants as Direct Controls of Insect Pests of Ornamentals ... 340
    Mass Trapping 340
    Mating Disruption 341
References 343

INTRODUCTION

The past 25 years have seen a tremendous increase in our understanding of how chemicals are used in insect communication and other aspects of their behavior. Throughout, an underlying interest in the field has been to learn how these chemicals might be used in managing insect pests in a more selective manner, minimizing effects from more ecologically disruptive controls such as use of insecticides.

Particular attention has been directed at insect **pheromones**, the chemicals used to communicate between members of the same species. Pheromones are involved in mate-finding, modifying egg-laying patterns, expressing alarm among colony members, and other behaviors in arthropods (Table 1).

Of these, the **sex pheromones**, which are used to find mates, have received the greatest development as potential controls for pest species of ornamental plants. Sex pheromones are widespread among the insects. Sex pheromones produced by some species of female moths can be detected by receptive males at concentrations of only a few molecules, making them among the most biologically active chemicals known. This has allowed their use as lures in traps or to disrupt mating in a variety of pest management strategies. The recent review by Inscoe et al.[1] lists almost 250 arthropod species for which pheromones were available in 1988.

In most insects, sex pheromones are produced by the female to attract the male. Among the Lepi-doptera, pheromones consist of various alcohols, aldehydes, and acetates. Terpenes are used by most bark beetles. Typically, the pheromone consists of a blend of various chemicals, and the relative proportions of each compound greatly affect its attractiveness. Furthermore, chemicals from the host plants may also be involved in attraction. For example, alpha-pinene is produced by trees wounded by female southern bark beetles (*Dendroctonus frontalis*) during the initiation of an attack. Combined with the weakly attractive sex pheromone produced by the female (frontalin) it creates a powerful attractant for male beetles.

Pheromones are also used to coordinate patterns of egg laying on host plants. Bark beetles, which usually must mass attack trees in high numbers to sufficiently overcome tree defenses, often use **aggregation pheromones** during early stages of attack, to draw in other beetles onto single trees. As attacks progress and the tree is successfully colonized, the beetles may then use anti-aggregation pheromones (also known as spacing or epideictic pheromones) which cause new beetles to avoid the colonized log. Pheromones with similar activity are also produced by fruit flies, such as the apple maggot (*Rhagoletis pomonella*), to deter additional females from egg laying so as to avoid having excess numbers of young try to develop in a single fruit.

Various **alarm pheromones** are produced by other insects. Social bees and wasps use alarm pheromones to coordinate attacks on intruders which

0-87371-350-8/94/$0.00+$.50

Table 1   Common types of pheromones used by insects and their general effects on behavior

Pheromone Type	Affected Behaviors	Associated Groups of Plant Pests
Sex pheromones	Long and short range location of mates	Most moths, bark beetles, scarab beetles, sawflies, some scales, spider mites
Aggregation pheromones	Coordinate mass attacks on plants for egg laying	Bark beetles
Anti-aggregation (spacing) pheromones	Space egg laying to avoid excess larval competition for food	Bark beetles, tephritid fruit flies
Alarm pheromones	Coordinate attacks in colony defense; warn siblings of natural enemy attacks	Social bees and wasps; some aphids
Trail-marking pheromones	Allow return to colony, plant, or other site	Ants, some caterpillars

threaten the hive. Some aphids produce alarm pheromones in response to attack by predators or parasites. These then have the effect of causing nearby aphids to move away or even drop from the plant.

In addition, there are a wide variety of insect attractants that are not pheromones but have been used in insect management. Most of these are food-based attractants or involve floral attractants. For example, the Japanese beetle, *Popillia japonica*, is commonly trapped by a blend of phenethyl proprionate, eugenol, and geraniol in a 3:7:3 ratio.[2] Methyl eugenol is the attractant used for managing the Oriental fruit fly (*Dacus dorsalis*), and hydrolyzed protein baits attract other tephritid fruit flies, such as the apple maggot or western cherry fruit fly (*Rhagoletis indifferens*). These attractants can sometimes be used in a similar manner as the pheromones for insect management.

USE OF PHEROMONES IN MANAGEMENT OF INSECT PESTS OF ORNAMENTALS

Pheromones and attractants have been used in a variety of strategies for managing insect pests affecting ornamental plants (Table 2). In general, these have included detection of exotic species, monitoring endemic species to improve control, and use as direct control agents. Several commercial pheromone products are currently marketed or are in development and can be used in insect management. Primary manufacturers include the Hercon Division of Health-Chem Corporation, Consep Membranes, Inc. (Bio-Lure™, Check-Mate™), Phero Tech, Inc., Scentry, Inc. (Scentry™), Shin-Etsu Chemical Company, Inc., and Trece, Inc.

(Pherocon™). Of the 21 pheromones currently registered in the U.S. by the EPA, 13 have potential application for control of insects and mites affecting ornamental plants.[3]

DETECTION OF EXOTIC SPECIES

Because they are so attractive, pheromones have proven to be highly sensitive lures to detect the presence of insect pests. This has been a tremendous boon to regulatory entomology efforts to track introduction and movement of new pest species. Currently, traps baited with pheromones or other attractants are regularly maintained at ports of entry to monitor for the possible introduction of new pest species into the U.S. This can then allow eradication efforts at an early stage of infestation, greatly increasing chances of success. For example, there have been several eradication programs to control various fruit flies, such as the oriental fruit fly and Mediterranean fruit fly (*Ceratitis capitata*), following their introductions in Florida and California. These have

Table 2   Ornamental insect pests[a] for which commercially available pheromone lures are useful for treatment timing

Peachtree borer[b]	Lilac/ash borer[b]
Oak borer[b]	Dogwood borer[b]
Rhododendron borer[b]	Banded ash borer[b]
Lesser peachtree borer	Nantucket pine tip
San Jose scale	moth

[a] In addition, Daterman[25] cites 12 additional instances where pheromones are used operationally for management of coniferous forest pests in the western U.S.; [b] Species trapped by the generic "clearwing borer" lure of Z,Z-3,13 octadecadienyl acetate.

prevented permanent establishment of these insects with tremendous savings in subsequent need for costly controls.

Pheromones and attractants are also used to monitor and further prevent the spread of species that are already locally established. Regulatory agencies such as nursery inspection services or state departments of agriculture rely heavily on such traps to detect movement of serious pest species, such as the Japanese beetle and gypsy moth (*Lymantria dispar*), that have the potential to spread and colonize other areas of North America. Traps can detect incipient infestations and enable eradication efforts. For example, the gypsy moth has been eradicated successfully from several locations in Colorado following accidental introductions during the late 1980s. Pheromone traps were not only used to detect and demarcate the initial infestations, but also were used to assess the effectiveness of control efforts. Failure to capture any male moths in pheromone traps for two consecutive seasons is generally considered an indication that the insect has been eradicated.[4]

MONITORING ENDEMIC SPECIES

Probably the most important use of pheromones in pest management for ornamentals has been for monitoring the seasonal biology of pest species that already are established in an area. Pheromone traps can be used to detect critical periods of insect activity, notably adult emergence and egg laying. This information can be important in appropriately timing control treatments. There has also been some use of attractant traps as tools to predict the need for treatments, i.e., as indicators of action thresholds.

Pheromones are most useful for monitoring pest species that tunnel into plant parts and that have egg hatch that follows soon after the onset of adult activity and egg laying. This is particularly true for clearwing borers, which are serious pests of many trees and shrubs. Many of the more damaging species are attracted to lures containing the attractant Z,Z-3,13 octadecadienyl acetate, commonly sold as the "clearwing borer lure." The value of monitoring clearwing borers has long been recognized,[5] and traps are widely used to detect flight activity of adult moths. Flight peaks can vary by several weeks over relatively small geographic areas; e.g., flights of the lilac/ash borer (*Podosesia syringae*) typically varied 3 weeks in the northern vs. southern Front Range area of Colorado.[6] Seasonal variation is also considerable with this species. Therefore, treatments can be timed much more accurately when using pheromone traps than with using calendar-scheduled sprays. Most state extension recommen-

dations suggest an application timing of 10 to 14 days after the capture of the first male moth, although a longer lag in treatment can also be effective with this species.[7]

Use of pheromones for monitoring insect activity has received most widespread attention among tree fruit crops, particularly apples. Many future uses in ornamental plant protection may be derived from experience with tree fruit crops. For example, pheromone trap captures can be linked with degree day models to more accurately predict egg laying. The BioFix model developed at Michigan State University predicts egg laying by the codling moth (*Cydia pomonella*). The model is initiated by pheromone trap captures that indicate onset of sustained adult flight and is designed to optimize the time of the first spray (13% egg hatch).[8]

There are some technical problems in pheromone monitoring of landscape insect pests. Although most pheromones are highly specific in their attractiveness, other species may be trapped accidentally and be misidentified. Furthermore, some attractant lures do show considerable cross attraction among related species, e.g., the "clearwing borer lure" may capture a half dozen or more species at a single site. Since most trap users are not highly trained in insect identification, assistance in distinguishing insects captured in pheromone traps is an important consideration in their effective use. Certain public perceptions involving pheromones also must be clarified with educational efforts. For example, many persons do not understand that attracting male moths to traps will not (at least in the case of Lepidoptera) result in greater pest injury.

Advances in chemical analysis techniques and bioassays have provided identification of the sex pheromones of most economically important Lepidoptera in North America.[1] Many pheromone suppliers will produce special orders of lures that are not normally advertised. However, complete identification of pheromones for many important pests of ornamentals is still lacking, notably several of the tip moths in the genus *Rhyacionia* or economically important *Dioryctria* species, such as the zimmerman pine moth (*D. zimmermani*) and *D. ponderosae*. Recently, promising progress toward identification of attractants for the latter has been reported.[9]

Use of pheromone trap captures as the basis for action thresholds of insect pests has been a less successful adaptation of pheromones for insect control. It has proved difficult to quantify insect populations based on trap captures alone, due to such factors as amount of competing pheromone produced by female insects, wind conditions, temperature conditions, trap location, and variable rate of pheromone released by the lures. For example, peak pheromone trap captures of the spotted tentiform leafminer (*Phyllonorycter blancardella*) were not correlated with peak emergence of the moths nor were changes in trap captures found to be reliable indicators of size of population change.[10] Furthermore, adult populations are also not reliable predictors of subsequent larval damage, since egg production and larval survival is so highly variable among insects. Early research attempts to use attractants associated with females of the southern masked chafer (*Cyclocephala lurida*) indicate that traps might be better predictors of subsequent populations when populations tend to be low, such as in home lawns, than in a golf course setting where high populations typically occur.[11] Fundamental to better quantification of insect populations with pheromone traps is an understanding of how insect behavior interacts with trap design and placement, a subject most recently reviewed by Muirhead-Thomson.[12]

Where pheromone trap captures have been used as action thresholds, the thresholds have been conservative. One of the few such uses in fruit tree crops involves the codling moth. In Michigan, trap captures of at least five moths per generation are needed as a minimum to justify treatments.[8] A forest example with potential for landscape applications is the use of pheromone traps to monitor Douglas-fir tussock moth (*Orygia pseudotsugata*). Pheromone trapping was found to provide a poor correlation with egg mass or larval densities the subsequent season. However, a threshold of 25 moths/trap (baited with a dilute 0.01% lure) provided a useful threshold for indicating a possibility of defoliation of the stand.[13] Pheromone trap captures (threshold of 25 moths/trap/day) have also been used to initiate sprays for cranberry girdler (*Chrysoteuchia topiaria*) in commercial fields of bluegrass.[14] Similarly, in landscape insect management very low trap captures might also be used as an indicator that treatment is not necessary.

USE OF PHEROMONES AND ATTRACTANTS AS DIRECT CONTROLS OF INSECT PESTS OF ORNAMENTALS

Several different approaches have been attempted using insect attractants as direct controls for pests. Some of the oldest uses have involved efforts to trap-out pests using lures and thus reduce the subsequent populations. As pheromone chemistry and formulation has advanced, more recent attention has focused on using insect pheromones to disrupt mating or host finding.

MASS TRAPPING

Nonpheromone attractants have been most widely used for trapping large numbers of pest insects for the purpose of effecting control. For several decades, traps baited with a floral lure have been used for control of Japanese beetles. The addition of the synthetic sex attractant of the Japanese beetle, Japonilure, makes an even more attractive lure.[2,15] Traps containing this lure have been sold under trade names such as Bag-a-Bug for over a decade. Although these traps are capable of trapping large numbers of Japanese beetles, they have little effect on reducing local beetle populations since the presence of favored host plants appears to be far more important in concentrating beetles.[15] Furthermore, studies have shown small-scale use of such traps may actually increase the risk of injury to adjacent vegetation.[16]

Use of insect sex pheromones in trap-out strategies have been even less successful. This is primarily related to the occurrence of highly attractive pheromones which are only active on males. Pest population reduction requires elimination of females before eggs are laid or reduction in mating success by annihilation of the male population before mating. For species where females produce the sex attractant, mating disruption is usually considered to be the more likely control approach using pheromones.[17] Roelofs et al.[18] demonstrated the obstacles to success with the latter, using a theoretical situation based on experience with the redbanded leafroller (*Argyrotaenia velutinana*), an important fruit pest. Given that a 95% reduction of the males is needed to reduce fertilization of eggs, a 2:1 pheromone trap:male ratio would be needed to provide substantial control in this example. Even higher theoretical trap ratios have been postulated for other species.[17] Furthermore, because of the effects of competing pheromone sources, expected control using attraction/annihilation techniques drops off dramatically as population density increases.[19]

However, for species where the females, or both sexes (e.g., bark beetles) can be attracted to a lure, chances for success increase. Trap-out strategies have been tested extensively for control of forest pests, notably the spruce bark beetle, *Ips typographus*, in Norway. One large-scale test involved use of over 600,000 traps. Individual traps captured an average of over 7400 beetles, but results were mixed. Evaluation of the trial suggested that the technique "probably will not suppress outbreaks in overmature stands after an outbreak has started…" However, there was still thought to be value in further reducing low-level populations of bark beetles enough so that the insects did not become abundant enough to overcome resistance mechanisms of the host trees.[20]

Mass trapping was also attempted for control of the European elm bark beetle (*Scolytus multistriatus*) in the Loveland-Fort Collins area of Colorado and in Evanston, IL, during the late 1970s. Again, although large numbers of beetles were captured, citywide use of the pheromone traps was not considered successful in suppressing incidence of infection by the fungus *Ceratocystis ulmi*, causal agent of Dutch elm disease.[21] The technique was thought to have greater potential for Dutch elm disease management in isolated groves. Suppression of twig feeding has been observed during trap-out trials involving *S. multistriatus*.[17]

Further, practical considerations of trap-out strategies involve their expense, difficulties in use, and effects on landscape appearance. Most traps are expensive, particularly at retail rates available to single homeowners. Many involve luring insects to sticky surfaces, which can be unpleasant to handle and to clean. Also, the traps can be conspicuous features on a landscape and may detract from the appearance of an ornamental planting.

MATING DISRUPTION

A substantially more promising use of sex pheromones for insect control is their use as mating disruptants. This tactic involves permeation of the atmosphere with pheromone so that the sexes are incapable of finding each other to mate. Although the exact means by which this is effected is rarely understood, several possible mechanisms include sensory adapation or habituation to the pheromone following high levels of exposure, competition between natural and synthetic sources of the pheromone, and camouflage of the natural plumes of pheromone.[22] Developments in formulation which allow consistent release rates of the pheromone and application techniques which provide even distribution have been critical to the success of this technique.

Effective use of mating disruption often requires that several conditions exist involving both the insect and the plant setting. Most important, the species must produce a powerful sex pheromone which can affect insect behavior over long ranges. Mating disruption is more likely to occur if adult females of the pest species are nonmigratory, so that mated females from outside a treated area are not a significant source of reinfestation. Because it is also important to permeate the area used by mating females, thorough distribution at a sufficiently high concentration is required. This makes isolated plantings more amenable to protection by mating disruption than plants that are widely used in local landscaping. Also, thorough permeation of pheromone may more easily be achieved on smaller plants, which present fewer difficulties in application.

The first successful uses of mating disruption involved pests of agricultural crops, often with a limited host range. These included artichoke plume moth (*Platyptilia carduidactyla*) in artichoke, pink bollworm (*Pectinophora gossypiella*) in cotton, grape berry moth (*Endopiza viteana*) in grapes, tomato pinworm (*Keiferia lycopersicella*) in processing tomatoes, and oriental fruit moth (*Grapholitha molesta*) in tree fruit crops. Commercial formulations of pheromones for use as mating disruptants are available for all these species, and some receive extensive use in pest management.

One limitation to the use of insect pheromones as direct control agents has been their classification as pesticides under the Federal Insecticide, Fungicide, and Rodenticide Act. Under Title 40 of the Code of Federal Regulations (40 CFR) they are included within the pesticide class of attractants. Such regulation is not conferred on pheromone uses which involve attraction of pest insects to traps where the pheromone is the sole active ingredient, but does apply to pheromone uses as mating disruptants. Pheromones which are identical to natural pheromones produced by insects are classified as semiochemical pesticides and may meet a reduced set of registration criteria. However, the majority of pheromone products are synthetically produced and often chemically different from the natural pheromones.[3]

Some of the first trials involving pheromones for mating disruption were targeted against the gypsy moth, an insect pest important in landscape plantings as well as forests. In forest settings, numerous trials have been conducted with variable success.[23] Applications to forest settings or to tall trees provide technical difficulties in applying the pheromone uniformly, given the great diversity of vegetation in forest systems. Also, gypsy moth males are capable of visually finding females, which occurs with greater success in high populations. The technique is considered to be useful for maintaining control where low gypsy moth populations occur. This could also include relatively small sites of landscape plants which are threatened by gypsy moth injury. However, most treatments for control of gypsy moth are made during outbreak conditions, which do not favor successful use of mating disruption.

Peachtree borer (*Synanthedon exitiosa*) and lesser peachtree borer (*S. pictipes*) are clearwing borers which occur as serious pests in both stone fruit orchards as well as in many ornamental *Prunus*. Snow[24] suppressed both species in Georgia peach orchards using an appropriately timed single application of the pheromones, at a rate of one lure per tree. He pointed out the need to treat all adjacent infested orchards, since mated females from these sites could move into the treated area. Similarly, control using mating disruption in landscape settings will be limited where nearby sources of the insect occur. Areawide use of the pheromone, involving entire neighborhoods in residential settings, will likely be needed to effect control in ornamental plant protection.

Mating disruption has also been attempted with the Douglas-fir tussock moth, a pest of forest as well as landscape plants. In aerial applications to large (16-ha) blocks of white fir/ponderosa pine forest areas, mating success as measured by egg production was reduced 74%, and there was a 68% reduction in larval populations the following season. Furthermore, numbers of male moths trapped the following season were very greatly reduced. The authors thought this latter result indicated that there was enough residual pheromone left to have measurable effects for a second season. Formulations that are effective for 2 years seem technically feasible for this species. No effects on the egg parasitoid *Telonomus californicus* were observed from pheromone treatment.[25]

Several tip moths and shoot borers also occur as pest problems in landscape plantings. In pine plantations, mating disruption has been attempted successfully for control of western pine shoot borer, *Eucosoma sonamana*.[26] This species showed considerable flexibility in what was required of a successful mating disruption dispenser. Treatments using a highly active lure that closely mimicked the natural pheromone blend typically suppressed infestation rates more than 75%. Suppression was slightly lower, but still acceptable (60 to 65%) with use of a low-cost blend with a substantially different isomeric ratio from the ideal. Daterman[26] pointed out that some "assistance may be necessary to encourage commercialization of pheromone products for pest species that are bound to be considered as minor marketing opportunities." He further noted that the U.S. Forest Service contributed basic toxicity testing in support of the registration of pheromones for control of the western pine shoot borer.

The pine tip moth *Rhyacionia zozona* has also been successfully managed in ponderosa pine plantations.[27] Pheromone lures applied at intervals of 10 m (100 dispensers/ha) provided nearly total disruption of mating, based on mating success monitored at traps. Larval populations were reduced 83.2%. In these trials, effects of mating disruptants on associated parasities of the tip moth were also monitored. Interestingly, although total parasitism levels were not affected by pheromone application, parasitism rates of individual species were affected. The authors found that incidence of a larval parasite (*Glypta zozonae*) declined in the pheromone-treated area, while a pupal parasite (*Mastrus aciculatus*) increased.[28] Host pheromones used in mating disruption trials can act as kairomones involved in host finding by their natural enemies. Because of this, pheromones used as mating disruptants may interact with the existing natural controls.

Based on capture of male moths visiting traps baited with virgin females, mating success of the cranberry girdler, a turfgrass pest, was reduced 98% by use of mating disruptant lures.[14] However, there has been little subsequent work on mating disruption of turfgrass insect pests. Because of the diverse species composition of potential turfgrass pests, as well as the absence of highly active pheromones among some of the "key" pest species such as billbugs and chinch bugs, successful adaptation of pheromones as mating disruptants in turfgrass pest management is unlikely. Furthermore, the relatively small area of most lawns allows for substantial migration of mated insects from adjacent, untreated turfgrass.

The use of mating disruption for control of a landscape pest was demonstrated by Klun et al.,[29] working with the bagworm (*Thyridopteryx ephemeraeformis*), a serious pest of arborvitae in much of the eastern U.S. They applied a racemic mixture of the pheromone in a tape formulation, which was wrapped spirally around the tree. Where high release rates were used (712 mg/tree/3 days) mating disruption averaged 87%. Mating suppression was independent of insect density, as was also found with Douglas-fir tussock moth.[25] Lower release rates, provided by a slow-release tape, did not suppress bagworm mating. The authors indicated that better formulation methods that could slowly release and maintain the high rates of pheromone throughout the 7-week insect mating period were needed to increase the practicality of this technique.

Although the tactic has been little explored in landscape plant protection, several important pest species affecting ornamental plants would appear to have potential for control by mating disruption using sex pheromones. For pests with highly active pheromone, mating disruption in isolated plantings could likely be achieved in most cases. Research is needed to identify the most active attractant compounds and to develop appropriate application methods.

Some insect pests of ornamentals seem particularly amenable to suppression by use of mating disruption. In particular are various species of Lepidoptera which are characterized by possessing both highly active sex pheromones and an adult female with limited mobility. Fall cankerworm (*Alsophila pometaria*), spring cankerworm (*Paleacrita vernata*), gypsy moth (*Lymantria dispar*), bagworm (*Thyridopteryx ephemeraeformis*), and Douglas-fir tussock moth (*Orgyia pseudotsugata*) are examples of important pest species possessing these traits. In addition, the latter two species also have a fairly restricted host range, and females occur in fairly limited areas around landscape plantings. This should greatly favor successful spot application of sex pheromones for mating disruption.

The limited market for pest control products designed for specific landscape pests would appear to be the most important restriction to adaptation of this technology. As long as registration criteria involve fairly high development costs, as currently required,[3] it will be difficult to economically develop pheromones as mating disruptants for landscape pests. Furthermore, there may prove to be some special application difficulties related to the sites where mating disruption would be used. For example, the diverse environment where landscape plants occur, involving buildings, irregular plant spacing and size, and considerable species diversity, may make it difficult to achieve uniform application of pheromone for mating disruption.

However, the associated safety features and ease of application particularly recommend use of this technology in ornamental plant protection. Potential nontarget effects from insecticide use are often of especially high concern in landscape settings. Furthermore, costs of applying pheromone lures to landscape plants should be very competitive compared to insecticide applications, which may cost over $100 per plant where large trees are involved. Their ease of use may also recommend them to applicators, who often have difficulties in scheduling treatments, particularly with notificiation requirements. Mating disruptant treatments should be much less sensitive to environmental factors during application, such as excessive wind speed or rainfall. Also, application timing may be less critical. Although lures should be in place at the onset of natural mating, lures typically release pheromone over the course of a month or more, so there is considerable flexibility as to when placement can be made.

REFERENCES

1. **Inscoe, M. N., Leonhardt, B. A., and Ridgway, R. L.,** Commercial availability of insect pheromones and other attractants, in *Behavior-Modifying Chemicals for Insect Management: Applications of Pheromones and Other Attractants*, Ridgway, R. L., Silverstein, R. M., and Inscoe, M. N., Eds., Marcel Dekker, New York, 1990, 631.

2. **Ladd, T. L., Klein, M. G., and Tumlinson, J. H.,** Phenethyl proprionate + eugenol + geraniol (3:7:3) and Japonilure: A highly effective joint lure for Japanese beetles, *J. Econ. Entomol.*, 74, 665, 1981.

3. **Tinsworth, E. F.,** Regulation of pheromones and other semiochemicals in the United States, in *Behavior-Modifying Chemicals for Insect Management: Applications of Pheromones and Other Attractants*, Ridgway, R. L., Silverstein, R. M., and Inscoe, M. N., Eds., Marcel Dekker, New York, 1990, 605.

4. **Leatherman, D. A.,** Colorado State Forest Service, personal communication, 1991.

5. **Neilsen, D. G.,** Sex pheromone traps: A breakthrough in controlling borers of ornamental trees and shrubs, *J. Arbor.*, 4(8), 181, 1978.

6. **Meyer, W. L., Cranshaw, W. S., and Eichlin, T. D.,** Flight patterns of clearwing borers in Colorado based on pheromone trap captures, *Southwest. Entomol.*, 13, 39, 1988.

7. **Bone, P. S. and Koehler, C. S.,** Study describes ash borer infestations, tests management method, *Calif. Agric.*, 45(5), 32, 1991.

8. **Johnson, J. W.,** Michigan State University, personal communication, 1991.

9. **Harrell, M. O.,** University of Nebraska, personal communication, 1991.

10. **Trimble, R. M.,** Assessment of a sex-attractant trap for monitoring the spotted tentiform leafminer, *Phyllonorycter blancardella* (Fabr.) (Lepidoptera: Gracillariidae): Relationship between male and female emergence and between trap captures and emergence, *Can. Entomol.*, 118, 1241, 1986.

11. **Potter, D. A.,** University of Kentucky, unpublished data, 1991.

12. **Muirhead-Thomson, R. C.,** *Trap Responses of Flying Insects: The Influence of Trap Design on Capture Efficiency*, Academic Press, New York, 1991, 304 pp.

13. **Shepherd, R. F., Gray, T. G., Chorney, R. J., and Daterman, G. E.,** Pest management of the Douglas-fir tussock moth, *Orygia pseudotsugata* (Lepidoptera: Lymantriidae): Monitoring endemic populations with pheromone traps to detect incipient outbreaks, *Can. Entomol.*, 117, 838, 1985.

14. **Kamm, J. A.,** Use of insect sex pheromones in turfgrass management, in *Advances in Turfgrass Entomology*, Niemczyk, H. D. and Joyner, B. G., Eds., ChemLawn Corp., Columbus, OH, 1982, 39.

15. **Klein, M. G.,** Mass trapping for suppression of Japanese beetles, in *Management of Insect Pests with Semiochemicals*, Mitchell, E. R., Ed., Plenum Press, New York, 1981, 183.

16. **Gordon, F. C. and Potter, D. A.,** Japanese beetle (Coleoptera: Scarabaeidae) traps: Evaluation of single and multiple arrangements for reducing defoliation in urban landscape, *J. Econ. Entomol.,* 79, 1381, 1986.

17. **Lanier, G. N.,** Principles of attraction-annihilation: Mass trapping, in *Behavior-Modifying Chemicals for Insect Management: Applications of Pheromones and Other Attractants*, Ridgway, R. L., Silverstein, R. M., and Inscoe, M. N., Eds., Marcel Dekker, New York, 1990, 25.

18. **Roelofs, W. L., Glass, E. H., Tette, J., and Comeau, A.,** Sex pheromone trapping for red-banded leaf roller control: Theoretical and actual, *J. Econ. Entomol.*, 63, 1162, 1970.

19. **Beroza, M. and Knipling, E. F.,** Gypsy moth control with sex attractant pheromone, *Science*, 177, 19, 1972.

20. **Lie, R. and Bakke, A.,** Practical results from the mass-trapping of *Ips typographus* in Scandanavia, in *Management of Insect Pests with Semiochemicals*, Mitchell, E. R., Ed., Plenum Press, New York, 1981, 175.

21. **Peacock, J. W., Cuthbert, R. A., and Lanier, G. N.,** Deployment of traps in a barrier strategy to reduce populations of the European elm bark beetle, and the incidence of Dutch elm disease, *Management of Insect Pests Using Semiochemicals*, Mitchell, E. R., Ed., Plenum Press, New York, 1981, 155.

22. **Carde, R. T.,** Principles of mating disruption, in *Behavior-Modifying Chemicals for Insect Management: Applications of Pheromones and Other Attractants*, Ridgway, R. L., Silverstein, R. M., and Inscoe, M. N., Eds., Marcel Dekker, New York, 1990, 47.

23. **Kolodny-Hirsch, D. M. and Schwalbe, D. P.,** Use of disparlure in the management of the gypsy moth, in *Behavior-Modifying Chemicals for Insect Management: Applications of Pheromones and Other Attractants*, Ridgway, R. L., Silverstein, R. M., and Inscoe, M. N., Eds., Marcel Dekker, New York, 1990, 363.

24. **Snow, J. W.,** Peachtree borer and lesser peachtree borer control in the United States, in *Behavior-Modifying Chemicals for Insect Management: Applications of Pheromones and Other Attractants*, Ridgway, R. L., Silverstein, R. M., and Inscoe, M. N., Eds., Marcel Dekker, New York, 1990, 241.

25. **Sower, L. L., Wenz, J. M., Dahlsten, D. L., and Daterman, G. E.,** Field testing of pheromone disruption on preoutbreak populations of Douglas-fir tussock moth (Lepidoptera: Lymantriidae), *J. Econ. Entomol.,* 83(4), 1487, 1990.

26. **Daterman, G. E.,** Pheromones for managing coniferous tree pests in the United States, with special reference to the western pine shoot borer, in *Behavior-Modifying Chemicals for Insect Management: Applications of Pheromones and Other Attractants*, Ridgway, R. L., Silverstein, R. M., and Inscoe, M. N., Eds., Marcel Dekker, New York, 1990, 317.

27. **Niwa, C. G., Daterman, G. E., Sartwell, C., and Sower, L. L.,** Control of *Rhyacionia zozana* (Lepidoptera: Tortricidae) by mating disruption with synthetic sex pheromone, *Environ. Entomol.,* 17(3), 593, 1988.

28. **Niwa, C. G. and Daterman, G. E.,** Pheromone disruption of *Rhyacionia zozana* (Lepidoptera: Tortricidae): Influence on the associated parasite complex, *Environ. Entomol.,* 18(4), 570, 1989.

29. **Klun, J. A., Neal, J. W., Jr., Leonhardt, B. A., and Schwarz, M.,** Suppression of female bagworm, *Thyridopteryx ephaemeraeformis*, reproduction potential with its sex pheromone, 1-methylbutyl decanoate, *Entomol. Exp. Appl.*, 40, 231, 1986.

Chapter 31

Life Cycles and Population Monitoring of Pest Mole Crickets

William G. Hudson, Department of Entomology, University of Georgia, Athens, GA

CONTENTS

Introduction 345
Life Cycles 345
    Egg Laying 346
    Nymphal Development 346
    Adults 346
    Implications for Control 346
Monitoring 347
    Sound Traps 347
    Grid-Square Rating 347
    Soil Flushing 348
    Pitfall Traps 348
    Other 348
References 348

INTRODUCTION

In most of the southeastern U.S., the primary pest species of mole crickets are the tawny mole cricket, *Scapteriscus vicinus* (TMC), and the southern mole cricket, *S. borellii* (SMC). A third species, the short-winged mole cricket, *S. abbreviatus* (SWC), does not fly and has not spread much beyond the vicinity of the Florida ports where it was introduced. Where they occur, mole crickets are far and away the most serious insect pests of turf and pasture grasses, causing more than $75 million in damage every year.

Mole crickets spend most of their lives underground. Both nymphs and adults come to the surface at night to feed, but they seldom leave the tunnel otherwise, unless forced out by flooding or predators. Food preferences of our pest species differ significantly.[1] As might be expected, TMC and SWC are vegetarians that feed on a wide variety of plants but with a preference for grasses. In contrast, SMC is largely predacious and does little direct damage to turfgrass. Mechanical damage to the root system can be significant as they tunnel along just below the soil surface foraging for earthworms and other small animals. In intensively managed turf areas such as golf putting greens, this tunneling alone is sufficient to require treatment to protect the grass surface.

LIFE CYCLES

Cycles of SWC are poorly understood, but it apparently breeds continuously with all stages present in the field at all times. Both TMC and SMC produce one generation per year, except that SMC has two generations per year in southern Florida.[2] Different life stages behave differently, and knowledge of the timing of local population cycles is critical for designing management strategies for particular sites. Unfortunately, detailed information about mole cricket cycles is available only for peninsular Florida, where long-term studies have been conducted for about 12 years. Entomologists in Alabama, Georgia, South Carolina, and North Carolina are collecting data on the cycles in their areas, but these studies are incomplete. The information presented here is based on Florida studies.

Mating and dispersal flights for mole crickets occur in the spring. Although these flight periods overlap for TMC and SMC, TMC tends to fly in greater numbers early, while SMC flights begin in earnest slightly later in the year and continue for a longer period of time.[3] Generally, TMC flights start in February and peak in March. They taper off sharply after mid-April, although a few may fly on any given night through May. In contrast, SMC flights begin in March and peak in early May, with considerable flight activity through June. In south

Florida, SMC has another flight in mid-summer, peaking in late July. There is a much smaller flight in the fall for both species, primarily in November for TMC and in October for SMC.

For both species, flights begin about 15 min after sunset and last about 60 to 90 min.[4] Local weather conditions determine whether crickets in a given area fly on a particular night. Cool (below approximately 65°F), rainy, or windy conditions will discourage flight activity. There have been few studies of individual flight behavior, but available data indicate a tendency for TMC females to fly just once, generally before beginning egg laying. Female SMC tend to fly more than once, and to fly between egg clutches.[5] Individual males have not been studied extensively, although males make up about 20% of the flying population.[6]

Flight typically ends in response to the calling song of a male mole cricket. This call is produced from a position at the base of a specially constructed acoustic horn that forms the end of the male's tunnel.[7] However, mate finding is not the only or even the main reason for a female to land near a particular male.[4] Only about 30% of the females attracted by a calling male enter the burrow of that male. Most have already mated and seem to be using the quality and intensity of the calling sound as an indicator of soil conditions, particularly moisture levels, that strongly influence the intensity of the call.[6] Both the crickets and their eggs are highly susceptible to desiccation, and adequate soil moisture is vital for successful oviposition.

EGG LAYING

After finding a suitable site, the female mole cricket spends about 2 weeks feeding and tunneling while her eggs mature. She then constructs a small chamber 4 to 12 in. below the soil surface where she lays a clutch of 25 to 60 (average of approximately 40) eggs.[5] After laying the eggs, she seals the entrance to the chamber and departs, leaving the eggs and young, which hatch in approximately 20 to 30 days (depending on temperature), to fend for themselves. Captive females of both species may lay as many as ten clutches of eggs in a lifetime, although this is probably extremely rare in nature.

In north Florida, egg laying by TMC begins as early as March but most eggs are laid from April through early June.[5] A few hatchlings may appear as early as April in some years, but most eggs hatch in May and June. By the first of July, virtually all TMC that are going to hatch have already done so.

Some SMC begin laying eggs in April, but most oviposition occurs in May and June with some continuing through July. Hatchlings begin to appear in April, but most emerge in May and June.[5] It is not unusual to see SMC hatchlings in the field as late as early September, although there are very few after late July.

NYMPHAL DEVELOPMENT

The exact number of nymphal instars for TMC is unknown. It is known that SMC is highly variable in its development, with a sample of five siblings reared under identical conditions to maturity in the laboratory passing through 8 (n = 2), 9 (n = 2), or 10 (n = 1) nymphal instars.[8] Presumably, the variability in the field is even greater since food supply, environmental conditions, and ancestry are much less uniform.

After spending the summer feeding and growing, TMC nymphs begin to molt to the adult stage in mid-September. By December, adults make up approximately 85% of the population and the rest are large nymphs.[2] Adult SMC also begin to appear in September, but the population matures much more slowly and by December only about 25% of SMC are adult, with the rest overwintering as nymphs.

ADULTS

Although mole crickets of both species mature as adults in the fall, there is little mating and no eggs are laid until the following spring.[5,9] Pitfall capture patterns suggest that adults of TMC do little tunneling in the fall. This is especially true for females, as pitfall captures are primarily males.[10] Flying TMC captured in sound traps in the fall include a higher proportion (>90%) of females than spring flights.[3] Apparently, female TMC are sedentary unless they fly, while males are more likely to move about on or near the soil surface.

IMPLICATIONS FOR CONTROL

Their extreme mobility and subterranean lifestyle make mole crickets the most difficult insect pest to control in turfgrass. There are, however, times when the crickets are more vulnerable to control efforts just as there are times when results are likely to be disappointing. Any IPM program must take into account the pest's seasonal cycles and differences in life stages if it is to offer optimal results. This is true for conventional chemical controls and for introduction of biological control agents as well.

Young nymphs are the most vulnerable stage for chemical control. They are smaller, less mobile, and spend most of their time in the surface layers of the soil where they are more likely to come in contact with insecticides. They appear at a time when the grass is growing vigorously, and their tunnels and feeding are often unnoticeable. Obviously, this is the optimal time to control mole crickets, before the turf suffers significant damage.

As the nymphs grow through the summer, the feeding damage increases. They tunnel more and the mechanical damage to the grass and disturbance of the turf surface become obvious, and they are more likely to escape hot or dry conditions by digging deeper in the ground where it is difficult to reach them by conventional means. Where irrigation is available, activity patterns can be manipulated by skipping one or more watering cycles and allowing the soil to dry out. Thorough irrigation will then bring the entire population to the surface the following night in a burst of activity that maximizes exposure to insecticides.

Once adults begin to appear, general surface activity decreases and becomes less predictable. Adults do not necessarily feed every night even if conditions are favorable, and so are less likely to be located where insecticides can reach them. This is especially true in the fall. There is also the possibility that a flight will occur after application of control materials, replacing those crickets killed with immigrants from outside the treated area. Although adults spend more time feeding at the surface in the spring, the probability of a flight replenishing the local supply of crickets after treatment is higher than in the fall. At either season, changeable weather conditions may bring on periods of cold or wet weather that reduce surface activity and make control difficult.

While adult mole crickets are less vulnerable to chemical insecticides, they are the stage that some promising biological control agents attack most successfully.[11] Obviously, there would be little point in releasing such an agent at a time when the population is comprised mostly of nymphs.

MONITORING

Several techniques are used by entomologists and turfgrass managers to monitor mole cricket populations. All have limitations, and the choice of a method for a particular situation depends on the type of information needed, life stage present, and time and manpower available to do the work. No one method will be appropriate for every site or time of year.

SOUND TRAPS

These traps attract flying adults to the highly amplified synthesized call of the male mole cricket. The trap set-up is simple, consisting of a caller positioned over some sort of catching device.[12] Most catching devices are funnels of some design with a bucket at the bottom where the crickets are collected in soil (for live capture) or a preservative such as alcohol if live crickets are not needed. A child's wading pool filled with water also works as a trap. The latest design of the callers was developed jointly by the University of Florida and Clemson University, and they are now produced on a custom-built basis by Eco-Sim Designs of Gainesville, FL.

Sound traps are effective only during the spring and fall flight seasons and capture only adults of the species whose call is broadcast. The numbers captured vary widely from year to year even in the absence of any control efforts.[3] Little is known about how far or how often individuals fly, or what proportion of the population might be flying at a given time. This uncertainty means that the sound traps sample an unknown proportion of the crickets from an unknown area, and so are of little value except as general indicators of cricket activity. However, they are the only method of collecting large numbers of live adults.

GRID-SQUARE RATING

This method of assessing mole cricket activity in an area employs a square, usually 2 or 3 ft on a side, that is divided into nine equal subdivisions. The grid is then placed on the turf surface, and each small square is rated as either a 0 (no mole cricket tunnels) or 1 (at least one tunnel). Sample values range from 0, for no activity, to 9 if there is one or more tunnels in each square.[13] The technique is very quick. With a little practice, many samples can be taken in a short time. Sampling should be done after a rain or irrigation so that only fresh activity will be counted.

The biggest advantage to grid-square sampling is that it is fairly sensitive to changes in low-density populations. Other solutions, such as soap flush (see below) require so many samples to overcome the variable results inherent at low population densities common on golf courses that time requirements become prohibitive. However, the grid-square method cannot differentiate between a rating of 9 produced by one tunnel per square and one caused by five tunnels per square and so should not be used at high densities. Surface tunneling by mole crickets is affected by many factors such as soil moisture, temperature, soil type, and turf quality. Grid-square samples from different areas should be compared only if conditions are similar among all areas. The relationship between surface activity and population density is not linear,[10] so a 50% reduction in grid-square rating does not necessarily reflect a 50% reduction in density. There is also a strong correlation between nymph size and surface tunneling, so comparisons over time are likely to be deceptive. Larger nymphs dig longer and more noticeable tunnels, and the grid-square rating for an area will go up over time even if there is no change in population. Because adults are less active on the surface than nymphs, the rating for an area will drop in late September as adults begin to appear.

SOIL FLUSHING

Mole crickets can be flushed from the soil by mixing detergent or some insecticides with water and pouring the solution on the ground. Soap and insecticide work equally well,[14] so soap is the material of choice. Both liquid and powdered soap are effective, but most research has been done with dishwashing liquid. The usual procedure is to mix $1/_2$ to 1 oz of the liquid in a gallon of water and pour the solution over an area 2 feet square. The area should be observed for about 3 min although most crickets will surface in the first minute.[15] The soap solution is effective at bringing nymphs to the surface but much less effective for adults.

Soap flushing is the only method of obtaining an estimate of population density. The relationship between numbers of nymphs present and numbers sampled by flushing is a function of soil moisture, and sample numbers can be converted to density estimates (number per square yard, for instance).[15] However, the estimates have very large confidence intervals and so require a great many samples to produce useful numbers. The effort is justified for research purposes but most turf managers can get equally useful information from the grid-square method described above at a much lower cost in time invested. Soap flushing is very useful for monitoring life stages present in the field and for determining when nymphs have hatched. It can also confirm that the problem in an area is, indeed, mole crickets when an infestation is just beginning.

PITFALL TRAPS

Linear pitfall traps[16] have been used to monitor local populations and represent one of the few sources of live juvenile mole crickets for research purposes. These traps are constructed by cutting a slot approximately 1 in. wide lengthwise in a section of 3-in.-diameter PVC pipe. The pipe is then buried in the ground so that the slot is at the level of the soil surface, with one end feeding into a 5-gal bucket and the other end capped. Up to four pipes can radiate out from a single bucket. The traps sample only from the immediate area around them, although at very high densities impressive numbers of nymphs can be taken continuously. The relationship between population density and

numbers captured is unknown, except that at low levels (below approximately 3 nymphs/square foot) very few crickets are captured.[10] Adults are less likely to be sampled by pitfalls than nymphs.

OTHER

Digging for mole crickets has also been used as a sampling technique, but it has not proven very effective. Tools as diverse as shovels and tree spades have been employed, but all are destructive to the turf and require substantial investments of time without providing any better information about the mole cricket population than the soap flush or grid-square sampling. Counting dead crickets on the surface has been used to compare insecticide treatments, but the numbers obtained are of little real value. Dead crickets on the surface are attractive to a variety of birds, and it is difficult to prevent loss of crickets from some or most plots before counts are made. Other studies have shown that some insecticides kill quickly enough that up to 65% of the dead crickets never make it to the surface.[17]

REFERENCES

1. **Matheny, E. L., Jr., Tsedeke, A., and Smittle, B. J.,** Feeding response of mole cricket nymphs (Orthoptera: Gryllotalpidae: *Scapteriscus*) to radio-labeled grasses with, or without, alternative food available, *J. Ga. Entomol. Soc.*, 16, 492, 1981.

2. **Walker, T. J., Ed.,** Mole crickets in Florida, *Univ. Fla. Agric. Exp. Sta. Bull.*, 846, 1984.

3. **Walker, T. J., Reinert, J. A., and Schuster, D. J.,** Geographical variation in flights of mole crickets, *Scapteriscus* spp. (Orthoptera: Gryllotalpidae), *Ann. Entomol. Soc. Am.*, 76, 507, 1983.

4. **Forrest, T. G.,** Mole crickets in Florida, *Univ. Fla. Agric. Exp. Sta. Bull.*, 846, 10, 1984.

5. **Forrest, T. G.,** Oviposition and maternal investment in mole crickets: effects of season, size, and senescence, *Ann. Entomol. Soc. Am.*, 79, 919, 1985.

6. **Forrest, T. G.,** Phonotaxis in mole crickets: Its reproductive significance, *Fla. Entomol.*, 63, 45, 1980.

7. **Nickerson, J. C., Snyder, D. E., and Oliver, C. C.,** Accoustical burrows constructed by mole crickets, *Ann. Entomol. Soc. Am.*, 72, 438, 1979.

8. **Hudson, W. G.,** Variability in development of *Scapteriscus acletus* (Orthoptera: Gryllotalpidae), *Fla. Entomol.*, 70, 403, 1987.

9. **Walker, T. J. and Nation, J. L.,** Sperm storage in mole crickets: fall matings fertilize spring eggs in *Scapteriscus acletus*, *Fla. Entomol.*, 65, 283, 1982.

10. **Hudson, W. G.,** Ecology of the Tawny Mole Cricket, *Scapteriscus vicinus* (Orthoptera: Gryllotalpidae): Population Estimation, Spatial Distribution, Movement, and Host Relationships, Ph.D. Dissertation, University of Florida, Gainesville, 1985, 78 pp.

11. **Hudson, W. G., Frank, J. H., and Castner, J. L.,** Biological control of *Scapteriscus* spp. mole crickets (Orthoptera: Gryllotalpidae) in Florida, *Bull. Entomol. Soc. Am.*, 13, 192, 1988.

12. **Walker, T. J.,** Sound traps for sampling mole cricket flights (Orthoptera: Gryllotalpidae: *Scapteriscus*), *Fla. Entomol.*, 65, 105, 1982.

13. **Cobb, P. P. and Mack, T. P.,** A rating system for evaluating tawny mole cricket, *Scapteriscus vicinus* scudder, damage (Orthoptera: Gryllotalpidae), *J. Entomol. Sci.*, 24, 142, 1988.

14. **Hudson, W. G.,** Field sampling of mole crickets (Orthoptera: Gryllotalpidae: *Scapteriscus*): A comparison of techniques, *Fla. Entomol.*, 71, 214, 1988.

15. **Hudson, W. G.,** Field sampling and population estimation of the tawny mole cricket (Orthoptera: Gryllotalpidae), *Fla. Entomol.*, 72, 337, 1989.

16. **Lawrence, K. O.,** A linear pitfall trap for mole crickets and other soil arthropods, *Fla. Entomol.*, 65, 376, 1982.

17. **Ulagaraj, S. M.,** Effects of insecticides on mole crickets (Orthoptera: Gryllotalpidae: *Scapteriscus*), *Fla. Entomol.*, 57, 414, 1974.

Chapter 32

Decision-Making Factors for Management of Fire Ants and White Grubs in Turfgrass

Beverly Sparks and S. Kristine Braman, *Department of Entomology, University of Georgia, Athens, GA*

CONTENTS

Red Imported Fire Ant: Introduction 351
Location of Mounds and Population Density 352
Selection and Timing of Control Measures 353
Management Programs for Recreational Turf Areas and Home Lawns 354
White Grubs: Introduction 354
Degree-Day Models...355
Influence of Rainfall and Soil Moisture 355
Life Cycle and Timing of Intervention 355
Other Factors Influencing Management Decisions 358
Summary 359
Acknowledgments 359
References 359

Fire ants and white grubs are among the most serious subterranean pests of managed turfgrasses. Here we discuss aspects of their identification, biology, and behavior, and factors affecting timing of control for both pest groups.

RED IMPORTED FIRE ANT: INTRODUCTION

The red imported fire ant, *Solenopsis invicta* Buren, an introduced species from South America, entered the U.S. in Mobile, Alabama, around 1930.[1,2] The species continues to spread into areas of the southern U.S. with a mild climate, adequate moisture, and food. Red imported fire ants (RIFAs) currently infest all of Florida and Louisiana and portions of Georgia, South Carolina, North Carolina, Tennessee, Alabama, Mississippi, Arkansas, Texas, and Oklahoma.

RIFAs are predators as well as scavengers and feed on a wide variety of foods. Due to their predaceous feeding behavior, the presence of fire ants should not always be considered detrimental. Fire ants primarily feed on other insects and in some situations are of great benefit. Pest species such as pecan weevils, boll weevils, bollworms, sugarcane borers, ticks, fly larvae, flea larvae, and cockroach eggs often serve as their prey. However, fire ants also feed on seeds, seedlings, fruits, vegetables, and will invade structures to feed on processed foods.

By far, the most significant problem associated with RIFAs is their stinging and biting behavior. They can sting repeatedly and will defensively and aggressively attack anything that disturbs their mounds or food sources. Symptoms of a RIFA sting include intense burning and itching. Often, a white pustule forms at the site where venom was injected beneath the skin. A few people are hypersensitive to the venom and may suffer chest pains, nausea, or lapse into a coma from even one sting.

The RIFA causes additional problems for man due to its mound building activities. RIFA mounds interfere with maintenance of pastures, recreational turf areas, and roadsides and can cause damage to mowers and other equipment. The mounds detract from the aesthetics of the land.

RIFAs build mounds in almost any type of soil, but prefer open, sunny areas such as pastures, parks, lawns, and cultivated fields. Mounds are occasionally located inside rotten logs, around stumps or trees, under structures, or inside electrical switch boxes. The size of the mound depends upon soil characteristics and the frequency with which the mound is disturbed. The aboveground portion of mounds is usually conical and can reach a height of 10 to 12 in. The tops of mounds are less well developed in very sandy soils or where land is frequently disturbed. Below ground, mounds are "V" shaped and may penetrate into the soil 3 to 4 ft (Figure 1).

A RIFA colony consists of the brood and three forms of adult ants (Figure 2): the black reproductive males, red-brown queens (reproductive females), and workers (sterile females). Egg-laying queens have no wings; unmated queens and males are winged. Workers are wingless and vary greatly in size. The brood includes the cream colored, globular eggs and grub-like larvae and pupae.

New fire ant colonies are not conspicuous until several months after the young queen begins egg laying. Upon completion of the mating flight, the newly mated queen constructs a small chamber in the soil and lays a cluster of a dozen or so eggs. The eggs hatch 7 to 10 days later and the immature ants are tended by the queen until they emerge from the pupae as adults in 15 to 25 days. These worker ants care for the queen, and she begins to lay up to 200 eggs per day. Worker ants gather food, defend the colony and care for the brood. The mature colony may contain 100,000 to 500,000 workers, the brood, and several hundred winged (reproductive) forms. Worker ants may live for 2 months, while queens can survive 5 years or more. Mating flights can occur at any time of year, but are most common from April through June. Males die soon after the mating flight; newly mated queens fall to the ground, shed their wings and begin searching for a suitable site to start a new fire ant colony.

LOCATION OF MOUNDS AND POPULATION DENSITY

Throughout the 1960s and 1970s attempts were made to eradicate the RIFA from the southeastern U.S.[3] These programs failed to eradicate the species and may actually have aided the spread of the

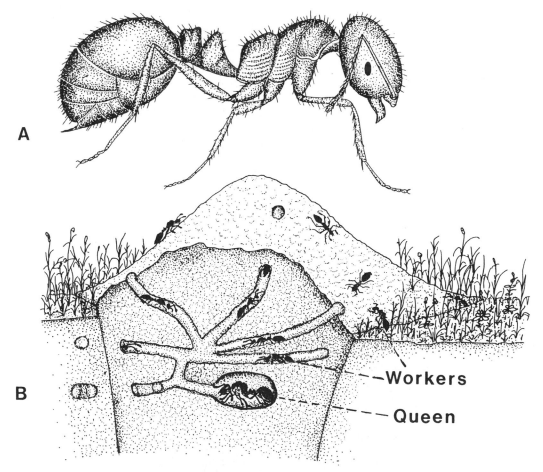

Figure 1 Worker of the red imported fire ant (A) and diagram illustrating mound construction, workers, and egg-laying queen (B). *(Illustration by Tong-Xian Liu.)*

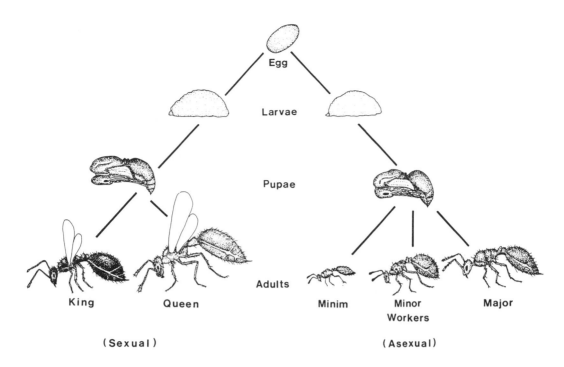

Egg

Larvae

Pupae

Adults

King   Queen   Minim   Minor   Major
                          Workers

(Sexual)   (Asexual)

Figure 2 Life stages of the red imported fire ant, *Solenopsis invicta. (Illustration by Tong-Xian Liu.)*

ants. Chemicals used in the eradication program also destroyed populations of native ant species.[4] In the absence of native ant species, RIFAs can quickly reinvade and subsequently prevent other ant species from reestablishing in an area.

With current technology, the elimination of RIFAs in fully infested areas is not technically, environmentally, or economically feasible.[5] It is possible, however, to control fire ants temporarily within a defined area with insecticides. The insecticide treatments must be applied periodically for as long as control is desired. If treatments are stopped, fire ant queens will reinvade, and the area will become reinfested. The decision to treat fire ants must be accompanied by a long-term commitment to continue periodic treatments.

In many agricultural settings, a long-term control program cannot be economically justified. In other areas, such as parks, playgrounds, recreational turf areas, and home lawns, the justification for a fire ant control program is more subjective. One must consider the potential health risk of fire ants as well as the environmental impact of chemical applications and then decide when ant populations reach intolerable levels and control measures are warranted.

SELECTION AND TIMING OF CONTROL MEASURES

Successful management programs make use of a combination of tactics to suppress RIFA popula-

tions. Methods used to control RIFA in recreational turf areas include broadcast applications of insecticides as well as individual mound treatments.

Several formulations of contact insecticides and bait formulated insecticides are available for broadcast application to control RIFA. Broadcast applications of contact insecticides are primarily successful in suppressing the activity of foraging ants and preventing small mounds from becoming established. As a single treatment, they are usually not effective in eliminating large, established mounds.

Broadcast treatment with bait products provides a gradual decline of RIFA populations. Elimination of mounds within a treated area is slow, requiring several weeks to 6 months or more. Broadcast treatment with baits is an effective and economical technique used to reduce large populations of RIFA when it is not feasible to individually treat mounds. Periodic broadcast treatments with baits will control colonies that reinvade the treated area, and will provide long-term suppression of RIFA populations.

The success of bait treatments is dependent on several factors. First, the bait must be accepted by the ants. Use products that are fresh or that have been properly stored. Commercially available RIFA baits are formulated on processed corn grit coated in soybean oil. Old or improperly stored bait may have become rancid and will not attract ants. Second, apply the bait when the ground and grass are dry and no rain (or irrigation) is expected for 24 hours. Make sure the ants are actively foraging for food when the baits are applied. Avoid application when soil temperatures are lower than 70°F, as foraging activity is reduced during cold periods. During hot summer days, apply baits in late afternoon or early evening as ants forage at night and are inactive during the day.

Treatment of individual mounds is a very effective, labor-intensive method of selectively eliminating RIFA colonies. Individual mounds may be treated using drenches, granular products, dusts, liquid fumigants, aerosols, or baits. Depending on the product that is used, individual mound treatments may take seconds (aerosols and liquid fumigants), days (drenches, granular products, dusts), or up to several weeks (baits) to have an effect on the colony.[6] In addition, the cost of individual mound treatments varies from pennies to dollars per mound depending on the product selected for use.[6]

The success of any individual mound treatment is based upon the ability to deliver the toxicant to the entire colony. Elimination of the mound requires the insecticide to come into contact with not only the workers but also the queen and her brood. Timing of the insecticide application for when the queen and her brood are located close to the soil surface provides access to the entire colony. The queen and brood are located closest to the soil surface in the spring and fall of the year and immediately following heavy rains. Care must be taken, however, not to drench saturated soil. During the winter and summer, the queen and brood are typically located deep within the mound where it is difficult to deliver contact insecticides. Failure to reach the queen and brood with the insecticide treatment often results in mound movement or relocation.

MANAGEMENT PROGRAMS FOR RECREATIONAL TURF AREAS AND HOME LAWNS

RIFA management programs must be designed for a specific site giving consideration to the combination of control tactics that will provide the desired level of control, be affordable, and the least harmful to the environment. Drees and Vinson[5] discuss three management programs for RIFA infestation in recreational turf and home lawns. The programs are designed for the following situations: (1) elimination of RIFA populations in small turf areas or home lawns; (2) long-term suppression of RIFA populations in large areas; and (3) elimination of RIFA in large areas.

Program 1 is designed for areas of 1 acre or less of ornamental turf. It can also be used in larger areas where preservation of native ant species is desired. This program selectively controls fire ants, but reinvasion of the ants will occur over time. This program requires regular monitoring for ant mounds and can be more labor intensive than Programs 2 and 3.

PROGRAM 1
1. Treat all unwanted RIFA mounds using an individual mound treatment of choice.
2. Selectively treat new or undesirable mounds as detected.

Program 2 is designed for long-term suppression of RIFAs in larger turf areas but will not eliminate all ant activity. Suppression of ant populations occurs over weeks or months. This is the most economical program to use in areas with large numbers of ants. This program should not be used for areas with large numbers of native ants and where RIFA mounds number less than 15 to 20 per acre.

PROGRAM 2
1. Make an annual or semi-annual broadcast application of a bait-formulated insecticide each spring and/or fall. If only one application can be made, research indicates that a fall application will suppress RIFA populations for the longest period of time.[7]
2. At least 2 days after the broadcast treatment with bait, begin treating individual mounds in sensitive areas with the individual mound treatment of choice.

Program 3 is designed to eliminate all mound building and foraging activity in the treated area. Effects of this program are more rapid than in Program 2, however, this program is more expensive and requires the use of more contact insecticides.

PROGRAM 3
1. Make an annual or semi-annual broadcast application of a bait-formulated insecticide in areas where there are more than 20 mounds per acre; or individually treat fire ant mounds in the area.
2. Routinely broadcast or spray a contact insecticide every 8 weeks or when new RIFA mounds are detected.

WHITE GRUBS: INTRODUCTION

White grubs are the immatures of several species of scarabaeid beetles. They are recognized as being one of the most difficult groups of turfgrass pests to manage.[8] Scarabaeid grubs feed on the roots of all species of commonly used turfgrasses causing irregular dead patches that can be lifted or rolled back like a carpet. Birds, moles, and skunks often cause further damage when they tear up the turf in search of grubs on which to feed. The subterranean habits of white grubs make them difficult to control because their presence often goes undetected until severe damage has already occurred and because surface-applied insecticides must penetrate the thatch layer to the root zone where grubs are actively feeding.

At least ten species of white grubs are pests of turfgrasses. Several of these are nonnative, e.g., the Japanese beetle, *Popillia japonica*, the European chafer, *Rhizotrogus majalis*, the oriental beetle, *Anomala orientalis*, and the asiatic garden beetle, *Maladera castanea*. The widely distributed native species, including northern and southern masked chafers, *Cyclocephala borealis* and *C. lurida*, the black turfgrass ataenius, *Ataenius spretulus*, and several species of *Phyllophaga* are also very destructive to turf. Green June beetle larvae, *Cotinus nitida,* may also cause significant damage to turf.

Most turfgrass-infesting grubs have annual (1-year) life cycles. Masked chafers, Japanese beetles, and green June beetles hatch from eggs deposited in June or July and cause their most serious damage late in the summer and fall before moving deeper into the soil to overwinter as large larvae (Figure 3).[9] Some species complete two generations per year in some areas (e.g., *A. spretulus*) and portions of populations of other species (e.g., European chafers and oriental beetle) or species at the northern limits of their geographic range (e.g., Japanese beetles) may require 2 years to complete development. *Phyllophaga* spp. may require from 1 to 4 years to complete their life cycle depending upon species and geographic location.

DEGREE-DAY MODELS

Insects develop at rates governed by temperature. Within a range of optimal temperatures most insects develop in direct relationship to temperature; the warmer the temperature the faster the rate of growth. Degree-day (DD) accumulations are a measurement of cumulative heat units, or physiological time, required for an event in the life cycle of an insect to occur. Daily temperatures can be converted to heat units and recorded, allowing the appearance of a particular life stage of a given pest to be predicted.

Models based on accumulated DD that allow prediction of phenology or adult emergence have been developed for the Japanese beetle[10] and masked chafers.[11] Timing of emergence, flight, or egg depo-

sition of European chafers and the black turfgrass ataenius has been related to dates of flowering of common indicator plants.[12,13] Potter[11] experimentally determined the lower developmental threshold for pupation and overwintering of the southern masked chafer as 10.8°C (approximately 51°F). Thermal unit accumulations in air and soil were closely correlated with first emergence of beetles, but were less useful for predicting the date of 50 and 90% flight. The ranges of DD accumulations required for first flight of northern and southern masked chafers were 898 to 905 and 1000 to 1110, respectively. Once emergence had begun, activity of chafers was more closely related to rainfall patterns than thermal accumulation.

Many turfgrass-infesting scarabaeid adults fly at night and are attracted to lights. Pheromones have been identified or demonstrated for several species (green June beetle, masked chafers, Japanese beetle). Light or pheromone traps may provide useful tools in addition to DD forecasting for detecting the timing and level of adult activity. Noninvasive means for relating trap captures to subsequent levels of the damaging stage remain elusive at present.

INFLUENCE OF RAINFALL AND SOIL MOISTURE

Temperature, soil type, and especially soil moisture influence oviposition, egg and larval survival, adult emergence and flight activity, and the expression of damage to turf.[11,14-18] Scarabaeid eggs must absorb a significant amount of moisture from the soil within a relatively short time period to survive and hatch.[19,20] Eggs desiccate and perish when soil moisture drops below critical thresholds. Soil moisture also influences vertical movement within the soil profile.[21]

Thermal unit DD accumulation alone was not an accurate method for predicting the 50 and 90% emergence of adult-masked chafers.[11] Following initial emergence, female flight activity was stimulated by heavy rainfall events. *Phyllophaga crinita* flights were also closely related to rainfall patterns.[14] However, oviposition activity and egg and larval survival were limited in soils containing excessive moisture.

LIFE CYCLE AND TIMING OF INTERVENTION

Life cycles of turfgrass-infesting scarabaeid larvae vary with species. Schematics of typical annual, 2- and 3-year life cycles are presented in Figures 3 and 4. Identification is often the key to effective

356

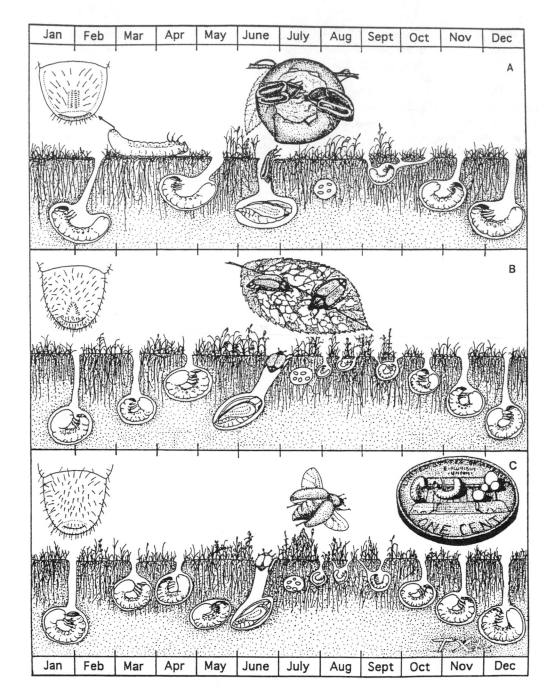

Figure 3 Life cycles and rastral patterns of common annual white grub species (A) green June beetles, (B) Japanese beetles, and (C) masked chafers (inset depicts relative size of eggs and newly hatched grubs). *(Illustration by Tong-Xian Liu.)*

control. The rastral pattern (arrangement of spines on the ventral surface of the last abdominal segment) is diagnostic. Rastral patterns for each species or genus (Phyllophaga) are indicated on the figures. Optimal timing of control efforts for white grubs in turf occurs after eggs have hatched but before the larvae have had time to become large and cause significant injury. Early instar larvae may also be more susceptible to insecticides than late stage larvae. Control failures often occur when insecticide applications are improperly timed: too early before eggs have hatched or too late when

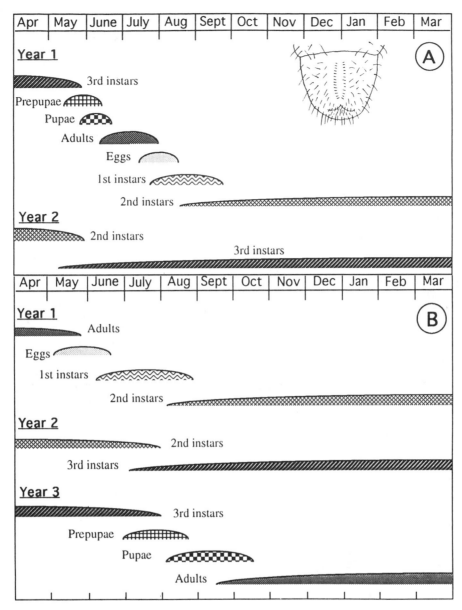

Figure 4 Schematic (A) 2- or (B) 3-year life cycles and rastral pattern for *Phyllophaga* spp. *(Illustration by Tong-Xian Liu.)*

larvae have already caused extensive damage and have left the root zone to spend the winter deeper in the soil.

Recent experiments evaluating efficacy and influence of timing of pesticide applications for southern masked chafers in central Georgia illustrate the variability in response to treatment that may occur depending upon time of application. Female chafers collected at light traps during June were confined in field microplots and allowed to lay eggs in KY 31 tall fescue for 2 weeks. Insecticide applications were made 2 months preoviposition (May 13), 1 month preoviposition (June 12), at oviposi-

tion (July 11), 1 month postoviposition (August 15), and 2 months postoviposition (September 11).

Control efforts early in the life cycle met with inconsistent success (Figure 5). Had oviposition occurred over a more extended period (as is often typical), pest reduction would likely have been even less successful. When applications were delayed until 2 months after oviposition, the high densities of grubs, equivalent to greater than 100/ft^2, killed the turf in the microplots decimating the food source and causing a reduction in population numbers. Unacceptable damage evident in the fifth trial period was avoided by applications delivered during the third and fourth trial periods equivalent to 2 to 6 weeks after peak adult activity. Treatments applied prior to the appearance of a pest problem, or prophylactic application of insecticide regardless of pest presence are wasteful and environmentally irresponsible. A combination of good record keeping, mapping of white grub "neighborhoods" (areas consistently infested), and monitoring of activity prior to judicious use of insecticides represent the use of IPM.

OTHER FACTORS INFLUENCING MANAGEMENT DECISIONS

IPM decision-making factors fall into two broad categories: pest population dynamics and management factors. Factors related to the pest's population dynamics may usually not be manipulated by the turfgrass professional, yet are important to decision making. These include such aspects as type of pest, type of damage, time of year, potential for population increase, influence of natural enemies, and damage potential of the pest for a particular turf use. Management activities more subject to manipulation include mapping, monitoring, and survey efforts; type of cultural practices employed

(mowing, fertilization, irrigation, cultivar selection, thatch management, top dressing); and control options which are influenced by economic and environmental costs, product efficacy, legal restrictions, and scheduling problems.

As understanding of management-related activities improves, especially in areas such as sampling and monitoring, we continue to be hampered by our lack of situation-specific thresholds for turfgrass pests causing aesthetic damage. Pest pressure requiring chemical intervention will vary depending upon turf use, turf species, cultural management practices, and public perception of tolerable turf injury. Rule-of-thumb thresholds for most turf pests are quite likely far too low. Healthy turf can tolerate considerable insect pressure, especially if adequate irrigation can be supplied at critical intervals.

Potter[22] discussed this issue in relation to southern masked chafer injury to Kentucky bluegrass turf. The effects of grub injury were compounded by moisture stress. Nonirrigated enclosures with 48 grubs per 0.1 m² produced only 57% as much growth as did irrigated, grub-free turf during a 2-week test period. However, regularly irrigated plots with initial infestation densities as high as 24 grubs

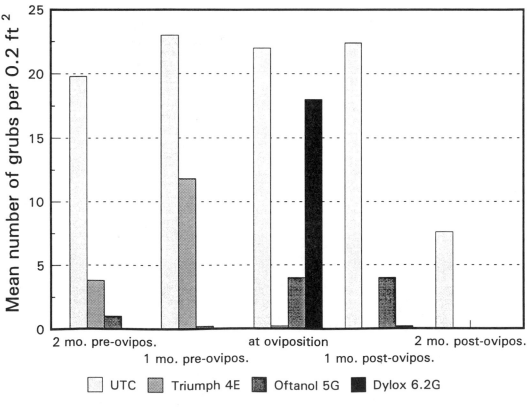

Figure 5 Efficacy and timing of application for white grub control.

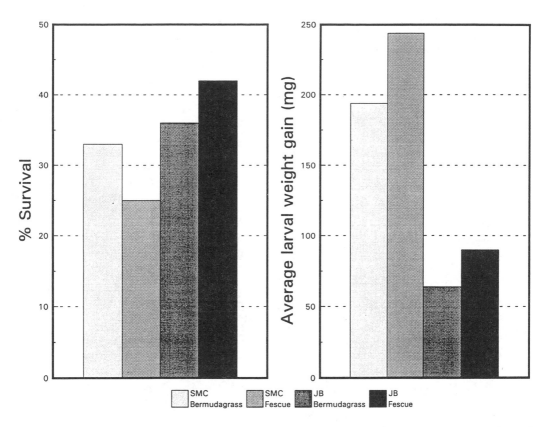

Figure 6 Survival and weight gain of southern masked chafers (SMC) and Japanese beetles (JB) on two grass species.

per 0.1 m² suffered little damage.

Recent studies evaluated the growth, survival, and damage relationships of white grubs in Georgia.[23] Third instar growth and survival, and the effect of larval density of the Japanese beetle and the southern masked chafer were determined on Tifway bermudagrass and KY-31 tall fescue in a 3-year field study. While winter survival of white grubs was sometimes favored in bermudagrass plots, larval weight gain was more often enhanced by feeding on fescue (Figure 6). Initial densities equivalent to 30/0.1 m² did not cause any measurable difference in aboveground turf quality in irrigated plots.

SUMMARY

Fire ant and white grub management provide challenges for the turfgrass professional. Current research on biological control of both groups suggests that viable alternatives to traditional means of chemical suppression are on the horizon. Host plant resistance research may provide additional options for white grub management. Successful pest management today requires considerable effort that is well founded in a knowledge of biology and life histories of problem pests. Applying all the principles of IPM: record keeping, host plant resistance, cultural practices, monitoring, and accurately targeting chemical and biological controls will provide effective pest suppression while minimizing excessive use of chemical controls.

ACKNOWLEDGMENTS

Technical contributions of S. Diffie, A. Pendley, M. Webb, and M. Gibson are appreciated.

REFERENCES

1. **Wilson, E. O.,** The fire ant, *Sci. Am.,* March, 36, 1958.
2. **Wilson, E. O. and Brown, W. L.,** Recent changes in the introduced population of the fire ant *Solenopsis saevissima* (Fr. Smith), *Evolution,* 12, 211, 1958.
3. **Collins, H.,** Control of imported fire ants. A review of current knowledge, U.S. Department of Agriculture, Animal and Plant Health Inspection Service, Tech. Bull., 1807, U.S. Department of Agriculture, Washington, DC, 1992.

4. **Summerlin, J. W., Hung, A. C. F., and Vinson, S. B.,** Residues in nontarget ants, species simplification and recovery of populations following aerial applications of mirex, *Environ. Entomol.,* 6, 193, 1977.

5. **Drees, B. M. and Vinson, S. B.,** Fire ants and their management, *Texas Ag. Ext. Bull.,* B-1536, 1991.

6. **Sparks, B. and Diffie, S.,** Imported Fire Ant Management: Results of Applied Research/Result Demonstrations 1987–1990 (A Cumulative Report of Statewide Field Trials), University of Georgia Extension, 90-015 (1), 1991.

7. **Collins, H., Callcott, A. M., Lockley, T. C., and Lander, A.,** Seasonal trends in effectiveness of hydramethylnon (Amdro) and fenoxycarb (Logic) for control of red imported fire ants (Hymenoptera: Formicidae), *J. Econ. Entomol.,* 85, 2131, 1992.

8. **Potter, D. A. and Braman, S. K.,** Ecology and management of turfgrass insects, *Annu. Rev. Entomol.,* 36, 383, 1991.

9. **Tashiro, H.,** *Turfgrass Insects of the United States and Canada,* Cornell University Press, Ithaca, NY, 19.

10. **Regniere, J., Rabb, R. L., and Stinner, R. E.,** *Popillia japonica:* Simulation of temperature-dependent development of the immatures, and prediction of adult emergence, *Environ. Entomol.,* 10, 290, 1981.

11. **Potter, D. A.,** Seasonal emergence and flight of northern and southern masked chafers in relation to air and soil temperature and rainfall patterns, *Environ. Entomol.,* 10, 793, 1981.

12. **Tashiro, H. and Gambrell, F. L.,** Correlation of European chafer development with the flowering period of common plants, *Ann. Entomol. Soc. Am.,* 56, 239, 1963.

13. **Wegner, G. S. and Niemczyk, H. D.,** Bionomics and phenology of *Ataenius spretulus,* *Ann. Entomol. Soc. Am.,* 74, 374, 1981.

14. **Gaylor, M. J. and Frankie, G. W.,** The relationship of rainfall to adult flight activity; and of soil moisture to oviposition behavior and egg and first instar survival in *Phyllophaga crinita, Environ. Entomol.,* 8, 591, 1979.

15. **Gyrisco, G., Whitcomb, W. H., Burrage, R. H., Logothetis, C., and Schwardt, H. H.,** Biology of European Chafer, *Amphimallon majalis* Razoumousky (Scarabaeidae), Cornell University Agricultural Experimental Station Memo No. 328, Cornell University Agricultural Experimental Station, Ithaca, NY, 1954.

16. **Ladd, T. L. and Burriff, C. R.,** Japanese beetle: influence of larval feeding on bluegrass yields at two levels of soil moisture, *J. Econ. Entomol.,* 72, 311, 1979.

17. **Potter, D. A.,** Influence of feeding by grubs on the quality and yield of Kentucky bluegrass, *J. Econ. Entomol.,* 75, 21, 1982.

18. **Regniere, J., Rabb, R. L., and Stinner, R. E.,** *Popillia japonica* (Coleoptera: Scarabaeidae): distribution and heterogeneous environments, *Can. Entomol.,* 115, 287, 1981.

19. **Potter, D. A. and Gordon, F. C.,** Susceptibility of *Cyclocephala immaculata* (Coleoptera: Scarabaeidae) eggs and immatures to heat and drought in turfgrass, *Environ. Entomol.,* 13, 794, 1984.

20. **Regniere, J., Rabb, R. L., and Stinner, R. E.,** *Popillia japonica:* Effect of soil moisture and texture on survival and development of eggs and first instar grubs, *Environ. Entomol.,* 10, 654, 1981.

21. **Villani, M. G. and Wright, R. J.,** Use of radiography in behavioral studies of turfgrass-infesting scarab grub species (Coleoptera: Scarabaeidae), *Bull. Entomol. Soc. Am.,* 34, 132, 1988.

22. **Potter, D. A.,** Influence of feeding by grubs of the southern masked chafer on quality and yield of Kentucky bluegrass, *J. Econ. Entomol.,* 75, 21, 1982.

23. **Braman, S. K. and Pendley, A. F.,** Growth, survival and damage relationships of white grubs in bermudagrass vs. tall fescue, *Int. Turfgrass Soc. J.,* 7, 370, 1993.

Chapter 33

Timing Controls for Insect Pests of Woody Ornamentals

James L. Hanula, U.S. Forest Service, Athens, GA (formerly with Connecticut Agricultural Experiment Station, New Haven, CT)

CONTENTS

Introduction 361
Examples 362
    Black Vine Weevil 362
        Insecticidal Control 362
        Biological Control 363
    Rhododendron Gall Midge 364
        Control 364
    Fletcher Scale 364
        Control 364
    Nantucket Pine Tip Moth 365
        Control 365
Conclusions 365
References 366

INTRODUCTION

Turfgrass represents a relatively uniform monoculture in the landscape when compared to ornamental trees and shrubs. Like most monocultures, lawns usually have a few recurring pests with occasional outbreaks of secondary or less common insects. In contrast, Raupp et al.[1] reported that 20 genera accounted for 77 to 89% of suburban landscape plants in Maryland. However, not all of those plants had perennial insect problems. Thus, the concept of key plants in the landscape as defined by Raupp et al.,[1] i.e., those that have recurring insect problems, can be used to greatly simplify ornamental IPM programs. Conversely, Nielsen[2] defined key plants as those that give the owner pleasure and are highly valued. These highly valued plants may not represent a large proportion of the local landscape, but when insect problems occur on them they require extra care.

Although the diversity of plants in urban landscapes does not have to be an impediment to utilizing IPM in this setting,[3] it is this diversity — and the variable value placed on landscape plants by their owners — that has made a broad IPM approach utilizing action thresholds, biological controls, etc., difficult. Currently, IPM for woody ornamentals relies heavily on cultural and pesticide treatments,[4] despite increasing interest in the development and utilization of biological control strategies. Again, because of the diversity of plants and associated insect problems, plants that are the most common and economically important will receive the greatest initial attention. Therefore, pesticides will remain a part of ornamental IPM until satisfactory alternatives can be found.

Although pesticide use will continue to be a part of landscape plant management, indiscriminate use of these chemicals is no longer tolerated. Laws requiring notification of neighbors and posting of warning signs have been enacted in a number of states. These laws not only increase the effort required in making an application of pesticide, but they also increase awareness of neighborhoods about the amount of pesticides being applied in their area. Consequently, it is essential to time applications of pesticides to maximize their efficacy, to reduce the probability of repeated applications for the same problem, and to minimize their effect on natural enemies. Whether using biological or insecticidal controls, achieving effective timing requires a knowledge of the pest insect biology so that the most susceptible stage is targeted and the number of applications per year can be kept to a minimum.

0-87371-350-8/94/$0.00+$.50
© 1994 by CRC Press Inc.

From the large number of insects that damage ornamental plants,[5] four have been selected to illustrate how targeting susceptible stages can provide effective control while minimizing pesticide usage, and in some cases, replace conventional pesticides with less toxic alternatives.

EXAMPLES

BLACK VINE WEEVIL

The black vine weevil, *Otiorhynchus sulcatus* (F.), is a flightless parthenogenetic insect that is capable of feeding on a large number of plants including many ornamentals. Adults feed on foliage, while larvae eat the roots and phloem tissue of belowground portions of the plant.

Adults begin to emerge from the soil at the end of May in Connecticut, and by mid-July, 85 to 90% of the population is in the adult stage. Following emergence, they feed on host foliage at night and hide in leaf litter during the day.[6] Most black vine weevil females do not disperse long distances, remaining within 10 m of their initial release points.[7,8]

Newly emerged females are not capable of laying eggs until their ovaries mature, a process that takes approximately 30 days at 20°C,[9-12] and 40 to 45 days under field conditions in eastern Pennsylvania[6] and Connecticut.[13] Following ovary maturation, females lay eggs at night, on or in the soil beneath suitable host plants.[10] Some females may oviposit at depths of 20 cm or more, although most eggs are probably laid near the surface of the soil under leaf litter or clods of soil.[14]

Females oviposit from mid July until early September,[6] but larvae hatching from eggs laid in early to mid August have the greatest likelihood of becoming established.[15] Eggs laid in the soil or in other moist locations are most likely to hatch.[7] As larvae mature, they move readily in the soil and may be found throughout the root system of their host.

Black vine weevils overwinter primarily as late instars or prepupae.[6] Adults may also survive the winter,[6,15] but they appear to contribute little to population levels the following year.[13]

Overwintered larvae begin feeding again when the soil temperatures reach 6°C (42°F).[6] It is during this short feeding period in the spring that the greatest damage occurs to woody ornamentals. Pupation occurs within the soil.

Insecticidal Control

Considerable effort has been expended on developing control strategies for the black vine weevil. Insecticides applied to the soil are popular with many nurserymen and landscape maintenance contractors because of the high level of control that was achieved with chlorinated hydrocarbon insecticides during the 1960s and 1970s. However, subsequent attempts to develop soil-applied larvicides have not been highly successful, and none are currently registered for general use on woody ornamentals.

Targeting larvae with conventional pesticides presents a number of problems. The extensive vertical distribution of larvae on the root systems requires large amounts of pesticides to achieve acceptable levels of mortality.[16-19] Late instar larvae are less susceptible to most insecticides than are early instars.[20,21] Finally, soil type or potting media can influence insecticide efficacy.[22,23]

Therefore, adults are considered to be the most suitable target,[20,21,24] since they occur above ground, are susceptible to foliar insecticide residues,[2,25] and they have an extended preoviposition period that provides a large treatment period. Adult emergence can be monitored effectively beneath shrubs with pitfall[20] or pan traps,[13] and the first foliar application should be scheduled 30 days after the first adults are captured. This delay allows a large proportion of the population to emerge as adults, but is short enough that no females reach reproductive maturity before the insecticide is applied.[13,26]

Although this strategy is effective on narrowleaf evergreens where adult feeding damage is not readily observed, or on deciduous hosts that lose damaged leaves each year, it may not be acceptable on broadleafed evergreens. On these species, the extensive leaf notching that can occur during the preoviposition period may preclude a long delay in insecticide applications since the aesthetic value or salability of the plant is reduced.

The number of applications required to control black vine weevil adults is dependent on the residual activity of the insecticide. One application of a short residual insecticide is not sufficient,[13] and three applications spaced 3 weeks apart are commonly recommended. Nielsen and Montgomery[25] examined the residual activity of a number of insecticides on *Taxus media* Rehder in a laboratory bioassay. Based on the data, Nielsen et al.[20] suggested that a single application of a long residual compound such as fenvalerate may be sufficient to control black vine weevil. Hanula[21] tested this hypothesis on field-grown *T. cuspidata* Sieb. & Zucc. and found that a single foliar spray, applied 21 days after the first adults were detected, was as effective in reducing the number of larvae in the next generation as two to four applications of fenvalerate throughout the growing season. Further tests showed that fluvalinate worked equally well.[13]

The type of host plant species also complicates control strategies for the black vine weevil. Despite the success of fenvalerate and fluvalinate in controlling the black vine weevil on *Taxus* spp., these compounds may not be as effective on other hosts. Shanks and Chamberlain[27] found that both compounds were more toxic to black vine weevil adults that were fed *T. media* foliage prior to being challenged, than to adults fed a number of other species. Therefore, host plant species must be considered when choosing a control strategy for this insect. However, in landscape plantings this may not be a serious drawback. Adults feed on a large number of hosts, but they preferentially oviposit on *T. cuspidata* after being exposed to it and will delay oviposition for up to 3 weeks when offered a less desirable host.[28] Consequently, *Taxus* spp. may be the foci of black vine weevil infestations in landscape plantings, so controls aimed at them may reduce the population to acceptable levels.

These data show that conventional control strategies for the black vine weevil that target the most accessible stage, take advantage of the long preoviposition period, and utilize longer residual compounds can result in reduced pesticide usage. For example, a nursery in Connecticut with 200 acres of *Taxus* spp. in cultivation applied 300 lb of acephate active ingredient in three applications spaced 3 weeks apart in 1988. In 1989, the nursery made a single application of fluvalinate at a higher volume per acre, resulting in only 52 lb active ingredient applied.[29] In comparison, a single application of carbofuran to control larvae at the labeled rate for black vine weevils in cranberries (currently not registered for use on ornamentals) would require 400 lb active ingredient per year.

Biological Control

Nematodes that kill insects have not been recorded naturally infecting black vine weevil, but the nematodes *Steinernema feltiae* (Filipjev), *S. carpocapsae* (Weiser), and *Heterorhabditis* sp. have been tested as control agents. In contrast to insecticides, nematodes are targeted at larvae and show promise as biological control agents for this insect. Larvae are less mobile and occupy a moist, protected habitat that favors the use of parasitic microorganisms.[30] Stimmann et al.[24] found that nematodes were most effective for the control of late instars, causing 70 to 90% mortality of larvae in container-grown nursery stock, while the same rates provide less than 60% control of early stage larvae. Delaying applications until late in their development allows black vine weevil larvae to cause significant damage, although no more than what would occur as a result of waiting until the adults emerge to apply an insecticide.

Another more serious problem with the use of nematodes is the ability of steinernematid and heterorhabditid nematodes to disperse from their point of release. Nematode dispersal is limited by pore size and soil moisture.[31] Georgis and Poinar[32,33] found that 90% of *S. feltiae* remained in the top 4 cm (1.5 in.) of columns containing loose sandy loam soil, and 90% of *H. heliothidis* (Khan, Brooks & Hirchmann) and *H. bacteriophora* remained in the top 6 cm (2.4 in.). Subsurface placement of heterorhabditid species improved their distribution.

Black vine weevil larvae are found throughout the root system of *T. cuspidata* plants to depths of 30 cm (12 in.), although the majority (90%) are found in the top 15 cm (6 in.) of soil.[34] Therefore, to be as effective as insecticide sprays for adults, nematodes must move to depths of 15 cm. Only *S. carpocapsae* and *S. feltiae* moved to that depth in sandy loam soils beneath field-grown plants in Connecticut.[34]

However, despite these apparent limitations, nematodes have shown great promise as biological control agents for black vine weevil when applied in the spring before adult emergence. For example, Dorschner et al.[35] reported a 70% reduction in adult populations following a late spring application of nematodes to the roots of hop plants, and similar results were reported for black vine weevil in a cranberry bog[36] and field-grown *T. cuspidata* plants.[34]

A number of fungi are naturally occurring pathogens of larval and adult black vine weevil, including *Beauveria bassiana* (Balsamo) Vuillemin, *Metarhizium anisopliae* var. *anisopliae* (Metshnikoff) Sorokin, *M. flavoviride* Gams and Rozsypal, and *Paecilomyces fumosoroseus* (Wize) Brown and Smith.[37] Poprawski et al.[38] found that *M. flavoviride* was an effective pathogen against first instar larvae, and Soares et al.[39] found *M. anisopliae* to be more effective against sixth instars. However, in both studies the fungi that were most effective were those that were originally isolated from black vine weevil. Again, larvae are the logical targets, since they are located in a moist habitat that favors the development of fungi. Some of these fungi also infect adult black vine weevils, but adults tend to avoid moist locations when resting, so they are less likely to become infected.[6] Zimmermann[40-42] has tested *M. anisopliae* extensively against black vine weevil in greenhouse trials and achieved a high degree of control (80 to 90%), but field trials were less successful.[42] As in the case of nematodes, the short vertical movement of conidia in field soils[43,44] will also have to be overcome to successfully utilize fungi for black vine weevil control. However, in contrast to nematodes, fungi may work well against early instars.[38]

RHODODENDRON GALL MIDGE

The rhododendron gall midge, *Clinodiplosis rhododendri* Felt, feeds primarily on the leaves of *Rhododendron catawbiense* Michaux and *R. maximum*.[45] Its biology has been described for field-grown *R. catawbiense* by Specker and Johnson[46] and for container-grown plants by Hanula.[47] Their results are summarized here.

Prepupae overwinter on or near the surface of the soil beneath the host plant that they developed on as larvae. Adult emergence in the spring coincides closely with bud break and the first vegetative growth flush. The short-lived females lay eggs on expanding vegetative buds and leaves. After hatching, first instars move into protected spaces between expanding leaves or within the rolled edges of new leaves to begin feeding. Larval feeding causes leaf distortion and may prevent leaves from unfolding. Although they do not produce distinct galls, distorted leaves provide a protected habitat where larvae are not as accessible to insecticides. Females will continue to oviposit on leaves where larvae are feeding as long as the leaf is actively growing. Mature larvae move down to the soil where they pupate. The process from egg to adult requires approximately 21 days at 25°C, although larvae may enter a period of dormancy during dry weather resulting in extended developmental times.[46]

The rhododendron gall midge completes at least one generation on each growth flush in container-grown plants[47] and probably on field-grown stock as well.[46] Only a small portion (approximately 20%) of the available buds are utilized during the first flush of growth in either location, while the second generation infests most of the new vegetative growth in the second flush.[46,47] Container-grown plants differ from landscape plantings in that they often undergo a third growth flush. The rhododendron gall midge completes a third generation on this additional growth, and possibly a fourth, since the last growth flush is not well synchronized and larvae are capable of developing on maturing foliage. All life stages of the midge are present on container plants from the beginning of the second growth flush until the end of the growing season.

Control

Foliar insecticide applications are effective against rhododendron gall midge during the first growth flush when midge emergence and oviposition is restricted to a relatively short period of time. However, to be effective these sprays must be directed at the adult, eggs, or young larvae before they move into protected habitats. The period from adult emergence to complete egg hatch is approximately 2 weeks, but the effective treatment period is probably much shorter. Thus, a landscape maintenance contractor with an extensive clientele, or a nursery with large numbers of rhododendron, would have difficulty treating all of the plants in their care during that time if the infestation is widespread. Foliar insecticide applications on later growth flushes are not likely to be effective, since the flushes are not well synchronized and all life stages of the midge may be present over an extended period of time.

In contrast, targeting the prepupae and pupae on the soil provides a large treatment period (October to May) and a high degree of efficacy. Prepupae are susceptible to a variety of insecticides, and treatment of the soil surface is sufficient to provide control.[47] In large-scale nursery trials, a single, early spring application was sufficient to suppress the population below damaging levels for the remainder of the growing season. Treatment can be restricted to only those locations where damaged plants are found, and since only a relatively small portion of the first growth flush is infested, subsequent infestations can be identified and treated before the second growth flush occurs, without extensive damage.

FLETCHER SCALE

Fletcher scale *Parthenolecanium fletcheri* (Cockerell) is a serious pest of *Taxus* spp. and an occasional pest of arborvitae (*Thuja* spp.). They have only one generation per year. Second instar nymphs overwinter and complete their development by May. Egg hatch occurs during June or early July depending on location, but usually happens during a relatively short period of time at any given site. Crawlers move a short distance before settling at a feeding site. Very little development or damage occurs during the remainder of the growing season with the majority of both occurring during the spring.[5]

Control

Control of Fletcher scale can be achieved by targeting crawlers with insecticide[5] or second instar nymphs with dormant oil in the fall.[48] Crawlers are present for a short period of time, so they provide a small treatment period. Conversely, targeting second instar nymphs expands the treatment period without sacrificing plant quality since the scale undergo very little development, and therefore cause little damage until the following year. McClure[48,49] demonstrated that fall or late winter dormant oil sprays provide almost 100% control of overwintering nymphs when plants were thoroughly wetted. Thus, delaying treatment to the fall can provide a high degree of control (without conventional insecticides), a large treatment period, and should help conserve natural enemies.

NANTUCKET PINE TIP MOTH

The examples above demonstrate that timing control measures to target susceptible stages does not always result in a short treatment period. However, for a number of ornamental insect pests this is not the case. Leaf miners, stem and twig borers, and bud and tip moths often have very narrow periods of time during which they are exposed to insecticide applications. This usually occurs between the time that adults emerge and lay eggs, and the time that larvae enter protected habitats. Although this may be a relatively short period for a given individual, control can be further complicated by protracted adult emergence and oviposition. Unfortunately, precise timing models are not readily available for many of these insects because of their relatively low economic importance. The Nantucket pine tip moth, *Rhyacionia frustrana* (Comstock), is an exception since it is a serious pest of young pine plantations and Christmas tree farms in the south, as well as a pest of ornamental hard pines throughout its range.

Berisford[50] recently reviewed the biology of *R. frustrana* so only a brief summary is provided here. Adult emergence coincides with host plant growth flushes. Eggs are laid singly on needles or shoots of susceptible pine species and hatch in approximately 14 days. First instars bore into needles or shoots,[51] and second instars feed at the needle or bud axils encased in a bubble constructed from silk and plant resin.[52] The remaining three instars feed inside the buds and shoots where they pupate after completing their development. *Rhyacionia frustrana* overwinter as pupae in the shoots, and two to five generations may be completed each year depending upon location.

Control

Prior to the development of spray timing methods based on pheromone trap captures and degree-day (DD) accumulations to predict optimum spray dates, control of *R. frustrana* was achieved by spraying trees on a biweekly or monthly schedule throughout the growing season, with limited success.[53] However, Berisford et al.[54] demonstrated that it was important to apply the insecticide dimethoate at the time when 70 to 80% of the eggs had hatched, i.e., when the population was primarily first and second instars. Although one well-timed application was effective for the first two generations in the Georgia piedmont, two applications were required on the third generation to achieve acceptable levels of control. Gargiullo et al.[53] developed predictive models for timing dimethoate and acephate sprays that utilized DD accumulations from the time that the first moth was captured in pheromone traps to 75% egg hatch. Their model was based on upper and lower developmental threshold temperatures of 9.3 and 33.5°C, respectively, with DD sums of 244 for the first generation, 348 for the second generation, and 267 and 467 for the third generation. Their technique was also applied to infestations in the coastal plain of Georgia where *R. frustrana* has four generations.[55] Although the DD sums were relatively close for the first three generations at both locations when dimethoate was used, the optimum time to spray with the pyrethroid fenvalerate was 1 to 2 weeks earlier than that for dimethoate. Thus, the type of insecticide influenced the timing of the application. Both insecticides provided good control, but only if applied at the right time for the given compound. Kudon et al.[56] refined the model for the third generation in the piedmont region and found that, although two applications were required to achieve a high degree of control, timing of the first application was more critical to overall success.

In California, Malinoski and Paine[57] utilized a slightly different and more simplified DD model. They used upper and lower threshold temperatures of 37.5 and 5.5°C, and accumulated DD from the date the first moth was captured to predict peak flight activity (575.2 DD). They then added 111 DD to that total to predict peak egg hatch. The same DD total was used for all four generations. Tests of the model were done with five pyrethroids and an insect growth regulator. All six compounds provided good control with only one application per generation; however, they did not determine if their model was equally effective with other classes of insecticide.

CONCLUSIONS

The examples above demonstrate that with adequate knowledge of an insect's life history and behavior; adequate monitoring of insect populations and life stages; a knowledge of the susceptibility of life stages to various control strategies; and a knowledge of the properties of insecticides and biological controls in the host habitat, one can often utilize that information to effectively time control measures to minimize their environmental impact. In addition, several of the above examples demonstrate that timing insecticides to susceptible stages does not always result in narrower treatment periods. However, it is also clear that considerable work is needed to define these parameters for the myriad of insects that feed on ornamental plants. This endeavor is particularly critical in the promising area of microbial controls, where this information is needed to improve the consistency of these materials so that they will be more widely accepted as alternatives to insecticides.

REFERENCES

1. **Raupp, M. J., Davidson, J. A., Holmes, J. J., and Hellman, J. L.,** The concept of key plants in integrated pest management for landscapes, *J. Arboric.,* 11, 317, 1985.
2. **Nielsen, D. G.,** Integrated pest management (IPM), *Plants Gardens, Brooklyn Bot. Garden Rec.,* 40, 70, 1983.
3. **Raupp, M. J. and Noland, R. M.,** Implementing landscape plant management programs in institutional and residential settings, *J. Arboric.,* 10, 161, 1984.
4. **Hellman, J. L., Davidson, J., and Holmes, J. J.,** Urban integrated pest management in Maryland, in *Advances in Turfgrass Entomology,* Niemczyk, H. D. and Joyner, B. G., Eds., Hammer Graphics, Picqua, OH, 1982, chap. 4.
5. **Johnson, W. T. and Lyon, H. H.,** *Insects That Feed on Trees and Shrubs,* Cornell University Press, Ithaca, NY, 1988.
6. **Smith, F. F.,** Biology and control of the black vine weevil, *USDA Tech. Bull.,* 325, 1932.
7. **Garth, G. S.,** Biology, Ecology and Behavior of the Black Vine Weevil, *Otiorhynchus sulcatus* (Fabricius), in Strawberry Fields in Western Washington, M.S. thesis, Washington State University, Pullman.
8. **Maier, C. T.,** Dispersal of adults of the black vine weevil, *Otiorhynchus sulcatus* (Coleoptera: Curculionidae), in an urban area, *Environ. Entomol.,* 7, 854, 1978.
9. **Cram, W. T. and Pearson, W. D.,** Fecundity of the black vine weevil, *Brachyrhinus sulcatus* (F.) fed on foliage of blueberry, cranberry, and weeds from peat bogs, *Proc. Entomol. Soc. Brit. Columbia,* 62, 25, 1965.
10. **Penman, D. R. and Scott, R. R.,** Adult emergence and egg production of the black vine weevil in Canterbury, *N. Zeal. J. Exp. Agric.,* 4, 385, 1976.
11. **Stenseth, C.,** Effect of temperature on development of *Otiorhynchus sulcatus, Ann. Appl. Biol.,* 91, 179, 1976.
12. **Nielsen, D. G. and Dunlap, M. J.,** Black vine weevil; reproductive potential on selected plants, *Ann. Entomol. Soc. Am.,* 74, 60, 1981.
13. **Hanula, J. L.,** Monitoring adult emergence, ovary maturation, and control of the black vine weevil (Coleoptera: Curculionidae), *J. Entomol. Sci.,* 25, 134, 1990.
14. **Montgomery, M. E. and Nielsen, D. G.,** Embryonic development of the black vine weevil, *Otiorhynchus sulcatus*: effect of temperature and humidity, *Entomol. Exp. Appl.,* 26, 24, 1979.
15. **Garth, G. S. and Shanks, C. H., Jr.,** Some factors affecting infestations of strawberry fields by the black vine weevil in western Washington, *J. Econ. Entomol.,* 71, 443, 1978.
16. **Nielsen, D. G. and Dunlap, M. J.,** Taxus, chemical control of black vine weevil larvae in field-grown plants, Lake County, Ohio, 1985, *Insect. Acar. Tests,* 13, 367, 1988.
17. **Saunders, J. L.,** Carbofuran drench for black vine weevil control on container-grown spruce, *J. Econ. Entomol.,* 63, 1698, 1970.
18. **Shanks, C. H., Jr.,** Granular carbofuran for black vine weevil control in nonflooded cranberry bogs, *J. Econ. Entomol.,* 72, 55, 1979.
19. **Halfhill, J. E.,** Larvicides for black vine weevil on woody ornamentals, *J. Agric. Entomol.,* 2, 292, 1985.
20. **Nielsen, D. G., Dunlap, M. J., and Boggs, J. F.,** Progress report on black vine weevil control, *Ohio Rep.,* 63, 41, 1978.
21. **Hanula, J. L.,** Field tests of fenvalerate for control of black vine weevil, *Conn. Agric. Exp. Sta. Bull.,* 860, 1988.
22. **Nielsen, D. G. and Boggs, J. F.,** Influence of soil type and moisture on the toxicity of insecticides to first-instar black vine weevil (Coleoptera: Curculionidae), *J. Econ. Entomol.,* 78, 753, 1985.
23. **Nielsen, D. G. and Roth, J. R.,** Influence of potting media on toxicity of bendiocarb and carbofuran to first instar black vine weevil (Coleoptera: Curculionidae), *J. Econ. Entomol.,* 78, 742, 1985.
24. **Stimmann, M. W., Kaya, H. K., Bulando, T. M., and Studdart, J. P.,** Black vine weevil management in nursery plants, *Calif. Agric.,* Jan/Feb, 25, 1985.
25. **Nielsen, D. G. and Montgomery, M. E.,** Toxicity and persistence of foliar insecticide sprays against black vine weevil adults, *J. Econ. Entomol.,* 70, 511, 1977.
26. **Hanula, J. L.,** unpublished data, 1991.
27. **Shanks, C. H., Jr. and Chamberlain, J. D.,** Effect of Taxus foliage and extract on the toxicity of some pyrethroid insecticides to adult black vine weevil (Coleoptera: Curculionidae), *J. Econ. Entomol.,* 81, 98, 1988.

28. **Hanula, J. L.,** Oviposition preference and host recognition by the black vine weevil, *Otiorhynchus sulcatus* (Coleoptera: Curculionidae), *Environ. Entomol.*, 17, 694, 1988.

29. **Hanula, J. L.,** unpublished data, 1989.

30. **Benz, G.,** Environment, in *Epizootiology of Insect Diseases,* Fuxa, J. R. and Tanada, Y., Eds., John Wiley & Sons, New York, 1987, chap. 8.

31. **Wallace, H. R.,** *The Biology of Plant Parasitic Nematodes,* Edward Arnold, London, 1963.

32. **Georgis, R. and Poinar, G. O., Jr.,** Effect of soil texture on the distribution and infectivity of *Neoaplectana carpocapsae* (Nematoda: Steinernematidae), *J. Nematol.*, 15, 308, 1983.

33. **Georgis, R. and Poinar, G. O., Jr.,** Vertical migration of *Heterorhabditis bacteriophora* and *H. heliothidis* (Nematoda: Heterorhabditidae) in sandy loam soil, *J. Nematol.*, 15, 652, 1983.

34. **Hanula, J. L.,** Vertical distribution of black vine weevil (Coleoptera: Curculionidae) immatures and infection by entomogenous nematodes in soil columns and field soil, *J. Econ. Entomol.*, 86, 340, 1993.

35. **Dorschner, K. W., Agudelo-Silva, F., and Baird, C. R.,** Use of Heterorhabditid and Steinernematid nematodes to control black vine weevil in hop, *Fla. Entomol.*, 72, 554, 1989.

36. **Shanks, C. H., Jr. and Agudelo-Silva, F.,** Field pathogenicity and persistence of Heterorhabditid and Steinernematid nematodes (Nematoda) infecting black vine weevil larvae (Coleoptera: Curculionidae) in cranberry bogs, *J. Econ. Entomol.*, 83, 107, 1990.

37. **Marchal, M.,** Fungi imperfecti isoles d'une population naturelle d'*Otiorhynchus sulcatus* Fabr. (Coleoptera: Curculionidae), *Rev. Zool. Agric. Pathol. Veg.*, 76, 101, 1977.

38. **Poprawski, T. J., Marchal, M., and Robert, P. H.,** Comparative susceptibility of *Otiorhynchus sulcatus* and *Sitonia lineatus* (Coleoptera: Curculionidae) early stages to five entomopathogenic Hyphomycetes, *Environ. Entomol.*, 14, 247, 1985.

39. **Soares, G. G., Jr., Marchal, M., and Ferron, P.,** Susceptibility of *Otiorhynchus sulcatus* (Coleoptera: Curculionidae) larvae to *Metarhizium anisopliae* and *Metarhizium flavoviride* (Deuteromycotina: Hyphomycetes) at two different temperatures, *Environ. Entomol.*, 12, 1887, 1983.

40. **Zimmermann, G.,** Gewachshausversuche zur Bekampfung des Gefurchten Dichmaulrusslers, *Otiorhynchus sulcatus* F., mit der pilz *Metarhizium anisopliae* (Metsch.) Sorok., *Nachr. Dtsch. Pflanzenschutzdientes*, 33, 103, 1981.

41. **Zimmermann, G.,** Weitere Versuche mit *Metarhizium anisopliae* (Fungi imperfecti, Moniliales) zur Bekampfung des Gefurchten Dickmaulrusslers, *Otiorhynchus sulcatus* F., a Topflanzen im Gewachshaus, *Nachr. Dtsch. Pflanzenschutzdientes*, 36, 55, 1984.

42. **Zimmermann, G.,** Ein schkagkraftiger pilz spart Insektizide, *Gb + Gw*, 85, 294, 1981.

43. **Storey, G. K. and Gardner, W. A.,** Vertical movement of commercially formulated *Beauveria bassiana* conidia through four Georgia soil types, *Environ. Entomol.*, 16, 135, 1987.

44. **Storey, G. K., Gardner, W. A., and Tollner, E. W.,** Penetration and persistence of commercially formulated *Beauveria bassiana* conidia in soil of two tillage systems, *Environ. Entomol.*, 18, 835, 1989.

45. **Gagne, R. J.,** *The Plant Feeding Gall Midges of North America,* Cornell University Press, Ithaca, NY, 1989.

46. **Specker, D. R. and Johnson, W. T.,** Biology and immature stages of the rhododendron gall midge, *Clinodiplosis rhododendri* Felt (Diptera: Cecidomyiidae), *Proc. Entomol. Soc. Wash.*, 90, 343, 1988.

47. **Hanula, J. L.,** Seasonal biology and control of the rhododendron gall midge, *Clinodiplosis rhododendri*, in container grown *Rhododendron catawbiense*, *J. Environ. Hortic.*, 9, 68, 1991.

48. **McClure, M. S.,** Fletcher scale control on Taxus using foliar sprays in autumn, 1986, *Insect. Acar. Tests,* 13, 365, 1988.

49. **McClure, M. S.,** Winter oil spray to control fletcher scale on Taxus, 1987, *Insect. Acar. Tests*, 13, 366, 1988.

50. **Berisford, C. W.,** The Nantucket pine tip moth, in *Dynamics of Forest Insect Populations: Patterns, Causes, Implications,* Berryman, A. A., Ed., Plenum Press, NY, 1988.

51. **Gargiullo, P. M. and Berisford, C. W.,** Life tables for the Nantucket pine tip moth, *Rhyacionia frustrana* (Comstock), and the pitch pine tip moth, *Rhyacionia rigidana* (Fernald) (Lepidoptera: Tortricidae), *Environ. Entomol.*, 12, 1391, 1983.

52. **Yates, H. O., III, Overgaard, N. A., and Koerber, T. A.,** Nantucket pine tip moth, Forest Insect and Disease Leaflet, No. 70, U.S. Department of Agriculture, Forestry Service, Washington, DC, 1981.

53. **Gargiullo, P. M., Berisford, C. W., Canalos, C. G., Richmond, J. A., and Cade, S. C.,** Mathematical descriptions of *Rhyacionia frustrana* (Lepidoptera: Tortricidae) cumulative catches in pheromone traps, cumulative egg hatching, and their use in timing of chemical control, *Environ. Entomol.*, 13, 1681, 1984.

54. **Berisford, C. W., Gargiullo, P. M., and Canalos, C. G.,** Optimum timing for insecticidal control of the Nantucket pine tip moth (Lepidoptera: Tortricidae), *J. Econ. Entomol.*, 77, 174, 1984.

55. **Gargiullo, P. M., Berisford, C. W., and Godbee, J. F., Jr.,** Prediction of optimal timing for chemical control of the Nantucket pine tip moth, *Rhyacionia frustrana* (Comstock) (Lepidoptera: Tortricidae) in the southeastern coastal plain, *J. Econ. Entomol.*, 78, 148, 1985.

56. **Kudon, L. H., Berisford, C. W., and Dalusky, M. J.,** Refinement of a spray timing technique for the Nantucket pine tip moth (Lepidoptera: Tortricidae), *J. Entomol. Sci.*, 23, 180, 1988.

57. **Malinoski, M. K. and Paine, T. D.,** A degree-day model to predict Nantucket pine tip moth, *Rhyacionia frustrana*, flights in southern California, *Environ. Entomol.*, 17, 75, 1988.

Chapter 34

A Case Study of the Impact of the Soil Environment on Insect/Pathogen Interactions: Scarabs in Turfgrass

Michael G. Villani, *Department of Entomology, Cornell University, Geneva, NY*

Stephen R. Krueger, *Boyce Thompson Institute for Plant Research, Ithaca, NY*

Jan Nyrop, *Department of Entomology, Cornell University, Geneva, NY*

(This chapter is a reprint of the authors' contribution to *Use of Pathogens in Scarab Pest Management*, Glare, T. R. and Jackson, T. A., Eds., published in 1992 by Intercept Press, Andover, Hampshire, UK. With permission.)

CONTENTS

Abstract 369
The Scarab Larvae/Turfgrass Complex 369
Behavior of Scarab Larvae in Turfgrass Ecosystems 370
Ecology of Scarab Pathogens in Soil 371
Integration and Conclusions 371
References 382

ABSTRACT

There is considerable interest in using microbial insecticides, including bacteria, entomophagous fungi, and entomogenous nematodes as biological control agents against the scarab pest complex. Because different classes of microbial control agents have different intrinsic modes of movement, invasion, and reproductive ability there is a need to understand more fully the interactions among the target scarab population, the specific control agent, and the soil environment. In this chapter we will integrate several important ecological characteristics of entomopathogenic nematodes, fungi and bacteria, as well as short residual insecticides, and Japanese beetle and European chafer larvae. Based or specific ecological characteristics, some suggestions are offered for treatment strategies that afford the best hope of consistent control for various combinations of larval phenological stage, control agent, and environmental conditions.

THE SCARAB LARVAE/TURFGRASS COMPLEX

In the northeastern United States, turfgrass is subject to intense feeding pressure from a complex of scarab beetles including the Japanese beetle, *Popillia japonica* (Newman) and European chafer, *Rhizotrogus majalis* (Razoumowsky). Insecticides have been the major control tactic used against these insects. However, because many of the plants attacked by the larvae are grown in urban or suburban areas, the potential for human exposure to insecticides, through application or environmental contamination, is great. Despite the efforts expended and costs incurred, the control of soil insects that damage turfgrass and horticultural commodities has been inconsistent since the loss of the cyclodiene insecticides.[1-3] Considering the impact of insecticide use in the urban environment, the importance of ecological research providing information leading to more effective management of scarab larvae is apparent. Information concerning the behavior of larvae is rudimentary. This is despite the fact that species and age-species differences of larval response to soil physical properties will affect both the stress that each species exerts on turfgrass and the efficacy of control tactics.

There is currently considerable interest in using microbial insecticides as biological control agents for the scarab complex in turfgrass. Evaluation of the agents has been equivocal; high levels of insect mortality are often observed in the laboratory while field releases are often disappointing. One reason for this inconsistency may be the lack of basic understanding of the interaction between the control agent, the target insect population, and the turfgrass environment. While each of these factors has been independently studied to varying degrees by basic and applied researchers, their interdependence has largely been ignored. If we assume that a combination of management tactics will be required to manage scarab larvae effectively and safely in turfgrass in the future, then the nature of the specific interactions between target insect populations and control agents in the turfgrass environment must be examined.

BEHAVIOR OF SCARAB LARVAE IN TURFGRASS ECOSYSTEMS

Much of the available literature relating to the impact of soil physical properties on the behavior of members of the scarab larvae complex has recently been reviewed by Tashiro.[4] Japanese beetle and European chafer beetles have somewhat similar seasonal patterns of vertical movement in response to gross temperature changes. After hatching, larvae feed at or near the soil surface from midsummer to late autumn. As surface temperatures cool, larvae migrate down into the soil and overwinter as late instar larvae below the frost line. In the spring, larvae migrate back up to the surface in response to temperature changes, complete feeding, and move down again into the soil profile to pupate and enclose. Field observations indicate that European chafer larvae remain at the soil surface later in the autumn and return to the soil surface earlier in the spring than do Japanese beetle larvae.

Superimposed on this yearly, temperature-mediated movement pattern there may be additional localized movement in response to changes in soil moisture. Various species of larvae have been reported to move deeper in the soil profile in response to decreasing soil moisture. Most of the observations concerning response to moisture are anecdotal.

Even under "favorable" conditions of soil moisture and temperature, field reports indicate that different larval species tend to be found at different soil levels. For example, European chafer larvae tend to feed further below the soil surface than Japanese beetle larvae. The differences in preferred feeding depths of different larval species have an effect on the types of plant tissues attacked. European chafer larvae, which feed for a longer period of time, have a greater potential effect per larvae than do Japanese beetle larva that leave the surface earlier in the autumn and return later in the spring.

Villani and Wright[5] radiographed soil blocks in the laboratory to study the response of three scarab species larvae: Japanese beetle, European chafer, and Oriental beetle (*Anomala orientalis*) to temperature flux. Temperature fluctuations had very little impact on the position of European chafer larvae. Population distribution of European chafer larvae within the soil profile was nearly identical in temperature regimes that were both stable (20°C throughout the profile over duration of experiment) and fluctuating (lowering of temperature from 20° to 2°C in 6°C weekly reductions, then returned to 20°C). This unresponsiveness to temperature conforms with field observations that indicate that European chafer larvae are often found in the upper root zone well into early winter and early spring (at times they feed in the upper root zone under snow if this zone is not frozen).[4] In contrast, the other two scarab species responded rapidly to shifting temperatures. Japanese beetle larvae fed in the upper root zone in the stable temperature regime whereas in the shifting temperature regime larvae moved from the upper root zone downward with the onset of cooling soil (14°C) and returned to the surface as temperature increased. Oriental beetle movement appeared to be more variable, but again larvae tended to remain in the upper root zone in the stable treatment and to respond to lower soil temperatures (8°C) by moving down in the soil profile. An increase in soil temperature moved a portion of the Oriental beetle population back to the upper root zone while not affecting the median population value. Similar movement behavior of Oriental beetle populations in response to warming has been observed in the field.[5]

Radiographs of soil blocks containing third instar Japanese beetle, Oriental beetle, European chafer, and Northern masked chafer (*Cyclocephala borealis*) larvae indicated that these species respond to simulated irrigation and drought.[5] Individual larvae of all species studied, moved upward after the addition of moisture in dry soils. However, median population response and variation around the median varied among species. European chafers showed the least sensitivity to decreased soil moisture; this fact may be related to their ability to escape rapidly from extreme conditions.

In study by Villani and Nyrop,[6] the movement patterns of Japanese beetle and European chafer larvae as influenced by gravity, host plant position, and external disturbances were studied in laboratory soil-turfgrass microcosms. Results demonstrated significantly different movement patterns between species and among age groups. The development stage of the larva has a large effect on Japanese beetle larva behavior and a measurable, but lesser effect on European chafers. Detailed behavior patterns of Japanese beetle and European chafer larvae are summarized in Tables 1 and 2.

ECOLOGY OF SCARAB PATHOGENS IN SOIL

The pathogens of scarab larvae considered here are the bacterium, *Bacillus papilliae*, nematodes (primarily *Heterorhabditis* spp.), and the fungi *Metarhizium anisopliae* and *Beauveria brongniartii*. The soil environment is the common habitat of each member of this group and each has been isolated from natural populations of larvae.[7] Control of Japanese beetle larvae with *B. papilliae* was first attempted by White and Dutky in the 1930s, and augmentation of natural levels of disease continues to be a long-term management option in many geographical regions.[8] Similarly, the nematode *Steinernema glaseri*, was first used in a Japanese beetle control program in New Jersey in the 1930s and 1940s but was discontinued. In recent years, renewed interest in controlling larval populations with nematodes and new initiatives with entomogenous fungi have occurred.[9-13]

Because of its high relative humidity, moderate temperature, and lack of UV radiation, soil is considered a promising habitat for microbial control of soil insects.[14,15] However, while protection from UV radiation is undoubtedly beneficial for all pathogen groups, the positive impact of high relative humidity and moderate temperature is less clear because their effects are mediated by the milieu of microorganisms that are present in the soil. High relative humidity enhances germination and sporulation of fungi and is critical for survival of nematodes. However, the relative humidity in soil at the permanent wilting point of plants is approximately 99%, and moisture is almost never a limiting factor in intensively managed turfgrass ecosystems.[16] An increase in bacterial activity is associated with high soil moisture.[17,18] In addition, the optimum temperature range for microbial decomposition of organic matter[19,20] encompasses the optimum temperature range for development of entomopathogenic fungi and nematodes.[21,22]

It is likely that the decreased survival of fungal propagules reported by Studdert et al.[23] and Krueger et al.[24,25] in soil environments at high temperature and moisture was caused by increased bacterial activity. Unfortunately there is little available information on the possible effect of microbial activity on survival of entomogenous nematodes. Properties of control agents that are used, or may be used, in turfgrass management programs for scarab larvae regulation are summarized in Table 3.

INTEGRATION AND CONCLUSIONS

Based on specific ecological characteristics we present some suggestions for treatment strategies that offer the best hope for consistent control for various combinations of larva phenological stage, control agents, and environmental conditions in Table 4 (Japanese beetles) and Table 5 (European chafer). Clearly, a number of factors must be considered when deciding the proper control agent, or group of agents, that is most appropriate for a particular soil insect pest population under specific environmental conditions. The flexibility of the system, in terms of temporal targeting of the pest population or manipulation of the soil environment to widen the treatment window of an agent, will increase the chance of effective control. To this end we offer several generalizations concerning the interaction of target soil insects, control agents, and the soil environment in the hope that management tactics based on the generalizations will be tested in the field:

1. Behavioral patterns of scarabs generally considered to have overlapping ecological niches may upon closer examination, be considerably different. These differences may enhance or detract from the efficacy of control tactics.

2. Screening of potential control agents should be done under a realistic range of environmental conditions with real target insects rather than representative species. The performance of a control agent will be modified by the environment's impact on the agents, including persistence, mobility, and reproductive ability; while susceptibility of an insect population will be modified by environmental stress, position in the soil profile, and phenological response to the environment. The persistence of the control agent under a variety of environmental conditions is extremely important, in order to realize insect control in the field, as it opens up the window of vulnerability of insect populations that have been shown to be susceptible in the laboratory.

Table 1   **Categorization of Japanese beetle behavior by phenological development**[4-6,9,26-30]

Stage	Depth in Soil	Feeding Behavior	Mechanism
Adult	Teneral adults may spend 2 to 14 days in pupal exuviae several centimeters down in the soil profile before climbing to the surface. Females will move several centimeters into the soil profile to oviposit. Typically several ovipositing bouts occur for each female.	Adult beetles feed on foliage and/or fruit of at least 300 plant species. Adults will not feed on common turfgrass but may feed on a variety of weed and pasture legumes in turfgrass systems.	Adult males are highly attracted to female pheromones. Both males and females are attracted to floral and fruit odors and are stimulated to feed by a number of simple and complex organic sugars.
Egg	Eggs are often oviposited 4 to 10 cm into the soil profile where they are protected from environmental extremes. Eggs hatch approximately 2 weeks after oviposition.	None	None
1st instar	Upon hatching, first instar larvae show little initial movement up to the soil surface. It is possible that they may feed at the hatching depth for several days or weeks before moving up to the soil thatch interface.	It is assumed that neonates will feed on organic material and fine roots in soil.	Little is known about the impact of the physical and biotic environment on the behavior of neonate larvae since they appear to have little mobility at this stage.
2nd instar	Most larvae feeding at the thatch soil interface <5 cm below soil surface. High temperatures and low moisture may cause larvae to move down temporarily into the soil profile.	2nd instar larvae feed on roots of grasses and weeds in soils. Some feeding in thatch layer (stems, rhizomes, and stolons) possible.	2nd instar larvae are attracted to or arrested by the presence of roots in soil. Innate response to disturbance is to burrow down into the soil profile. Unfavorable temperature or moisture may cause larvae to move down into soil profile.
Early 3rd instar		Most larvae feed on the roots of grasses and weeds in soil. Some feeding in thatch layer possible. Larvae found down in the profile appear unresponsive to turfgrass roots.	Larvae show innate response to gravity identical to 2nd instar larvae. Attraction or arrestment by host plants may be diminished. Unfavorable temperature or moisture may cause larvae to move down the soil profile.
Pre-winter 3rd instar	Larvae have begun to move downward into the soil profile. Approximately 60% of Japanese beetle larvae were found in the upper 5 cm of soil while the remainder had moved below this depth. Studies indicate that a soil temperature below 10°C will initiate downward movement.	Larvae, even those in the upper root zone, appear to do little or no feeding.	Larvae appear unresponsive to root material. In the field, larvae move down in the soil profile as soil temperatures fall. Laboratory studies indicate that larval movement is regulated, at least in part, by the physiological state of the organism and is not merely a direct response to changes in the environment.

Early-winter 3rd instar	<5% of larvae can be found in the first 5 cm of soil while approximately 62% can be found at the 5 to 10 cm level and 33% below the 10 cm depth.	Diapausing
Late-winter 3rd instar	Laboratory studies of larvae suggest feeding behavior identical to 2nd instar larvae.	Laboratory studies indicate response similar to 2nd instar larvae. Position in field appears to be in response to cold soil temperatures at surface.
Post-winter 3rd instar	Larvae are moving from overwintering depths to the soil surface to resume spring feeding. Depth of larvae is determined by soil temperature at site. Studies indicate that a soil temperature >10°C will initiate upward movement.	Laboratory data suggests that Japanese beetles show an innate negative geotropic response with arrestment in sod. In addition, if larvae do not locate sod, they commence a random search behavior throughout the soil profile.
Pre-pupae	Majority of larvae are found at the 5 to 10 cm depth in the soil profile.	After cessation of feeding, larvae move downward in the profile. Larvae are unresponsive to host plants and appear motivated primarily by positive geotropic movement. It is possible that larvae form pupal sites with favorable soil moisture and temperature conditions.
Pupae	None	None

Table 2 Categorization of European chafer behavior by phenological development[4-6,9,31-35]

Stage	Depth in Soil	Feeding Behavior	Mechanism
Adult	Teneral adults may spend 2 to 14 days in pupal exuviae several centimeters down in the soil profile before climbing to surface. Oviposition occurs over a short period of time at depth similar to ovipositing Japanese beetle.	There is little feeding by adult European chafers.	Males and females emerge from turfgrass at dusk and are strongly attracted to lights and vertical structures, such as trees. Mating will begin in the trees but coupled pairs will drop to ground to complete mating. Females will oviposit close to mating aggregation site.
Eggs	Eggs are often oviposited 4 to 10 cm into the soil profile in small chambers created by ovipositing females. Eggs hatch approximately 2 weeks after oviposition.	None	None
1st instar	Neonates show strong movement to soil/thatch interface in turfgrass systems.	Feeding occurs on roots of grasses and weed in soil. Some feeding in thatch layer possible.	Larvae are attracted to, or arrested by, the presence of roots in soil. Innate response to disturbance is to burrow down into soil profile. Unfavorable temperature or moisture may cause larvae to move down into soil profile but larvae appear less sensitive to environmental conditions than Japanese beetle.
2nd instar	Most larvae feeding at the thatch soil interface <5 cm below soil surface. However, there appears to be a greater amount of "trivial" movement, both vertical and horizontal, with European chafer larvae when compared with Japanese beetle larvae at a similar stage. High temperatures and low moisture may cause larvae to move temporarily down into the soil profile.		
Early 3rd instar			
Pre-winter 3rd instar			Larvae show innate response to gravity identical to early instar. Attraction or arrestment by host plant may be diminished. Unfavorable temperature or moisture may cause larvae to move down into soil profile but larvae appear less sensitive to environmental conditions than Japanese beetle.
Overwinter 3rd instar	Most larvae are down in the soil profile in response to low temperatures at the soil surface. Larvae will move to the soil/thatch interface if several days of warming occurs.	Larvae down in the soil profile will be quiescent. However, larvae in the root zone will feed if the soil environment warms.	

Stage		
Post-winter 3rd instar	Larvae are moving from overwintering depths to the soil surface to resume spring feeding. Depth of larvae is determined by soil temperature at soil.	Most larvae feed on the roots of grasses and weeds in soil. Some feeding in thatch layer possible.
Pre-pupae	A majority of larvae are found at the 5 to 10 cm depth in the soil surface.	None — After cessation of feeding, larvae move downward in the profile. Larvae are unresponsive to host plants and appear motivated primarily by positive geotropic movement. It is possible that larvae form pupal chambers in sites with favorable soil moisture and temperature conditions.
Pupae	None	

Table 3 Compilation of attributes of control agents that have shown field or laboratory efficacy against Japanese beetle or European chafers in turfgrass[7,8,22,36-40,43-45]

Pathogens	Host Range	Primary Mode of Entry	Pathogenicity[a]	Latent Period[b]	Primary Mode of Transmit	Mobility of Transmit Stage	Persistence in Soil	Environmental Factors Affecting Efficacy
Toxins[c]	Variable	Contact	High	None	None	Passive	Short	Variable
Milky[d]	Narrow	Ingestion	Low	Long	Horizontal	Passive	Extended	Temperature
Fungi[e]	Broad	Contact	Moderate	Moderate	Horizontal	Passive	Variable	Temperature
Nematodes[f]	Relatively broad	Contact	High	Moderate	Horizontal	Active	Variable	Temperature

[a] High: mortality <48 hours postinfection; medium: 48 hours to 2 weeks postinfection; low: >2 weeks postinfection; [b] Time period from infection until production of transmission stage; [c] Biological and short residual synthetics; [d] *Bacillus papilliae* (commercial); [e] *Metarhizium anisopliae* and *Beauveria brongniartii*; [f] *Heterorhabditis* sp.

Table 4 Prediction of stage-specific results of control agents based on integration of ecology of Japanese beetle (Table 1) and properties of control agents (Table 3)

Stage	Toxins	Milky Disease	Fungi	Nematodes
Adult	Treatment of aggregation sites will reduce feeding damage to valued plants but has not been shown to reduce larval populations.	Although adult beetles may be susceptible to milky disease, infected adults have never been reported in nature. Treatment of adults with spores for transport to oviposition site is of questionable value because it is doubtful whether an individual could transport challenging dose of pathogen to a site.	Possible strategies would entail treating feeding sites with conidia or blastospores to infect aggregating females. Reduction in damage may occur through direct mortality of adults at feeding sites. Deposition of inoculum into the soil in close proximity to egg may occur via surface-contaminated or infected ovipositing females. Neonate larvae may contact and become infected by this deposited inoculum.	Aboveground habitats are relatively hostile to heterorhabditid nematodes making treatment of adults to reduce larval populations unattractive.
Egg	Relatively high mortality at egg and neonate stage make this tactic wasteful. Control agent must be in contact with eggs at the site of oviposition for any chance of spatial overlap. Control agents discussed here are not known to infect Japanese beetle larvae at the egg stage in the field although laboratory manipulation may cause mortality through atypical introduction of past chorion.			
1st instar	Treatments for control of 1st instar larvae might be applied at the oviposition zone for control of neonates at egg hatch or, alternatively, agent may be targeted for the soil/thatch interface, the ultimate feeding site for majority of late 1st instar or early 2nd instar larvae. As an overall strategy, the relatively high mortality of neonate larvae may make the targeting of this stage with a bioinsecticide unwise particularly if monitoring of populations will be a component of an overall management strategy. The persistence and mobility of the control agent will determine the placement of the agent if first larvae are targeted.			
	Short residual insecticides placed at the soil/thatch interface at egg hatch may not have sufficient residual to control late 1st instars that reach this zone. Insecticides placed on the turf crown will	First instar larvae actively feeding in soil above 20°C may become infected with milky disease. Soil temperatures at the zone of oviposition will rarely be high enough to allow for pathogen replication in the insect. Spores	The lack of an effective method of depositing fungi to the oviposition zone and the relative lack of movement of neonates within this zone reduces the probability of agent/target contact, thereby adversely impacting this placement	Laboratory studies have suggested that 1st instar Japanese beetle larvae are extremely susceptible to heterorhabditid nematodes in small arenas where nematodes are placed in direct contact with target larvae. Under field conditions nematodes

Stage				
	not move through thatch to reach 1st instar in the oviposition zone.	ingested by 1st instars at this level may multiply as larvae move to warmer soil layers. Larvae infected at this stage will add relatively little to the pool of pathogen in the profile.	strategy. Efficacy of pathogen placed at the soil/thatch interface will depend upon the persistence of the product and the ability to predict the ascent of the population of this zone in the profile.	may have difficulty locating relatively small larvae down in the profile.
2nd instar	Control agents should be targeted for the soil/thatch interface. Agents may reach this position through active or passive transport. Larvae at this stage may move vertically and horizontally in the profile before settling into semipermanent feeding chambers and may have the greatest chance of contacting infective particles.			
	Except for periods of extreme drought 2nd instar larvae should be actively moving and feeding in the upper soil profile. Larvae would contact insecticides both dermally and topically. Efficacy of most insecticides is inversely related to body weight. Relatively rapid activity of toxins at this stage reduces the chance of subsequent damage by larvae.	Second instar larvae have the greatest chance of ingesting pathogen under environmental conditions conducive to pathogen multiplication.	Under suitable conditions, fungal applications at this time will allow replication and further infection of population before larvae have moved down in the soil profile as winter approaches.	Nematodes placed in the upper thatch zone with sufficient irrigation may move through thatch to infect larvae at the interface. Under suitable conditions, nematode applications will allow replication and further infection of population before larvae have moved down in soil profile as winter approaches.
Early 3rd instar	Control agents should be targeted for the soil/thatch interface. Agents may reach this position through active or passive transport.			
	Larvae are at a physiological stage where a small portion of the populations have stopped feeding and begun to move down the profile. Toxins that require ingestion for activity may show reduced efficacy.	Soil temperatures have dropped in the field to levels below 20°C. Spores ingested by larvae may remain in the hemolymph and multiply as temperatures warm in spring.	Treatment of soil/thatch interface to infect larvae moving down profile. Change in soil moisture and temperature may move larvae out of treated zone and/or affect the persistence of pathogen.	Larvae are relatively high in the soil profile and may provide a better target for actively-searching nematodes than earlier instars. High pathogenicity makes nematodes a good choice when larval damage has occurred or is imminent.

Table 4 (Continued) **Prediction of stage-specific results of control agents based on integration of ecology of Japanese beetle (Table 1) and properties of control agents (Table 3)**

Stage	Toxins	Milky Disease	Fungi	Nematodes
Pre-winter 3rd instar	No treatment is recommended at this time. Falling soil temperatures and larvae at a physiological stage where a large % of population has stopped feeding and begun to move down in the profile makes this a less than ideal time to treat. Many soil insecticides are less effective in cool soils.	General lack of feeding by larvae makes ingestion of pathogen unlikely. Soil temperatures make replication of the pathogen within larvae minimal.	Larvae in general are at, or near, the soil/thatch interface but sudden cold temperatures may drive larvae deeper into profile thereby reducing target/agent spatial overlap.	Declining soil temperatures may reduce the searching ability of nematodes while driving larvae down into the soil profile. Differing response to fluctuating environmental conditions may reduce spatial overlap of larval population and nematodes. Moderate latent nematodes through the larval population.
Over-winter 3rd instar	Larvae are down in the profile. Chances for control with insecticides application is remote, regardless of placement.	Lack of feeding and larval depth in soil profile make ingestion of pathogen unlikely. Soil temperatures make replication of pathogen within larvae minimal.	Do not treat at this stage. Larvae are down in the profile and not moving thereby reducing probability of agent/target contact even if pathogen is placed at level of diapausing larvae.	Soil temperatures and position of larvae in the profile preclude the use of nematodes at this stage.
Post-winter 3rd instar	Due to lack of transmission of short residual products, use will be of limited value in reducing damage from current generation and numbers of subsequent generations.	Larvae ingesting pathogen may complete development before chronic nature of disease causes death. Spores ingested earlier may continue to develop and cause high mortality. Larval mortality would do little to reduce damage by larval feeding from current generation but would have the largest impact in	Variable temperatures and high moisture from snowmelt may reduce pathogen efficacy and persistence. Moderate pathogenicity may result in extensive feeding damage by larvae. Large sporulating larvae may provide additional inoculum for infection of next generation if pathogen persistence	If soil temperatures are above threshold for searching behavior, nematodes placed in the soil/thatch interface or at the surface with sufficient irrigation may provide control. Fluctuating soil conditions may reduce overlap of nematodes and larval populations. Large larvae may provide the best target for

Control agents should be targeted for the soil/thatch interface. Agents may reach this position through active or passive transport. Larvae are moving back to this zone to feed for short period before entering pupal stage. Heavy feeding, high moisture, and movement of larvae through treated zone increases the probability that larvae will contact the control agent.

searching nematodes cuing in on larval respiration or waste products, and greatest resource for nematode reproduction.

increasing pathogen titer for control of subsequent generations.

is high.

Pre-pupae and pupae — Treatment at this stage is not recommended as larvae are down in the soil profile in pupation chambers, making contact with pathogen unlikely.

Lack of feeding depth in soil profile make ingestion of pathogen unlikely.

Although pupae may be susceptible to pathogen, the enclosure in pupal cavity and the relatively greater depth of the pre-pupae and pupae at these stages than at earlier stages may make targeting of this stage impractical.

Due to the mobility of adult beetles reduction in pupal populations on a localized basis would have little impact on treated turf the following autumn.

Table 5 Prediction of stage-specific results of control agents based on integration of ecology of European chafer (Table 2) and properties of control agents (Table 3)

Stage	Toxins	Milky Disease	Fungi	Nematodes
Adult	Treatment of aggregation sites may reduce number of larvae in soil. This may be feasible because females will often oviposit near aggregation sites.	Although chafer larvae are susceptible to endemic milky disease, commercial milky disease is not infective to European chafers.	Possible strategies would entail treating feeding sites with conidia or blastospores to infect aggregating females. Reduction in damage may occur through direct mortality of adults at aggregation sites. Deposition of inoculum into the soil in close proximity to eggs may occur via surface-contaminated or infected ovipositing females. Neonate larvae may contact and become infected by this deposited inoculum.	Aboveground habitats are relatively hostile to heterorhabditid nematodes making treatment of adults to reduce larval populations unattractive.
Egg	Relatively high mortality at egg and neonate stage make this tactic wasteful. Control agent must be placed in contact with eggs at the site of oviposition for any chance of spatial overlap. Control agents discussed here are not known to infect European chafer larvae at the egg stage in the field although laboratory manipulation may cause mortality through atypical introduction of past chorion.			
1st through pre-winter 3rd instar	Except for periods of extreme drought chafer larvae should be actively moving and feeding in the upper soil profile. Larvae would contact insecticides both dermally and topically.	Although chafer larvae are susceptible to endemic milky disease, commercial milky disease is not infective to European chafers.	Under suitable conditions, fungal applications at this time will allow replication and further infection of additional individuals in the population before larvae have moved down in soil profile as winter approaches.	Nematodes placed in the upper thatch zone with sufficient irrigation may move through thatch to infect larvae at the interface.

Control agents should be targeted for the soil/thatch interface. Agents may reach this position through active or passive transport. Larvae may move vertically and horizontally in the profile before settling into semipermanent feeding chambers and may have the greatest chance of contacting infective particles. High level of "trivial" movement when compared to Japanese beetles makes encounters with control agents more likely but also decreases the probability that larvae will remain in contact with any particular particle for a long period of time unless the agent physically adheres to the chafer's integument.

Overwinter 3rd instar	Although larvae may be found at the soil/thatch interface in periods of warm weather there is rarely sufficient time for control agents to be effective.			
Post-winter 3rd instar	Control agents should be targeted for the soil/thatch interface. Agents may reach this position through active or passive transport. Larvae are moving back to this zone to feed for a short period before entering pupal stage. Heavy feeding, high moisture and movement of larvae through treated zone increases the probability that larvae will contact the control agents.			
	Due to the lack of transmission of short residual products, use will be of limited value in reducing damage from current generation and reducing numbers of subsequent generations.	Although chafer larvae are susceptible to endemic milky disease, commercial milky disease is not infective to European chafers.	Variable temperatures and high moisture from snowmelt may reduce pathogen efficacy and persistence. Moderate pathogenicity may result in extensive feeding damage by larvae. Large sporulating larvae may provide additional inoculum for infection of next generation if pathogen persistence is high.	If soil temperatures are above threshold for searching behavior, nematodes placed in the soil/thatch interface or at the surface with sufficient irrigation may provide control. Fluctuating soil conditions may reduce spatial overlap of nematodes cuing in on larval respiration or waste products and greatest resource for nematode reproduction.
Pre-pupae and pupae	Treatment at this stage is not recommended as larvae are down in soil profile in pupation chambers, making contact with pathogen unlikely.			
	Due to the mobility of adult beetles, reduction in pupal populations on a localized basis would have little impact on treated turf the following autumn.	Although chafer larvae are susceptible to endemic milky disease, commercial milky disease is not infective to European chafers.	Although pupae may be susceptible to pathogen, the enclosure in the pupal cavity and the relatively greater depth of the prepupae and pupae at these stages than at earlier stages may make targeting this stage impractical.	

3. In general, the most effective target zone for toxins, microbials, and nematodes is the soil/thatch interface; control agents may reach this zone either by passive or active transport. Sampling and monitoring of soil insect populations will serve to optimize the placement of control agents in the field.

REFERENCES

1. **Harris, C. R.,** Factors influencing the toxicity of insecticides in soil, *Annu. Rev. Entomol.,* 17, 177, 1972.
2. **Harris, C. R.,** *Advances in Turfgrass,* Niemczyk, H. D., and Joyner, B. D., Eds., Entomology Chemlawn Corporation, 1982.
3. **Baker, P. B.,** Responses by Japanese and Oriental beetle grubs (Coleoptera: Scarabaeidae) to bendiocarb, chloropyrifos and isofenphos, *J. Econ. Entomol.,* 79, 452, 1986.
4. **Tashiro, H.,** *Turfgrass Insects of the United States and Canada,* Cornell University Press, Ithaca, NY, 1987.
5. **Villani, M. G. and Wright, R. J.,** Use of radiography in behavioral studies of turfgrass-infesting scarab grub species (Coleoptera: Scarabaeidae), *Bulletin Entomological Society of America,* 34(3), 132, 1988.
6. **Villani, M. G. and Nyrop, J.,** Age dependent movement patterns of Japanese beetle and European chafer grubs in soil-turfgrass microcosms, *J. Environ. Entomol.,* in press, 1990.
7. **Fleming, W. E.,** Biological control of the Japanese beetle, *U.S. Department of Agriculture Technical Bulletin* No. 1383, 1968.
8. **Klien, M. G.,** Pest management of soil inhabiting insects with microorganisms, *Ag., Ecosystems Environ.,* 24, 337, 1988.
9. **Villani, M. G. and Wright, R. J.,** Entomogenous nematodes as biological control agents of European chafer and Japanese beetle (Coleoptera: Scarabaeidae), *J. Econ. Entomol.,* 81, 484, 1988.
10. **Wright, R. J., Villani, M. G., and Agudelo-Silva, F.,** Steinernematid and Heterorhabditid nematodes for control of larval European chafer and Japnese beetles (Coleoptera: Scarabaeidae) in potted yew, *J. Econ. Entomol.,* 81, 152, 1988.
11. **Ferron, P.,** Etude en laboratoire des conditins ecologiques favorisant le development de la mycose a *Beauveria tenella* du ver blanc, *Entomophaga,* **12**, 257, 1967.

12. **Keller, S., Keller, E., and Auden, L. A. L.,** Ein Grossversuch zur Bekampfung des Maikafers (*Melolontha melolontha* L.) mit dem Pilz *Beauveria brongnartii* (Sacc.) Petch, *Bull. Societe Entomolgique Suissse,* 59, 47, 1986.
13. **Samuels, K. D. Z., Pinnock, D. E. and Bull, R. M.,** Scarabaeid larvae control in sugarcane using Metarhizium anisopliae, *J. Invert. Pathol.,* 55, 135, 1990.
14. **Wraight, S. P. and Roberts, D. W.,** Insect control efforts with fungi, *J. Ind. Microbiol.,* (Suppl. 2), 77, 1987.
15. **Roberts, D. W.,** World picture of biological control of insects by fungi, *Mem. Inst. Oswaldo cruz, Rio de Janeiro,* 84(Suppl. 3), 89, 1989.
16. **Griffin, D. M.,** Soil moisture and the ecology of soil fungi, *Biological Rev.,* 38, 141, 1963.
17. **Wilson, J. M. and Griffin, D. M.,** Water potential and the respiration of microorganisms in the soil, *Soil Biol. Biochem.,* 7, 199, 1975.
18. **Griffin, D. M.,** Water potential as a selective factor in the microbial ecology of soils, in *Water Potential Relations in Soil Microbiology,* Soil Science Society of America, Special Publication 9, 141, 1981.
19. **Widung, R. E., Garlang, T. R., and Bushbom, R. L.,** The interdependent effects of soil temperature and water content on soil respiration rate in plant root decomposition in arid grassland soils, *Soil Biol. Biochem.,* 7, 373 , 1975.
20. **Nyhan, J. W.,** Influence of soil temperature and water tension on the decomposition rate of carbon-14 labelled herbage, *Soil Sci.,* 121, 288, 1976.
21. **Ferron, P.,** Biological control of insect pests by entomogenous fungi, *Ann. Rev. Entomol.,* 23, 409, 1978.
22. **Kaya, H. K.,** Diseases caused by nematodes, in *Epizootiology of Insect Diseases,* Fuxa, J. R. and Tanda, Y., Eds., John Wiley & Sons, New York, 1987, 453.
23. **Studdert, J. P., Kaya, H. K., and Duniway, J. M.,** Effect of water potential, temperature, and clay coat on survival of *Beauveria bassiana* conidia in a clay and peat soil, *J. Invert. Pathol.,* 55, 417, 1990.
24. **Krueger, S. R., Villani, M. G., Nyrop, P. J., and Roberts, D.,** Effect of soil environment on the efficacy of fungal pathogens against scarab grubs in laboratory bioassays, *J. Biol. Control,* (submitted), 1991a.

25. **Krueger, S. R., Nechols, J. R., and Ramoska, W. A.,** Infection of chinch bug, *Blissus leucopterus leucopterus* (Hemiptera: Lygaeidea), adults from *Beauveria bassiana* (Deuteromycotina: Hyphomycetes) conidia in soi under controlled temperature and moisture conditions, *J. Invert. Pathol.*, 56, in press, 1991b.

26. **Hartzell, A. and McKenna, G. F.,** Vertical migration of Japanese beetle larvae, *Contribution. Boyce Thompson Institute,* 11, 87, 1939.

27. **Hawley, I. M.,** Notes on the biology of the Japanese beetle, *U.S. Department of Agriculture, Bureau of Entomology and Plant Quarantine,* E 615, 1944.

28. **Regniere, J., Rabb, R. L., and Stinner, R. E.,** *Popillia japonica*: Seasonal history and associated Scarabaeidae in eastern North Carolina, *Environ. Entomol.*, 10, 297, 1981a.

29. **Regniere, J., Rabb, R. L., and Stinner, R. E.,** *Popillia japonica*: Simulation of temperature dependant development of immatures, and prediction of adult emergence, *Environ. Entomol.*, 10, 290, 1981b.

30. **Regniere, J., Rabb, R. L., and Stinner, R. E.,** *Popillia japonica*: Effect of soil moisture and texture on survival and development of eggs and first instar grubs, *Environ. Entomol.*, 10, 654, 1981c.

31. **Villani, M. G. and Wright, R. J.,** Managing the scarab grub complex in turfgrass: some ecological considerations, in *Integrated Pest Management for Turfgrass and Ornamentals,* Leslie, A. and Metcalf, R., Eds., U.S. Environmental Protection Agency, Washington, DC, 1989, 127.

32. **Villani, M. G. and Wright, R. J.,** Environmental influences on soil macroarthropod behavior in agricultural systems, *Annu. Rev. Entomol.*, 35, 249, 1990a.

33. **Villani, M. G. and Wright, R. J.,** Environmental considerations in soil insect pest management, in *Handbook of Pest Management in Agriculture*, Vol. 1, Pimentel, D., Ed., CRC Press, Boca Raton, FL, 1990b.

34. **Tashiro, H., Gyrisco, G. G., Gambrell, F. L., Flori, B. J., and Brietfeld, H.,** Biology of the European chafer, *Amphimallon majilis* Razoumowsky (Scarabaeidae) in northeastern United States, *New York State Agricultural Experiment Station Bulletin*, No. 1366, 1969.

35. **Gyrisco, G. G., Whitcomb, W. H., Burrage, R. H., Logothetis, C., and Schwardt, H. H.,** Biology of European chafer, *Amphimallon majalis* Razoumowsky (Scarabaeidae), *Cornell University Agricultural Experiment Station Memo* No. 328, Ithaca, NY, 1954.

36. **Krieg, A.,** Diseases caused by bacteria and other procaryotes, in *Epizootiology of Insect Diseases,* Fuxa, J. R. and Tanda, Y., Eds., John Wiley & Sons, New York, 1987, 323.

37. **Carruthers, R. I. and Soper, R. S.,** Fungal disease, in *Epizootiology of Insect Diseases,* Fuxa, J. R. and Tanda, Y., Eds., John Wiley & Sons, New York, 1987, 357.

38. **Klein, M. G.,** Suppression of white grubs with microorganisms and attractants, in *Integrated Pest Management for Turfgrass and Ornamentals,* Leslie, A. and Metcalf, R., Eds., U.S. Environmental Protection Agency, Washington, DC, 1989, 273.

39. **Gaugler, R.,** Ecological considerations in the biological control of soil-inhabiting insects with entomopathogenic nematodes, *Ag. Ecosystems Environ.*, 24, 351, 1988.

40. **Georgis, R. and Poinar, G., Jr.,** Field effectiveness of entomophilic nematodes Neoaplectana and Heterhabditis, in *Integrated Pest Management for Turfgrass and Ornamentals,* Leslie, A. and Metcalf, R., Eds., U.S. Environmental Protection Agency, Washington, DC, 1989, 215.

41. **Fleming, W. E.,** The Japanese beetle in the United States, *U.S. Department of Agriculture Handbook* No. 236, 1962.

42. **Fleming, W. E.,** Biology of the Japanese beetle, *U.S. Department of Agriculture Technical Bulletin* No. 1449, 1972.

43. **Storey, G. K. and Gardner, W. A.,** Soil profiles of applied aqueous suspensions of commercially-formulated *Beauveria bassiana,* in *Fundamental and Applied Aspects of Invertebrate Pathology,* Samson, R. A., Vlak, J. M., and Peters, D., Eds., Wageningen, The Netherlands, 1986, 257.

44. **Storey, G. K. and Gardner, W. A.,** Vertical movement of commercially formulated *Beauveria bassiana* conidia through four Georgia soil types, *Environ. Entomol.*, 16, 178, 1987.

45. **Storey, G. K. and Gardener, W. A.,** Movement of and aqueous spray of *Beauveria bassiana* into the profile of four Georgia soils, *Environ. Entomol.*, 17, 135, 1988.

Ahmad, S., Streu, H. T., and Vasvary, L. M., The Japanese beetle: a major pest of turfgrass, *Am. Lawn Appl.*, 4(2), 2, 1983.

Fleming, W. E., Integrating control of the Japanese beetle — A historical review, *U.S. Department of Agriculture Technical Bulletin* No. 1545, 1978.

Integrated Management of Disease

Lab tests for biocontrol of fungus
disease on grass

Turfgrass field days — student
presentation on turf disease
experiments

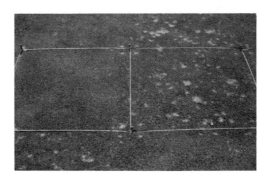

Turf plots showing biocontrol of
dollar spot on creeping bentgrass with *Fusarium
heterosporum* (photo courtesy of Lee Burpee)

Chapter 35

Disease Management for Warm-Season Turfgrasses

Monica L. Elliott, University of Florida, Research and Education Center, Fort Lauderdale, FL

CONTENTS

Introduction 387
Cultural Control Practices 387
    Turfgrass Selection 388
    Mowing 388
    Irrigation 388
    Plant Nutrition 389
    Soil and Thatch Management 389
Chemical Control Practices 389
Biological Control Practices 390
Diagnostic Features and Controls of Warm-Season Turfgrass Diseases 391
Symptoms and Controls for Nematode Infestations of Warm-Season Turfgrasses 394
References 394

INTRODUCTION

Although we can divide turfgrasses into cool- and warm-season types based on plant physiology and even designate the areas of the country where each should be grown, it is people who determine the grass grown in any particular landscape. The result is that we find bentgrass grown in Florida and bermudagrass grown in the central U.S. To confuse the issue even further, bermudagrass putting greens are overseeded with ryegrass or bentgrass during the winter months in the southern states. No matter where turfgrass is grown, the primary goal is to have attractive turfgrass areas. Diseases, of course, can quickly blemish this picturesque setting.

While turfgrass injuries or disorders may look like diseases, they are not diseases and should not be treated as such. An *injury* to turfgrass is a one-time occurrence such as a lightning strike or damage from car tires. A turfgrass *disorder* is an interaction between the plant and its environment caused by the imbalance (too much or too little) of certain physical or chemical factors, such as moisture, temperature, and nutrients. Disorders are not caused by pathogens and so are not infectious. Examples would be nutritional deficiencies, drought, herbicide injury, cold stress, or heat stress.

A *disease* is primarily an interaction between the plant and a pathogen and consists of three components: turfgrass host, pathogen, and the environment in which the host and pathogen interact. In most turfgrass situations, the environment is the key factor for disease development since the turfgrass host and turfgrass pathogens are virtually always present. Thus, the environment may predispose the plant to infection by the pathogen, but the environmental conditions are not the *direct* cause of the disease observed. While turfgrasses may be affected by diseases all year long, individual turf diseases are prominent for only a few months each year, usually due to weather patterns and the resulting environmental effects. Therefore, disease control recommendations are aimed at (1) suppressing the pathogen and (2) altering the environment so it is less favorable for disease development. An integrated disease management program that includes cultural, chemical, and biological control methods is the key to preventing and controlling turfgrass diseases because it uses all available tools to manage diseases below an economic or aesthetic threshold level.

CULTURAL CONTROL PRACTICES

Cultural practices should promote an environment that is *not* conducive to pathogen infection and disease development. If a disease should affect the

0-87371-350-8/94/$0.00+$.50

turfgrass, these practices should be implemented first or, at the very least, implemented at the same time fungicides are applied. If a turfgrass area has a history of developing a particular disease at a particular time of year, then it makes sense to implement cultural practices to prevent this yearly recurrence. Explain to the club membership or homeowner your reasons for altering a practice. Provide them with records indicating disease outbreak, cost of fungicide applications, etc. Explain the potential benefits for altering a maintenance practice in both economical and ecological terms. Discussed below are cultural practices designed to alter the turfgrass environment to prevent diseases or at least lessen their severity.

TURFGRASS SELECTION

The best method for controlling diseases is to prevent their occurrence. Selection of turfgrass species and cultivars should be based on your geographical location and on how the turf will be used and maintained. Selections that are not suited for a particular area will be continually stressed and more susceptible to diseases, thus requiring increased maintenance costs in terms of labor and pesticides. For example, nonirrigated areas would be satisfactory for bahiagrass but not St. Augustinegrass. Soils with a high pH (>7) would be suitable for St. Augustinegrass but not centipedegrass. Sometimes turfgrass is not the appropriate plant material selection for a particular landscape. For example, most warm-season turfgrasses do not thrive in heavily shaded areas.

Certain cultivars of some turf species are resistant to specific diseases. Selection of disease-resistant cultivars is especially important for control of viral diseases, in part because there are no chemicals to control these diseases. The one virus associated with warm-season turfgrasses, Panicum mosaic virus, is most frequently observed in centipedegrass and susceptible St. Augustinegrass cultivars. There are no resistant cultivars of centipedegrass, but the virus is normally a mild pathogen of this grass. Although it is a severe pathogen of St. Augustinegrass, there are a number of St. Augustinegrass cultivars that are resistant to this virus. Therefore, *before* you plant any grass, consult with your local experts to determine the most appropriate turfgrass to plant and then make sure the correct material is installed. This simple task will save you time and money and allow you to be a proper steward of the environment. If it is not practical to replace a poorly adapted turfgrass, then other cultural practices must be initiated to prevent disease occurrence.

MOWING

Mowing is the most common turf maintenance operation. Naturally, it can have a large influence on disease development and prevention. A healthy plant is better able to resist infection. However, turfgrasses used in the landscape that are cut *below* their optimum height will be stressed and more susceptible to some diseases. In addition, every time a mower removes leaf tissue, a wound is created through which a pathogen may enter the plant. To minimize the extent of wounding, always use a sharp mower blade so the leaf blade is cut and not torn or shredded. If at all possible, turf with active disease areas should be mowed last, as mowers may actually spread the pathogen from one location to another. Likewise, clean the equipment between jobs. A thorough rinse with water is sufficient to remove clippings and debris which may carry plant pathogens. This is especially important for lawn care services and golf courses.

For golf course putting greens with active disease areas, raising the height of cut is one of the first cultural practices that should be implemented. Plants produce their energy by a process called photosynthesis. Over the past few years, the height of cut on greens has been reduced substantially with $^3/_{16}$ in. or lower as the standard on bermudagrass putting greens. The low height of cut reduces photosynthesis. In addition, diseases eventually reduce the leaf canopy, and photosynthesis is decreased even more. Raising the height of cut increases the green plant tissue needed for photosynthesis resulting in more energy for turfgrass growth and development and, subsequently, recovery from disease.

In the past, it has been suggested that leaf clippings should be collected when a leaf disease is active. Because these clippings are often disposed of in landfills, this is no longer ecologically acceptable and has become illegal in some states already. In general, do *not* collect leaf clippings unless you have an acceptable method for recycling the material. One method would be composting of clippings. Properly composted organic waste material will have no pathogens, so you will not infect a turfgrass area by using this material in the landscape. In addition, recent studies suggest that using a mulching lawnmower blade with a closed mower deck may help to limit leaf diseases when it is necessary to return clippings to a turf area.

IRRIGATION

While irrigation is essential to prevent drought damage, the amount of water and the timing of its application can prevent or contribute to disease development. Most fungal pathogens require free water or very high humidity to initiate the infection process. Dew, and more importantly, the length of the dew period, which is dependent on temperature and humidity, is a critical factor for disease development. Irrigating in the evening before dew forms or in the morning after the dew evaporates extends the dew period. Therefore, irrigate when dew is already present, usually in the pre-dawn hours. In addition, this will dilute or remove the guttation fluid that can accumulate at the cut leaf tip, and that may provide a food source for some pathogens.

When you do irrigate, apply enough water *each* time to adequately moisten the entire root system. Irrigate to the depth of the roots, but not below them. Shallow irrigations will require you to irrigate more frequently and thus increase the chances for pathogen infection and pathogen movement, since the leaves will remain wet for longer periods of time. Appropriate landscape pruning and planting can enhance air movement and speed the evaporation of water from the turf surface. In general, following good, basic water management principles will enhance your disease management skills.

PLANT NUTRITION

Many diseases are also influenced by the nutritional status of the turfgrass, especially nitrogen. One of the advantages, especially in areas with substantial rainfall, of a slow-release, synthetic nitrogen fertilizer is the constant release of the nitrogen over a specific period of time. This is in contrast to rapid-release (soluble) nitrogen fertilizers which release the nitrogen quickly into the turf environment. Unless the rapid-release nitrogen fertilizer is applied on a regular basis, a cycle can result in which the turfgrass is subjected to excess nitrogen availability followed by a nitrogen deficit. Apply only the amount of nitrogen recommended for the turfgrass being grown. Remember, it is easier to add nitrogen to the soil than to remove it. Excessive nitrogen applications encourage Rhizoctonia diseases, gray leaf spot, "Helminthosporium" leaf spot, and Pythium blight. If nitrogen must be applied when these diseases are active, use a slow-release, synthetic nitrogen source to prevent a sudden flush of leaf growth that is favorable for disease development. Low nitrogen levels encourage diseases such as dollar spot, rust, and anthracnose. If these diseases occur, a rapid-release nitrogen source may be useful to stimulate leaf growth and allow the plant to "outgrow" the disease.

SOIL AND THATCH MANAGEMENT

Physical and chemical soil properties may not affect disease development directly, but they do affect turfgrass health. Soil pH is an important factor for growth of warm-season turfgrasses. For example, centipedegrass and bahiagrass require an optimum soil pH between 5 and 6 whereas zoysiagrass and St. Augustinegrass prefer soil with pH between 6 and 7. A soil pH greatly above or below these optimum values results in turfgrass that is constantly stressed and more susceptible to turfgrass pathogens. Soils that are compacted and poorly drained result in stressed turf also, especially the root system. Installation of drainage or routine aerification of these chronically stressed areas help to reduce disease development.

Thatch is the tightly bound layer of living and dead stems and roots that develops between the zone of green vegetation and soil surface. It is a *natural* component of a turfgrass ecosystem. An *excessive* thatch accumulation indicates that plant tissue is being produced more quickly than it is being decomposed. Factors that impede microbial decomposition of thatch materials are excessively wet or dry conditions, very high or low thatch pH, inadequate or excessive nitrogen levels, and repeated use of chemical pesticides that reduce the level of microorganisms responsible for decomposition. Thatch accumulation is probably most severe with zoysiagrass and does require periodic renovations.

Many of the fungi that cause turfgrass diseases do not simply disappear when a disease is no longer active. Some pathogens remain dormant when environmental conditions are not conducive for their growth. Other pathogens continue to grow and survive by decomposing organic matter in the thatch layer. In fact, these and other fungi are important components in the thatch decomposition process. In both cases, the fungi are not causing any harm to the turf.

CHEMICAL CONTROL PRACTICES

Except for St. Augustinegrass decline, caused by the Panicum mosaic virus, diseases of warm-season turfgrasses are caused by fungi. Chemical control of these fungal diseases is accomplished by using fungicides. The primary misconception turf managers have concerning fungicides is that these chemicals are fungicidal. In fact, most fungicides are simply fungistatic. Therefore, fungicides do *not* eliminate the pathogens from the turfgrass area but primarily *suppress* the growth of the fungal pathogens to prevent them from infecting the plant during the time period when the environment is conducive for disease development. Turfgrass fungicides can be divided into two broad categories based on the location of their activity: (1) contact fungicides, and (2) systemic fungicides, which include true systemic compounds and local-systemic compounds.

Contact fungicides are generally applied to the leaf and stem surfaces of turfgrasses and act as protective compounds. These fungicides remain on the plant surface and do not penetrate into the plant. They remain active only as long as the fungicide remains on the plant in sufficient concentration to inhibit fungi. Leaves which emerge after the fungicide has been applied will not be protected. In addition, fungicide on the plant surface will be gradually lost due to mowing, irrigation, rainfall, and decomposition. Consequently, they are only effective for short durations, usually 7 to 14 days. To obtain optimum protection, it is important that contact fungicides evenly coat the entire leaf surface and are allowed to dry completely before irrigating or mowing. Ideally, the turf area should be mowed and irrigated prior to a fungicide application to allow a maximum time interval between fungicide application and the next turfgrass maintenance operation.

Contact fungicides are normally used to control foliar diseases and not root/crown diseases. The exceptions would be chloroneb and ethazol when used to control Pythium root rot. Some contact fungicides have a broad spectrum of disease control activity and have been used extensively in the turf industry for a number of years. Today, they are often mixed with systemic fungicides to obtain a fungicide mixture with broad spectrum control or to enhance the activity of the systemic fungicide.

True systemic fungicides are chemicals that do penetrate plant surfaces and are then translocated (moved) within the plant, either in the xylem or phloem tissue. Xylem tissue is formed from plant water-conducting cells, so compounds in the xylem move primarily in an upward direction with the water stream. Phloem tissue is formed from tubes which move photosynthates (plant carbohydrates and sugars) from their site of production (source) to other plant organs (sinks) in upward and downward directions. With turfgrasses, this usually means fully expanded green leaves are sources and roots are sinks, so most photosynthates move downward. Except for fosetyl-Al which is translocated in both xylem and phloem (primarily phloem) tissue, systemic fungicides are xylem-limited and move primarily upward. Therefore, protection of root and crown tissue with systemic fungicides often requires that these fungicides be watered in after application to allow for absorption by roots and upward movement in the plant.

In general, systemic fungicides have both curative and protective properties with extended residual activity. Because systemic fungicides are absorbed by the plants, they "work" inside the plant to (1) control pathogenic fungi that have already entered the plant and initiated a disease (curative action) and (2) inhibit fungi that enter the plant from initiating a disease (preventative action). Their residual activity is also due to the fact that they are absorbed by the plant. Once a systemic fungicide is inside the plant, it will not be removed by rain or irrigation, and newly emerged leaves may contain sufficient concentrations of the fungicide to protect them from fungal infection. Therefore, systemic fungicides do not need to be applied as often as contact fungicides; usually, 15- to 30-day intervals are adequate. Systemic fungicides usually have a very specific mode of action and do not have as broad a spectrum of disease control as contact fungicides.

Local-systemic fungicides are capable of penetrating the plant surface but only move very short distances within the plant and usually not within the xylem or phloem tissue. The majority of fungicide remains on or near the plant surface. Included in this group of fungicides are iprodione and vinclozolin. These fungicides are primarily protective whereas the true systemic compounds have both curative and protective properties.

You would not think of giving a family member any medication without reading the instructions first. Turfgrass fungicides deserve the same amount of respect. In addition to rates and intervals for application, labels provide information concerning the use or nonuse of surfactants with the material, compatibility with other pesticides or fertilizers, amount of water to use in the application process, posting or reentry restrictions, etc. Take the time to read label instructions completely at least once each year as they do change frequently.

BIOLOGICAL CONTROL PRACTICES

Biological control of turfgrass diseases is a new area of disease management that is still in the experimental research phase for warm-season turfgrasses. Experiments are in progress concerning the use of nonpathogenic fungi and bacteria for disease control. However, the most active area of research involves the use of natural organic fertilizers for disease suppression. Some of these products are thought to stimulate the growth of microorganisms which antagonize turfgrass pathogens. In both cases, further testing is required to substantiate their value in the consistent control of turfgrass diseases and the proper methods for their use. When that is accomplished, biological control will be routinely incorporated into an integrated turfgrass management program.

DIAGNOSTIC FEATURES AND CONTROLS OF WARM-SEASON TURFGRASS DISEASES

Listed below are the most common diseases of warm-season turfgrasses. An asterisk (*) is used to denote the most susceptible turfgrass species for each disease. Cultural control recommendations are brief since more detailed information on these practices was discussed previously.

Disease: Anthracnose

Causal Agent(s): *Colletotrichum graminicola*

Susceptible Grasses: Bahiagrass, bermudagrass and centipedegrass*; there are no resistant cultivars.

Occurrence: Primarily in the spring when the weather is warm and very humid or raining frequently.

Symptoms/Signs: Reddish-brown to brown lesions with yellow halos expand to cause yellowing of entire leaf blade. Tiller infection results in stem girdling. Small, irregular patches of dead plants are observed. Fruiting bodies are dark cushion-like bodies with small black spines that can be observed with a hand lens.

Controls: Avoid nitrogen deficiency and N:K fertility imbalances, alleviate compaction, improve drainage, and remove excessive thatch. Stress due to insects or nematodes should be eliminated. Raise mowing height. Irrigate deeply and infrequently. Chemical controls include benomyl, chlorothalonil, fenarimol, mancozeb, propiconazole, thiophanate methyl, and triadimefon.

Disease: Bermudagrass Decline

Causal Agent(s): *Gaeumannomyces graminis* var. *graminis*

Susceptible Grasses: Bermudagrass*; there are no resistant cultivars.

Occurrence: Primarily observed in Florida during those months when temperatures are above 75 to 80°F (including night) with high humidity and frequent rainfall. This disease is only observed on putting greens due to the stress imposed by the low cutting height, with outside edges exhibiting the symptoms first.

Symptoms/Signs: Initial symptoms of this root rot disease are irregular, yellow (chlorotic) patches (a few inches to a few feet in diameter) that

will continue to thin out to bare ground. Roots with beginning symptoms are usually off-white in color with numerous isolated black lesions. Eventually, roots become short, thin, and rotted (black color). Black strands of fungi (runner hyphae), visible with a microscope, are present on surface of diseased stolons and roots.

Controls: Since this disease is very difficult to control once it is established, *preventative* measures are necessary to control or decrease the damage. Aerify greens frequently during the late spring and summer, especially greens that are compacted. Topdress frequently with the appropriate root-zone mix. Balance nitrogen applications with equal or increased amounts of potassium. Raise mowing height, especially during stressful growth periods. Do not add material that will increase soil pH. The fungicides benomyl, fenarimol, propiconazole, thiophanate methyl, and triadimefon have been useful in controlling similar patch diseases as is indicated on their labels, but only when used preventively.

Disease(s): Rhizoctonia Blight (Brown Patch) Rhizoctonia Leaf and Sheath Spot

Causal Agent(s): *Rhizoctonia solani* *Rhizoctonia zeae*, *Rhizoctonia oryzae*

Occurrence: Rhizoctonia blight is most frequently observed in the spring and fall, especially when humidity is high or rainfall is frequent. The leaf and sheath spot symptoms are observed in the summer when the temperature and humidity are very high (>85°F).

Susceptible Grasses: Bahiagrass, bermudagrass, centipedegrass*, St. Augustinegrass*, and zoysiagrass*

Symptoms/Signs: Rhizoctonia blight begins as small, circular light green patches that turn yellow and then brown or straw-colored. Patches expand to several feet in diameter. Turf at outer margin of patch may appear dark and wilted (smoke ring). Whole leaf fascicles pull up easily due to a soft, dark-color rot at the base of the leaf sheath. Eventually, entire tillers will easily separate from the stolons.

The leaf and sheath spot symptoms include distinct light-brown foliar lesions and, more common, total leaf blight resulting in a reddish-yellow coloration of the leaves.

Controls: Avoid excessive use of nitrogen, especially quick-release nitrogen sources. Irrigate deeply and infrequently. Chemical controls include chlorothalonil, fenarimol, iprodione, maneb, mancozeb, PCNB, propiconazole, thiram, and triadimefon. Use benomyl and thiophanate methyl *only* for Rhizoctonia blight as *R. zeae* and *R. oryzae* are insensitive to these fungicides.

Disease: Dollar Spot

Causal Agent(s): *Sclerotinia homoeocarpa* (*Lanzia* spp. and *Moellerodiscus* spp.)

Susceptible Grasses: Bahiagrass*, bermudagrass*, centipedegrass, St. Augustinegrass, zoysiagrass*. There are no resistant cultivars.

Occurrence: Most frequently observed in the spring and fall and only on turf with low nitrogen levels.

Symptoms/Signs: On close-cut grass such as bermudagrass putting greens, small (< 2 in.) bleached patches of dead grass will form. On higher-cut grass, patches will be larger and more diffuse. These patches do not expand but do coalesce with other spots to form large patches. Leaves have irregular, light tan lesions with distinct brown borders. White, cottony mycelium may be observed in early morning when dew is present.

Controls: Avoid nitrogen deficiency. Fertilize with soluble nitrogen sources when symptoms first appear. Irrigate deeply and infrequently. Chemical controls include benomyl, chlorothalonil, fenarimol, iprodione, mancozeb, maneb, propiconazole, thiophanate methyl, thiram, triadimefon, and vinclozolin.

Disease: Cercospora Leaf Spot

Causal Agent(s): *Phaeoramularia* (*Cercospora*) *fusimaculans*

Susceptible Grasses: St. Augustinegrass*; bitter-blue cultivars (e.g., Floratam) are less susceptible than yellow-green cultivars.

Occurrence: Most frequently observed in the summer months.

Symptoms/Signs: Small, dark brown or purple lesions develop on leaf blades and sheaths. Spots elongate and become tan as the disease becomes more severe. High disease severity results in leaf death and turf areas that thin-out.

Controls: Avoid nitrogen deficiency. Irrigate deeply and less frequently. Contact fungicides such as chlorothalonil, iprodione, or mancozeb may provide disease suppression.

Disease: Fairy Ring

Causal Agent(s): *Chlorophyllum, Marasmius, Lepiota, Lycoperdon,* and other basidiomycete fungi

Susceptible Grasses: All warm-season turfgrasses.

Occurrence: Rings are primarily observed in the spring and summer.

Symptoms/Signs: The most common symptoms for lawn grasses are circular to semicircular bands of dark green turf with or without mushrooms present in the band. Some fairy rings are bands of dead turf that may be associated with bands of dark green turf. Rings expand each year. On new putting greens, the rings are often associated with hydrophobic areas.

Controls: If necessary for aesthetic purposes, mask dark green fairy ring symptoms with nitrogen fertilizers. Irrigate deeply if turf area is susceptible to dryness or the fungi have created a hydrophobic area. Addition of a wetting agent may also be useful. Remove mushrooms as some are poisonous. Before planting, eliminate large sources of organic matter such as tree stumps, wood building materials, etc. as these provide a food base for the fungi that cause fairy rings. No fungicides are currently registered. The only way to eliminate fungi causing the rings is to fumigate and replant.

Disease: Gray Leaf Spot

Causal Agent(s): *Pyricularia grisea*

Susceptible Grasses: St. Augustinegrass*; yellow-green cultivars are less susceptible than blue-green/bitter-blue cultivars. However, the yellow-green cultivars are more susceptible to chinch bug damage.

Occurrence: Most frequently observed during the summer months during warm, humid weather. St. Augustinegrass treated with the herbicide atrazine is more susceptible to infection due to herbicide-induced stress.

Symptoms/Signs: Lesions begin as small, brown spots that expand into oval lesions with tan centers and dark purple or brown margins. Yellow halos may be present. During warm humid weather, lesions are covered with gray velvet mat of mycelium. Leaves wither resulting in a scorched appearance.

Controls: Avoid excessive use of nitrogen. Use slow-release nitrogen sources. Irrigate deeply and infrequently. Chemical controls include chlorothalonil, propiconazole (sod only), and thiophanate methyl combined with mancozeb.

Disease: "Helminthosporium" Leaf Spot/ Melting-Out

Causal Agent(s): *Bipolaris, Drechslera, Exserohilum* spp. (previously *Helminthosporium* fungi), and *Curvularia* spp.

Susceptible Grasses: Bermudagrass*, St. Augustine-grass, zoysiagrass. There are no resistant cultivars.

Occurrence: At any given time of the year, at least one species within this group of fungi can be isolated from symptomatic grass. As a general rule, the leaf spot phase is most severe during mild periods in the spring and fall, but the root rot or "melting-out" phase is most likely to occur during the summer months.

Symptoms/Signs: Leaf spot symptoms vary with each specific pathogen/host combination from small, solid brown or purple color lesions to expanded lesions with bleached centers that girdle the leaf blade. Severely infected leaves turn reddish-brown to straw color. There will be an overall appearance of thinning and chlorosis. "Melting-out" occurs under severe infection as turf areas thin and die due to extensive rot of stems, crowns, and roots.

Controls: Avoid excessive use of nitrogen and, more importantly, balance fertility components. Irrigate deeply and less frequently. Avoid thatch accumulation. Raise mowing height during disease outbreaks. Chemical controls include chlorothalonil, iprodione, maneb, mancozeb, propiconazole, and vinclozolin. However, overuse of certain fungicides (benomyl) and herbicides (MCPP, 2,4-D and dicamba) have been shown to enhance disease development on cool-season grasses. Note if there is a correlation between disease outbreaks and the use of certain pesticides, then avoid using that pesticide, or, if that pesticide must be used, treat preventatively for "Helminthosporium" disease control prior to its use.

Disease: Pythium Root Rot

Causal Agent(s): *Pythium* spp. These species are probably different from those that cause Pythium blight which is a rare disease on warm-season turfgrasses.

Susceptible Grasses: All warm-season turfgrasses. There are no resistant cultivars.

Occurrence: Symptoms may appear at any time of the year, but they will always be associated with wet conditions — either from too much precipitation or too much irrigation. Poor drainage conditions will compound this problem. Root damage from nematodes may contribute to this disease also.

Symptoms/Signs: Symptoms are typically a nonspecific decline in turf quality with a general turf chlorosis and thinning observed. Roots appear thin with few root hairs and have

a general discoloration. Turf does not respond to fertilizer applications.

Controls: Improve drainage, aerate, and reduce irrigation. When symptoms are present, irrigate only as needed to prevent permanent wilt. Increase mowing height to reduce stress and promote root growth. The following fungicides *may* be effective: chloroneb, ethazol, metalaxyl, propamocarb, and a mixture of metalaxyl with mancozeb. These fungicides should be irrigated into the root zone. Reduction of high nematode populations may be useful also.

Disease: Rust

Causal Agent(s): *Puccinia* spp.

Susceptible Grasses: Bermudagrass, St. Augustine-grass, and zoysiagrass*. There are no resistant cultivars.

Occurrence: Varies with location, pathogen species, and turfgrass species.

Symptoms/Signs: Orange to reddish-brown pustules on leaves. Severe infections cause yellowing of leaves and thin turf.

Controls: Avoid all practices that stress the turf — nutrient and water deficiencies, mowing too short, etc. Irrigate deeply and less frequently. Chemical controls include maneb, mancozeb, propiconazole, and triadimefon.

Disease: Spring Dead Spot

Causal Agent(s): *Leptosphaeria korrae*, *Ophios-phaerella herpotricha*, or *Gaeumannomyces graminis* var. *graminis*

Susceptible Grasses: Bermudagrass*. There are no resistant cultivars.

Occurrence: Observed in those areas where bermudagrass undergoes dormancy due to cold weather. The pathogenic fungi actually infect the grass in the fall, but aboveground symptoms are most prevalent the following spring.

Symptoms/Signs: Circular patches of bleached, sunken, dead grass become visible as dormant turf resumes growth in spring. Root, crown, and stolon rot is evident on plants in these patches.

Controls: Avoid low mowing heights, thatch, compaction, and excessive use of nitrogen fertilizers during the late summer. The use of ammonium-nitrogen fertilizers balanced with potassium help to alleviate the disease over time. Chemical controls include benomyl, fenarimol, and propiconazole.

Disease: St. Augustinegrass Decline

Causal Agent(s): Panicum Mosaic Virus

Susceptible Grasses: Centipedegrass and St. Augustinegrass*. There are no resistant centipedegrass cultivars. Resistant St. Augustinegrass cultivars include Floratam, Floralawn, Raleigh, and Seville.

Symptoms/Signs: Initially, a chlorotic (yellow) mosaic or mottle on leaf blades is observed that gradually becomes more extensive until area appears uniformly yellow and thin. Death may eventually occur with susceptible St. Augustinegrass cultivars.

Controls: Do not plant susceptible cultivars. No chemical will control this disease because it is a virus that is primarily spread by mechanical sap transmission.

SYMPTOMS AND CONTROLS FOR NEMATODE INFESTATIONS OF WARM-SEASON TURFGRASSES

Causal Agent(s): Sting, Ring, Stubby-Root, Lance, Root-Knot, Spiral, Dagger, and Cyst (see Chapter 44 for more information.)

Susceptible Grasses: All warm-season turfgrasses are affected by one or more of the above nematodes.

Occurrence: Most of the nematodes listed are widely distributed throughout the areas where warm-season turfgrasses are grown. The population levels required for damage vary based on soil type, geographic location, host-nematode interactions, and the general health of the turf affected.

Symptoms: Irregular patches of turf, due to degradation of the root system, will decline resulting in chlorosis and necrosis. Affected plants will wilt easily and be slow to recover from the wilt. Plants in these patches will have short roots with little branching, as compared to healthy roots of the same turfgrass species. Roots will appear discolored, but not black, with or without obvious lesions. Root tips may be swollen. The root system can be so reduced that it is unable to hold soil together when plugs or cores are lifted. It is quite likely that root pathogens will also be present, since the nematodes have weakened the root system making it more susceptible to pathogen infection.

Sampling: Do *not* consider treatment for nematodes without submitting soil samples for analysis. Sample from the *edge* of a symptomatic area and not from the center. As with root diseases, nematodes are often blamed for loss in turf vigor and quality without a proper diagnosis.

This results in unnecessary applications of nematicides. It is not uncommon to observe a "greening" response to a nematicide, even if few nematodes were present. This is due to the release of nitrogen and other nutrients from organisms killed by the nematicide placed in the soil.

Cultural Controls: Irrigate to promote a healthy, deep root system. Avoid excessive use of nitrogen fertilizers, balance nitrogen applications with potassium applications, and correct any nutrient deficiencies that stress the turf. Avoid any cultural practice that will stress the turf, e.g., mowing too short. You cannot eliminate the nematodes, but you can reduce their number or prevent their number from increasing to damaging levels.

Chemical Controls: Never apply a nematicide without obtaining a soil sample first to determine if there is a justification for using a nematicide. Nematicides will only *temporarily* affect the nematode population. Also, nematodes vary in their susceptibility to nematicides. Consult with your local experts before selecting a nematicide. Unless cultural practices are initiated at the same time, the nematode populations will likely increase to prenematicide levels, sometimes rather rapidly. Products currently registered for nematode control are fenamiphos, ethoprop, and isazophos. However, these compounds are severely restricted in their use, and *all* must be applied by certified pesticide applicators. None is registered for use on home lawns. Do not overuse nematicides, as enhanced biodegradation, already documented for fenamiphos in Florida, could develop.

REFERENCES

1. **Burpee, L. and Martin, B.,** Biology of *Rhizoctonia* species associated with turfgrasses, *Plant Dis.*, 76(2), 112, 1992.
2. **Colbough, P. F., Hipp, B. W., and Knowles, T.,** Influence of clippling recycling on disease incidence in three turfgrass species, *Phytopathology*, 82(Abstr.), 1109, 1992.
3. **Couch, H. B., Lucas, L. T., and Haywood, R. A.,** The nature and control of Rhizoctonia blight, *Golf Course Manage.*, 58(6), 48, 1990.
4. **Freeman, T. E.,** Diseases of Southern Turfgrasses, Agricultural Experimental Station Bulletin 713, University of Florida, Gainesville, 1967.

5. **Holcomb, G. E., Liu, T.-Y. Z., and Derrick, K. S.,** Comparison of isolates of panicum mosaic virus from St. Augustinegrass and centipedegrass, *Plant Dis.*, 73(4), 355, 1989.

6. **Lyr, H., Ed.,** *Modern Seletive Fungicides*, John Wiley & Sons, New York, 1987.

7. **McCarty, L. B., Black, R. J., and Ruppert, K. C.,** Florida Lawn Handbook, SP-45, Florida Cooperative Extension Service, Gainesville, 1991.

8. **Smiley, R. W., Dernoeden, P. H., and Clarke, B. B.,** *Compendium of Turfgrass Diseases*, 2nd ed., APS Press, St. Paul, MN, 1992.

9. **Smith, J. D., Jackson, N., and Woolhouse, A. R.,** *Fungal Diseases of Amenity Turf Grasses*, 3rd ed., E. & F.N. Spon, London, 1989.

Chapter 36

Use of Disease Models for Turfgrass Management Decisions

William W. Shane, *Southwest Michigan Research and Extension Center, Michigan State University, Benton Harbor, MI*

CONTENTS

Introduction 397
Current Models for Turfgrass Diseases 397
    Anthracnose Foliar Blight 397
    Pythium Blight 398
    Dollar Spot 400
    Brown Patch 401
    Assumptions of Predictive Models 401
Delivery Systems 401
    Electronic Predictors 401
    Integration with Irrigation Control Systems 402
    Prediction Based on Regional Weather Stations 402
Summary 402
References 403

INTRODUCTION

Maintenance of turfgrass quality on golf courses requires the use of fungicides because genetic resistance and cultural management strategies do not provide adequate control of many diseases. Ideally, fungicides are targeted to stop the infection, internal spread, and/or reproduction of pathogens. In actuality, the increase of fungal pathogen populations is difficult to monitor because of their microscopic size. Pathogens infect and colonize plant tissue before symptoms are evident to the unaided eye. Temperature, moisture, and humidity have profound effects on pathogen activity and on the effectiveness of applied pesticides. It is difficult, however, to judge the effects of weather on turfgrass pathogens and to always use fungicides most efficiently.

Predictive models for guiding management actions have proven useful with several plant diseases. Many farmers rely on predictive models in controlling scab on apples,[9] leaf spot diseases on peanuts,[12] and Cercospora leaf spot on sugar beets.[13] These crops have a high value per hectare, and multiple fungicide applications are made in single growing seasons. Predictive systems for these diseases use weather data such as temperature, relative humidity, and leaf wetness to determine the potential for infection and disease development.

Anthracnose, Pythium blight, dollar spot, and brown patch are diseases of turfgrasses for which predictive models have been developed. Most golf courses are sprayed routinely to prevent damage from one or more of these diseases. This chapter will describe predictive models for these diseases, weather data collection systems, and future prospects for the use of disease models for turfgrass management.

CURRENT MODELS FOR TURFGRASS DISEASES

ANTHRACNOSE FOLIAR BLIGHT

Anthracnose, caused by *Colletotrichum graminicola*, is a problem on annual bluegrass and bentgrass. Roots, stolons, and leaves can become infected by the causal fungus. During hot weather, leaf blighting is produced and appears as round to elongate reddish-brown lesions, with characteristic microscopic spines called setae. Under favorable conditions, abundant conidia are produced on blighted leaves. *C. graminicola* is considered to be a weak pathogen because the disease usually develops following or during periods of environmental stress. Field trials

0-87371-350-8/94/$0.00+$.50
© 1994 by CRC Press Inc.

have shown, however, that fungicide treatment can effectively reduce anthracnose damage.

A model predicting anthracnose foliar blight, based on growth chamber studies, was developed at Michigan State University.[2,3] The model generates an anthracnose severity index (ASI) using the maximum temperature and number of hours of leaf wetness averaged over three consecutive days (Table 1). An ASI value of two is designated as the minimum condition for infection.[3] Greater damage from anthracnose was associated with higher ASI values. The model assumes that adequate inoculum and a susceptible host are present. This model was developed for anthracnose foliar blight occurring during hot weather. The anthracnose pathogen also causes a basal rot of annual bluegrasses and bentgrasses during cool weather;[17] however, the model is not designed to predict Colletotrichum basal rot.

No independent evaluation of the anthracnose model has been published, although the authors report it shows good accuracy.[3,19] In validation trials, the model sometimes predicted disease development in early May, but none occurred.[3] This failure was attributed by the authors to a lack of anthracnose spores, as determined with the use of live trap plants. Guidelines for the use of the ASI

to schedule fungicide applications have not been specified, beyond the suggestion that the onset of days with ASI greater than two would signal the need for fungicide applications in those sites with a perennial anthracnose problem.[3] Temperature is the most important factor in the anthracnose model (Table 1). Essentially, pathogen activity is predicted when the average daily temperature for the 3-day window exceeds 22°C (72°F). Under Ohio conditions, ASI values of two or more are very common (Figure 1), indicating that this threshold may predict disease too often.

PYTHIUM BLIGHT

Pythium blight, caused by *Pythium* spp. is an important disease on highly managed turf during hot, humid weather. *Pythium aphanidermatum* is the primary cause of Pythium blight in central Ohio, but other species may be involved in other areas of the U.S.[16] Pythium blight appears as irregular patches of blighted turf that are distinct on close-cut areas, but diffuse on high-cut turfgrasses. Individual leaves are dark-colored, watersoaked, and fade to a tan brown when dry.

In climates with cool to cold winters, overwintering oospores in thatch germinate to begin new infection cycles.[8] Onset of blight symptoms usually occurs during mid-summer, and these are most severe at temperatures above 30°C.[5] Under conditions of high temperature and humidity, *Pythium* can quickly damage and kill succulent turfgrass.

Golf course managers apply fungicides preventively for Pythium blight control on tees and greens during periods of hot, humid weather. Costs for control of Pythium blight on the large areas of turfgrass in fairways (12 to 24 ha for an 18-hole course) can be considerable. Fungicides for Pythium blight control are an added expense, since these chemicals are ineffective against most other turfgrass diseases. The need for fungicides to control Pythium blight varies considerably from year to year, and various guidelines have been used to guide fungicide applications. An old guideline for prediction of Pythium blight is the "150 rule." This rule states that if relative humidity and temperature sum to 150 or more, then there is risk of Pythium blight. The "150 rule" for Pythium blight is a low-risk model, predicts blight too frequently, and thus results in unnecessary fungicide applications.

Two models were developed during the 1980s to predict Pythium blight occurrence. Both models were based on regression analyses of weather data and visual symptoms from golf courses. The model by Hall et al.[7] is based on temperature only (Table 2). The model by Nutter et al.[11] (NCS model) uses

Table 1 **Anthracnose severity index (ASI) values for selected average temperatures and average number of hours leaves are wet**

Hours Leaves Wet	ASI for Average Air Temperature[a]						
	18	20	22	24	26	28	30
2	0.9	1.4	2.1	2.9	3.8	5.0	6.2
4	1.0	1.5	2.3	3.1	4.2	5.4	6.7
6	1.1	1.7	2.5	3.4	4.5	5.7	7.2
8	1.1	1.8	2.6	3.6	4.8	6.1	7.6
10	1.2	1.9	2.8	3.9	5.1	6.5	8.0
12	1.2	2.0	3.0	4.1	5.4	6.8	8.4
16	1.3	2.2	3.3	4.5	6.0	7.5	9.2
20	1.3	2.4	3.6	4.9	6.5	8.1	10.0
24	1.3	2.4	3.8	5.3	6.9	8.7	10.7

[a] The anthracnose model generates an ASI for each day based on the average daily Celsius temperature, T, for a 3-day period and the average hours of leaf wetness, L, per day for the same period. $ASI = 4.0233 - 0.2283 L - 0.5308 T - 0.0013 L^2 + 0.0197 T^2 + 0.0155 LT$.

From Danneberger, T. K., Vargas, J. M., and Jones, A., A model for weather-based forecasting of anthracnose on annual bluegrass, *Phytopathology,* 74, 448, 1984. With permission.

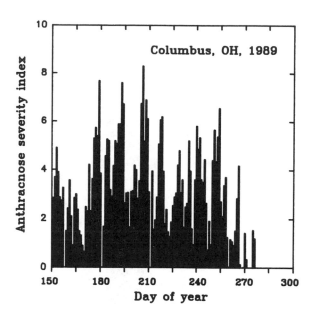

Figure 1 Anthracnose severity indices predicted for a turfgrass site in Columbus, OH during the summer of 1989.

Table 2 **Pythium blight forecasting model of Hall et al.**

For the past 24 h

Hours that temperature is equal or greater than 21°C (70°F)	>=18 h	
	Minimum temperature during last 24 h	
<18 h	<20°C (68°F)	>=20°C (68°F)
Index = 0, No risk	Index = 1, Moderate risk	Index = 2 High risk

From Hall, T. J., Larsen, P. O., Madden, L. V., and McCormick, D. J., Forecasting turfgrass diseases — an update, *Golf Course Manage.*, 53, 74, 1985. With permission.

relative humidity in addition to temperature data (Table 3). The index for the Hall model is 0, 1, or 2; relatively indicating that the risk of Pythium blight is none, moderate, or high for the day in question. The NCS model generates a 0 or 1 index, indicating no or high risk of Pythium blight. The NCS model has been the most widely used of the two models because it has been deployed in a commercial electronic disease predictor. The authors reported that the model had good predictive power — false positive predictions (blight predicted when it did not occur) was 6.0%, and false negative predictions (blight was not predicted but it occurred) was 0.77% for 2-year data.

Experience with the NCS model in Ohio suggests that it is risky.[14] This model predicts Pythium

blight too infrequently. Examination of Pythium blight episodes in Ohio indicates that the NCS model has too stringent a requirement for long periods of high relative humidity (Table 4). The NCS model states that a day favorable for blight requires at least 14 hours of ≥90% relative humidity. The NCS model correctly identified all 50 blight-negative days. The NCS model predicted 4, 3, and 2 days were favorable for blight in the monitored periods in 1987, 1988, and 1989, respectively. In contrast, the NCS model correctly identified only 2 of 10 onset days for Pythium blight. Seven out of ten onset days were incorrectly classified by the NCS model as low-risk days because during these days the number of hours with a relative humidity ≥90% and temperatures ≥20°C was less than 14 (Table 4).

Table 3 Pythium blight forecasting model of Nutter et al.

For the past 24 h

| Maximum daily temperature greater than 30°C (86°C) followed by at least 14 h of relative humidity >90% with minimum temperature>20°C (68°F) |

Yes   Index = 1, High risk

No

Index = 0,
No risk

From Nutter, F. W., Cole, H., Jr., and Schein, R. D., Disease forecasting system for warm weather Pythium blight of turfgrass, *Plant Dis.*, 67, 1126, 1983. With permission.

Our work in Ohio has resulted in a model that appears to have an improved ability to predict Pythium blight. This model, PY2, predicts blight activity on any day (noon to noon) when the maximum air temperature is ≥28°C, the minimum air temperature ≥20°C, and ≥9 hours of relative humidity ≥90% or leaves wet. The PY2 model has a less stringent requirement for long periods of moisture than the NCS model (9 hours instead of 14 hours). Analysis of data from 3 years indicates that PY2, in comparison to NCS, offered a lower false-negative error rate at the price of a somewhat higher false positive rate. Care should be taken in the use of either NCS and PY2 to time fungicide sprays until they are tested further.

DOLLAR SPOT

Dollar spot, caused by the fungi *Lanzia* and *Moellerodiscus* spp., is a common foliar disease on most turfgrasses. On golf courses, more fungicide applications are made to control dollar spot than

Table 4 Environmental conditions and *Pythium* model predictions associated with the onset day of ten Pythium blight epidemics on The Ohio State University golf course fairways and tees during 1987–1989[a]

Year and Julian Day	Model Indices NCS[b]	PY2[c]	Hours with RH ≥90% and Temp. ≥20°C	Temp. ≥20°C	Temperature (°C) Max	Min
1987						
162	0	0	0	0	24.8	11.8
173	1	1	20	24	30.5	21.9
187	1	1	20	24	30.8	21.5
203	0	1	11	24	34.8	21.9
208	0	1	8	19	31.5	19.5
1988						
189	0	1	0	21	37.4	20.3
214	0	1	12	20	29.3	20.4
218	0	1	10	24	33.9	24.5
225	0	1	12	24	34.6	24.1
1989						
205	0	1	10	24	31.4	21.7

[a] Blight-onset day was defined as a 24-hour period (noon of the previous day to 1 PM of the current day) in which (a) blight symptoms have appeared following at least 1 day with no symptoms at any site; or (b) blight symptoms have persisted from the previous day and a significant increase (1-tailed t test, $p = 0.05$) in an antibody assay reading for *Pythium* since the previous sampling date was detected;[14,15] [b] NCS model: 0 = no risk; 1 = high risk if during 24-hour period maximum air temperature ≥30°C and at least 14 hours when both the RH is ≥90% and air temperature ≥20°C; [c] PY2 model: 0 = no risk; 1 = high risk if during 24-hour period maximum air temperature ≥28°C and minimum temperature ≥20°C and relative humidity ≥90% or leaves wet ≥9 hours.

any other disease. Initial damage associated with dollar spot may be superficial, but prolonged pathogen activity can result in turfgrass death and destruction of the putting surfaces on greens.

The fungi that cause dollar spot have simple life cycles. Sclerotia in the thatch layer germinate and infect plant tissue in the spring. Fungal mycelia grow from plant to plant, and are spread great distances by mowers, shoes, and other mechanical means. Dollar spot is favored by dew and nutrient-rich guttation droplets on grass foliage during periods of warm days and cool nights. In Ohio, dollar spot damage is most severe in mid- to late-summer, but the disease may become a problem as early as May.

Two models have been developed for scheduling fungicide sprays for control of dollar spot — the R. Hall model[6] and the Mills and Rothwell model.[10] The R. Hall model states that pathogen activity can be expected when there have been two consecutive days with rain and a mean temperature >22°C, or three consecutive days with rain and a mean temperature >15°C. The Mills and Rothwell model predicts disease activity if the maximum temperature is greater than 77°F and the relative humidity is >90% during any 3 in 7 days.

Burpee and Goulty evaluated both models in the field over two growing seasons.[1] They related the predictions of the two models to increases in the numbers of dollar spot infection centers. They concluded that the R. Hall model predicted disease too infrequently, and the Mills and Rothwell model predicted disease activity too often. Of the two, the Mills and Rothwell model appeared to provide better fungicide timing information for control of dollar spot.[1] More field evaluation is needed, however, before the reliability of these two models is certain.

BROWN PATCH

Brown patch is a common turfgrass disease, caused by *Rhizoctonia solani* Kuhn. *Rhizoctonia zeae* and *R. oryzae* are also associated with brown patch pathogen development in warmer areas of the U.S.; however, *R. solani* is the species of concern to most turfgrass managers in the country during hot, humid weather.

On close-cut turfgrass, brown patch appears as an irregular to round patch, up to 1 m in diameter. On taller-cut grass, the patches may appear larger, more irregular, and less obvious to the eye. Brown patch is primarily a foliar disease, but can result in death of entire plants if allowed to progress unchecked.

The association of brown patch with warm and wet conditions has been recognized for many years.[4] Temperatures of 21 to 37°C and high humidity are commonly associated with the disease. Prediction of brown patch activity has been the focus of research at the University of Massachusetts, where a preliminary model has been developed.[18] The model uses precipitation, relative humidity, and air and soil temperature to predict pathogen activity and disease development.

ASSUMPTIONS OF PREDICTIVE MODELS

Most current predictive models for turfgrass diseases do not allow for factors such as inoculum pressure, soil type, cultivar resistance to pests, soil fertility, or irrigation. These factors can have a profound effect on disease epidemics. For example, anthracnose problems are worse on certain biotypes of *Poa annua*, thus, the potential for anthracnose problems is greater at a site where these susceptible biotypes predominate. Pathogens such as *Pythium* and *Rhizoctonia* have limited mobility but survive well in the soil or thatch. The potential for outbreaks of Pythium or Rhizoctonia blight at any site is greater if the disease has been a problem in the past. Most predictive models assume that a pathogen is present at the monitored site in sufficient amount to produce disease if appropriate environmental parameters are met. If the pathogen is not present, such models overpredict disease risk and result in unnecessary use of fungicide. The development of rapid diagnostic techniques, such as antibody assays, will aid in the refinement of predictive models.[15]

DELIVERY SYSTEMS

A major obstacle in the development of useful models has been the inability to obtain detailed weather data. The critical environmental factors are air and soil temperature, relative humidity, precipitation, irrigation, and leaf wetness. In the past, golf superintendents have experimented with thermometers and chart-recording temperature and relative humidity instruments for calculating disease indices.[10] Cost, reliability of such instruments, and usefulness of the resultant information have all been at issue.

ELECTRONIC PREDICTORS

Recently, electronic weather data collection instruments with the capability to calculate indicies using various models have become available. The first such electronic instrument adapted for turfgrass was the Reuter-Stokes Predictor (Reuter-Stokes, Twinsburg, OH). This instrument is no longer being sold. The Neogen EnviroCaster (Neogen Food Tech Corporation, Lansing, MI) currently has a Pythium blight model, an anthracnose foliar blight model, and a model for predicting annual bluegrass seedhead emergence. Neogen has plans to incorporate into their EnviroCaster models for dollar spot, summer patch, and brown patch. A somewhat similar electronic weather data instrument/predictor, Metos, has been recently developed for the turfgrass system by Gottfried Pessl, Austria.

INTEGRATION WITH IRRIGATION CONTROL SYSTEMS

Increasingly, electronic controllers and monitors are being used to guide turfgrass management decisions. Irrigation on many golf courses is now being controlled by microcomputers linked to automatic weather stations. As an example, the Toro Company has added the capacity for degree day calculations to their Network 8000 automatic irrigation system (The Toro Company, Irrigation Division, Riverside, CA). The incorporation of models for disease, insect, and weed prediction into such systems will be a natural follow-on development.

PREDICTION BASED ON REGIONAL WEATHER STATIONS

Some states maintain online weather station systems. For example, Ohio currently has eleven electronic weather stations distributed throughout the state. Weather data from all stations is automatically sent every morning to a mainframe computer at Wooster. At Ohio State University, we have linked turfgrass disease index models to this database, using it to predict disease based on known optimum temperature ranges for particular diseases. For example, the optimum temperature for development of Drechslera leaf spot (*Drechslera poae*) on Kentucky bluegrass is 14 to 19°C. A simple predictive index for this disease can be constructed by counting the number of hours in the day that fall within this temperature range (Figure 2A). Potential disease activity for different years and locations can be easily compared. Indices can be calculated for other diseases in the same fashion, such as

for brown patch (*Rhizoctonia solani*) with an optimum temperature range of 21 to 32°C (Figure 2B). Although an on-site weather station would provide the best disease predictions for a specific location, such a statewide system can provide an acceptable guide for predicting disease activity. A regional weather network linked to predictive models would likely prove to be of great benefit to the lawn care industry. Daily or weekly updates on predicted disease activity could be used to guide management decisions for diseases such as Drechslera leaf spot, brown patch, and dollar spot.

SUMMARY

With a reliable weather data collection system, a disease prediction model can provide an objective guide that is not subject to the frailties of human memory. Development of predictive models for plant diseases is a time-consuming process. A minimum of 4 years of field data is needed for model formulation and testing. Data from growth chamber experiments are useful for initial development of models, but extensive field testing under a variety of environmental conditions is required.

Models developed for turfgrass diseases may prove to be most useful as general indices for predicting disease activity during a season, rather than as tools to precisely time fungicide applications. A turf manager learns through experience how many fungicide applications are needed to manage diseases in an average year at a particular site. Disease indices provide a prediction of pathogen activity. A disease index for the current year can be compared to the historical average, so that timing and numbers of fungicide applications can be adjusted accordingly.

Use of predictive models for turfgrass diseases will become more feasible as costs decrease for equipment for collecting and processing weather data. Information from regional weather stations is becoming increasingly available throughout the country. Considerable work is still needed to establish historical databases for weather data and to accumulate field observations that are required for evaluation and improvement of current disease-predicting models for turfgrasses.

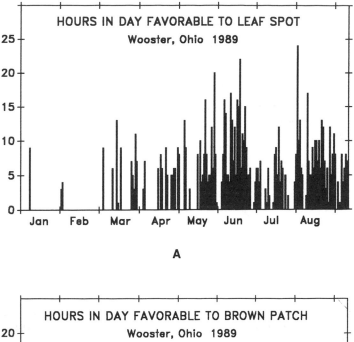

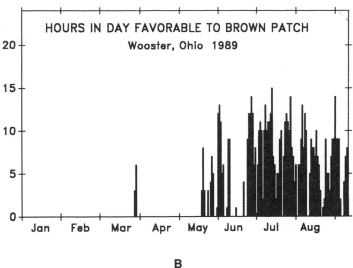

Figure 2 Disease indices predicted for (A) leaf spot and (B) brown patch based on data from the regional weather station at Wooster, OH.

REFERENCES

1. **Burpee, L. L. and Goulty, L. G.,** Evaluation of two dollarspot forecasting systems for creeping bentgrass, *Can. J. Plant Sci.*, 66, 345, 1986.
2. **Danneberger, T. K.,** Epidemiology and Control of Anthracnose Incited by *Colletotrichum graminicola* (Ces.) Wils. on Annual Bluegrass, Ph.D. dissertation, Michigan State University, East Lansing, 1982, 106 pp.
3. **Danneberger, T. K., Vargas, J. M., and Jones, A. L.,** A model for weather-based forecasting of anthracnose on annual bluegrass, *Phytopathology*, 74, 448, 1984.
4. **Dickinson, L. S.,** The effect of air temperature on pathogenicity of *Rhizoctonia solani* parasitizing grasses on putting green turf, *Phytopathology*, 20, 597, 1930.
5. **Freeman, T. E.,** Effect of temperature on cottony blight of ryegrass, *Phytopathology*, 50 (Abstr.), 575, 1960.

6. **Hall, R.,** Relationship between weather factors and dollar spot of creeping bentgrass, *Can. J. Plant Sci.*, 64, 167, 1984.

7. **Hall, T. J., Larsen, P. O., Madden, L. V., and McCormick, D. J.,** Forecasting turfgrass diseases — an update, *Golf Course Manage.*, 53(3), 74, 1985.

8. **Hall, T. J., Larsen, P. O., and Schmitthenner, A. F.,** Survival of *Pythium aphanidermatum* in golf course turfs, *Plant Dis.*, 64, 1100, 1980.

9. **Jones, A. L., Fisher, P. D., Seems, R. C., Kroom, J. C., and Van de Motter, P. J.,** Development and commercialization of an in-field microcomputer delivery system for weather-driven predictive models, *Plant Dis.*, 68, 458, 1984.

10. **Mills, S. G. and Rothwell, J. D.,** Predicting diseases — the hygrothermograph, *Greenmaster*, 18(4), 14, 1982.

11. **Nutter, F. W., Cole, H., Jr., and Schein, R. D.,** Disease forecasting system for warm weather Pythium blight of turfgrass, *Plant Dis.*, 67, 1126, 1983.

12. **Parvin, D. W., Jr., Smith, D. H., and Crosby, F. L.,** Development and evaluation of a computerized forecasting method for cercospora leafspot of peanuts, *Phytopathology*, 64, 385, 1974.

13. **Shane, W. W. and Teng, P. S.,** Cercospora infection model presented, *Sugar Producer*, 10, 14, 1984.

14. **Shane, W. W.,** Evaluation of Pythium blight prediction models using an antibody-aided detection technique, *Phytopathology*, 78 (Abstr.), 1612, 1988.

15. **Shane, W. W.,** Prospects for early detection of Pythium blight by antibody-aided monitoring of Pythium blight on turfgrass, *Plant Dis.*, 75, 921, 1991.

16. **Smiley, R. W.,** *Compendium of Turfgrass Diseases*, American Phytopathological Society, St. Paul, MN, 1992, 98 pp.

17. **Smith, J. D., Jackson, N., and Woolhouse, A. R.,** *Fungal Diseases of Amenity Turf Grasses*, 3rd ed., Routledge, Chapman and Hall, New York, 1989.

18. **Vallencourt, L. and Schumann, G. L.,** Preliminary model for predicting brown patch (*Rhizoctonia* spp.) on creeping bentgrass, *Phytopathology*, 81 (Abstr.), 125, 1990.

19. **Vargas, J. M., Jr., Roberts, D., Danneberger, T. K., Otto, M., and Detweiler, R.,** Biological management of turfgrass pests and the use of prediction models for more accurate pesticide applications, in *Integrated Pest Management for Turfgrass and Ornamentals*, Leslie, A. R. and Metcalf, R. L., Eds., U.S. Environmental Protection Agency, Washington, DC, 1989, chap. 11.

Chapter 37

Managing Cool-Season Lawn Grasses to Minimize Disease Severity

Peter H. Dernoeden, *Department of Agronomy, University of Maryland, College Park, MD*

The relationship among environmental conditions, turfgrass plants, and pathogens as they interact with one another are the key factors involved in disease development. The environmental factor is the most important determinant of a disease outbreak. For serious disease problems to occur, the environment must simultaneously exert pressures on both host and pathogen whereby environmental conditions are conducive for growth of the pathogen while at the same time they stress plants. For example, snow mold pathogens can only be destructive when low temperatures slow or stop growth of the turf. As long as there is sufficient moisture and favorable temperatures (32 to 55°F), snow mold fungi can actively grow and parasitize grass plants. The intensity and nature of turfgrass management, however, also greatly influences the severity, and sometimes the types of diseases that may occur. Without question, diseases are much more severe on golf courses, and fungicide use on these sites is common. This is due not only to traffic, but more importantly, to low and frequent mowing, frequent irrigation, and other intensive management practices routinely performed on golf courses. Because of greater flexibility in maintenance tactics, many lawn diseases can be managed with little input of fungicides. The key to an effective lawn disease management program begins with planting regionally adapted, disease-resistant species and cultivars of turfgrasses. Planting improved cultivars alone does not eliminate disease problems from lawns. However, proper mowing, fertility, and irrigation practices can greatly minimize the severity of numerous fungal diseases of lawn grasses.

A turf lawn is normally composed of plants representing a single or a few species. Planting a single cultivar of highly apomictic Kentucky bluegrass or planting vegetatively propagated warm-season grasses results in stands comprised of mostly genetically identical individuals. Where a mono-culture exists, little or no genetic variation in disease resistance occurs. Therefore, if one plant is susceptible to a particular disease, all plants are susceptible. Except where warm-season grasses are grown, monoculture is not a chief factor in disease susceptibility because the common practice of blending and mixing turfgrass species and cultivars provides a wide genetic base, which reduces the probability of a disease epidemic. Hence, blending cultivars of cool-season grasses, particularly Kentucky bluegrass, is an extremely important first step in managing lawns for potential disease problems.

Disease susceptibility in turf is primarily governed by environmental and cultural factors associated with turfgrass management. Weather conditions obviously cannot be manipulated, but cultural practices can be adapted to reduce disease problems. The management factors that influence lawn diseases most include mowing, irrigation, fertilization, thatch depth, and soil pH.

Mowing favors infection by creating wounds through which a pathogen can enter plants easily. Mowing also spreads spores and fungal mycelium of some pathogens. Hence, when foliar diseases are active, especially brown patch, dollar spot and Pythium blight, it is wise to mow when leaves are completely dry. Clipping removal from greens, and poling or dragging of greens or fairways to speed leaf drying, may help to alleviate some disease. For lawns, however, clipping removal probably would not affect disease severity. Indeed, it could be argued that clippings provide nutrients that promote turf vigor and thereby indirectly reduce some diseases or help speed recovery of turf from injury. Preliminary research suggests that clipping removal or nonremoval does not affect disease development as much as environmental stresses.

Height of cut is a major factor in disease susceptibility. Close mowing exacerbates most turf diseases, particularly Helminthosporium diseases, rust, summer patch, anthracnose, dollar spot, and injury

0-87371-350-8/94/$0.00+$.50

caused by plant parasitic nematodes. Cultivars of bluegrass that are apparently resistant to summer patch may be predisposed to this disease by low mowing. It is important to note that the term "resistance" means only that a plant is less susceptible to a disease. Resistance is a relative concept that does not imply that a cultivar is immune to a particular disease. There are few examples of turfgrass immunity to common fungal diseases.

The continuous removal of the youngest, most photosynthetically productive tissues, below recommended heights, causes depletion of carbohydrate reserves in the grass plant. These reserves play a key role in active disease resistance processes in plants, and these reserves also are needed by plants to recover from injury. Increasing mowing height increases leaf area, resulting in an increase in carbohydrate synthesis. High mowing also helps to moderate soil temperature. Hence, one of the most important management recommendations for most disease problems is to increase mowing height when turf injury is first observed. For example, studies have shown that Kentucky bluegrass was damaged much less by summer patch when maintained at 3.0 in. as compared to 1.5 in. When bluegrass turf maintained at 1.5 in. was allowed to grow to 3.0 in. after being severely damaged, it recuperated more rapidly from summer patch than turf maintained continuously at the 1.5-in. height.

Irrigation provides moisture critical to spore germination and fungal development. The timing, duration, and frequency of irrigation may greatly affect disease intensity. Light, frequent irrigations discourage root development and predispose turf to injury when extended periods of drought occur. Light and frequent irrigations that keep soils moist during periods of high temperature stress greatly increase summer patch of Kentucky bluegrass and creeping red fescue. This is because the fungus that causes summer patch (*Magnaporthe poae*) grows much more rapidly along roots when soils are moist and very warm (>85°F in the root zone). Studies also have shown that summer patch was much less severe when 3.0-in.-high turf (4% injury) was subjected to deep and infrequent irrigation when compared to 3.0-in.-high turf (16% injury) or 1.5-in.-high turf (36% injury) subjected to light and frequent irrigation. The light and frequent irrigation treatments in this study kept soil moist for extended periods, which favored root infection by *M. poae* during the warm summer months. Frequent cycles of wetting and drying of thatch enhance spore production by Heiminthosporium leaf spot (i.e., *Drechslera* and *Bipolaris* spp.) pathogens. Irrigating deeply helps to encourage bacteria that keep leaf spot pathogens from producing enormous spore populations in thatch. Sometimes, however, light and frequent irrigation is recommended in situations where a disease or an insect pest has caused extensive damage to roots. For example, studies conducted at Michigan State University revealed that the severity of necrotic ring spot, another root disease of Kentucky bluegrass, is reduced by light daily watering.

Subjecting turf to extreme drought stress appears to enhance the level of injury in stands affected by Helminthosporium diseases, stripe smut, powdery mildew, dollar spot, necrotic ring spot, and fairy rings. Conversely, excessive irrigation restricts root development and encourages disease. Turfgrasses grown under wet conditions develop succulent tissues and thinner cell walls, which presumably are more easily penetrated by fungal pathogens. Algae, mosses, and the germination of weed seeds are encouraged by wet or waterlogged soils, particularly where stand density is poor. Early morning or afternoon irrigation is often recommended during summer to ensure that plant tissues are dry by nightfall. This practice reduces the number of nighttime hours when leaves are wet and thereby helps to minimize the intensity of foliar blighting diseases such as Pythium blight and brown patch.

Proper soil fertility improves plant vigor and the ability of plants to resist or recover from disease. Excessive use of nitrogen has been reported to encourage Helminthosporium leaf spot, brown patch, and Pythium blight. Excessive use of nitrogen promotes tissue succulence and thinner cell walls which, as previously mentioned, are more easily penetrated by fungal pathogens. Excessively applied nitrogen may also divert carbohydrates or important biochemicals, synthesized by plants to resist pathogen attack, into growth processes devoted to leaf production. Incorporation of these plant chemicals into developing tissues, rather than for purposes of plant defense, may result in shortages that adversely affect the capacity of the plant to resist fungal invasion. Conversely, turfgrasses grown in nutrient-poor soils lack vigor and are more prone to severe damage from dollar spot, red thread, and rust diseases. Application of nitrogen to diseased turf under these conditions stimulates foliar growth at a rate that exceeds the capacity of these fungi to colonize new tissues.

In general, slow-release nitrogen fertilizers are recommended where chronic disease problems occur. The controlled release of these nitrogen sources results in a moderate rather than rapid or excessive growth rate. Selected composted animal waste or sludge nitrogen sources show promise in managing diseases such as dollar spot, red thread, brown patch, and necrotic ring spot. While slow release nitrogen sources are preferred for disease management, there are some exceptions to this generalization. Rapid release, water soluble nitrogen sources may be recommended during autumn months to help speed recovery of turf injured by disease or environmental stress. Furthermore, water soluble nitrogen from ammonium sulfate and ammonium chloride have been shown to reduce the severity of take-all of bentgrass and spring dead spot of bermudagrass, and ammonium sulfate can reduce pink snow mold and summer patch damage.

There is little documented evidence that phosphorus, potassium, or micronutrients alone impact significantly on turf diseases. However, deficiencies in these nutrients can weaken plants and predispose them to more disease injury than normally would have occurred. For example, take-all disease of bentgrass is frequently more severe in putting greens where phosphorus is deficient. More importantly, N + K and N + P + K fertility have been shown to reduce the severity of take-all, stripe smut, spring dead spot, red thread, and Fusarium patch when compared with nitrogen alone.

Many turfgrass pathogens survive as resting structures or as saprophytes (organisms living on dead organic matter) in thatch. Thatch also provides fungi with moisture. As previously noted some pathogens, such as *Helminthosporium* spp., produce enormous populations of spores in thatch, particularly when the thatch is subjected to frequent wetting and drying. Stripe smut, Helminthosporium melting-out, Fusarium blight, and summer patch are diseases that appear to be favored by excessive thatch accumulation.

Traffic, like mowing, produces wounds that are easily invaded by some fungal pathogens. Compaction caused by heavy traffic impedes air and water movement into soil and eventually restricts root func-

tion, causing a decline in plant vigor and disease resistance. Summer patch, Helminthosporium leaf spot, and anthracnose tend to be more damaging on highly trafficked or compacted soil sites. Hence, core aerification can alleviate compaction and reduce the severity of these diseases. Conversely, plant parasitic nematode population densities tend to be higher in light-textured, sandy loam soils than in clays or compacted soils.

Soil pH also may affect disease development in turfgrasses. In general, most turfgrass pathogens are able to grow at any pH encountered by turf. High soil pH, however, favors take-all and pink snow mold on golf course putting greens. Factors that encourage take-all under alkaline soil conditions remain imperfectly understood. It is likely, however, that this pathogen achieves a competitive advantage over antagonistic microorganisms in alkaline soils. Acidifying nitrogen fertilizers have been shown to reduce the severity of take-all, pink snow mold, summer patch, and spring dead spot. For most diseases, however, soil pH does not impact directly on disease, but it is prudent to maintain or adjust soil pH to a range of 6.0 to 7.0.

Adherence to sound cultural practices is basic to reducing disease severity in turf. The turfgrass environment, however, is not static and managers must continually modify cultural practices to encourage vigorous growing conditions. To maintain these conditions, the turfgrass manager must routinely monitor the nutrient and pH status of soil, adjust mowing heights, judiciously irrigate, overseed with disease resistant cultivars, manipulate different sources of nitrogen fertilizer, and control thatch and compaction. The basic principles of cultural disease management of cool-season grasses are outlined in Table 1.

Table 1 Basic principles of managing cool-season lawn grasses to minimize damage caused by common fungal pathogens

1. Plant disease resistant species and cultivars. Recommendations for regionally adapted cultivars can be obtained from most county cooperative extension service offices.
2. Use balanced N-P-K fertilizer at the appropriate times. In the absence of a soil test, N-P-K should be applied at least once annually in a ratio of 3:1:2. Most of the total annual nitrogen (about 75%) should be applied to cool-season grasses during autumn months.
3. At least 50% of all nitrogen applied per year should be from a slow release nitrogen source. Some examples of slow release nitrogen sources include: isobutylidene diurea (IBDU), methylene urea, sulfur coated urea, and composted sludges or animal waste products.
4. Maintain a high mowing height within a species' adapted range. Raise mowing height during periods of environmental stress or disease outbreak. For reducing disease damage, it is best to maintain Kentucky bluegrass or perennial ryegrass lawns at 2.5 to 3.0 in., and fescue species at 3.0 to 3.5 in. in height.
5. Irrigate deeply to wet soil to a 4- to 6-in. depth when turf first exhibits signs of wilt. Avoid frequent and light applications of water.
6. Test soil every 2 or 3 years for phosphorus and potassium levels as well as soil pH. Adjust soil pH to a range of 6.0 to 7.0.

408

REFERENCES

Colbaugh, P. F. and Endo, R. M., Drought stress — an important factor stimulating the development of *Helminthosporium sativum* on Kentucky bluegrass, in *Proc. 2nd Int. Turfgrass Res. Conf.,* Roberts, E. C., Ed., American Society of Agronomy, Madison, WI, 1974, 328.

Davis, D. B. and Dernoeden, P. H., Summer patch and Kentucky bluegrass quality as influenced by cultural practices, *Agron. J.,* 83, 670, 1991.

Shurtleff, M. C., Fermanian, T. W., and Randell, R., *Controlling Turfgrass Pests*, Prentice-Hall, Englewood Cliffs, NJ, 1987.

Smiley, R. W., Dernoeden, P. H., and Clarke, B. B., Eds., *Compendium of Turfgrass Diseases*, 2nd ed., American Phytopathological Society, St. Paul, MN, 1992.

Smith, J. D., Jackson, N., and Woolhouse, A. R., *Fungal Diseases of Amenity Turf Grasses*, 3rd ed., Routledge, Chapman and Hall, New York, 1989.

Thompson, D. C., Clarke, B. B., Heckman, J. R., and Murphy, J. A., Influence of nitrogen source and soil pH on summer patch development in Kentucky bluegrass, *Int. Turfgrass Soc. Res. J.,* 7, 317, 1993.

Vargas, J. M., Jr., *Management of Turfgrass Diseases*, rev. ed., Lewis Publishing, Boca Raton, FL, 1993.

Chapter 38

Biological Control of Turfgrass Diseases

Eric B. Nelson, *Department of Plant Pathology, Cornell University, Ithaca, NY*

Lee L. Burpee, *Department of Plant Pathology, University of Georgia, Griffin, GA*

Mark B. Lawton, *Monsanto Canada, Mississauga, Ontario, Canada*

CONTENTS

Introduction 409
Approaches to Biological Control 410
The Use of Composts for the Biological Control of Turfgrass Diseases 411
    The Composting Process 412
    Disease Suppression with Composts 413
    The Need for Predictably Suppressive Composts 415
Turfgrass Disease Control with Suppressive Soils 417
The Use of Microbial Inoculants for Turfgrass Disease Control 417
Suppression of Typhula Blight: A Model for Biological Control 419
    Enhancing the Efficacy of *Typhula phacorrhiza* 420
    The Mechanism of Disease Suppression by *T. phacorrhiza* 421
    Methods for Applying *T. phacorrhiza* to Turf 421
Future Perspectives 423
References 424

INTRODUCTION

Disease management represents a significant challenge for turfgrass managers. This is made particularly demanding by the perennial nature of turfgrass plantings as well as that of the disease-causing organisms. Since most, if not all, of the fungal pathogens of turfgrass are always present in turfgrass plantings, rarely is the pathogen's presence or population level limiting for disease development.[1] As a result, the principal factors determining the incidence and severity of turfgrass diseases are environmental factors and plant stresses that influence not only the activity of pathogens, but the susceptibility of turfgrass plants. This is particularly true for some root pathogens with which turfgrass plants remain infected year round.[2] Since in many cases these factors cannot be manipulated adequately to minimize losses from fungal diseases, turfgrass managers rely largely on applications of fungicides for the management of fungal diseases.

Of the materials currently being used for turfgrass disease control, most are broad-spectrum systemic fungicides. Many problems have arisen from the repeated and prolonged use of such chemicals. Included among these are the development of fungicide-resistant pathogen populations,[3,4] deleterious effects on nontarget organisms, particularly those involved in carbon and nitrogen cycling,[5] enhancement of nontarget diseases,[6] and the selection of fungicide-degrading microorganisms.[7] In an effort to reduce this fungicide dependency and to prevent many of the undesirable biological and environmental effects of excessive fungicide use, alternative management practices are being explored.

One of several exciting alternative management strategies being developed is the use of biological controls in which individual or mixtures of microorganisms are deployed to either reduce the activities of pathogens or enhance the tolerance of plants to disease.[8] This approach to disease control has been used successfully on an experimental as well as a commercial basis for the control of plant pathogens on several crop plant species.[8]

APPROACHES TO
BIOLOGICAL CONTROL

Most turfgrass managers are familiar with the negative aspects of soil microorganisms since some are pathogenic and can damage a turfgrass stand. However, in addition to pathogens, the soil harbors a variety of nonpathogenic microorganisms that actually improve plant health. These soil bacteria and fungi are responsible for increasing the availability of plant nutrients and forming symbiotic associations with turfgrass roots,[9-11] for producing substances stimulatory to plant growth,[12] and for protecting plants against infection from pathogenic fungi.[8]

The practice of biological control attempts to take advantage of all the above-mentioned microbial attributes in order to minimize damage from plant pathogens. For example, the application of composts or other sources of organic matter to turf may introduce large populations of antagonistic microorganisms[13] which may reduce disease by interfering with the activities of pathogenic fungi. Similarly, cultural management techniques such as core aeration, fertilization, or the application of lime, may reduce disease development by altering the soil and thatch microbial communities within which pathogens must function. In such cases, cultural practices may indirectly affect disease severity by changing the environment to favor the antagonistic microflora to the detriment of pathogen populations.

Biological control may be achieved either through the application of introduced antagonists or through the manipulation of native antagonists present in disease-suppressive soils and on plant parts.[8] In either case, the goal is to reduce or eliminate pathogen activities by reducing pathogen inoculum in soil, by protecting plant surfaces from infection, or by inducing natural defense mechanisms within the plant.

Biological control of pathogen inoculum may occur by the microbial destruction of pathogen propagules and the prevention of inoculum formation through the action of mycoparasites (i.e., fungi parasitic on other fungi). In addition, antibiotic-producing antagonists may displace pathogens in decaying plant residues such as thatch and reduce their populations in soil. Some nonpathogenic soil microorganisms are able to effectively colonize aboveground as well as belowground plant parts[14] and in so doing, protect these tissues from infection by pathogens. It is also apparent that some biological control agents can induce natural defense mechanisms in plants; a phenomenon referred to as cross protection or induced resistance.[15,16]

Some of the more commonly studied antagonists include fungi in the genera *Fusarium, Gliocladium, Laetisaria, Penicillium, Sporidesmium, Talaromyces, Trichoderma,* and *Verticillium* and bacteria in the genera *Bacillus, Enterobacter, Erwinia,* and *Pseudomonas.* Research has shown that these microorganisms can interfere with pathogen populations in a number of ways. Mycoparasites such as *Trichoderma* and *Sporidesmium* may parasitize pathogen propagules and mycelium.[17,18] Other antagonists, particularly *Pseudomonas, Bacillus, Enterobacter, Erwinia,* and *Gliocladium* produce antibiotics inhibitory to pathogen growth.[19-24] Some strains of *Pseudomonas* and *Enterobacter* species are efficient competitors for essential nutrients and other growth factors, thereby reducing pathogen germination, growth, and plant infection.[25-27]

Antagonists of turfgrass pathogens can be found in a variety of sites.[28] They are particularly abundant in turfgrass soils and thatch as well as in decaying organic substrates.[13] Nelson and Craft[28] found a greater percentage of antagonists of *Pythium aphanidermatum* associated with thatch than with the rhizosphere soil of turfgrasses in both low and high maintenance sites. Also, those associated with thatch were generally more effective in suppressing Pythium blight. Among the bacteria screened, those members of the Enterobacteriaceae were significantly more efficacious than other general heterotrophic bacteria and *Pseudomonas* species.[28]

To predictably and successfully manipulate biological control agents, their biology and ecology in turfgrass ecosystems must be understood. Biological control agents differ fundamentally from chemical fungicides in that they must grow and proliferate to be effective. Therefore, effective antagonists must be able to become established in turfgrass ecosystems and remain active against target pathogens during periods favorable for plant infection. The two factors most important in determining antagonist establishment and growth are (1) the environmental conditions, particularly temperature, moisture, nutrients, and pH and (2) their ability to compete with the existing soil and plant microbial communities. Just as some organisms are antagonists of pathogens, antagonists have their own antagonists as well.

Biological control agents must also be compatible with other management inputs. In particular, biological control agents must be tolerant of fungicides, insecticides, herbicides, and fertilizers currently used in management programs. Their

Table 1 **Known examples of turfgrass disease biological control**

Disease (Pathogen)	Antagonists	Location	Ref.
Brown patch	*Rhizoctonia* spp.	Ontario, Canada	39, 69, 70, 77
(*Rhizoctonia solani*)	*Laetisaria* spp.	North Carolina	
	Compost microbes	New York; Maryland	
Dollar spot	*Enterobacter cloacae*	New York	38, 71, 72, 78, 79
(*Sclerotinia homoeocarpa*)	*Fusarium heterosporum*	Ontario, Canada	
	Gliocladium virens	South Carolina	
	Compost microbes	New York	
Necrotic ringspot	Native soil microbes	Michigan	87
(*Leptosphaeria korrae*)			
Pythium blight	*Pseudomonas* spp	Illinois; Ohio	28, 65, 66, 80–82
(*Pythium aphanidermatum*)	*Trichoderma* spp.	Ohio	
	Trichoderma hamatum	Colorado	
	Enterobacter cloacae	New York	
	Various bacteria	New York; Pennsylvania	
	Compost microbes	Pennsylvania	
Pythium root rot	*Enterobacter cloacae*	New York	40
(*Pythium graminicola*)	Compost microbes	New York	
Red thread	Compost microbes	New York	42
(*Laetisaria fuciformis*)			
Southern blight	*Trichoderma harzianum*	North Carolina	83
(*Sclerotium rolfsii*)			
Summer patch	Various bacteria	New Jersey	86
(*Magnaporthe poae*)			
Take-all patch	*Pseudomonas* spp.	Colorado; France	55, 56, 88, 89
(*Gaeumannomyces*	*Gaeumannomyces* spp.	Australia	
graminis var. *avenae*)	*Phialophora radicicola*	Australia	
	Microbial mixtures	Australia	
Typhula blight	*Typhula phacorrhiza*	Ontario, Canada	41, 67, 84
(*Typhula* spp.)	*Trichoderma* spp.	Massachusetts	
	Compost microbes	New York	

activities must also not be discouraged by cultural practices used in turfgrass maintenance. Just as pathogens are influenced by environmental conditions, so too are biological control agents. Therefore, biological control strategies must be employed primarily to control pathogens, but at the same time to maintain the associated antagonistic microflora.

Although few in-depth studies on the biological control of turfgrass diseases have been conducted, promising results have been obtained using complex mixtures of microorganisms and individual antagonists as tools for managing fungal diseases of golf course turf (Table 1). While individual organisms isolated from many different environments can be suitable for use as biological control agents, composts and organic fertilizers are perhaps the best sources of complex mixtures of antagonistic microorganisms.[13]

THE USE OF COMPOSTS FOR THE BIOLOGICAL CONTROL OF TURFGRASS DISEASES

To reestablish the microbiological balance of soils on which intensively managed turfgrasses are grown, sufficient organic matter must be introduced into the soil/plant system to support microbial growth and activity.

Some of the best sources of both organic matter and populations of antagonistic microorganisms are composted materials.[13] Fortunately, composts are currently available, in many cases at no charge, and they can be applied as a top dressing without the need for elaborate and expensive equipment. Golf greens and tees are top-dressed several times a season with a mixture of sand and some type of organic matter (usually peat) or soil. Most sphagnum peats used in top dressings, however, have little or no disease-suppressive properties.[29] It should be relatively easy, therefore, to replace the peat with composted manures, sludges, and/or food and agricultural wastes that are readily available and inherently disease-suppressive.

THE COMPOSTING PROCESS

Composting has been defined as the "biological decomposition of organic constituents in wastes under controlled conditions."[30] Since composting relies exclusively on microorganisms to decompose the organic matter, the process has biological as well as physical limitations. During composting, the environmental parameters (i.e., moisture, temperature, aeration) must be stringently controlled. This is necessary to maintain adequate rates of decomposition and to avoid the production of decomposition by-products that may be harmful to plant growth. To maintain proper temperatures, the composting mass must be large enough to be self-insulating but not so large that compaction results in reduced air exchange. The composting mass must be moist enough to support microbial activity but not excessively moist so that air exchange is limited. The particle size of the material must be small enough to provide proper insulation but not too small to limit air exchange.

Composting involves successions of both mesophilic (moderate-temperature) and thermophilic (high-temperature) microorganisms during various phases of organic matter decomposition.[31-33] Each microbial community makes an important contribution to the nature of the composted material. Failure to maintain environmental conditions favorable for adequate microbial activity could jeopardize the quality of the final product.

When all the environmental and physical conditions are optimized, composting should proceed through three distinct phases (Figure 1). During the initial phase lasting one to several days, depending on the type of starting material, temperatures in the internal portions of the composting mass rise as a result of the growth and activity of the indigenous mesophilic microorganisms associated with the starting organic material. During this self-heating phase, most of the soluble, readily degradable materials are broken down by these naturally occurring microorganisms, precluding the need for additional inoculum. At this stage of composting, populations of microorganisms increase in magnitude and activity.

As temperatures increase above 40°C, the mesophilic populations are replaced by thermophilic populations capable of degrading most of the resistant polymers such as cellulose and

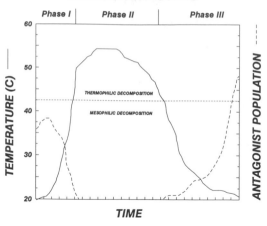

AEROBIC COMPOSTING

Figure 1 Schematic diagram of the composting process. During Phase I, initial heating takes place and readily soluble components are degraded. During Phase II, cellulose and hemicellulose are degraded under high temperature (thermophilic) conditions. This is accompanied by the release of water, carbon dioxide, ammonia, and heat. Finally, during Phase III, curing and stabilization are accompanied by a drop in temperatures and increased humification of the material. Recolonization of the compost by mesophilic microorganisms occurs during phase III. Included in these microbial communities are populations of antagonists.

hemicellulose. During this stage of decomposition, microbial diversity decreases and only a few species of *Bacillus* are active in decomposition processes.[32,34,35] The thermophilic phase may last several months, depending on the cellulose content of the material and the temperatures maintained during this period. Generally, the higher the cellulose content, the longer the thermophilic phase. Temperatures required for thermophilic decomposition range from 35 to 70°C. However, the highest rate of microbial activity and decomposition occurs at the lower end of the thermophilic range at temperatures of 35 to 55°C.[36] Increases in temperature above 55°C can be self-limiting to the decomposition process.[35] To overcome these constraints, most composts need to be aerated either through repeated pile inversions or through forced air ventilation. Prior to placing in windrows, many composts are started in aerated vessel systems where temperatures can be precisely regulated[37] and uniform decomposition can be established.

Since composting consumes much oxygen, aeration serves to keep the composting mass aerobic instead of anaerobic. Should composts become

Figure 2 Biological suppression of dollar spot on a creeping bentgrass/annual bluegrass putting green 32 days after application of selected composts and organic fertilizers. (A) The plot on the left was untreated while the one on the right was treated with ~10 lb/1000 ft² of an organic fertilizer. (B) The plot on the bottom was left untreated while the one on the top was treated with a poultry litter/cow manure compost mixture at the rate of ~10 lb/1000 ft².

anaerobic, many toxic microbial metabolites can accumulate resulting in detrimental effects on plants coming in contact with such material.[30] Additionally, undesirable odors are present in uncontrolled anaerobic composts. Most composts produced in a proper aerobic environment should have little or no odor associated with the decomposing mass. Aeration also serves as a means of drying the material making it more suitable for handling and transport.

As the cellulose and hemicellulose components are exhausted, the compost enters a curing or stabilization phase where temperatures decline, decomposition rates decrease, and the thermophilic microbial populations are again replaced by mesophilic populations. In general, the longer the curing period, the more diverse the colonizing mesophilic microflora. It is this recolonizing by mesophilic microflora that is most important in suppressing turfgrass diseases since a large proportion of the recolonizing mesophilic microflora is composed of microbial antagonists that render the compost disease-suppressive.[13] Unfortunately, there is no reliable way to predict the disease-suppressive properties of composts since the nature of the recolonizing antagonist population is left to chance and determined largely by the microflora present at the composting site.

DISEASE SUPPRESSION WITH COMPOSTS

Applications of some composted materials can suppress a number of turfgrass diseases. Monthly applications of top dressings composed of as little as 10 lb of suppressive compost/1000 ft² have been shown to be effective in suppressing diseases such as dollar spot[38] (Figure 2), brown patch,[39] Pythium root rot,[40] Typhula blight,[41] and red thread[42] (Table 2). Reductions in severity of Pythium blight, summer patch, and necrotic ringspot have also been observed in sites receiving periodic applications of composts.[43] Excellent control of *Pythium graminicola*-incited root rot on creeping bentgrass putting greens has also been obtained with root-zone amendments of various composts (20% compost:80% sand; v:v) (Figure 3).[85] Perhaps the most important benefit of compost use on established turfgrasses is its impact on root-rotting pathogens in soil. Populations of soilborne *Pythium* species are generally not suppressed following traditional chemical fungicide applications[2] (Figure 4A) but can be reduced on putting greens receiving continuous compost applications in the absence of any chemical fungicide applications[2] (Figure 4B). Additionally, heavy applications of certain composts (~200 lb/1000 ft²) to putting greens in late fall can be effective, not only in suppressing winter diseases such as Typhula blight, but in protecting putting surfaces from winter ice and freezing damage.[43]

Composts prepared from different starting materials as well as those at different stages of decomposition vary in the level of disease suppression[2,44] and in the spectrum of diseases that are controlled (Table 3). This is primarily a result of the microbial variability among different composts and among the different qualities of organic matter present in any one compost at various stages of decomposition.[13] Although microbial activity is necessary for disease-suppressive properties to be expressed in most composts,[13] the specific nature of disease suppressiveness is, in general, unknown.

Table 2 **Turfgrass diseases for which composts have been suppressive**

Disease (Pathogen)	Mode of Application	Turfgrasses
Brown patch (*Rhizoctonia solani*)	Top dressings[a]	Creeping bentgrass/annual bluegrass Tall fescue
Dollar spot (*Sclerotinia homoeocarpa*)	Top dressings	Creeping bentgrass/annual bluegrass
Necrotic ringspot (*Leptosphaeria korrae*)	Top dressings	Kentucky bluegrass
Pythium blight (*Pythium aphanidermatum*)	Top dressings	Perennial ryegrass
Pythium root rot (*Pythium graminicola*)	Top dressings and heavy fall applications[b] Root-zone amendments[c]	Creeping bentgrass/annual bluegrass
Red thread (*Laetisaria fuciformis*)	Top dressings	Perennial ryegrass
Typhula blight (*Typhula* spp.)	Heavy fall applications	Creeping bentgrass/annual bluegrass

[a] Applied at the rate of ~10 lb/1000 ft^2; [b] Applied at the rate of ~200 lb/1000 ft^2; [c] Incorporated into sand at the rate of 20% compost, 80% sand (v,v).

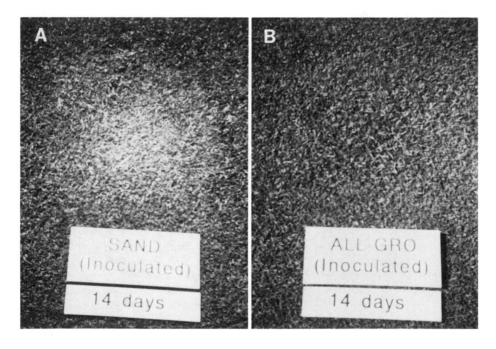

Figure 3 Suppression of Pythium root rot (*Pythium graminicola*) on a creeping bentgrass, sand-based putting green. (A) 100% sand root-zone mix; (B) root-zone mix consisted of 80% sand, 20% brewery waste compost (v/v). Both plots were inoculated with *P. graminicola* 14 days earlier.

The microbiology of disease-suppressive composts has not been extensively studied. Fungal and bacterial antagonists suppressive to a number of plant pathogens have been recovered from hardwood bark and sewage sludge composts.[44-49] Relationships between microbial activity and *Pythium* suppression in composts have also been described.[50-52] Nelson et al.[44] found *Trichoderma hamatum, T. harzianum,* and *Gliocladium virens* from suppressive bark composts to be the most

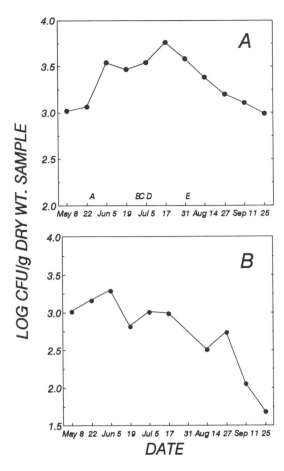

Figure 4 Populations of soilborne *Pythium* spp. in creeping bentgrass/annual bluegrass putting greens. (A) Populations of *P. graminicola/P. torulosum* in a green receiving standard management inputs. Letters A–E indicate various pesticide applications; A = propamocarb @ 3 oz/1000 ft², B = fosetyl Al @ 4 oz/1000 ft², C = propiconazole @ 2 oz/1000 ft², D = isophenphos @ 3 oz/1000 ft², and E = chlorothalonil @ 6 oz/1000 ft². (B) Populations of *P. graminicola/P. torulosum* in a green receiving no pesticide applications. Green received compost applications for the previous 2 years.

important fungi in the suppression of *R. solani*, whereas Kwok et al.[46] identified a number of bacterial species effective against *R. solani* in bark composts. However, strains of *Enterobacter cloacae, Flavobacterium balustinum, Xanthomonas maltophila,* and various fluorescent *Pseudomonas* spp. were more effective in suppressing *R. solani* when combined with an isolate of *Trichoderma hamatum.*

Antagonists of other plant pathogens have been studied in less detail. Various oligotrophic *Pseudomonas* species from composts are effective root colonists and antagonists of *Pythium ultimum.*[53] In some sewage sludge composts, thermophilic strains of *Bacillus subtilis* have been shown to be effective in inducing suppression to a number of soilborne plant pathogens.[48,49]

Although a wide variety of microbial antagonists can be found in composted substrates, the predominant species and their relative contributions to disease suppression remain unknown. However, those microorganisms that are rapid and aggressive colonizers of organic matter are more likely to contribute the most to disease suppression in composts.[13]

THE NEED FOR PREDICTABLY SUPPRESSIVE COMPOSTS

The use of top dressings and root-zone mixes amended with disease-suppressive composts is being accepted by turfgrass managers as an attractive disease control alternative. In the few cases that have been examined, substantial reductions in fungicide use have accompanied the adoption of these strategies.[54] Many composted materials and other organic fertilizers are commercially available and can be found in many garden centers. Others are available from municipal waste treatment facilities. Research has shown that the use of composts and organic fertilizers for turfgrass disease control is economically and technologically practical and, in some instances, can provide levels of control equivalent to that currently attained with fungicides.[38,39]

One of the principal problems associated with the use of composts for disease control is that a given compost may not be predictably suppressive from year to year, batch to batch, and from one site to the next. Turfgrass managers and compost producers agree that the future success of these materials in commercial turfgrass management depends upon the abilities of producers to provide material with predictable levels of disease control. Gross variations in disease-suppressive qualities of composts cannot be tolerated because end-users need to be assured that every batch of compost used specifically for disease control will work every time. Unfortunately, it is not known how to predict the suppressive properties of certain composts without actually testing them in field situations.

A number of assays have been developed to determine compost maturity and degree of stabilization for the purpose of reducing the variability in physical and chemical properties.[55] However, none

Table 3 **Biological suppression of various turfgrass diseases with compost-amended top dressings**[a]

Treatments	Spots/Plot Dollar Spot	% Plot Area Diseased Brown Patch	Red Thread	Typhula Blight	Pythium Root Rot
Untreated	19.8	72	47	33	38
Organic fertilizer CP	5.2*	18*	20	55*	22
Organic fertilizer GR	6.8*	24*	43	45	20
Turkey litter compost	13.8	18*	10*	28	18*
Endicott sludge	13.0	42*	40	10*	16*
IPS cow manure	16.9	54	43	—	—
Baltimore sludge	17.3	60	23	—	—
AB brewery waste	17.8	54	30	10*	24
Endicott leaf compost	18.9	44*	53	—	—
MH manure compost	20.2	72	53	15	—
Schenectady sludge	21.4	66	57	—	—
Spent mushroom	21.8	54	53	—	—
Fungicide standard[b]	0.6*	8*	—	22	22

Note: Numbers followed by (*) are significantly ($p = 0.05$) different from the untreated check.

[a] Determined 30, 13, 27, and 19 days post-inoculation for dollar spot, brown patch, red thread, and *Pythium* root rot, respectively. Gray snow mold evaluated in the spring (April), 6 months after the last fall application;

[b] Fungicide standard for all diseases except *Pythium* root rot consisted of propiconazole applied at the rate of 174 mg a.i./m^2. For Pythium root rot, metalaxyl was applied at the rate of 0.75 mg a.i./m^2.

have been designed to directly assess microbiological aspects of maturity and disease suppressiveness. As composts decompose and mature, changes in the quality of the organic matter are apparent. During composting the levels of carbohydrates decrease while aromatic carbon compounds increase.[56] Readily available carbon compounds and cellulose can support the activities of plant pathogens such as *Pythium* and *Rhizoctonia*.[51-57] However, as these compounds are reduced during composting, saprophytic growth of such pathogens is dramatically reduced. Antagonists also are affected by these changes in organic matter quality. For example, *Rhizoctonia*-suppressive strains of *Trichoderma hamatum* and *T. harzianum* are unable to suppress Rhizoctonia damping-off in immature composts but are extremely effective when introduced into mature composts.[44,57] Spectroscopic analyses of the organic matter in composts is now being used as a means to predict both compost maturity and the disease suppressiveness of composts.[56,58] However, this technology is not widely available and is cost prohibitive for routine compost analyses.

Another approach for predicting disease-suppressive properties in composts involves determinations of levels of microbial activity and microbial biomass. In *Pythium*-suppressive composts and peats, direct relationships have been established between both the level of general microbial activity and the amount of microbial biomass and the degree of *Pythium* suppression.[50-52,59] While this technique is useful for diseases caused by *Pythium* spp. and perhaps other nutrient-dependent pathogens, its utility in predicting disease suppressiveness for other nutrient-independent pathogens is uncertain. Additionally, because a limited number of composts are disease-suppressive due to nonmicrobiological mechanisms,[2,13] this approach is not useful for predicting suppressiveness in these composts.

Despite the requirement of microbial activity for the expression of disease-suppressive properties in most composts used in turfgrass applications,[2] we know little of the specific microorganisms involved in imparting disease-suppressive properties. Identification of specific organisms with biological control activity in composts will be a key factor in understanding how composts suppress diseases. This knowledge has proven to be important in developing hardwood bark composts for use in the production of container-grown ornamentals.[13]

Several aspects of the ecology of key compost-inhabiting antagonists in turfgrasses will be important in developing more effective biological control strategies with compost-based materials. For example, the ability of antagonists to establish and survive in turfgrass ecosystems is necessary for biological control to occur. The interactions of antagonists with other soil organisms and the soil and plant factors affecting optimum biological control activity will be important in developing strategies with compost-based materials. In addition, these organisms may serve as indicators of how long to compost a material before it can be certified to be disease-suppressive. Research aimed at understanding the fate of antagonistic organisms in soils and on plants following compost applications will aid in understanding why composts fail at certain times and in certain locations but not at others. Such research should also help to predict the compatibility of composts and their resident antagonists with pesticides and other cultural practices commonly used in turf management.

As a means of making composts more predictably disease-suppressive, it may be possible to introduce antagonistic organisms with known biological control properties into composting materials at key stages in the curing process.[13] This strategy has been used successfully to produce composts more predictably disease-suppressive[13] and more highly suppressive to a number of plant pathogens.[48,49] This approach should enable compost producers to ultimately produce predictably suppressive biological control materials.

Over the past 5 years, a large number of composts have become available for turfgrass applications. Some are properly composted and formulated and of high quality, whereas others are not. In the past, quality control was of less concern when composts were used primarily as fertilizers. However, for the purposes of disease management, quality control is extremely important. Some organic materials that are improperly composted can be extremely phytotoxic. Other improperly composted materials can even accentuate the development of some diseases.[13]

TURFGRASS DISEASE CONTROL WITH SUPPRESSIVE SOILS

Most soils vary in their ability to support plant disease development. Soils in which a pathogen cannot establish, persist, or cause significant disease despite the presence of an adequate population are referred to as suppressive soils.[8,60] Like composts, the microbial activity present in these soils is responsible, in most cases, for their disease-suppressive properties. Many of the microorganisms responsible for making soils suppressive can be found among many of the genera of bacteria and fungi previously listed.

Some suppressive soils have been used to control turfgrass diseases in a similar manner to that of composts. As little as a 5-mm top-dressing application of a take-all suppressive soil has been used to completely control take-all patch (*G. graminis* var. *avenae*) on creeping bentgrass turf in greenhouse experiments.[61,62] Although disease suppression appears to be a result of the complex microflora present in these types of soils, various fluorescent *Pseudomonas* species and *Phialophora radicicola* are particularly effective in suppressing take-all patch.

THE USE OF MICROBIAL INOCULANTS FOR TURFGRASS DISEASE CONTROL

Microbial inoculants used for disease control consist of preparations of living microorganisms that are inhibitory to plant pathogens. In the development and use of microbial inoculants, beneficial microorganisms commonly found in nature are isolated from the environment (usually from soils or plant tissues) and their populations are artificially increased. In some instances, they may be culturally or genetically improved in the laboratory, and then they are introduced back into the environment in the form of an inoculant.

Unlike traditional synthetic chemical fungicides, more careful consideration must be made of various aspects of the storage and application of microbial inoculants. Of particular importance is the shelf life of microbial inoculants since the organisms present in such products may not remain viable for extended periods of time. One also needs to consider that, for any microbial-based inoculant to be effective, the organism(s) present in such a product must become established in turfgrass plantings and must remain active throughout the period when disease pressure is greatest. Additionally, the organisms present in these types of products must be compatible with other agrichemicals used in management systems. For example, whereas bacterial preparations should generally be tolerant of most chemical fungicides used in disease management programs,[63] fungal preparations may or may not be as suitable as bacterial preparations.

Through the past couple of decades, it has become apparent that the use of microbial inoculants is not without problems. This is primarily due to the lack of knowledge about how to adequately produce, formulate, and handle living organisms. However, through continued evaluation in agronomic and horticultural systems, it has become evident that microbial inoculants may have an important place in commercial plant production and realistically offer important disease-control alternatives in plant health management. They can provide levels of disease control that, in many cases, facilitate reduced applications of fungicides and, in a few cases, may eliminate the need for fungicide applications altogether. In addition, microbial inoculants are a potentially important tool in managing fungicide resistance among pathogen populations. Furthermore, the success of sustainable plant production is largely dependent upon the integration of biological and other nonchemical means of control into disease management strategies. Recent developments in integrated pest management (IPM) are a direct result of the awareness of the importance of biological controls in holistic approaches to plant health management.

Although the biological control of turfgrass diseases is still very much in the early developmental stages, the future of microbial inoculants for turf disease control is extremely bright. It is encouraging that a number of chemical pesticide companies are now funding biological control research and have made commitments to the development of microbial inoculants. Strategies by which some companies are developing microbial-based inoculants for turfgrass disease control have been described.[64]

The search for candidate strains of bacterial and fungal antagonists has been promising based on laboratory and greenhouse tests. Antagonists suppressive to take-all patch caused by *Gaeumannomyces graminis* var. *avenae*,[61,62] to dollar spot caused by *Sclerotinia homoeocarpa*,[78] and to Pythium blight caused by *Pythium aphanidermatum*[28,65,66] have been described. Nelson and Craft[28] used Pythium blight as a model to develop a rapid and miniaturized assay to screen soil bacteria for the suppression of various turfgrass diseases (Figure 5). Using this assay, they observed that thatch-inhabiting bacteria from both low- and high-maintenance turfgrass areas were the most effective antagonists against Pythium blight. Among these strains, enteric bacteria were particularly efficacious. Strains of *Enterobacter cloacae* were as effective as metalaxyl in suppressing Pythium blight of creeping bentgrass (Figure 6).

Promising results have also been obtained from field studies with biological control agents. Preparations of *T. phacorrhiza* Fr. applied to creeping bentgrass swards effectively suppressed Typhula blight caused by *Typhula incarnata* and *T. ishikariensis*.[67,68] On creeping bentgrass putting greens, isolates of binucleate *Rhizoctonia* spp. and *Laetisaria arvalis* were suppressive to brown patch caused by *Rhizoctonia solani*[69,70] whereas strains of *Enterobacter cloacae* and *Gliocladium virens* have been shown to be effective in suppressing dollar spot.[71,72]

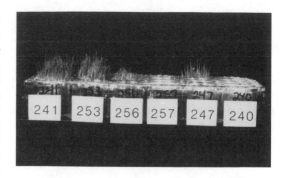

Figure 5 Effect of various soil bacteria on the suppression of Pythium blight of creeping bentgrass in tissue culture plate assays. Each row of four wells contains the same test bacterium. Wells with healthy turf indicate an effective bacterial antagonist. Strain numbers are indicated below each row of wells. *(From Nelson, E. B. and Craft, C. M., Plant Dis., 75, 510, 1991. With permission.)*

When antagonists are applied at the proper time and in an appropriate manner, they can establish high population levels in bentgrass putting greens and can be as effective as some of the newest chemical fungicides in controlling turfgrass diseases. Nelson and Craft[71] were able to establish populations of *Enterobacter cloacae* in creeping bentgrass/annual bluegrass turf at levels of 10^8 to 10^9 cells/g thatch (Figure 7). After 13 weeks, populations had declined to 10^5 to 10^6 cells/g thatch, but these levels were more than adequate to achieve disease control.

The future use of antagonists as microbial inoculants will come only from a better understand-

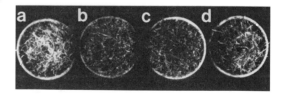

Figure 6 Effect of various strains of *Enterobacter cloacae* and the fungicide metalaxyl on the suppression of Pythium blight of perennial ryegrass in growth chamber experiments. Disease severity was rated on a scale of 1–5 for which 1 = no foliar blight and 5 = 100% foliar blight. (a) Nontreated; (b) drenched with metalaxyl (750 fg a.i./ml); (c) treated with *E. cloacae* strain EcCT-501; and (d) treated with *E. cloacae* strain E6. *(From Nelson, E. B. and Craft, C. M., Phytopathology, 82, 206, 1992. With permission.)*

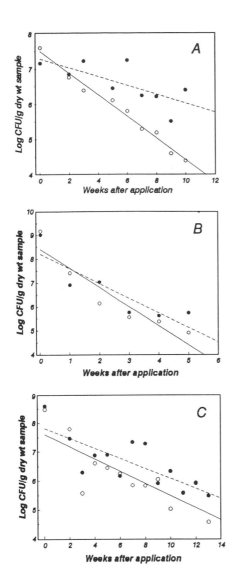

Figure 7 Population dynamics of introduced strains of *Enterobacter cloacae* in a creeping bentgrass/annual bluegrass putting green. (A) Spring application (1988) of strains E6-R8 (--●--) and E1-R6 (—O—), (B) Fall application (1988) of E6-R8 (--●--) and E1-R6 (—O—), (C) Spring application (1989) of strains EcCT-501-R3 (--●--) and EcH-1-R8 (—O—). *(From Nelson, E. B. and Craft, C. M., Plant Dis., 75, 510, 1991. With permission.)*

ing of how antagonists function and how they interact with other turfgrass management inputs. Recent developments in molecular biology have tremendously increased our ability to answer some of these questions.[73] These advances have been one of the principal reasons that biological control of

fungal plant pathogens has become a more viable option for turfgrass disease management than it was just a few years ago.

SUPPRESSION OF TYPHULA BLIGHT: A MODEL FOR BIOLOGICAL CONTROL

Typhula blight (gray snow mold) caused by the fungi *Typhula incarnata* and *T. ishikariensis* is a serious disease of turfgrasses in regions where snow-cover persists for 90 days or longer. The causal organisms grow at the interface of snow and the soil surface where they infect and parasitize numerous species of perennial plants and winter annuals, including turfgrasses and winter cereals. All cool-season species of turfgrass are susceptible to *Typhula* spp., but Typhula blight is usually most severe on high-maintenance grasses such as creeping bentgrass, perennial ryegrass, and annual bluegrass maintained at mowing heights <2.5 cm (Figure 8). On these grasses, acceptable control of the disease is achieved most commonly through preventive application(s) of fungicides in late fall or early winter, prior to the first snowfall.

A logical alternative to chemical control of Typhula blight would be to formulate a low-temperature-tolerant, disease-suppressive microbe as a preventive inoculant. Such a microbe was isolated in 1983 from Kentucky bluegrass (*Poa pratensis* L.) near Cambridge in Ontario, Canada. The organism *Typhula phacorrhiza* Fr. is closely related to the fungi that cause Typhula blight but it is not pathogenic to turfgrasses,[67] and specific isolates are highly disease-suppressive (Figure 9).

Isolates of *T. phacorrhiza* can be grown in the laboratory on a number of different substrates. For field experiments in Ontario, the fungus was grown on a mixture of heat-killed (autoclaved) grain that was dispensed by hand onto plots of creeping bentgrass in late November prior to the first snowfall. Results of studies conducted from 1985 through 1988 indicated an inverse relationship between the concentration of inoculum of *T. phacorrhiza* applied to turf in November and the intensity of Typhula blight observed the following March (Figure 10). The biocontrol agent, applied at a rate of 7.0×10^3 colony-forming-units (CFU)/m^2 of turf, provided control of Typhula blight equal to that achieved with the fungicide pentachloronitrobenzene (PCNB) applied at 30 kg a.i./ha (Table 4).

The prospect of residual disease suppression over several months or seasons makes biocontrol particularly attractive. Evidence from studies

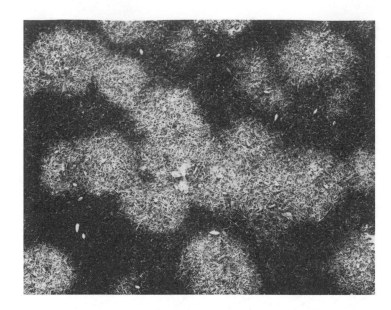

Figure 8 Typhula blight of creeping bentgrass.

Figure 9 Suppression of Typhula blight of creeping bentgrass by *Typhula phacorrhiza* (isolate T011). Grain infested with isolate T011 was applied, prior to snowfall, to the plot on the left. *(From Burpee, L. L., Kaye, L. M., Goulty, L. G., and Lawton, M. B., Plant Dis., 71, 97, 1987. With permission.)*

conducted with *T. phacorrhiza* suggests that the fungus survives and reproduces in turfgrass thatch and soil, and suppresses Typhula blight up to 16 months after two annual applications of infested grain.[68] Residual suppression of disease is a significant advantage that biocontrol agents may have over chemical agents.

ENHANCING THE EFFICACY OF *TYPHULA PHACORRHIZA*

Prior to 1985, experiments on biocontrol of Typhula blight were conducted with only two isolates of *T. phacorrhiza*, designated T011 and T016. These isolates provided good disease control, but only at relatively high rates of application. Therefore, a study was conducted to determine the potential of selecting isolates more highly disease suppressive than the original isolates.

In the spring of 1986, 33 isolates of *T. phacorrhiza* were collected from various locations in southern Ontario. The fungus colonizes corn stover and other crop debris under snow, and sclerotia can be found on the debris after snow melt. The potential of *T. phacorrhiza* isolates to suppress Typhula blight was assessed in field plots of creeping bentgrass during the winter of 1986 to 1987. Of the isolates, 22 provided significantly ($\alpha = 0.05$) greater control of Typhula blight than the standard isolate T011 (Figure 11). These results indicate that disease suppressive potential is a variable trait that can be enhanced by selection. The possibility exists that more suppressive isolates can be found, and the breeding of strains through sexual and/or somatic hybridization may lead to further improvements in the efficacy of biocontrol.

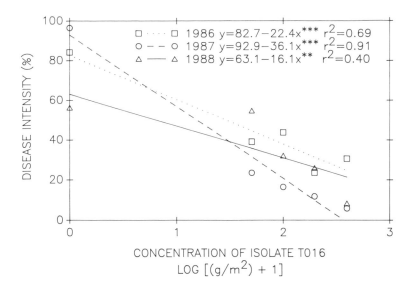

Figure 10 Relationship between intensity of Typhula blight of creeping bentgrass and concentration of inoculum of the biocontrol agent *Typhula phacorrhiza*. (From Lawton, M. B. and Burpee, L. L., Phytopathology, *80, 70, 1990.*)

Table 4 **Suppression of Typhula blight on creeping bentgrass by isolate T011 of** *Typhula phacorrhiza*

Treatment	Rate	Disease Suppression (%)[a]
Heat-killed inoculum	200 g/m^2	−23.57 a[b]
Heat-killed inoculum + cellulose	50 g/m^2 + 5% w/v	−16.73 a
Cellulose	5% w/v	−7.85 a
Heat-killed inoculum	50 g/m^2	19.23 a
Grain inoculum + cellulose	50 g/m^2 + 5% w/v	28.39 a
Grain inoculum	50 g/m^2	52.02 b
Grain inoculum[c]	200 g/m^2	88.87 b
Quintozene (PCNB)	30 kg a.i./ha	95.35 b

[a] Mean of four values calculated as a percentage of disease in an untreated plot in each block recorded on 30 March 1986; [b] Values followed by same letter are not significantly different at $p = 0.05$ according to cluster analysis; [c] Application rate equivalent to 7.0×10^3 colony-forming-units/m^2.

THE MECHANISM OF DISEASE SUPPRESSION BY *T. PHACORRHIZA*

Laboratory studies indicate that isolates of *T. phacorrhiza* do not produce antibiotics, and the fungus is not parasitic on other species of *Typhula*. A plausible hypothesis is that some form of competition for nutrients and/or space may function in the biocontrol observed in field studies. The competition hypothesis is supported by evidence of a significant ($\alpha = 0.05$) reduction in growth of isolates of *T. incarnata* and *T. ishikariensis* on media previously exposed to *T. phacorrhiza*.[67] Applications of incremental increases in nutrients to the media, after exposure to *T. phacorrhiza*, resulted in significant (p = 0.001) linear increases in colony diameters of *T. incarnata* and in dry weights of sclerotia of *T. incarnata* and *T. ishikariensis*. These results suggest that, during saprophytic phases of growth, isolates of *T. phacorrhiza* may competi-

tively utilize nutrients essential for the growth and development of pathogenic species of *Typhula*.

METHODS FOR APPLYING *T. PHACORRHIZA* TO TURF

To be a commercial success, microbial inoculants must be developed as economical formulations in which microbes can be delivered to a specific target, such as the surface of plants or soil. Recently, several researchers have concentrated on incorporating biocontrol agents into sodium alginate pellets.[74,75] The biocontrol agent, in the form of cells or tissue fragments, is mixed with a suspension of sodium alginate in water. A bulking agent of clay or organic matter is usually added to ensure that the pellets are uniform in size and shape. An organic bulking agent may also serve as a nutrient source for the biocontrol agent. The pellets are formed by dripping the alginate suspension into a gellant,

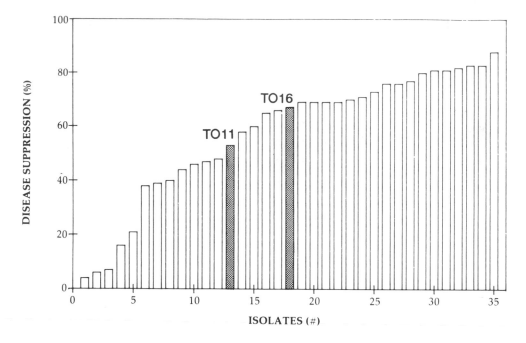

Figure 11 Differences in disease suppression potential among 35 isolates of *T. phacorrhiza*.

usually calcium chloride or calcium gluconate. After drying, the pellets are about the size of granules of commercial fertilizer, 3 to 4 mm in diameter, and can be spread with conventional fertilizer spreaders. Pellets distributed onto soil or plant surfaces serve as sites from which the biocontrol agent can grow and develop.

The alginate pellet technique has been investigated as a means of formulating isolates of *T. phacorrhiza* for application to turf.[74] Pellets were produced that contained mycelium of the fungus and either wheat endosperm or cornmeal as a bulking agent (Figure 12). The viability of the pellets was 96% after 32 weeks of storage at –10°C (Figure 13), but it was reduced significantly after 8 weeks of storage at room temperature. Field studies are required to assess the efficacy of the pellets as a system for applying *T. phacorrhiza* to swards of turfgrasses.

The studies with *T. phacorrhiza* serve as an example of how a potential biocontrol agent can be selected, assessed, and formulated as a microbial inoculant. However, many questions need to be answered prior to commercialization of a product. For example, (1) Is the pellet formulation effective under field conditions? (2) What effects do com-

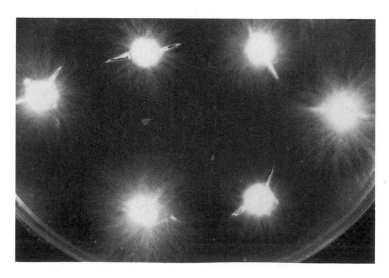

Figure 12 Growth of mycelium of *T. phacorrhiza* from sodium alginate pellets.

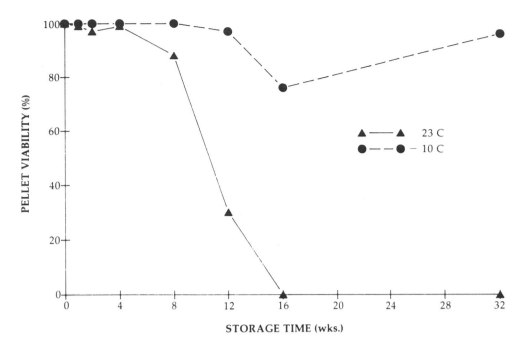

Figure 13 Effect of storage temperature on viability of *T. phacorrhiza* in sodium alginate pellets.

mon turfgrass chemical and cultural practices have on the viability of *T. phacorrhiza*? and (3) What effect does *T. phacorrhiza* have on diseases other than Typhula blight?

FUTURE PERSPECTIVES

Biological control of turfgrass diseases is still very much in the developmental stages. Although there is a number of biological control products available for disease control on other commodities, none is available specifically for turfgrass disease control. Despite the past lack of emphasis on biological control research, the last 5 to 10 years have seen tremendous advances in our efforts to understand and develop biological control strategies for turfgrass diseases. As the need to reduce fungicide dependency and provide sound environmental stewardship becomes more critical, the greater the need will be to develop safe, effective, and environmentally sound alternative control strategies.

The potential of composts to suppress turfgrass diseases is clear. At present, applications of these types of materials provide excellent alternatives to the use of fungicides on turf and may, in the long term, provide the only means of reducing soil populations of pathogens in turfgrass plantings. As we learn more about composting and the benefits of composted materials to plant health, there will undoubtedly be a greater demand from turfgrass managers for high quality disease-suppressive

composts. Composted products for use in turfgrass applications are becoming increasingly available. In general, compost producers are committed to providing the highest quality materials to turfgrass managers at an equivalent cost of disease control far below that of traditional fungicides. In addition to providing effective disease control, the use of composts will help ease the burden on our nation's landfills and foster a commitment from turfgrass managers to sound environmental stewardship.

Because microbial inoculants used for disease control are relatively new to the marketplace, it is not yet clear, particularly in the U.S., whether they will compete well with chemical fungicides and be acceptable to federal and state regulatory agencies. Although it is encouraging that more biological control products are becoming available, time will tell whether the beneficial properties of such materials can augment or replace traditional fungicides. It is critical that some of the initial biological control products consistently perform comparable to conventional fungicides if they are to find their way into the marketplace and gain widespread acceptance. Biological control systems such as the *T. phacorrhiza*-Typhula blight system have provided useful models for the study of turfgrass disease biological control. As our search for more effective antagonists of turfgrass pathogens expands, suitable bacterial and fungal antagonists will provide a pool from which these organisms can be developed into microbial inoculants. Biological control is on the verge of a new era of discovery and commercialization. The benefits of biological controls, once realized, may ultimately change the way in which disease control is approached.

REFERENCES

1. **Smiley, R. W.,** *Compendium of Turfgrass Diseases*, APS Press, St. Paul, MN, 1983.
2. **Nelson, E. B. and Craft, C. M.,** unpublished data, 1991.
3. **Sanders, P. L.,** Failure of metalaxyl to control pythium blight on turfgrass in Pennsylvania, *Plant Dis.,* 68, 776, 1984.
4. **Sanders, P. L., Coffey, M. D., Greer, G. D., and Soika, M. D.,** Laboratory-induced resistance to fosetyl-Al in a metalaxyl-resistant field isolate of *Pythium aphanidermatum, Plant Dis.,* 74, 690, 1990.
5. **Vyas, S. C.,** *Nontarget Effects of Agricultural Fungicides*, CRC Press, Boca Raton, FL, 1988.
6. **Smiley, R. W.,** Nontarget effects of pesticides on turfgrasses, *Plant Dis.,* 65, 17, 1981.
7. **Walker, A. and Welch, S. J.,** Enhanced biodegradation of dicarboximide fungicides in soil, in *Enhanced Biodegradation of Pesticides in the Environment*, Racke, K. D. and Coats, J. R., Eds., ACS Symposium Series # 426, American Chemical Society, 1990, 53.
8. **Cook, R. J. and Baker, K. F.,** *The Nature and Practice of Biological Control of Plant Pathogens,* The American Phytopathological Society, St. Paul, MN, 1983.
9. **Haahtela, K., Laasko, T., Nurmiaho-Lassila, E., Ronkko, R., and Korhonen, T.,** Interactions between N$_2$-fixing enteric bacteria and grasses, *Symbiosis*, 6, 139, 1988.
10. **Haahtela, K., Laasko, T., Nurmiaho-Lassila, E., Ronkko, R., and Korhonen, T.,** Effects of inoculation of *Poa pratensis* and *Triticum aestivum* with root-associated N$_2$-fixing *Klebsiella, Enterobacter and Azospirillum, Plant Soil*, 106, 239, 1988.
11. **Smith, J. D., Jackson, N., and Woolhouse, A. R.,** *Fungal Diseases of Amenity Turfgrasses*, 3rd ed., Routledge, Chapman & Hall, New York, 1989.
12. **Arshad, M. and Frankenberger, W. T., Jr.,** Microbial production of plant hormones, in *The Rhizosphere and Plant Growth*, Keister, D. L. and Cregan, P. B., Eds., Kluwer Academic Publishers, Dordrecht, 1991, 327.
13. **Hoitink, H. A. J. and Fahy, P. C.,** Basis for the control of soilborne plant pathogens with composts, *Ann. Rev. Phytopathol.*, 24, 93, 1986.
14. **Nelson, E. B.,** Current limits to biological control of fungal phytopathogens, in *Handbook of Applied Mycology, Vol 1. Soil and Plants*, Arora, D. K., B., Rai, Mukerji, K. G., and Knudsen, G. R., Eds., Marcel-Dekker, New York, 1990, 327.
15. **Mandeel, Q. and Baker, R.,** Mechanisms involved in biological control of Fusarium wilt of cucumber with strains of nonpathogenic *Fusarium oxysporum, Phytopathology*, 81, 462, 1991.
16. **van Peer, R., Niemann, G. J., and Schippers, B.,** Induced resistance and phytoalexin accumulation in biological control of Fusarium wilt of carnation by *Pseudomonas* sp. strain WCS417r, *Phytopathology*, 81, 728, 1991.
17. **Adams, P. B.,** The potential of mycoparasites for biological control of plant diseases, *Annu. Rev. Phytopathol.*, 28, 59, 1990.
18. **Papavizas, G. C.,** Trichoderma and Gliocladium: biology, ecology, and potential for biocontrol, *Annu. Rev. Phytopathol.*, 23, 23, 1985.
19. **Howell, C. R., Beier, R. C., and Stipanovic, R. D.,** Production of ammonia by *Enterobacter cloacae* and its possible role in the biological control of Pythium pre-emergence damping-off by the bacterium, *Phytopathology*, 78, 1075, 1988.
20. **Howell, C. R. and Stipanovic, R. D.,** Gliovirin, a new antibiotic from *Gliocladium virens* and its role in the biological control of *Pythium ultimum, Can. J. Microbiol.*, 29, 321, 1983.
21. **Gueldner, R. C., Reilly, C. C., Pusey, P. L., Costello, C. E., Arrendale, R. F., Cox, R. H., Himmelsbach, D. S., Crumley, F. G., and Cutler, H. G.,** Isolation and identification of iturins as antifungal peptides in biological control of peach brown rot with *Bacillus subtilis, J. Agric. Food Chem.*, 36, 366, 1988.
22. **Thomashow, L. S. and Weller, D. M.,** Role of a phenazine antibiotic from *Pseudomonas fluorescens* in biological control of *Gaeumannomyces graminis* var. *tritici, J. Bacteriol.*, 170, 3499, 1988.
23. **Thomashow, L. S., Weller, D. M., Bonsall, R. F., and Pierson, L. S.,** Production of the antibiotic phenazine-1-carboxylic acid by fluorescent *Pseudomonas* species in the rhizosphere of wheat, *Appl. Environ. Microbiol.*, 56, 908, 1990.

24. **Gupta, V. K. and Utkhede, R. S.,** Factors affecting the production of antifungal compounds by *Enterobacter aerogenes* and *Bacillus subtilis*, antagonists of *Phytophthora cactorum, J. Phytopathol.,* 117, 9, 1989.

25. **Nelson, E. B.,** Exudate molecules initiating fungal responses to seeds and roots, *Plant Soil*, 129, 61, 1990.

26. **Loper, J. E.,** Role of fluorescent siderophore production in biological control of *Pythium ultimum* by a *Pseudomonas fluorescens* strain, *Phytopathology*, 78, 166, 1988.

27. **Paulitz, T. C.,** Effect of *Pseudomonas putida* on the stimulation of *Pythium ultimum* by seed volatiles of pea and soybean, *Phytopathology*, 81, 1282, 1991.

28. **Nelson, E. B. and Craft, C. M.,** A miniaturized and rapid bioassay for the selection of soil bacteria suppressive to Pythium blight of turfgrasses, *Phytopathology*, 82, 206, 1992.

29. **Boehm, M. J.,** *Suppression of Pythium Root Rot of Poinsettia in Canadian Sphagnum Peat and Compost-Amended Container Media,* M. S. thesis, Ohio State University, Columbus, OH, 1990.

30. **Golueke, C. G.,** *Composting: A Study of the Process and its Principles*, Rodale Press, Emmaus, PA, 1972.

31. **Bagstam, G.,** Population changes in microorganisms during composting of spruce bark. II. Mesophilic and thermophilic microorganisms during controlled composting, *Eur. J. Appl. Microbiol. Biotechnol.*, 6, 279, 1979.

32. **Finstein, M. S. and Morris, M. L.,** Microbiology of municipal solid waste composting, *Adv. Appl. Microbiol.*, 19, 113, 1975.

33. **Hardy, G. E. St. J. and Sivasithamparam, K.,** Microbial, chemical and physical changes during composting of a eucalyptus (*Eucalyptus calophylla* and *Eucalyptus diversicolor*) bark mix, *Biol. Fertil. Soils*, 8, 260, 1989.

34. **Strom, P. F.,** Identification of thermophilic bacteria in solid-waste composting, *Appl. Environ. Microbiol.*, 50, 906, 1985.

35. **Strom, P. F.,** Effect of temperature on bacterial species diversity in thermophilic solid-waste composting, *Appl. Environ. Microbiol.*, 50, 899, 1985.

36. **McKinley, V. L. and Vestal, J. R.,** Biokinetic analyses of adaptation and succession: Microbial activity in composting municipal sewage sludge, *Appl. Environ. Microbiol.*, 47, 933, 1984.

37. **Kuter, G. A., Hoitink, H. A. J., and Rossman, L. A.,** Effects of aeration and temperature on composting of municipal sludge in a full-scale vessel system, *J. Water Pollut. Control. Fed.*, 57, 309, 1985.

38. **Nelson, E. B. and Craft, C. M.,** Suppression of dollar spot on creeping bentgrass and annual bluegrass turf with compost-amended top-dressings, *Plant Dis.*, 76, 954, 1992.

39. **Nelson, E. B. and Craft, C. M.,** Suppression of brown patch with top-dressings amended with composts and organic fertilizers, 1989, *Biol. Cult. Tests Control Plant Dis.*, 6, 90, 1991.

40. **Nelson, E. B. and Craft, C. M.,** Suppression of Pythium root rot with top-dressings amended with composts and organic fertilizers, 1991, *Biol. Cult. Tests Control Plant Dis.*, 7, 104, 1992.

41. **Nelson, E. B. and Craft, C. M.,** Suppression of Typhula blight with top-dressings amended with composts and organic fertilizers, 1991, *Biol. Cult. Tests Control Plant Dis.*, 7, 107, 1992.

42. **Nelson, E. B. and Craft, C. M.,** Suppression of red thread with top-dressings amended with composts and organic fertilizers, 1989, *Biol. Cult. Tests Control Plant Dis.*, 6, 101, 1991.

43. **Nelson, E. B.,** personal observation, 1990.

44. **Nelson, E. B., Kuter, G. A., and Hoitink, H. A. J.,** Effects of fungal antagonists and compost age on suppression of Rhizoctonia damping-off in container media amended with composted hardwood bark, *Phytopathology*, 73, 1357, 1983.

45. **Kuter, G. A., Nelson, E. B., Hoitink, H. A. J., and Madden, L. V.,** Fungal populations in container media amended with composted hardwood bark suppressive and conductive to Rhizoctonia damping-off, *Phytopathology*, 73, 1450, 1983.

46. **Kwok, O. C. H., Fahy, P. C., Hoitink, H. A. J., and Kuter, G. A.,** Interactions between bacteria and *Trichoderma hamatum* in suppression of Rhizoctonia damping-off in bark compost media, *Phytopathology*, 77, 1206, 1987.

47. **Tunlid, A., Hoitink, H. A. J., Low, C., and White, D. C.,** Characterization of bacteria that suppress Rhizoctonia damping-off in bark compost media by analysis of fatty acid biomarkers, *Appl. Environ. Microbiol.*, 55, 1368, 1989.

48. **Phae, C. G., Sasaki, M., Shoda, M., and Kubota, H.,** Characteristics of *Bacillus subtilis* isolated from composts suppressing phytopathogenic microorganisms, *Soil Sci. Plant Nutr.*, 36, 575, 1990.

49. **Phae, C. G. and Shoda, M.,** Expression of the suppressive effect of *Bacillus subtilis* on phytopathogens in inoculated composts, *J. Ferment. Bioeng.*, 6, 409, 1990.

50. **Chen, W., Hoitink, H. A. J., and Madden, L. V.,** Microbial activity and biomass in container media for predicting suppressiveness to damping-off caused by *Pythium ultimum, Phytopathology*, 78, 1447, 1988.

51. **Chen, W., Hoitink, H. A. J., Schmitthenner, A. F., and Tuovinen, O. H.,** The role of microbial activity in suppression of damping-off caused by *Pythium ultimum, Phytopathology*, 78, 314, 1988.

52. **Mandelbaum, R. and Hadar, Y.,** Effects of available carbon source on microbial activity and suppression of *Pythium aphanidermatum* in compost and peat container media, *Phytopathology*, 80, 794, 1990.

53. **Sugimoto, E. E., Hoitink, H. A. J., and Tuovinen, O. H.,** Oligotrophic pseudomonads in the rhizosphere: suppressiveness to Pythium damping-off of cucumber seedlings (*Cucumis sativus* L.), *Biol. Fertil. Soils*, 9, 231, 1990.

54. **Skorulski, J. E.,** Development of Integrated Pest Management Strategies for Turfgrass Pests of Golf Courses, MPS thesis, Cornell University, Ithaca, NY, 1990, 68 pp.

55. **Inbar, Y., Chen, Y., Hadar, Y., and Hoitink, H. A. J.,** New approaches to compost maturity, *BioCycle*, 31, 64, 1990.

56. **Inbar, Y., Chen, Y., and Hadar, Y.,** Solid-state carbon-13 nuclear magnetic resonance and infrared spectroscopy of composted organic matter, *Soil Sci. Soc. Am. J.*, 53, 1695, 1989.

57. **Chung, Y. R., Hoitink, H. A. J., Dick, W. A., and Herr, L. J.,** Effects of organic matter decomposition level and cellulose amendment on the inoculum potential of *Rhizoctonia solani* in hardwood bark media, *Phytopathology*, 78, 836, 1988.

58. **Inbar, Y., Chen, Y., and Hadar, Y.,** Humic substances formed during the composting of organic matter, *Soil Sci. Soc. Am. J.*, 54, 1316, 1990.

59. **Inbar, Y., Boehm, M. J., and Hoitink, H. A. J.,** Hydrolysis of fluorescein diacetate in sphagnum peat container media for predicting suppressiveness to damping-off caused by *Pythium ultimum, Soil Biol. Biochem.*, 23, 479, 1991.

60. **Schneider, R. W., Ed.,** *Suppressive Soils and Plant Disease*, The American Phytopathological Society, St. Paul, MN, 1982.

61. **Wong, P. T. W. and Baker, R.,** Suppression of wheat take-all and Ophiobolus patch by fluorescent pseudomonads from a Fusarium-suppressive soil, *Soil Biol. Biochem.*, 16, 397, 1984.

62. **Wong, P. T. W. and Siviour, T. R.,** Control of Ophiobolus patch in *Agrostis* turf using avirulent fungi and take-all suppressive soils in pot experiments, *Ann. Appl. Biol.*, 92, 191, 1979.

63. **Smiley, R. W. and Craven, M. M.,** Microflora of turfgrass treated with fungicides, *Soil Biol. Biochem.*, 11, 349, 1979.

64. **Woodhead, S. H., O'Leary, A. L., O'Leary, D. J., and Rabatin, S. C.,** Discovery, development, and registration of a biocontrol agent from an industry perspective, *Can. J. Plant Pathol.*, 12, 328, 1990.

65. **Wilkinson, H. T. and Avenius, R.,** The selection of bacteria antagonistic to *Pythium* spp. pathogenic to turfgrass, *Phytopathology*, 74(Abstr.), 812, 1984.

66. **O'Leary, A. L., O'Leary, D. J., and Woodhead, S. H.,** Screening potential bioantagonists against turf pathogens, *Phytopathology*, 78(Abstr.), 1593, 1988.

67. **Burpee, L. L., Kaye, L. M., Goulty, L. G., and Lawton, M. B.,** Suppression of gray snow mold on creeping bentgrass by an isolate of *Typhula phacorrhiza, Plant Dis.*, 71, 97, 1987.

68. **Lawton, M. B. and Burpee, L. L.,** Effect of rate and frequency of application of *Typhula phacorrhiza* on biological control of Typhula blight of creeping bentgrass, *Phytopathology*, 80, 70, 1990.

69. **Burpee, L. L. and Goulty, L. G.,** Suppression of brown patch disease of creeping bentgrass by isolates of nonpathogenic *Rhizoctonia* spp., *Phytopathology*, 74, 692, 1984.

70. **Sutker, E. M. and Lucas, L. T.,** Biocontrol of *Rhizoctonia solani* in tall fescue turfgrass, *Phytopathology*, 77(Abstr.), 1721, 1987.

71. **Nelson, E. B. and Craft, C. M.,** Introduction and establishment of strains of *Enterobacter cloacae* in golf course turf for the biological control of Dollar Spot, *Plant Dis.*, 75, 510, 1991.

72. **Haygood, R. A. and Mazur, A. R.,** Evaluation of *Gliocladium virens* as a biocontrol agent of dollar spot on bermudagrass, *Phytopathology*, 80(Abstr.), 435, 1990.

73. **Nelson, E. B. and Maloney, A. P.,** Molecular approaches for understanding biological control mechanisms in bacteria: studies of the

interaction of *Enterobacter cloacae* with *Pythium ultimum*, *Can. J. Plant Pathol.*, 14, 106, 1991.

74. **Fravel, D. R., Marois, J. J., Lumsden, R. D., and Connick, W. J., Jr.,** Encapsulation of potential biocontrol agents in an alginate-clay matrix, *Phytopathology*, 75, 774, 1985.

75. **Walker, H. L. and Connick, W. J., Jr.,** Sodium alginate for production and formulation of mycoherbicides, *Weed Sci.*, 31, 333, 1983.

76. **Lawton, M. B.,** Studies on the Control of Snow Molds on Turfgrass and Winter Wheat in Ontario, Ph.D. thesis, University of Guelph, Guelph, Ontario, Canada, 1989.

77. **O'Neill, N. R.,** Plant pathogenic fungi in soil/compost mixtures, in *Research for Small Farms*, Kerr, H. W., Jr. and Knutson, L., Eds., U.S. Department of Agriculture, ARS Misc, Publ. No. 1422, Goverment Printing Office, Washington, D. C., 1982, 285.

78. **Goodman, D. M. and Burpee, L. L.,** Biological control of dollar spot disease of creeping bentgrass, *Phytopathology*, 81, 1438, 1991.

79. **Goodman, D. M. and Burpee, L. L.,** Isolate pba7 of *Fusarium* as an antagonist of *Sclerotinia homoeocarpa*, in *The Guelph Turfgrass Institute 1988 Annual Report*, Burpee, L. L., Ed., Guelph Turfgrass Institute, Guelph, Ontario, Canada, 1988, 71.

80. **Rasmussen-Dykes, C. and Brown, W. M., Jr.,** Integrated control of Pythium blight on turf using metalaxyl and *Trichoderma hamatum*, *Phytopathology*, 72(Abstr.), 974, 1982.

81. **Sanders, P. L. and Soika, M. D.,** Biological control of Pythium blight, 1990, *Biol. Cult. Tests Control Plant Dis.*, 6, 99, 1991.

82. **Soika, M. D. and Sanders, P. L.,** Effect of various nitrogen sources, organic amendments and biological control agents on turfgrass quality and disease development, 1990, *Biol. Cult. Tests Control Plant Dis.*, 6, 91, 1991.

83. **Punja, Z. K., Grogan, R. G., and Unruh, T.,** Comparative control of *Sclerotium rolfsii* on golf greens in Northern California with fungicides, inorganic salts and *Trichoderma* spp., *Plant Dis.*, 66, 1125, 1982.

84. **Harder, P. R. and Troll, J.,** Antagonism of *Trichoderma* spp. to sclerotia of *Typhula incarnata*, *Plant Dis. Rep.*, 57, 924, 1973.

85. **Thurn, M.,** unpublished data, 1992.

86. **Thompson, D. C. and Clarke, B. B.,** Evaluation of bacteria for biological control of summer patch of Kentucky bluegrass caused by *Magnaporthe poae*, *Phytopathology*, 82, 1123, 1992.

87. **Melvin, B. P., Vargas, J. M., and Berndt, W. L.,** Biological control of necrotic ringspot, *Phytopathology*, 78, 1503, 1988.

88. **Sarniguet, A. and Lucas, P.,** Evolution of bacterial populations related to decline of take-all patch on turfgrass, *Phytopathology*, 81, 1202, 1991.

89. **Sarniguet, A. and Lucas, P.,** Etude preliminaire sur l'effcacite de differents moyens de lutte contre le pietin-echaudage du gazon (*Gaeumannomyces graminis* var. *avenae*), *Agronomie*, 12, 187, 1992.

Chapter 39

IPM for Tree and Shrub Diseases

George W. Hudler, *Department of Plant Pathology, Cornell University, Ithaca, NY*

CONTENTS

Introduction 429
Some Thoughts About Diagnosis 429
Integrating Disease Management in New Plantings 430
    Plant Selection 430
    Site Selection 430
    Careful Inspection of Planting Stock 433
Integrating Disease Management for Established Landscapes 434
    Evaluation and Documentation 434
    Reduction of Primary Inoculum via Sanitation 434
    Maintaining Host Vigor 434
    Pruning with Care and Common Sense 435
    Use of Pesticides 435
        Surface Active, Broad Spectrum Fungicides 435
        Surface Active and Locally Systemic, Broad Spectrum Fungicides 436
        Systemic Fungicides 436
Biological Control of Plant Pathogens on Landscape Trees and Shrubs 437
IPM in the Landscape — The Future 437
References 438

INTRODUCTION

Managers of amenity plants in landscapes face unique challenges in the area of pest management. Foremost of these is that plants being grown for ornamental purposes must not only be alive, but by their very nature must be pleasing to the eye. Thus, many diseases which pose little or no threat to plant health *per se*, but which cause significant loss of aesthetic appeal, may command an inordinate amount of attention. Use of chemical pesticides is one of several strategies successfully employed to prevent disease occurrence. However, it is *only* one and as the use of chemicals comes under increased scrutiny from environmental activists and the public at large, successful pest managers will have to do a better job of informing their clients of alternatives and of using them.

In the discussion that follows, IPM strategies for disease control in landscapes are presented in three parts. First, actions that can be taken to prevent or minimize disease occurrence in *new* plantings are described. Second, strategies for managing plants so as to minimize diseases in *established* plantings are discussed. And third, innova-

tions to strengthen IPM programs for the future are presented. But first...

SOME THOUGHTS ABOUT DIAGNOSIS

The landscape plant manager cannot adequately manage a plant disease unless the cause of the problem has been correctly identified. This crucial first step in integrated pest management is often overshadowed by more "newsworthy" endeavors, but it remains the foundation for the rest of the program. Thus, time spent honing one's skills in this area is every bit as important as anything else the pest manager does.

One source of continual difficulty in diagnosis of plant diseases is to distinguish between problems caused by infectious agents such as viruses, bacteria, fungi, and mycoplasma-like organisms (MLOs) and those caused by noninfectious agents such as freezing temperatures, drought, and toxic chemicals. Clues to help distinguish between the two groups of agents often only surface after rigorous "interrogation" but here are some general observations to look for:

1. Symptoms caused by noninfectious agents are more likely to occur on more than one species of plant. Few infectious diseases attack plants of different species or genera.

2. Symptoms caused by noninfectious agents are more likely to affect all members of one species in a site. Infectious diseases rarely affect all individuals of a single species in a single year. However, factors such as late frost or toxic chemical drift may affect all or nearly all plants within a given area in a short period of time.

3. Symptoms caused by noninfectious agents are more likely to occur in a systematic pattern (i.e., only close to the road, only on the west side of the crown). Infectious diseases usually occur in a more or less random fashion, certainly not so regular as to occur only in certain rows.

A second issue of concern with regard to diagnosis is to distinguish between those diseases most likely to occur on plants predisposed by one or more adverse environmental conditions and those caused by primary pathogens. Diseases associated with predisposed plants are far more common in the landscape than they are on agronomic crops, and it is important for the diagnostician to be aware of them so as to avoid prescribing treatment for a disease while leaving its root cause(s) unattended.

Insofar as basic mechanics of diagnosis are concerned, one is strongly advised *not* to approach the task by holding a diseased specimen in one hand and flipping through picture books with the other in hopes of finding something that looks like the specimen in question. Such an approach may occasionally work, but far more often it will lead to frustration and erroneous conclusions.

A more satisfactory strategy is to consult one or more host indices to find out what pathogens are actually known to occur on the plant in question and where in the world one might expect to find them. Books by Hepting,[1] Pirone,[2] Horst,[3] and Farr et al.[4] and compendia published by the American Phytopathological Society (APS)[5-7] will help to expedite this part of the task.

Then, with some verbal description in hand and with some notion of what one is looking for, microscopic work, culturing, or other tests may be in order. This portion of the process may have to be done in a well-equipped laboratory. However, kits such as those manufactured by Agri-Diagnostics Associates (Cinnamonson, NJ), Neogen (Lansing, MI) and AgDia (Mishawaka, IN), utilizing differences in serological reactions to distinguish among taxa of plant pathogens, show great promise in aiding diagnosticians trying to identify suspect pathogens in the field.

Finally, comparison with pictures in comprehensive treatises such as Sinclair et al.[8] or in more specialized books like the APS compendia is in order. Obviously, the plant pest diagnostician needs to invest a considerable sum in a personal library in order to diagnose diseases accurately.

INTEGRATING DISEASE MANAGEMENT IN NEW PLANTINGS

PLANT SELECTION

Landscape designers and others responsible for choosing plant materials usually select plants for particular sites based on factors such as shape, color, texture, and size at maturity. Little concern is given to pest resistance even though considerable effort has been expended to develop and/or identify disease-resistant cultivars of some species. Table 1 lists some representatives of selected genera where disease resistance has either been documented via observation of specimens in service, or inoculation and evaluation in the context of a controlled experiment. Trees listed therein were not necessarily immune to the disease in question, but infection levels were sufficiently low that applications of pesticides to them were not deemed necessary by the observer. If one were planning to use representatives of any of the genera listed in Table 1, one would be well advised to consider resistant individuals.

SITE SELECTION

Many pathogens of woody ornamental plants, especially those that attack the living bark, cortex, and vascular cambium, are facultative parasites. Fungi in the genera *Botryosphaeria, Endothia, Nectria, Sphaeropsis, Steganosporium, Thyronectria,* and *Valsa* are typical examples. They pose little or no threat to vigorously growing plants but are quick to take advantage of individuals under stress from factors such as defoliation, insufficient water, and noxious chemicals in the soil or air. In one case, *Pachysandra terminalis,* a shade-loving plant, was predisposed to stem canker caused by *Volutella pachysandricola* when grown in full sun.[9] It is likely that other plants are also predisposed to opportunistic pathogens when moved to environments that are too different from those in which they have evolved.

The nature of the predisposing action has been the source of considerable research in recent years, and one recurring theme is that stressed plants are slower to repair bark wounds. Most pathogens of

Table 1 **Some trees known to be resistant to selected diseases**

Resistant Trees	Resistant To	Ref.	Comments
Acer (maple) *A. platanoides* cv. 'Jade Glen', 'Parkway'	Verticillium wilt (*Verticillium dahliae*)	41	Cvs. 'Crimson King' and 'Greenlace' were most susceptible in this study of controlled inoculations. Cvs. 'Globosum', 'Royal Red', 'Cleveland', 'Summer Shade', 'Emerald Queen', 'Silver Variegated', Schwedleri', 'Superform', 'Columnare Compacta' were intermediate in symptom expression.
Aesculus (horsechestnut, buckeye) *A. arguta, A. glabra* var. *monticola, A. glabra* var. *sargentii, A parviflora, A. parviflora* var. *serotina*	Leaf blotch (*Guignardia aesculi*)	42	
Albizzia (mimosa) *A. julibrssin* cv. 'Union'	Wilt (*Fusarium oxysporum* f.sp. *pernicosum*)	43	Cultivar has not been widely planted, thus resistance has not had extensive testing in the field.
Celtis (hackberry) *C. sinensis C. jessoensis*	Witches'-broom (*Sphaerotheca phytoptophyla* + an eriophyid mite)	44, 45	Based on cursory observations of established street trees, there seem to be resistant individuals of *C. occidentalis*, as well.
Chamaecyparis (falsecypress) *C. pisifera* cvs. 'Filifera', 'Aurea variegata', 'Plumosa aurea', 'Plumosa lutescens', and 'Squarrosa sulfurea'	Twig blight (*Phomopsis juniperovora*)	44	
Cornus (dogwood) *C. kousa*	Anthracnose (*Discula destructiva*)	46	*Cornus florida* is highly susceptible to this disease and many specimens in the eastern U.S. have been killed by the pathogen.
Crataegus (hawthorn) *C. crus-galli* (Cockspur thorn)	Leaf blight (*Diplocarpon mespili*) Rust (*Gymnosporangium* spp.)	44	Hawthorns are susceptible to a large number of insects and diseases. *C. crus-galli* is highly susceptible to leaf miners. *C. laevigata* cv. Paul's Scarlett is highly susceptible to leaf blight.

Table 1 (Cont.) **Some trees known to be resistant to selected diseases**

Resistant Trees	Resistant To	Ref.	Comments
Juniperus (juniper, red cedar) *J. chinensis* cvs. 'Femina', 'Keteleeri', 'Pfitzeriana aurea', 'Robusta', 'Sargentii', 'Sargentii glauca', 'Shoosmith' *J. communis* cvs. 'depressa', 'Repanda', 'saxatalis', 'suecica', *J. sabina* cv. 'Skandia' *J. squamosa* cvs. 'Prostrata', 'Pumila'	Blight (*Phomopsis juniperovora*) Rust (*Gymnosporangium* spp.)	44	Plants injured by deicing salt or stressed by drought will be more susceptible to disease, even if listed as resistant at left. New cultivars are being tested.
Malus (flowering crabapple) *Malus* cvs. 'Beverly', 'Bob White', 'David', 'Dolgo', 'Selkirk', 'Sentinel', 'Strawberry parfait', 'Sugar Tyme', 'White Angel' *Malus floribunda* cvs. 'Golden Hornet', 'Golden Gem', *Malus* x *micromalus*, *M. robusta*, *M. sargentii*, *M.* x *zumi* 'Calocarpa'	Scab (*Venturia inaequalis*) Rust (*Gymnosporangium* spp.) Powdery mildew (*Sphaerotheca* spp.) Fireblight (*Erwinia amylovora*)	47	Many other cultivars are susceptible to one or more of the major crabapple diseases but are still deemed worth planting. Also, new cultivars of flowering crabapples are introduced annually. Check with your Cooperative Extension Service for up-to-date ratings.
Platanus *P.* x *acerifolia* cvs. 'Columbia', 'Liberty'	Anthracnose (*Gnomonia platani*)	48	This tree is a hybrid between American sycamore and Oriental plane. Several crops of nursery grown seedlings thought to be *P.* x *acerifolia* but really *P.* x *acerifolia* x *P. occidentalis* with lost resistance to anthracnose were produced and distributed about 25 years ago, but currently available trees should be true *P.* x *acerifolia*.
Populus (Hybrid poplar) *Populus* 'Assiniboine'	Canker (*Septoria musiva*) Rust (*Melampsora* spp.)	49	Cultivar is susceptible to lime-induced chlorosis and poplar petiole gall aphid.
Pyrus (pear) *P. calleryana*	Fire blight (*Erwinia amylovora*)	45	Most cultivars of *P. calleryana* are resistant to fireblight.
Ulmus (elm) *Ulmus* cvs. 'Dynasty', 'Homestead', 'Jacan', 'Pioneer Regal',	Dutch elm disease (*Ophiostoma ulmi*) Elm yellows, caused by a mycoplasm-like organism	44	All elms native to North America, including Dutch elm disease-resistant cv. 'Liberty', are susceptible to elm yellows.

Table 1 (Cont.) **Some trees known to be resistant to selected diseases**

Resistant Trees	Resistant To	Ref.	Comments
'Sapporo Autumn Gold', 'Thompson', 'Urban'			
U. glabra (Scots elm)			
U. hollandica cv. 'Christine Buisman'			
U. parvifolia (Chinese elm)			
U. pumila (Siberian elm)			

live bark can only gain access to their hosts when the outer layer of dead bark is ruptured and inoculum is introduced into the wound. A nonstressed host forms a barrier of impervious tissue and, eventually, new periderm in living cells subjacent to the wound and does so quickly enough to limit advance of the pathogen. The stressed host, though erecting the same barriers, does so too slowly or in some other ineffectual manner such that growth of the pathogen is not contained.[9,10]

Obviously, it is crucial as one develops a landscape plan with an eye to minimizing plant stress to ensure that the site is well surveyed and characteristics of soil, drainage, air flow, and traffic patterns for pedestrian and vehicular traffic are known. When planting actually begins, workers must take care to employ needed soil modifications, maintain the integrity of the root systems, and ensure that the plants will receive adequate water for at least 1 year after planting.

CAREFUL INSPECTION OF PLANTING STOCK

It is unlikely that a reputable nursery would knowingly distribute diseased stock, but the characteristics of some pathogens and/or the diseases they cause are such that detection in the nursery may be difficult if not impossible.

An example is Ploioderma needlecast of Austrian (*Pinus nigra*), pitch (*P. rigida*), and (occasionally) mugo (*P. mugo*) pines in the northeast U.S. The pathogen, *Ploioderma lethale,* produces ascospores on needles infected the previous year, and those spores infect newly developing needles in the spring. However, there is virtually no evidence that infection of the new needles has occurred until the following spring when those needles, now 1-year-old and fully expanded, suddenly turn a straw brown color. If infected trees were taken from the nursery and transplanted early enough in the spring, while still dormant, overt symptoms of the disease may not have shown up,

and the disease will have been introduced to a new site.[8]

Other diseases with extended latent periods between infection and symptom expression include Rhabdocline and Swiss needlecasts of Douglas-fir, Lophodermium needlecast of Scots pine, crown gall on a wide variety of deciduous hosts, MLO-caused ash yellows and lilac witches'-broom, and Eutypella canker on maple.[8]

In another case where inadvertent introduction of a pathogen may occur, the pathogen survives on the dormant host as microscopic resting structures. With the onset of warmer weather and the growth of new tissues, the resting structures either become or produce infectious propagules to cause disease on new twig and leaf tissue. Fungi in the genus *Taphrina*[11] and most powdery mildews[12] are notorious in this regard. *Erwinia amylovora* (cause of fire blight), *Diplocarpon roseae* (cause of black spot of rose), and *Discula destructiva* (cause of dogwood anthracnose) are typical examples of other pathogens which overwinter on their hosts as minute, unobtrusive bodies which could be spread to new sites on those hosts.

Pathogens could also be introduced in soil or plant debris associated with roots or root balls. Root pathogens such as *Phytophthora cinnamomi* on ericaceous plants and *Verticillium dahliae* on a wide variety of deciduous hosts survive as resting structures in the soil until stimulated to germinate by exudates from host roots. No matter how careful one is, transplanting does cause injury to roots, and those wounds often provide the stimulus for soilborne pathogens to become active.

Because disease control is more likely to be effective when the population of a pathogen is low, rigorous inspection of newly planted stock is an oft-overlooked but valuable strategy in an IPM program. It is incumbent upon those who are doing the planting to be aware of diseases that occur on material they are installing and to be alert for symptoms and signs of those diseases. Reinspection of new stock for up to 1 year after installation will help the pest manager to identify problems which might have originated in the nursery. Quick action can be taken on the site as needed, and suppliers can be made aware of problems within their production areas to minimize chances for continued introduction of the pathogen(s) to new sites.

INTEGRATING DISEASE MANAGEMENT FOR ESTABLISHED LANDSCAPES

Established landscapes present an additional array of problems for pest managers because the plant material has usually become an integral part of the overall landscape. More often than not, the plants are too large to simply be removed and replaced with others of equal stature if errors in species or site selection were made. Furthermore, as species of plants age they become more sensitive to problems associated with maturation and old age, especially decay of stem and root wood.

Political boundaries also emerge as important factors in plant disease management in established landscapes. While relatively innocuous strategies may contain low levels of a particular pathogen, those same strategies may be totally ineffectual if plants are overwhelmed by inoculum from unmanaged adjoining properties.

EVALUATION AND DOCUMENTATION

In all but a few cases, once plant pathogens infect their hosts, they cannot be eradicated. Thus, the prevailing theme of a successful management program is to prevent disease before it occurs rather than to act afterwards. The pest manager becomes better at predicting the need for action as (s)he becomes more familiar with the plants and the site in question. To this end there is no substitute for periodic, well-documented inventories of the plant materials being managed and their problems. A map of the site showing locations of all plants accompanied by a list with a history of each specimen, updated at least once per year, quickly becomes the most important tool in the pest manager's arsenal.

REDUCTION OF PRIMARY INOCULUM VIA SANITATION

Many foliage diseases overwinter in the duff on leaves infected the previous year. Apple scab (*Venturia inaequalis*), tar spot (*Rhytisma* spp.), honeysuckle leaf blight (*Insolibasidium deformans*), and horsechestnut leaf blotch (*Guignardia aesculi*) are typical examples. It is reasonable to expect that if primary inoculum of these and similar pathogens could be significantly reduced by collecting and destroying infected leaves in the autumn, primary infection the following spring would also be reduced.

Similarly, it has been suggested that incidence of many bark pathogens causing twig blights and cankers could be reduced by removing diseased twigs and branches and thus lessening inoculum. Such a strategy is suspected to be effective for control of some diseases such as fire blight (*E. amylovora*), pine-pine gall rust (*Endocronartium harknessii*), and black knot of *Prunus* spp. (*Apiosporina morbosa*).

MAINTAINING HOST VIGOR

No matter how carefully sites were chosen at the time of planting, landscape trees often seem to find themselves with limited space for root growth as they get larger and older. Roots of those with adequate room may suffer if soil around them has become compacted by repeated mowing of turf and/or excessive traffic. The consequences of these actions over time are that older trees are more likely to suffer from drought stress and may be deficient in nutrients.

The predisposing effects of various stress-inducing factors on the incidence of bark diseases in newly planted stock was discussed previously. For the most part, these diseases are of greater consequence on younger trees because they are likely to girdle the main stem. Older trees with rough bark seem to be less threatened by such diseases, but Valsa canker of spruce — especially Colorado blue spruce (*Picea pungens*) — caused by *Valsa kunzei* is a notable exception. This disease is uncommon on trees less than about 20 years old but often destroys older, more valuable specimens.

One additional group of facultative parasites of significant importance on older trees are fungi in the genus *Armillaria*. This group was once lumped together as a single species, *Armillaria mellea*, but it is now suspected to be comprised of as many as a dozen species.[13] These fungi kill trees that have been defoliated or have been attacked by sapsucking insects, and they are suspected of being able to kill trees weakened by any number of other agents, including those associated with "old age."[14] Inasmuch as the primary site for colonization by *Armillaria* spp. is the living bark and vascular cambium in roots, it is suspected that the pathogens act in a manner similar to canker diseases, and that trees will respond with defense reactions similar to those created to slow other bark pathogens. However, the fungi also rot wood, and so must contend with and overcome another array of defense responses in the xylem. Nonstressed hosts are rarely reported to fall prey to *Armillaria*.

Another issue of concern is the relationship, if any, between reduced vigor and the ability of a tree to cope with wounds to the xylem and subsequent decay. In view of the increased threat to people and property as rotten stems and branches get older and larger (further off the ground!), the prudent landscape manager will do everything possible to ensure that defense reactions in the xylem have the best chance at slowing the decay process. It is assumed, though poorly documented, that trees suffering from water stress, inadequate nutrition, defoliation, or any of a number of other adverse environmental conditions are more likely to sustain losses from wood decay.

Any action taken to improve host vigor should optimize defense responses to both bark and xylem pathogens. Inasmuch as the most common and pervasive problem seems to be water stress, efforts to improve the soil in the rhizosphere by way of aeration and watering are worthy endeavors. Conventional equipment adapted from the turf industry has been used with some success. However, innovations like the Grow Gun® (Grow Gun Inc., Arvada, CO) show even greater promise in fracturing compacted soil and in providing a means to add water, nutrients, and other amendments to the rhizosphere.

PRUNING WITH CARE AND COMMON SENSE

Landscape plants often have to be pruned in order to reduce their size, improve their shape, or ensure their safety. Because pruning cuts are wounds, and wounds are known to be sites for origination of canker diseases and wood decay, the prudent landscape manager will make them no more often than is absolutely necessary and will do so at a time of year when the plant has the best chance to repair the damage.

Studies of wound repair in bark of several different species of woody plants indicate that barrier zones in live bark are formed most rapidly at times of the year when other growth is also occurring.[9,15,16] No wound repair occurs when trees are dormant. Not surprisingly, bark disease incidence is also higher on trees wounded while dormant.[17-19] Thus, pruning from early spring through midsummer would seem to be the best time for such action and pruning in the beginning of the dormant season, i.e., November and December, is to be avoided as much as possible. A notable exception to this suggestion is where oak wilt threatens the lives of oaks. Insect vectors of this disease are attracted to fresh wounds in oaks in the spring and early summer, at the same time the pathogen is sporulating. Wounds of any kind on oaks at this time of year are to be avoided.

The mechanics of how to best prune branches and the use of wound dressings to protect wounds from colonization by noxious organisms are points of some contention among tree care professionals. Shigo's dissections of thousands of trees (pruned, butchered, or otherwise abused) and his re-creations of their histories suggest that (1) once touted flush cuts do not minimize decay, (2) discoloration and decay do not stop when wounds "callus over"; thus, speed of closure is not a measure of incidence of decay, (3) the proper line of cut is at the natural junction between stem and branch — the "branch collar," and (4) dressings used to cover wounds do nothing to prevent discoloration and decay and may actually impede optimum functioning of wound repair processes.[20]

More recent experiments by Neely,[21] however, challenge Shigo's assertions about the speed of wound repair and its relationship to incidence of discoloration. And, results of experiments by Peterson and Helmer[22] and by Biggs[10] indicate that properly formulated wound dressings may have a place in reducing disease incidence in stems and branches of woody plants. That questions about pruning, the single most researched and discussed arboricultural practice in the past 20 years, remain unresolved is testament to the complex nature of woody plants and their care.

USE OF PESTICIDES

Despite the foregoing array of nonchemical means for preventing diseases on landscape plants, they still do occur. And, there may be no alternatives for management except to either apply pesticides or remove the plants. Fortunately, those chemicals currently registered for disease control in landscapes have been subjected to rigorous tests by various government and industry laboratories. So far as is known, they pose minimal threats to "nontarget" organisms, including people and pets, when used according to label directions.

Most pesticides for disease control are fungicides and these can act in one or more ways to achieve desired results:

Surface Active, Broad Spectrum Fungicides

Compounds in this group are applied to plant surfaces, primarily foliage, where they act to inhibit survival of germinating spores, thus preventing infection and colonization by pathogens. They are effective against a broad array of fungi, mostly Ascomycetes and Deuteromycetes, and remain effective for at least 10 to 14 days after application.

Most have relatively high LD_{50} values for acute toxicity to test animals and humans and pose little threat to nontarget organisms. Representative compounds in this group include:

Compound	Trade Names	Comments
Captan	Captec 4L	For control of foliage diseases and cutting
	Ortho garden fungicide	rots. Also used as seed treatment.
Chlorothalonil	Daconil 2787	Especially useful for control of needlecasts
	Bravo 720	of conifers. Bravo 720 has excellent retention on hard-to-wet surfaces.
Copper hydroxide	Kocide 101	May be toxic to foliage of some deciduous trees and shrubs.
Mancozeb	Dithane M-45	General protectant fungicide. Use with spreader-sticker for maximum effectiveness.
Funginex	Triforine	General protectant fungicide including activity against several rust diseases

Surface Active and Locally Systemic, Broad Spectrum Fungicides

Chemicals included in this group prevent spores from germinating and/or penetrating host tissue. They are also absorbed into the plant through epidermal cells and thus may kill established infections.

Most of the fungicides included in this group are members of a new generation of chemicals known as "azoles." Azoles kill fungi by inhibiting their ability to synthesize complex organic chemicals called "sterols" which play crucial roles in many normal life processes. Although all living organisms need sterols for growth and development, some of those occuring in fungi do not occur in other organisms. Thus, chemicals which target those sterols unique to fungi should pose little threat to other types of organisms. Some compounds in this group are:

Compound	Trade Names	Comments
Fenarimol	Rubigan	Registered for control of scab, powdery mildew, and rusts on apples, cherries, and pears. Expect more uses for ornamentals in future years.
Propiconazole	Banner	Registered for control of many leaf diseases on turf,
	Alamo	bedding plants, and broadleaved and coniferous trees and shrubs.
Triadimefon	Bayleton	Effective against at least 65 diseases on trees and shrubs.

Systemic Fungicides

Compounds in this group may be applied via foliar spray, trunk injection, or soil drench. They are readily mobile in plants and are relatively long-lived because within plant tissue they are protected from rainfall or sunlight. Those chemicals applied via injection or drench pose the least threat for exposure of nontarget sites.

Compound	Trade Names	Comments
Fosetyl Al	Alliette	For control of foliage and root diseases caused by Oomycetes. Especially effective for control of downy mildews.
Metalaxyl	Subdue	For control of Phytophthora and Pythium root diseases. May be applied as spray or drench; best results are with drench.
Thiabendazole	Arbotect 20S	For control of Dutch elm disease. Applied via injection to prevent infection in healthy trees and to eradicate disease from trees with less than 5% crown loss.
Thiophanate methyl	Cleary's 3336 Topsin M	Systemic fungicide for control of many leaf and twig blights. May be a suitable replacement for benomyl in many cases.

Despite the effectiveness and relative safety of the aforementioned materials, public sentiment against the use of chemical pesticides continues to grow and successful landscape managers will have to aggressively explore alternative chemical strategies if they are to remain competitive.

One strategy which bears continual scrutiny is the potential use of materials which have little or no toxicity to nontarget organisms and have little chance for accumulation in the food chain. Soaps and oils have captured attention as insecticides because they meet both of these criteria. More recently, sodium bicarbonate (baking soda) has proven to have little or no toxicity to nontarget organisms but a high degree of efficacy in controlling certain leaf diseases. The value of this material as a fungicide was first demonstrated by Homma et al.[23] who used it to control powdery mildew (*Sphaerotheca fuliginea*) on cucumber. More recently, however, Horst et al.[24] demonstrated its efficacy in preventing powdery mildew and black spot on rose. In the latter experiments, addition of 1% ultrafine spray oil (Sunspray 6E, Sun Oil Co, Philadelphia, PA) significantly reduced the incidence of both diseases.

Finally, film-forming polymers such as Vapor Gard® and Wilt Pruf®, chemicals usually used as antitranspirants, have been proven effective in reducing the incidence of powdery mildew on several plants.[25-27] Although they have not been thoroughly tested for efficacy as disease control agents on ornamental plants (except for roses), there is little reason to think they will not be effective, at least in reducing incidence of some powdery mildews.

BIOLOGICAL CONTROL OF PLANT PATHOGENS ON LANDSCAPE TREES AND SHRUBS

Despite tremendous interest by the mass media and the research community on the potential use of microorganisms to control plant diseases and a deluge of publications on the subject, only two have actually found their way into "mainstream" agriculture. Both are used to control diseases of trees and shrubs.

Galltrol-A® (also Norbac® and Agrocin 84®), a strain of nonpathogenic *Agrobacterium radiobacter,* is applied as a root dip or drench to prevent colonization of potential hosts by *A. tumefaciens,* the cause of crown gall. Treatment with this microorganism has become standard practice in many nurseries throughout the world.[28-30] In many countries, although not in North America, the fungus *Peniophora gigantea* is applied to freshly cut pine stumps to protect against colonization by the root decay fungus, *Heterobasidion annosum.*[31]

Several innovations in biological control with considerable promise for nonchemical management of landscape tree diseases have appeared in recent years. Binab® T, a formulation of two species of fungi in the genus *Trichoderma,* is purported to prevent canker diseases and wood decay when applied to fresh wounds. Implantation of Binab® pellets into elms was also suspected to prevent Dutch elm disease in one experiment.[32]

Trichoderma spp. have also shown efficacy in protecting pine stumps from colonization by *H. annosum,*[31] apple fruit from *Botrytis cinerea,*[33] and apple and pear stems from *Chondrostereum purpureum*[34,35] and *Nectria galligena.*[36] These positive results from field tests coupled with many more in laboratory experiments continue to buoy enthusiasm within the research community for eventual discovery of more practical uses of this amazing group of microbes.

The future for using microbial agents to prevent leaf and twig diseases seems to be not quite so promising as that for preventing wood decay, but some observations warrant continued exploration. In laboratory and growth chamber experiments, and to a lesser extent in the field, saprophytic fungi in the genus *Chaetomium* reduced incidence of apple scab,[37] nonpathogenic *Erwinia herbicola* prevented fire blight,[38] and another nonpathogen, *Pseudomonas fluorescens,* suppressed the incidence of Scleroderris canker on pines caused by *Ascocalyx abietina.*[39] Young and Andrews[40] successfully reduced primary inoculum of the apple scab pathogen by "inoculating" leaf litter in an orchard with the mycoparasite *Athelia bombasina.* Such treatments with appropriate microbes could also be used to reduce primary inoculum of other leaf pathogens overwintering in/on leaves infected the previous year.

IPM IN THE LANDSCAPE — THE FUTURE

Future generations will have to manage trees and shrubs that are being planted right now. There is little reason to think that public aversion toward chemical pesticides will improve in the years to come, and producers and managers of landscape plants might just as well come to grips with that fact. From the array of IPM strategies presented above, three areas to receive increased emphasis emerge.

First is that the nursery industry, supported by research from universities, agricultural experiment stations, arboreta, etc., must increase its efforts to identify, produce, and market disease-resistant plants. Many such trees and shrubs already dot the landscape. They stand disease-free or almost so next to severely-infected relatives, and for all intents and purposes have already proven their ability to resist a particular pathogen. It should be a relatively straightforward matter to identify these plants, propagate them either vegetatively or by grafting, test them for resistance to known doses of inoculum and an array of isolates of the pathogen, and get the best of them into people's hands. For the longer term, it will be prudent for developers of disease resistant trees and shrubs to keep abreast of emerging technologies in the area of gene transfer so as to be able to add those to the arsenal as needs dictate. But, there is no need to wait for those technologies. Selection can begin now with what is already growing. At the same time, the industry must aggressively discourage the use of trees and shrubs known to be highly susceptible to disease.

Second, people who design landscapes and/or specify plants for particular sites, those who produce them, those who install them, and those who manage them must become more sensitive to the needs of those plants. With each passing month, research results illustrate how absolutely crucial site quality is in sustaining healthy plants. Poor choices anywhere along the line will lead to unthrifty, disease- and insect-prone specimens requiring undue (chemical intensive?) maintenance for the rest of their lives. The heightened awareness for those already in the business is most expeditiously obtained by inclusion of speakers to address such topics in professional society or association meetings. Development and inclusion of courses on plant health management in curricula for future generations of students is a must.

Third, it is time for plant health care professionals to become more proactive in educating their customers about how plants grow and what they will and will not tolerate. At the beginning of this chapter, I suggested that disease management in ornamental plants is especially challenging because aesthetic appeal is such an overriding issue. If those who ultimately call the shots had a better sense of the kinds of problems that (do or do not) threaten plant health, perhaps they would view some of the more insignificant maladies with more tolerance. Foliage problems are particularly visible, and it is likely that more resources go into prevention of foliage problems than into those on any other portion of an ornamental plant. Yet, trees and shrubs can endure substantial amounts of damage to foliage with little or no impact to overall health. People ought to know that, along with a lot more about tree biology, as they make choices about care of their plants.

Landscape plant disease management is a dynamic field. The sheer numbers of varieties and cultivars of species involved and the wide variety of sites on which they are grown compound the problems of pest management so much that one never lacks for a challenge in caring for them.

REFERENCES

1. **Hepting, G. H.,** *Diseases of Forest and Shade Trees of the United States, USDA Agricultural Handbook No. 386*, U.S. Department of Agriculture, Washington, DC, 1971.
2. **Pirone, P. P.,** *Diseases and Pests of Ornamental Plants*, 5th ed., John Wiley & Sons, New York, 1978.
3. **Horst, R. K.,** *Westcott's Plant Disease Handbook*, 5th ed., Van Nostrand Reinhold, New York, 1990.
4. **Farr, D. F., Bills, G. F., Chamuris, G. P., and Rossman, A. Y.,** *Fungi on Plants and Plant Products in the United States*, APS Press, St. Paul, MN, 1989.
5. **Coyier, D. L. and Roane, M. K.,** *Compendium of Rhododendron and Azalea Diseases*, APS Press, St. Paul, MN, 1986.
6. **Horst, R. K.,** *Compendium of Rose Diseases*, APS Press, St. Paul, MN, 1983.
7. **Stipes, R. J. and Campana, R. J.,** *Compendium of Elm Diseases*, APS Press, St. Paul, MN, 1981.
8. **Sinclair, W. A., Lyon, H. H., and Johnson, W. T.,** *Diseases of Trees and Shrubs*, Comstock Publishing, Ithaca, NY, 1987.
9. **Hudler, G. W., Neal, B. G., and Banik, M. T.,** Effects of growing conditions on wound repair and disease resistance in *Pachysandra terminalis*, *Phytopathology*, 80, 272, 1990.
10. **Biggs, A. R.,** Managing wound associated diseases by understanding wound healing in the bark of woody plants, *J. Arboric.*, 16, 108, 1990.
11. **Mix, A. J.,** A monograph of the genus Taphrina, *Univ. Kansas Sci. Bull.*, 23(1), 1949.
12. **Gadoury, D. M. and Pearson, R. C.,** Initiation, development, dispersal, and survival of cleistothecia of *Uncinula necator* in New York vineyards, *Phytopathology*, 78, 1413, 1988.
13. **Watling, R., Kile, G. A., and Burdsall, H. H., Jr.,** Nomenclature, taxonomy, and identification, in *Armillaria Root Disease, USDA Agricultural Handbook 691*, Shaw, C. G. and Kile, G. A., Eds., U.S. Department of Agriculture Forestry Service, Washington, DC, 1991, chap. 1.

14. **Wargo, P. M. and Harrington, T. C.,** Host stress and susceptibility, in *Armillaria Root Disease, USDA Agricultural Handbook 691,* Shaw, C. G. and Kile, G. A., Eds., U.S. Department of Agriculture Forestry Service, Washington, DC, 1991, chap 7.

15. **Marshall, R. P.,** *The Relation of Season of Wounding and Shellacking to Callus Formation in Tree Wounds, USDA Tech. Bull No. 246,* U.S. Department of Agriculture Forestry Service, Washington, DC, 1931.

16. **Mullick, D. B. and Jensen, G. D.,** Rates of nonsuberized impervious tissue development after wounding at different times of the year in three conifer species (studies of periderm, VII), *Can. J. Bot.,* 54, 881, 1976.

17. **Bertrand, P. F. and English, H. E.,** Virulence and seasonal activity of *Cytospora leucostoma* and *C. cincta* in French prune trees in California, *Plant Dis. Rep.,* 60, 106, 1976.

18. **Wood, F. A. and Skelly, J.,** Relation of time of year to canker initiation by *Fusarium solani* in sugar maple, *Plant Dis. Rep.,* 53, 753, 1969.

19. **Wensley, R. N.,** Rate of healing and its relation to canker of peach, *Can. J. Plant Sci.,* 46, 257, 1966.

20. **Shigo, A. L.,** *A New Tree Biology,* Shigo & Trees Associates, Durham, NH, 1986.

21. **Neely, D.,** Branch pruning and wound closure, *J. Arboric.,* 17, 205, 1991.

22. **Peterson, J. L. and Helmer, D. B.,** A wound treatment system to suppress cankers and wood rot in trees, *J. Arboric.,* 18, 155, 1992.

23. **Homma, Y., Arimoto, Y., and Misato, T.,** Effect of sodium bicarbonate on each growth stage of cucumber powdery mildew fungus (*Sphaerotheca fuliginea*) in its life cycle, *J. Pest. Sci.,* 6, 201, 1981.

24. **Horst, R. K., Kawamoto, S. O., and Porter, L. L.,** Effect of sodium bicarbonate and oils on the control of powdery mildew and black spot of roses, *Plant Dis.,* 76, 247, 1992.

25. **Elad, Y., Ziv, O., Ayish, N., and Katan, A.,** The effect of film-forming polymers on powdery mildew of cucumber, *Phytoparasitica,* 1 (7), 279, 1989.

26. **Hagiladi, A. and Ziv, O.,** The use of antitranspirants for the control of powdery mildew of roses in the field, *J. Environ. Hort.,* 4, 69, 1986.

27. **Ziv, O. and Hagiladi, A.,** Control of powdery mildew with antitranspirants, *HortScience,* 19, 708, 1984.

28. **Du Plessis, H. J., Hattingh, M. J., and Van, V. H. J.,** Biological control of crown gall in South Africa by *Agrobacterium radiobacter* strain K84, *Plant Dis.,* 69, 302, 1985.

29. **Kerr, A.,** Biological control of crown gall through production of Agrocin 84, *Plant Dis.,* 64, 24, 1980.

30. **Ryder, M. H. and Jones, D. A.,** Biological control of crown gall using *Agrobacterium* strains K84 and K1026, *Aust. J. Plant Physiol.,* 18, 571, 1991.

31. **Lundberg, B. and Unestam, T.,** Antagonism against *Fomes annosus.* Comparison between different methods in vitro and in vivo, *Mycopathologica,* 70, 107, 1980.

32. **Ricard, J. L.,** Field observations on the biocontrol of Dutch elm disease with *Trichoderma viride* pellets, *Eur. J. For. Pathol.,* 13, 60, 1983.

33. **Tronsmo, A. and Ystaas, J.,** Biological control of *Botrytis cinerea* on apple, *Plant Dis.,* 64, 1009, 1980.

34. **Corke, A. T. K.,** The prospect for biotherapy of trees infected by silver leaf, *J. Hort. Sci.,* 49, 391, 1974.

35. **Dubos, B. and Ricard, J. L.,** Curative treatment of peach trees against silver leaf disease (*Stereum purpureum*) with *Trichoderma* preparations, *Plant Dis. Rep.,* 58, 147, 1974.

36. **Corke, A. T. K. and Hunter, T.,** Biocontrol of *Nectria galligena* infection of pruning wounds on apple shoots, *J. Hort. Sci.,* 54, 47, 1979.

37. **Boudreau, M. and Andrews, J. H.,** Factors influencing antagonism of *Chaeotmium globosum* to *Venturia inaequalis:* a case study in failed biocontrol, *Phytopathology,* 77, 1470, 1987.

38. **Beer, S. V. and Rundle, J. R.,** Recent progress in development of biological control for fire blight — a review, *Acta Hortic.,* 51, 195, 1984.

39. **Knudsen, G. R. and Hudler, G. W.,** Interactions between epiphytic bacteria and conidia of *Gremmeniella abietina,* in *Scleroderris Canker of Conifers,* Manion, P. D., Ed., Nijhoff and Junk, The Hague, 1984, 217.

40. **Young, C. S. and Andrews, J. H.,** Inhibition of pseudothecial development of *Venturia inaequalis* by the basidiomycete *Athelia bombacina* in apple leaf litter, *Phytopathology,* 80, 536, 1990.

41. **Townsend, A. M., Schreiber, L. R., Hall, T. J., and Bentz, S. E.,** Variation in response of Norway maple cultivars to *Verticillium dahliae, Plant Dis.,* 74, 44, 1990.

42. **Neely, D. R.,** Additional *Aesculus* species and subspecies susceptible to leaf blotch, *Plant Dis. Rep.,* 55, 37, 1971.

43. **Gill, D. L.,** 'Union' mimosa, *HortScience,* 14, 644, 1979.

44. **Hudler, G. W.,** Disease and nematode control for trees and shrubs, in *1992 Pest Management Recommendations for Commercial Production and Maintenance of Trees and Shrubs,* Cornell University, Ithaca, NY, 1992, chap. 2.

45. **Dirr, M. A.,** *Manual of Woody Landscape Plants,* Stipes Publications, Champaign, IL, 1977.

46. **Holmes, F. W. and Hibben, C. R.,** Field evidence confirms *Cornus kousa* dogwood's resistance to anthracnose, *J. Arboric.,* 15, 290, 1989.

47. **Smith, E. L and Treaster, S.,** Evaluation of flowering crabapple susceptibility to apple scab in Ohio in 1989, in *OARDC Special Circular #135,* Ohio Research & Development Center, Columbus, OH, 1990.

48. **Santamour, F. S.,** 'Columbia' and 'Liberty' planetrees, *HortScience,* 19, 901, 1984.

49. **Schroeder, W. H. and Lindquist, C. H.,** 'Assiniboine' poplar, *Can. J. Plant Sci,* 68, 644, 1989.

An Inventory of Biological Controls, Available and Under Development

Aphid lion (*Chrysopa rufilabris*, or lacewing larva) eating an aphid, a common pest of ornamental plants (photo courtesy of Max Badgely, Biological Photography)

A *Scapteriscus* mole cricket (a serious turf pest in the South) under attack by a natural enemy, the digger wasp, *Larra bicolor* (photo courtesy of James L. Castner, University of Florida)

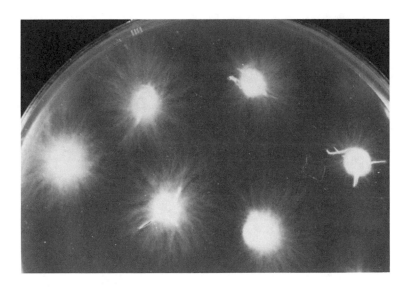

Biocontrol agent for Typhula blight of creeping bentgrass, *Typhula phacorrhiza*, growing from alginate pellets (photo courtesy of Lee Burpee)

Chapter 40

Beneficial Insects and Mites

Richard Weinzierl and Tess Henn, Office of Agricultural Entomology, University of Illinois, Champaign, IL

(This chapter was excerpted from the authors' *Alternatives in Insect Management: Biological and Biorational Approaches,* published in 1991 by the North Central Regional Extension Services in cooperation with the U.S. Department of Agriculture.)

CONTENTS

Introduction 443
Natural Enemies of Pest Species 443
    Predators 443
    Parasitoids 444
    Competitors 444
Types of Biological Control 444
    Classical Biological Control 445
    Conservation 445
    Augmentation 446
Some Common Introduced or Naturally Occurring Beneficial Insects and Mites 447
    Predators 447
    Parasitoids 450
Some Basic Principles of Biological Control 451
    Natural Enemies Survive Better in Stable Environments 451
    Natural Enemies Usually Leave a Moderate Pest Residue 451
    Control by Natural Enemies Takes Time 451
    Natural Enemies Are Products of Evolution, Not Manufacturing 452
Summary 452
Acknowledgments 452
Selected References 452

INTRODUCTION

Synthetic chemical insecticides have played important and beneficial roles in the control of agricultural pests and the reduction of insect-borne diseases for nearly 50 years. Their use will remain essential for many more years. Nonetheless, insecticides also pose real hazards. Some leave undesirable residues in food, water, and the environment. Low doses of many insecticides are toxic to humans and other animals, and some insecticides are suspected to be carcinogens. As a result, many researchers, farmers, and homeowners are seeking less hazardous alternatives to conventional synthetic insecticides.

Many insects and related arthropods perform functions that are directly or indirectly beneficial to humans. They pollinate plants, contribute to the decay of organic matter and the cycling of soil nutrients, and attack other insects and mites that are considered to be pests. Only a very small percentage of over 1 million known species of insects are pests. Although all the remaining nonpest species might be considered beneficial because they play important roles in the environment, the beneficial insects and mites used in pest management are natural enemies of pest species. A natural enemy may be a predator, a parasitoid, or a competitor.

NATURAL ENEMIES OF PEST SPECIES

PREDATORS

Predaceous insects and mites function much like other predaceous animals. They consume several to many prey over the course of their development, they are free living, and they are usually as big as or bigger than their prey. Some predators, including certain syrphid flies and the common green lacewing, are predaceous only as larvae; other

444

lacewing species, lady beetles, ground beetles, and mantids are predaceous as immatures and adults. Predators may be generalists, feeding on a wide variety of prey, or specialists, feeding on only one or a few related species. Common predators include lady beetles, rove beetles, many ground beetles, lacewings, true bugs such as *Podisus* and *Orius*, syrphid fly larvae, mantids, spiders, and mites such as *Phytoseiulus* and *Amblyseius*.

PARASITOIDS

Parasitoid means parasitelike. Although parasitoids are similar to true parasites, they differ in important ways. True parasites are generally much smaller than their hosts. As they develop, parasites usually weaken but rarely kill their hosts.

In contrast, many parasitoids are almost the same size as their hosts, and their development always kills the host insect. Although parasitoids are sometimes called parasites or parasitic insects, these terms are not completely accurate. In contrast to predators, parasitoids develop on or within a single host during the course of their development.

The life cycles of parasitoids are quite unusual. In general, an adult parasitoid deposits one or more eggs into or onto the body of a host insect or somewhere in the host's habitat. The larva that hatches from each egg feeds internally or externally on the host's tissues and body fluids, consuming it slowly; the host remains alive during the early stages of the parasitoid's development. Late in development, the host dies and the parasitoid pupates inside or outside of the host's body. The adult parasitoid later emerges from the dead host or from a cocoon nearby (see Figure 1).

Most parasitoids are highly host-specific, laying their eggs on or into a single developmental stage of only one or a few closely related host species. They are often described in terms of the host stage(s) within which they develop. For example, there are egg parasitoids, larval parasitoids, larval-pupal parasitoids (eggs are placed on or into the larval stage of the host, and the host pupates before it dies), pupal parasitoids, and a few species that parasitize adult insects.

The vast majority of parasitoids are small to minute wasps that do not sting humans or other animals. Certain species of flies and beetles also are parasitoids. *Bathyplectes, Trichogramma, Encarsia, Muscidifurax, Spalangia,* and *Bracon* are some of the more important parasitoids studied or used in agricultural systems.

COMPETITORS

Competitors are often overlooked in discussions of natural enemies, perhaps because many competitors of common crop pests also are pests themselves. Competitors can be beneficial, however, in instances where they compete with a nondamaging stage of a pest species. For example, dung beetles in the genera *Onthophagus* and *Aphodius* break up cow pats in pastures as they prepare dung to feed their larvae. This action speeds the drying of dung and makes it less suitable for the development of the larval stages of horn flies, face flies, and other pest flies. Some nonpest flies also develop in pasture dung and compete with pest species for the resources it provides. Despite these and a few other examples, the use of competitors in pest management is not common.

TYPES OF BIOLOGICAL CONTROL

Biological control, sometimes referred to as biocontrol, is the use of predators, parasitoids, competitors, and pathogens to control pests. In biological control, natural enemies are released, managed, or manipulated by humans. Without human intervention, however, natural enemies exert some degree of control on most pest populations. This ongoing, naturally occurring process is termed biotic natural control. Applied biological control produces only a small portion of the total benefits provided by the many natural enemies of pests.

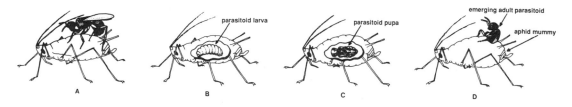

Figure 1 Generalized life cycle of an aphid parasitoid. (A) Adult parasitoid wasp injects an egg into a live aphid. (B) The parasitoid larva feeds within the apihid; late in the parasitoid's development the aphid dies. (C) The parasitoid pupates within the enlarged, dry shell of the dead aphid. (D) The new adult parasitoid cuts an exit hole in the back of the aphid and flies away, leaving behind the empty "aphid mummy."

There are three basic approaches to the use of predators, parasitoids, and competitors in insect management. These approaches are (1) classical biological control — the importation and establishment of foreign natural enemies; (2) conservation — the preservation of naturally occurring beneficials; and (3) augmentation — the inundative or inoculative release of natural enemies to increase their existing population levels. Broad definitions of biological control sometimes include the use of *products* of living organisms (such as purified microbial toxins, plant-derived chemicals, pheromones, etc.) for pest management. Although these products are biological in origin, their use differs considerably from that of traditional biological control agents.

CLASSICAL BIOLOGICAL CONTROL

Importing natural enemies from abroad is an important step in pest management in part because many pest insects in the U.S. and elsewhere were originally introduced from other countries. Accidental introductions of foreign pests have occurred throughout the world as a result of centuries of immigration and trade. Although the foreign origins of a few recently introduced pests such as the Asian tiger mosquito, Russian wheat aphid, and Mediterranean fruit fly are often noted in news stories, many insects long considered to be serious pests in this country are also foreign in origin. Examples of such pests include the gypsy moth, European corn borer, Japanese beetle, several scale insects and aphids, horn fly, face fly, and many stored-product beetles. In their native habitats, some of these pests cause little damage because their natural enemies keep them in check. In their new habitats, however, the same set of natural enemies does not exist, and the pests pose more serious problems. Importing and establishing their native natural enemies can help to suppress populations of these pests.

Importation typically begins with the exploration of a pest's native habitat and the collection of one or several species of its natural enemies. These foreign beneficials are held in quarantine and tested to ensure that they themselves will not become pests. They are then reared in laboratory facilities and released in the pest's habitat until one or more species become established. Successfully established beneficials may moderate pest populations permanently and at no additional cost if they are not eliminated by pesticides or by disruption of essential habitats.

Importation of natural enemies has produced many successes. An early success was the introduction of the Vedalia beetle, *Rodolia cardinalis,* into California in 1889 for the control of cottony cushion scale on citrus. For over 100 years this predaceous lady beetle from Australia has remained an important natural enemy in California citrus groves. In Illinois, populations of *Coccinella septempunctata*, an introduced lady beetle, have become increasingly widespread in recent years. This beetle feeds on a variety of aphids, including such pests as the green peach aphid and the pea aphid. *Bathyplectes curculionis* and *Bathyplectes anurus* are introduced parasitoids that help to regulate alfalfa weevil populations in many regions. Efforts to introduce and establish natural enemies of several important Midwestern pests are ongoing.

Although the importation of new natural enemies is important to farmers, gardeners, and others who practice pest management, the scope of successful introduction projects (involving considerable expertise, foreign exploration, quarantine, mass rearing, and persistence through many failures) is so great that only government agencies commonly conduct such efforts. Introducing foreign species is not a project for the commercial farmer or home gardener.

CONSERVATION

Conserving natural enemies is often the most important factor in increasing the impact of biological control on pest populations. Conserving or encouraging natural enemies is important because a great number of beneficial species exist naturally and help to regulate pest densities. Among the practices that conserve and favor increases in populations of natural enemies are the following:

Recognizing beneficial insects. Learning to distinguish between pests and beneficial insects and mites is the first step in determining whether or not control is necessary. This circular provides general illustrations of several predators and parasitoids. Picture sheets that feature common pests of many crops and sites can be obtained by writing Extension entomologists or county Extension offices in most states. Insect field guides are also useful for general identification of common species (see Borror and White 1970).

Minimizing insecticide applications. Most insecticides kill predators and parasitoids along with pests. In many instances natural enemies are more susceptible than pests to commonly used insecticides. Treating gardens or crops only when pest populations are great enough to cause appreciable damage or when levels exceed established economic thresholds minimizes unnecessary reductions in populations of beneficial insects.

Using selective insecticides or using insecticides in a selective manner. Several insecticides are toxic only to specific pests and are not directly harmful to beneficials. For example, microbial insecticides containing different strains of the bacterium *Bacillus thuringiensis (Bt)* are toxic only to caterpillars, certain beetles, or certain mosquito and black fly larvae. Other microbial insecticides offer varying degrees of selectivity.

Other insecticides that function as stomach poisons, such as the plant-derived compound ryania, do not directly harm predators or parasitoids because these compounds are toxic only when ingested along with treated foliage. Insecticides that must be applied directly to the target insect or that break down quickly on treated surfaces (such as natural pyrethrins or insecticidal soaps) also kill fewer beneficials. Leaving certain areas unsprayed or altering application methods can also favor survival of beneficials. For example, spraying alternate middles of orchard rows, followed by treating the opposite sides of the trees a few days later, allows survival and dispersal of predatory mites and other natural enemies and helps to maintain their impact on pest populations.

Maintaining ground covers, standing crops, and crop residues.
Many natural enemies require the protection offered by vegetation to overwinter and survive. Ground covers supply prey, pollen, and nectar (important foods for certain adult predators and parasitoids), and some degree of protection from weather. Most studies show greater numbers of natural enemies in no-till and reduced tillage cropping systems. In addition, some natural enemies migrate from woodlots, fencerows, and other noncrop areas to cultivated fields each spring. Preserving such uncultivated areas contributes to natural biological control.

Maintaining standing crops also favors the survival of natural enemies. Where entire fields of alfalfa are cut, natural enemies must emigrate or perish. Alternate strip cutting (with time for regrowth between the alternate cutting dates) allows dispersal between strips, so natural enemies remain in the field and help to limit outbreaks of pests.

Providing pollen and nectar sources or other supplemental foods.
Adults of certain parasitic wasps and predators feed on pollen and nectar. Plants with very small flowers (such as some clovers, Queen Anne's lace, and other plants in the family Umbelliferae) are the best nectar sources for small parasitoids and are also suitable for larger predators. Seed mixes of flowering plants intended to attract and nourish beneficial insects are sold at garden centers and through mail-order catalogs. Although no published data document the effectiveness of particular commercial mixes, these flower blends probably encourage a variety of natural enemies. The presence of flowering weeds in and around fields may also favor natural enemies.

Artificial food supplements containing yeast, whey proteins, and sugars may attract or concentrate adult lacewings, lady beetles, and syrphid flies. As adults these insects normally feed on pollen, nectar, and honeydew (the sugary, amino acid-rich secretions from aphids or scale insects), and they may require these foods for egg production. Lady beetles are predaceous as adults, but some species eat pollen and nectar when aphids or other suitable prey are unavailable. The proteins and sugars in artificial foods provide enough nutrients for some species to produce eggs in the absence of abundant prey. Wheast, BugPro, Bug Chow, and PredFeed are a few of the artificial foods available from suppliers of natural enemies.

The practices listed above must be judged according to their impacts on pest populations as well as their effects on natural enemies. Practices that favor natural enemies may or may not lessen overall pest loads or result in acceptable yields. For example, reduced tillage favors beneficials but also contributes to infestations of such pests as the common stalk borer and European corn borer in corn. Moreover, tillage decisions may be influenced more by soil erosion and crop performance concerns than by impacts on pests or natural enemies. Flower blends and flowering weeds can serve as nectar sources for moths (the adult forms of cutworms, armyworms, and other caterpillar pests) as well as beneficials. The ultimate goal of conserving natural enemies is to limit pest problems and damage to crops, rather than simply to increase numbers of predators or parasitoids. Pest densities and crop performance are factors that must be included in any evaluation of the effectiveness of natural enemy conservation efforts.

AUGMENTATION
Augmentation involves releasing natural enemies into areas where they are absent or exist at densities too low to provide effective levels of biological control. The beneficial insects or mites used in such releases are usually purchased from a commercial insectary (insect-rearing facility) and shipped in an inactive stage (eggs, pupae, or chilled adults) ready for placement into the habitat of the target pest. Augmentation is broadly divided into two categories — inoculative releases and inundative releases.

Inoculative releases involve relatively low numbers of natural enemies and are intended to inoculate or "seed" an area with beneficial insects that will reproduce. As the natural enemies increase in number, they suppress pest populations for an extended period. They may limit pest populations over an entire season (or longer) or until climatic conditions or a lack of prey results in population collapse. Generally only one or two inoculative releases are made in a single season.

In contrast, inundative releases involve large numbers of natural enemies that are intended to overwhelm and rapidly reduce pest populations. Such releases may or may not result in season-long establishment of natural enemies in the release area. Inundative releases that do not result in season-long establishment are the most expensive way to employ natural enemies because the costs of rearing and transporting large numbers of insects produce only short-term benefits. Such releases are usually most appropriate against pests that undergo only one or two generations per year.

The distinction between inoculative and inundative releases is not absolute. Many programs attempt to blend long-term establishment with short-term results. In addition, conservation and augmentation may be used together in a variety of ways to produce the best results.

Beneficial insects and mites (Figures 2–10) may be purchased from insectaries or gardening and farming supply outlets. For more information on such sources, see "Commercial and Cooperative Insectaries" in Chapter 41 of this book. Mahr and Ridgway (1993) summarize the methods involved in purchasing and using benefical insects.

SOME COMMON INTRODUCED OR NATURALLY OCCURRING BENEFICIAL INSECTS AND MITES

Few guidelines exist for monitoring populations of natural enemies and determining their likely impacts on pest infestations. Nonetheless, recognizing the beneficials that are present in any situation and understanding their roles are useful steps in deciding on appropriate pest management practices. Some common, naturally occurring species are described below.

PREDATORS

Lady Beetles (Family Coccinellidae). Beetles in the family Coccinellidae are known as lady beetles, though they are commonly referred to as "ladybugs." There are more than 400 species of lady beetles in North America, ranging in color from the familiar orange with black spots to various

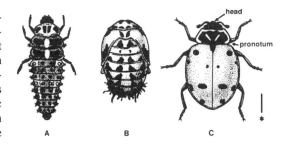

Figure 2 The convergent lady beetle, *Hippodamia convergens*. (A) Larva. (B) Pupa. (C) Adult.

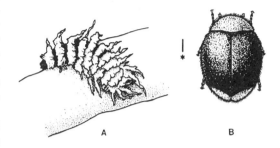

Figure 3 The mealybug destroyer, *Cryptolaemus montrouzieri*. (A) Larva. (B) Adult.

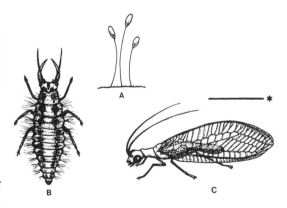

Figure 4 The common green lacewing, *Chrysoperia carnea*. (A) Eggs. (B) Larva, commonly known as an "aphid lion." (C) Adult.

shades of red and yellow to jet black. The vast majority of lady beetles are beneficial predators of soft-bodied insects (aphids and scale insects in particular), mites, and insect eggs. In each species, adults and larvae consume similar prey and generally can be found together where their prey is abundant. Most species of lady beetles are not available for purchase and release, but many of them provide significant levels of pest control if they are not eliminated by insecticides, tillage, or other land-use practices.

The Convergent Ladybeetle, *Hippodamia convergens.* The convergent lady beetle is one of the best known of all insect natural enemies. The adult beetle has orange wing covers, usually with six small black spots on each side. The beetle's pronotum (the shieldlike plate often mistaken for the head) is black with white margins and two diagonal white dashes. These "convergent" dashes give this lady beetle its common name. The immature beetle is a soft-bodied, alligator-shaped larva. It is grey and orange and is covered with rows of raised black spots (Figure 2).

Larval and adult convergent lady beetles feed primarily on aphids. Where aphids are not available, they may feed on scale insects; other small, soft-bodied insect larvae; insect eggs; and mites. Adults also feed occasionally on nectar, pollen, and honeydew (the sugary secretions of aphids, scales, and other sucking insects). Development from egg to adult takes 2 to 3 weeks, and adults live for several weeks to several months, depending on location and time of year.

The convergent lady beetle occurs naturally throughout much of North America. In the Midwest, adult beetles overwinter in small groups beneath bark or in other protected sites. In California, adult beetles overwinter in huge aggregations in the foothills of the central and southern mountain ranges. These California beetles are harvested from their overwintering sites, stored at cool temperatures to maintain their dormant state, and shipped to customers in the spring and summer for release in gardens or crops.

Coccinella septempunctata, referred to as "C-7" for its seven spots, is a Eurasian lady beetle that was introduced into the U.S. several times in the 1970s and 1980s. Now common in several Mid-

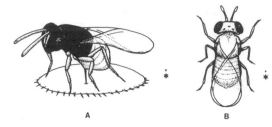

Figure 6 (A) An adult *Encarsia formosa* attacking greenhouse whitefly. (B) *Trichogramma.*

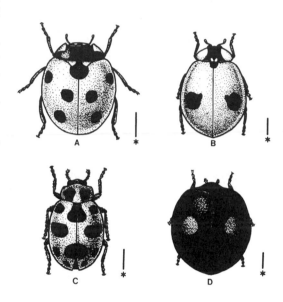

Figure 7 Some important Illinois lady beetles. (A) C-7, *Coccinella septempunctata.* (B) The two-spotted lady beetle, *Adalia bipunctata.* (C) The spotted lady beetle, *Coleomegilla maculata.* (D) The twice-stabbed lady beetle, *Chilocorus stigma.*

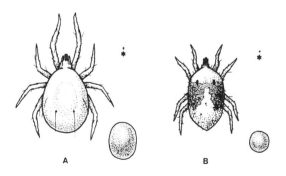

Figure 5 (A) A predatory mite, *Phytoseiulus persimilis,* adult and egg. (B) The two-spotted spider mite, *Tetranychus urticae,* adult and egg.

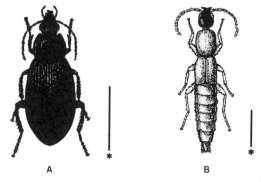

Figure 8 (A) A ground beetle, Family Carabidae. (B) A rove beetle, Family Staphylinidae.

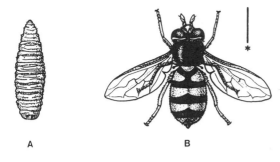

Figure 9 A syrphid fly. (A) Larva. (B) Adult.

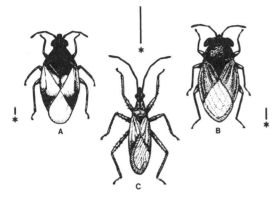

Figure 10 Some important predaceous bugs. (A) The minute pirate bug, *Orius insidiosus*. (B) A big-eyed bug, *Geocoris* species. (C) The common damsel bug, *Nabis americoferus*.

west states, it is a significant natural enemy of several important aphid species, including the pea aphid and the green peach aphid (see Figure 7A).

Other common aphid-feeding lady beetles found in the north-central states include the two-spotted lady beetle (*Adalia bipunctata,* Figure 7B), and the spotted lady beetle (*Coleomegilla maculata,* Figure 7C). *Coleomegilla maculata* may play a key role in limiting European corn borer populations by feeding on corn borer egg masses. The twice-stabbed lady beetle (*Chilocorus stigma,* Figure 7D) is a predator of many species of scale insects.

Stethorus punctum — an important predator of spider mites in apple orchards throughout Michigan, Pennsylvania, and western New York — is found sporadically in Illinois. Adults are tiny (3 mm), round, shiny, black beetles. Both adults and larvae feed on mites, and they are most often found where spider mite populations are high (15 or more mites per leaf). Adults overwinter in debris at the base of apple trees or in fields or wooded areas near orchards. Although *Stethorus punctum* is suscep-

tible to standard rates of most orchard insecticides and miticides, adults are mobile and can fly into orchards after spray residues have declined.

Ground Beetles (Family Carabidae) and Rove Beetles (Family Staphylinidae). Adult and larval ground beetles and rove beetles prey on a wide range of insects and are especially important as predators of caterpillars and other soft-bodied insects in field crops, forests, and many other habitats. Together these two families of beetles include nearly 5000 species that are widely distributed throughout North America (see Figure 8).

Both ground beetles and rove beetles are commonly found under plant debris and beneath the soil surface. Many species are nocturnal (active at night) and as a result are not as apparent as other natural enemies. Some of the larger species of ground beetles can be found in trees, where they prey on various caterpillar pests, including tent caterpillars, tussock moth larvae, and gypsy moth larvae. Ground beetles and rove beetles, along with spiders, are the most common predators found in many field crops.

The Green Lacewings, *Chrysoperla* (formerly *Chrysopa*) Carnea and *Chrysoperla Rufilabris*. Green lacewings occur naturally throughout North America and are widely available for purchase and release. Adult green lacewings have delicate, light green bodies; large, clear wings; and bright golden or copper-colored eyes. The larvae are small, grayish brown, and elongate and have pincerlike mandibles. Green lacewing eggs are found on plant stems and foliage. They are laid singly or in small groups on top of fine, silken stalks which reduce predation and parasitism by keeping the eggs out of reach (Figure 6).

Green lacewing larvae are generalist predators of soft-bodied insects, mites, and insect eggs, but they feed primarily on aphids and are commonly known as "aphid lions." Lacewing larvae are also cannibalistic, feeding readily on other lacewing eggs and larvae if prey populations are low. Larvae feed for about 3 weeks before pupating inside silken cocoons that are usually attached to the undersides of leaves.

Although adults of some lacewing species are predaceous, *Chrysoperla carnea* adults feed only on nectar, pollen, and aphid honeydew. *Chrysoperla carnea* females cannot produce eggs if these foods are not available. Green lacewing adults make long dispersal flights soon after emerging from the pupal stage; this dispersal takes place regardless of the availability of food when the adults emerge. Lacewings are night fliers and may travel many miles before mating and starting to produce eggs.

Females are mobile throughout their egg-laying period, although they concentrate where nectar and honeydew are abundant. They tend to lay eggs wherever they land to feed or rest.

Artificial foods such as Bug Chow, BugPro, Wheast, or PredFeed can be used in place of natural foods (nectar and honeydew) to attract and concentrate adult lacewings. The presence of artificial foods does not keep newly emerged adults from dispersing, but such foods may attract older adults that are in the area. Food sprays are useful only when a substantial population of lacewings is present in the area.

Syrphid, Flower, or Hover Flies (Family Syrphidae). Syrphid flies are common in many habitats. The small, wormlike larvae of many species are found on foliage, where they prey on aphids. Adult syrphid flies feed on pollen and nectar. The adults of many species closely resemble bees or wasps but do not sting or bite (see Figure 9).

True Bugs (Order Hemiptera) Many species of true bugs are predaceous, and several play important roles in the control of turf and ornamental pests. The minute pirate bug (*Orius insidiosus*, Figure 10A) feeds on the eggs of caterpillar pests in corn and other crops; it also feeds on many other small soft-bodied insects. The big-eyed bugs (*Geocoris* species, Figure 10B) also prey on caterpillar eggs and other small insects. Damsel bugs (*Nabis* species, Figure 10C) are common in gardens and crops, where they feed on aphids and other pests.

The Spined Soldier Bug, *Podisus Maculiventris*. The spined soldier bug is a predaceous "true bug" available for purchase. It occurs naturally throughout the Midwest and is one of the most common predatory stink bugs throughout much of the country. The adult spined soldier bug is grayish brown with sharply pointed "corners" on its pronotum. Nymphal soldier bugs are various shades of orange with black markings. They are round bodied and wingless. Nymphs and adults stab their prey with long, pointed "beaks" that are held folded under their bodies while not feeding. Although the spined soldier bug is sold mainly as a predator of Mexican bean beetle (*Epilachna varivestis*) larvae, it is a generalist that feeds readily on many soft-bodied insects and larvae.

Praying Mantids. Although several mantid species occur naturally in the southern U.S., many of the species found commonly in the Midwest were originally introduced form the tropics. In the fall, adult female mantids produce egg cases that may contain up to two hundred eggs. These eggs remain dormant until early summer when tiny mantid nymphs hatch and begin to search for prey. Only one generation of mantids develops each year.

Mantid nymphs and adults are indiscriminate generalist predators that feed readily on a wide variety of insects, including many beneficial insects and other mantids. Older mantids feed on medium-sized insects such as flies, honey bees, crickets, and moths. They are not effective predators on aphids, mites, or most caterpillars. Most of the mantids that hatch from an egg case die as young nymphs as a result of starvation, predation, or cannibalism. In addition, mantids are territorial, and by the end of the summer often only one adult is left in the vicinity of the original egg case.

Predators of Spider Mites. Mites in the genera *Phytoseiulus* and *Amblyseius* are fast-moving, pear-shaped predators with short life cycles (from 7 to 17 days, depending on temperature and humidity) and high reproductive capacities. They are pale to reddish in color and can be distinguished from two-spotted spider mites by their long legs, lack of spots, and rapid movement when disturbed. The eggs of predatory mites are elliptical and larger than the spherical eggs of spider mites (see Figure 5). Predatory mite nymphs feed on spider mite eggs, larvae, and nymphs. Adult predators feed on all developmental stages of spider mites.

Several species of predatory mites are sold by U.S. distributors, but the only species that has been studied extensively for use on a commercial scale is *Phytoseiulus persimilis*. This mite develops, reproduces, and preys on spider mites most effectively in a temperature rage of 21 to 27°C (70 to 80°F), with relative humidities of 60 to 90%. Above and below these ranges, *Phytoseiulus persimilis* is less able to bring two-spotted spider mite populations under control.

PARASITOIDS

Parasitoids are very common as naturally occurring biological control agents, with more than 8500 species of parasitic wasps and flies occurring in North America. Despite their prevalence and importance, parasitoids go largely unnoticed because of their small size and inconspicuous behavior. A detailed discussion of individual, naturally occurring parasitoid species is beyond the scope of this chapter. For useful background information and specific details on selected parasitoids, see Askew (1971), Clausen (1940), and De Bach (1964).

Encarsia Formosa, a Parasitoid of the Greenhouse Whitefly. *Encarsia formosa*, a tiny parasitic wasp, has been used to control greenhouse whiteflies on tomatoes and cucumbers in Europe for over 50 years. *Encarsia* adults lay their eggs into the scalelike third and fourth nymphal stages of whiteflies (see Figure 6A). Parasitized whitefly nymphs blacken and die as the parasitoid larva develops inside. Adult wasps provide additional whitefly control by feeding directly on early and late nymphal stages.

Encarsia performs best when greenhouse temperatures are maintained between 21 and 26°C (70 and 80°F), with relative humidities of 50 to 70%. In these conditions, *Encarsia* reproduces much faster than the whitefly. At lower temperatures the whitefly reproduces more rapidly than the parasitoid, and *Encarsia* may not provide adequate control. In addition, *Encarsia* requires bright light for optimum reproduction and development. This dependence on light intensity further limits the parasitoid's effectiveness during winter months when day length is shorter and light intensities are lower.

***Trichogramma* Wasps, Egg Parastioids.** The *Trichogramma* wasps are the most commonly used parasitoids worldwide. They occur naturally and are reared and released extensively in Europe and Asia for the control of many species of caterpillar pests in various crops. *Trichogramma* wasps are very small, averaging about 0.7 mm long as adults (the size of the period at the end of this sentence; see Figure 6B).

Most *Trichogramma* species lay their eggs into the eggs of moths and butterflies. A few species parasitize eggs of other kinds of insects. *Trichogramma* larvae develop within host eggs, killing the host embryos in the process. Instead of a caterpillar hatching from a parasitized egg, one or more adult *Trichogramma* wasps emerge. Because the caterpillar pests are killed in the egg stage, no feeding damage occurs.

SOME BASIC PRINCIPLES OF BIOLOGICAL CONTROL

The actions of natural enemies can be manipulated by the timing and magnitude of releases and by habitat management. Unlike insecticides, however, natural enemies are living organisms with biological needs and behavioral traits that may conflict with the goals or constraints of pest management programs. Success with natural enemies depends on understanding and accommodating their biological and evolutionary limitations.

NATURAL ENEMIES SURVIVE BETTER IN STABLE ENVIRONMENTS

Many pests colonize annual crops or temporary habitats rapidly. In contrast, most natural enemies colonize such habitats slowly. Where crop residues are frequently removed and soil is disturbed by tillage, or where insects are nearly eliminated by the use of insecticides, the rebuilding of natural enemy populations tends to lag behind the regrowth of pest populations. Consequently, naturally occurring predators and parasitoids are more prevalent and more easily maintained in perennial crops (such as alfalfa, pasture grasses, and orchard crops) than annual crops. Providing a more stable environment by using reduced tillage, ground covers, or strip harvesting contributes to the survival of a range of natural enemies, both naturally occurring and introduced. These practices may also favor certain pests, however, and each must be evaluated according to its overall benefits and drawbacks.

NATURAL ENEMIES USUALLY LEAVE A MODERATE PEST RESIDUE

As natural enemies reduce pest population densities, it becomes increasingly difficult for them to locate and attack the few surviving pests. When pest population densities become too low, natural enemies often leave the pest residue (the remaining pest population) and disperse in search of more abundant hosts or prey. In the absence of predators or parasitoids, the remaining pest population rebuilds, providing hosts or prey for subsequent generations of natural enemies. This cycling of populations allows the natural enemies and their hosts to avoid extinction. Exceptions to this rule occur in greenhouses, for example, where natural enemies are confined with their hosts or prey and cannot disperse when pest population densities fall.

The occurrence of moderate pest residues may or may not limit the use of biological control. Where nearly 100% control is necessary, natural enemies alone usually do not provide sufficient control. This is the case with certain pests of commercial fruits and vegetables, cut flowers, ornamentals, and other commodities for which current cosmetic or grade standards are very strict. In many situations, however, moderate levels of pest infestation are acceptable. This is true for most pests of gardens, lawns, and field crops. In these settings natural enemies may provide acceptable levels of control.

CONTROL BY NATURAL ENEMIES TAKES TIME

Unlike insecticides, which are uniformly applied and have nearly immediate effects on pest populations, natural enemies require time to disperse from release sites, search for hosts or prey, and handle (consume or lay eggs into) each host or prey individually. In some cases, the natural enemies that are released must reproduce before a significant degree of control occurs. Consequently, where commodities must be rapidly disinfected or protected from a pest that is already causing serious losses, predators and parasitoids do not (or only rarely) provide sufficiently rapid control.

Determining the correct timing for releases is one of the most important steps in the implementation of biological control. Because natural enemies do not provide immediate control, they usually must be released before severe damage is imminent. Preventive releases, however, are almost never appropriate or effective because natural enemies die or disperse in the absence of hosts or prey. In general, releases should begin when pest populations are substantial enough to sustain the natural enemy but low enough to allow the natural enemy time to catch up and provide an adequate degree of control. Knowledge of pest development and careful monitoring of pest populations are key factors in determining when to make releases.

NATURAL ENEMIES ARE PRODUCTS OF EVOLUTION, NOT MANUFACTURING

In evolutionary terms, the success of a natural enemy is defined by its ability to survive as a species. This survival depends on the continued availability of the natural enemy's host or prey. In pest management terms, however, the success of a natural enemy is defined by the degree to which it controls its host or prey. Where high levels of control are required, naturally selected behaviors that result in moderate pest residues and guarantee species survival may prevent predators and parasitoids from "succeeding" in pest management terms. Rearing and releasing greater numbers of natural enemies does not always overcome these limitations.

Traits such as host or prey preferences, host-finding behaviors, dispersal thresholds, climatic dependencies, habitat preferences, and sensitivity to insecticides are genetically determined. Some of these traits — climatic tolerances and insecticide resistance, in particular — can be manipulated through selective breeding. In addition, some trait selection may occur unintentionally during mass rearing. Continuous rearing of natural enemies in insectaries can select for undesirable traits such as a preference for hosts other than the target pests or a lack of vigor under field conditions. Many traits, however, are difficult to alter and must be accommodated when developing biological control programs.

Rearing and releasing natural enemies is much more complicated than spraying an insecticide. Consequently, attempts to use natural enemies without understanding their behavior often result in failure and disappointment. However, when natural enemies are used in ways that accommodate their strengths and weaknesses, they can become effective components of integrated pest management programs.

SUMMARY

Biological control is a complex subject that can be presented only superficially in a chapter of this length. Successful use of natural enemies in pest management requires detailed understanding of insect biology and pest management techniques. In addition, it requires realistic expectations. The possibilities are not endless; there are real limitations that result from biological constraints and from current agricultural production and marketing practices. Nonetheless, biological control utilizing beneficial insects and mites represents an effective alternative for insect management in some situations. In almost all settings, encouraging or conserving naturally occurring populations of beneficial insects and mites is possible. Conservation may be aided greatly by the development and use of more selective, rapidly degrading insecticides and by the use of insecticides in a more selective manner. The greatest promise for biological control may lie in such conservation efforts.

ACKNOWLEDGMENTS

The following reviewers contributed to this chapter: Charles Helm, William Ruesink, and Audrey Hodgins, Illinois Natural History Survey; John Obrycki, Iowa State University; and Daniel Mahr, University of Wisconsin.

SELECTED REFERENCES

Askew, R. R., 1971. *Parasitic Insects*, American Elsevier, New York, 316 pp.

Borror, D. J. and R. E. White, 1970. *A Field Guide to the Insects of America North of Mexico*, Houghton Mifflin, Boston, 404 pp.

Canard, M., Semeria, Y., and T. R. New, Eds., 1984. *Biology of Chrysopidae*, Dr. W. Junk, The Hague, 294 pp.

Clausen, C. P., 1940. *Entomophagous Insects*, McGraw-Hill, New York, 688 pp.

Croft, R. A., 1975. *Integrated Control of Apple Mites*, Extension Bulletin E-825, Cooperative Extension Service, Michigan State University, East Lansing, 12 pp.

Davis, D. W., S. C. Hoyt, J. A. McMurtry, and M. T. AliNiazee, Eds., *Biological Control and Insect Pest Management*, Bulletin 1911. Agricultural Experiment Station, University of California, Division of Agriculture and Natural Resources, Berkeley, 102 pp.

De Bach, P., 1964. *Biological Control of Insect Pests and Weeds*, Reinhold, New York, 844 pp.

Hagen, K. S., 1962. Biology and ecology of predaceous Coccinellidae, *Annu. Rev. Entomol.*, 7, 289.

Hoy, M. A. and D. C. Herzog, Eds., 1985. *Biological Control in Agricultural IPM Systems*, Academic Press, Orlando, FL, 589 pp.

Hussey, N. W. and N. Scopes, 1985. *Biological Pest Control: The Glasshouse Experience*, Cornell University Press, Ithaca, NY, 240 pp.

Mahr, D. L. and N. M. Ridgway, 1993. Biological Control of Insects and Mites. NCR 481, Cooperative Extension Service, University of Wisconsin, Madison.

Osborne, L. S., L. E. Ehler, and J. R. Nechols, 1985. *Biological Control of the Twospotted Spider Mite in Greenhouses*, Bulletin 853, Institute of Food and Agricultural Services, University of Florida, Gainesville, 40 pp.

Papavizas, C. C., Ed., 1981. *Biological Control in Crop Production,* Allanheld, Osmun Publishers, Granada, 461 pp.

Steiner, M. Y. and D. P. Elliott, 1983. *Biological Pest Management for Interior Plantscapes,* Alberta Environmental Centre, Vegreville, Alberta, 30 pp.

van Driesche, R. G., R. Prokopy, and W. Coli, 1989. *Using Biological Control in Massachusetts: Spider Mites in Apples,* Cooperative Extension Service, University of Massachusetts, Amherst, 6 pp.

Weinzierl, R., and T. Henn, 1991. Alternatives in Insect Management: Biological and Biorational Approaches. NCR 401, Cooperative Extension Service, University of Illinois at Urbana-Champaign.

Chapter 41

Commercial Biological Controls for Insect and Mite Pests of Ornamentals

Carol S. Glenister, IPM Laboratories, Inc., Locke, NY

CONTENTS

Definition of Biological Controls 455
Sources of Beneficial Insects, Mites, and Nematodes 456
    Naturally Occurring Beneficials 456
    Commercial and Cooperative Insectaries 456
        Suppliers of Beneficials 456
        Commercial Availability 457
    Public Sector Introductions of New Species 457
Status of Research and Development on Using Beneficials in Ornamentals 458
Commercially Available Beneficial Species 458
    Natural Enemies of Aphids 458
    Natural Enemies of Caterpillars 459
    Natural Enemies of Mealybugs 459
    Natural Enemies of Scales 460
        Expected Developments — Natural Enemies of Euonymus Scale 460
    Natural Enemies of Mites 460
    Natural Enemies of Thrips 461
    Natural Enemies of Whiteflies 462
    Biological Control of Soil Pests and Woodborers with Beneficial Nematodes 462
Optimizing Biological Control 462
    Keeping the Environment Hospitable 463
        Manage Pesticides 463
        Provide Food and Habitat 463
    Choose Beneficials Appropriate to the Pest 463
    Synchronize Life Cycles 463
    Release Rates 463
    Monitor and Maintain Quality 463
    Take Time to Experiment 464
What Lies Ahead 464
References 464

DEFINITION OF BIOLOGICAL CONTROLS

Biological controls are living organisms that limit or reduce pest populations. Also referred to as natural enemies and beneficials, biological controls of pest insects and mites are comprised of beneficial insects and mites as well as insect-attacking bacteria, fungi, viruses, protozoa, and nematodes. Vertebrates such as birds and mice also serve as biological controls of insects and mites. (Scientists have observed that gypsy moth numbers were higher where family cats minimized mouse predation on gypsy moth caterpillars and pupae.)[1] In this chapter, after a brief discussion of the various sources of beneficials, I will concentrate my discussion on commercially available insects, mites, and beneficial nematodes that prey upon and/or parasitize insect and mite pests of ornamentals.

455

SOURCES OF BENEFICIAL INSECTS, MITES, AND NEMATODES

There are three major sources of beneficials. Nature is the single largest producer, followed by commercial and cooperative insectaries, and then by federal, state, and university laboratories.

NATURALLY OCCURRING BENEFICIALS

Despite their enormous numbers, naturally occurring beneficials do their work unnoticed and typically are appreciated only after they fail. IPM practitioners know from experience that every plant in the natural summer landscape is being protected in some way by beneficial species. In *Insect Pest Management Guidelines for California Landscape Ornamentals*, Koehler recommends that a miticide be included with carbaryl sprays for controlling Cooley spruce gall aphid, since the carbaryl will kill off the predatory mites that normally keep the spider mites in check.[2]

Natural enemies' failures have been dramatized the most by devastating outbreaks of scales and mites in the landscape following municipal spray programs with broad spectrum pesticides. In a residential area of San Diego, the purple scale (*Lepidosaphes beckii*) had been controlled by an imported parasite, *Aphytis lepidosaphes*. However, when carbaryl was used in 1973 and 1974 to eradicate Japanese beetle in an area of about 100 residential blocks, the treatment killed most of the scale parasites in the eradication area, allowing the scale to damage more than 60% of the 50 ornamental citrus trees studied in the eradication zone within two seasons. Scale densities were so thick on some of the trees that they completely covered the bark and caused acute defoliation. Nearby, outside the eradication zone, 50 comparable trees supported harmless numbers of live purple scale that were 12,000-fold lower than in the eradication zone.[3]

Similarly, a weekly fogging of malathion for mosquito control in the city of South Lake Tahoe killed most of the native natural enemies (four parasites and two predators) of the pine needle scale (*Chionaspis* (*Phenacaspis*) *pinifolia*).[4] The 5-year mosquito control program only managed to free the city of mosquitos for a few hours following treatment each week, but was toxic to the beneficial insects for 5 days following each application. This resulted in a heavy infestation of pine needle scale throughout natural stands of Jeffrey and lodgepole pine. Beginning with the cessation of fogging in 1969, the natural enemy populations grew enough within two seasons to cause a dramatic decline in most of the infested area. By the end of the third season, all remaining scales were heavily parasitized. In this case, the predators were responsible for the major decline of the infestations while the parasites prevented low scale populations from increasing.[4]

Important practices for conserving naturally occurring beneficial species in the landscape are the same as for purchased beneficials reviewed below in the "Keeping the Environment Hospitable" section.

COMMERCIAL AND COOPERATIVE INSECTARIES

Commercial and cooperative insectaries have grown in response to markets for beneficials in agriculture. One of the oldest insectaries in California is a citrus-grower cooperative that has been producing parasites and predators of scales and mealybugs since 1928.[5] Billions of beneficials are also produced for grapes, cotton, field corn, strawberries, almonds, and greenhouse crops, with smaller quantities of beneficials applied to other crops. Many of these beneficial species can also be used in landscapes and they are listed and described below in the "Commercially Available Beneficial Species" section.

Suppliers of Beneficials

Several lists of beneficial suppliers have been compiled. Below are three of the most comprehensive and regularly updated lists. Each list contains entries not found in the other lists.

The BioIntegral Resource Center (BIRC) provides a list of suppliers of beneficial organisms that is updated annually and can be obtained by writing or calling:

BioIntegral Resource Center
PO Box 7414
Berkeley, CA 94707
Tel. (510) 524-2567

The 1992 BIRC list includes 34 suppliers from the U.S. and Canada, and 20 suppliers from the Eurasian continent and Australia. The list keys each of 60 beneficial species to its suppliers in addition to listing all the offerings of each supplier.

Several state agencies and sustainable agriculture organizations keep lists of beneficial suppliers. One such list is provided by the California Environmental Protection Agency (Department of Pesticide Regulation, Environmental Monitoring, and Pest Management) entitled *Suppliers of Beneficial Organisms in North America*, and is updated approximately every 2 years. It is free and can be obtained by writing or calling:

Department of Pesticide Regulation
EM & PM Branch
Attn: Beneficial Suppliers Booklet
1020 N Street, Room 161
Sacramento, CA 95814
Tel. (916) 324-4101

The 1992 California list notes the beneficials supplied by each of 95 companies and includes 134 beneficial species, including 26 for weed control.

The Association of Natural Bio-control Producers (ANBP) provides a free list of its members that supply beneficials. This list can be obtained by writing or calling:

Maclay Burt, Executive Director
Association of Natural Bio-control Producers
10202 Cowan Heights Drive
Santa Ana, CA 92705
Tel./FAX (714) 544-5295

The 1992 ANBP list contains the telephone numbers and addresses of 40 beneficial suppliers throughout the U.S. The objectives statement of the organization includes a set of guidelines for quality assurance procedures to ensure that beneficial products reaching the consumer are healthy and effective.

Commercial Availability

Most commonly used beneficials, such as Trichogramma, spider mite predators, lacewings, Encarsia, Aphidoletes, thrips predators, and Cryptolaemus, are in adequate supply throughout the year. However, heavy demand in the spring may put orders on a first-come, first-serve basis. Occasionally, production crashes make individual species unavailable while production is renewed.

New beneficial species enter the market as a result of demand. If enough customers request an individual species, suppliers will find a way to fill the need.

PUBLIC SECTOR INTRODUCTIONS OF NEW SPECIES

Federal and state agencies, along with their university cooperators, are primarily involved in what is termed classical biological control: the importation and colonization of foreign species of beneficials. These importations usually target foreign insects that have become pests without the natural enemies that suppressed them in their native land. Ideally, scientists examine the invading pest in its native land and collect parasites and predators that they find attacking the pest. These parasites and predators are cultured in federal quarantine laboratories and observed to be sure that they are free of their own natural enemies. Once they have passed quarantine requirements, state, federal, and university research staff cooperate in colonizing the natural enemies in outbreak areas of the target pest.

Classical biological control has been extremely successful against several pests of landscape plants such as scales, sawflies, and mealybugs. Over all of agriculture and forestry and horticulture in general, the most consistently dramatic successes in the U.S. in the last century have been scale control beginning with the introduction of the Vedalia beetle. A century ago, the cottony cushion scale was covering California's citrus groves and threatening that region's citrus industry. Pesticide applications had little effect. Following the importation of the Vedalia beetle from Australia, the little black and orange lady beetle brought the scale population under control within the year. Outbreaks are now usually due to the use of insecticides that kill the scales' natural enemies.[6]

The most recent dramatic example of classical biological control in the urban landscape is that of the ash whitefly, *Siphoninus phillyreae*. This exotic whitefly probably arrived in California in 1986 or 1987, but was first discovered in high densities in Los Angeles in 1988 on a wide range of fruit and ornamental broadleaf trees.[7] The whitefly is not generally considered a pest in Europe but in California, free of its native natural enemies, this whitefly became so numerous that it it could be counted in terms of percent leaf underside covered by immatures (whitefly scales). Flying adults were thick enough to discourage outdoor activities. The insects defoliated ornamental and shade trees, producing so much honeydew that cars and sidewalks were blackened and furniture and carpets got sticky from honeydew carried in from outside.[7]

The first ash whitefly parasites, *Encarsia partenopea* from Israel, were released in October 1989. By September 1991, whitefly numbers were 1,000- to 10,000-fold lower than the peak numbers observed in 1990, and 80 to 98% of ash whitefly scales were found to be parasitized. In January 1992, it was difficult to find an ash whitefly in southern California.

The U.S. Department of Agriculture Agricultural Research Service maintains a database entitled *Releases of Beneficial Organisms* in the United States (ROBO). This database maintains records on importations and releases from 1981 on. For more information, write:

Dr. Jack R. Coulson
ARS Biological Control Documentation Center
Beneficial Insects Laboratory, BBII, ARS, USDA
Building 264, BARC-East
Beltsville, MD 20705

STATUS OF RESEARCH AND DEVELOPMENT ON USING BENEFICIALS IN ORNAMENTALS

The majority of biological control research is conducted on agricultural crops and forests; relatively little attention is paid to landscape ornamentals. However, ornamentals share many pests with agriculture and forests, so much is being done simply by applying general knowledge to landscape problems. Scientists that specialize in pest management on ornamentals especially need financial and moral support to justify their research in the 1990s economic climate.

COMMERCIALLY AVAILABLE BENEFICIAL SPECIES

NATURAL ENEMIES OF APHIDS

The aphid midge, *Aphidoletes aphidimyza* is a nonpredaceous gnat-like fly that produces 70 to 100 eggs in its 1-week life-span.[8] The predaceous larvae are tiny orange maggots that kill aphids by injecting a paralyzing toxin into the aphids' leg joints and then sucking out the body fluids. They can kill 10 to 50+ aphids during their 7- to 14-d development and are known to attack at least 60 different aphid species.[9] *A. aphidimyza* can be an important natural predator in greenhouses, field vegetables, and apple orchards.[8,10,11] Aphidoletes adults are such efficient searchers that they can locate one aphid-infested brassica plant among 75 aphid-free plants.[12]

The environment has an important influence on their predatory activities. The adults prefer shady, humid areas, so eggs are more likely to be laid in the shade than in bright sunlight. The predators stop development under short daylengths and overwinter as cocoons in the soil.[9] This suspended state, known as diapause, can be prevented by nightlong, low-intensity lighting.[13] Recommended release rates for greenhouses range from one to five aphid midge cocoons per 10 ft^2 at the first sign of aphids, with one to three repeat releases at 1- to 2-week intervals.[9,14] On apple trees, Applied Bionomics recommends releasing up to five midges per tree at the first sign of aphids.

The aphid parasite, *Aphidius matricariae*, is a tiny wasp that inserts single eggs into young green peach aphids.[15] As the aphid matures, the wasp larva feeds internally on the aphid which continues to look normal. However, the parasite eventually consumes the entire contents of the by now adult aphid, which dies and leaves a bloated, brown and papery shell called an aphid mummy. The adult wasp exits the mummy after chewing a circular, hinged "door" in the top of the mummy. Subsequently adult females mate and "sting" new aphids. Aphidius spp. and close relatives occur naturally where there are aphids, sometimes surviving the presence of harsh pesticides.[16] The effectiveness of *Aphidius matricariae* is frequently hindered in the summer by its own parasites (termed hyperparasites) so it is recommended most often for use in winter greenhouses against its host, green peach aphid.[14]

The lady beetle family includes the shiny red to orange beetles with black spots that are well known for their voracious appetite for aphids. Other members of the lady beetle family have feeding preferences ranging from spider mites to scales. Hagen tabulated 25 species of aphid-predatory lady beetles in the U.S.[17] People are less familiar with the black and orange alligator-shaped lady beetle larvae with large mandibles that also feeds on aphids. Although aphids may be their optimum food, most of these lady beetles also consume insect eggs, small caterpillars, small beetle larvae, scales, mealybugs, and other slow moving, small insects.

The seasonal habits of the convergent lady beetle, *Hippodamia convergens*, have made it particularly convenient for commercialization during most of the 20th century. Each year when aphid-prey get scarce and temperatures hot, they migrate to mountains and woods, and gather in large clusters of clinging beetles which can be conveniently collected by the gallons between May and February.[17] Beetles in these aggregations survive the winter in diapause, feeding little and living off their fat reserves. Collected in this state, beetles are well-adapted to cold-storage until sale. Mating is delayed until the following spring, and egg laying waits until after their dispersal flight. These latter traits cause great controversey regarding the usefulness of these lady beetles. Once released from cold-storage (or winter), their habit is to fly straight up to a temperature gradient of about 55°F where they travel for miles on wind currents until nightfall. Dispersal continues until abundant aphids slow them down.[17] During and between migrations, each beetle can consume thousands of aphids.[18] Testimonials about the effectiveness of purchased lady beetles in greenhouses, fields, and landscapes indicate that in some cases, the lady beetles must manage to consume aphids before dispersing. Pest managers try to overcome lady beetle disappearance by weekly releases.

Lacewing larvae, *Chrysoperla carnea* and *C. rufilabris*, are alligator-shaped with large, sickle-shaped, hollow mandibles with which they impale their prey and suck out the body juices. Also known as aphid lions, they can consume 200 to 400 aphids during their 2-week existence. Although they prefer aphids, they will also feed on all kinds of insect eggs, small caterpillars, scales including whitefly scales, mites, and thrips.[19,20] Larvae pass through three instars before spinning an ovoid, silken cocoon, reminiscent of many spider egg cases.

The lacewing is named for its shimmering, delicately veined wings which overshadow a slender, soft abdomen. They can live for 2.5 to 3 months under ideal conditions.[20] Adults lay several hundred eggs, but egg viability declines with the mother's age.[19]

Different lacewing species have specific habitat preferences. The Taubers recommend deploying *C. rufilabris* in more humid environments and *C. carnea* in drier environments.[21] In general, lacewing larvae are hindered by waxes and strong pubescence on leaves. Optimal temperatures for *C. carnea* development lie within the range of 61 to 79°F.[19] Since adults feed only on pollen, nectar, and aphid honeydew, a diverse landscape with season-long availability of nectar-bearing flowers will attract adults. Egg laying is induced by aphid honeydew, but commercially available mixtures of protein and sugar can be sprayed or dabbed onto plants to induce egg laying. Use these products with caution because they can be phytotoxic.

Lacewing eggs are commercially available in enormous numbers. The more expensive, larger, pre-fed lacewing larvae have at least three advantages over the eggs: (1) they can immediately eat many more aphids than newly hatched larvae, (2) pre-fed larvae have better defenses against ants and other predators, and (3) pre-fed larvae can search better than newly hatched larvae.

From results of greenhouse studies in the Soviet Union, one *C. carnea* egg for 1.3 aphids, or one larva for 5 to 30 aphids is recommended.[22] Releases of *C. carnea* eggs in the crowns of pear trees in the U.S. resulted in substantial reduction of fruit infestation by the grape mealybug, *Pseudococcus maritimus*.[23] Releases of 10,000 larvae per acre of cotton have yielded substantial reductions of bollworm and tobacco budworm.

NATURAL ENEMIES OF CATERPILLARS

The most commonly used biological control for caterpillars is a bacterium, *Bacillus thuringiensis* (Bt), that has been marketed as a sprayable pesticide in the U.S. for nearly half a century. The second most common biological controls are *Trichogramma* spp., tiny wasps that parasitize the caterpillar eggs. These parasites are necessarily miniscule, as they complete their entire development within the eggs. Other natural enemies of caterpillars are generalist predators such as lady beetles, lacewings, and Orius spp. discussed elsewhere in this chapter. In addition, some egg and larval parasites of gypsy moth are available on a contract basis.

Different *Trichogramma* species are available for different habitats. *Trichogramma pretiosum* is used against caterpillar eggs in field crops and vegetables and *Trichogramma minutum* is used against caterpillar eggs in orchards and ornamentals. *T. minutum* has been reported in high numbers from pest species ranging from redbanded leafroller to spruce budworm and even spruce sawfly.[24] A few other *Trichogramma* species are available for specific applications. For example, *T. platneri* is produced for the control of the Amorbia and Avocado Looper in avocado groves.

Trichogramma can only use eggs less than 2- to 3-d old. Thus releases must be made as soon as moths are detected and repeated throughout the moth flight. Pheromone and light traps are essential aids in determining moth flights. During moth flights, pest managers make weekly blanket releases to ensure that *Trichogramma* wasp adults are available to attack newly laid caterpillar eggs.

Purchased *Trichogramma* are shipped while they are still maturing within the host eggs which are glued to cards densely enough to resemble grains of sand in sandpaper. Young *Trichogramma* adults emerge within a few days after delivery. Environmental conditions have an enormous impact on the usefulness of *Trichogramma*. Parasite emergence and health will decline outside the range of 40 to 80% relative humidity and at or above 95°F. Predators such as ants can consume large proportions of the parasitized eggs that are set out on cards. Free water can also reduce egg hatch or impede adult activity. The optimum release strategy is to hold the maturing *Trichogramma* in uncrowded conditions at 85°F and 60% relative humidity until the adults have emerged and are flying.[25]

NATURAL ENEMIES OF MEALYBUGS

Four mealybug species plagued the citrus orchards of southern California in the early 1900s. Three out of the four species were successfully controlled by imported parasites that are relatively specific to the species in question. The fourth species, the citrus mealybug, is controlled on tens of thousands of acres by periodic releases of the mealybug destroyer *Cryptolaemus montrouzieri* ("Crypts"), which is produced in private and grower-cooperative insectaries.

Crypt larvae and adults voraciously consume all stages of many species of mealybugs and often surprise first-time users with their fast cleanup. The adult is an eighth-inch long lady beetle with an orange head and an orange abdomen that is just visible under hard black wing covers. Crypt larvae resemble mealybugs except that their white waxy coating has thicker, longer filaments than on mealybugs, and mature larvae are almost twice the size of citrus mealybugs.

Development from egg to adult takes about 30 days at temperatures between 74 and 80°F. Adults can lay an average of 439 eggs over a 51-day period at 76 to 78°F.[5] The optimum conditions for egg laying and larval development are 72 to 84°F and 70 to 80% relative humidity.[26] Temperatures below 68°F are inadequate for beetle development. Failures are often reported in greenhouses during the winter.[27] Citrus growers compensate for this problem by releasing beetles by hand into the orchards in late spring and early fall at rates of between 5 and 20 predators per tree.

Since ants will protect mealybugs for the honeydew that they secrete, ant control will improve the crypts effectiveness.[26]

NATURAL ENEMIES OF SCALES

The citrus industry currently supports the production of two of the many scale parasite species that control scale on citrus. *Aphytis melinus* is a tiny black and yellow wasp that was introduced into California from Pakistan and India in 1956 and 1957.[28] It lays eggs by piercing the scale covering with its ovipositor and inserting them under the shell of the California red scale, *Aonidiella aurantii*, Oleander scale, *Aspidiotus nerii*, yellow scale, *Aonidiella citrina*, and dictyospermum scale, *Chrysomphalus dictyospermi*. *Metaphycus helvolus* is a tiny black and yellow wasp that parasitizes black scale, *Saissetia oleae*, hemispherical scale, *Saissetia coffeae*, nigra scale, *Parasaissetia nigra*, and brown soft scale, *Coccus hesperidum*.[26] Both parasite species take 2 to 3 weeks to mature within the scale host, after which the new adult parasite emerges by chewing a tiny hole in the scale shell. As the only available scale parasites, they are being deployed against a wide variety of scale species despite the advice of scale biocontrol researchers that scale parasites are fairly specific in their choice of hosts.

Expected Developments — Natural Enemies of Euonymus Scale

Euonymus scale is a serious pest of ornamental trees and shrubs in the U.S., partially because it is an exotic species with few natural enemies to keep it under control. Since 1984, a lady beetle predator that was imported from the Republic of Korea, *Chilocorus kuwanae*, has been released into heavy scale populations in the U.S. by the USDA Agricultural Research Service and their cooperators.[29] Scale populations on release plants were drastically reduced by the third year after predator introduction. Scale predators continue to be collected from release sites and transferred to new sites for colonization. They are commercially available from at least two suppliers.

NATURAL ENEMIES OF MITES

Pest mites are commonly controlled by their natural enemies, provided that the natural enemies have survived the local pesticide regimen. McMurtry et al. have compiled a list of 40 species of predatory mites in the world literature that protect agricultural crops from spider mites.[30,31] Of these, five species are commercially available in the U.S.: *Phytoseiulus persimilis, Mesoseiulus longipes, Neoseiulus californicus, Galendromus occidentalis,* and *Amblyseius fallacis*. Two additional species, *Neoseiulus cucumeris* and *Neoseiulus barkeri* are used for control of cyclamen mite and broad mite, respectively.

Pest managers must be able to distinguish predatory mites from pest mites so that they can monitor ratios of predators to prey. All these predatory species belong to the phytoseiid family and can easily be distinguished from pest mites despite their tiny size. The phytoseiids' bodies are tear drop-shaped and appear shiny and smooth under low magnification. Their legs are noticeably longer than their spider mite prey and the two front legs are commonly extended forward like feelers. The predatory mites run in a circular fashion searching for food while their prey usually move slowly and erratically. These movements are easily distinguished without a hand lens. *P. persimilis* is a distictive, bright orange. The slightly elongated eggs are noticeably larger than prey mite eggs and can be recognized with minimal practice.

The phytoseiid mite's life cycle consists of an egg, three immature stages, and the adult. Time from egg to adult varies from 5 days to 2 weeks or more depending upon temperature, available food, and predator species. The developmental rates of *P. persimilis, M. longipes, G. occidentalis,* and *A. fallacis* are faster than the other two species, at optimum temperatures taking only 5 d to produce the next generation. By combining this fast generation time with an egg-laying rate of almost three eggs per day under optimum conditions, these mites can reproduce faster than their spider mite prey.[32]

Under optimum conditions, *P. persimilis* will control two-spotted spider mites the fastest, because it eats more spider mites daily. *P. persimilis* can consume between 14 and 23 eggs per day while *G. occidentalis* and *A. fallacis* can consume about eight eggs per day.[31] However, high consumption rates are not necessarily an indication of the predator's effectiveness in any given habitat. The key is that each predator has a unique set of optimum conditions. *P. persimilis* successfully controls the two-spotted spider mite at relative humidities of 60 to 90% while it fails at high temperatures and 40% relative humidity. In one experiment at 81°F, only 7.5% of the eggs hatched at 40% relative humidity whereas 99.7% of the eggs hatched at 80% relative humidity.[32] Comparing relative humidity tolerance, *G. occidentalis* exhibits the best tolerance for lower relative humidity, *P. persimilis* and *A. fallacis* require high relative humidity, and *M. longipes* and *N. californicus* fall in between.

Plant structure should also be considered in choosing predators. *P. persimilis* has not performed well in North American orchards, whereas *G. occidentalis* has controlled mites in western orchards and *A. fallacis* has controlled mites in the more humid eastern orchards.[30,31] In California, *G. occidentalis* has been found occurring naturally on Acer, Aesculus, Citrus, Convolvulus, Cynodon, Dactylon, Fragaria, Juglans, Magnolia, Malus, Prunus, Quercus, Salix, Sambucus, Vitis, and Rubus.[33] At first glance, *N. californicus* does not seem to be a very useful predator, however, it excels where the other predators fail. A relatively slow feeder, *N. californicus* will stay on the crop even when spider mite prey are scarce whereas the other predators will disappear. *N. californicus* is particularly useful in preventing outbreaks from spider mite populations that are too low to support the other predators.

NATURAL ENEMIES OF THRIPS

Five species are currently available for biological control of thrips. Two species of phytoseiid mites, *Neoseiulus cucumeris* and *N. barkeri*, are similar in appearance to spider mite predators, but consume thrips as well as a wide range of mites. A third predatory mite, *Hypoaspis miles* is a soil-dwelling generalist predator that is used primarily for fungus gnat and shorefly control and secondarily for thrips that pupate in the soil, such as western flower thrips and onion thrips. The most recently utilized thrips predators are *Orius insidiosus* and *O. tristicolor*, two species of minute pirate bugs.

Neoseiulus cucumeris and *N. barkeri* are typical phytoseiids with pear-shaped bodies, two front legs held forward like feelers, fast movements, and the same developmental stages. *N. cucumeris* adults are usually lighter tan, while *N. barkeri* adults are more brown-orange. Further identification can only be done under a microscope.

Pest managers that use the considerably larger spider mite specialist, *P. persimilis*, must be careful not to have the same expectations. Thrips predators feed much more slowly, consuming 1 to 2 young thrips per day. In absence of prey they can subsist on spider mites and/or pollen. In contrast, *P. persimilis* consumes all stages of spider mites, killing 20 spider mite eggs or five adults daily. The solution is to release thrips predators in much higher numbers than spider mite predators. Fortunately, purchase of large quantities is feasible since they are much less expensive to produce than spider mite predators.

It is important to introduce the predators before or at the first sign of thrips, and to apply regularly at weekly or biweekly intervals. They are shipped and distributed over the crop in bran. Alternatively, the bran is available in teabag-like packets called sachets that can be hung on plants. The thrips predators move out of the sachets for 3 to 6 weeks.

Diapause in both species is induced by various combinations of short day-length, scarce prey, and cool temperatures.[34,35] Thus, control from October to April in the greenhouse depends on maintaining summer-like light and temperature regimes.

Minute pirate bugs are tiny (about $1/_{10}$ in.) black, brown, and silver bugs with beaks that pierce prey and suck them dry. The wingless orange to brown nymphal stages look very different from the adult stage. Adult females lay two to four eggs per day, and development from egg to adult at 75°F takes 20 days. They prey on all stages of thrips, butterfly and moth eggs, spider mites, aphids, whitefly pupae, and other beneficial insects. They also cannibalize their own kind. Adult *Orius insidiosus* can eat 5 to 20 thrips larvae per day. Pollen and plant juices will also sustain these predators, an important factor in keeping high predator populations to suppress sudden jumps in thrips numbers. Their feeding has never been considered injurious to plants.[36]

The best control strategy is to release a few *Orius* early in the crop at or before the first detection of thrips. Predator reproduction will follow and overtake the thrips numbers. Releases as low as one *Orius* per plant have controlled thrips for the entire season in cucumber.

Commercially available *Orius* species are not recommended from October through April except where summer-like conditions are maintained since they enter reproductive diapause induced by short daylength and low temperatures. They also exhibit plant preferences: *Orius insidiosus* is extremely useful for thrips control on cucumber and tomato but deserted roses in greenhouse experiments soon after release.[37] Prefences on other plant species need to be determined. Our general knowledge of *Orius* spp. is that they prefer flowering plants, so they should be supported by a diverse landscape.

NATURAL ENEMIES OF WHITEFLIES

The greenhouse whitefly parasite, *Encarsia formosa*, was first identified in England in 1926 and became the first beneficial insect used in greenhouses. It quickly spread to Australia, Canada, and continental Europe in the 1930s.[38] *Encarsia formosa* will also control the sweet potato whitefly, *Bemisia tabaci*, but control is usually not as dramatic as with the greenhouse whitefly.

The adult *Encarsia formosa* is so tiny that it appears like a crawling speck. Magnified observation will distinguish a dark head and thorax, and a light-colored abdomen covered by shining, white wings. The female lays eggs within immature whiteflies (the whitefly scales found on leaf undersides) by drilling a hole with her ovipositor. At 75°F, the offspring take about 3 weeks to mature to adulthood inside the whitefly scale. Half-way through their development, they turn the greenhouse whitefly scales black, making it easy to distinguish parasitized from healthy scales. Parasitized sweet potato whiteflies are not as distinctive, as they turn an orange-brown.

Parasite activity is slowed at low temperatures, low light intensities, and on plants with hairy leaves (for example, on cucumber). Large quantities of honey dew also reduce parasitization.[36]

Tactics used in the greenhouse to bring greenhouse whitefly and sweet potato whitefly under biological control within 4 to 8 weeks include: parasite introduction at the first sign of whitefly, multiple releases to get overlapping generations of parasites, and use of pesticides to reduce whitefly numbers that are growing too fast for the *Encarsia*.

The tiny ($1/_{16}$th in.), brown-black lady beetle, *Delphastus pusillus*, became commercially available in the U.S. in 1990. Both the larvae and adults prey on all stages of many species of whiteflies. They are found from Massachusetts and California to South America in a range of crops from field crops to citrus groves.[39,40] Adults consume an average of 153 sweet potato whitefly (SPW) eggs per day or 11 large SPW scales per day at 77 to 80°F.[41] The adult female requires a minimum of 100 SPW eggs per day to support egg-laying.

D. pusillus has been targeted as a possible solution to the massive 1991 outbreaks of SPW in California crops.[42] Several species of wasps also under investigation for SPW control are *Encarsia transvena*, *E. deserti*, *E. tabacivora*, *E. near formosa*, and *Eretmoceros californicus*.

BIOLOGICAL CONTROL OF SOIL PESTS AND WOODBORERS WITH BENEFICIAL NEMATODES

Steinernema carpocapsae is the most readily available nematode for control of soil pests and woodborers. It has carried many variations of this name over the last several years, but nearly everything on the market is some strain of the above species. Also available in limited supply is *Heterorhabditis heliothidis*, a nematode that is highly effective against grubs in turf, but its relatively short shelf-life has hindered its development.

Juvenile steinernematid nematodes called infectives enter through natural openings on the insect body,[43] often through the mouth, and rapidly kill the insect. After the nematode completes one or more generations, the insect cadaver becomes crowded with nematode offspring, and juvenile nematodes grow into "infective" nematodes, a nonfeeding, survival stage with closed mouths, collapsed gut, and a thick skin that can even withstand some slow drying. For steinernematids, the infective stage is the only stage that can attack and kill the insect. Infectives can survive for several years in the soil, "awakening" only to wriggle toward chemical cues from insects.

Although nematodes kill a very wide range of insect species, the best results have been in moist environments: against black vine weevil and other soil insects in pots and cranberry bogs, against grubs in well-watered lawns, against wood and cane borers in their tunnels, and against fungus gnats in greenhouses.[44,45]

The nematodes can be applied through a sprayer or watering can, or in a granular mixture. Although infectives can withstand controlled drying, they are harmed by sudden drying, so the application method must get them into a moist environment quickly. Nematodes are also extremely sensitive to ultraviolet light. For this reason, we recommend that applications be made early in the morning or in the evening.

OPTIMIZING BIOLOGICAL CONTROL

Although there is a large body of scientific literature on specific aspects of many natural enemies, there are few instances where knowledge of a single beneficial species is complete enough to provide precise and fool-proof instructions on use. For most applications, the most useful knowledge continues to come from practice: trial and error in the habitat in question. The most knowledgeable handlers of individual species are the people that grow them in the commercial insectaries and public laboratories. These insect farmers have learned the optimum conditions to support growth and reproduction, and the types of conditions that kill or debilitate the natural enemies that they grow.

Successful field users act as if they are trying to manage a field insectary. As insectary managers, they provide a hospitable environment, choose natural enemies capable of attacking the pest, synchronize life cycles, release beneficials at appropriate rates, and monitor the quality of the beneficials that they release, as described below.

KEEPING THE ENVIRONMENT HOSPITABLE
Manage Pesticides
Avoid long-residual pesticides. Generally these are toxic to beneficial insects longer than to the pest itself. For example, the pyrethroid, Decis, is toxic to the predatory mite, *P. persimilis*, and the whitefly parasite, *E. formosa*, for more than 30 d after application in the greenhouse.[26] Similarly, pyrethroids used in orchards are extremely disruptive to naturally occurring spider mite controls. Pesticides that allow some survival of biological controls include the insect growth regulators such as Dimilin and Enstar, neem, and the insecticidal soaps and oils.

Provide Food and Habitat
Beneficial insects and mites cannot survive without food and shelter appropriate to their species. Food sources include nectar and pollen in addition to prey. The easiest way to maintain food appropriate to various beneficials is to maintain diverse plantings that provide blooms throughout the season. Umbellifers are most often cited for providing nectar and pollen to beneficial species, but professional horticulturists have an enormous variety of other plants to choose from as well. The diverse structure in these plantings will automatically provide a diverse habitat for shelter of various beneficial species.

CHOOSE BENEFICIALS APPROPRIATE TO THE PEST
Proper identification of the pest is crucial to appropriate choice of predators and parasites. Many beneficial species are very particular in choosing what pest species to attack. For example, *Encarsia formosa*, the greenhouse whitefly parasite, was not an appropriate choice for controlling ash whitefly in California. Similarly, we do not yet have a predatory mite species to recommend for the control of spruce spider mite.

SYNCHRONIZE LIFE CYCLES
Some beneficial species only attack certain life stages of their prey and are short-lived enough to miss attacking pests that are not in the appropriate life-stage. For example, the whitefly parasites only parasitize medium and large whitefly scales successfully. If the whitefly parasites are released against the egg stage, they will perish before they can find whiteflies to parasitize. Presently, the solution is to make weekly or biweekly releases of beneficials until synchrony is attained. At the same time, multiple releases provide overlapping generations of beneficials so that adults are constantly available and laying eggs. Without overlapping generations, there will be times when all the parasites are immature and there are no adults attacking pests.

RELEASE RATES
The appropriate number of predators or parasites to release depends on the number of prey available. For example, the predatory mite *P. persimilis* can kill 20 spider mite eggs a day under warm temperatures. However, figuring out a release ratio is not easy because the spider mite population is constantly changing: adults are laying fresh eggs, immatures are maturing to adults, and old adults are dying. The predator population is also constantly changing. Experience has shown that a safe release rate is one predator for ten spider mites under moderate temperatures and high relative humidity. Control of the spider mites is likely within 2 weeks. When *P. persimilis* is released at a lower rate, the predators will eventually catch up with the spider mites, but not before the spider mite population has increased substantially.

Recommendations based on number of plants, square feet, or acres are automatically assuming certain numbers of pests per unit. Usually the pest density is assumed to be very low, and it is important to start the beneficial releases at the very first sign of the pest.

MONITOR AND MAINTAIN QUALITY
Reputable suppliers guarantee live and healthy delivery of their beneficial insects and mites. It is important that you inspect the deliveries immediately on arrival and get them onto the plants as soon as possible. If the beneficials seem low in number or many dead individuals are obvious, call the supplier immediately for a replacement.

Timely release is important. The vigor of the release will depend on the age and nutritional status of the individuals. If they are starved for too long, they will lose the ability to lay eggs, which is crucial to successful colonization. You should work out a delivery time that will allow release within a day after arrival. Most beneficial species can be stored in a cool, dark place until release. Release instructions should give storage recommendations.

TAKE TIME TO EXPERIMENT

New users should recognize that it will take time to develop skill in applying beneficial insects. They should take the time to experiment on one or two small areas, learning from successes and failures. After several seasons, the pest manager will have enough experience to safely judge which beneficials have been useful in the specific area.

WHAT LIES AHEAD

Production and use of beneficial insects, mites, and nematodes is gaining momentum made possible by adoption of integrated pest management, a management system where pests are closely monitored, pesticides are carefully managed to minimize impact on natural enemies, and cultural practices are finely tuned to minimize pest levels. More massive than the output from commercial insectaries, the environment's natural production of beneficial insects and mites is rebounding with the decreased use of toxic, long-residual pesticides. Growing use of IPM's pest monitoring will continue to increase awareness of the environment's natural potential and improve pesticide selection to conserve beneficial species.

The Environmental Protection Agency's 1992 move to evaluate ways to support the registration of less toxic, more environmentally friendly chemicals for pest control, suggests that in the future it may be possible to use more selective pesticides to suppress individual pest outbreaks without concurrently killing the natural enemies of the ten other pests that are not in outbreak phase. For example, it may be possible to spray for lacebug without killing the spider mite predators and inducing a spider mite outbreak.

The increased momentum of the biological control industry is already yielding an increase in the species of biological control agents produced. It is safe to speculate that by the year 2000 there will be close to 100 beneficial species produced in massive numbers for pest control, given current trends. The species will be selected on the basis of market demand, potential for successful control, and the motivation level of the commercial insectaries for developing new products.

REFERENCES

1. **U.S. Department of Agriculture,** The homeowner and the gypsy moth: guidelines for control, USDA Combined Forest Pest Research and Development Program, Home and Garden Bulletin No. 227, U.S. Department of Agriculture, Washington, DC, 1979.

2. **Koehler, C. S.,** Insect Pest Management Guidelines for California Landscape Ornamentals, Public. 3317, Cooperative Extension, University of California, 1987.

3. **DeBach, P. and Rose, M.,** Environmental upsets caused by chemical eradication, *Calif. Agric.*, 31(7), 8, 1977.

4. **Roberts, F. C., Luck, R. F., and Dahlsten, D. L.,** Natural decline of a pine needle scale population at South Lake Tahoe, *Calif. Agric.*, 27(10), 10, 1973.

5. **Fisher, T. W.,** Mass culture of *Cryptolaemus* and *Leptomastix* — natural enemies of the citrus mealybug, *CA Ag. Exp. Stn. Bull.*, 797, 1963.

6. **Simmonds, F. J., Franz, J. M., and Sailer, R. I.,** History of biological control, in *Theory and Practice of Biological Control, Huffaker,* C. B. and Messenger, P. S., Eds., Academic Press, New York, 1976, chap. 2.

7. **Bellows, T. S., Paine, T. D., Gould, J. R., Bezark, L. G., and Ball, J. C.,** Biological control ash whitefly: a success in progress, *Calif. Agric.*, 46(1), 24, 1992.

8. **Bouchard, D., Tourneur, J. C., and Paradis, R. O.,** Bio-ecologie d'*Aphidoletes aphidimyza* (Rondani) (Diptera: Cecidomyiidae) predateur du puceron du pommier, Aphis pomi DeGeer (Homoptera: Aphididae), *Ann. Soc. Entomol. Que.*, 26(2), 119, 1981.

9. **Markkula, M. and Tiittanen, K.,** Biology of the midge Aphidoletes and its potential for biological control, in *Biological Pest Control: The Glasshouse Experience*, Hussey, N. W. and Scopes, N., Eds., Cornell University Press, Ithaca, NY, 1985, chap. 2.8.

10. **Farrar, C. A., Perring, T. M., and Toscano, N. C.,** A midge predator of potato aphids on tomatoes, *Calif. Agric.*, 40(11/12), 9, 1986.

11. **Tracewski, K. T., Johnson, P. C., and Eaton, A. T.,** Relative densities of predaceous Diptera (Cecidomyiidae, Chamaemyiidae, Syrphidae) and their aphid prey in New Hampshire, U.S.A., apple orchards, *Protection Ecol.*, 6(3), 199, 1984.

12. **El-Titi, A.,** Zur Ausloesung der Eeiablage beider aphidophagen Gallmuecke *Aphidoletes aphidimyza* (Diptera: Cecidomyiidae), *Entomol. Exp. Appl.*, 17, 9, 1974.

13. **Gilkeson, L. A. and Hill, S. B.,** Diapause prevention in *Aphidoletes aphidimyza* (Diptera: Cecidomyiidae) by low-intensity light, *Environ. Entomol.*, 15(5), 1067, 1986.

14. **Applied Bionomics Ltd.,** Biological Control Agents Price List and Application Rates, Sidney, British Columbia, Canada, 1991.

15. **Rabasse, J. M. and Wyatt, I. J.,** Biology of aphids and their parasites in greenhouses, in *Biological Pest Control: The Glasshouse Experience*, Hussey, N. W. and Scopes, N., Eds., Cornell University Press, Ithaca, NY, 1985, chap. 2.7.

16. **Wyatt, I. J.,** The distribution of *Myzus persicae* (Sulz.) on year-round chrysanthemums. II. Winter season: the effect of parasitism *Aphidius matricariae* Hal., *Annu. Appl. Biol.*, 65, 31, 1970.

17. **Hagen, K. S.,** Biology and ecology of predaceous Coccinellidae, *Annu. Rev. Entomol.*, 7, 289, 1962.

18. **Hagen, K. S. and Sluss, R. R.,** Quantity of aphids required for reproduction by *Hippodamia* spp. in the laboratory, in *Ecology of Aphidophagous Insects*, Hodek, I., Ed., Academia, Prague, 1966.

19. **New, T. R.,** The biology of Chrysopidae and Hemerobiidae (Neuroptera), with reference to their usage as biocontrol agents: a review, *Trans. R. Entomol. Soc. London*, 127(2), 115, 1975.

20. **Canard, M. and Principi, M. M.,** *Biology of Chrysopidae*, Canard, M., Semeira, Y., and New, T. R., Eds., Dr. W. Junk, The Hague, 1984, chap. 4.

21. **Tauber, M. J. and Tauber, C. A.,** Life history traits of *Chrysopa carnea* and *Chrysopa rufilabris* (Neuroptera: Chrysopidae): influence of humidity, *Annu. Entomol. Soc. Am.*, 76, 282, 1983.

22. **Tulisalo, U.,** Biological control in the greenhouse, in *Biology of Chrysopidae*, Canard, M., Semeira, Y., and New, T. R., Eds., Dr. W. Junk, The Hague, 1984, chap. 8.3.

23. **Ridgway, R. L. and Murphy, W. L.,** Biological control in the field, in *Biology of Chrysopidae*, Canard, M., Semeira, Y., and New, T. R. Eds., Dr. W. Junk, The Hague, 1984, chap. 8.2.

24. **U.S. Forest Service,** Insects of Eastern Forests, Misc. Publ. 1426, U.S. Department of Agriculture, Washington, DC, 1985, 608 pp.

25. **Gross, H. R.,** Effect of temperature, relative humidity, and free water on the number and normalcy of *Trichogramma pretiosum* Riley (Hymenoptera: Trichogrammatidae) emerging from eggs of *Heliothis zea* (Boddie)) Lepidoptera: Noctuidae), *Environ. Entomol.*, 17(3), 470, 1988.

26. **Steiner, M. Y. and Elliott, D. P.,** *Biological Pest Management for Interior Plantscapes*, 2nd ed., Alberta Environmental Center, AECV87-E1, Vegreville, AB, 1987, 32 pp.

27. **Copland, M. J. W., Tingle, C. C. D., Saynor, M., and Panis, A.,** Biology of glasshouse mealybugs and their predators and parasitoids, in *Biological Pest Control: The Glasshouse Experience*, Hussey, N. W. and Scopes, N., Eds., Cornell University Press, Ithaca, NY, 1985, chap. 2.9.

28. **Bennett, F. D., Rosen, D., Cochereau, P., and Wood, B. J.,** Biological control of pests of tropical fruits and nuts, in *Theory and Practice of Biological Control*, Huffaker, C. B. and Messenger, P. S., Eds., Academic Press, New York, 1976, chap. 15.

29. **Drea, J. J. and Carlson, R. W.,** The establishment of *Chilocorus kuwanae* (Coleoptera: Coccinellidae) in eastern United States, *Proc. Entomol. Soc. Wash.*, 89(4), 821, 1987.

30. **McMurtry, J. A., Huffaker, C. B., and van de Vrie, M.,** Ecology of tetranychid mites and their natural enemies: a review. I. Tetranychid enemies: their biological characters and the impact of spray practices, *Hilgardia*, 40(11), 331, 1970.

31. **McMurtry, J. A.,** The use of Phytoseiids for biological control: progress and future prospects, in *Recent Advances in Knowledge of the Phytoseiidae*, Hoy, M. A., Ed., University of California Division of Agricultural Sciences Publication 3284, 1982, chap. 2.

32. **Tanigoshi, L. K.,** Advances in knowledge of the biology of the Phytoseiidae, in *Recent Advances in Knowledge of the Phytoseiidae*, Hoy, M. A., Ed., University of California Division of Agricultural Sciences Publication 3284, 1982, chap. 1.

33. **Flaherty, D. L. and Huffaker, C. B.,** Biological control of Pacific mites and Willamette mites in San Joaquin Valley vineyards. I. Role of *Metaseiulus occidentalis*, *Hilgardia*, 40 (10), 267, 1970.

34. **Gilkeson, L. A. and Hill, S. B.,** Diapause prevention in *Aphidoletes aphidimyza* (Diptera: Cecidomyiidae) by low-intensity light, *Environ. Entomol.*, 15(5), 1067, 1986.

35. **Glenister, C. S.,** Personal observation of *N. barkeri.*

36. **Malais, M. and Ravensberg, W. J.,** *Knowing and Recognizing,* Koppert, B. V., Ed., Berkel en Rodenrijs, The Netherlands, 1992.

37. **Oetting, R.,** personal communication, 1991.

38. **Hussey, N. W.,** History of biological control in protected culture: Western Europe, in *Biological Pest Control: The Glasshouse Experience,* Hussey, N. W. and Scopes, N., Eds., Cornell University Press, Ithaca, NY, 1985, chap. 1.1.

39. **Dowell, R. V., Fitzpatrick, G., Zumstein, R., Johnson, M., and Fiore, J.,** Integrating biological control of citrus blackfly and current Florida citrus spray programmes, *Trop. Agric. (Trinidad),* 63(4), 301, 1986.

40. **Gold, C. S., Altieri, M. A., and Bellotti, A. C.,** The effects of intercropping and mixed varieties of predators and parasitoids of cassava whiteflies (Hemiptera: Aleyrodidae) in Colombia, *Bull. Entomol. Res.,* 79, 115, 1989.

41. **Hoelmer, K. V.,** personal communication, 1990.

42. **Parella, M. P., Bellows, T. S., Gill, R. J., Brown, J. K., and Heinz, K. M.,** Sweetpotato whitefly: prospects for biological control, *Calif. Agric.,* 46(1), 25, 1992.

43. **Poinar, G. O.,** *Nematodes for Biological Control of Insects,* CRC Press, Boca Raton, FL, 1979, 277 pp.

44. **Gaugler, R.,** Biological control potential of *Neoaplectanid* nematodes, *J. Nematol.,* 13(3), 241, 1981.

45. **Kaya, H. K.,** Princes from todes, *Am. Nurseryman,* September 1, 1988, 63.

Chapter 42

Inoculative Biological Control of Mole Crickets

J. Howard Frank, Entomology & Nematology Department, University of Florida, Gainesville, FL

CONTENTS

Abstract 467
Immigrant Mole Crickets in the U.S., Puerto Rico, and the U.S. Virgin Islands 467
    Distribution and Origin 467
    Control Strategies and the IFAS Mole Cricket Program 468
The Strategy of Inoculative Biological Control 469
Inoculative Biological Control Agents 469
    Larra polita and *Larra bicolor* 469
    Steinernema scapterisci 470
    Ormia depleta 470
    Pheropsophus aequinoctialis 470
    Current Level of Control in Florida 471
    Availability of Inoculative Biocontrol Agents 473
Inoculative Biological Control and IPM 473
Acknowledgments 473
References 474

ABSTRACT

Six species of mole crickets have become pests in parts of the U.S. (including Hawaii) and its Caribbean territories (Puerto Rico and the U.S. Virgin Islands). All are of exotic origin and arrived without the specialist natural enemies that help to keep them in check in their homelands. They can be killed with chemical pesticides, but the cost of treatment is recurrent and very high. Importation and establishment of their specialist natural enemies might reduce their populations permanently to a lower level. Such inoculative (or classical) biological control was attempted in the 1920s in Hawaii against *Gryllotalpa orientalis* Burmeister, and in the 1930s in Puerto Rico against *Scapteriscus didactylus* (Latreille), but the level of success was not measured. Introductions of inoculative biocontrol agents into Florida in the 1980s against *Scapteriscus vicinus* Scudder, *S. borellii* Giglio-Tos, and *S. abbreviatus* Scudder, are now beginning to result in areawide reduction of populations of these pests. This should reduce the use of pesticides, and result in substantial economic savings. It should also reduce risk to the environment of contamination with chemicals. Three such biocontrol agents were released and established by personnel of the University of Florida's Institute of Food and Agricultural Sciences (IFAS) research program. They are *Larra bicolor* Fabricius (a digger wasp), *Steinernema scapterisci* Nguyen and Smart (an entomopathogenic nematode), and *Ormia depleta* (Wiedemann) (a tachinid fly). There is evidence of reduced populations of *S. vicinus* and *S. borellii* in areas where the fly and the nematode are established.

IMMIGRANT MOLE CRICKETS IN THE U.S., PUERTO RICO, AND THE U.S. VIRGIN ISLANDS

DISTRIBUTION AND ORIGIN

Gryllotalpa orientalis Burmeister is a four-clawed mole cricket related to the native four-clawed *Neocurtilla* mole crickets of continental North America. It is native to Asia and it probably immigrated to Hawaii as a stowaway (Table 1). There, it caused sufficient damage that its control by a parasitoid wasp was attempted.[1]

Two-clawed mole crickets of the genus *Scapteriscus* are immigrants to the southern U.S.

Table 1   **Distribution of pest mole crickets in the U.S., Puerto Rico, and the U.S. Virgin Islands**

Species	Origin	Distribution
G. orientalis[a]	Asia	HI
S. abbreviatus	Southern SAm	FL, GA, Puerto Rico, St. Croix
S. borellii[b]	Southern SAm	FL, GA, SC, NC, AL, MS, LA, TX, AZ-CA
S. vicinus[c]	Southern SAm	FL, GA, SC, NC, AL, MS, LA, TX
S. didactylus[d]	Northern SAm	Puerto Rico, St. Thomas, St. John
S. imitatus	Northern SAm	Puerto Rico

Note: G. = Gryllotalpa, S. = Scapteriscus, SAm = South America.

[a] Reported until recently to be *G. africana* Palisot, but now believed to be *G. orientalis*;[36]   [b] *S. acletus* was the name used until recently in the U.S. for this species;[37]   [c,d] Generally considered to be the most harmful in the southeastern U.S.[c] and the Caribbean,[d] respectively, by combination of large populations and phytophagous diet.

and are important pests of vegetable seedlings and of pasture and turf grasses. They arrived from southern South America at the end of the 19th century in ships' ballast. Disembarking at port cities in the southeastern U.S., their populations grew rapidly, probably due to absence of specialist natural enemies. Three of them are now recognized: the short-winged mole cricket (*S. abbreviatus* Scudder), the southern mole cricket (*S. borellii* Giglio-Tos, formerly known as *S. acletus* Rehn & Hebard), and the tawny mole cricket (*S. vicinus* Scudder).[2] The short-winged mole cricket, incapable of flight, still is largely restricted to coastal areas in Florida and Georgia. The tawny mole cricket, a strong flier, now occurs as far north as North Carolina, and as far west as Texas. The southern mole cricket, another strong flier, has expanded its population still farther north and west; it has appeared in North Carolina and at the Arizona/California border at Yuma. The destruction wrought by *Scapteriscus* generally is attributed in the popular press to "the mole cricket" so that all members of the mole cricket family (Gryllotalpidae) now have a bad reputation, deserved or not. However, the only native mole cricket in the southern coastal U.S., *Neocurtilla hexadactyla* (Perty), is rarely, if ever, a pest.

Three immigrant species occur in Puerto Rico. They are *S. abbreviatus*, *S. didactylus* (Latreille), and *S. imitatus* Nickle & Castner.[3] *Scapteriscus didactylus* appears to have arrived in Puerto Rico by 1797,[4] and was noted as "the most serious pest of general agriculture."[5] The earliest record of *S. abbreviatus* dates from 1917,[6] whereas that of *S. imitatus* dates from 1938.[3] *S. didactylus* and *S. imitatus* are native to northern South America, but the distribution of *S. imitatus* extends farther southward in that continent. There are no native mole cricket species in Puerto Rico or in the U.S. Virgin Islands.

S. didactylus occurs on St. Thomas and St. John, whereas *S. abbreviatus* occurs on St. Croix,[7] where it causes minor damage.[8] Reports of the "changa or West Indian mole cricket" (*S. didactylus*) in Georgia and Florida, and of *S. vicinus* in the West Indies, are due to misidentification.

CONTROL STRATEGIES AND THE IFAS MOLE CRICKET PROGRAM

Early control strategies in the southern U.S., as in Puerto Rico, were based on chemicals. Arsenic and cyanide were incorporated into baits and applied in plantings of vegetables and on turf.[9,10] By 1940, damage to vegetables in Florida had become so severe that federal help was given to Florida vegetable growers. It took the form of a U.S. Department of Agriculture program which, in the year 1940, provided an additional 1258 tons of arsenic-based baits.[11]

The inadequacy of such chemical control methods prompted investigation of biological control methods, first in Puerto Rico, and then in Florida. The Puerto Rican program had some measure of success (see below) but the attempt in Florida was abandoned when use of chlorinated hydrocarbons became widespread in the late 1940s. Chlordane was such an inexpensive, persistent and effective material, that vegetable growers and turf managers alike based their control strategy upon it. For the first time, cattle ranchers had a material that they could use to control mole crickets in pastures.

With the banning of the use of chlordane in the 1970s, it was cattle ranchers who were most concerned about damage caused by mole crickets. Vegetable growers and turf managers could afford to use the more expensive carbamate and organophosphorous pesticides, but economics of the cattle industry would not permit use of many of these materials, even if they were not potentially toxic to cattle. Political pressure by Florida ranchers caused the state legislature in 1978 to authorize funding to be spent on research on mole crickets at the IFAS, which is the agricultural college of the University of Florida.[12]

The IFAS mole cricket program undertook basic research on taxonomy, physiology, biochemistry, behavior, ecology, sampling methods, and biological control of pest mole crickets, none of which was being investigated elsewhere. It did not ignore chemical control methods, though these were being researched elsewhere. It sponsored work by several students who obtained MS or PhD degrees and, by 1992, has produced over 150 publications on mole crickets. Its primary emphases have been to develop an ecological basis for management of pest mole crickets, and to achieve biological control. Unfortunately, much of the rest of the country has perceived mole crickets as being a problem unique to Florida, and external funding has been difficult to obtain. Although it seems true that the problem in Florida is worse than in any other state, the research being performed in Florida has potential application in many states.

THE STRATEGY OF INOCULATIVE BIOLOGICAL CONTROL

Many animals kill *Scapteriscus* mole crickets. Amphibians (such as frogs and toads), birds (such as egrets and sandhill cranes), mammals (such as armadillos and foxes), insects (such as tiger beetles, ground beetles, and assassin bugs), and spiders have been seen to kill them. Perhaps as many as a third of all *Scapteriscus* nymphs in Florida pastures are killed by such predators,[13] but the number of mole crickets killed simply is not enough to prevent damage. Fungal pathogens of the genera *Sorosporella, Beauveria,* and *Metarhizium* add to the total mortality. Four things differ in managed turf: (1) the tolerance of damage is less than in pastures, (2) more chemical pesticides are used per acre, (3) there usually are fewer mole crickets per unit area, and (4) there are fewer natural enemies.

The various natural enemies mentioned above are not specific to *Scapteriscus* mole crickets — they have a varied diet. Specific natural enemies — those which will feed only on *Scapteriscus* mole crickets and nothing else — are likely to be found in the homelands of *Scapteriscus*, where there has been a long history of co-evolution.

Biological control is the deliberate attempt to control pests by the use of natural enemies. Two of its major forms are inoculative (=classical) biocontrol and inundative use of biopesticides.

Inoculative biocontrol is used mainly against immigrant pest species (such as *Scapteriscus* mole crickets) and requires search for and importation of specific natural enemies from the area of origin of the pest. The natural enemies generally are predators, parasites, parasitoids, or competitors. It is then expected that the imported natural enemy can be established and will reduce the population of the pest to a nondamaging level permanently. This method of biocontrol generally does not result in a product that can be marketed by a commercial company, so research costs have to be met by the beneficiaries and/or by the government.

Biopesticides also are biocontrol agents, but generally are pathogens. They may be of native or foreign origin, but produce no permanent effect on the pest population. Thus, a marketable product is produced, and research therefore is supported often by commercial companies which can benefit from sales. Biopesticides are applied in much the same way as chemical pesticides but are safer to many nontarget organisms. They need repeated application just like chemical pesticides. The economics of use of biopesticides therefore resemble those of chemical pesticides, and are very different from those of inoculative biocontrol.

INOCULATIVE BIOLOGICAL CONTROL AGENTS

LARRA POLITA AND LARRA BICOLOR

Digger wasps of the genus *Larra* (Hymenoptera: Sphecidae) are specialized natural enemies of mole crickets.[1,14] *Larra analis* Fabricius is a parasitoid of *Neocurtilla hexadactyla* in the southern U.S.[15] Examination of tens of thousands of *Scapteriscus* mole crickets in Florida by personnel of the IFAS mole cricket program has failed to show even one of them bearing an egg or larva of *L. analis*, attesting to the specialization of this parasitoid. Laboratory observation in Florida likewise has failed to show successful attack on *N. hexadactyla* by the introduced *L. bicolor* Fabricius.

L. polita (Smith) subspecies *luzonensis* Rohwer was imported from the Philippines into Hawaii in 1925 to combat *Gryllotalpa orientalis*, and it became established.[1] *L. bicolor* was imported from Amazonian Brazil into Puerto Rico to combat *Scapteriscus didactylus*, and it became established by 1941.[16,17] I have discovered no analysis of the effects of either of these introductions. A short-lived attempt was made in the 1940s to import *L. bicolor* into Florida. Failure to obtain living *Larra* from Puerto Rico or northern South America delayed the program, which then was cancelled by

1949 on discovery of the efficacy of chlordane for killing mole crickets.[18]

A renewed attempt was made to import *L. bicolor* from Puerto Rico into Florida in the late 1970s. Planning included preparation of five release sites with plantings of a weed (*Spermacoce verticillata* (L.), a preferred nectar source for the parasitoid) and a laboratory comparison of attack by *L. bicolor* on adults of several *Scapteriscus* species.[19] Culminating in 1981, several dozen adult *L. bicolor* from Puerto Rico were released at each of the sites, but establishment occurred only at a site in Broward County.[18] Study of the established population in 1986 to 1987 failed to show that *L. bicolor* was attacking southern and tawny mole crickets, and revealed only a small proportion of field-collected short-winged mole crickets attacked by it.[20] This raises questions of which species (*Scapteriscus abbreviatus, S. didactylus, S. imitatus*) is attacked by *L. bicolor* in Puerto Rico, and whether this parasitoid regulates populations of any of them; no answer is evident in Puerto Rican literature. Some effort has been put into study of other South American *Larra* species relative to their importation into Florida as biocontrol agents for *Scapteriscus*.[21]

STEINERNEMA SCAPTERISCI

Nematodes of the genus *Steinernema* (Rhabditida: Steinernematidae) are all entomopathogenic, carrying in their guts a complement of bacteria of the genus *Xenorhabdus* which destroy the host insect. The best-known species is *Steinernema carpocapsae* (Weiser), which is widespread and uses various hosts. It can be reared in the laboratory on wax moth (*Galleria mellonella* (L.)) larvae, and also can be reared on artificial diets. It has been found effective as a biopesticide against various pest insects. It will kill mole crickets, but was not found in mole crickets trapped in Florida.[22]

Exploration for natural enemies of mole crickets in South America revealed a nematode which initially was believed to be a strain of *S. carpocapsae*. Imported from Uruguay into Florida in 1985, it proved ultimately to be a new species, *S. scapterisci* Nguyen and Smart.[23] It reproduced well in adults and large nymphs of *Scapteriscus borellii* and *Scapteriscus vicinus,* but poorly in *Galleria mellonella.*[22] Releases of this nematode in small plots in pastures in Alachua County, Florida, in 1985 showed suppression of *Scapteriscus* mole cricket populations,[13] persistence of nematode populations at the sites of release for at least 5 years,[24] and lack of effect on native *Neocurtilla* mole cricket populations.[24] The nematode also was shown to disperse from release points via flight of infected mole crickets.[25]

Experimental releases of this nematode subsequently were hindered by need to produce progeny *in vivo*, but *in vitro* methods were developed by personnel of biosys (Palo Alto, CA). Provision by biosys of large numbers of nematodes reared *in vitro* allowed experimental releases in Florida. As of late 1991, populations of the nematode are established in 13 Florida counties, and are being monitored. Provision of samples of this nematode to researchers in other states (in addition to Florida) by biosys have led to its evaluation as a biopesticide. This nematode is capable of killing insects other than *Scapteriscus* mole crickets, but it can reproduce only in *Scapteriscus* and in *Acheta* (house crickets), attesting to its safety for widespread release.[22]

ORMIA DEPLETA

Investigation in the mid-1980s in South America revealed that gravid adult females of a parasitic fly (*Ormia depleta* (Wiedemann), Diptera: Tachinidae) are attracted at night to the song of male *Scapteriscus* mole crickets.[26] This species of parasitic fly had been encountered by Puerto Rican entomologists in Amazonian Brazil, but attempts to introduce it into Puerto Rico were not successful.[17,18,27] Female flies are attracted to, and lay larvae on or near *Scapteriscus* males calling from their acoustical burrows. This behavior makes them specific to *Scapteriscus* mole crickets, though their larvae can develop in at least two other genera of crickets. After about 3 years needed to accomplish laboratory rearing of the fly,[28] flies were released in Florida first in Alachua County in April 1988, in Manatee County in October 1988, in Osceola County in April 1989, and in Dade County in March 1990. Populations of flies became established at three of the sites, but there is yet no evidence of establishment in Osceola County. In 1990 to 1991, flies were released on 28 golf courses and a sod farm in a program sponsored by the Florida Turfgrass Research Foundation. In 1991, flies were provided to the North Carolina Department of Agriculture for release on a golf course in North Carolina.

PHEROPSOPHUS AEQUINOCTIALIS

Larvae of the South American bombardier beetle *Pheropsophus aequinoctialis* (L.) (Coleoptera: Carabidae) appear to be specialist predators of *Scapteriscus* eggs.[18] They may become useful inoculative biocontrol agents for these mole crickets, but research on host-specificity required before their release is at an impasse due in part to problems in rearing the beetles.[29]

CURRENT LEVEL OF CONTROL IN FLORIDA

The inoculative biocontrol agents mentioned above are at various stages of development. *Larra bicolor* has been released in Puerto Rico (where its effect has not been recorded) and Florida (where its effect appears to be negligible). *Pheropsophus aequinoctialis* has not been released in Florida (or anywhere else) pending completion of host-specificity trials. *Steinernema scapterisci* and *Ormia depleta* have both been released in Florida, the former first in 1985, the latter first in 1988. Evidence for their effect on *Scapteriscus* mole cricket populations is presented below for four areas of Florida.

Dade County

A population of *Ormia depleta* was established in 1990 on a golf course in urban Miami by placing a box of 220 puparia in sand at a protected site in March 1990. Recovery of the box later showed that 203 adult flies had emerged successfully. Trapping for flies at monthly intervals beginning in May 1990 suggested an increasing population of the fly, with numbers trapped peaking in late autumn in 1990 and 1991. The superintendent's report that mole cricket damage was less in 1991 than in 1990 may or may not have been due to action by the fly.

Broward County

Approximately 5 billion *Steinernema scapterisci* were applied on each of three golf courses during May 1990 at 200,000 infective juveniles per square meter on edges of fairways of three holes per course. Populations of the nematode became established at two courses. Systematic monthly sampling at these courses over almost 2 years indicated reduction in mole cricket populations on treated vs. untreated fairways.[30]

Alachua/Gilchrist Counties

Two trapping stations for mole crickets have been operated nightly since 1979 in southwestern Alachua County. Each of the stations, approximately 3 km apart, has a pair of traps, one baited with synthetic sound of *S. vicinus*, and the other with sound of *S. borellii*. Total annual catch has fluctuated from year to year.

Ormia depleta puparia were placed at one of the stations in April and June 1988, using the same method that later was used in Dade County. A population of the fly became established. In July 1989, flies appeared at the other trapping station. In 1991, flies appeared at temporary traps some 15

km east; later in 1991, flies appeared at a temporary trap at the northern border of Alachua County, and at another some 20 km west of the western border, in Gilchrist County.[31] It can be assumed that a population of flies was well-established during much of 1991 in much of Alachua County.

Steinernema scapterisci was first released on small plots in pastures in northern and eastern Alachua County in summer 1985. In 1988, *S. scapterisci* was found in mole crickets trapped at one of the trapping stations, and in 1989 at the other, though the stations are >20 km from the sites of release, indicating that the nematode can be dispersed by flying mole crickets before they sicken and die.[25]

About five billion nematodes were released on each of three golf courses in Alachua County in October 1989 and April 1990. Routine monthly sampling on the golf courses, as in Broward County, showed establishment of nematode populations and reduction of mole cricket populations on treated fairways vs. untreated fairways.[30]

Records of mole crickets at both trapping stations during 1991 showed that numbers of *S. vicinus* were lower than at any time in the previous 8 years, while numbers of *S. borellii* were lower than at any time in the entire history of the stations.[32] It is difficult to partition the effects of the fly and of the nematode at the trapping stations. The golf courses, too, are now within the area colonized by the fly. The golf courses now within that area had less mole cricket damage in 1991 than in previous years, resulting in less pesticide usage. The evidence suggests reduction of mole cricket populations over a considerable area of Alachua County in 1991 as a result of actions of inoculative biocontrol agents.

Manatee/Sarasota/Hardee Counties

The IFAS mole cricket program has maintained a trapping station for mole crickets since 1980 at the IFAS research station in Manatee County. Its operation is the same as for the two stations in Alachua County. *Ormia depleta* puparia were placed at this station in October and November 1988 and February 1989, using the same method that later was used in Dade County. By spring 1989 a population clearly was established.

No *Steinernema scapterisci* were released in Manatee County. However, *O. depleta* were released also at two golf courses in Sarasota County (the county immediately south), together with small numbers of *S. scapterisci*, early in 1990. The exercise was repeated at two more golf courses in Sarasota County early in 1991. Additionally, the same routine was applied at two golf courses in

Hillsborough County (immediately north of Manatee County) early in 1990.

Temporary traps operated in June 1990 in Manatee County in concentric rings around the point of release, showed that the fly population had spread. It had extended 16 km in all directions from the point of release, crossing the southern border of Manatee County into Sarasota County and reaching the northern border of Manatee County.[33] In May 1991, a temporary trap line was extended east of the point of release and showed that the population had spread a dramatic 72 km east into Hardee County.[31] Meanwhile, in autumn 1990, a mole cricket parasitized by fly larvae was captured at a golf course in Sarasota County, and more were captured at another course in the same county in 1991.[30]

Events in Manatee County seemed to allow the effects of the fly on mole cricket populations to be evaluated. Numbers of S. vicinus trapped in 1991 were about as low as during any year since the trapping station was established in 1980. However, numbers of S. borellii were very high. Possibly this suggests that the fly has a much greater effect on S. vicinus than it does on S. borellii. This would be a welcome effect since S. vicinus is by far the most damaging species. (Incidentally, Larra bicolor seems to specialize on S. abbreviatus,[20] and Steinernema scapterisci infects a greater percentage

of sound-trapped S. borellii than of S. vicinus.[25]) In addition, mole crickets infected with Steinernema scapterisci were captured at the trapping station in mid-1991, which will make further evaluation of effects of the fly more difficult.

The fly is now widely distributed in Manatee/Sarasota/Hardee counties, and the nematode is established at least at one locality in Manatee County. The nematode is not known to be established in Sarasota County, but is established at three localities at least in Hillsborough County (to the north), from one of which it is presumed to have been carried to Manatee County by flying mole crickets. Against this background, a survey form was sent in November 1991 to golf course superintendents throughout Florida, to ask them to compare mole cricket damage in 1991 with that in 1990. Survey forms were mailed from, and received by, the Florida Turfgrass Association. Results are shown in Table 2.[34]

The ratios of less/same/more are almost identical in the south and west, so may be combined. In these counties, there were 56 courses reporting equal or less damage, 50 reporting more. Thus, the random probability of a course having equal or less damage is $56/(56 + 50) = 0.53$, and it takes six courses within one of these counties (or within contiguous counties) all showing equal or less damage to achieve significance ($p < 0.05$). But in

Table 2   **Results of survey of Florida golf course superintendents asking them whether damage by mole crickets was less/same/more in 1991 than in 1990**

Counties	Damage — Numbers Responding			
	Less or Much Less	Same	More or Much More	Total
South				
Dade/Broward/Palm Bch./Martin/ Collier/Lee/DeSoto/Charlotte	24	14	34	72
Manatee/Sarasota[a]	6	5	0	11
West				
Hillsborough/Pinellas/Polk/ Pasco/Hernando/Citrus/Sumter/ Leon	12	6	16	34
East				
St. Lucie/Indian R./Brevard/ Volusia/Flagler/St. Johns/ Duval/Nassau/Highlands/ Lake/Seminole/Orange/Baker	21	16	15	52
Panhandle				
Gadsden/Okaloosa	2	1	0	3
Total:	65	42	65	172

[a] These two counties are contiguous; they are north of Charlotte County and south of Hillsborough County. *Ormia depleta*, an inoculative biological control agent from South America, has been established in them.

Manatee and Sarasota counties (which are between the other two groups) all 11 superintendents responding reported equal or less damage; the probability of this being a chance occurrence is $0.53^{(11-1)}$ = 0.0017, or less than 1 in 500. It is surely no coincidence that in Manatee/Sarasota counties, where the fly is widely established, all 11 superintendents reported equal or less damage in 1991 than in 1990, in highly significant contrast to all other counties of the south and west. However, assistance from the nematode, at least in Manatee County, may have contributed there. No golf courses reported from Hardee, Alachua, or Gilchrist counties.

AVAILABILITY OF INOCULATIVE BIOCONTROL AGENTS

People needing methods to control pests are used to paying for a marketable product, but inoculative biocontrol agents do not fit the concept of a marketable product. The services of *Ormia depleta* are free in areas where this biocontrol agent is now established in Florida. Populations of the fly eventually will spread to all areas of Florida (and neighboring states) in which it can live. Current research is aimed at measuring the fly's effectiveness in supressing mole cricket populations. Future research will try to demonstrate seasonal and latitudinal variation in nectar sources used by the adult flies; it may be possible to build up populations of the flies in localized areas (specific golf courses, for example) by selecting landscape plants that are useful to the flies as nectar sources.

No research is now in progress on *Larra* or *Pheropsophus*. The reason for this is simply that Florida government funds allotted to mole cricket research will cover the salary of only one researcher whose efforts are concentrated for the present on *Ormia*. There is room in the program for graduate students wishing to undertake research on some aspect of biocontrol of mole crickets toward an M.S. or Ph.D. degree, but no funds to pay them stipends nor to pay the salaries of additional full-time researchers. Through the University of Florida, the program has a mechanism of accepting tax-exempt donations or grants toward research on *Ormia, Larra,* or *Pheropsophus*.

Small numbers of *Ormia depleta* have been supplied to researchers in North Carolina and Georgia. No attempt will be made to mass-produce the fly in Florida for distribution in Florida or elsewhere because it is best suited as an inoculative biocontrol agent, not as a biopesticide.

Steinernema scapterisci has a dual role as an inoculative biocontrol agent and as a biopesticide. It is likely to be marketed as a biopesticide in the near future by industry. Methods of industrial-scale production of it have been developed by biosys, which can thus produce it in bulk at a cost several orders of magnitude lower than can the IFAS mole cricket program.

INOCULATIVE BIOLOGICAL CONTROL AND IPM

The fundamental concept of IPM is based upon knowledge of the dynamics of the target pest population and of the action of its existing natural enemies. Knowledge of the dynamics of pest mole cricket populations gained by studies which were part of the IFAS mole cricket program have allowed the more precise use of chemical pesticides in Florida and elsewhere.[2] The value of native natural enemies, especially in pastures, has been observed, and efforts can be made to preserve their populations. For, example, though chemical sprays are likely to kill nontarget insects and beneficial insect predators, these are unlikely to be attracted to toxic baits. Development of a bait containing malathion, by a graduate student participating in the IFAS mole cricket program,[35] was a step toward protection of nontarget organisms.

Replacement of chemical pesticides by biopesticides aimed at insects should eliminate the risk to vertebrates entirely, as well as eliminating risk of pollution of ground-waters by chemicals. Further, it might eliminate some of the risk to nontarget insects, including beneficial predators, dependent upon the method of application and the kind of biopesticide.

Biopesticides are unlikely to be less costly than chemical pesticides. The only realistic way of reducing the cost of control is to reduce recurrent costs by employing inoculative biological control. This is precisely the current objective of the IFAS mole cricket research program, which will first develop inoculative biological control with the intent to reduce mole cricket populations over most of Florida, and second develop biopesticides for use in remaining areas of damage. Work on biopesticides by members of the IFAS mole cricket program has not yet progressed beyond the stage of laboratory evaluation.

ACKNOWLEDGMENTS

Members of the IFAS mole cricket research program are especially grateful to members of the Florida Turfgrass Association who have contributed funds through the Florida Turfgrass Research Foundation. The Caribbean Basin Advisory Group

(USDA-CSRS) provided funds in 1983 to 1986 under a grant which allowed initial exploration in South America for natural enemies of *Scapteriscus* mole crickets. A small stock of *Steinernema scapterisci* was supplied to biosys (Palo Alto, CA), which then developed commercial-scale *in vitro* rearing methods; billions of the nematodes, kindly supplied by that company, were then used for trials as an inoculative biocontrol agent in Florida. W. G. Hudson (University of Georgia), and T. J. Walker and J. P. Parkman (University of Florida) kindly reviewed manuscript drafts.

REFERENCES

1. **Williams, F. X.**, Studies in tropical wasps — their hosts and associates (with descriptions of new species), *Hawaii. Sugar Plant. Assoc. Exp. Stn. Entomol. Ser. Bull.*, 19, 1, 1928.
2. **Walker, T. J., Ed.**, Mole crickets in Florida, *Univ. Fla. Agric. Exp. Stn. Bull.*, 846, 1, 1985.
3. **Nickle, D. A. and Castner, J. L.**, Introduced species of mole crickets in the United States, Puerto Rico, and the Virgin Islands (Orthoptera: Gryllotalpidae), *Annu. Entomol. Soc. Am.*, 77, 450, 1984.
4. **Frank, J. H., Woodruff, R. E., and Nuñez, C. A.**, *Scapteriscus didactylus* (Orthoptera: Gryllotalpidae) in the Dominican Republic, *Fla. Entomol.*, 70, 478, 1987.
5. **Zwaluwenburg, R. H. van**, The changa, or West Indian mole cricket, *Puerto Rico Agric. Exp. Stn. Bull.*, 23, 1, 1918.
6. **Wolcott, G. N.**, "Insectae Borinquenses," a revision of "Insectae Portoricensis," *J. Agric. Univ. Puerto Rico*, 20, 1, 1936.
7. **Ivie, M. A. and Nickle, D. A.**, Virgin Islands records of the changa, *Scapteriscus didactylus* (Orthoptera: Gryllotalpidae), *Fla. Entomol.*, 69, 760, 1986.
8. **Frank, J. H. and Keularts, J. C.**, unpublished data, 1991.
9. **Worsham, E. L. and Reed, W. V.**, The mole cricket (*Scapteriscus didactylus*), *Ga. Exp. Stn. Bull.*, 101, 251, 1912.
10. **Thomas, W. A.**, The control of the Puerto Rican mole-cricket, or changa, on golf courses, *U.S. Golf Assoc. Green Sec. Bull.*, 6, 197, 1926.
11. **Schroeder, H. O.**, Mole cricket project, Plant City, Florida. Report on Control Activities September 1–December 31, 1940, *USDA Bur. Entomol. Pl. Quarantine*, 1, 1941.
12. **Koehler, P. G., Short, D. E., and Barfield, C. B.**, Mole crickets: IFAS research project, *Fla. Cattleman*, June, 91, 1979.
13. **Hudson, W. G., Frank, J. H., and Castner, J. L.**, Biological control of *Scapteriscus* spp. mole crickets (Orthoptera: Gryllotalpidae) in Florida, *Bull. Entomol. Soc. Am.*, 34, 192, 1988.
14. **Castner, J. L.**, Biology of the mole cricket parasitoid *Larra bicolor* (Hymenoptera: Sphecidae), in *Advances in Parasitic Hymenoptera Research*, Gupta, V. K., Ed., Brill, Leiden, 1988, 423.
15. **Smith, C. E.**, *Larra analis* Fabricius, a parasite of the mole cricket *Gryllotalpa hexadactyla* Perty, *Proc. Entomol. Soc. Wash.*, 37, 65, 1935.
16. **Wolcott, G. N.**, The introduction into Puerto Rico of *Larra americana* Saussure, a specific parasite of the "changa," or Puerto Rican mole cricket, *Scapteriscus vicinus* Scudder, *J. Agric. Univ. Puerto Rico*, 22, 193, 1938.
17. **Wolcott, G. N.**, The establishment in Puerto Rico of *Larra americana* Saussure, *J. Econ. Entomol.*, 34, 53, 1941.
18. **Frank, J. H.**, Mole crickets and other arthropod pests of turf and pastures, in *Classical Biological Control in the Southern United States*, Habeck, D. H., Bennett, F. D., and Frank, J. H., Eds., *Southern Coop. Ser. Bull.*, 355, 131, 1990.
19. **Castner, J. L.**, Suitability of *Scapteriscus* spp. mole crickets as hosts of *Larra bicolor* (Hymenoptera: Sphecidae), *Entomophaga*, 29, 323, 1984.
20. **Castner, J. L.**, Evaluation of *Larra bicolor* as a Biological Control Agent of Mole Crickets, Ph.D. dissertation, University of Florida, Gainesville, 1988.
21. **Bennett, F. D.**, personal communication, 1990.
22. **Nguyen, K. B.**, A New Nematode Parasite of Mole Crickets: Its Taxonomy, Biology, and Potential for Biological Control, Ph.D. dissertation, University of Florida, Gainesville, 1988.
23. **Nguyen, K. B. and Smart, G. C.**, *Steinernema scapterisci* n. sp. (Rhabditida: Steinernematidae), *J. Nematol.*, 22, 187, 1989.
24. **Parkman, J. P., Hudson, W. G., Frank, J. H., Nguyen, K. B., and Smart, G. C.**, unpublished data, 1992.
25. **Parkman, J. P. and Frank, J. H.**, Infection of sound-trapped mole crickets, *Scapteriscus* spp., by *Steinernema scapterisci*, *Fla. Entomol.*, 75, 163, 1992.
26. **Fowler, H. G. and Kochalka, J. N.**, New record of *Euphasiopteryx depleta* (Diptera: Tachinidae) from Paraguay: attraction to

broadcast calls of *Scapteriscus acletus* (Orthoptera: Gryllotalpidae), *Fla. Entomol.*, 68, 225, 1985.

27. **Sabrosky, C. W.**, Taxonomy and host relations of the tribe Ormiini in the western hemisphere (Diptera: Larvaevoridae), *Proc. Entomol. Soc. Wash.*, 55, 167, 1953.

28. **Wineriter, S. A. and Walker, T. J.**, Rearing phonotactic parasitoid flies (Diptera: Tachinidae, Ormiini, *Ormia* spp.), *Entomophaga*, 35, 621, 1990.

29. **Frank, J. H. and Cicero, J. M.**, unpublished data, 1991.

30. **Parkman, J. P. and Frank, J. H.**, unpublished data, 1992.

31. **Amoroso, J. and Walker, T. J.**, personal communication, 1991.

32. **Walker, T. J.**, Acoustic methods of monitoring and manipulating insect pests and their natural enemies, in *Biological Control and IPM: The Florida Experience*, Rosen, D., Capinera, J. L., and Bennett, F. D., Eds., Intercept Press, in press.

33. **Amoroso, J. and Walker, T. J.**, personal communication, 1990.

34. **Frank, J. H. and Yount, R. A.**, unpublished data, 1992.

35. **Kepner, R. L. and Yu, S. J.**, Development of a toxic bait for control of mole crickets (Orthoptera: Gryllotalpidae), *J. Econ. Entomol.*, 80, 659, 1987.

36. **Townsend, B. C.**, A revision of the Afrotropical mole crickets (Orthoptera: Gryllotalpidae), *Bull. Br. Mus. Nat. Hist. Entomol.*, 46, 175, 1983.

37. **Nickle, D. A.**, *Scapteriscus borellii* Giglio-Tos: the correct species name for the southern mole cricket in southeastern United States, *Proc. Entomol. Soc. Wash.*, 94, 524, 1992.

Chapter 43

Nematodes as Bioinsecticides in Turf and Ornamentals

Ramon Georgis, biosys, Palo Alto, CA

George O. Poinar, Jr., Department of Entomology, University of California/Berkeley

CONTENTS

Introduction 477
Product Characteristics 477
    Mode of Action 477
    Host Spectrum and Safety 479
    Persistence and Dispersal 480
        Abiotic Factors 481
        Biotic Factors 482
Commercial Development 482
    Production 482
    Formulation 482
    Field Efficacy 482
    Compatibility with Chemical Pesticides 484
    Standardization and Quality Control 484
Future Prospects 485
References 487

INTRODUCTION

Environmental concerns associated with chemical pesticide usage are forcing the industry to search for less toxic pest management methods. Entomopathogenic nematodes in the genera, *Steinernema* and *Heterorhabditis*, and their associated bacteria *Xenorhabdus* spp. have received considerable attention recently as biological insecticides,[1,2] particularly because of their ability to search for and kill their host rapidly, their wide host range, and their safety to mammals and plants and exemption from federal and state requirements in most countries.[3] Recent developments in production through liquid fermentation and in formulation stability have led to the introduction of a number of products against insect pests in soil and cryptic environments[4,5] (Table 1).

Despite these advantages, the widespread usage of nematode products will not be achieved unless they become competitive with chemical pesticides on the basis of field efficacy, cost, and ease of use.[6] This chapter addresses the characteristics of nematode-based products and the significant progress achieved in the commercialization of these nematodes.

PRODUCT CHARACTERISTICS

Over the last 10 years, companies developing nematode products have expended considerable effort researching entomopathogenic nematodes. The research has consisted of in-house experimental research programs as well as grants to universities and independent researchers. Within the framework of insect control, there is no product that cannot be improved upon or replaced with a better one. This reasoning is responsible for the current three-stage product development pathway (Table 2) adopted by industries involved in commercializing nematodes. The overall characteristics of the commercially available entomopathogenic nematode-based products in turf and ornamentals are summarized in Table 3.

MODE OF ACTION

Steinernematid and heterorhabditid nematodes are mutualistically associated with the bacteria

478

Table 1 **Products commercially available in turf and ornamentals**

Nematode Species	Product	Market Segment	Company
S. carpocapsae[a]	BioSafe®	Home lawn & garden	Ortho Monsanto, USA
		Turf	SDS Biotech, Japan
	Exhibit®	Turf & Ornamentals	Ciba-Geigy, USA
		Ornamentals	Ciba-Geigy, (W. Europe)
	Boden-Nützlinge	Home lawn & garden	Celeflor, Germany
	Sanoplant®	Home lawn & garden	Dr. R. Maag, Switzerland
S. feltiae	Nemasys	Ornamentals	Agricultural Genetics Co., UK
	—	Ornamentals	biosys, USA
S. glaseri	Vector®-WG	Turf	biosys, USA
S. scapterisci	Proact	Turf	BioControl, USA
S. riobravis	Vector-MC	Turf	biosys, USA
H. bacteriophora	Otinem	Ornamentals	Ecogen, USA
H. megidis	Nemasys-H	Ornamentals	Agricultural Genetics Co., UK

[a] Produced by biosys, Palo Alto, CA.

Table 2 **Normal product development pathway used by companies involved with research on steinernematid and heterorhabditid nematodes**

Product Development Path	Function
Stage 1: Basic research	Identification and characterization of nematodes and symbiont bacteria
	DNA based identification
Stage 2: Applied research	Host range
	Impact on nontarget organisms
	Activity spectrum at various temperatures
	In vitro production efficiency
	Formulation research
	Application techniques
	Efficacy and field evaluation
	Quality control
Stage 3: Commercialization	Production scale-up
	Formulation scale-up
	Standardization and quality assurance
	Demonstration trials with growers and distributors
	Marketing

belonging to the genus *Xenorhabdus* spp.[7,8] The free-living, infective-stage nematode is the invasive form which locates insects, initiates infection, and is the only stage in the nematode's life cycle that survives outside the insect in the soil[2]. The infective-stage nematode enters the insects via natural opening (mouth, anus, or spiracles) and penetrates mechanically into the hemocoel where it releases the bacterium. In addition, heterorhabditids have the ability to enter the hemocoel of certain insects by penetrating through the intersegment.[7] The bacteria proliferate, cause a septicemic death of the insect within 24 to 72 hours, and establish favorable conditions for nematode reproduction by providing nutrients and inhibiting the growth of many foreign microorganisms. The nematodes feed on multiplying bacteria and dead host tissue, passing through several generations. Eventually, infective stage nematodes, carrying the mutualistic bacteria in their gut, emerge from the depleted insect cadaver. Depending on the species, at 18 to 28°C, it takes 8 to 20 days for the nematodes to complete their life cycle in most insects.

The relationship between the nematode and the bacterium is considered mutualistic because the bacterium cannot enter into an insect's hemocoel without the nematode and the nematode cannot reproduce without the bacterium.

Table 3 **Characteristics of steinernematid- and heterorhabditid-based products in turf and ornamentals**

Character	Data	Ref.
Active ingredient	Third-stage infective juveniles and their symbiotic bacteria	7, 8
Mode of entry	Primary insect natural openings	7
Mode of action	Bacteria proliferate in insect hemocoel, cause septicemic death within 24 to 72 hours	8
Field efficacy and application		
Target insects and stage	Soil, cryptic and greenhouse-inhabiting insect larvae	4, 5
Dosage	2.5×10^9 to 12.5×10^9 nematodes/ha	4, 48
Timing	Evening or early morning to avoid UV light	3, 48
Temperature	12 to 30°C range for activity	21, 37
Moisture	Adequate moisture is needed for nematode movement, survival, and infection	21, 27
Spray volume	1500 to 2400 I/ha	48
Equipment	Most common sprayers and nozzles and pressures of up to 2068 KPu (300 lb/in.^{2})	48
Persistence	In general 3 to 6 weeks half field life, but few epizootics have been documented	21
Dispersal	In all soil types, majority of nematodes remain close to the application site	21
Compatibility	Compatible with many chemical pesticides	12, 48
Production	Liquid fermentation and solid media culture	47
Formulation	Immobilization or partial desiccation	47
Carrier	Polyacrylimide gel, alginate gel, and clay	48
Storage	Up to 5 months (room temperature) and 12 months (refrigeration) for steinernematids. Up to 3 months (refrigeration) for heterorhabditids	48
Quality control	LT_{50} of LD_{50} standards throughout all stages of product development	48
Safety	Exempted from registration requirements in most countries	3
Mammalian toxicity	No infection symptoms or mortality by oral, intradermal subcutaneous, and interperitonal inoculation	7, 9
Bird toxicity	No mortality or disease symptoms by subcutaneous inoculation	9
Fish toxicity	No disease symptoms or mortality in inoculated water tank	9
Honey bees	Hive temperature lethal to nematodes	12
Nontarget invertebrates	No impact or detrimental effect including earthworms and predatory insects	10–12

HOST SPECTRUM AND SAFETY

Under laboratory conditions, nearly 300 insect species from ten orders are reported to serve as hosts for these nematodes.[7] Moreover, exposure of representative invertebrates (Gastropoda, Symphyla, Arachnida, Crustacea, and Diplopoda) to high concentration of nematodes results in in-fection and has expanded the host range beyond the class Insecta.[9]

However, under natural conditions, these nematodes have been used mainly against the larval stages of certain soil and cryptic-inhibiting insects.[4,5] Soil insects are a logical target since the nematodes naturally occur in soil and may have advantages

Table 4 **Comparison data showing the nondetrimental effect of entomopathogenic nematodes on nontarget arthropods compared to the chemical pesticide in two golf courses**

Nontarget Arthropods	Average Number of Nontarget Arthropods 28 Days After Treatment			
	Golf Course 1[a]		Golf Course 2[b]	
	S. carpocapsae	Ethoprop	H. bacteriophora	Isofenphos
Predatory insects[c]	15.0	6.3[d]	31.1	14.7[d]
Predatory mites[e]	47.3	24.1[d]	46.1	19.7[d]
Collembola	5.0	1.3	17.0	5.5[d]

[a] Tawny mole cricket-infested golf course; [b] Japanese beetle-infested golf course; [c] Representing Carabidae, Staphylinidae, and Gryllidae; [d] Significant differences between the nematode and the insecticide treatments, but not between the nematodes and the control treatments ($p > 0.05$) using Kruskal-Wallis Test;[60] [e] Representing Gamasida, Actinedida, and Oribatida.

Modified from *Georgis, R., Kaya, H. K., and Gaugler, R., Environ. Entomol., 20, 815, 1991.*

over pesticides and microbial pathogens by virtue of their ability to seek out the insect host.[3] Akhurst[10] has reviewed the available data on the effect on nontarget beneficial soil organisms and concluded that these nematodes have no negative impact. In two field releases of entomopathogenic nematodes against the Japanese beetle *Popillia japonica* and tawny mole cricket *Scapteriscus vicinus*, nematodes showed no detrimental effect on beneficial predatory insects and mites, whereas the effect of chemical pesticides was significant[11] (Table 4).

Attempts to control aboveground insects have been discouraging, with low host mortality, insignificant population reduction, or inadequate crop protection.[12] UV light and desiccation were singled out as major factors limiting nematode survival on foliage. According to Kaya,[12] nematodes survived up to 60 min on plants during the day, but up to 10 h when applied at night. Similarly, lack of oxygen and the inability of nematodes to move and locate the hosts in aquatic environments has precluded their use as control agents of aquatic insects.[5] No impact on beneficial organisms is expected with nematodes in these environments.[11]

Tests conducted on rats, mice, chicks, rabbits, pigs, lizards, turtles, and fish showed no disease symptoms or mortality caused by nematodes and their associated bacterium.[9] In two published reports,[13,14] newly hatched tadpoles exposed to a high concentration of *S. carpocapsae* became infected and died after 48 hours. Death was not due to the nematodes, but to foreign bacteria that entered the tadpoles through the penetration holes made by the nematodes. The infection decreased markedly with age, which was probably correlated with a thickening of the epithelial gut wall, thus making it more difficult for the nematodes to penetrate.[14] Since the

nematodes have a short lifespan in aquatic environments,[11] detrimental effect on tadpoles is unlikely in nature.

The application of entomopathogenic nematodes to soil may have positive indirect effects on soil organisms. For example, Ishibashi and Kondo[15,16] and Smitley et al.[17] showed that applications of steinernematids or heterorhabditids increased native rhabditid nematode species and decreased plant parasitic soil nematodes. Bird and Bird[18] also demonstrated that repetitive applications of *S. glaseri* reduced the ability of 2-week-old second-stage root knot nematodes, *Meloidogyne javanica*, to attack plant roots. When *S. glaseri* was applied at a high concentration, the numbers and reproductive capacity of *M. javanica* were significantly reduced, resulting in larger plants. *S. glaseri* is attracted to the growing root tips of the plants, probably by attraction to carbon dioxide,[19] and competitively displaces the smaller and less active plant parasitic nematodes.

PERSISTENCE AND DISPERSAL

Steinernematids and heterorhabditids are widespread and have been isolated from every inhabited continent and many islands. Few epizootics have been documented.[20-22]

In most field studies, nematode persistence as indicated by insect mortality over time, declined significantly 3 to 6 weeks after their release.[11,12] However, few reports showed the ability of the nematodes to achieve long-term control.[20]

The source of variation may be related to application methods, nematode species and strain, host species, stage and availability, insect defense mechanisms, and the biotic and abiotic soil environment.

Because of the nematode's moderate persistence,

its field application has stressed inundative releases to reduce pest population to a significant level.

The bacterium *Xenorhabdus*, which is associated with the nematodes, has never been recovered in nature except in the body cavities of insects invaded by the nematodes and inside the intestine of these nematodes. The bacteria are protected inside the nematode since the cells cannot survive in soil and water.[23]

An understanding of the relationship between the soil environment and the survival of steinernematids and heterorhabditids is essential to developing an effective biological insecticide.[3]

Abiotic Factors

There are many abiotic factors that may limit nematode success in soil, including: soil texture (soil type), pore size, moisture content, aeration, temperature, and soil chemistry.[24]

Soil type significantly affects nematode persistence in the laboratory.[25] The persistence of *S. carpocapsae* decreased with an increase in the clay content of soil and differed between *S. carpocapsae* and *S. glaseri* in each of four different soil types tested (sandy loam, sand, clay loam, and clay). Little information is available on the effects of soil type on the persistence of heterorhabditid nematodes in soil. Soil texture also affects the ability of nematodes to infect host. Infection by *S. glaseri, S. carpocapsae,* and *H. bacteriophora* was lower in a clay loam soil with a high clay content than in a sandy soil.[26,27] Soil texture also affects vertical and horizontal dispersal of nematodes, since movement is impeded considerably in soils with a higher clay content.[21,28-30] It is generally believed that with the exception of *S. glaseri,* steinernematid nematodes do not disperse far from the placement site.[21,29] *S. glaseri* is known to move downward in soil readily, whereas other steinernematids do not.[21,29] However, Choo et al.[31] and Jansson et al.[32] noted that heterorhabditids readily move downward in search of hosts.

Soil water potentials significantly affect the ability of various nematodes to infect hosts.[27] In general, very wet or very dry soil reduces nematode infection; however, the range in water potentials in which infection occurred varies among different soil types. Also, the larger *S. glaseri* is less effective than the smaller *Heterorhabditis* sp. in clay soil with a small pore diameter. This data indicates that soil type, soil moisture, and nematode species can significantly affect the ability of nematodes to find and infect a potential host. Georgis and Gaugler[33] noted a remarkable improvement in field efficacy of *H. bacteriophora* HP88

against the Japanese beetle as irrigation frequency increased from 7- to 10-day to 1- to 4-day intervals. Furthermore, they concluded that it is important that nematode treatments are applied to moist soil and recommended 1 to 2 cm of irrigation immediately after treatment. Similarly, Shetlar et al.[34] reported that field application of infective juveniles to dry turfgrass followed by 0.64 cm irrigation after treatment and normal rainfall resulted in grub reductions comparable with a standard insecticide, but applications without irrigation achieved less suppression.

S. carpocapsae can survive some desiccation if it occurs slowly.[35,36] Kung et al.[37] found that steinernematids became inactive and survived best in soils with a moisture content between 2 and 4%. They suggested that these nematodes were able to enter into a state similar to anhydrobiosis, because when the soil was rehydrated to 16% moisture, these nematodes regained their pathogenicity. Because steinernematids can survive for long periods in soil with little moisture, moisture levels that occur normally in natural soils may not be limiting for these nematodes; however, infectivity is dependant on nematode activity and a certain moisture content is necessary for activity. Thus, Georgis and Gaugler[33] concluded that the effective uses of nematodes in soil may be restricted to sites where irrigation is available or to geographical areas with frequent rainfall.

Soil temperature can significantly affect survivorship, host finding, and infectivity of the nematodes.[37,38] The efficacy of *H. bacteriophora* HP88 against the Japanese beetle *P. japonica* was invariably poor in spring trials.[33] By contrast, nematode efficacy was comparable with the chemical pesticides in fall trials. The overall mean of soil temperature was $12.8 \pm 0.5°C$ in the spring trials, whereas the overall mean temperature was $20.2 \pm 0.2°C$ in the fall.

Kung et al.[39] showed that survivorship was significantly affected by soil pH. Steinernematids survived poorly in soils with a pH of 10. Survivorship also decreased with a decrease in soil pH from 8 to 4. When steinernematids were kept in soil with a pH of 10 for 1 week they lost the ability to cause lethal infection. However, nematode pathogenicity and persistence did not decline when exposed in soil with a pH between 4 and 8 for 4 weeks. The pH range is similar to that needed for turf and ornamental growth. They also found that survivorship decreased with a decrease in soil oxygen content. Other factors, such as an increase in clay content, an increase in soil moisture and/or an increase in soil organic matter can affect soil aeration (oxygen content) and concomitantly affect nematode survivorship, infectivity, and fitness.

Biotic Factors

Survival of *S. carpocapsae* and *S. glaseri* was greater in sterilized and nonsterilized bark compost and sterilized sandy soil than in nonsterile sandy soil.[16] Kaya et al.[40] also observed better survival of *S. carpocapsae* in sterilized soil as compared with nonsterilized soil. These nematodes are probably susceptible to a number of antagonists, particularly bacteria, fungi, and predatory invertebrates. To date, the only discovered naturally occurring disease of a steinernematid is a microsporidan.[41] Under laboratory conditions, the entomopathogenic nematodes are infected with parasitic nematophagous fungi,[42,43] predacious nematode-trapping fungi,[44] and preyed upon by mononchid and dorylaimid nematodes,[45] mites,[45,46] collembolans,[45] and tardigrades.[45] These observations suggest that some of these antagonists may cause significant mortality of entomopathogenic nematodes in the soil.

COMMERCIAL DEVELOPMENT

The successful commercialization of nematode products for insect control depends heavily on the ability to produce sufficient quantities of the product to supply full-scale pest management programs. Without industrialization, the benefit of scientific research might never become available to the public. If the user cannot be assured of a standardized product that offers a determinable set of characteristics and if this product cannot also be attained in sufficient quantity as needed, then no practical success in the marketplace can be expected.

PRODUCTION

Currently, steinernematids are produced consistently and effectively in fermentors up to 80,000 L. Ox-kidney homogenate-yeast extract or a medium containing soy flour, yeast extract, corn oil, and egg yolk have been shown to support a yield as high as 100,000 infective juveniles/mL.[7] As a result, the current cost of application of nematode products is 10 to 20% higher than the standard insecticides. Optimum aeration, the nematode's sensitivity to shear, and an understanding of the nematode-bacterium interaction are the most significant issues that need to be considered for production efficiency in liquid culture[47] and are undoubtedly some of the factors causing inconsistent production of viable heterorhabditid nematodes in liquid culture. At the present time, the production of heterorhabditids is more efficient in solid media culture.[20]

FORMULATION

The development of quality control standards by industry is important in maintaining a high quality product in a form that is easy to store and handle.[48]

To maintain nematode virulence, current technology utilizes gel polymers (e.g., polyacrylamide and alginate) or clay to immobilize or partially desiccate the nematodes. These formulations reduce nematode metabolism and improve their tolerance to temperature extremes. Most significantly, large scale application of the nematodes is feasible due to the ease of extracting, mixing, and applying these formulations.[6,48]

The package design and size depends on the formulation type and the nematode species. To achieve a viable product, the oxygen and moisture requirements for each nematode species and their compatibility with the selected formulation type must be assessed.[48] For example, 250×10^6 *S. carpocapsae* formulated on an alginate sheet enclosed in a 4-L container will achieve 3 months storage at room temperature or 6 months under refrigeration. Cold temperature storage is recommended for heterorhabditids, probably because of their high oxygen demand.

Prior to formulation, infective nematodes produced from *in vitro* culture are stored for 2 to 3 months at 4 to 10°C in large aerated aqueous suspension tanks.

FIELD EFFICACY

Steinernematids and heterorhabditids have been proven efficacious against various soil and cryptic-inhabiting insects of ornamentals and turf[4,5] (Table 5). These nematodes differ in virulence to specific host, tolerance to adverse environmental conditions, ability to seek out host, and behavior in the soil.[49,50] Based on these characteristics, efforts made in recent years have led to selecting the proper strain or species in a particular habitat or against a particular insect species (Tables 6 and 7).

Many factors affect our ability to place quantities of nematodes on or in close proximity to the target host in order to produce optimal results at the lowest possible cost.[48] To overcome the impact of abiotic and biotic factors on nematode efficacy and persistence, the inundative application of high concentration of a specific nematode species (approximately 2.5×10^9 to 12.5×10^9 infectives/ha) has been used as the primary control strategy to ensure that sufficient nematodes will come in contact with the target insect.[5,48] Such a strategy has produced inconsistent results. Georgis and Gaugler[33] demonstrated that biological control attempts using nematodes without careful consideration of optimal strain and soil parameter, risk a high probability of failure.

Table 5 **Major target insects[a] for steinernematid- and heterorhabditid-based products in turf and ornamentals**

Latin Name	Common Name	Country
Turfgrass		
Agrotis ipsilon	Black cutworm	U.S., Japan, Canada
Spodoptera depravata	Japanese lawn cutworm	Japan
Pseudaletia unipuncta	Armyworm	U.S.
Parapediasia teterrella	Bluegrass sod webworm	U.S., Japan
Herpetogramma phaeopteralis	Tropical sod webworm	U.S.
Sphenophorus spp.	Billbugs	U.S., Japan
Hyperodes spp.	Annual bluegrass weevil	U.S.
Scarabaeidae[b]	White grubs	U.S., Japan, Canada, W. Europe
Scapteriscus spp.	Mole crickets	U.S.
Tipula paludosa	European crane fly	U.S., Canada, W. Europe
Ornamentals		
Otiorhynchus sulcatus	Black vine weevil	U.S., Canada, W. Europe
O. ovatus	Strawberry root weevil	U.S., Canada, W. Europe
Bradysia spp	Fungus gnats	U.S., Canada, W. Europe
Scatella sp	Shore flies	U.S.
Scarabaeidae[c]	White grubs	U.S.
Liriomyza trifolii	Leaf miner	U.S., W. Europe
Spodoptera exigua	Beet armyworm	U.S.
A. ipsilon	Black cutworm	U.S.
Sesiidae	Stem borers	U.S., W. Europe
Chrysoteuchia topiaria	Cranberry girdler	U.S.
Pachneus litus	Blue green weevil	U.S.
Opogona sp.	Banana moth	U.S., W. Europe

[a] Immature stages. All soil applications[5,60] except for leaf miner[61] and beet armyworm[5] (foliar applications; cryptic environment); [b] Major insects are Japanese beetle (*Popillia japonica*), European chafer (*Rhizotrogus majalis*), masked chafers (*Cyclocephala* spp.), June beetles (*Phyllophaga* spp.), and soybean beetle (*Anomala rufocuprea*); [c] Major insects are Japanese beetle, European chafer, and oriental beetle (*Anomala orientalis*).

They emphasized the importance of understanding the underlying reasons for failure through an analysis of multiple trials. In an examination of entomopathogenic nematodes used against Japanese beetle larvae in inundative releases on turfgrass, Georgis and Gaugler[33] analyzed 380 treatments from 82 field trials performed between 1984 and 1988 using a standard protocol. The results showed that most test failures could be explained on the basis of unsuitable nematode strains or environmental conditions. Most significantly, data showed that the incidence of nematode failure could be reduced by optimizing application strategy. Abiotic factors for example, can often be manipulated to maximize nematode-insect contact. Providing irrigation and adjusting the time of application to avoid low soil temperature enhanced the performance of nematodes to a level comparable with that of standard insecticides. In these studies, *H. bacteriophora* HP88 strain at 2.5×10^9 infectives/ha was the optimum nematode under the following conditions: soil temperature above 20°C (fall generation), frequent irrigation (1- to 4-day intervals), silty clay soil type, and thatch layer (organic matters between soil and turfgrass foliage) less than 10 mm.

The development of stable formulations and adopting a quality control program by industry has ensured the application of standardized and viable products.[6,48]

One of the most important elements in the commercialization of nematode products is their compatibility with common existing agrichemical equipment, including pressurized sprayers, helicopters, and irrigation systems. Successful applications through irrigation systems have increased the willingness of the growers to use nematode products for reasons related to minimal labor requirements, automatic provision of sufficient moisture for the nematodes, availability of permanently installed equipment, and flexibility in the timing of application.[48]

Table 6　**Field efficacy comparison between steinernematid and heterorhabditid nematodes and chemical pesticides against selected turfgrass insects**[a] [6,33,60]

Treatment	Application Rate/ha	No. Trials	% Reduction[b]
Popillia japonica, **Japanese Beetle**			
H. bacteriophora HP88	2.5×10^9	32	74.8 ± 6.9
S. feltiae SN	2.5×10^9	14	52.4 ± 13.7
S. glaseri	2.5×10^9	11	73.7 ± 6.1
S. carpocapsae All	2.5×10^9	18	43.9 ± 17.3
Isofenphos	2.5 kg ai	34	81.4 ± 7.6
Bendiocarb	4.0 kg ai	13	87.1 ± 8.5
Scapteriscus vicinus, **Tawny Mole Cricket**			
S. scapterisci	2.5×10^9	18	70.8 ± 9.4
S. carpocapsae All	2.5×10^9	16	57.7 ± 16.2
Isofenphos	2.5 kg ai	8	77.2 ± 7.8
Acephate	3.4 kg ai	22	68.4 ± 8.5
Agrotis ipsilon, **Black Cutworm**			
S. carpocapsae All	2.5×10^9	19	94.5 ± 7.2
H. bacteriophora HP88	2.5×10^9	6	62.1 ± 12.8
Chlorpyrifos	1.1 kg ai	3	99.1 ± 5.3
Sphenophorus purvulus, **Bluegrass Billbug**			
S. carpocapsae All	2.5×10^9	10	78.4 ± 7.4
H. bacteriophora HP88	2.5×10^9	4	74.1 ± 8.0
Isazophos	2.3 kg ai	8	83.6 ± 5.9

[a] Study conditions and methodology detailed in Georgis[6] and Georgis et al.[60]; [b] Each value is the mean ± standard error. Significant differences between treated and control plots ($p > 0.05$) using the Kruskal-Wallis test.[62]

COMPATIBILITY WITH CHEMICAL PESTICIDES

It should be emphasized that nematodes are particularly attractive as applied or introduced components of integrated management systems, because they have little or no impact on predacious or parasitic arthropods and other fauna that influence populations of pest species.[10,11] Additionally, since more than 90% of insect species spend at least part of their life cycle in soil,[51] most can be considered candidates for suppression by steinernematids and heterorhabditids. Other important considerations for successful pest management programs are the compatibility of nematodes with chemical pesticides and fertilizers in tank mixes and with chemical residues.

Steinernematids and heterorhabditids are compatible with commercial preparations of *Bacillus thuringiensis*,[52] insect growth regulators, and many chemical pesticides including insecticides, miticides, fungicides, herbicides, and fertilizers.[12,48,52] Certain pesticides can adversely effect nema-todes.[53-57] Removal or dilution of the pesticides with water usually results in nematode recovery, infection of the insect host, and normal development.[56,58] Surprisingly, in some studies, nematode viability was used as the only measurement to determine their compatibility with pesticides. However, recent studies showed that compatibility should be based on nematode pathogenicity to the insect host along with its ability to tolerate the effect of the pesticides.[53-56] Although nematodes and certain pesticides are not compatible, they can possibly be used together if they are separated spatially or temporally[58] (Table 8).

STANDARDIZATION AND QUALITY CONTROL

Maintaining a high quality of the final product is the most important factor determining the successful market introduction of a product. For this reason, the development of a quality control program is an essential step in the commercialization of nematodes.[48]

Table 7 **Efficacy comparison between steinernematid and heterorhabditid nematodes and chemical pesticides against selected soil-inhabiting insects of potted ornamentals, shrubs, and flowers[a]**

Treatment	Application Rate[b]	No. Trials	% Reduction[c]
Otiorhynchus sulcatus, Black Vine Weevil			
S. carpocapsae All	7.5×10^9	24	81.7 ± 10.9
H. bacteriophora HP88	7.5×10^9	11	87.4 ± 12.3
Bendiocarb	2.6 kg	13	90.1 ± 9.7
Acephate	1.4 kg	7	77.6 ± 10.6
Popillia japonica, Japanese Beetle			
S. carpocapsae All	7.5×10^9	7	61.4 ± 16.4
S. glaseri	7.5×10^9	3	85.7 ± 7.3
S. feltiae SN	7.5×10^9	5	62.9 ± 5.6
Isofenphos	4.0 kg	2	88.3 ± 3.8
Bradysia spp., Fungus Gnats			
S. carpocapsae All	7.5×10^9	5	71.2 ± 10.7
S. feltiae SN	7.5×10^9	4	87.3 ± 15.4
Gnatrol	10.0 I	2	80.9 ± 8.8

[a] Study conditions and methodology detailed in Georgis[6] and Georgis et al.[60]; [b] Expressed as number of infective juveniles for nematode treatments/ha and the ai/ha or dose/1000 L water for insecticides; [c] Each value is the mean ± standard error. Significant differences between treated and control plots ($p > 0.05$) using the Kruskal-Wallis test.[62]

Presently, LC_{50}, LD_{50}, or LT_{50} (i.e., the concentration, dose, or time needed to kill 50% of test insects) standards have been determined for each nematode species or strain with commercial prospects. Maintaining these performance standards throughout all the stages of product development and between products made at various times is essential to assure consistent field performance.

The first step in standardization is aimed at obtaining reliable and consistent nematode production. Inoculum batches for _in vitro_ culture are produced from stocks of nematode strains that are stored by cryopreservation to minimize variation in nematode pathogenicity between various production lots.

Differences in pathogenicity and behavior between products made from various lots have been insignificant in current formulations (i.e., clay, polyacrylamide, and alginate gels). By immobilizing or partially desiccating the nematodes, these materials maintain nematode survivorship and pathogenicity at room temperature and refrigeration.

FUTURE PROSPECTS

The intense interest in steinernematid- and heterorhabditid-based products is a reflection on impressive characteristics of an ideal insecticide.

More significant is the spectrum of their activity and compatibility with commonly used conventional sprayers. Certainly, progress made in liquid fermentation processes, formulation stability, and application strategy has allowed the nematodes to become competitive with chemical insecticides in the market, not only in effectiveness in protecting the crop, but also in the cost/benefit ratio. However, further advancement in formulation (i.e., shelflife, ease of use) and application technology (i.e., irrigation, spray volume) are needed to expand the usage of current products in the marketplace. The recent introduction of water-dispersable granule formulation into the turfgrass market is a significant step toward expanding the market share of nematode-based products. Furthermore, preliminary investigations have demonstrated that genetic manipulation through selective breeding and gene manipulation has potential to improve field efficacy, formulation, stability, and production of current products.[2,49]

It is important to emphasize that the successful penetration of nematode products will require a change in the attitudes and behavior of the technical advisers and the end users. For example, managers will need to focus more attention on monitoring of pest populations and timing applications.

Since steinernematids and heterorhabditids differ in efficacy against a particular insect species

Table 8 Chemicals that can be used with _Steinernema carpocapsae_ in turf and ornamentals[63]

Compounds	Chemical Class	Trade Name
	Tank Mix	
Biopesticides	Azatin	Margo-san
	Bacillus thuringiensis	M-One, Dipel
	Fatty acids	Safer soap
Insect growth regulators	Diflubenzurion	Dimilin
	Fenoxycarb	Logic
	Kinoprene	Enstar
	Methroprene	Apex
Insecticides	Acephate	Orthene
	Bifenthrin	Talstar
	Carbaryl	Sevin
	Cyfluthrin	Tempo
	Cythion	Malathion
	Diazinon	Knox-out
	Endosulfan	Thiodan
	Esfenvalerate	Asana
	Etridiazole	Terrazole
	Isofenphos	Oftanol
	Methidathion	Supracide
	Trichlorfon	Dylox
Fungicides	Benomyl	Benlate
	Bromine-chlorine	Agribrom
	Chlorothalonil	Daconil
	Copper hydroxide	Kocide
	Fosethyl-Al	Aliette
	Iprodione	Chipco 26019
	Metalaxyl	Subdue
	Oryzalin	Surflan
	Oxazoidinedione	Ornalin
	Pentachloronitrobenzene	Terraclor
	Thiophanate-methyl	Zyban
	Triademefon	Bayleton
Herbicides	Chlorthal dimethyl	Dacthal
	Glyphosate	Roundup
Miticides	Dienochlor	Pentac
Fertilizers	Most fertilizers are compatible with nematodes	
	Use 1 Week After Nematodes[a]	
Insecticides	Bendiocarb	Turcam
	Chlorpyrifos	Dursban
Fungicides	Anilazine	Dyrene
	Dimethyl benzyl ammonium chloride	Physan 20
	Fenarimol	Rubigan
	Mercurous chloride	Calo-Clor
Herbicides	2,4-D	2,4-D
	Triclopyr	Turflon
	Use 2 Weeks After Nematodes[a]	
Insecticides	Ethoprop	Mocap
	Isazophos	Triumph
Nematicides	Fenamiphos	Nemacur

[a] Laboratory bioassays. Days needed to assure that the survival and the pathogenicity of the nematodes are not affected by pesticides at recommended field dosages, using a modified method reported in Rovesti et al.[53] and Forschler et al.[55]

Table 9 Host preference of selected entomopathogenic nematodes[a]

Nematode Species	Host Preference[b]
S. carpocapsae	Cutworms, armyworms, sod webworms, billbugs, black vine weevil, stem borers, fungus gnats, leaf miner
S. feltiae	Fungus gnats, leaf miner
S. scaptersci	Mole crickets
S. glaseri	Black vine weevil, billbugs, white grubs
H. bacteriophora	Black vine weevil, billbugs, white grubs

[a] Georgis,[60] Kaya and Gaugler,[2] Lewis et al.,[50] and Parwinder et al.[64,65];

[b] Scientific names reported in Table 5.

(Table 9), product optimization will require the introduction of various species to the market. However, to achieve this, it is important to increase our understanding on the fate and the behavior of these nematodes in the soil and their interaction with the target insect.[2,59] Most of these issues are currently investigated by Dr. Harry Kaya (University of California/Davis) and Dr. Randy Gaugler (Rutgers University), and a number of publications have been generated from both laboratories.

REFERENCES

1. **Gaugler, R. and Kaya, H. K.,** *Entomopathogenic Nematodes in Biological Control*, CRC Press, Boca Raton, FL, 1990.
2. **Kaya, H. K. and Gaugler, R.,** Entomopathogenic nematodes, *Annu. Rev. Entomol.*, 38, 181, 1993.
3. **Gaugler, R.,** Ecological considerations in the biological control of soil-inhabiting insects with entomopathogenic nematodes, *Agric. Ecosys. Environ.*, 24, 351, 1988.
4. **Georgis, R. and Hague, N. G. M.,** Nematodes as biological insecticides, *Pestic. Outlook*, 2, 29, 1991.
5. **Georgis, R.,** Present and future prospects for entomopathogenic nematode products, *BioCont. Sci. Tech.*, 2, 83, 1992.
6. **Georgis, R.,** Commercialization of steinernematid and heterorhabditid entomopathogenic nematodes, *Brighton Crop Protection Conference, Pests and Diseases*, 1, 275, 1990.
7. **Poinar, G. O., Jr.,** Taxonomy and biology of Steinernematidae and Heterorhabditidae, in *Entomopathogenic Nematodes in Biological Control*, Gaugler, R. and Kaya, H. K., Eds., CRC Press, Boca Raton, FL, 1990, chap. 2.
8. **Akhurst, R. J. and Boemare, N. E.,** Biology and taxonomy of *Xenorhabdus*, in *Entomopathogenic Nematodes in Biological Control*, Gaugler, R. and Kaya, H. K., Eds., CRC Press, Boca Raton, FL, 1990, chap. 4.
9. **Poinar, G. O., Jr.,** Non-insect host for the entomogenous rhabditoid nematodes *Neoaplectana* (Steinernematidae) and *Heterorhabditis* (Heterorhabditidae), *Rev. Nematol.*, 12, 423, 1989.
10. **Adhurst, R. J.,** Safety to non-target invertebrates of nematodes of economically important pests, in *Safety of Microbial Insecticides*, Laird, M., Lacey, L. A., and Davidson, E. W., Eds., CRC Press, Boca Raton, FL, 1990, chap. 16.
11. **Georgis, R., Kaya, H. K., and Gaugler, R.,** Effect of steinernematid and heterorhabditid nematodes (Rhabditida: Steinernematidae and Heterorhabditidae) on non target arthropods, *Environ. Entomol.*, 20, 815, 1991.
12. **Kaya, H. K.,** Entomogenous nematodes for insect control in IPM systems, in *Biological Control in Agricultural IPM Systems*, Hoy, M. A. and Herzog, D. C., Eds., Academic Press, New York, 1985, 283.
13. **Kermarrec, A. and Mauleon, H.,** Nocuité potentielle du nematode entomoparasite *Neoaplectana carpocapsae* Weiser pour le crapaud *Bufo marinus* aux Antilles, *Meded. Fac. Landbouw. Rijksuniv. Gent.*, 50, 831, 1985.
14. **Poinar, G. O., Jr. and Thomas, G. M.,** Infection of frog tadpoles (Amphibia) by insect parasitic nematodes (Rhabditida), *Experientia*, 44, 528, 1988.
15. **Ishibashi N. and Kondo, E.,** *Steinernema feltiae* (DD-136) and *S. glaseri*: persistence in soil and bark compost and their influence on native nematodes, *J. Nematol.*, 18, 310, 1986.
16. **Ishibashi, N. and Kondo, E.,** Dynamics of the entomogenous nematode, *Steinernema feltiae*, applied to soil with and without nematicide treatment, *J. Nematol.*, 19, 404, 1987.
17. **Smitley, D. R., Warner, F. W., and Bird, G. W.,** Influence of irrigation and *Heterorhabditis bacteriophora* on plant-parasitic nematodes in turf, *J. Nematol.*, 24, 637, 1992.
18. **Bird, A. F. and Bird, J.,** Observations on the use of insect parasitic nematodes as a means of biological control of root-knot nematodes, *Int. J. Parasitol.*, 16, 511, 1986.

19. **Gaugler, R., LeBeck, L., Nakagaki, B., and Boush, G. M.,** Orientation of the entomogenous nematode, *Neoaplectana carpocapsae*, to carbon dioxide, *Environ. Entomol.*, 8, 658, 1980.

20. **Hominick, W. M. and Reid, A. P.,** Perspectives on entomopathogenic nematology, in *Entomopathogenic Nematodes in Biological Control*, Gaugler, R. and Kaya, H. K., Eds., CRC Press, Boca Raton, FL, 1990, chap. 17.

21. **Kaya, H. K.,** Soil ecology, in *Entomopathogenic Nematodes in Biological Control*, Gaugler, R. and Kaya, H. K., Eds., CRC Press, Boca Raton, FL, 1990, chap. 5.

22. **Georgis, R. and Hague, N. G. M.,** A neoaplectanid nematode in the web-spinning larch sawfly *Cephalcia iariciphila*, *Ann. Appl. Biol.*, 99, 171, 1981.

23. **Poinar, G. O., Jr.,** *Nematodes for Biological Control of Insects*, CRC Press, Boca Raton, FL, 1979.

24. **Wallace, H. R.,** *The Biology of Plant Parasitic Nematodes*, Edward Arnold, London, 1963.

25. **Kung, S., Gaugler, R., and Kaya, H. K.,** Soil type and entomopathogenic nematode persistence, *J. Invertebr. Pathol.*, 55, 401, 1990.

26. **Geden, C. J., Axtell, R. C., and Brooks, W. M.,** Susceptibility of the lesser mealworm, *Alphitobius diaperinus* (Coleoptera: Tenebrionidae) to the entomogenous nematodes *Steinernema feltiae*, *S. glaseri* (Steinernematidae) and *Heterorhabditis heliothidis* (Heterorhabditidae), *J. Entomol. Sci.*, 20, 331, 1985.

27. **Molyneux, A. S. and Bedding, R. A.,** Influence of soil texture and moisture on the infectivity of *Heterorhabditis* sp. D1 and *Steinernema glaseri* for larvae of the sheep blowfly, *Lucillia cuprina*, *Nematologica*, 30, 358, 1984.

28. **Georgis, R. and Poinar, G. O., Jr.,** Effect of soil texture on the distribution and infectivity of *Neoaplectana carpocapsae* (Nematoda: Steinernematidae), *J. Nematol.*, 15, 308, 1983.

29. **Georgis, R. and Poinar, G. O., Jr.,** Effect of soil texture on the distribution and infectivity of *Neoaplectana glaseri* (Nematoda: Steinernematidae), *J. Nematol.*, 15, 219, 1983.

30. **Georgis, R. and Poinar, G. O., Jr.,** Vertical migration of *Heterorhabditis bacteriophora* and *H. heliothidis* (Nematoda: Heterorhabditidae) in sandy loam soil, *J. Nematol.*, 15, 652, 1983.

31. **Choo, H. Y., Kaya, H. K., Burlando, T. M., and Gaugler, R.,** Entomopathogenic nematodes: host-finding ability in the presence of plant roots, *Environ. Entomol.*, 18, 1136, 1989.

32. **Jansson, R. K., LeCrone, S. H., Gaugler, R., and Smart, G. C., Jr.,** Potential of entomopathogenic nematodes as biological control agents of sweetpotato weevil (Coleoptera: Cruculionidae), *J. Econ. Entomol.*, 83, 1818, 1990.

33. **Georgis, R. and Gaugler, R.,** Predictability in biological control using entomopathogenic nematodes, *J. Econ. Entomol.*, 84, 713, 1991.

34. **Shetlar, D. J., Suleman, P. E., and Georgis, R.,** Irrigation and use of entomogenous nematodes *Neoaplectana* spp. and *Heterorhabditis heliothidis* (Rhabditida: Steinernematidae and Heterorhabditidae) for control of Japanese beetle (Coleoptera: Scarabaeidae) grubs in turfgrass, *J. Econ. Entomol.*, 81, 1318, 1988.

35. **Womersley, C. Z.,** Dehydration survival and anhydrobiotic potential, in *Entomopathogenic Nematodes in Biological Control*, Gaugler, R. and Kaya, H. K., Eds., CRC Press, Boca Raton, FL, 1990, chap. 6.

36. **Simons, W. R. and Poinar, G. O., Jr.,** The ability of *Neoaplectana carpocapsae*, (Steinernematidae: Nematoda) to survive extended periods of desiccation, *J. Invertebr. Pathol.*, 22, 228, 1973.

37. **Kung, S., Gaugler, R., and Kaya, H. K.,** Effects of soil temperature, moisture and relative humidity on entomopathogenic nematode persistence, *J. Invertebr. Pathol.*, 57, 242, 1991.

38. **Gaugler, R.,** Biological control potential of neoaplectanid nematodes, *J. Nematol.*, 13, 241, 1981.

39. **Kung, S. P., Gaugler, R., and Kaya, H. K.,** Influence of soil pH and oxygen on entomopathogenic nematode persistence, *J. Nematol.*, 21, 574, 1990.

40. **Kaya, H. K., Mannion, C., Burlando, T. M., and Nelsen, C. E.,** Escape of *Steinernema feltiae* from alginate capsules containing tomato seeds, *J. Nematol.*, 19, 287, 1987.

41. **Poinar, G. O., Jr.,** A microsporidian parasite of *Neoaplectana glaseri* (Steinernematidae: Rhabditida), *Rev. Nematol.*, 11, 359, 1988.

42. **Timper, P. and Kaya, H. K.,** Role of the second-stage cuticle of entomogenous nematodes in preventing infection by nematophagous fungi, *J. Invertebr. Pathol.*, 54, 314, 1989.

43. **Poinar, G. O., Jr. and Jansson, H. B.,** Infection of *Neoaplectana* spp. and *Heterorhabditis heliothidis* to the endoparasitic fungus *Drechmeria coniospora*, *J. Nematol.*, 18, 225, 1986.

44. **Poinar, G. O., Jr. and Jansson, H. B.,** Infection of *Neoaplectana* and *Heterorhabditis*

(Rhabditida: Nematoda) with the predatory fungi, *Monacrosporium elipsosporum* and *Arthrobotrys oligospora* (Moniliales: Deuteromycetes), *Rev. Nematol.*, 9, 241, 1986.

45. **Ishibashi, N., Young, F. Z., Nakashima, M., Abiru, C., and Haraguchi, N.,** Effects of application of DD-136 on silkworm, *Bombyx mori*, predatory insect, *Agriosphodorus dohrni*, parasitoid, *Trichomalus apanteloctenus*, soil mites, and other non-target soil arthropods, with brief notes on feeding behavior and predatory pressure of soil, tardigrades, and predatory nematodes on DD-136 nematodes, in *Recent Advances in Biological Control of Insect Pests by Entomogenous Nematodes in Japan*, Ishibashi, N., Ed., Ministry of Education, Japan, Grant No. 59860005, 1987, 158.

46. **Epsky, N. D., Walter, D. E., and Capinera, J. L.,** Potential role of nematophagous microarthropods as biotic mortality factors of entomogenous nematodes (Rhabditida: Steinernematidae, Heterorhabditidae), *J. Econ. Entomol.*, 81, 821, 1988.

47. **Friedman, M. J.,** Commercial production and development, in *Entomopathogenic Nematodes in Biological Control*, Gaugler, R. and Kaya, H. K., Eds., CRC Press, Boca Raton, FL, 1990, chap. 8.

48. **Georgis, R.,** Formulation and application technology, in *Entomopathogenic Nematodes in Biological Control*, Gaugler, R. and Kaya, H. K., Eds., CRC Press, Boca Raton, FL, 1990, chap. 9.

49. **Gaugler, R., McGuire, T., and Campbell, J. F.,** Genetic variability among strains of the entomopathogenic nematode *Steinernema feltiae*, *J. Nematol.*, 21, 247, 1989.

50. **Lewis, E., Gaugler, R., and Harrison, R.,** Entomopathogenic nematode host finding: relevance of host contact cues to cruise and ambust foragers, *Parasitology*, 105, 309, 1992.

51. **Akhurst, R. J.,** Controlling insects with entomopathogenic nematodes, in *Fundamental and Applied Aspects of Invertebrate Pathology*, Samson, R. A., Vlak, J. M., and Peters, D., Eds., Proc. 4th Int. Coll. Invertebr. Pathol., 1986, 265.

52. **Poinar, G., Jr., Thomas, G. M., and Lighthart, B.,** Bioassay to determine the effect of commercial preparations of *Bacillus thuringiensis* on entomogenous rhabditid nematodes, *Agric. Ecosys. Environ.*, 30, 195, 1990.

53. **Rovesti, L., Heinzpeter, E. W., Tagliente, F., and Deseo, K. V.,** Compatibility of pesticides with the entomopathogenic nematode *Heterorhabditis bacteriophora*, Poinar

(Nematoda: Heterorhabditidae), *Nematologica*, 34, 462, 1988.

54. **Rovesti, L., Fiorini, T., Bettini, G., Heinzpeter, E. W., and Tagliente, F.,** Compatibilita di *Steinernema* spp. e *Heterorhabditis* spp. con fitofarmaci, *Inform. Fito.*, 9, 55, 1990.

55. **Forschler, B. T., All, J. N., and Gardner, W. A.,** *Steinernema feltiae* activity and infectivity in response to herbicide exposure in aqueous and soil environment, *J. Invertebr. Pathol.*, 55, 375, 1990.

56. **Kaya, H. K. and Burlando, T. M.,** Infectivity of *Steinernema feltiae* in fenamiphostreated sand, *J. Nematol.*, 21, 434, 1989.

57. **Zimmerman, R. J. and Cranshaw, W. S.,** Compatibility of three entomogenous nematodes (Rhabditida) in aqueous solutions of pesticides used in turfgrass maintenance, *J. Econ. Entomol.*, 83, 97, 1990.

58. **Hara, A. J. and Kaya, H. K.,** Toxicity of selected organophosphate and carbamate pesticides to infective juveniles of the entomogenous nematode *Neoaplectana carpocapsae* (Rhabditida: Steinernematidae), *Environ. Entomol.*, 12, 496, 1983.

59. **Klein, M. G. and Georgis, R.,** Persistence of control of Japanese beetle (Coleoptera: Scarabaeidae) larvae with steinernematid and heterorhabditid nematodes, *J. Econ. Entomol.*, 85, 727, 1992.

60. **Georgis, R., Redmond, C. T., and Martin, W. R.,** *Steinernema* B-326 and B-319 (Nematoda): new biological soil insecticides, *Brighton Crop Protection Conference, Pests and Diseases*, 2, 73, 1992.

61. **Harris, M. A., Begley, J. W., and Warkentin, D. L.,** *Liriomyza trifolii* (Diptera: Agromyzidae) suppression with foliar applications of *Steinernema carpocapsae* (Rhabditida: Steinernematidae) and Abamectin, *J. Econ. Entomol.*, 83, 2380, 1990.

62. **Hollander, M. and Wolfe, O. A.,** *Nonparametric Statistical Methods*, John Wiley & Sons, New York, 1973.

63. **Hom, A. and Georgis, R.,** unpublished data, 1990.

64. **Parwinder, S. G., Gaugler, R., and Lewis, E. E.,** Host recognition behavior by entomopathogenic nematodes during contact with gut contents and its adaptive significance, *J. Parasitol.*, 79, 495, 1993.

65. **Parwinder, S. G., Gaugler, R., and Selvan, M. S.,** Host recognition by entomopathogenic nematodes: behavioral response to contact with host faeces, *J. Parasitol.*, 19, 1219, 1993.

Chapter 44

Biological Control for Plant-Parasitic Nematodes Attacking Turf and Ornamentals

George O. Poinar, Jr., *Department of Entomology, University of California/Berkeley*

Ramon Georgis, *biosys, Palo Alto, CA*

CONTENTS

Introduction 491
Types of Plant-Parasitic Nematodes 491
Nematode Genera Commonly Found on Turf and Ornamentals 492
    Parasites of Aerial Plant Parts 492
    Root Parasites 492
Plant Damage by Nematodes 492
Control Measures 494
Pathogen Groups Attacking Plant-Parasitic Nematodes on Turf and Ornamentals 494
    Nematophagous Fungi 495
        Predaceous Fungi 495
        Endoparasitic Fungi 495
        Other Fungi 496
    Bacteria 497
Other Biological Control Agents 497
Selection of Bio-Control Agents for Turf and Ornamental Nematodes 498
Availability of Bio-Control Agents 499
Future Goals 500
References 500

INTRODUCTION

There are many types of nematodes; some 20,000 have been described and some 480,000 await descriptions. Nematodes occur in all conceivable habitats (e.g., soil, plants, invertebrates, vertebrates, freshwater, saltwater, polar ice) and all nematodes obtain nourishment from living cells; thus, they are either predators or parasites. Those that feed on microorganisms are called microbotrophs and are usually free-living in soil or water. Those that feed on filamentous fungi and higher plants are known as plant parasites. Those that obtain nourishment from animals comprise the vertebrate and invertebrate parasites.

In the present work, we are concerned with plant-parasitic nematodes, especially those attacking turf and ornamentals. Known diseases of these nematodes will be discussed in light of establishing a biological control program which could be used to reduce populations of these nematodes.

TYPES OF PLANT-PARASITIC NEMATODES

Plant-parasitic nematodes all obtain their nourishment from multicellular plants, including algae, fungi, and vascular plants. Most of them are less than 1 mm in length and all possess a stylet or spear. This hard and usually hollow structure, located in the head of the nematode, is used to penetrate the plant cell wall and remove the plant cell sap, in a manner similar to the functioning of a hypodermic syringe.[1]

Most plant-parasitic nematodes attacking turf and ornamentals are ectoparasites, characterized by feeding on plant parts from outside the plant. Those nematodes which enter inside plant tissues are called endoparasites. In both groups, there are representatives which are sendentary (remaining in one locality on the plant throughout most of their life cycle) and migratory (moving from one feeding site to another). Most are root parasites but a few occur in stems, leaves, buds, and flowers of some plants.

The life cycle of most active plant-parasitic nematodes is relatively short (a month or less). However, under adverse weather conditions (winter, drought) many can become quiescent and survive in the soil or plant parts for a much longer period. Species of *Anguina* can survive as juveniles in dried grass florets for 28 years.[2]

In contrast to the short life span of most root parasites, representatives of the Longidoridae which contain the genera *Longidorus, Paratrichodorus, Trichodorus,* and *Xiphinema* on turf and ornamentals can remain actively feeding and reproducing for several years. All nematodes undergo four molts from egg to adult and during this time pass through four juvenile stages. For further information on the biology of plant parasitic nematodes, the books of Poinar[1] and Dropkin[3] can be consulted.

Figure 1 Leaf galls on grass caused by *Anguina* nematodes.

NEMATODE GENERA COMMONLY FOUND ON TURF AND ORNAMENTALS

PARASITES OF AERIAL PLANT PARTS

These nematodes cause leaf blotching and other deformities on plants. One of the most common aerial species attacking ornamentals is the chrysanthemum foliar nematode, *Aphelenchoides ritzemabosi*. These nematodes crawl up the stems and enter the leaves through the stomata. As they feed, the damaged mesophyll cells darken resulting in black leaf spots. The nematodes can remain viable inside fallen dried leaves for several years. Other nematodes which feed on the above portion of grass plants (Figure 1) belong to the genera *Anguina* and *Paranguina*. These are the seed and leaf gall nematodes which are often highly specialized in their selection of plant host and plant part attacked. They form galls on stems, flower heads, and roots of a number of turfgrasses.

ROOT PARASITES

Most nematodes attacking turf and ornaments feed on the roots. As mentioned earlier, root parasites include those which feed from the outside on the outer root cells (ectoparasites), those which enter the roots and feed from the inside (endoparasites), those which move constantly from one site to another (migratory), and those which remain and feed from one location (sedentary). All combinations of the above are possible. As can be seen from Table 1, root ectoparasites predominate. The most commonly encountered forms include representatives

of the genera *Tylenchorhynchus* (stylet nematodes), *Hoplolaimus* (lance nematodes), *Belonolaimus* (sting nematodes), *Trichodorus* (stubby-root nematodes), *Longidorus* (needle nematodes), *Paratylenchus* (pin nematodes), *Xiphinema* (daggar nematodes), *Helicotylenchus* (spiral nematodes), and *Criconemella* (ring nematodes), which are mostly migratory ectoparasites. The eggs of these forms are deposited in the soil, and the juveniles hatch and begin feeding on root hairs and rootlets, moving from location to location as they grow and enter the adult stage. Damage consists of injury to the roots, resulting in less hardy plants which are susceptible to secondary pathogens.

Lesion nematodes of the genus *Pratylenchus* are migratory root endoparasites which move about in the roots, feeding at various locations. Members of this genus are found on a large number of host plants, including turf and ornamentals.

Sedentary root parasites of turf and ornamentals include representatives of the genera *Meloidogyne* (root knot) and *Heterodera* (cyst nematodes). In these forms the juveniles remain in one location in the plant root as they mature, eventually swelling up into sedentary sac-like bodies. Although these two genera represent the most economically important of all plant parasitic nematodes, relatively few species occur on grasses; a preference for the dicots seems to exist.

PLANT DAMAGE BY NEMATODES

Estimates indicate that about 7 to 15% of agricultural crops in the U.S. are destroyed annually by

Table 1 **Commonly used turfgrasses, their nematode parasites, and location of damage**[4,22]

Grass	Nematode Genera	Location of Damage
Agrostis spp.	*Anguina*	Seed, leaf, and stem galls
(Bent grasses)	*Ditylenchus*	Leaf galls
	Longidorus	Root ectoparasites
	Meloidogyne	Root knot
	Paratrichodorus	Root ectoparasites
	Pratylenchus	Ecto- and endoparasites
	Subanguina	Crown and root galls
Cynodon spp.	*Anguina*	Seed, leaf, and stem galls
(Bermuda and	*Belonolaimus*	Root ectoparasites
Bradley grass)	*Criconemella*	Root ectoparasites
	Dolichodorus	Root ectoparasites
	Hoplolaimus	Root ectoparasites
	Meloidogyne	Root knot
	Paratrichodorus	Root ectoparasites
	Rotylenchus	Root ecto- and endoparasites
	Trichodorus	Root knot
Eremochloa ophiuroides	*Criconemella*	Root ectoparasites
(Munro)	*Pratylenchus*	Root ecto- and endoparasites
(Centipede grass)	*Trichodorus*	Root ectoparasites
Festuca spp. (Fescue)	*Anguina*	Seed and leaf galls
	Helicotylenchus	Root ecto- and endoparasites
	Heterodera	Cyst nematodes
	Meloidogyne	Root knot
	Pratylenchus	Root endoparasites
	Trichodorus	Root ectoparasites
Lolium spp.	*Longidorus*	Root ectoparasites
(Ryegrass)	*Meloidogyne*	Root knot
	Pratylenchus	Root endoparasites
	Subanguina	Crown and root galls
	Trichodorus	Root ectoparasites
	Xiphinema	Root ectoparasites
Poa spp.	*Anguina*	Seed and leaf galls
(Meadow grass)	*Criconemella*	Root ectoparasites
	Ditylenchus	Root galls
	Helicotylenchus	Root ectoparasites
	Heterodera	Cyst nematodes
	Meloidogyne	Root knot
	Longidorus	Root ectoparasites
	Paratylenchus	Root ectoparasites
	Rotylenchus	Root ectoparasites
	Trichodorus	Root ectoparasites
	Xiphinema	Root ectoparasites
Stenotaphrum secundatum	*Belonolaimus*	Root ectoparasites
(Walt)	*Criconemella*	Root knot
(St. Augustinegrass)	*Ditylenchus*	Root ecto- and endoparasites
	Hoplolaimus	Root ecto- and endoparasites
	Meloidogyne	Root knot
	Pratylenchus	Root endoparasites
	Rotylenchus	Root ecto- and endoparasites
	Trichodorus	Root ectoparasites
Zoysia spp.	*Criconemella*	Root ectoparasites
(Zoysia grass)	*Meloidogyne*	Root knot
	Pratylenchus	Root ecto- and endoparasites
	Trichodorus	Root ectoparasites

Table 2   **Genera of plant-parasitic nematodes reported from ornamentals, the location of damage, and representative plant hosts**[23]

Nematode	Damage	Representative Plant Hosts
Aphelenchoides spp.	Buds and leaves	Chrysanthemum, ferns, begonia, aster, dahlia, gloxinia, delphinium, phlox, verbena, zinnia, African violet
Belonolaimus spp.	Root ectoparasite	Chrysanthemum
Criconemoides spp.	Root ectoparasite	Carnation
Ditylenchus spp.	Leaf galls	Tulips, narcissus, iris
Helicotylenchus spp.	Root ectoparasite	Azalea, rhododendron
Heterodera spp.	Cyst nematode	Cacti
Longidorus spp.	Root ectoparasite	Rose
Meloidogyne spp.	Root knot	Chrysanthemum, carnation, cacti
Pratylenchus spp.	Root lesion	Rose, ferns
Trichodorus spp.	Root ectoparasite	Chrysanthemum, azalea, rhododendron
Tylenchorhynchus spp.	Root ectoparasite	Azalea, rhododendron
Xiphinema spp.	Root ectoparasite	Rose

nematodes, totaling a loss of $4 billion in 1976.[1] Undoubtedly this cost is higher today, and it is impossible to say whether that percentage has remained constant or whether it still applies today for turf and ornamental pests.

Nematodes have been recognized as important pests of turfgrasses and can cause considerable damage to golf course greens, lawns, and turf nurseries.[3] Most of the nematode parasites of commonly occurring turfgrasses are root ectoparasites; however, some are endoparasites and a few cause galls on the inflorescences, leaves, and roots (Table 1). Ornamental plants, defined here as any plant that is used to ornament homes, gardens, parks, etc. also are attacked by both ectoparasites and endoparasites (Table 2). (Normally, neither the plant nor its parts or product is eaten, but this is not always the case.) Ornamentals as a group include flowers, shrubs, and trees and are a bit subjective, since what one person calls an ornamental may not be tolerated by another. A list of the commonly occurring nematode genera found on ornamentals is presented in Table 2.

Nematode infestations by root ectoparasites result in general plant symptoms associated with root damage. In most cases it is not possible to determine the cause of these symptoms without examining the roots or surrounding soil for pathogens. Methods for extracting nematodes from soil are summarized elsewhere.[1] In some instances, nematode cysts or root galls can be spotted, giving an indication that the aboveground symptoms are the result of nematodes. Such symptoms include poor growth, chlorosis, wilting, dieback, and discoloration (bronzing, reddening, blackening) of plant parts.

CONTROL MEASURES

Although many types of control measures for plant parasitic nematodes have been proposed,[5] nematocides are by far the most common control used today. Even in 1970, over $60 million was spent for nematocides to treat some 1.7 million acres, and in 1989 over 1 billion pounds of pesticides were applied to the environment in the U.S. alone.[6] More nematocides were probably used for controlling nematode pests of ornamentals than of turf. Nemacure is used in California to control *Anguina radicola* on annual bluegrass (*Poa annua*) in golf greens. Other chemicals that have been used for the control of turf nematodes include Phenamiphos, DBCP, Dasanit, Metham-sodium, Carbofuran, Aldicarb, Fensulfothion, isazophos (Triumph), ethoprop (Mocap), and Phenamiphos. Environmental concern dictates the need for safe, specific biological control programs for nematode parasites.

PATHOGEN GROUPS ATTACKING PLANT-PARASITIC NEMATODES ON TURF AND ORNAMENTALS

Pathogens of plant-parasitic nematodes include viruses, protozoa, fungi (Figure 2), and bacteria. Representatives of the former two groups are obligate parasites (cannot be grown on artificial media) which are not practical at this time for controlling plant parasitic nematodes. However, it is likely that virus and protozoan diseases of plant-parasitic nematodes do exist and could be used in control programs. Very little time and effort has been involved in searching for pathogens of these groups.

Figure 2 Nematode attacked by fungus.

By far, most of the pathogens which show promise for nematode control belong to the fungi and one genus of bacteria.

NEMATOPHAGOUS FUNGI

Nematophagous fungi fall into two broad groups. The first group are commonly called predaceous or nematode trapping fungi and are characterized by an extensive hyphal network in the environment. These fungi produce trapping devices along the hyphae which serve to catch and hold nematodes. The prey is then invaded by hyphae which penetrate through the nematode's body wall and enter its tissues. The second group, often called endoparasites, produce small conidia or zoospores but very little or no mycelium in the environment. The spores arise from conidiophores emerging from diseased nematodes and infect new prey directly through the body surface or by ingestion. Reviews on nematophagous fungi have been prepared by Gray,[7] Morgan-Jones and Rodriguez-Kabana,[8] Jansson and Nordling-Hertz,[9] Stirling,[10] and Alam.[11]

Predaceous Fungi

All predaceous fungi produce some type of trap for catching and infecting nematodes. The traps are usually fixed in position and consist of adhesive hyphae, adhesive nets, adhesive knobs, or constricting and nonconstricting rings (Figures 3 to 6). Sometimes the adhesive portion breaks off and is carried some distance by the nematode before infection occurs, but usually the nematode is kept stationary. Once a nematode is caught, an infection hypha grows out from the adhesive body, penetrates the host's cuticle, forms assimilative hyphae, and absorbs the nematode's body contents.

Conidiophores bearing conidia or chlamydospores are formed on the surface or inside the body of the diseased nematode. A list of the commonly occurring genera of nematophagous predaceous fungi are presented in Table 3. Those with adhesive structures normally attack mobile vermiform nematodes, while those that infect with normal hyphae usually attack stationary eggs or cysts.

Endoparasitic Fungi

Most of the nematophagous fungi in this group infect nematodes with conidia which rest in the soil. Hyphae are produced only inside the host's body, and only evacuation tubes or conidiophores and conidia are produced externally. Infection can occur per os (spores are ingested) or through the

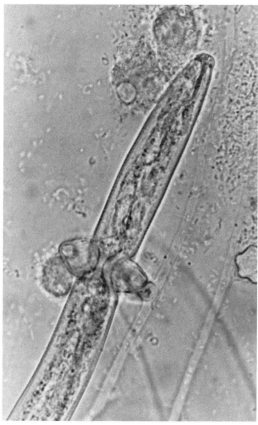

Figure 3 Nematode caught in vise-like grip of a ring trap of *Arthrobotrys anchonia*. *(Photo courtesy of R. Mankau.)*

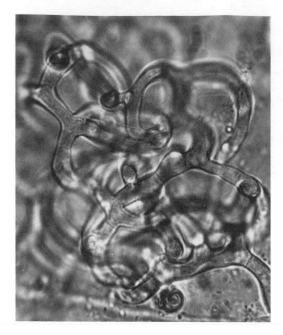

Figure 4 Net-like trap of the predatory fungus, *Arthrobotrys cnoides. (Photo courtesy of R. Mankau.)*

body wall (spores are adhesive or encyst on the nematode cuticle) (Figure 7). Since the conidia of endoparasites do not require large food reserves for the production of vegetative hyphae, they are usually smaller in size than those of predaceous fungi. Endoparasitic fungi can be grouped according to the nature of the conidia. Those which encyst on the nematode's cuticle include members of the Chytridiomycetes and Oomycetes (Zygomycetes), while adhesive and ingested conidia are produced by Deuteromycetes and Basidiomycetes. A list of commonly occurring genera of nematophagous endoparasitic fungi is presented in Table 4. It may be noted that some genera contain species with different life cycles and are therefore included as predators and endoparasites.

Other Fungi

There are many other soil fungi which appear to have antagonistic effects on plant parasitic nematodes. One of these is the collagenolytic fungus *Cunninghamella elegans*. This and other such fungi are able to digest collagen as a single source of carbon and nitrogen. The fungi produce chitanese and collagenase, and these enzymes have a deleterious effect on eggs and second stage juveniles of root knot nematodes (*Meloidogyne javanica*), and reduce the motility of *Rotylenchulus reniformis* and *Xiphinema index*.[12]

One potential problem with the use of nematophagous fungi is the application of fungicides for

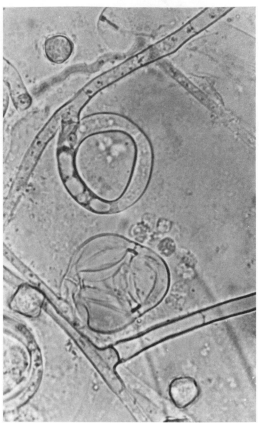

Figure 5 Triggered ring trap of *Arthrobotrys dactyloides. (Photo courtesy of R. Mankau.)*

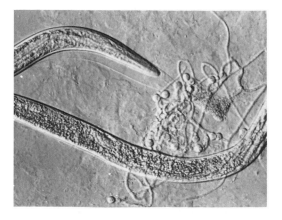

Figure 6 Sticky knobs of *Monacrosporium ellipsosporum* hold these nematodes prior to fungal penetration.

the control of plant parasitic fungi. It has been shown that the fungicide benomyl stopped the growth of *Verticillium chlamydosporium* and adversely affected several other nematophagous fungi.[13]

Table 3 **Some common genera of nematophagous predatory fungi, their infectious units and important plant-parasitic genera attacked**

Genera[a]	Infectious Units	Important Plant-Parasitic Nematode Genera Attacked[b]
Arthrobotrys (D)	Adhesive hyphae, adhesive nets, constricting rings	*Meloidogyne, Heterodera, Ditylenchus*
Cylindrocarpon (D)	Hyphae	*Heterodera*
Cystopage (Z)	Adhesive hyphae	?
Dactylaria (D)	Adhesive hyphae, adhesive knobs, constricting rings, nonconstricting rings	*Heterodera*
Dactylella (D)	Adhesive bunches, adhesive knobs, constricting rings	*Meloidogyne, Heterodera*
Exophiala (D)	Hyphae	*Heterodera*
Fusarium (D)	Hyphae	*Heterodera, Meloidogyne*
Gliocladium (D)	Hyphae	*Heterodera, Meloidogyne*
Hirsutella (D)	Hyphae	*Heterodera, Criconemella*
Monacrosporium (D)	Adhesive knobs	*Meloidogyne, Heterodera*
Nematoctonus (B)	Adhesive knobs	*Heterodera*
Nematophthora (B)	Hyphae	*Heterodera*
Paecilomyces (D)	Hyphae	*Heterodera, Meloidogyne*
Phoma (B)	Hyphae	*Heterodera*
Stylopage (Z)	Adhesive hyphae	?
Verticillium (D)	Hyphae	*Heterodera*

[a] Z = Zygomycetes; D = Deuteromycetes; B = Basidiomycetes; [b] ? = Host nematode genera not documented.

Figure 7 Spores of the endoparasitic fungus *Meria coniospora* attached to the head of a nematode prior to fungal penetration.

BACTERIA

One very unusual bacterial genus attacking nematodes is *Pasteuria*. The spores of the *Pasteuria penetrans* group are sticky and adhere to the cuticle of susceptible nematodes as they move through the soil (Figure 8). These endospores germinate by forming a germ tube which penetrates the nematode cuticle. In the host's body cavity the germ tube develops into a septate mycelia-like body which eventually breaks up into sporangia. Each spo-

rangium yields a single endospore (Figure 9). The diseased nematode dies and liberates the spores into the soil. A number of nematodes belonging to different groups (plant-parasites, microbotrophic free-living soil forms, predaceous forms) have been reported as hosts of *Pasteuria* bacteria. However, much interest has centered on the species recovered from diseased root knot (*Meloidogyne* spp.) nematodes.[14] Aside from *Meloidogyne* spp., other genera of plant parasitic nematodes found on turf and ornamentals which are susceptible to *Pasteuria* include *Belonolaimus, Criconemella, Ditylenchus, Dolichodorus, Helicotylenchus, Heterodera, Hoplolaimus, Longidorus, Paralongidorus, Pratylenchus, Rotylenchus, Trichodorus, Tylenchorhynchus, Tylenchulus,* and *Xiphinema*.

OTHER BIOLOGICAL CONTROL AGENTS

Aside from pathogens, there are other candidates for the biological control of plant-parasitic nematodes. One of these is invertebrate predators, some of which are listed in Table 5. As is the case with predators, they are nonspecific and attack other invertebrates in soil besides nematodes. Also, when attacking nematodes, they do not select just plant parasites but other soil dwelling nematodes as well. Promising candidates include predatory nematodes

Table 4 **Genera of nematophagous endoparasitic fungi, type of spores produced and plant-parasitic nematode genera attacked**

Genus[a]	Type of Spore	Plant-Parasitic Genera Attacked
Acrostalagmus (D)	Adhesive conidia	?[b]
Catenaria (Z)	Encysting zoospores	*Xiphinema, Heterodera*
Cephalosporium (D)	Adhesive conidia	?
Harposporium (D)	Ingested conidia	?
Meria (D)	Adhesive conidia	*Meloidogyne*
Myzocytium (Z)	Encysting zoospores	*Xiphinema*
Nematoctonus (B)	Adhesive conidia	?
Spicaria (D)	Adhesive conidia	?
Verticillium (D)	Adhesive conidia	?

[a] D = Deuteromycetes; B = Basidiomycetes; Z = Zygomycetes;
[b] ? = Host nematode genera not documented.

which belong to several groups. One group includes the mononchoids which swallow smaller nematodes (Figure 10). Another group includes dorylaimoids which puncture their prey with a stylet and suck out the body contents (Figure 11).

Antagonistic plants which produce products which serve as naturally occurring nematicides are not uncommon,[15] especially in representatives of some of the tribes of Compositae. There are examples of protecting desirable plants from nematodes by planting them next to marigolds. Compounds given off by marigold roots will repel some plant parasitic nematodes. Extracts of the Indian Neem tree (*Azadirachta indica*: Meliaceae) are nematicidal. When these plants are grown together with nematode-infested tomato and eggplants, the naturally produced and liberated extract can protect the food plants from nematodes (*Meloidogyne* and *Rotylenchus*). Neem extracts have also been used for root dip and seed dressing treatments of nematode-susceptible crops with favorable results.[16]

Some interesting results have been obtained by applying one species of nematode to the soil in order to reduce the populations of a pest plant-parasitic nematode. Thus, when a million *Aphelenchus avenae* were applied to soil 3 days before adding *Meloidogyne* juveniles, the number of nematode-induced galls was reduced by 50%. It doesn't seem to matter what nematode is applied since galls were reduced when the entomopathogenic nematode, *Steinernema carpocapsae*, was also applied.[17] It is possible that the increase in nematode populations stimulates the growth of microorganisms which have deleterious effects on all nematodes. Further research is needed to explain these results.

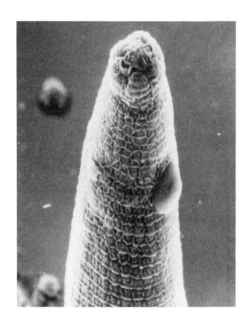

Figure 8 Spore of the bacterium *Pasteuria penetrans* attached to the cuticle of a nematode. *(Photo courtesy of R. M. Sayre.)*

SELECTION OF BIO-CONTROL AGENTS FOR TURF AND ORNAMENTAL NEMATODES

The selection of a nematode bio-control agent depends on many factors, both theoretical and practical. The selected agents should be effective in reducing plant-parasitic nematodes below the level at which plant damage occurs. The selected agent should be readily available. If a living agent, it should be readily culturable with the production of infectious units, whether those be spores or mycelia (with pathogens) or multicellular units (in the case of predators). These units should be storable, shipable, and capable of being applied to the environment. The lists of pathogens and predators cited in Tables 3 through 5 will be considerably reduced when we eliminate those that don't meet the above criteria. Even if they are effective in killing plant parasitic nematodes, if they can't be cultured, stored, applied, or persist in a viable form, they cannot be considered.

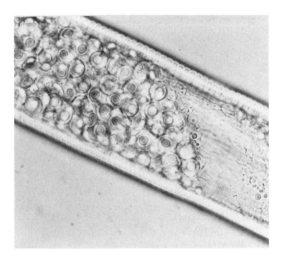

Figure 9 Developing spores of the bacterium *Pasteuria penetrans* inside the body cavity of a nematode. *(Photo courtesy of R. M. Sayre.)*

In reviewing the nematophagous fungi, most of the obligate pathogens will be ruled out because of the difficulty of culturing them. This leaves the predatory types which can be effective and thus do have potential as nematode bio-control agents. However very little field data exists regarding their effectiveness against nematode parasites of turf and ornamentals.

Once a predatory fungus has been found which is culturable and infective for the target nematode, then greenhouse and field experiments should be conducted, together with controls, followed by an attempt to re-isolate diseased nematodes and the fungus from treated soils. Other considerations that should be taken into account when considering biological control agents of nematodes are discussed by Kerry.[18]

AVAILABILITY OF NEMATODE BIO-CONTROL AGENTS

At this time, there are no commercially available nematode biological control agents produced in North America. Likely candidates which may soon be available are predatory fungi belonging to the genera *Verticillium, Monacrosporium Nematophthora, Hirsutella, Paecilomyces,* and *Arthrobotrys* and bacteria belonging to the *Pasteuria penetrans* complex. Two commercial preparations of *Arthrobotrys* have been available in France. These preparations (called "Royal 300" and "Royal 350") were used

Table 5 Representative invertebrate predators and plant-parasitic nematode genera attacked

Predator Group	Type of Predator	Plant-Parasitic Nematode Genera Attacked
Insects (Insecta)	Spring tails (Collembola)	*Tylenchulus, Tylenchorhynchus, Meloidogyne*
	Rove beetle (Staphylinidae)	*Heterodera* cysts
Mites (Acari)	*Pergamascus*	*Heterodera* cysts
	Gaeolaelaps	*Xiphinema*
	Hypoaspis	*Meloidogyne, Heterodera*
Nematodes (Nematoda)	Dorylaims (*Actinolaimus, Dorylaimus, Labronema*)	*Meloidogyne, Heterodera, Anguina*
	Mononchids (*Mononchus, Clarkus*)	*Heterodera, Anguina*
	Seinura	*Ditylenchus*
	Diplogasterids (*Butlerius, Mononchoides, Mylonchulus*)	*Hoplolaimus, Meloidogyne, Tylenchulus, Pratylenchus Tylenchorhynchus, Xiphinema, Hoplolaimus*

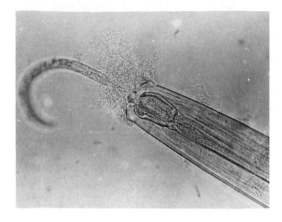

Figure 10 A predatory mononchoid nematode attacking a smaller nematode. *(Photo courtesy of R. Mankow.)*

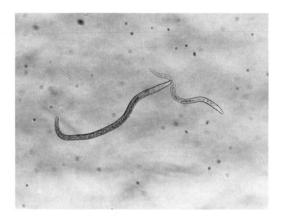

Figure 11 A predatory dorylaimoid nematode (*Thornia* sp.) attacking a smaller plant-parasitic nematode. (*Photo courtesy of R. Mankau.*)

for controlling *Ditylenchus myceliophagus* on mushrooms and root knot (*Meloidogyne* spp.) juveniles on tomatoes and other vegetables.[19,20] A product containing *Paecilomyces lilacinus* is being commercially marketed under the name of Biocon in the Philippines. This fungus, produced by Asiatic Technologies, Inc. in Manila, is sold in 10-g foil pouches. Each pouch contains about 8 billion spores of *P. lilacinus*, sells for 150 pesos (about $7.50) and covers a half hectare. The product can be used as a root drench or mixed with organic fertilizers which are then applied to the field.[21]

FUTURE GOALS

Nematode biological control agents are badly needed and will be used once effective products are commercially available. The phasing out of chemical nematicides in attempts to improve the environment will continue, and replacements are necessary.

Nematode bio-control agents can be used by themselves if effective. However, until a wide selection of agents is available, it may be necessary to use them in an integrated control program. The goal should be to produce a range of fungi or bacterial pathogens which will be effective over a range of temperatures and soil types against several genera of plant parasitic nematodes.

The need for biological control alternatives for nematode parasites of turf and ornamentals is especially desirous since applications of chemical nematicides against nematode parasites of these plants almost always exposes homeowners, their families, and pets to potentially dangerous chemicals. Let us hope that funds, researchers, and industry will assist in the production of safe, effective nematode biological agents in the near future.

REFERENCES

1. **Poinar, G. O., Jr.,** *The Natural History of Nematodes*, Prentice-Hall, Englewood Cliffs, NJ, 1983.
2. **Fillding, M. J.,** Observations on the length of dormancy in certain plant infecting nematodes, *Proc. Helminth. Soc. Wash.*, 18, 110, 1950.
3. **Dropkin, V. H.,** *Introduction to Plant Nematology*, John Wiley & Sons, New York, 1980.
4. **Eriksson, K. B.,** Nematode diseases of pasture legumes and turfgrasses, in *Economic Nematology*, Webster, J. M., Ed., Academic Press, London, 1972, chap. 4.
5. **Mai, W. F.,** *Control of Plant-Parasitic Nematodes*, Publication 1696, National Academy of Sciences, Washington, D.C., 1968.
6. **Ferris, H.,** Biological approaches to the management of plant parasitic nematodes, in *Beyond Pesticides*, Division of Agriculture and Natural Resources Publications, University of California, Oakland, 1992, 68.
7. **Gray, N. F.,** Fungi attacking vermiform nematodes, in *Diseases of Nematodes,* Vol. 2, Poinar, G. O., Jr. and Jansson, H.-B., Eds., CRC Press, Boca Raton, FL, 1988, chap. 1.
8. **Morgan-Jones, G. and Rodriguez-Kabana, R.,** Fungi colonizing cysts and eggs, in *Diseases of Nematodes*, Vol. 2, Poinar, G. O., Jr. and Jansson, H.-B., Eds., CRC Press, Boca Raton, FL, 1988, chap. 2.

9. **Jansson, H.-B. and Nordbring-Hertz, B.,** Infection events in the fungus-nematode system, in *Diseases of Nematodes*, Vol. 2, Poinar, G. O., Jr. and Jansson, H.-B., Eds., CRC Press, Boca Raton, FL, 1988, chap. 3.

10. **Stirling, G. R.,** Biological control of plant-parasitic nematodes, in *Diseases of Nematodes*, Vol. 2, Poinar, G. O., Jr. and Jansson, H.-B., Eds., CRC Press, Boca Raton, FL, 1988, chap. 5.

11. **Alam, M. M.,** Nematode destroying fungi, in *Nematode Bio-Control*, Jairajpuri, M. M., Alam, M. M., and Ahmad, I., Eds., CBS Publishers, Delhi, India, 1990, chap. 6.

12. **Galper, S., Cohen, E., Spiegel, Y., and Chet, I.,** A collagenolytic fungus, *Cunninghamella elegans* for biological control of plant-parasitic nematodes,. *J. Nematol.*, 23, 269, 1991.

13. **Meyer, S. L. F., Sayre, R. M., and Huettel, R. N.,** Benomyl tolerance of ten fungi antagonistic to plant parasitic nematodes, *J. Nematol.*, 23, 402, 1991.

14. **Sayre, R. M. and Starr, M. P.,** Bacterial diseases and antagonists of nematodes, in *Diseases of Nematodes*, Vol. 2, Poinar, G. O., Jr. and Jansson, H.-B., Eds., CRC Press, Boca Raton, FL, 1988, chap. 5.

15. **Gommers, F. J. and Bakker, J.,** Physiological diseases induced by plant responses or products, in *Diseases of Nematodes*, Vol. 1, Poinar, G. O., Jr. and Jansson, H.-B., Eds., CRC Press, Boca Raton, FL, 1988, chap. 1.

16. **Alam, M. M.,** Neem in nematode control, in *Nematode Bio-Control*, Jairajpuri, M. M., Alam, M. M., and Ahmad, I., Eds., CBS Publishers, Delhi, India, 1990, chap. 5.

17. **Ishibashi, N. and Choi, D.-R.,** Biological control of soil pests by mixed application of entomopathogenic and fungivorous nematodes, *J. Nematol.*, 23, 175, 1991.

18. **Kerry, B. R.,** An assessment of progress toward microbial control of plant-parasitic nematodes, *J. Nematol. (Suppl.)*, 22, 621, 1990.

19. **Cayrol, J. C. and Frankowski, J. P.,** Une methode de lutte biologique contre les nematodes a galles des racines appartenant au genre *Meloidogyne, Pepinieristes, Horticulteurs, Maraichers-Revue Horticole,* 193, 15, 1979.

20. **Cayrol, J. C., Frankowski, J. P., Laniece, A., d'Hardemare, G., and Talon, J. P.,** Contre les nematodes en champignoniere. Mise au point d'une methode de lutte biologique a l'aide d'un Hyphomycete predateur: *Arthrobotyrs robusta* Souche 'antipolis' (Royal 300), *Pepinieristes, Horticulteurs, Maraichers-Revue Horticole,* 184, 23, 1978.

21. **Timm, M.,** 'Biocon' controls nematodes biologically, *Bio/Technology,* 5, 772, 1986.

22. **Griffin, G. D.,** Nematode parasites of alfalfa, cereals and grasses, in *Plant and Insect Nematodes*, Nickle, W. R., Ed., Marcel Dekker, New York, 1984, chap. 8.

23. **Hague, N. G. M.,** Nematode diseases of flower bulbs, glasshouse crops and ornamentals, in *Economic Nematology*, Webster, J. M., Ed., Academic Press, London 1972, chap. 17.

24. **Small, R. W.,** Invertebrate predators, in *Diseases of Nematodes* Vol. 2, Poinar, G. O., Jr. and Jansson, H.-B., Eds., CRC Press, Boca Raton, FL, 1988, chap. 4.

Chapter 45

Microbial Control of Insect Pests of Landscape Plants

Whitney S. Cranshaw, Department of Entomology, Colorado State University, Fort Collins, CO

Michael G. Klein, Horticultural Insects Research Laboratory, USDA-Agricultural Research Service, Wooster, OH

CONTENTS

Strategies for Use of Microbes for Control of Landscape Insect Pests 504
Bacteria 505
    Bacillus thuringiensis 505
        Advantages of Bt Insecticides...506
        Limitations of Bt Insecticides 507
    Milky Diseases 508
    Serratia spp. 509
Viruses 509
    Gypsy Moth Baculovirus 510
    Sawfly Baculovirus 511
Protozoa 511
    Nosema locustae 512
    Nosema fumiferanae 512
Fungi 513
    Beauveria bassiana 513
        Use of *B. bassiana* for Control of Landscape Insect Pests 514
    Metarhizium anisopliae 515
        Use of *M. anisopliae* for Control of Landscape Insect Pests 515
    Verticillium lecanii 515
Acknowledgments 516
References 516

Naturally occurring insect disease organisms (entomopathogens) provide a potentially rich source of control options for use in plant protection for ornamental and landscape plants. Diseases are widespread in the insect world, involving various bacteria, viruses, fungi, and protozoa. They can effect spectacular changes in insect populations during outbreaks (epizootics), and more often occur at chronic levels with more subtle effects in population regulation. As stated in one of the first comprehensive reviews of insect pathology,[1] these microogranisms should not be envisioned as a panacea for insect control, yet they show potential, and in some cases proven, utility as pest control options.

Among the more obvious advantages of entomopathogens for ornamental plant protection is their specificity of effects. None of the insect pathogens considered for development as applied controls has been demonstrated to have serious effects on humans, mammals, or other vertebrates.[2-4] Furthermore, most show activity that is limited to various insect taxa, minimizing deleterious effects on beneficial insects also involved in natural control of pest species. As such, microbial controls are widely perceived as being particularly safe to use, even though the underlying assumption that these organisms are without harm is based on circumstantial evidence for several entomopathogens that have not been rigorously tested.[3] These safety features are of especial benefit in turfgrass and ornamental plant protection, where high human traffic around treated sites and applicator exposure can create substantial risks from pesticide exposure.

Microbial insecticides kill insects by a different mode of action than do most synthetically derived insecticides; this allows them to be considered for applications where resistant pest strains have developed. Pesticide resistance is an acute concern among many pest species affecting ornamental (particularly greenhouse) and certain agricultural commodities which are intensively treated with pesticides. Although resistance to microbial insecticides can occur, and has already with *Bacillus thuringiensis* in vegetable crops,[5] the typically limited use of microbial insecticides in ornamental settings is unlikely to provide sufficient selective pressure to rapidly induce resistant strains.

Many microbial insecticides reproduce in susceptible hosts and are capable of spread. This can allow secondary spread of the disease-producing organisms following the original application, providing more persistent control effects.

Despite the long recognition of insect diseases as important natural controls, successful adaptation of them for use in applied control programs has been relatively modest. Although there are some successful uses of microbes for insect control (e.g., *Bacillus thuringiensis*, *Bacillus popilliae*), it has been difficult to bring marketable products to pest managers. This fact has its roots in many causes. Most of the microorganisms (i.e., viruses, bacteria, protozoa) must be ingested to cause infection, and do not exhibit the contact activity common to most synthetically derived insecticides. They require thorough application coverage, which causes difficulty in getting sufficient exposure by susceptible insects, particularly for species that tunnel into plant parts.

Production and distribution difficulties also are obstacles common to many biological control organisms. Many insect pathogens are obligate parasites, requiring a living host for reproduction. Perishability (e.g., "shelf life") is also a concern. Viable, infective organisms are required to produce most insect diseases, and viability can be lost during extended storage or poor handling. In addition, pathogenicity can be lost by attenuation during the production process.

Activity of microbial agents as pest control agents also differs from currently used insecticides. Infection of insects requires days, sometimes weeks, to develop symptoms, and noticeable effects on insect suppression are delayed. Microbial agents also may be more sensitive to environmental factors, such as ultraviolet radiation or moisture, which can decrease persistence or infectivity. As such their application can require substantially more sophisticated techniques for effective use.

Finally, microbial insecticides, unlike nematodes or "higher" organisms such as predatory or parasitic insects (e.g., lady beetles, lacewings), are regulated as pesticides by the U.S. Environmental Protection Agency (EPA). As such, development and registration costs can be substantial. Since most insect pathogens only can infect a small number of pest species, this limits potential markets, which in turn reduces profitability.

STRATEGIES FOR USE OF MICROBES FOR CONTROL OF LANDSCAPE INSECT PESTS

Microbes can be used in a variety of pest management strategies as biological control agents, based on their epidemiological characteristics. Most involve either introduction and colonization of the pathogen at a site for long-term or permanent establishment; supplemental augmentation of existing pathogens to provide higher levels of insect control; mass production and release for immediate control; or conservation and enhancement of existing natural controls through manipulation of the environment.[6,7]

The original use of the Japanese beetle (*Popillia japonica*) strain of *Bacillus popilliae* is an example of introducing an insect pathogen for permanent establishment, as an inoculative release. Through various state, federal, and now private efforts, the milky spore pathogen was produced and distributed throughout the northeastern and mid-Atlantic states, where it has subsequently reproduced and spread. Control was not expected in the year of release, but establishment has helped suppress pest populations for several years, sometimes with permanent effects.

In contrast, use of the *Bacillus thuringiensis* insecticides is designed to provide immediate suppression of a pest population. They normally do not reproduce within the affected insect, and direct effects are short-lived. Indeed, some current commercial formulations are now merely derived from *Bacillus thuringiensis* products, notably the active toxins that the bacteria produces, and living organisms are not applied. Use characteristics closely resemble those of standard insecticides.

Somewhat intermediate are those pathogens that are endemic among insect populations, but not at levels which provide adequate control. For example, the gypsy moth (*Lymantria dispar*) baculovirus and the muscardine fungi are widespread among populations of their insect hosts. However, to be more useful in applied pest management, formulations are produced and distributed to supplement the existing levels of the disease. These may then also spread and provide long-term pest suppression benefits.

The area which has received the least attention in ornamental plant protection, as well as in agriculture, is manipulations which conserve or enhance naturally occurring levels of an insect pathogen.[8] Environmental modifications, such as altering humidity to induce epizootics of fungi, have been utilized in greenhouse operations using microbial control agents,[9] and watering is very important in successful use of insect parasitic nematodes, discussed elsewhere in this publication. There are no examples, however, where plant protection programs for landscape plants have been developed that consciously augment existing pathogens through environmental manipulations.

This chapter will discuss the microbial agents which have received most research and development as microbial controls for turfgrass and ornamental insects. This review is by no means exhaustive for the subject of potential entomopathogens. Many other organisms exist which exert some level of biological suppression, even if they do not lend themselves to production as applied microbial control agents. Readers interested in further information on these and other microorganisms impacting insect and mite pest populations can find additional information among the numerous references on this subject, many of which are cited at the end of the chapter.

BACTERIA

BACILLUS THURINGIENSIS

Bacillus thuringiensis (or Bt) is the most widely used microbial control agent for ornamental insects affecting landscape plants. It occurs naturally as a common saprophytic bacteria in many soils, and several strains have been found which can kill insects. It can be manufactured easily and can be formulated into products which have use features similar to standard insecticides, yet it retains many of the advantages of microbial control agents. Furthermore, Bt products, notably the delta endotoxin, are being introduced into other insect control formulations through genetic engineering techniques.[10,11]

The insecticidal activity of *B. thuringiensis* was first discovered in 1911, but it was not developed commercially until the late 1950s. In recent years there has been increased interest in *B. thuringiensis* as an insecticide, and several new products have been developed. Most Bt products are designed to be mixed with water and sprayed on leaves to be eaten by pest insects, released into water, or used as a soil drench, depending on the feeding habits of the target pest.

The active insecticides involved in *B. thuringiensis* are various proteins produced by the bacteria. In particular, compounds known as the delta-endotoxins, which form crystals inside the bacteria, kill insects.[12] These crystals must be ingested, then activated in the gut of susceptible insects to cause disease. The delta-endotoxin then disrupts the cells along the lining of the insect gut (epithelial cells) within minutes after contact. Cells lining the gut are killed after a few hours, paralyzing its function, at which time the insect stops or greatly reduces feeding. Death of the insect often takes several days, usually caused by starvation.

Unlike most other entomopathogens, *B. thuringiensis* usually does not reproduce within the affected insect. (In some large caterpillars the bacteria may enter the hemolymph and reproduce, but this is uncommon.) Typically, Bt insecticides are used in a manner very similar to chemical insecticides. The active ingredients are the crystalized delta-endotoxins, which provide direct, but short residual, control of existing pest populations.

Chemistry of the insect gut is very important for the delta-endotoxins to affect insects. For example, not all caterpillars, including many important agricultural pests in the family Noctuidae, are susceptible to the commonly used *kurstaki* strain of Bt because of differences in gut pH. Furthermore, changes in gut pH can occur due to diet, and also affect susceptibility to *B. thuringiensis*.[13] Bacteriostatic and bacteriocidal effects can also occur from host plant chemistry, which can affect persistence of *B. thuringiensis*.[14]

Some strains of *B. thuringiensis* produce other toxins that are insecticidal.[15] One of these, thuringiensin (alpha-exotoxin), is currently being evaluated as an insecticide/miticide. However, thuringiensin is highly toxic to certain warm-blooded animals, such as rabbits, and its development is separately regulated. Currently registered *B. thuringiensis* insecticides cannot contain thuringiensin.

Although there are numerous strains of *B. thuringiensis*, only a few have been commercially developed. Each strain shows specific activity, being effective against a limited spectrum of insects. At present the following strains have been developed for the U.S. market (see Table 1).

The most widely used strain, *kurstaki*, has activity against a broad spectrum of leaf-feeding Lepidoptera. Although it also has extensive applications in agriculture, it is most widely used for control of species affecting forest and shade trees, primarily for control of gypsy moth and various budworms (*Choristoneura* species). Ornamental insects specifically labeled as target pests of *kurstaki*-strain *B. thuringiensis* products sold in the U.S. are such important species as tent caterpillars (*Malacosoma*

Table 1 Commercially developed strains of *Bacillus thuringiensis*

B. thuringiensis Strain	Trade Name(s)	Susceptible Insect Species
kurstaki	Thuricide, Biobit, Dipel, Caterpillar Attack, Javelin, MVP,[b] Steward, etc.	Most caterpillars
aizawa	Certan	Caterpillars of the wax moth
tenebrionis[a]	Trident, Novodor, M-Trak,[b] M-One	Leaf feeding beetles (e.g., Colorado potato beetle, elm leaf beetle)
tenebrionis/kurstaki	Foil[c]	Leaf beetles, caterpillars
israelensis	Vectobac, Gnatrol, Teknar, Mosquito Attack, Bactimos	Mosquito larvae, blackfly larvae, fungus gnat larvae

[a] M-One and M-Trak formulations, produced by Mycogen Corporation, were originally identified as containing genetic material of the strain *san diego*. In a recent court settlement[16] *san diego* was acknowledged to be identical to *tenebrionis*, which held a prior patent; [b] Formulation contains the *B. thuringiensis* delta-endotoxin produced in *Pseudomonas fluorescens* through genetic transfer; [c] Formulation contains the delta endotoxin of both the *tenebrionis* and *kurstaki* strains, produced through genetic transfer techniques.

species), fall webworm (*Hyphantria cunea*), cankerworms (*Alsophila pometaria, Paleacrita vernata*), fruittree leafroller (*Archips argyrospila*), redhumped caterpillar (*Schizura concinna*), bagworm (*Thyridopteryx ephemeraeformis*), spruce budworms (*Choristoneura* species), and the gypsy moth.

The *tenebrionis* strain of *B. thuringiensis* has effects limited largely to leaf beetles (Family: Chrysomelidae). It is among the most recently identified strains of Bt,[17] and commercial products containing the strain were originally developed for control of the Colorado potato beetle (*Leptinotarsa decemlineata*). However, shade tree pests included on current U.S. labels include elm leaf beetle (*Xanthogaleruca luteola*), cottonwood leaf beetle (*Chrysomela scripta*), and elm calligrapha (*Calligrapha scalaris*). Additional activity has recently been demonstrated against larvae of the Japanese beetle and northern masked chafer (*Cyclocephala borealis*).[18]

(Note: Previously, a beetle-active Bt strain known as *B. thuringiensis* subsp. *san diego* was developed by Mycogen Corporation. This strain shared many common characteristics[19] with *tenebrionis* and recently was acknowledged in a court settlement to be genetically identical to the *tenebrionis* strain.[16] The *tenebrionis* strain held prior patent rights.)

The *israelensis* strain is effective against larvae of certain flies in the suborder Nematocera. Most use is for biting flies which develop in water — the mosquitoes and black flies. Formulations, however, are also sold for control of fungus gnats (*Bradysia* species) in ornamental plants. These are applied as a soil drench. Recent studies have also shown this strain has good activity when applied against larvae of certain *Tipula* sp. craneflies.[20] However, registration of these Bt strains for control of pest *Tipula* species, such as the European crane fly (*T. paludosa*), has not been granted in the U.S.

One of the newest strains identified is *japonensis*. This has insecticidal activity limited to larvae of various scarab beetles, including the Japanese beetle.[21] Because of the importance of this group of insects as pests of turf and ornamental plants, research is in progress to adapt this bacteria for insect control in field applications.

B. thuringiensis does not need to be applied as a living organism to kill susceptible insects. Indeed, some new products contain only the Bt delta-endotoxin which is produced in a different species of bacteria, transformed by inserting the genes for toxin production. For example, Mycogen Corporation recently began marketing M-Trak®, which contains the bacteria *Pseudomonas fluorescens* genetically transformed to produce the toxin of the *tenebrionis* (= *san diego*) strain of *B. thuringiensis*. In this Cell-Cap® process the bacteria are killed, fixing the cell wall to provide a more persistent coating.[10]

Advantages of Bt Insecticides

The selective activity of *B. thuringiensis*-based insecticides is their primary advantage for use in ornamental settings. Each strain is capable of only affecting a specific group of insects (e.g., certain leaf-feeding caterpillars, larvae of leaf beetles) and only when ingested. It has no direct adverse effects on desirable vertebrate species such as fish, wildlife, or pets,[2] a highly desirable feature in landscape plant protection where concerns about nontarget effects of pesticide applications are acute.

B. thuringiensis is also considered essentially nontoxic to humans, which is an important feature for applicators and postapplication users of the treat site. Many formulations of *B. thuringiensis* subsp. *kurstaki* are registered on a wide range of food crops and do not even specify that an interval elapse between application and harvest. This can be a very useful characteristic where drift during application to ornamental plants results in inadvertent residues on garden plants grown on or near treated properties.

Bt insecticides have only modest effects on beneficial insects. Typical insect natural enemies (e.g., lady beetles, parasitic wasps) are not directly susceptible to effects of *B. thuringiensis*. There are some indirect effects, because Bt-killed insects are removed and do not allow reproduction of these other species. However, Bt does not normally reproduce following application, and the potential indirect effects on these other beneficial insects is less likely to be serious than with other microbial organisms which cause epizootics. Compared to other insecticides, *B. thuringiensis* conserves and integrates well with these natural controls, a central feature of IPM programs.

Certain use characteristics of Bt insecticides have also recommended their use by plant managers. Formulations are mixed and applied by conventional application equipment. Results of treatments can be evaluated within a few days of application, and they require less attention to environmental variables (e.g., humidity) than do other insect pathogens considered for use in plant protection. Most use features closely parallel use of synthetic insecticides, making Bt insecticides easily adapted by applicators.

B. thuringiensis can be produced at costs competitive with standard chemical insecticides. Commercial production typically involves fermentation processes that can be highly efficient for production of both *B. thuringiensis* spores and the toxic crystals of delta-endotoxin.[22] Alternative production strategies are available, derived from the ability to clone delta-endotoxin producing genes to other organisms, such as *Pseudomonas fluorescens*.[10]

Limitations of Bt Insecticides

One important limitation is that *B. thuringiensis* acts strictly as a stomach poison. In order to be effective, the insect must eat sufficient amounts to be affected. Hence, feeding habits are an important consideration. Insects which tunnel into plants are not well controlled, even though they may be susceptible to the Bt toxin. This may vary by plant. For example, the tobacco budworm, *Heliothis virescens*, was controlled with *B. thuringiensis* subsp. *kurstaki* on petunia, where the insect feeds on exposed flower petals; control was not acceptable on geranium, where it tunnels unopened buds and feeds on the developing ovaries.[23]

The short persistence of Bt insecticides also is a significant limitation. *B. thuringiensis* occurs naturally as a soil bacterium and is poorly adapted to leaves or other surfaces exposed to sunlight, particularly ultraviolet light.[24] Cranshaw et al.,[25] for example, showed a sharp decline in activity of *B. thuringiensis* subsp. *tenebrionis* (= *san diego*) within 24 hours after application to elm leaves. Since developing insects often stop feeding for considerable periods during molting, short-lived applications may not be effective when insects resume feeding. Also, insects which have not yet hatched will not be controlled, so timing of application becomes more critical.

Various techniques have been used to extend persistence of Bt insecticides and other microbial agents that are sensitive to sunlight. Several adjuvants that absorb ultraviolet light or otherwise provide increased protection on leaf surfaces can extend persistence of microbial agents.[8] Genetic engineering adaptations have also been developed. For example, the Cell-Cap® technology of Mycogen Corporation incorporates the Bt-toxin genes in a bacteria more adapted to survival on leaf surfaces, and produces the Bt toxin in a form with greater persistence.[10,11] Persistence can also be affected by site of application. For example, when applied to more sheltered sites which avoid UV breakdown, such as shaded lower leaves, *B. thuringiensis* can persist for longer periods.

As reviewed by Klein,[26] *B. thuringiensis* is also poorly suited for soil applications, as indigenous soil microorganisms quickly degrade the toxic parasporal crystal. In addition, application to a turfgrass surface provides difficult challenges for effective use of any insecticide which must be ingested, because the surface area of the treated area is so extensive, greatly diluting the applied product. Despite this, some *B. thuringiensis* products make claims for control of surface feeding turfgrass pests such as sod webworms, and may be sold specifically for this purpose (e.g., Sod Webworm Attack®). Although sod webworms and other Lepidoptera found in turfgrass are undoubtably susceptible to the *kurstaki* strain of *B. thuringiensis*, particularly early instars, evidence of their effectiveness under field conditions is another question. For instance, Cranshaw has never observed suppression of sod webworms with *B. thuringiensis* in Colorado trials he has conducted.

B. thuringiensis shows considerable variation in activity against different insect life stages. Older larvae are typically much less susceptible than are young larvae. For example, there was almost a six-fold range in LC_{50} (lethal concentration) values for a *B. thuringiensis* subsp. *tenebrionis* (= *san diego*) formulation (M-One®) applied against elm leaf beetle;[27] similarly, LC_{50} values for second instar cottonwood leaf beetle were 20 times lower than for third instar larvae.[28] Hence, treatments typically must target younger larvae, as noted in label use directions. This can require increased care in application timing, based on surveys of egg hatch and larval development, compared to treatments with greater activity against all damaging life stages. The combination of short persistence and activity limited to early instars requires that multiple applications will be needed if egg hatch is extended.

It is important to note that formulation changes may also affect activity characteristics. The original *Bacillus thuringiensis* subsp. *tenebrionis* (= *san diego)* shows much greater activity against adult beetles than does the formulation involving *Pseudomonas fluorescens*-encapsulated Bt toxin, produced through genetic engineering.[27]

The selectivity of Bt insecticides, considered an advantage in one context, is also a disadvantage in some situations. They can only control a limited number of potential pest species (e.g., leaf beetle larvae, caterpillars) requiring applicators to make pesticide purchases and tank mixes for only a single species. This can cause complications in application scheduling and pesticide ordering, resulting in increased costs to applicators. Multiple pest complexes can occur on landscape plants, which creates situations where a second insecticide may be necessary, negating much of the benefit of the Bt treatment. This can, in part, be overcome by increasing the spectrum of activity by combining different Bt strains through genetic recombination. Such a development was announced by the Ecogen Corporation, with the product Foil®, which contains two strains active against beetles and caterpillars.

Bt insecticides can be more perishable than conventional insecticides. Stored properly, most formulations of *B. thuringiensis* can remain effective for over a year, particularly dry formulations. For example, the wettable powder formulations of Dipel® show little reduction in effectiveness for 2 to 3 years, when stored dry and not exposed to sunlight. Liquid formulations using water (aqueous formulations) have a considerably shorter effective shelf life, less than 1 year. *B. thuringiensis* that is allowed to be diluted with water or wetted degrades much more rapidly than the original formulated material.

MILKY DISEASES

Larvae of at least 29 species of scarabs have been found in the field infected by various strains or varieties of *Bacillus popilliae* Dutky, which produce the "milky diseases."[29] The term is descriptive of the condition of larvae in an advanced state of infection, which turn milky white as a result of buildup of spores and parasporal bodies in the hemolymph. Symptomology is similar among the different grub species infected by milky disease bacteria, and all are considered to be strains of *B. popilliae* by many authors. Each strain, however, has a very narrow host range and is only pathogenic under field conditions against a single scarab genus.

Milky spore diseases are slow-acting and may require a month or more to kill the host insect. Infection occurs following ingestion of the spores by the grubs. The exact cause of death from *B. popilliae* infection is not understood. The bacteria reproduce in very high numbers within the insect hemolymph, resulting in release of 1 to 2 billion spores following infection of the Japanese beetle.[30]

The only commercially developed *B. popilliae* produces the milky disease in the Japanese beetle. This was the first microbial insecticide registered in the U.S., with a history of use of over 55 years. In the early development, state and federal agencies were very heavily involved in production and dessemination of the milky spore bacteria, as a central part of the Japanese beetle biological control program. It is presently being commercially produced and distributed under the trade names of Doom® and Japidemic®.

B. popilliae applications are typically made not to control the white grub populations during the immediate season, but rather to inoculate the lawn with the disease to provide long-term suppression over several subsequent seasons. The recommended application rate of Japanese beetle milky spore powder is about 10 lb/acre, and it is usually applied in spots spaced at intervals of 4 to 10 ft. Such spot applications have been found more effective than broadcast treatments, even when higher rates of spores are used in broadcast treatment.[31]

Once established in an area, the bacteria reproduce within infected grubs and are released back into the soil. Within a few years, natural spread of the bacteria usually occurs throughout the treated area. Where larval populations are high and soil temperatures are favorable for infection, occurrence of the disease can increase rapidly, often following a classic pattern.[32] However, many factors limit the rapid buildup and dispersal of *B. popilliae* in field populations. Milky disease organisms are best thought of as agents that provide general population suppression in an area, rather than as a microbial insecticide that will control scarab larvae at a particular location.

Persistence of the milky spore bacteria remains an open question, and studies suggest individual spores may last 2 to more than 10 years. Soil from New Jersey and Delaware treated with milky spore 25 to 30 years earlier, however, still had active populations of *B. popilliae* capable of infecting grubs, possibly due to periodic replenishment from newly infected grubs.[33]

Perhaps the biggest limitation to further use of the Japanese beetle milky disease has been its inability to be cultured successfully *in vitro*.[34] Only the spore form of the bacteria can be used to produce infection in grubs, and spore production is very low in artificial culture. Since the first field trials by Dutky in 1937, the only source of the Japanese beetle milky spore has been from field collection, which is the source of both the Doom and Japidemic formulations. Such a procedure is expensive, and it is difficult to produce a standardized product. An apparent breakthrough in *in vitro* production of the spores resulted in a patent on the process in April 1989, and products made by this process were manufactured and sold under the Grub Attack® trade name for several years. The bacteria produced by the manufacturing process described in the patent have been subsequently identified as other *Bacillus* species, not *B. popilliae*, and lack infectivity for Japanese beetle larvae. Subsequently Grub Attack® products have been withdrawn from the market.

The specificity of activity of *B. popilliae* strains is also considered a limitation where mixed populations of grub species occur. Commercial preparations are not effective against other species of white grubs, notably the masked chafers, *Cyclocephala* spp.[33] *B. popilliae* is also sensitive to temperature, requiring minimal temperatures of 16°C for development of the disease, with an optimum of 21°C, which has raised questions as to its effectiveness in cooler soils.[26] Furthermore, there remain concerns about attenuation of the bacteria or resistance to it by Japanese beetle.[35] These limitations have caused most extension personnel in the Northeast to no longer recommend milky disease bacteria as a control for white grubs. In addition, serious questions have been raised about the development and spread of milky disease bacteria in field populations of Japanese beetle in Kentucky,[36] Ohio,[37] and the Azores.[38]

Several other *B. popilliae* strains produce milky diseases in white grubs, although they have not been developed commercially. For example, milky disease affects several species of *Cyclocephala*, the chafers, several of which are important turfgrass pests. In addition, at least three morphologically different spore forms of *Bacillus* infect the black turfgrass ataenius, *Ataenius spretulus*,[39] producing symptoms nearly identical to those found in Japanese beetle. Milky disease is widespread among black turfgrass ataenius and causes increasing infection rates as the season progresses. Klein[26] points out that introduction of the spores earlier in the season might have the effect of producing greater impact on populations, particularly with the early season generation.

SERRATIA SPP.

There has been a long history of interest in using species of *Serratia*, particularly the red pigment-producing bacterium *S. marcescens,* for control of insect pests.[40] However, despite the development of spectacular epizootics in field and laboratory insect populations, concerns about safety and efficacy have negated attempts at commercial development of the bacteria as microbial insecticides.

Other species of *Serratia* have now been identfied as pathogens of the grass grub (*Costelytra zealandica*) in New Zealand, and a commercial product (Invade®) based on these bacteria has been marketed.[29] *S. entomophila* was isolated from amber-colored larvae in declining field populations. Further examination revealed the bacteria colonized the gut of the insect, causing an immediate cessation of feeding and subsequent death of the larvae by a combination of toxic products and starvation. Bacteria can be produced by fermentation and applied in the field by subsurface injection. This process speeds up the natural development of the pathogen in grass grub populations and causes population collapse before damage occurs in the pastures.

Scarab larvae with symptoms similar to the New Zealand amber disease have been observed in the U.S. In addition, bacteria similar to *S. entomophila* have been isolated by one of the authors (MGK) from Japanese beetle and masked chafer larvae. However, considerable effort is needed to develop these bacteria for larval control, including safety and other testing required for registration.

VIRUSES

Several important groups of plant pests are attacked by virus diseases. Among the most important are the baculoviruses (also called nuclear polyhedrosis viruses or NPVs) which attack many leaf-feeding caterpillars and sawflies that are damaging to ornamental and forest plants. Baculoviruses produce the "wilt diseases" which cause the infected insects to hang limply, later rupturing in death to spill virus polyhedra. They are important in the natural control of such major pest species as the gypsy moth, Douglas-fir tussock moth, and red-headed pine sawfly. For some of these species, there has been some commercial development of their viruses as microbial controls.

The biggest advantage of using baculoviruses for insect control is their specificity of action. Most only are active against a single species or small group of closely related insects. As such, they are among the most selective microbial diseases considered for use in insect management. This family of viruses also has no biochemical or physical similarities to viruses which affect plants or vertebrates, a situation that is fairly unique among viruses considered for use in insect management.[41] Finally, the baculoviruses replicate in the insect host, so that initial applications can spread and have potential long-term effects once they are introduced among a pest population.

Development of baculoviruses as microbial insecticides has been limited by several factors. Their very specificity of action, considered an advantage in one context, also limits their potential market. Furthermore, the potential market for virus diseases affecting gypsy moth, Douglas-fir tussock moth, and many other leaf-feeding caterpillars directly competes with another, already widely adapted microbial control agent, *Bacillus thuringiensis*. As a result, much of the development of baculoviruses has been done by governmental agencies. Presently, only the gypsy moth baculovirus is currently being pursued commercially.

There are also some important limitations regarding the effectiveness of baculovirus applications. Baculoviruses must be ingested, and young larvae are by far the more susceptible stage to infection. The viruses are unstable, losing most of their activity within a few days, or even hours. Since egg hatch and feeding activity of a plant pest, such as gypsy moth, is typically staggered, it is difficult to obtain adequate control with a single application. In addition, infected insects may not die for 7 to 14 days following application, and may do significant feeding injury in the interim.[42]

Persistence of viruses can be improved by various techniques. Applications made late in the day may reduce degradation,[43] or materials may be added to retard UV breakdown and extend virus persistence.[8,44] Viruses can be mixed with feeding stimulants, which can concentrate feeding on virus-treated surfaces.[45] These adjuvants, however, can add significant costs to virus treatments.

Production of viral insecticides poses some difficult challenges.[46] Viruses only reproduce within a living cell, so production in living host insects is currently the only means for commercially producing viral insecticides. This is an expensive procedure, and the amount of harvestable virus is dependent on a large number of factors involved in rearing procedures. Other organisms, such as certain bacteria with potential safety problems, commonly contaminate virus-killed insects. Regulations governing viral insecticides require that commercial formulations must be fully categorized, which is difficult when manufacture varies.

These problems could be largely circumvented if virus production were possible through *in vitro* cell culture methods. Very promising progess has been made in developing methods for cell culture of insect viruses, although costs remain too high.[47] There have been problems with virus attenuation in these *in vitro* systems, with decreased production of infectious virus particles following production of successive generations of the virus in cell culture lines. This requires production to be batched.

GYPSY MOTH BACULOVIRUS

Development of baculoviruses for control of gypsy moth has been the most extensive among insects affecting forest and shade trees. Currently, research involving Gypcheck, a baculovirus strain registered by the U.S. Forest Service, is being pursued actively. Although originally developed for aerial application to large forested tracts, it may also have promise for ground application to smaller areas, including protection of trees around homes.[48,49] Lewis[50] summarized the advantages of gypsy moth baculovirus use as having "no demonstrable effects on beneficial forms of life. It is a natural component of the gypsy moth ecosystems that does not affect other natural mortality factors. It protects foliage and is a sound pest management tool because of its safety and compatibility with other forms of insect control…"

Gypsy moths sustain high mortality from infection by baculovirus infections under natural conditions, and epizootics commonly cause collapse of outbreak populations. Several strains of the virus occur, with variable activity against gypsy moth.[51] One strain (the Hamden strain) is used as the standard and is the parent source of the registered product Gypchek.[50] Gypsy moth larvae used for virus production can be reared on an artificial diet.

Susceptibility of gypsy moth to infection by its baculovirus is also related to many environmental factors.[8] For example, larvae feeding on acidic oak foliage sustained lower mortality than larvae feeding on aspen, which had less effect on lowering midgut pH.[52]

Persistence of gypsy moth baculovirus has been the focus of several studies. Foliage treated with Gypchek was capable of causing infection for at least 14 days following application.[49] Even longer persistence has been shown for naturally occurring virus on bark and other surfaces,[53] allowing transmission between seasons. Egg masses commonly become infected on trunks and branches by contamination from virus-killed late instar caterpillars.[53] However, it does not appear to be transmitted from adults to larvae via infected eggs, as has been often postulated.[54]

In field trials, ground applications of Gypcheck have been able to increase incidence of infection by the gypsy moth baculovirus in plots involving small numbers of specimen trees around house lots[48] as well as larger plots.[49] Late-season baculovirus outbreaks sometimes, but not always, followed these applications, and these combined infections provided suppression of egg mass production at most sites. However, current-season defoliation was not significantly decreased by baculovirus application,[49] due in part to the relatively slow course of infection with this virus.

Perhaps the most important limitation to adaptation of gypsy moth baculovirus in landscape pest management is that it competes with a microbial control that users are more familiar with, *B. thuringiensis*. Lewis[50] makes the point that "it is tempting to compare the use of *B.t.* vs. baculovirus for control of the gypsy moth since both are microbial entities. However, they differ in their effects on gypsy moth in the following ways: (1) Baculovirus is more specific; (2) *B.t.* acts quicker and can achieve foliage protection without large population reduction; (3) Baculovirus can exhibit carryover effects, *B.t.* cannot; (4) *B.t.* can affect alternate parasite hosts whereas baculovirus will not; (5) *B.t.* is less affected by UV radiation than baculovirus; (6) *B.t.* is more nearly analogous to a synthetic pesticide than is baculovirus." These raise several issues (specifically points 2, 5, 6) that would adversely impact operational use of gypsy moth baculovirus in landscape plant protection, compared to existing practice of Bt application.

SAWFLY BACULOVIRUS

Over 25 sawfly species are affected by baculoviruses,[55] including such important species as the red-headed pine sawfly (*Neodiprion lecontei*) and the European pine sawfly (*N. sertifer*). Their potential as biological control agents has been well recognized since the European spruce sawfly (*Gilphinia*

hercyniae) was substantially suppressed as a pest species following the accidental introduction of its baculovirus in the 1930s.[56] This virus continues to provide long-term control of the European spruce sawfly.

As a general rule, the sawfly baculoviruses show excellent ability to spread among sawfly populations, enabling them to be used as inoculations which self-replicate and can provide long-term control. Dosages needed to induce outbreaks are much lower than those needed for caterpillars which also attack conifers, even though virus production is also much lower per individual insect when compared to viruses affecting caterpillars.

Sawfly baculoviruses have had a considerable history of development, although much occurred before 1975.[57] Due to their rapid activity and ease of spread through insect populations, they are considered to be the most efficacious of the insect viruses in ability to suppress insect pest populations.[42] The virus of the European pine sawfly was used extensively by the Forest Service until 1971, when attempts were made to develop it commercially. Testing of this product (Neochek-S) has been successful and it has been produced in Canada and England. Similarly, Lecontvirus, the baculovirus affecting the redheaded pine sawfly, has been developed in Canada.

At present, formulations of neither of these viruses are commercially developed in the U.S., nor are there plans to do so. Although they potentially can be highly effective for control of several sawfly pest species, the limited markets they represent, and problems with their production in live hosts are likely to preclude their future use in ornamental plant protection unless their development is subsidized substantially by government agencies.

PROTOZOA

Many different species of protozoans naturally infect a wide range of insects. Of these, the most studied and adapted to biological control are microsporidia in the genus *Nosema*. Among insects which affect woody and herbaceous ornamental plants, grasshoppers, several leafrollers, and budworms suffer natural infections by various species of *Nosema*. In turfgrass, *Nosema* infection is also widespread among many sod webworms[58] and white grubs.[33] A recent review of protozoan pathogens of insects identified several species which have potential importance for control of forest or horticultural crops.[59] Of these, two species, *Nosema locustae* and *N. fumiferanae*, have received the most research attention and development.

Nosema infections are not considered to produce highly virulent effects in susceptible insects, and rarely cause rapid death or incapacitation. Instead, they act as chronic diseases that reduce the fitness of infected insects by weakening and/or affecting reproduction.

NOSEMA LOCUSTAE

N. locustae is presently the only protozoan which has been commercially developed for biological control in the U.S. It is used for suppression of grasshoppers, primarily certain *Melanoplus* species. Currently there are two basic manufacturers of the infectious spores that sell under the trade names NoLo Bait®, Semaspore®, and Grasshopper Attack®. Most sales of *N. locustae* have been to suppress populations on rangeland, particularly where insecticide applications are not desirable due to potential nontarget impacts. In many areas of the western U.S., grasshopper outbreaks also greatly affect ornamental plantings, and *N. locustae* is frequently applied to protect gardens and landscapes.

Once established in a grasshopper population, *N. locustae* spread can occur through cannibalism of infected individuals. Under natural conditions in the western U.S., typically less than 1% of the population is infected, although infection can occur at much higher rates.[60]

Commercial *N. locustae* products are sold as living, infectious spores, formulated on a food-base carrier, typically wheat bran. *N. locustae* causes infection when the spores are eaten by susceptible grasshoppers. The fat body is the most common reproduction site, though other tissues may be infected. Infection slows development, reduces feeding, and may produce less viable offspring.[61]

N. locustae was originally developed to be applied during early stages of outbreaks, and to provide another means to suppress grasshopper populations between outbreak cycles. In practice, most applications have been made to effect short-term control during outbreaks, and often have not met user expectations.[61]

The appropriate use of *N. locustae* in ornamental plant protection is as a grasshopper management tool, for long-term suppression. Effectiveness is much more limited when used as a control agent during acute phases of a grasshopper outbreak. As such, it should be used as part of a maintenance program with applications made to grasshopper breeding areas around landscape plantings when surveys indicate increasing pest populations.

In practice, this can be difficult to achieve. Such management practice requires long-term contracts at a landscape site, or at least cooperation between contractors. Breeding populations of grasshoppers need to be monitored at a site over the course of several years, so that *N. locustae* can be applied during the appropriate period of incipient outbreak. At least as important, the greatest suppression costs are incurred during periods prior to outbreaks, which may only occur at intervals of a decade or more. Property managers are often unwilling to provide funds for such maintenance practices except when an imminent threat to property damage exists, which often is too late for *N. locustae* use. Furthermore, applications often need to be made to breeding areas which may not be on the property of the client, although nonchemical grasshopper suppression during severe outbreaks rarely invites publicly stated environmental concerns.

NOSEMA FUMIFERANAE

N. fumiferanae is a protozoan disease largely restricted to insects in the genus *Choristoneura*, particularly the spruce budworm, *C. fumiferana*.[62] Surveys indicate that it is widespread in the field, commonly infecting more than a third of the larvae, with higher infection levels developing during outbreaks. Under laboratory conditions some other caterpillars have been found infected, which does not appear to occur under natural conditions.

N. fumiferanae is a spore-producing microsporidian and infections follow ingestion of the spores by susceptible insects. It primarily infects cells of the midgut, and there are no external signs of infection. Larval and pupal development is retarded, and adults have a shortened lifespan, with females being particularly affected by infection. Early instars (I and II) are much more susceptible to infection than are older stages. Like many protozoan diseases of insects, transmission can occur through the egg (transovarial transmission) as well as by ingestion of the spores.

A system for mass-production of the spores has been published.[62] Applications of spores in field trials were shown to increase the incidence of the disease among spruce budworm larvae when applied against late instar larvae which feed on needles, but not against the needlemining early instars. Also, infection rates were higher on balsam fir vs. white spruce, perhaps due to differences in onset of bud break affecting larval exposure. Since *N. fumiferanae* acts only as a debilitating agent on these later stages, differences in mortality were not observed following application. However, levels of infection were advanced in treated populations after 2 to 3 years. Although Wilson[62] pointed out its potential in integrated control of spruce budworm, there has been no subsequent field testing and development.[61]

FUNGI

Fungi have long been recognized as insect pathogens, due to their conspicuous symptoms and the spectacular effects they can have on insect populations during epizootics. For example, *Beauveria bassiana*, the causal organism of the "white muscardine" disease of many insects, was the first microbe recognized as an insect pathogen, by Agostino Bassi in 1835. Fungi were also involved in the first attempts to use insect pathogens as control agents, during the late 1880s and 1890s.[1]

Over 750 species of fungi are known to be pathogenic to insects,[63] providing a rich source of potential biological control agents. Interest and research involving fungi as biological control organisms has accelerated greatly in the last 20 years.[64] Commercial development has never been successful in North America, with no entomopathogenic fungi having ever been marketed. At present, at least five organisms are currently being considered for commercial development (*B. bassiana, Metarhizium anisopliae, Verticillium lecanii, Paecilomyces fumosa-rosea, Hirsutella thompsonii*), with several others in earlier stages of testing as potential microbial pesticides.[64]

Of the various entomopathogens considered for insect control, fungi are the only ones which do not require ingestion by the insect. Indeed, ingested spores are often excreted by the insect.[65] Instead, most fungi invade the insect by penetrating the exoskeleton. This gives them activity that is more analogous to contact insecticides vs. the stomach poison activity of other microbes.

Entomopathogenic fungi may occur naturally in several different stages including spores of conidia, blastospores, resting spores, mycelia, or hyphae. Primarily, conidia are the stage which are purposefully applied for insect control, being more persistent than blastospores.[66,67] Alternative stages may also be useful for application, such as the use of dried mycelium preparations.[68,69]

Successful infection of insects by fungi is highly dependent on environmental conditions. High humidity, required for spore germination and infection, sensitivity to ultraviolet degradation, and deleterious effects from competing organisms can all greatly affect the performance of fungi used as mycoinsecticides.[70] Erratic performance has marked field applications of fungi for insect control since the first attempts around the turn of the century, and it continues to be a major limitation in the development of fungi as entomopathogens. Pest species which occur in soil, water, or in greenhouses are seen as having the best potential for control, because environmental factors are more optimal or are easier to control.[64] Several pests of turf and ornamentals occur in such sites. Furthermore, resting sites of foliar-feeding insects may provide sufficient humidity for infection. For example, infection of chinch bug (*Blissus leucopterus leucopterus*) was higher in drier soil, since soil cracking occurred, which provided microenvironments of suitable humidity.[71]

BEAUVERIA BASSIANA

B. bassiana is perhaps the best known of the naturally occurring fungus diseases of insects, and is the species that has been most fully exploited for insect control. It attacks a very wide range of insects, including many pest species. For example, in his review of turfgrass pests Tashiro[39] cites reports of it naturally infecting cranberry girdler, Japanese beetle, imported fire ants, and chinch bugs. It is often the major natural control of the latter group of important turfgrass pests. Adult billbugs are also infected. Root weevils, aphids, and greenhouse whitefly are ornamental pests cited by McCoy[64] as species which have recently been controlled in field trials.

B. bassiana can infect insects either through the cuticle, or possibly, through the mouthparts. For example, Ferron[65] reported that most natural infections of the cockchafer grub, *Melonontha melonontha*, by the related fungus *B. brongniartii*, occurred either through the membranes between the head and thorax, or the segments of appendages.

Conidia, the stage of the fungus that is typically applied to initiate new infections, require at least 92% relative humidity to germinate. However, the microclimate around a soil insect may be sufficient to allow infection even when soil moisture is below this.[72] The germinated conidia then produce hyphae which penetrate the insect cuticle both through mechanical and enzymatic means. If the insect molts during early stages of infection, an unsuccessful infection occurs as the fungi are cast off. Once inside the body cavity of the insect, toxins (beauvericin) are produced which destroy insect tissues. After killing the insect, the fungus grows throughout the infected insect during the course of about 2 days, producing a red pigmented antibiotic to inhibit competing bacteria. The dead insects become stiff, and when high humidity conditions occur, a white covering of the conidiophores is produced.[65]

Optimal growth of *B. bassiana* is around 23 to 25°C, and the occurrence of natural outbreaks is usually related to temperature. There are differences, however, in temperature sensitivity among strains of *B. bassiana*, and it may be possible to select strains with temperature requirements more closely matched to sites of use.[65]

B. bassiana also occurs as a saprophyte in soil.[73] This is reflected in numerous studies which show an active soil growth phase after application of conidia that effectively extends the range of infection of this species. The extent of hyphal growth is affected by soil conditions. For example, colonies forming on pupae of the beet armyworm (*Spodoptera exigua*) extended 4.9 cm in a test soil, but only 1.3 to 1.6 cm in soils high in organic matter.[74] Hyphae from infected pecan weevils (*Curculio caryae*) extended through a volume of up to 327.26 cm², sometimes infecting and killing adjacent weevil larvae.[75]

Diet of the insect pest can greatly affect the pathogenicity and spread of *B. bassiana*. Ramoska and Todd[76] reported that chinch bugs fed on corn or sorghum were less susceptible to infection by *B. bassiana* than those reared on other diets. Furthermore, fungal development was inhibited in insects fed the corn or sorghum diet, which the authors attributed to a fungal inhibitor associated with the plants. Hare and Andreadis[77] also found diet suitability to be negatively correlated with susceptibility to *B. bassiana* in the Colorado potato beetle. It may be that use of *B. bassiana* in pest management of landscape pests would be optimized by integration with host plant resistance.

Pesticides used in ornamental plant protection may also affect *B. bassiana*. Anderson et al.[78] found that none of five insecticides tested significantly affected *B. bassiana* growth rate. Furthermore, combinations of the insecticides with *B. bassiana* were consistently more toxic to the Colorado potato beetle. However, an evaluation of four fungicides found that mancozeb was highly deleterious to conidia survival in both the laboratory and field. In the same test, the fungicides chlorothalonil and metalaxyl were not inhibitory to the fungus and would therefore be compatible in use.[79] Similarly, effects of herbicides on growth of *B. bassiana* can vary considerably.[80] It may therefore be important to better understand these interactions with pesticides in landscape settings where *B. bassiana* is to be integrated with chemical control of plant pathogenic fungi, weeds, or other pest species.

B. bassiana is fairly nonselective as a entomopathogen and has potential adverse effects on desirable nontarget species including honeybees,[81] lady beetles,[82] and green lacewings.[83] As with pest species, diet can apparently affect susceptibility to *B. bassiana*. Green lacewings that were starved or raised on a poor diet were much more susceptible to infection.[83]

B. bassiana infection can interfere with development of entomogenous nematodes. Although co-infections of both the nematodes and fungi can reduce the time of hosts to succumb to lethal infection, dually infected hosts cannot support both entomopathogens.[84]

Experimental evidence indicates that humans and other vertebrates are not hosts of *B. bassiana*. However, there are concerns that continued exposure to spores or mycelia may cause allergenic reactions in some individuals.[70]

Use of *B. bassiana* for Control of Landscape Insect Pests

B. bassiana has not been widely tested specifically against insect pests of landscape plants, with much of the more recent work involving agricultural pests, notably the Colorado potato beetle.[85] Suppression of cottonwood borer (*Plectodera scalator*), an imported pest of the *Populus* species, has been reported in nurseries.[86]

Although many landscape pest species are highly susceptible to *B. bassiana*, attempts to use the fungus for control of turfgrass pests in field applications have been disappointing.[87] Much of this is due to short persistence of the fungus in such sites. Furthermore, it is difficult for applied fungal preparations to reach areas where many turfgrass pest species occur, e.g., subterranean species such as scarab beetle larvae.

Fungistatic effects of soil to survival of conidia have been observed repeatedly and are related primarily to effects on competing microorganisms.[88,89] Conidia survival is inversely related to both temperature and soil water, both of which stimulate soil microflora. High organic matter is also deleterious to conidia survival,[90] as was the use of nitrogen fertilizers.[88] Stimulation of antagonistic microflora by deposition of plant debris has also been recognized as important in conidia survival.[88] The turfgrass site, so rich in organic matter and decomposition processes, would therefore be highly deleterious to *B. bassiana*. Application sites with a more sterile soil enviornment, such as around the base of trees or shrubs and in nursery stock production, would likely be more advantageous.

Application of conidia to sites where pest species occur also is a problem. Conidia applied to the surface persist poorly, although soil incorporation can greatly improve persistence.[91,92] Surface applications do not readily move through soil, due to mechanical filtration effects. Movement is greatest in coarse soils.[93]

Formulation differences may also improve the persistence of *B. bassiana*. Clay coating can substantially lengthen conidial survival,[74] though increased mortality of insects did not result.[94] The use of alginate encapsulated mycelium also can protect the fungus from ultraviolet degradation and increase storage life.[69]

Successful use of *B. bassiana* in landscape plant protection may require alternative application methods. For example, conidia were found to persist much longer (more than 3 weeks) when applied within tubers attractive to leaf beetles.[95] Such trap sites may be designed for other species, using food, floral, or pheromone lures. Optimal use of *Beauveria* may be as a long-term suppressor of pest populations, rather than for short-term control. For example, *B. brongniartii* was found to be effective for control of the European cockchafer, a serious pasture pest. However, population reductions were not observed until the generation following the treated one. Breeding areas became enriched with the fungus and subsequently allowed epizootics which controlled at least two later generations.[67]

METARHIZIUM ANISOPLIAE

The history of research with *Metarhizium anisopliae* rivals that of *Beauveria bassiana*. Known as the "green muscardine" disease because of the color of sporulating conidiophores that cover infected insects, *M. anisopliae* has a very wide distribution and host range, and has been reported infecting over 100 species from several insect orders.[70] For example, among turfgrass insect pests Tashiro[39] cites incidence of field infection observed among mole crickets (*Scapteriscus* spp.), red imported fire ants (*Solenopsis invicta*), and a wide variety of scarab beetle larvae including green June beetle (*Cotina nitida*), May/June beetles (*Phyllophaga* spp.), Japanese beetle, and European chafer (*Rhizotrogus majalis*).

Infection usually occurs through penetration of the insect cuticle by germinating spores. However, oral infection may also be an alternate route of infection with some species. For example, adults of the pales weevil (*Hylobius pales*) were very susceptible to oral infection by *M. anisopliae*.[65] Death of infected larvae occurs in conjunction with the production of toxins (destruxins) by the fungus,[70] with lethal effects typically taking several weeks, depending on temperature.

M. anisopliae does not have a soil saprophytic habit and does not grow through the soil, as does *B. bassiana*. For example, following infection of larvae of the pecan weevil, spores were largely restricted to larval cadavers and the inside of larval chambers.[75]

Optimum growth of *M. anisopliae* is around 27 to 28°C[65] and insect susceptibility to infection drops with lower temperatures. However, differences in strains exist. For example, the black vine weevil

was less susceptible to *M. anisopliae* when exposed at 15°C than at 20°C. Temperature sensitivity, however, was much less acute in the best performing strain,[88] indicating that strain selection would be very important for successful use of the entomopathogen in cool temperature environments.

Use of *M. anisopliae* for Control of Landscape Insect Pests

The recent review of field trials with *M. anisopliae*[67] include some with pest species of importance to landscape plants. For example, in several trials conducted for control of black vine weevil control exceeded 80%. Also, successes using *M. anisopliae* for applied insect control in field settings have been reported among insects that have habits similar to landscape pests, including white grubs in sugarcane,[97] leaf beetles on coconut,[98] and froghoppers in sugarcane.[99] Some of the most successful trials using *M. anisopliae* have involved control of larvae of the rhinoceros beetle, *Oryctes rhinoceros*. These involved treatment of larval breeding sites with spores of the fungus grown on oats. A high level of infection was achieved, and spores persisted for 2 years.[100]

VERTICILLIUM LECANII

The fungus disease that has received most development as a foliar treatment vs. soil use is *Verticillium lecanii*. This fungus is widely distributed in the tropics and subtropics.[101] It is somewhat less common in temperate regions, but can be locally important. For example, it has been recognized as an important natural control of pear thrips (*Taeniothrips inconsequens*) in Vermont maple forests[102] and it is recognized as an important natural control of many aphids affecting ornamentals.[103]

V. lecanii has a very wide host range. Most commonly recorded hosts are scales, aphids, and other Homoptera, but various Diptera, Hymenoptera, and Lepidoptera and eriophid mites are also hosts.[101] On infected insects the fungus produces a cottony, whitish-yellow colony. It also is saprophytic on various foodstuffs (including honeydew) and can even be hyperparasitic on other fungi.[104] This adaptability allows it to be easily cultured artificially for commercial development.

There has been some commercial interest in developing the fungus, particularly in England. Products containing *V. lecanii* have been marketed in the past under the trade names Vertalec and Mycotol for use in control of aphids and whiteflies.[64]

Interest in *V. lecanii* has been concentrated in greenhouse production rather than in landscape pest management. Optimal temperature requirements are moderate (20 to 25°C) but the fungus has a high humidity requirement, 85 to 90%, which must be maintained at least 10 to 12 hours daily.[104] In England[101,105] and in the U.S.[106] it has been used most successfully for control of aphids on chrysanthemums.

V. lecanii is considered to be highly promising in integrated control programs in greenhouses.[101,104] However, environmental conditions in outdoor plantings are considered to be too variable for its effective use as a mycoinsecticide in temperate climates. Specialized use for application to moist microhabitats or in tropical or subtropical zones with high humidity appears much more promising.[101]

ACKNOWLEDGMENTS

We wish to acknowledge the substantial contributions made by the reviewers that have greatly improved this paper. In particular we would like to recognize Dr. Pat Vittum of the University of Massachusetts, Dr. Dennis Knudsen of Colorado State University, and Amy Downing of the Chemlawn Chemical Research and Development Division.

REFERENCES

1. **Steinhaus, E. A.,** *Principles of Insect Pathology*, McGraw-Hill, New York, 1949, 757 pp.
2. **Heimpel, A. M.,** Safety of insect pathogens for man and invertebrates, in *Microbial Control of Insects and Mites*, Burges, H. D. and Hussey, N. W., Eds., Academic Press, New York, 1971, 469.
3. **Burges, H. D.,** Safety, safety testing and quality control of microbial insecticides, in *Microbial Control of Pests and Plant Diseases 1970–1980*, Burges, H. D., Ed., Academic Press, New York, 1981, chap. 40.
4. **Laird, M., Lacey, L. A., and Davidson, E. W.,** *Safety of Microbial Insecticides*, CRC Press, Boca Raton, FL, 1990, 259 pp.
5. **Tabashnik, B. E., Cushing, N. L., Finson, N., and Johnson, M. W.,** Field development of resistance to *Bacillus thuringiensis* in diamondback moth (Lepidoptera: Plutellidae), *J. Econ. Entomol.*, 83, 1671, 1990.
6. **Lacey, L. A. and Harper, J. D.,** Microbial control and integrated pest management, *J. Entomol. Sci.*, 21, 206, 1986.
7. **Fuxa, J. R.,** Ecological considerations for the use of entomopathogens in IPM, *Annu. Rev. Entomol.*, 32, 225, 1987.
8. **Jaques, R. P.,** Manipulation of the environment to increase effectiveness of microbial agents, in *Microbial Control of Insect Pests: Future Strategies in Management Systems*, Allen, G. E., Ignoffo, C. M., and Jaques, R. P., Eds., 1978, 72.
9. **Dedryver, C. A.,** Initiating an epizootic in a glasshouse with *Entomophthora fresenii* on *Aphis fabae* through inoculum introduction and control of relative humidity, *Entomophaga*, 24, 443, 1979.
10. **Feitelson, J. S., Quick, T. C., and Galertner, F.,** Alternative hosts for *Bacillus thuringiensis* delta-endotoxin genes, in *New Directions in Biological Control: Alternatives for Suppressing Agricultural Pests and Diseases*, Baker, R. R. and Dunn, P. E., Eds., Alan R. Liss, New York, 1990, 561.
11. **Leemans, J., Reynaerts, A., Hofte, H., Peferoan, M., Van Mellaert, M., and Joos, H.,** Insecticidal crystal proteins from *Bacillus thuringiensis* and their use in transgenic plants, in *New Directions in Biological Control: Alternatives for Suppressing Agricultural Pests and Diseases*, Baker, R. R. and Dunn, P. E., Eds., Alan R. Liss, New York, 1990, 573.
12. **Cooksey, K. E.,** The protein crystal of *Bacillus thuringiensis*: Biochemistry and mode of action, in *Microbial Control of Insects and Mites*, Burges, H. D. and Hussey, N. W., Eds., Academic Press, New York, 1971, 247.
13. **Yendol, W.,** Manipulation of the environment: bacteria, in *Microbial Control of Insect Pests: Future Strategies in Management Systems*, Allen, G. E., Ignoffo, C. M., and Jaques, R. P., Eds., 1978, 91.
14. **Smirnoff, W. A. and Hutchinson, P. M.,** Bacteriostatic and bacteriocidal effects of extracts of foliage from various plant species on *Bacillus thuringiensis* var. *thuringiensis* Berliner, *J. Inv. Pathol.*, 7, 273, 1965.
15. **Dulmage, H. T.,** Insecticidal activity of isolates of *Bacillus thuringiensis* and their potential for pest control, in *Microbial Control of Pests and Plant Diseases 1970–1980*, Burges, H. D., Ed., Academic Press, New York, 1981, chap. 11.
16. **Thayer, A.,** Suit over biopesticide patent rights settled, *Chem. Eng. News*, 70(35), 9, 1992.
17. **Kreig, A., Huger, A. M., Langenbruch, G. A., and Schnetter, W.,** *Bacillus thuringiensis* var. *tenebrionis*: ein neuer, gegen Larven con Coleopteren witsamer Pathotype, *Z. Ang. Entomol.* 96, 500, 1983.
18. **Downing, A. S.,** personal communication, 1991.

19. **Kreig, A., Huger, A. M., and Schnetter, W.,** *Bacillus thuringiensis* var "San Diego" Stamm M-& ist identisch mit dem zuvor in Deutschland isolierten kaferwirksamen *B. thuringiensis* subsp. *tenebrionis* Stamm Bl 256-82, *J. Appl. Entomol.,* 104, 417, 1987.

20. **Smits, P. H. and Vlug, H. J.,** Control of tipulid larvae with *Bacillus thuringiensis* var. *israelensis,* in *Proceedings and Abstracts, 5th International Colloquium on Invertebrate Pathology and Microbial Control,* Glen Osmond, Australia, 1990, 343.

21. **Ohba, M., Iwahana, H., Asano, S., Suzuki, N., Sato, R., and Hori, H.,** A unique isolate of *Bacillus thuringiensis* with high larvicidal activity specific for scarabaeid beetles, *Let. Appl. Microbiol.,* 14, 54.

22. **Talmage, H. T. and Rhodes, R. A.,** Production of pathogens on artificial media, in *Microbial Control of Insects and Mites,* Burges, H. D. and Hussey, N. W., Eds., Academic Press, New York, 1971, 530.

23. **Cranshaw, W. S.,** The tobacco budworm on bedding plants in Colorado, *Colo. Greenhouse Growers Assoc. Res. Bull.,* 467, 1989.

24. **Ignoffo, C. M. and Garcia, C.,** UV-photoactivation of cells and spores of *Bacillus thuringiensis* and effects of peroxidase on inactivation, *Environ. Entomol.,* 7, 270, 1978.

25. **Cranshaw, W. S., Day, S. J., Gritzmacher, T. J., and Zimmerman, R. J.,** Field and laboratory evaluations of *Bacillus thuringiensis* strains for control of elm leaf beetle, *J. Arboricul.,* 15(2), 31, 1988.

26. **Klein, M. G.,** Biological suppression of white grubs in turf, in *Integrated Pest Management for Turfgrass and Ornamentals,* Leslie, A. R. and Metcalf, R. L., Eds., U.S. Environmental Protection Agency, Office of Pesticide Programs, Washington, DC, 1989, 297.

27. **Zimmerman, R. J.,** Contributions Towards an Integrated Pest Management Program for Elm Leaf Beetle, *Xanthogaleruca luteola* (Muller) (Coleoptera: Chrysomelidae) in Colorado, Ph.D. dissertation, Colorado State University, Fort Collins, 1991.

28. **Bauer, L. S.,** Response of the cottonwood leaf beetle, (Coleoptera: Chrysomelidae) to *Bacillus thuringiensis* var. *san diego, Environ. Entomol.,* 19(2), 428, 1990.

29. **Klein, M. G. and Jackson, T. A.,** Bacterial diseases of scarabs, in *Use of Pathogens in Scarab Pest Management,* Jackson, T. A. and Glare, T. R., Eds., Intercept Books, London, 1992, 43.

30. **Fleming, W. E.,** Biological control of the Japanese beetle, *USDA Tech. Bull.,* 1383, 78, 1968.

31. **Klein, M. G.,** Biological suppression of turf insects, in *Advances in Turfgrass Entomology,* Niemcyzk, H. D. and Joyner, B. G., Eds., Chemlawn Corp., Columbus, OH, 1982, 91.

32. **Klein, M. G.,** Use of *Bacillus popilliae* in Japanese beetle control, in *Use of Pathogens in Scarab Pest Management,* Jackson, T. A. and Glare, T. R., Eds., Intercept Books, London, 1992, 179.

33. **Klein, M. G.,** Pest management of soil-inhabiting insects with microorganisms, *Agric. Ecosyst. Environ.,* 24, 337, 1988.

34. **Stahly, D. P. and Klein, M. G.,** Problems with in vitro production of spores of *Bacillus popilliae* for use in biological control of the Japanese beetle, *J. Inv. Pathol.,* in press.

35. **Hanula, J. L. and Andreadis, T. G.,** Parasitic microorganisms of the Japanese beetle (Coleoptera: Scarabaeidae) and associated Scarabaeidae larvae in Connecticut soils, *Environ. Entomol.,* 17, 709, 1988.

36. **Redmond, C. T.,** Evaluation of *Bacillus popilliae,* Causal Agent of Milky Disease, for the Biological Control of Japanese Beetle Grubs, M.S. Thesis, University of Kentucky, Lexington, 1990.

37. **Shetlar, D. J.,** White grubs in turfgrass, OCES/OSU HYG Fact Sheet 2500-91, 1991.

38. **Lacey, L. A.,** personal communication, 1991.

39. **Tashiro, H.,** *Turfgrass Insects in the United States and Canada,* Cornell University Press, Ithaca, NY, 1987.

40. **Steinhaus, E. A.,** *Insect Pathology an Advanced Treatise,* Vol. 1 and 2, Academic Press, New York, 1963.

41. **Tinsley, T. W. and Kelly, D. C.,** Taxonomy and nomenclature of insect pathogenic viruses, in *Viral Insecticides for Biological Control,* Maramorosch, K. and Sherman, K. E., Eds., Academic Press, New York, 1985.

42. **Podgwaite, J. D.,** Strategies for field use of Baculoviruses, in *Viral Insecticides for Biological Control,* Maramorosch, K. and Sherman, K. E., Eds., Academic Press, New York, 1985, 775.

43. **Smirnoff, W. A.,** The effect of sunlight on the nuclear polyhedrosis virus of *Neodiprion swainei* with measurement of solar energy received, *J. Inv. Pathol.,* 19, 179, 1972.

44. **Jaques, R. P.,** Stability of insect viruses in the environment, in *Viral Insecticides for Biological Control,* Maramorosch, K. and Sherman, K. E., Eds., Academic Press, New York, 1985, 285.

45. **Luttrell, R. G., Yearian, W. C., and Young, S. Y.,** Effect of spray adjuvants on *Heliothis zea* (Lepidoptera: Noctuidae) nuclear polyhedrosis virus efficacy, *J. Econ. Entomol.,* 76, 162, 1983.

46. **Sherman, K. E.,** Considerations in the large-scale and commercial production of viral insecticides, in *Viral Insecticides for Biological Control,* Maramorosch, K. and Sherman, K. E., Eds., Academic Press, New York, 1985, 757.

47. **Hink, W. F.,** Production of *Autographa californica* nuclear polyhedrosis virus in cells from large-scale suspension cultures, in *Microbial and Viral Pesticides,* Kurstak, E., Ed., Marcel Dekker, New York, 1982, 493.

48. **Webb, R. E., Podgwaite, J. D., Shapiro, M., Tatman, K. M., and Douglass, L. W.,** Hydraulic spray application of Gypchek as a homeowner control tactic against gypsy moth (Lepidoptera: Lymantriidae), *J. Entomol. Sci.,* 25(3), 383, 1990.

49. **Podgwaite, J. D., Reardon, R. C., Kolodny-Hirsch, D. M., and Walton, G. S.,** Efficacy of ground application of the gypsy moth (Lepidoptera: Lymantriidae) nucleopolyhedrosis virus product, Gypchek, *J. Econ. Entomol.,* 84(2), 440, 1991.

50. **Lewis, F. B.,** Control of the gypsy moth by a baculovirus, in *Microbial Control of Pests and Plant Diseases 1970–1980,* Burges, H. D., Ed., Academic Press, New York, 1981, chap 18.

51. **Shapiro, M. and Robertson, J. L.,** Natural variability of three geographic isolates of gypsy moth (Lepidoptera: Lymantriidae) nuclear polyhedrosis virus, *J. Econ. Entomol.,* 84(1), 71, 1991.

52. **Keating, S. T., Schultz, J. C., and Yendol, W. G.,** The effect of diet on gypsy moth (*Lymantria dispar*) larval midgut pH, and its relationship with larval susceptibility to a baculovirus, *J. Inv. Pathol.,* 56, 317, 1990.

53. **Woods, S. A., Elkinton, J. S., and Shapiro, M.,** Factors affecting the distribution of a nuclear polyhedrosis virus among gypsy moth (Lepidoptera: Lymantriidae) egg masses and larvae, *Environ. Entomol.,* 19, 1330, 1991.

54. **Murray, K. D., Shields, K. S., Bruan, J. P., and Elkinton, J. S.,** The effect of gypsy moth metamorphosis on the development of nuclear polyhedrosis virus infection, *J. Inv. Pathol.,* 57, 352, 1991.

55. **Martignoni, M. E. and Iwai, P. J.,** A catalogue of viral diseases of insects, mites, and ticks, in *Microbial Control of Pests and Plant Diseases 1970–1980,* Burges, H. D., Ed., Academic Press, New York, 1981, Appendix 2.

56. **Bird, F. T. and Burk, J. M.,** Artificially disseminating virus as a factor controlling the European spruce sawfly, *Diprion hercyniae* (Htg.), in the absence of introduced parasites, *Can. Entomol.,* 93, 228, 1961.

57. **Cunningham, J. C. and Entwistle, P. F.,** Control of sawflies by baculovirus, in *Microbial Control of Pests and Plant Diseases 1970–1980,* Burges, H. D., Ed., Academic Press, New York, 1981, chap 19.

58. **Banerjee, A. C.,** Microsporidia diseases of sod webworms in bluegrass lawns, *Ann. Entomol. Soc. Am.,* 61, 544, 1968.

59. **Brooks, W. M.,** Entomogenous protozoa, in *Handbook of Natural Pesticides,* Ignoffo, C. M., Ed., CRC Press, Boca Raton, FL, 1988, chap. 1.

60. **Henry, J. E. and Oma, E. A.,** Pest control by *Nosema locustae,* a pathogen of grasshoppers and crickets, in *Microbial Control of Pests and Plant Diseases 1970–1980,* Burges, H. D., Ed., Academic Press, New York, 1981, chap. 30.

61. **Henry, J. E.,** Control of insects by protozoa, in *New Directions in Biological Control: Alternatives for Suppressing Agricultural Pests and Diseases,* Baker, R. R. and Dunn, P. E., Eds., Alan R. Liss, New York, 1990, 161.

62. **Wilson, G. G.,** *Nosema fumiferanae,* a natural pathogen of a forest pest: potential for pest management, in *Microbial Control of Pests and Plant Diseases 1970–1980,* Burges, H. D., Ed., Academic Press, New York, 1981, chap. 32.

63. **Carruthers, R. I. and Hural, K.,** Fungi as naturally occurring pathogens, in *New Directions in Biological Control: Alternatives for Suppressing Agricultural Pests and Diseases,* Baker, R. R. and Dunn, P. E., Eds., Alan R. Liss, New York, 1990, 115.

64. **McCoy, C. W.,** Entomogenous fungi as microbial pesticides, in *New Directions in Biological Control: Alternatives for Suppressing Agricultural Pests and Diseases,* Baker, R. R. and Dunn, P. E., Eds., Alan R. Liss, New York, 1990, 139.

65. **Ferron, P.,** Pest control by the fungi *Beauveria* and *Metarhizium,* in *Microbial Control of Pests and Plant Diseases 1970–1980,* Burges, H. D., Ed., Academic Press, New York, 1981, 465.

66. **Watt, B. A. and LeBrun, R. A.,** Soil effects of *Beauveria bassiana* on pupal populations of the Colorado potato beetle (Coleoptera: Chrysomelidae), *Environ. Entomol.,* 13, 15, 1984.

67. **Keller, S. and Zimmerman, G.,** Mycopathogens of soil insects, in *Insect-Fungus Interactions*, Wilding, N., Collins, N. M., Hammond, P. M., and Webber, J. F., Eds., Academic Press, London, 1989, 239.

68. **Pereira, R. M. and Roberts, D. W.,** Dry mycelium preparations of entomopathogenic fungi, *Metarhizium anisopliae* and *Beauveria bassiana*, *J. Inv. Pathol.*, 56, 39, 1990.

69. **Pereira, R. M. and Roberts, D. W.,** Alginate and cornstarch mycelial formulations of entompathogenic fungi, *Beauveria bassiana* and *Metarhizium anisopliae*, *J. Econ. Entomol.*, 84, 1657, 1991.

70. **McCoy, C. W., Samson, R. A., and Boucias, D. G.,** Entomogenous fungi, in *Handbook of Natural Pesticides*, Ignoffo, C. M., Ed., CRC Press, Boca Raton, FL, 1988, chap. 2.

71. **Krueger, S. R., Nechols, J. R., and Ramoska, W. A.,** Infection of chinch bug, *Blissus leucopterus leucopterus* (Hemiptera: Lygaeidae), adults from *Beauveria bassiana* (Deuteromycotina: Hyphomycetes) conidia in soil under controlled temperature and moisture conditions, *J. Inv. Pathol.*, 58, 19, 1991.

72. **Ramoska, W. A.,** The influence of relative humidity on *Beauveria bassiana* infectivity and replication in the chinch bug, *Blissus leucopterus*, *J. Inv. Pathol.*, 433, 389, 1984.

73. **de Hoog, D. S.,** The genera *Beauveria, Isaria, Tritirachium* and *Acrodontium* gen. nov., *Studies in Mycol.*, 1972, vol. 1.

74. **Studdert, J. P. and Kaya, H. K.,** Water potential, temperature, and clay-coating of *Beauveria bassiana* conidia: effect on *Spodoptera exigua* pupal mortality in two soil types, *J. Inv. Pathol.*, 56, 327, 1990.

75. **Gottwald, T. R. and Tedders, W. L.,** Colonization, transmission, and longevity of *Beauveria bassiana* and *Metarhizium anisopliae* (Deuteromycotina: Hyphomycetes) on pecan weevil larvae (Coleoptera: Curculionidae) in the soil, *Environ. Entomol.*, 13, 557, 1984.

76. **Ramoska, W. A. and Todd, T.,** Variation in efficacy and viability of *Beauveria bassiana* in the chinch bug (Hemiptera: Lygaeidae) as a result of feeding activity on selected host plants, *Environ. Entomol.*, 14, 146, 1985.

77. **Hare, J. D. and Andreadis, T. G.,** Variation in the susceptibility of *Leptinotarsa decemlineata* (Coleoptera: Chrysomelidae) when reared on different host plants to the fungal pathogen in the field and laboratory, *Environ. Entomol.*, 12, 1892, 1983.

78. **Anderson, T. E., Hajek, A. E., Roberts, D. W., Preisler, K. K., and Robertson, J. L.,** Colorado potato beetle (Coleoptera: Chrysomelidae): effects of combinations of *Beauveria bassiana* with insecticides, *J. Econ. Entomol.*, 82, 83, 1989.

79. **Loria, R., Galaini, S., and Roberts, D. W.,** Survival of the entomopathogenic fungus *Beauveria bassiana* as influenced by fungicides, *Environ. Entomol.*, 12, 1724, 1983.

80. **Gardner, W. A. and Storey, G. K.,** Sensitivity of *Beauveria bassiana* to selected herbicides, *J. Econ. Entomol.*, 78, 1275, 1985.

81. **Vandenberg, J. D.,** Safety of four entomopathogens for caged adult honey bees (Hymenoptera: Apidae), *J. Econ. Entomol.*, 83, 755, 1990.

82. **Magalhaes, B. P., Lord, J. C., Wraight, S. P., Daoust, R. A., and Roberts, D. A.,** Pathogenicity of *Beauveria bassiana* and *Zoopththora radicans* to the coccinellid predators *Coleomegilla maculata* and *Eriopis connexa*, *J. Inv. Pathol.*, 52, 471, 1988.

83. **Donegan, K. and Lighthart, B.,** Effect of several stress factors on the susceptibility of the predatory insects, *Chrysoperla carnea* (Neuroptera: Chrsyopidae), to the fungal pathogen *Beauveria bassiana*, *J. Inv. Pathol.*, 54, 79, 1989.

84. **Barbercheck, M. E. and Kaya, H. K.,** Interactions between *Beauveria bassiana* and the entomogenous nematodes, *Steinernema feltiae* and *Heterorhabditis heliothidis*, *J. Inv. Pathol.*, 55, 225, 1989.

85. **Roberts, D. W., LeBrun, and Semel, M.,** Control of the Colorado potato beetle with fungi, in *Advances in Potato Pest Management*, Lashomb, J. H. and Casagrande, R. A., Eds., Hutchinson Ross, Stroudsburg, PA, 1982, 119.

86. **Forschler, B. T. and Nordin, G. L.,** Impact of *Beauveria bassiana* on the cottonwood borer, *Plectodera scalator* (Coleoptera: Cerambycidae), in a commercial cottonwood nursery, *J. Entomol. Sci.*, 24, 186, 1989.

87. **Shetlar, D. J.,** personal communication, 1992.

88. **Lingg, A. J. and Donaldson, M. D.,** Biotic and abiotic factors affecting stability of *Beauveria bassiana* conidia in soil, *J. Inv. Pathol.*, 38, 191, 1981.

89. **Groden, E. and Lockwood, J. L.,** Effects of soil fungistasis on *Beauveria bassiana* and its relationship to disease incidence in the Colorado potato beetle, *Leptinotarsa decemlineata*, in Michigan and Rhode Island soils, *J. Inv. Pathol.*, 24, 57, 7, 1991.

90. **Studdert, J. P., Kaya, H. K., and Duniway, J. M.,** Effect of water potential, temperature, and clay-coating on survival of *Beauveria bassiana* conidia in a loam and peat soil, *J. Inv. Pathol.,* 55, 417, 1989.

91. **Gaugler, R., Costa, S. D., and Lashomb, J.,** Stability and efficacy of *Beauveria bassiana* soil inoculations, *Environ. Entomol.,* 18, 412, 1989.

92. **Storey, G. K., Gardner, W. A., and Tollner, E. W.,** Penetration and persistence of commercially formulated *Beauveria bassiana* conidia in soil of two tillage systems, *Environ. Entomol.,* 18, 835, 1989.

93. **Storey, G. K. and Gardner, W. A.,** Movement of an aqueous spray of *Beauveria bassiana* into the profile of four Georgia soils, *Environ. Entomol.,* 17, 135, 1988.

94. **Studdert, J. P. and Kaya, H. K.,** Water potential, temperature, and soil type on the formation of *Beauveria bassiana* soil colonies, *J. Inv. Pathol.,* 56, 380, 1990.

95. **Daoust, R. A. and Pereira, R. M.,** Stability of entomopathogenic fungi *Beauveria bassiana* and *Metarhizium anisopliae* on beetle-attracting tubers and cowpea foliage in Brazil, *Environ. Entomol.,* 15, 1237, 1986.

96. **Soares, G. G., Jr., Marchal, M., and Ferron, P.,** Susceptibility of *Otiorhynchus sulcatus* (Coleoptera: Curculionidae) larvae to *Metarhizium anisopliae* and *Metarhizium flavoviride* (Deuteromycotina: Hyphomycetes) at two different temperatures, *Environ. Entomol.,* 12, 1887, 1983.

97. **Samuels, K. D. Z., Pinnock, D. E., and Bull, R. M.,** Scarabaeid larvae control in sugarcane using *Metarhizium anisopliae, J. Inv. Pathol.,* 55, 135, 1990.

98. **Liu, S. D., Lin, S. C., and Shiau, J. F.,** Microbial control of coconut leaf beetle (*Brontispa longissma*) with green muscardine fungus, *Metarhizium anisopliae* var. *anisopliae, J. Inv. Pathol.,* 53, 307, 1989.

99. **Allard, G. B., Chase, C. A., Heale, J. B., Isaac, J. E., and Prior, C.,** Field evaluation of *Metarhizium anisopliae* (Deuteromycotina: Hyphomycetes) as a mycoinsecticide for control of sugarcane froghopper, *Aeneolamia varia saccharina* (Hemiptera: Cercopidae), *J. Inv. Pathol.,* 55, 41, 1990.

100. **Latch, G. C. M. and Falloon, R. E.,** Studies in the use of *Metarhizium anisopliae* to control *Oryctes rhinoceros, Entomophaga,* 21, 39, 1976.

101. **Hall, R. A.,** The fungus *Verticillium lecanii* as a microbial insecticide against aphids and scales, in *Microbial Control of Pests and Plant Diseases 1970–1980,* Burges, H. D., Ed., Academic Press, New York, 1981, chap. 25.

102. **Skinner, M., Parker, B. L., and Bergdahl, D. R.,** *Verticillium lecanii* isolated from larvae of pear thrips, *Taeniothrips inconsequens,* in Vermont, *J. Inv. Pathol.,* 58, 157, 1991.

103. **Vehrs, S. L. and Parrella, M. P.,** Aphid problems increase on ornamentals, *Calif. Agric.,* 45, 28, 1991.

104. **Samson, R. A. and Rombach, M. C.,** Biology of the fungi *Verticillium* and *Aschersonia,* in *Biological Pest Control: The Glasshouse Experience,* Hussey, N. W. and Scopes, N., Eds., Cornell University Press, Ithaca, NY, 1985, 30.

105. **Sopp, P. I., Gillespie, A. T., and Palmer, A.,** Comparison of ultra-low-volumne electrostatic and high-volume hydraulic application of *Verticillium lecanii* for aphid control on chrysanthemums, *Crop Prot.,* 9, 177, 1990.

106. **Gardner, W. A., Oetting, R. D., and Storey, G. K.,** Scheduling of *Verticillium lecanii* and benomyl applications to maintain aphid (Homoptera: Aphididae) control on chrysanthemums, *J. Econ. Entomol.,* 77, 514, 1984.

Chapter 46

The Role of Endophytes in Integrated Pest Management for Turf

Melodee L. Fraser, *Pure Seed Testing, Inc., Rolesville, NC*

Jane P. Breen, *Center for Turfgrass Science, Cook College, New Brunswick, NJ*

CONTENTS

Introduction 521
Biology of Endophytes 521
Benefits of Endophytes in Turf 522
Development and Availability of Endophyte-Infected Turfgrasses 523
    Tall Fescue 524
    Perennial Ryegrass 525
    Fine Fescue 525
    Creeping Bentgrass 525
    Kentucky Bluegrass 526
Conclusion 526
References 526

INTRODUCTION

The development, production, and culture of genetically improved turfgrass cultivars have increased substantially in recent years. Turfgrass breeding has grown into a discipline encompassing many dedicated individuals in the public and private sectors of many countries. Through the efforts of turfgrass breeders, dramatic improvements have been made in the characteristics of several species.

The intensive management of turfgrasses for particular uses has also increased. Turfgrass managers show a growing interest in utilizing turfgrasses with distinct characteristics for specific situations. Well-maintained turfgrasses are not only aesthetically pleasing, but also contribute to improving the environment.

With an increasing emphasis on reducing the use of synthetic agricultural chemicals, many turfgrass managers are developing IPM programs. There is a high demand for quality turfgrasses that can persist and form functional, attractive turf with fewer management inputs. Consequently, turfgrass breeders are selecting plants with characteristics that enhance pest resistance and improve persistence. One of the biggest contributions turfgrass breeders have made to integrated pest management is the discovery and utilization of *Acremonium* endophytes in the development of elite turfgrass cultivars.

BIOLOGY OF ENDOPHYTES

Fungal endophytes form infections within leaves and stems of healthy plants but have limited, if any, pathogenic effects.[1] Endophytic fungi have been defined as fungi that are contained in or grow entirely within the substrate plant and spend all or nearly all of their life cycle in the host.[2] Fungal endophytes have been known in grasses for over 100 years[3] and have been found in numerous groups as diverse as sedges and pine trees. The considerable biological and agricultural significance of fungal endophytes, however, has only recently been discovered.

Most endophytes are highly host-specific and infect a single genus or species, but others have a relatively wide host range. The relationship between the fungus and its host may be mutalistic or parasitic, depending on the species involved[4] and environmental conditions. Many endophytes have not been known to produce pathogenic effects, while similar types may become pathogenic under certain conditions. For instance, some endophytes produce a choke disease on turfgrass hosts. This disease arises when the fungal fruiting structure or stromata suppresses emergence of the grass inflorescence, thereby decreasing seed yield. Environmental conditions affect the ability of fungal isolates to produce this sexual stage. Field data indicate that high soil fertility reduces the ability of *Acremonium typhinum* to produce choke in Chewings fescue, *Festuca rubra* L. subsp. *commutata* Gaud.[5]

Basic information on endophyte biology and endophyte-host interactions is limited. However, the presence of these organisms in widely diverse plant taxa, and their increasingly apparent biological significance, suggest that fungal endophytes may prove to be as common and as important as the mycorrhizal fungi.[1]

Fungal endophytes occur in a wide range of grasses. The traits these fungi share are (1) intercellular mycelia; (2) systemic nature, with concentration in the leaf sheath, culm, meristem, and seed; (3) asymptomatic infections; and (4) persistent infection through the life of the host.

The *Acremonium* endophytes are obligate seed-borne fungi that have long been recognized as a botanical curiosity in species such as ryegrass, *Lolium*,[3,6] and fescue.[7,8] The biological and agricultural significance of the *Acremonium* endophytes became clear when their presence was associated with ryegrass staggers disease in livestock[9-12] and resistance to the Argentine stem weevil, *Listronotus bonariensis* Kuschel, a major insect pest of perennial ryegrass, *L. perenne* L., in New Zealand.[13]

The taxonomy of *Acremonium* endophytes is being debated. Morgan-Jones and Gams[14] created a section in the genus, *A. coenophialum* Morgan-Jones and Gams, for the endophyte of tall fescue, *F. arundinacea* Schreb. They also named the endophyte of a fine fescue, *A. typhinum* Morgan-Jones and Gams. Other endophytes were subsequently placed in this section. To date, five species of *Acremonium* endophytes have been identified. *A. lolii* Latch, Christensen, and Samuels and *A. coenophialum* infect perennial ryegrass and tall fescue, respectively. *A. typhinum* infects a range of turf and conservation grasses, six species of fine-leaved fescue,[15] and big bluegrass, *Poa ampla* Merr.[16] *A. huerfanum* White infects *F. arizonica* Vasey.[17] *A. chisosum* White infects *F. arizonica*, *Bromus anomalus* Rupr., *F. obtusa* Bieler, and *F. subulata* Trin.[18] The *Acremonium* endophytes differ markedly from other *Acremonium* species. Latch et al.[19] suggested that they may be species of *Epichloe* that rarely form visible stroma.

Acremonium endophytes in grasses are relatively common. Bacon et al.[20] and White[17] reported widespread distribution of endophytes both geographically and taxonomically. In the subfamily Festucoideae, for instance, 25% of the 157 species examined were infected with fungal endophytes. Saha et al.[15] found *Acremonium*-type endophytic fungi in 19 of 83 seed lots of fine fescue cultivars and selections. Sun et al.[21] observed a low frequency of endophyte infection, 3.2% in fine fescue plants collected from old turfs in the U.S.

Acremonium endophytes are transmitted maternally. As the grass plant matures and produces an inflorescence, the fungal hyphae move into the pith of the reproductive culm and then into the developing ovule and seed.[2,20,22] Approximately 2 to 3 weeks after the seed germinates, during sheath differentiation, the hyphae penetrate the leaf tissue.[23] Within the plant, the fungal hyphae of the *Acremonium* endophytes occur intercellularly in leaf and stem tissue. No *Acremonium* endophyte mycelium has been isolated from turfgrass roots.

Within a seed lot or a turf stand, the endophyte level is often measured as the percent infected seeds or plants. Mycelia concentration, however, varies within a plant. During vegetative growth, almost half of the mycelia of *A. lolii* are in the leaf sheath, with older leaves having higher concentrations of mycelia than young leaves.[24] Most of the remaining mycelia are evenly distributed between the crown and pseudostem.[25] During reproductive growth, Musgrave and Fletcher[25] found the highest endophyte concentration in the crown (34%). The flowering stem and leaf sheath contained 25 and 26% of the total, repectively. During plant dormancy, the endophyte is found within the crown and rhizomes. The within-plant distribution for *A. coenophialum* in tall fescue is similar to that of *A. lolii* in perennial ryegrass.

BENEFITS OF ENDOPHYTES IN TURF

One of the most important characteristics conveyed by *Acremonium* endophyte infection in turf is enhanced insect resistance. Endophytic fungi were first reported to be associated with insect resistance in the early 1980s.[13] Since this observation, well-documented situations of turfgrass resistance to aboveground-feeding insects have been attributed to endophyte infection.

Funk et al.[26] reported that endophyte-infected perennial ryegrasses demonstrated increased resistance to sod webworms, *Crambus* spp. Johnson-Cicalese and White[27] found endophytes improved resistance to four species of billbug, *Sphenophorus* spp., in New Jersey. Chinch bugs, *Blissus leucopterus hirtus* Montandon, have been shown to cause less damage on endophyte-infected fine fescue[15,28] and perennial ryegrass.[29] Breen[30] found endophyte infection increased resistance to three aphid species in perennial ryegrass; tall fescue; hard fescue, *F. longifolia* Thuill.; Chewings fescue; and blue fescue, *F. glauca* Lam. Fall armyworms, *Spodoptera frugiperda* J. E. Smith, were adversely affected by feeding on endophyte-infected perennial ryegrass, hard fescue, and Chewings fescue.[31]

Endophytes have been observed to occasionally deter some root-feeding pests; however, this association has not been demonstrated as conclusively as their effect on foliar-feeding herbivores. Researchers in Kentucky,[32] New Jersey, and Rhode Island have found variable effects of endophytes on white grubs depending upon grub age, grass species, and environmental conditions. Patterson et al.[33] found alkaloids produced by the endophyte of tall fescue deter Japanese beetle, *Popilla japonica* Newman, grubs feeding under laboratory conditions.

There are also reports of decreased nematode populations on roots of endophyte-infected tall fescue. Pedersen et al.[34] found reduced numbers of stubby root, *Paratrichodorus minor* (Colbran) Siddigi, and spiral nematodes, *Helicotylenchus dihystera* (Cobb) Sher, associated with endophyte infection.

Acremonium endophytes are considered either undesirable or beneficial, depending upon the herbivore. Their association with toxicosis in livestock has resulted in an effort to produce endophyte-free tall fescue pastures in the U.S. Improved insect resistance is a desirable characteristic in endophyte-infected turfgrasses that can be an important component of an IPM program. The use of these grasses provides an alternative to applications of synthetic chemical insecticides and reduces management inputs.

Research is being conducted to determine whether endophyte infection imparts disease resistance to the host plant. Isolates of *A. lolii* and *A. coenophialum* have been shown to produce compounds *in vitro* that hinder the growth of certain plant-pathogenic fungi.[34,35] Few incidents of endophyte-enhanced disease resistance have been documented, however. Tajimi[37] reported improved resistance to leaf spot, *Cladosporium phlei* (Gregory) De Vries, and stem rust, *Puccinia graminis* Pers. f. sp. *phlei-pratensis* Stak. et Piem., in endophyte-infected timothy, *Phleum pratense* L.

Researchers at Rutgers University have observed decreased dollar spot, *Sclerotinia homoeocarpa* Bennett, incidence on endophyte-infected Chewings fescue, hard fescue, and strong creeping red fescue,

F. rubra L. subsp. *rubra*.[37] These same researchers, however, have also observed increased *Pythium* blight infection on tall fescue plots containing certain strains of endophytes.[37]

Endophyte-enhanced disease resistance would be particularly valuable in turf and could decrease many management inputs. Chemical fungicide use would decrease in an integrated pest management program that utilized these grasses. Herbicide applications to control weeds that often invade turf as a result of disease damage would also decrease. Disease resistance can also lower the requirements of plants for other inputs such as water and fertilizers. Endophytes have been shown to modify their host plant's physiology and their utilization of nitrogen, water, and light (see Bacon[4]). Such modifications may decrease water and fertilizer applications.

DEVELOPMENT AND AVAILABILITY OF ENDOPHYTE-INFECTED TURFGRASSES

Turfgrass breeders are increasingly aware of the benefits of *Acremonium* endophytes. They and other scientists are developing methods to incorporate beneficial endophytes into elite cultivars. The predominant methods used to date have been modifications of conventional breeding methods, such as backcrossing and recurrent selection. Because *Acremonium* endophytes are transmitted maternally, and most cool-season turfgrasses are largely cross-pollinated, these techniques have been successful. Cultivars with high levels of endophyte infection have been developed in tall fescue, perennial ryegrass, and some species of fine fescue. Research programs continue to build upon these successes and work toward incorporating endophytes into other important species where *Acremonium* endophytes may not occur naturally, such as creeping bentgrass, *Agrostis palustris* Huds., and Kentucky bluegrass, *P. pratensis* L.

Other methods of incorporating endophytes into turfgrasses are being developed. Seedling inoculation[39] is the most common. Tissue culture, which introduces the endophyte *in vitro* to somatic embryos,[40] has been successfully used to incorporate the fungus. Other techniques involving manipulation of both endophyte and host at the cellular level are being investigated.

In the development of endophyte-infected turfgrasses, it is important to consider not only the genotype of the host, but also the genotype of the endophyte and its possible interaction with the host plant. *Acremonium* endophytes vary not only among species, but among strains of the same species. The performance of a host infected with one strain can be different from that of the same host infected with a strain possessing other characteristics.

Differences among fungal isolates grown *in vitro*, in production of allelochemicals,[20] or in optimal growth temperature suggest that endophytes may be selected for specific traits. Phenotypic differences in tall fescue plants were associated with endophyte infection when a number of genetically identical infected and free pairs were compared for differences in growth, morphology, and ergovaline production.[41] Forage yield, tiller number, and total nonstructural carbohydrates varied among endophyte-infected and free genotype pairs.

Levels of insect resistance have also varied within species of *Acremonium*. Endophyte-enhanced perennial ryegrass resistance to greenbug, *Schizaphis graminum* Rondani, and fall armyworm varied among four endophyte-host combinations.[31,42] This variation was due to differences in endophyte, host genotype, and endophyte-grass interactions. Hill et al.[43] found strains of *A. coenophialum*, isolated from tall fescue, varied in production of ergopeptine alkaloids.

Seasonal differences in degrees of insect resistance, endophyte concentration, lolitrem B concentration, and ryegrass staggers indicated that temperature may affect the expression of endophyte effects.[25,44,45] Breen[42] found that endophyte concentration in four genotypes of perennial ryegrass was higher at 14 and 21°C than at 7 and 28°C. A seasonal variation in mycelial concentration was also noted in the field, with highest concentrations of *A. lolii* in Repell perennial ryegrass recorded during spring and fall in central New Jersey. Endophyte concentration, however, varied more among genotypes than among temperature treatments. Higher endophyte concentration in the field conveyed greater resistance to greenbug.

Endophyte concentration may also be affected by soil fertility levels[46] and soil pH.[47] A number of abiotic factors could affect the expression of resistance and influence techniques in seed production and turf management.

Further investigation into interactions of genetic and endophytic resistance are necessary. The initial suitability of the host plant and its responses to injury by insects may affect the potential of endophytes to confer resistance. Plants that have some genetic resistance or are less suitable hosts to particular insects may be more likely to benefit from *Acremonium* endophyte infection.[48,49] In a comparison of endophyte-enhanced resistance to fall armyworm in two perennial ryegrasses with the same maternal parent, and thus the same source of endophyte, one endophyte-infected grass that was genetically more susceptible was less resistant than the other ryegrass selection which had a higher level of genetic resistance.[30] Fall armyworms have been able to avoid some genetic resistance by feeding on more nutritious plant parts or highly fertilized plants.[50] If genetic resistance is already high, an endophyte might not increase resistance. Although differences are often dramatic, endophytes should not be expected to confer immunity from insect pests.

Increased knowledge of the genetic variation within endophytes, their interactions with various hosts, and responses to environmental conditions will be needed to gain the greatest possible benefits from endophyte-enhanced performance. Ideally, endophytes possessing characteristics that convey different benefits to their host will be identified. Funk et al.[49] suggested that optimum performance may require a combination of genetic and endophytic resistance. They proposed that the most efficient technique will be to select the "best" endophytes and host plants separately and incorporate the endophytes into compatible elite turfgrasses. Breeding programs using these techniques should be useful in developing cultivars that require fewer management inputs.

As previously mentioned, endophyte-infected cultivars of certain major cool-season turfgrass species are commercially available. The consumer must be aware, however, of the importance of proper seed storage in maintaining endophyte viability. Seed of these cultivars must be kept in cool, dry conditions and used as soon as possible. Under hot, humid storage conditions, viability may decline rapidly.[49] Also, endophyte viability is likely to decrease with increasing age of stored seed. With proper storage and utilization, however, endophyte-infected cultivars can be very useful for turf and play an important role in IPM programs. A brief description of the development and availability of endophyte-infected cultivars, in the major cool-season turfgrass species, follows.

TALL FESCUE

Tall fescue is an important species used widely for turf, pastures, and conservation in many parts of the U.S., particularly the transition zone. Tall fescue breeding is a relatively young discipline. The first improved cultivar, Kentucky 31, was released in 1943 and is still the most widely used. Most lots of Kentucky 31 seed are highly infected with endophyte. This cultivar is well suited as a pasture or conservation grass. The presence of the endophyte, however, is detrimental to grazing animals. Kentucky 31 is also widely used for lawns, particularly in the transition zone and parts of the southeast where the endophyte contributes to its persistence.

Many improvements have been made in tall fescue cultivars for turf. The first turf type, Rebel, was released in 1980.[51] Since its release, continued dramatic improvements have been made in color, leaf texture, growth habit, and seed yield. While these improvements have resulted in the availability of many tall fescue cultivars that are superior, for turf, to the original turf-types, there are still problems with persistence in the areas of this species' highest use. The primary characteristics that need improvement are problems with disease and insect resistance. Newer cultivars have better resistance to net blotch, *Drechslera dictyoides* (Drechs.) Shoemaker, crown rust, *P. coronata* Corda., and stem rust. Their resistance to brown patch, *Rhizoctonia solani* Kuhn., and *Pythium* blight is still poor. Evaluation programs in tall fescue's primary use areas continue to work on these problems.

Only a few turf-type tall fescue cultivars have high levels of endophyte infection. Two of these are Shenandoah and Titan. A number of other cultivars contain lower levels of endophyte. The primary reasons that endophytes have not been incorporated into many tall fescue cultivars are their wide use in pastures and the use of tall fescue straw for feeding livestock. Endophyte infection may play an important role in tall fescue turf persistence in the transition zone and southeastern U.S. The identification of endophyte strains that are most beneficial in these situations will be an important factor in future tall fescue improvement programs.

PERENNIAL RYEGRASS

There has been dramatic progress in the development of perennial ryegrasses for turf in recent years. Darker color, lower growth habit, improved disease resistance and stress tolerance, and better mowing quality are some of the improvements that have been made in turf-type cultivars. As a result, many attractive, wear-tolerant cultivars are available and widely used for permanent turfs in temperate regions of the U.S. They are also used extensively throughout the southeastern U.S. and southern California for the winter overseeding of dormant warm-season turfgrasses.

Acremonium endophytes have been incorporated into many perennial ryegrass cultivars including Advent, Affinity, All*Star, APM, Assure, BrightStar, Citation II, Dandy, Dasher II, Gettysburg, Legacy, Manhattan II (E), Palmer II, Pennant, Pinnacle, Prelude II, Regal, Repell, Repell II, Saturn, Seville, Sherwood, SR 4000, SR 4100, SR 4200, and Yorktown III. Many other endophyte-infected cultivars are being developed.

FINE FESCUE

The fine fescues are a group of *Festuca* species with fine, bristlelike leaves. They possess a number of characteristics that make them desirable for certain turf situations. They are quite shade tolerant and well adapted to acidic, low-fertility soils. They perform especially well under low maintenance. The primary species used for turf are blue; Chewings; hard; sheeps, *F. ovina* L.; slender creeping red, *F. rubra* subsp. *litoralis* (Meyer) Auguier; and strong creeping red fescue.

Blue fescues are primarily used for ornamental purposes. Their characteristic bluish leaf color does not mix well with other turfgrasses. They are used in wildflower mixes or in ornamental situations where this color is attractive and desirable. *Acremonium* endophytes have been found to exist in blue fescues[15] and have been associated with insect resistance.[30,31] The potential exists to develop cultivars with high levels of endophyte infection.

Chewings fescues are becoming more widely used for turf. They are bright green, low growing, dense turfgrasses that are used in mixes with Kentucky bluegrass and perennial ryegrass for lawns in temperate regions of the U.S. They are also used in overseeding mixtures with other turfgrasses.

Cultivars of Chewings fescue with high endophyte infection are available. A source of endophyte found in plants growing in an old lawn in Cambridge, MA, has been used to develop these cultivars. Some of the available cultivars containing this endophyte are Banner II, Jamestown II, and SR 5000. Other cultivars are being developed.

Hard fescues produce a bright green, dense, low-growing turf that often retains its color under drought stress. An *Acremonium* endophyte found in two European cultivars is being used in hard fescue breeding programs. SR 3000 is one commercially available cultivar containing high levels of endophyte. Other cultivars are being developed and should be available soon.

Acremonium endophytes have been detected in slender creeping red, strong creeping red, and blue fescues.[15] Improved cultivars will be developed with these endophytes.

CREEPING BENTGRASS

An endophyte-infected creeping bentgrass would potentially be very valuable to turfgrass managers. Bentgrasses form some of the most intensively managed turf surfaces, including golf course greens, tees, and fairways. Cultivars containing the proper beneficial endophytes could lower management inputs. Endophytes have been detected in some bentgrass species.[17] Researchers are searching for useful endophytes in many species and are incorporating such fungi into creeping bentgrass germ plasm.[52]

KENTUCKY BLUEGRASS

To date, *Acremonium* endophytes have not been detected in Kentucky bluegrass. Researchers are searching for endophytes in Kentucky bluegrass and other species of *Poa*. The discovery of an endophyte that conveys beneficial characteristics to a Kentucky bluegrass host will be very useful in the development of new cultivars. Because Kentucky bluegrass is a highly apomictic species, a high percentage of the plants in a cultivar are genetically identical to each other. For this reason, only one plant would need to be successfully inoculated to produce a cultivar with a high level of endophyte infection.

CONCLUSION

Acremonium endophytes are very useful to turfgrass managers. They convey such benefits as herbivore resistance, resistance to some diseases, and improved stress tolerance. Turfgrass breeders and other scientists successfully incorporate beneficial endophytes into elite turfgrass germplasm to develop new cultivars. Endophyte-infected cultivars of perennial ryegrass, tall fescue, and some fine fescue species are commercially available. With proper storage and use, such cultivars can be valuable components of IPM programs for turf.

REFERENCES

1. **Carroll, G. C.,** The biology of endophytism in plants with particular reference to woody perennials, *Microbiology of the Phylloplane*, Fokkema, N. J. and Van der Heavel, J., Eds., Cambridge University Press, Cambridge, UK, 1986, 205.
2. **Siegel, M. R., Johnson, M. C., Varney, D. R., Nesmith, W. C., Buckner, R. C., Bush, L. P., Burrus, P. B., Jones, T. A., and Boling, J. A.,** A fungal endophyte in tall fescue: incidence and dissemination, *Phytopathology*, 74, 932, 1984.
3. **Vogl, A. E.,** Mehl und die anderen Mehlprodukte der Cereallen und Leguminosen, *Nahrungsm. Unters. Hyg. Warenk.*, 12, 25, 1898.
4. **Bacon, C. W.,** Procedure for isolating the endophyte from tall fescue and screening isolates for ergot alkaloids, *Appl. Environ. Microbiol.*, 54, 2615, 1988.
5. **Sun, S., Clarke, B. B., and Funk, C. R.,** Effect of fertilizer and fungicide applications on choke expression and endophyte transmission in Chewings fescue, in *Proc. Int. Symp. on Acremonium Grass Interactions*, Quisenberry, S. S. and Joost, R. E., Eds., Louisiana Agricultural Experimental Station, Baton Rouge, LA, 1990, 62.
6. **Freeman, E. M.,** The seed fungus of *Lolium tementulum* L., the darnel, *Philos. Trans.*, 196, 1, 1904.
7. **Neill, J. C.,** The endophyte of ryegrass (*Lolium perenne*), *N.Z. J. Sci. Tech.*, 21, 280, 1940.
8. **Sampson, K.,** The systemic infection of grasses by *Epichloe typhina* (Pers.), *Tul. Trans. Br. Mycol. Soc.*, 18, 30, 1933.
9. **Fletcher, L. R. and Harvey, I. C.,** An association of a *Lolium* endophyte with ryegrass staggers, *N.Z. Vet. J.*, 29, 185, 1981.
10. **Gallagher, R. T., Campbell, A. C., Hawkes, A. D., Holland, P. T., McGaveston, D. A., and Pansler, E. A.,** Ryegrass staggers: the presence of lolitrem neurotoxins in perennial ryegrass seed, *N.Z. Vet J.*, 30, 183, 1982.
11. **Gallagher, R. T., Hawkes, A. D., Steyen, P. S., and Vleggar, R.,** Tremorgenic neurotoxins from perennial ryegrass causing ryegrass staggers disorder of livestock: structure elucidation of lolitrem B in perennial ryegrass by high performance liquid chromatography with fluorescence detection, *J. Chromatogr.*, 321, 217, 1984.
12. **Mortimer, P. H., Young, P. W., and diMenna, M. E.,** Perennial ryegrass staggers research: an overview, *Proc. N.Z. Soc. Animal Prod.*, 44, 181, 1984.
13. **Prestidge, R. A., Pottinger, R. P., and Barker, G. M.,** An association of *Lolium* endophyte with ryegrass resistance to Argentine stem weevil, *Proc. 35th N.Z. Weed Pest Cont. Conf.*, 119, 1982.
14. **Morgan-Jones, G. and Gams, W.,** Notes on Hyphomycetes XLI. An endophyte of *Festuca arundinacea* and the anamorph of *Epichloe typhina*, new taxa in one of the two new sections of *Acremonium*, *Mycotaxon*, 15, 311, 1982.
15. **Saha, D. C., Johnson-Cicalese, J. M., Halisky, P. M., Van Heemstra, M. I., and Funk, C. R.,** Occurrence and significance of endophytic fungi in the fine fescues, *Plant Dis.*, 71, 1021, 1987.
16. **Sun, S. and Breen, J. P.,** Inhibition of *Acremonium* endophyte by Kentucky bluegrass, *Proc. 2nd Intern. Symp. on Acremonium-Grass Interactions*, Hume, D. E., Latch, G. C. M., and Easton, H. S., Eds., Palmerston North, New Zealand, 19, 1993.

17. **White, J. F.,** Widespread distribution of endophytes in the Poaceae, *Plant Dis.*, 71, 340, 1987.

18. **White, J. F. and Morgan-Jones, G.,** Endophyte-host associations in forage grasses. VI. *Acremonium chisosum,* a new species isolated from *Stipa eminens* in Texas, *Mycotaxon*, 28, 179, 1987.

19. **Latch, G. C. M., Christensen, M. J., and Samuels, G. J.,** Five endophytes of *Lolium* and *Festuca* in New Zealand, *Mycotaxon*, 20, 535, 1984.

20. **Bacon, C. W., Lyons, P. C., Porter, J. K., and Robbins, J. D.,** Ergot toxicity from endophyte-infected grasses: a review, *Agron. J.*, 78, 106, 1986.

21. **Sun, S., Clarke, B. B., Huff, D. R., Betts, L. L., Smith, D. A., and White, J. F.,** Enhanced performance and new sources of *Acremonium typhinum* in fine fescues, *Proc. 2nd Intern. Symp. on Acremonium-Grass Interactions*, Hume, D. E., Latch, G. C. M., and Easton, H. S., Eds., Palmerston North, New Zealand, 23, 1993.

22. **Latch, G. C. M. and Christensen, M. J.,** Ryegrass endophyte incidence and control, *N.Z. J. Agric. Res.*, 25, 443, 1982.

23. **Lyons, P. C. and Bacon, C. W.,** Growth of the tall fescue endophyte into seedlings as related to temperature and sheath differentiation, *Phytopathology*, 75, 501, 1985.

24. **Breen, J. P.,** unpublished data, 1989.

25. **Musgrave, D. R. and Fletcher, L. F.,** The development and applications of ELISA detection of *Lolium* endophyte in ryegrass staggers research, *Proc. N.Z. Soc. Animal Prod.*, 44, 185, 1984.

26. **Funk, C. R., Halisky, P. M., Johnson, M. C., Siegel, M. R., Stewart, A. V., Ahmad, S., Hurley, R. H., and Harvey, I. C.,** An endophytic fungus and resistance to sod webworms: Association in *Lolium perenne* L., *Bio/Technology*, 1, 189, 1983.

27. **Johnson-Cicalese, J. M. and White, R. H.,** Effect of *Acremonium* endophytes on four species of billbug found on New Jersey turfgrass, *J. Am. Soc. Hortic. Sci.*, 115, 602, 1990.

28. **Funk, C. R., Halisky, P. M., Ahmad, S., and Hurley, R. H.,** How endophytes modify turfgrass performance and response to insect pests in turfgrass breeding and evaluation trials, in *Proc. 5th Int. Turf. Res. Conf.*, Lemaire, F. Ed., International Turfgrass Society, Avignon, France, 1985, 132.

29. **Mathias, J. K., Ratcliffe, R. H., and Hellman, J. L.,** Association of an endophytic fungus in perennial ryegrass and resistance to the hairy chinch bug (Hemiptera:Lygaeidae), *J. Econ. Entomol.*, 83, 1640, 1990.

30. **Breen, J. P.,** Enhanced resistance to three species of aphids in *Acremonium* endophyte infected turfgrasses, *J. Econ. Entomol.*, 86, 1279, 1993.

31. **Breen, J. P.,** Enhanced resistance to fall armyworm in *Acremonium* endophyte infected turfgrasses, *J. Econ. Entomol.*, 86, 621, 1993.

32. **Potter, D. A., Patterson, C. G., and Redmon, C. T.,** Feeding ecology of Japanese beetle and southern masked chafer grubs (Coleoptera:Scarabaeidae): influence of turfgrass species and tall fescue endophyte, *J. Econ. Entomol.*, 85, 900, 1992.

33. **Patterson, C. G., Potter, D. A., and Fannin, F. F.,** Feeding deterrency of alkaloids from endophyte-infected grasses to Japanese beetle grubs, *Entomol. Exp. Appl.*, 61, 285, 1992.

34. **Pedersen, J. F., Rodriguez-Kabana, R., and Shelby, R. A.,** Ryegrass cultivars and endophyte in tall fescue affect nematode in grass and succeeding soybean, *Agron. J.*, 80, 811, 1988.

35. **Bayaa, B. O., Halisky, P. M., and White, J. F.,** Inhibitory interactions between *Acremonium* spp., and the mycoflora from seeds of *Festuca* and *Lolium*, *Phytopathology*, 77, 115, 1987.

36. **White, J. F. and Cole, G. T.,** Endophyte-host associations in forage grasses. III. *In vitro* inhibition of fungi by *Acremonium coenophialum*, *Mycologia*, 77, 487, 1985.

37. **Tajimi, A.,** Useful and harmful effects of endophyte in timothy (*Phleum pratense* L.), *Annu. Rep. Hokkaido Nat. Ag. Exp. Sta.*, 34, 1990.

38. **Funk, C. R.,** personal communication, 1991.

39. **Latch, G. C. M. and Christensen, M. J.,** Artificial infection of grasses with endophyte, *Ann. Appl. Biol.*, 107, 17, 1985.

40. **Kearny, J. F., Parrott, W. A., and Hill, N. S.,** Infection of somatic embryos of tall fescue with *Acremonium coenophialum*, *Crop Sci.*, 31, 979, 1991.

41. **Hill, N. S., Stringer, W. C., Rottinghaus, G. E., Belesky, D. P., Parrott, W. A., and Pope, D. D.,** Growth, morphological and chemical component responses of tall fescue to *Acremonium coenophialum*, *Crop Sci.*, 30, 156, 1990.

42. **Breen, J. P.,** Temperature and seasonal effects on expression of *Acremonium* endophyte-enhanced resistance to *Schizaphis graminum* Homo-ptera:Aphididae, *Environ. Entomol.*, 21, 68, 1992.

43. **Hill, N. S., Parrott, W. A., and Pope, D. D.,** Ergopeptine alkaloid production by endophytes in a common tall fescue genotype, *Crop Sci.*, 31, 1545, 1991.

44. **Fletcher, L. R.,** Effects of presence of *Lolium* endophyte on growth rates of weaned lambs, growing onto hoggets on various ryegrasses, *Proc N.Z. Grassland Assoc.,* 47, 99, 1983.

45. **Prestidge, R. A., diMenna, M. E., Van der Zijpp, S., and Baden, D.,** Ryegrass content, *Acremonium* endophyte and Argentine stem weevil in pastures in the volcanic plateau, *Proc. 38th N.Z. Weed Pest Cont. Conf.,* 38, 38, 1985.

46. **Stewart, A. V.,** Plant Breeding Aspects of Ryegrasses (*Lolium* spp.) Infected with Endophytic Fungi, Ph.D. dissertation, Lincoln College, University College of Agriculture, Canturbury, New Zealand, 1987.

47. **Breen, J. P. and Duell, R. W.,** unpublished data, 1989.

48. **Clay, K.,** Fungal endophytes of grasses: a defensive mutualism between plants and fungi, *Ecology,* 69, 10, 1988.

49. **Funk, C. R., Clarke, B. B., and Johnson-Cicalese, J. M.,** Role of endophytes in enhancing the performance of grasses used in conservation and turf, *IPM for Turfgrass and Ornamentals,* Leslie, A. R. and Metcalf, R. L., Eds., U.S. Environmental Protection Agency, Washington, DC, 1989, 203.

50. **Wiseman, B. R., Leuck, D. B., and McMillan, W. W.,** Effect of crop fertilizer on feeding of larvae of fall armyworm on excised leaf sections of corn foliage, *J. Ga. Entomol. Soc.,* 8, 136, 1973.

51. **Funk, C. R., Engel, R. E., Dickson, W. K., and Hurley, R. H.,** Registration of Rebel tall fescue, *Crop Sci.,* 21, 632, 1981.

52. **Lee, L.,** personal communication, 1992.

Chapter 47

Stress Tolerance of Endophyte-Infected Turfgrass

Michael D. Richardson, *Turf Merchants, Inc., Aurora, OR*

Charles W. Bacon, *Toxicology and Mycotoxin Research Unit, R.B. Russell Research Center, USDA-ARS, Athens, GA*

CONTENTS

Introduction 529
Nitrogen Nutrition 530
Drought and Flooding 531
    General Stress Tolerance Mechanisms 531
    Osmotic Adjustment 532
    Photosynthesis and Stomatal Conductance 532
Adaptive Metabolic Strategy to Abiotic Stresses 533
Conclusions 534
References 535

INTRODUCTION

The culture and maintenance of high-quality turf and ground cover is often inhibited by environmental constraints associated with water availability, mineral nutrition, and biotic predators such as insects and disease organisms. Traditional approaches used to confront abiotic stresses such as inadequate water and fertility include supplementary irrigation and soil mineral additives. Although these approaches can be successful in reducing stress damage, economic and environmental problems associated with these types of supplementation preclude their long-term use. The exploitation of a mutualistic symbiosis between fungal endophytes and certain turfgrass species offers a new approach in reducing damage which results from abiotic stress.[1,2] Interest in these fungus-grass associations was stimulated after their identification in pasture cultivars of tall fescue (*Festuca arundinacea* (Schreb.) and perennial ryegrass (*Lolium perenne* L.) as a causal agent in toxicities associated with grazing animals.[3,4] At the time of this discovery, the ecological significance of the symbiosis to plant survival was largely unknown, and it was only after manipulation of the symbiosis that a mutualism between plant and fungus became apparent.[5]

Ecologically and economically significant symbiotic associations are widespread in turf crops and ground cover. Parasitic symbionts such as bacterial, fungal, and viral pathogens are extensive and cause economic losses in most aspects of turf utilization. On the other hand, habitation of leguminous ground cover species by nitrogen-fixing, bacterial symbionts is a common mutualism which augments the nitrogen needs of the host plant. Mycorrhizal fungi are also notable mutualistic symbionts of turf species, inhabiting the root epidermal area and enhancing the plant's ability to scavenge the rhizosphere for water and nutrients. The fact that these and other mutualisms occur throughout the plant kingdom and the seemingly constitutive nature of each relationship suggests that the association is vitally important to the ecological well-being of the partners.

Infection of grasses by systemic fungal endophytes of the Balansiae tribe is extensive, with estimates of over 80 genera of grasses serving as hosts for these organisms.[6,7] The grass host range is physiologically and geographically diverse, with endophytes infecting both warm-season and cool-season grasses.[8] Although most grass endophytes identified thus far have been classified in the Balansiae tribe, there are other genera whose classification is based on their reproductive structure produced in agar after isolation.[7] A distinct sexual reproductive cycle which gives rise to meiotic spores is characteristic of endophytes classified in the genera *Atkinsonella*, *Balansia*, *Myriogenospora*, and *Epichloë*. These endophytes typically form external fertile stromata on the host plant and, in the process, often prevent a host from producing fertile inflorescences.[9]

0-87371-350-8/94/$0.00+$.50

Endophytic fungi which inhabit economically important turf species such as the fescues and ryegrasses belong to the genus *Acremonium* Section Albo-lanosa. These are characterized by the absence of a sexual reproductive cycle, but their anamorphic state resembles that of *Epichloë typhina*.[10] These endophytes complete their entire life cycle within the vegetative and reproductive tissues of the host plant. However, the absence of external fruiting bodies capable of sexual reproduction does not appear to have interfered with the evolution and survival strategy of the endophyte. The life cycle, taxonomy, and hosts of several commonly encountered *Acremonium* species have recently been reviewed by White et al.[11]

If a mutualistic symbiosis indeed exists between grasses and endophytes, what benefit does the fungal partner offer to the plant in this relationship? The initial observation which linked the endophyte to toxic animal disorders provided the first clue to this very complex relationship. As noted by Siegel and Schardl,[8] those persons concerned with animal husbandry only saw the fungus as a detriment to these grasses and failed to realize the ecological significance of the anti-herbivore characteristics. Enhanced insect resistance and livestock toxicities associated with endophyte-infected grasses have received considerable attention, and comprehensive reviews on the topic have recently appeared.[1,8,12-14]

Aside from the well-established facts of herbivore resistance associated with endophytes, observations that infected grasses are more tolerant of abiotic stresses such as drought and low fertility are equally as common.[2,15,16] The physiological mechanisms by which endophytes improve plant growth and persistence under abiotic stress conditions are probably ubiquitous, but diverse. In an attempt to initiate understanding of these mechanisms, the following review will focus on the adaptive nature of endophyte-infected tall fescue and ryegrass to abiotic stresses associated with nitrogen nutrition, drought, flooding, and light. Although this review is relative to turf-type grasses, most of the material reviewed will be the results of research conducted with pasture selections of tall fescue and perennial ryegrass. It also is important to distinguish between studies describing genetically diverse populations of infected and uninfected materials vs. experiments in which genetically identical, cloned plant material was utilized for comparison.

NITROGEN NUTRITION

The involvement of grass endophytes with host plant nitrogen metabolism was evident in early research which demonstrated a positive relationship between nitrogen fertilization and severity of ruminant toxicosis.[17] Although the interaction of nitrogen fertilization with herbivore performance problems is an important aspect of this symbiosis, it is outside the scope of this text. Other soil fertility parameters such as acidity or micronutrients also might be involved with endophyte-plant associations, but the only nutritional studies presently available have compared plant response to various forms and rates of nitrogen fertilization.

Grasses respond to nitrogen fertilization by increasing leaf expansion and tiller growth.[18] Unfortunately, this macronutrient is quickly removed from soils in high-producing systems, and supplemental fertilizers are costly. In tall fescue, the endophyte has been shown to increase shoot biomass in several cloned grasses, especially where nitrogen was in excess.[15,19] Moreover, one infected genotype also was found to utilize inadequate soil nitrogen more efficiently during regrowth after defoliation.[15] In a population study, infected plants also outyielded noninfected plants, but again, the response was restricted to high nutrient levels.[20] Besides increased herbage growth, nitrogen rate may also influence leaf morphology and development of certain clones. Infected plants grown under high nitrogen levels had thicker and narrower blades than their endophyte-free partners, and air pockets occurred within the sheath tissue mesophyll.[15] Arachevaleta et al.[15] reported that leaves of infected plants matured faster, a response that we suspect would stimulate new leaf initiation from existing tillers. Rapid leaf maturation followed by leaf initiation might explain why shoot biomass is often increased in infected plants even when number of tillers remains the same.[15]

The interaction of fertility and endophyte-infection on perennial ryegrass is not well defined. Many of the controlled studies to date found that infected plants were no more productive than endophyte-free plants under all levels of nitrogen fertilization,[20] suggesting that the energy cost to the host plant in this particular association may outweigh advantages to nitrogen efficiency. However, when CO_2 concentrations were increased approximately twofold higher than normal atmospheric levels, infected grasses were more responsive to nitrogen fertilization by increasing leaf area, shoot dry weight, and root dry weight compared to noninfected plants.[21] The improved efficiency of C_3 plants under high CO_2 may offset the proposed energy deficits associated with the endophyte of perennial ryegrass. Although these studies suggest no advantage for infected grasses under normal conditions, Latch et al.[22] found certain clones of endophyte-infected perennial ryegrass to be much more productive than uninfected materials. Future work with these grasses and their endophytes might define any role of fertility with increased fitness.

The differential growth response of endophyte-infected grasses to nitrogen fertilization, as well as the abundance of nitrogen-rich intermediates arising from fungal metabolism,[23-25] suggests that basic nitrogen metabolism in the host plant is altered by this symbiosis. A comprehensive study of nitrogen utilization in tall fescue demonstrated that leaf blades of infected plants had significant increases in glutamine synthetase activity, a key enzyme involved in the reassimilation of ammonia lost during processes such as photorespiration.[26] Ammonia and total amino acid pools were increased in sheath tissues and leaf blades of infected plants, while nitrate levels were reduced. Levels of certain amino acids also increased in infected plants, with the most pronounced effect on levels of asparagine and glutamine in the sheath material. Although it is still impossible to delineate mechanisms or effects which are of either plant or fungal origin, it is obvious from this study that the plant nitrogen status is influenced by the endophyte in all parts of the host plant, including those uninhabited by the fungus.

Nitrogen nutrition also plays a role in seedling growth of endophyte-infected tall fescue and perennial ryegrass. Infected seedlings were less vigorous than endophyte-free seedlings at low nitrogen levels, but more vigorous when nitrogen was in excess.[20] These data further emphasize that metabolic costs to the host plant may outweigh beneficial aspects of the symbiosis and that, in addition to environmental adversity, certain ecological advantages may only be observed in situations where nutrients and energy are excessive. The possibility of other factors such as hormonal imbalances during seedling development should not be ruled out, as the plant growth regulator indole acetic acid has been produced *in vitro* by the tall fescue endophyte.[27]

DROUGHT AND FLOODING

GENERAL STRESS TOLERANCE MECHANISMS

The widespread use of tall fescue as a turf and pasture species is a direct result of its persistence under water stress. The involvement of the endophyte with this adaptability was realized when researchers concerned with animal toxicities discovered that endophyte-free tall fescue was not as persistent under severe drought stress.[16] Although their experimental conditions were not controlled, the results demonstrated that drought tolerance in tall fescue was partially due to the endophyte. This report was also the stimulus for further controlled experiments by others to identify mechanism(s) involved in this adaptability.

Plants endure or survive drought with a variety of escape and tolerance mechanisms, all of which serve to improve water utilization or water conservation.[28] Escape mechanisms are primarily associated with annual species and are characterized by rapid phenological development during periods of adequate soil moisture, and plants with such mechanisms undergo a dormancy period during severe drought. The perennial nature of endophyte-infected grasses suggests that drought escape is not an alternative, so tolerance mechanisms are utilized. Numerous morphological and physiological features have been associated with drought tolerance in endophyte-infected tall fescue, including increased root growth, enhanced leaf rolling, stomatal closure, and osmotic adjustment.

Increased root growth of infected plants subjected to drought stress has been observed both at the population level and in experiments with cloned material.[21,29,30] Grasses often respond to a water deficit by increasing the allocation of photosynthates to root growth and subsequently improving the opportunity for water absorption.[31] Although reduced nematode feeding on infected tall fescue roots has been one explanation for improved rooting,[32] other factors must also be involved since root growth of infected plants was increased when plants were subjected to drought under parasite-free conditions.[22,29] It is likely that other physiological processes such as photosynthesis or osmotic adjustment are the actual mechanisms altered by endophyte-infection while increased root growth is only an indirect consequence of these effects.

The regulation of water loss in endophyte-infected grasses by leaf rolling has been reported in several studies comparing cloned genotypes of tall fescue.[15,33] The phenomenon of leaf rolling is also a common drought tolerance mechanism among grass species, and is the result of subepidermal bulliform cells collapsing.[34] The main advantage in leaf roll is a reduction in leaf radiant heat, thereby decreasing transpirational demand and subsequent water loss. However, the direct effect of leaf rolling on stomatal function is unclear, as evidence suggests that a microenvironment of high humidity and low CO_2 is created within the rolled leaf,[35] conditions which should cause stomates to remain open. Although morphological distinctions such as subcellular air spaces have been identified in infected leaves,[15] there is no information comparing the anatomy or distribution of bulliform cells in infected and uninfected grasses.

Although the ability to withstand drought is a significant survival feature of tall fescue, the facility to tolerate extended flooded conditions is equally widespread and has also been studied in endophyte-infected cloned materials.[36] Infected plants were found to yield more dry matter than uninfected plants[36] and a stoloniferous growth habit was observed in infected plants after extended inundation.[61] The excessive stolons produced shortly after inundation by infected plants could be interpreted as a means for escaping the flooded condition.

OSMOTIC ADJUSTMENT

A drought tolerance mechanism which has emerged at the forefront of endophyte research is the ability of infected plants to accumulate solutes such as sugars, amino acids, or inorganic ions in response to water deficits, a phenomenon generally referred to as osmotic adjustment. The rationale for osmotic adjustment is to decrease the water potential of a cell, such that a downward gradient of water potential is generated from outside to inside the cell. This gradient causes water to move into the cell and prevents the cell from dehydrating under water stress. West et al.[37] were the first to document osmotic adjustment in developing tissues of endophyte-infected tall fescue, although a similar response was observed in perennial ryegrass before the association with an endophyte was considered.[38] Studies on tall fescue indicate that osmotic adjustment in infected plants occurs primarily in developing leaves and tiller bases rather than mature tissues.[39] Although the maintenance of turgor in these young tissues may allow growth to occur, these authors suggest that osmotic adjustment and increased turgor are more likely related to plant survival than growth per se. The data of Arachevaleta et al.[15] would support this contention, as no differences in herbage yield were observed under severe stress, but survival and regrowth potential of infected plants was increased dramatically over noninfected clones.

Typical solutes which plants may accumulate in response to drought or saline environments are normal cellular constituents such as carbohydrates, ions, or amino acids.[40] In cool-season grasses, high concentrations of mono- and oligosaccharides make up a large portion of the cellular osmotic potential, but these are rapidly utilized for growth and development.[41,42] As leaf elongation and tiller development are early casualties to water deficit, carbohydrates accumulate at greater rates than they are utilized, leading to an increase in osmotic pressure at the growing point.[43] In a study on infected tall fescue exposed to water stress, the leaf sheath of infected plants accumulated fructose and glucose in greater amounts than noninfected tissues.[44] These data suggest that either infected plants maintain a greater supply of photosynthate to the growing point, or energy sinks in the form of fungal metabolism enhance deposition of carbohydrates to infected areas. In many parasitic symbioses, high concentrations of soluble carbohydrates are deposited at infection sites, altering the osmotic and water status of those tissues.[45]

Sugar alcohols, or polyols, are major components of the carbohydrate pool in many fungi, and the accumulation of these compounds as cellular osmoticum is a common phenomenon in fungi, algae, lichens, and some higher plants.[46] Mannitol and arabitol have been shown to be present in infected tall fescue leaf sheaths under well-watered conditions, while arabitol concentrations increased when plants were exposed to water stress.[44] The effect these compounds exert on plant water potential is unknown, but it is interesting to speculate that their presence in this system would enhance cellular osmotic potential. Mannitol also has been identified in tissues of perennial ryegrass,[47] and although this study was conducted without knowledge or interest in a fungal symbiont, it is likely that the polyol was the result of infection by an endophyte.

In addition to carbohydrates and sugar alcohols, the amino acids proline and betaine are often utilized as osmoticum in several plant species.[40] The interactions of endophyte and nitrogen nutrition suggests that amino acid metabolism in infected plants may influence drought tolerance. A greater accumulation of proline in infected tissue was observed by Lyons et al.[26] in plants grown under high nitrogen fertilization, but it is unknown how drought might affect this process. Drought-induced accumulations of proline have been reported earlier in perennial ryegrass[48] and tall fescue[49] but, again, these studies were conducted without addressing the possible endophyte influence on this response.

PHOTOSYNTHESIS AND STOMATAL CONDUCTANCE

The effect of endophytes on photosynthesis of tall fescue has been investigated only briefly, although extensive research into this area is in progress.[19,50] Because endophytes are energetically classified as biotrophs, the host plant must meet those energy needs through the products of photosynthesis. Plant products which may be metabolized by fungi to meet their energy needs include carbohydrates, lipids, and amino acids.[51] In many plant-fungal associations, especially parasitic relationships, the host plant increases photosynthetic output presumably to satisfy both plant and fungal demands, as the fungus is an additional sink for photosynthate. The studies comparing photosynthesis of infected and noninfected plants have not shown increases in photosynthetic rate due to endophyte. Conversely, a slight reduction in photosynthesis of infected grasses has been demonstrated in several genotypes, implying that the endophyte is not a significant sink for photosynthates.[19,50]

The published studies on photosynthetic rates of selected clonal lines of infected grasses have used light intensity as a factor in plant response, but the data have been conflicting. Belesky et al.[19] showed that infected plants had reduced photosynthetic rates compared to noninfected plants and the differences were greater under highest light intensities. In a study examining diurnal responses of photosynthesis, Richardson et al.[50] also found decreased photosynthesis in certain infected clones, but differences were noted only at low light intensities. The disparity between results may reflect differences in genetic material or even experimental conditions. In either case, the small differences between infected and noninfected plants demonstrate that the vital process of photosynthesis is not significantly affected by the infection. This strengthens the proposal that this symbiosis is not parasitic.

The intercellular localization of endophytes, and a lack of any haustoriate-type structures,[52] indicate that the fungus derives nutrients exclusively from the apoplast by a simple diffusion process. *In vitro* studies of endophyte nutritional requirements have shown that the choice of carbon and nitrogen sources for growth and development is diverse,[53,54] suggesting that the endophyte could persist on a number of compounds depending on availability within the apoplast. The slow growth of the endophyte in culture[54,55] may also be suggestive of its growth in the plant, a situation which would not require a high nutrient uptake. However, a nonhaustoriate confinement to the apoplasm does not prevent the possibility of an extracellular-directed and driving movement of nutrients from the grass cell to the apoplasm and fungus. This certainly warrants research and an increase in intercellular sap nutrients may be indicative of this mechanism. No research has been attempted to determine the makeup of intercellular sap in uninfected and infected grasses or to determine the effects environmental factors may have on the contents of the sap.

Photosynthetic studies of tall fescue have suggested that reduced photosynthesis of infected plants may result from decreased stomatal conductance.[19,50] Stomatal conductance also plays an important role in drought tolerance by regulating the amount of water lost during the process of transpiration. The hormone abscisic acid (ABA) figures centrally in the mechanisms by which stomatal closure is regulated. A preliminary report[56] indicates that in the presence of precursors, ABA is produced by *Acremonium* endophytes *in vitro* and ABA concentrations in infected plants were greater than in uninfected material. Because ABA is translocated through the vascular system,[57] movement from the site of fungal colonization to active sites at stomatal guard cells may be possible, although any direct control by the fungus to regulate stomatal closure is unknown. Both drought and flooding can initiate either increased or decreased ABA synthesis, respectively; therefore, stomatal closure under these conditions may be more closely regulated in infected plants.

ADAPTIVE METABOLIC STRATEGY TO ABIOTIC STRESSES

The focus of attention in the area of endophyte-grass relationships is the potential for manipulating or transferring stress resistances to other grasses. This exciting concept is only possible if the mechanism is responsive to the environment and is regulatable. We believe that there is a regulated adaptive metabolic response to all stresses imposed on endophyte-infected grasses. Such a strategy is essential and is expected to fit within the overall energy economy of the symbiosis. Further, it is essential that these strategies must be linked to specific environmental triggers. This would mean that resistances to stresses are not simply secondary results of an ineffective metabolism operating under stress but rather are regulated adjustments responding quantitatively to less than normal growing conditions. If and when such plant metabolic strategies are identified, they may then be subjected to breeding and selection for further improvements.

Some of the benefits attributed to endophyte-infected grasses are the results of secondary metabolites, although, as discussed before, primary metabolites may exert an indirect effect. Identifying a biochemical basis of any stress mechanism will be difficult. Currently, a stress metabolite in endophyte-infected grasses has yet to be identified. Stress resistance is probably due to an interaction, since most endophyte-infected grasses contain several biologically active compounds.[58] Also, optimum conditions necessary for growth of the symbiosis are probably not the conditions necessary for optimum concentrations of these compounds, since stress compounds are expected to accumulate under conditions less favorable for growth. A further difficulty is the fact that most of these compounds are of fungal origin and we must look for an interaction between two systems.

The available evidence indicating effects of abiotic stress on possible metabolic strategies of endophyte-infected grasses is indirect and limited to a few studies. We do have information related to two classes of compounds considered by many as indicators of stress: amino acids and sugar alcohols. While ergopeptide alkaloids probably respond directly to herbivory, i.e., biotic stress, they nevertheless show some response to abiotic factors that are not stressing. These are discussed for comparison.

The accumulation of proline under conditions of stress is a complex sequence of events which may be directly or indirectly influenced by the moisture or nutritional status of an endophyte-infected grass. As indicated earlier, this amino acid is considered important in several aspects of stress mechanisms. Proline accumulates in endophyte-infected grasses without stress,[26] but the effect of drought stress on this accumulation pattern has not been determined. If any effects from this amino acid are directly related to stress, proline accumulation may be considered to have an adaptive role in drought and would have application as an indicator of stress tolerance in grass breeding.

The recent finding of Richardson et al.[44] that sugar alcohols, especially arabitol, accumulate in plants under water stress is an additional indicator that there is a directed mechanism related to abiotic stresses. Since arabitol is probably produced by the fungus, this metabolic strategy might be a response of the fungus to moisture stress. This response is a common mechanism in many microorganisms, including fungi, and as a consequence of the fungal response, the grass is protected as well. This is one more suggestion that the grass-endophyte relationship is a mutualism.

Lyons et al.[26] reported that, under low nitrogen fertilization, endophyte-infected grasses accumulated essentially the same amount of total amino acids as uninfected grasses. While an endophyte effect was seen under high soil nitrogen, this effect is not viewed as stress-related. However, plants grown under high nitrogen did produce significantly more of the herbivore deterrents, ergopeptide alkaloids. In another study,[36] a similar conclusion was reached using clonal material. In infected plants, ammonia nitrogen was effective in stimulating an increase in alkaloid synthesis only at a high concentration whereas nitrate nitrogen was effective in stimulating this increase at low and high concentrations.[26,36] These studies suggest that secondary metabolism may be regulated by both the form and concentration of soil nitrogen. Since increased alkaloid levels followed an increase in soil nitrogen and nitrogen availability, this also suggests that the induction of ergopeptide alkaloids may involve nitrogen precursors, and that alkaloid synthesis is related to stimulatory plant and fungus growth. This may be contradictory to studies on the *in vitro* synthesis of ergopeptide alkaloids which indicated that ergopeptide alkaloids accumulated only after fungal growth had ceased and when the culture had aged.[6,25]

Under high soil nitrogen and moderate drought (–0.50 MPa) or low soil nitrogen and moderate moisture (–0.05 MPa), both conditions where there is less plant growth, endophyte-infected tall fescue contained a higher concentration of ergopeptide alkaloids compared to controls.[36] It seems likely that the regulatory mechanisms discussed above are not the only ones involved in the controls of ergopeptide synthesis, and that undoubtedly additional ways exist in the grass to control the flow of metabolites associated with the pathway for the synthesis of ergopeptide alkaloids. Certainly the grass is expected to exert some control over ergopeptide alkaloid synthesis, and the data of Hill et al.[59] suggest such a control. These data, nevertheless, support our contention that this class of compounds responds directly to conditions (biotic and/or abiotic) favoring excess growth, but is indirectly related to a specific stress, i.e., herbivory.

In addition to the ergopeptide class of metabolites, endophyte-infected grasses also produce an insect deterrent, peramine, which is a nonergot alkaloid. However, there are no studies to indicate how abiotic stresses influence its accumulation pattern.

CONCLUSIONS

Natural, symbiotic endophytes of the genus *Acremonium* could have a positive impact in the development and management of future turfgrass systems. We have shown in this review that endophyte-infected plants are generally more drought-tolerant, flood-tolerant, and nitrogen-efficient than plants which are not infected by these fungi, especially in stressful environments. Although relatively little information is currently available regarding the influence of endophytes in plants subjected to typical turf cultural practices,[12] there is reason to assume that infected grasses managed as a turf will continue to outperform uninfected plants, especially when environmental stress allows the expression of these characteristics. So far, the major obstacle to utilizing endophytes in pasture grasses has been the potential animal toxicities associated with infected plants. However, when endophyte-infected grasses are used as a turf, animal toxicity can be ignored and the positive benefits associated with drought stress, nitrogen utilization, and insect resistance may be fully realized.

As breeders of endophyte-infected pasture grasses attempt to reduce or eliminate the toxic components of these grasses, turfgrass breeders should take an opposite approach and search for endophyte-infected germplasm which is highly toxic and subsequently more resistant to predators. Although there is no clear relationship between toxicity of endophyte-infected grasses and enhanced drought stress, the selection of infected grasses with superior drought-tolerance characteristics can also be approached without concern for the effects those selections may have on animal toxicity. Finally, methods of reintroducing endophytes into endemic grass hosts and methods of introducing endophytes into grasses that are not natural hosts for these organisms are in various stages of development.[60] If endophytes isolated from plants with superior stress-tolerance characteristics are able to transfer those traits to other germ plasm or other grass species, these fungi could become a powerful biocontrol option for the turfgrass industry.

REFERENCES

1. **Clay, K.,** Fungal endophytes of grasses: a defensive mutualism between plants and fungi, *Ecology*, 69, 10, 1988.

2. **Bacon, C. W.,** Abiotic stress tolerance (moisture, nutrients) and photosynthesis in endophyte-infected tall fescue, *Agric. Ecosystems Environ.*, 44, 123, 1993.

3. **Bacon, C. W., Porter, J. K., Robbins, J. D., and Luttrell, E. S.,** *Epichloe typhina* from toxic tall fescue grasses, *Appl. Environ. Microbiol.*, 34, 576, 1977.

4. **Fletcher, L. R. and Harvey, I. C.,** An association of *Lolium* endophyte with ryegrass staggers, *N.Z. Vet. J.*, 29, 185, 1981.

5. **Bacon, C. W. and Siegel, M. R.,** The endophyte of tall fescue, *J. Prod. Agric.*, 1, 45, 1988.

6. **Bacon, C. W. and De Battista, J. P.,** Endophytic fungi of grasses, in *Handbook of Applied Mycology, Vol. 1, Soils and Plants*, Arora, D. K., Ed., Marcel Dekker, New York, 1990, 231.

7. **Clay, K.,** Clavicipitaceous fungal endophytes of grasses: coevolution and the change from parasitism to mutualism, in *Coevolution of Fungi with Plants and Animals*, Pirozynski, K. A. and Hawksworth, D. L., Eds., Academic Press, New York, 1988, 79.

8. **Siegel, M. R. and Schardl, C. L.,** Fungal endophytes of grasses: detrimental and beneficial associations, in *Microbial Ecology of Leaves*, Andrews, J. H. and Hirano, S. S., Eds., Springer-Verlag, New York, 1990, 198.

9. **Bacon, C. W., Porter, J. K., and Robbins, J. D.,** Toxicity and occurrence of *Balansia* on grasses from toxic fescue pastures, *Appl. Microbiol.*, 29, 553, 1975.

10. **Morgan-Jones, G. and Gams, W.,** Notes on hyphomycetes. XLI. An endophyte of *Festuca arundinacea* and the anomorph of *Epichloe typhina*, new taxa in one of two new sections of *Acremonium*, *Mycotaxon*, 15, 311, 1982.

11. **White, J. F., Morgan-Jones, G., and Morrow, A. C.,** Taxonomy, life cycle, reproduction, and detection of *Acremonium* endophytes, *Agric. Ecosystems Environ.*, 44, 13, 1993.

12. **Funk, C. R., Clarke, B. B., and Johnson-Cicalese, J. M.,** Role of endophytes in enhancing the performance of grasses used for conservation and turf, in *Integrated Pest Management for Turfgrass and Ornamentals*, Leslie, A. R. and Metcalf, R. L., Eds., U.S. Environmental Protection Agency, Washington, DC, 1989, 203.

13. **Garner, G.,** Fescue foot — the search for the cause continues, in *Missouri Cattle Backgrounding and Feeding Seminar*, University of Missouri Press, Columbia, 1984, 62.

14. **Latch, G. C. M.,** Physiological interactions of endophytic fungi and their hosts. Biotic stress tolerance imparted to grasses by endophytes, *Agric. Ecosystems Environ.*, 44, 143, 1993.

15. **Arechavaleta, M., Bacon, C. W., Hoveland, C. S., and Radcliffe, D. E.,** Effect of the tall fescue endophyte on plant response to environmental stress, *Agron. J.*, 81, 83, 1989.

16. **Read, J. C. and Camp, B. J.,** The effect of the fungal endophyte *Acremonium coenophialum* in tall fescue on animal performance, toxicity, and stand maintenance, *Agron. J.*, 78, 848, 1986.

17. **Stuedemann, J. A. and Hoveland, C. S.,** Fescue endophyte: history and impact on animal agriculture, *J. Prod. Agric.*, 1, 39, 1988.

18. **MacAdam, J. W., Volenec, J. J., and Nelson, C. J.,** Effects of nitrogen on mesophyll cell division and epidermal cell elongation in tall fescue leaf blades, *Plant Physiol.*, 89, 549, 1989.

19. **Belesky, D. P., Devine, O. J., Pallas, J. E., Jr., and Stringer, W. C.,** Photosynthetic activity of tall fescue as influenced by a fungal endophyte, *Photosynthetica*, 21, 82, 1987.

20. **Cheplick, G. P., Clay, K., and Marks, S.,** Interactions between infection by endophytic fungi and nutrient limitation in the grasses *Lolium perenne* and *Festuca arundinacea*, *New Phytol.*, 111, 89, 1989.

21. **Marks, S. and Clay, K.,** Effects of CO_2 enrichment, nutrient addition, and fungal endophyte-infection on the growth of two grasses, *Oecologia*, 84, 207, 1990.

22. **Latch, G. C. M., Hunt, W. F., and Musgrave, D. R.,** Endophytic fungi affect growth of perennial ryegrass, *N.Z. J. Agric. Res.*, 28, 165, 1985.

23. **Lyons, P. C., Plattner, R. D., and Bacon, C. W.,** Occurrence of peptide and clavine ergot alkaloids in tall fescue grass, *Science*, 232, 487, 1986.

24. **Porter, J. K., Bacon, C. W., and Robbins, J. D.,** Ergosine, ergosinine, and chanoclavine I from *Epichloe typhina*, *J. Agric. Food Chem.*, 27, 595, 1979.

25. **Bacon, C. W., Porter, J. K., and Robbins, J. D.,** Laboratory production of ergot alkaloids by species of *Balansia*, *J. Gen. Microbiol.*, 113, 119, 1979.

26. **Lyons, P. C., Evans, J. J., and Bacon, C. W.,** Effects of the fungal endophyte *Acremonium coenophialum* on nitrogen accumulation and metabolism in tall fescue, *Plant Physiol.*, 92, 726, 1990.

27. **De Battista, J. P., Bacon, C. W., Severson, R., Plattner, R. D., and Bouton, J. H.,** Indole acetic acid production by the fungal endophyte of tall fescue, *Agron. J.*, 82, 878, 1990.

28. **Hsiao, T. C.,** Plant responses to water stress, in *Annual Review of Plant Physiology*, Annual Reviews, Inc., Palo Alto, CA, 1973, 519.

29. **Belesky, D. P., Stringer, W. C., and Hill, N. S.,** Influence of endophyte and water regime upon tall fescue accessions. I. Growth characteristics, *Ann. Bot.*, 63, 495, 1989.

30. **De Battista, J. P., Bouton, J. H., Bacon, C. W., and Siegel, M. R.,** Rhizome and herbage production of endophyte-removed tall fescue clones and populations, *Agron. J.*, 82, 651, 1990.

31. **Passioura, J. B.,** Water collection by roots, in *The Physiology and Biochemistry of Drought Resistance in Plants*, Paleg, L. G. and Aspinall, D., Eds., Academic Press, New York, 1981, 39.

32. **West, C. P., Izekor, E., Oosterhuis, D. M., and Robbins, R. T.,** The effect of *Acremonium coenophialum* on the growth and nematode infestation of tall fescue, *Plant Soil*, 112, 3, 1988.

33. **Hill, N. S., Stringer, W. C., Rottinghaus, G. E., Belesky, D. P., Parrott, W. A., and Pope, D. D.,** Growth, morphological, and chemical component responses of tall fescue to *Acremonium coenophialum*, *Crop Sci.*, 30, 156, 1990.

34. **Bittman, S. and Simpson, G. M.,** Drought effect on leaf conductance and leaf rolling in forage grasses, *Crop Sci.*, 29, 338, 1989.

35. **O'Toole, J. C. and Cruz, R. T.,** Response of leaf water potential, stomatal resistance, and leaf rolling to water stress, *Plant Physiol.*, 65, 428, 1980.

36. **Arechavaleta, M., Bacon, C. W., Plattner, R. D., Hoveland, C. S., and Radcliffe, D. E.,** Accumulation of ergopeptide alkaloids in symbiotic tall fescue grown under deficits of soil water and nitrogen fertilizer, *Appl. Environ. Microbiol.*, 58, 857, 1992.

37. **West, C. P., Oosterhuis, D. M., and Wullschleger, S. D.,** Osmotic adjustment in tissues of tall fescue in response to water deficit, *Environ. Exp. Bot.*, 30, 149, 1989.

38. **Lawlor, D. W.,** Plant growth in polyethylene glycol solutions in relation to the osmotic potential of the root medium and the leaf water balance, *J. Exp. Bot.*, 20, 895, 1969.

39. **Elmi, A. A., West, C. P., Turner, K. E., and Oosterhuis, D. M.,** *Acremonium coenophialum* effects on tall fescue water relations, in *Proceedings of International Symposium on Acremonium/Grass Interactions*, Quisenberry, S. S. and Joost, R. E., Eds., New Orleans, 1990, 137.

40. **Borowitzka, L. J.,** Solute accumulation and regulation of water activity, in *The Physiology and Biochemistry of Drought Resistance in Plants*, Paleg, L. G. and Aspinall, D., Eds., Academic Press, New York, 1981, 97.

41. **Schnyder, H. and Nelson, C. J.,** Growth rates and carbohydrate fluxes within the elongation zone of tall fescue leaf blades, *Plant Physiol.*, 85, 548, 1987.

42. **Spollen, W. G. and Nelson, C. J.,** Characterization of fructan from mature leaf blades and elongation zones of developing leaf blades of wheat, tall fescue, and timothy, *Plant Physiol.*, 88, 1349, 1988.

43. **Horst, G. L. and Nelson, C. J.,** Compensatory growth of tall fescue following drought, *Agron. J.*, 71, 559, 1979.

44. **Richardson, M. D., Chapman, G. W., Jr., Hoveland, C. S., and Bacon, C. W.,** Sugar alcohols in endophyte-infected tall fescue under drought, *Crop Sci.*, 32, 1060, 1992.

45. **Smith, D. E., Muscatine, L., and Lewis, D. H.,** Carbohydrate movement from autotrophs to heterotrophs in parasitic and mutualistic symbiosis, *Biol. Rev.*, 44, 17, 1969.

46. **Lewis, D. H. and Smith, D. C.,** Sugar alcohols (polyols) in fungi and green plants. I. Distribution, physiology, and metabolism, *New Phytol.,* 66, 143, 1967.

47. **Harwood, V. D.,** Analytical studies on the carbohydrates of grasses and clovers. VII. The isolation of D-mannitol from perennial rye-grass (*Lolium perenne*), *J. Sci. Food Agric.,* 5, 453, 1954.

48. **Kemble, A. R. and MacPherson, H. T.,** Liberation of amino acids in perennial ryegrass during wilting, *Biochem. J.,* 58, 46, 1954.

49. **Belesky, D. P., Wilkinson, S. R., and Pallas, J. E., Jr.,** Response of four tall fescue cultivars grown at two nitrogen levels to low soil water availability, *Crop Sci.,* 22, 93, 1982.

50. **Richardson, M. D., Bacon, C. W., and Hoveland, C. S.,** The effect of endophyte removal on gas exchange in tall fescue, in *Proceedings of International Symposium on Acremonium/Grass Interactions,* Quisenberry, S. S. and Joost, R. E., Eds., New Orleans, 1990, 189.

51. **Mayer, A. M.,** Plant-fungal interactions: a plant physiologist's viewpoint, *Phytochemistry,* 28, 311, 1989.

52. **Hinton, D. M. and Bacon, C. W.,** The distribution and ultrastructure of the endophyte of toxic tall fescue, *Can. J. Bot.,* 63, 36, 1985.

53. **Kulkarni, R. and Nielsen, B. D.,** Nutritional requirements for growth of a fungus endophyte of tall fescue grass, *Mycologia,* 78, 781, 1986.

54. **Davis, N. D., Clark, E. M., Schrey, K. A., and Diener, U. L.,** In vitro growth of *Acremonium coenophialum,* an endophyte of toxic tall fescue grass, *Appl. Environ. Microbiol.,* 52, 888, 1986.

55. **Pope, D. D. and Hill, N. S.,** Effects of various culture, media, antibiotics, and carbon sources on growth parameters of *Acremonium coenophialum,* the endophyte of tall fescue, *Mycologia,* 83, 110, 1991.

56. **Bunyard, B. and McInnis, T., Jr.,** Evidence for elevated phytohormone levels in endophyte-infected tall fescue, *International Symposium on Acremonium/Grass Interactions,* (Abstract), 1990.

57. **Zhang, J., Schurr, U., and Davies, W. J.,** Control of stomatal behaviour by abscisic acid which apparently originates in the roots, *J. Exp. Bot.,* 38, 1174, 1987.

58. **Siegel, M. R., Latch, G. C. M., Bush, L. P., Fannin, F. F., Rowan, D. D., Tapper, B. A., Bacon, C. W., and Johnson, M. C.,** Fungal endophyte-infected grasses: alkaloid accumulation and aphid response, *J. Chem. Ecol.,* 16, 3301, 1990.

59. **Hill, N. S., Parrott, W. A., and Pope, D. D.,** Ergopeptine alkaloid production by endophyte in a common tall fescue genotype, *Crop Sci.,* 31, 1545, 1991.

60. **Kearney, J. F., Parrott, W. A., and Hill, N. S.,** Infection of somatic embryos of tall fescue with *Acremonium coenophialum, Crop Sci.,* 31, 979, 1991.

61. **Hoveland, C. S.,** personal communication.

An Inventory of New Generation Chemical Controls: Alternative Inorganics, Botanicals, New Generation Insecticides

Landscape maintenance worker spraying shrubbery
(photo courtesy of *Grounds Maintenance* magazine)

Spray Oil for Fruit Trees

Active ingredient:
   Superior Petroleum Oil...98.6%
Inert Ingredients..1.4%
                    Total.................100.0%
Superior Petroleum Oil Minimum Unsulfonated Residue not less than 92%.
Petroleum Oil Classification: Unclassified

EPA Reg. No. 11111-1
EPA Est. 11111-II-1

ACME CHEMICAL COMPANY
Box 111, San Diablo, CA 90404
700/828–0736

CAUTION
KEEP OUT OF REACH OF CHILDREN
Statement of Practical Treatment
If swallowed, do not induce vomiting. If in eyes, flush with plenty of water. If on skin, wash with soap and water.

See Below for Additional Precautionary Statements

COMPATIBILITY: Acme Spray Oil for Fruit Trees is compatible with most commonly used insecticides and fungicides. Do not use in combination with or immediately before or after spraying with dinitro compounds, fungicides such as Captan, Folpet, Dyrene, Karathene, Morestan or any other product containing sulfur. Also do not use with Sevin or dimethoate.
TIMING THE TREATMENT: You must determine the precise timing to fit local growth and climatic conditions.

Specifications from a typical label for a fictitious spray oil (reproduction courtesy of
IPM Education & Publications, University of California/Davis)

Chapter 48

Botanical Insecticides and Insecticidal Soaps

Richard Weinzierl and Tess Henn, Office of Agricultural Entomology, University of Illinois, Champaign, IL

(This chapter was excerpted from the authors' *Alternatives in Insect Management: Biological and Biorational Approaches*, published in 1991 by the North Central Regional Extension Services in cooperation with the U.S. Department of Agriculture.)

CONTENTS

Advantages of Botanical Insecticides and Insecticidal Soaps 542
Disadvantages of Botanical Insecticides and Insecticidal Soaps 542
What Are Botanical Insecticides and Insecticidal Soaps? 542
     Synergists 543
The Botanical Insecticides 543
    Pyrethrum and Pyrethrins 543
    Rotenone 547
    Sabadilla 548
    Ryania 549
    Nicotine 549
    Citrus Oil Extracts: Limonene and Linalool 550
    Other Essential Plant Oils: Herbal Repellents and Insecticides 550
    Neem 551
Insecticidal Soaps 553
Summary 554
Terms to Understand 554
Acknowledgments 555
References 555

The ideal insecticide should control target pests adequately and should be target-specific (able to kill the pest insect but not other insects or animals), rapidly degradable, and low in toxicity to humans and other mammals. Two classes of insecticides that exhibit some of these characteristics are the botanical insecticides and the insecticidal soaps. Botanical insecticides, sometimes referred to as "botanicals," are naturally occurring insecticides derived from plants. Insecticidal soaps are soaps that have been selected and formulated for their insecticidal action.

Botanical insecticides and insecticidal soaps are promising alternatives for use in insect management. However, like conventional synthetic insecticides, botanicals and insecticidal soaps have advantages and disadvantages and should be judged accordingly. Each compound should be evaluated in terms of its toxicity, effectiveness, environmental impacts, and costs.

The strengths and weaknesses of botanical insecticides and insecticidal soaps are briefly summarized in this chapter. Each compound is discussed in terms of its mode of action, mammalian toxicity, and practical uses. General information on the history and development of botanicals, insecticide toxicology, and state registration requirements is also presented. Insecticide toxicity is stressed throughout this chapter. Even though botanicals and insecticidal soaps are naturally derived and most are less toxic than many conventional insecticides, they are poisons and should be handled with the same caution as synthetic chemicals.

ADVANTAGES OF BOTANICAL INSECTICIDES AND INSECTICIDAL SOAPS

Many compounds with diverse chemical structures and different modes of action are classified as botanical insecticides or insecticidal soaps. It is therefore difficult to present a detailed list of advantages or disadvantages that apply to all of the compounds included in this category. Some general advantages shared by most of these compounds are:

Rapid degradation. Botanicals and insecticidal soaps degrade rapidly in sunlight, air, and moisture and are readily broken down by detoxification enzymes. This is very important because rapid breakdown means less persistence in the environment and reduced risks to nontarget organisms. Soaps and many botanicals may be applied to food crops shortly before harvest without leaving excessive residues.

Rapid action. Botanicals and soaps act very quickly to stop feeding by pest insects. Although they may not cause death for hours or days, they often cause immediate paralysis or cessation of feeding.

Low mammalian toxicity. Most botanicals and insecticidal soaps have low to moderate mammalian toxicity. There are exceptions, however; see "The Toxicity of Insecticides" for a general discussion of insecticide toxicities.

Selectivity. Although most botanicals have broad-spectrum activity in standard laboratory tests, in the field their rapid degradation and the action of some as stomach poisons make them more selective in some instances for plant-feeding pest insects and less harmful to beneficial insects.

Low toxicity to plants. Most botanicals are not phytotoxic (toxic to plants). Insecticidal soaps and nicotine sulfate, however, may be toxic to some ornamentals.

DISADVANTAGES OF BOTANICAL INSECTICIDES AND INSECTICIDAL SOAPS

The following disadvantages do not preclude the effective use of botanicals or insecticidal soaps, but do call attention to certain factors that must be considered when using these insecticides.

Rapid degradation. Rapid breakdown of botanicals, although desirable from an environmental and human health standpoint, creates a need for more precise timing, more frequent insecticide applications, or both.

Toxicity. Although the insecticidal soaps and most botanicals are the lesser of many "evils" in terms of general pesticide toxicities, they are toxins nonetheless. All toxins used in pest control pose some hazard to the user and to the environment. In addition, toxins are useful only when incorporated into a conscientious program of pest management that includes sanitation, cultural control, crop rotation, and use of resistant plant varieties. No insecticides, natural or synthetic, should be used as the sole means of defense against pest insects.

Cost and availability. Botanicals tend to be more expensive than synthetics, and some are not as widely available. In addition to problems of supply, the potency of some botanicals may differ from one source or batch to the next.

Lack of test data. Data on effectiveness and long-term (chronic) toxicity are unavailable for some botanicals. Tolerances for residues of some botanicals on food crops have not been established.

State registration. Several botanicals are registered by the U.S. Environmental Protection Agency (EPA) and are available by mail order, but are not registered for legal sale in specific states. (See "Registration of Botanical Insecticides" for more information on this problem.)

WHAT ARE BOTANICAL INSECTICIDES AND INSECTICIDAL SOAPS?

Botanicals are naturally occurring insecticides derived from plant sources. They are processed into various forms.

Preparations of the crude plant material. These are dusts or powders made from ground, dried plant parts that have not been extracted or treated extensively. They are marketed either full strength or diluted with carriers such as clays, talc, or diatomaceous earth. Examples include dusts or wettable powders of cubé roots (rotenone); pyrethrum flowers; sabadilla seeds; ryania stems; or neem leaves, fruits, or bark.

Plant extracts or resins. These are water or solvent extracts that concentrate the insecticidal components. Such extracts or resins are formulated into liquid concentrates or are impregnated onto dusts or wettable powders. Botanicals in this form include pyrethrins, cubé resins (rotenone), citronella and other essential oils, and neem seed extracts or oils.

Pure chemicals isolated from plants. These are purified insecticidal compounds that are isolated and refined by a series of extractions, distillations, or other processes and are formulated into concentrates. Included in this category are nicotine, d-limonene, and linalool.

Insecticidal soaps are specially formulated soaps that contain the potassium or sodium salts of insecticidal fatty acids. Soaps marketed under the Safer and M-Pede labels are most widely known; they contain the potassium salt of oleic acid, a fatty acid found in certain vegetable oils.

Crude botanical insecticides have been used for centuries and were known in tribal or traditional cultures long before being introduced into Europe or the U.S.. Botanicals with long histories of traditional use include neem in India, rotenone in East Asia and South America, and pyrethrum in Persia (Iran).

From the late 1800s to the 1940s, botanicals and insecticidal soaps were used extensively on certain crops. Soaps and nicotine-based insecticides were important before the turn of the century, whereas pyrethrum, rotenone, sabadilla, and ryania were popular in the 1930s and early 1940s. During that time, research on botanicals focused on efficacy, development of new formulations, and the use of synergists (compounds that enhance insecticidal action). With the development of synthetic insecticides in the mid-1940s, the use of botanicals was largely abandoned in commercial agriculture; the new synthetic compounds were less expensive, more effective, and longer lasting.

From 1945 to the early 1970s, the only botanicals remaining in widespread use were pyrethrins (used as household and industrial sprays and aerosols) and nicotine (used in greenhouses and orchards). Home gardeners continued to use rotenone on a small scale. Use of sabadilla and ryania was virtually abandoned, and for years these compounds were nearly unavailable in the U.S.

In the past 10 or 15 years, interest in botanicals has increased, primarily as a result of concerns about environmental contamination and pesticide residues in foods. In a few instances, botanical insecticides have come into commercial use because a key pest has become resistant to most classes of synthetic insecticides, for example, the use of rotenone for control of the Colorado potato beetle on Long Island in New York.

Botanicals and insecticidal soaps are not widely used in conventional commercial agriculture, but small-scale organic growers and home gardeners are using them more extensively. In addition, state and federal certification programs for organic farming generally allow the use of insecticidal soaps and most botanical insecticides. As a result, botanical insecticides that for years were unavailable have

been reregistered and are being produced and marketed in limited quantities. Several new botanical insecticides, such as the citrus oil derivatives and new formulations of neem also are available or are under development.

SYNERGISTS

Insects, like humans and other animals, possess enzymes that are capable of breaking down a wide variety of toxic substances. Certain natural insecticides — pyrethrins in particular, as well as several other botanicals — are readily attacked and degraded by these enzymes once inside the insect's system. In some cases, degradation occurs so rapidly that the insecticide is not active long enough to kill the insect. The insect is temporarily stunned, but then recovers. Synergists are compounds that enhance insecticidal action by inhibiting certain detoxification enzymes. The synergists used in commercial insecticides block a system of enzymes known as the multifunction oxidases (MFOs); consequently, these synergists increase the effective toxicity of those insecticides that are easily degraded by the MFOs. Insecticides that are not readily detoxified by the MFOs are not synergized by these compounds. The synergists themselves are low in toxicity, have little or no inherent insecticidal activity, and are not persistent.

The most commonly used synergist is piperonyl butoxide (PBO). PBO is found in most products that contain pyrethrins. It is also an effective synergist for rotenone, sabadilla, ryania, and the citrus oil derivatives, as well as some synthetic pyrethroids and carbamate insecticides. MGK 264 (N-octyl bicycloheptene dicarboximide) is another synergist that is sometimes used in livestock and animal shelter insecticides. Both of these compounds have low mammalian toxicity (see Table 1). They are usually formulated with insecticides in ratios of from 2:1 to 10:1 (synergist:insecticide). High cost is the major factor limiting more widespread use of synergists. In addition, some organic certification programs do not allow the use of PBO.

THE BOTANICAL INSECTICIDES

A general discussion of insecticide toxicity and a table of toxicity estimates are presented in the section, "Toxicity of Insecticides." Refer to that box and to the glossary of terms, "Terms to Understand," for further explanations of the toxicological information presented in this section.

PYRETHRUM AND PYRETHRINS
Source. Pyrethrum is the powdered, dried flower

TOXICITY OF INSECTICIDES

Although many household and industrial products — including cleaning and polishing agents, degreasers, paints, and solvents — are toxic, insecticides are among only a few compounds used intentionally as poisons. The nature of insecticide toxicity and the manner in which insecticides are applied can make these compounds particularly hazardous. Because of this, it is important that insecticide users understand the basics of insecticide toxicology.

Insecticides are, by definition, compounds intended to kill insects. Most insecticides attack basic physiological processes such as nerve transmission or cellular respiration, processes that are common to insects, humans, and other animals. Whether natural or synthetic in origin, most insecticides can poison many forms of animal life. The risk of accidental poisoning may be heightened if a pesticide user perceives an insecticide to be harmless and consequently overapplies or misuses it. Such a problem can result from the widely held misconception that naturally derived insecticides are basically nontoxic to humans and other animals and are therefore "safe" to use in a careless manner. While most botanicals pose fewer hazards than many synthetic insecticides, their toxicity is still a factor to be considered.

INSECTICIDE TOXICITY

The toxicity of any chemical is usually evaluated in terms of both acute and chronic effects. Acute toxicity refers to the effects of a single dosage or exposure. Chronic toxicity refers to the effects of repeated doses or exposures over time.

The acute toxicity of an insecticide is generally described in terms of an LD_{50}, the dose required to kill 50% of the animals in a test. An LD_{50} is therefore a "median lethal dose." LD_{50} values are usually expressed as milligrams of toxicant per kilogram of body weight of a test animal (mg/kg). Consequently, *a lower LD_{50} indicates a more toxic compound*. For example, an insecticide with an LD_{50} of 60 mg/kg is much more toxic than one with an LD_{50} of 5,000 mg/kg.

The most common measures of an insecticide's toxicity to mammals are its oral and dermal LD_{50} values. Testing is generally performed on laboratory animals such as mice, rats, guinea pigs, or rabbits. Test animals are exposed to a range of single doses — a certain number of animals at each dose — to determine the insecticide's acute oral or dermal toxic effects. Estimates of an insecticide's acute human toxicity are derived from this type of animal testing.

Although LD_{50} values are useful indicators of toxicity, they do not provide a full picture of all of the risks associated with insecticide contact. For example, LD_{50} values fail to indicate toxic effects other than death. These effects may include eye injury, throat and lung irritation, chemical burns, neurological damage, and many other forms of injury. In addition, LD_{50} values generally indicate the acute toxicity of pure, unformulated insecticidal compounds rather than the diluted, formulated products actually used by consumers. Formulation (the process of turning a pure active ingredient into a finished insecticidal product) usually reduces poisoning risks because it involves diluting the active compounds with various carriers and additives. In some instances, however, formulation may increase poisoning risks; this occurs when carriers or other ingredients include toxic solvents or solvents that speed the entry of the active compounds into the body.

In comparison with questions about acute toxicity, questions about the chronic effects of repeated exposures to lower doses of pesticides are much more difficult to answer clearly. To investigate chronic effects, a compound is administered to laboratory animals at a range of doses (usually including the maximum dose that the animal can survive) for an extended period of time. The most common means of administering compounds is in combination with the animal's food, but compounds are sometimes administered dermally or by injection or inhalation. Following chronic exposures over a normal life span, the test animals are killed and examined for tumor production and other changes in major tissues and organs. A key purpose for chronic toxicity testing is to identify probable carcinogens (cancer-causing agents). Investigations of reproductive effects (birth rate, birth weight, incidence of birth defects) might also be conducted during each study.

Acute and chronic toxicity tests give indications of the immediate and long-term effects of human exposure to pesticides, but such tests provide only limited information. For instance, because different species react to insecticides in different ways, tests that measure rodent responses may not always accurately predict human responses. In addition, single-compound tests do not measure the effects of real-life human exposures to multiple compounds. Chronic, high-dose tests used to identify possible carcinogens may or may not yield results that are relevant for the prediction of cancer risks associated with the infrequent, low-level exposures that most humans are expected to encounter. Yet, despite these limitations, toxicological testing has produced results that help to describe the risks associated with pesticide exposure and that allow some meaningful comparisons of synthetic and botanical insecticides.

BOTANICALS VS. SYNTHETICS

Natural compounds are not inherently less toxic to humans than synthetic ones. Some of the most deadly, fast-acting toxins and some potent carcinogens occur naturally. *Despite the claims presented in some advertising materials, "natural" does not necessarily mean safe or nontoxic, and it certainly does not mean nonchemical.*

The LD_{50} values presented in Table 1 illustrate the fact that botanical insecticides range from practically nontoxic (such as neem or insecticidal soap) to very toxic (such as nicotine). Most are only slightly to moderately toxic. LD_{50} values for some common organophosphate, carbamate, and pyrethroid insecticides are also listed in Table 1 for comparison.

In evaluating the toxicities listed in Table 1, it is important to consider the situations that might lead to pesticide exposures and poisoning. Human exposure to pesticides usually results from careless contact during application or from contacting or eating residues that remain on treated materials or foods. Because botanical insecticides and insecticidal soaps break down rapidly in the environment, they rarely pose any risk to consumers. Many synthetic insecticides are much more stable, and problems associated with persistent residues present a more realistic concern. As a result, where a short-lived insecticide (a botanical, a soap, or certain synthetics such as resmethrin) and a more persistent insecticide (certain synthetics such as permethrin or chlorpyrifos) are characterized by similar LD_{50} values, the persistent product is much more likely to pose some risk to persons other than the applicator.

For applicators, however, certain botanical insecticides and synthetic insecticides can pose similar risks. Rotenone and ryania, for example, are similar to carbaryl and malathion in acute toxicity; applicators may be poisoned by careless exposure to any of these products. The important concept to remember is that although the environmental safety of all botanical insecticides is enhanced by their rapid degradation, several botanical insecticides can readily poison the careless applicator. Users should always wear protective clothing and avoid insecticide exposure.

head of the pyrethrum daisy, *Chrysanthemum cinerariaefolium* (Figure 1). Most of the world's pyrethrum crop is grown in Kenya. The term "pyre*thrum*" is the name for the crude flower dust itself, and the term "pyre*thrins*" refers to the six related insecticidal compounds that occur naturally in the crude material. The pyrethrins constitute 0.9 to 1.3% of dried pyrethrum flowers. They are extracted from crude pyrethrum dust as a resin that is used in the manufacture of various insecticidal products.

Mode of Action. Pyrethrins exert their toxic effects by disrupting the sodium and potassium ion exchange process in insect nerve fibers and interrupting the normal transmission of nerve impulses. Pyrethrin insecticides are extremely fast-acting and cause an immediate "knockdown" paralysis in insects. Despite their rapid toxic action, however, many insects are able to metabolize (break down) pyrethrins quickly. After a brief period of paralysis, these insects may recover rather than die. To

Table 1 **Estimates of acute toxicity of botanical insecticides, insecticidal soap, and selected synthetic insecticides**

Generic Name (Trade Name)	Class	Oral LD_{50}	Dermal LD_{50}	Signal Word[a]
Insecticidal soap (Safer)	Soap	16,500	—	Caution
Neem	Botanical	13,000	—	N/A
Piperonyl butoxide (PBO)	Synergist	>7,500	7,500	Caution
d-Limonene (VIP)	Botanical	>5,000	—	Caution
Sabadilla (Red Devil)	Botanical	4,000	—	Caution
MGK 264	Synergist	2,800	—	Caution
Resmethrin	Pyrethroid	>2,500	3,000	Caution
Linalool (Demize)	Botanical	2,440 to 3,180	3,578 to 8,374	Caution
Pyrethrins	Botanical	1,200 to 1,500	>1,800	Caution
Malathion	Organophosphate	885 to 2,800	4,100	Caution
Carbaryl (Sevin)	Carbamate	850	>4,000	Warning/caution
Ryania (Ryan 50)	Botanical	750 to 1,200	4,000	Caution
Permethrin (Pounce, Ambush)	Pyrethroid	430 to 4,000[b]	>2,000	Danger/warning
Chlorpyrifos (Lorsban)	Organophosphate	135 to 163	2,000	Warning/caution
Rotenone	Botanical	60 to 1,500[b]	940 to 3,000	Caution
Nicotine (Black-Leaf 40)	Botanical	50 to 60	50	Danger
Carbofuran (Furadan)	Carbamate	8 to 14	>2,500	Danger/warning
Terbufos (Counter)	Organophosphate	2 to 5	7	Danger

[a] See Table 2 for explanation of signal words; [b] Toxicity varies greatly depending on type of solvent used as a carrier.

prevent insects from metabolizing pyrethrins and recovering from poisoning, most products containing pyrethrins also contain the synergist PBO. Without PBO the effectiveness of pyrethrins is greatly reduced.

Mammalian Toxicity. Pyrethrins are low in mammalian toxicity (see Table 1), and few cases of human poisonings have ever been reported. Cats, however, are highly susceptible to poisoning by pyrethrins, and care must be taken to follow label directions closely when using products containing pyrethrins to treat cats for fleas.

When ingested, pyrethrins are not readily absorbed from the digestive tract, and they are rapidly hydrolyzed under the acid conditions of the gut and the alkaline conditions of the liver. Pyrethrins are more toxic to mammals by inhalation than by

Figure 1 Stages of the pyrethrum daisy, the source of natural pyrethrins.

ingestion because inhalation provides a more direct route to the bloodstream. Exposure to high doses may cause nausea, vomiting, diarrhea, headaches, and other nervous disturbances. Repeated contact with crude pyrethrum dusts may cause skin irritation or allergic reactions. The allergens that cause these reactions are not present in products containing refined pyrethrins. Tests indicate that chronic exposure to pyrethrins does not cause genetic mutations or birth defects (see Casida, 1973 and Hayes, 1982).

There is no single antidote for acute pyrethrin poisoning. Treatment of poisoning is symptomatic, that is, the various symptoms of poisoning are treated individually as they occur because there is no way to counteract the source of the poisoning directly.

Uses. Pyrethrins are contact poisons that have almost no residual activity in most applications. They break down very rapidly in sunlight, air, and moisture. Degradation is accelerated under acid or alkaline conditions, and for this reason pyrethrins should not be mixed with lime or soap solutions for application. Formulated products containing pyrethrins are stable in storage for long periods if not diluted, but powders made directly from ground pyrethrum flowers may lose up to 20% of their potency in a single year. Synergism by PBO or MGK 264 is essential for obtaining full effectiveness from pyrethrins.

Pyrethrins are used against a broad range of pests. Products containing synergized pyrethrins (pyrethrins plus PBO) are registered for use on pets and livestock to control fleas, flies, and mosquitoes. They are also registered as indoor household sprays, aerosols, and "bombs" for the control of various flying insects, fleas, and (less effectively) ants and roaches. Formulations containing microencapsulated pyrethrins may provide some residual activity for indoor use. Synergized pyrethrins are used in food-processing plants and food warehouses to control stored-product pests (flour beetles, Indianmeal moth, and others). Pyrethrins are also formulated with rotenone and ryania or copper for general use in gardens; rotenone and ryania are slower acting compounds to which pyrethrins are added for their rapid knockdown effect.

Pyrethroid Insecticides. Pyrethroids are not botanical insecticides. They are synthetic compounds that are based on the chemical structure and physiological action of the natural pyrethrins, but they are more toxic to insects and generally more persistent in the environment. A few pyrethroids, such as resmethrin (used for household insect control), are very low in mammalian toxicity and degrade quickly. Others, such as cypermethrin (used

on cotton, vegetables, and fruits), are moderately toxic and more persistent. For many of the pyrethroids, acute toxicity and hazard vary greatly depending on the kind of solvent that is used as a carrier for the formulated product. Pyrethroids are generally effective at very low concentrations and are used at much lower application rates than most other synthetic insecticides.

ROTENONE

Source. Rotenone is an insecticidal compound found in the roots of *Lonchocarpus* species in South America, *Derris* species in Asia, and several other related tropical legumes. Commercial rotenone was once produced from Malaysian *Derris*. Currently the main commercial rotenone source is Peruvian *Lonchocarpus*, often referred to as cubé root.

Rotenone is extracted from cubé roots in acetone or ether. Extraction produces a 2 to 40% rotenone resin, which contains several related but less insecticidal compounds known as rotenoids. The resin is used to make liquid concentrates or to impregnate inert dusts or other carriers. Most rotenone products are made from the complex resin rather than from purified rotenone itself. Alternatively, cubé roots may be dried, powdered, and mixed directly with an inert carrier to form an insecticidal dust.

Mode of Action. Rotenone is a powerful inhibitor of cellular respiration, the process that converts nutrient compounds into energy at the cellular level. In insects rotenone exerts its toxic effects primarily on nerve and muscle cells, causing rapid cessation of feeding. Death occurs several hours to a few days after exposure. Rotenone is extremely toxic to fish, and is often used as a fish poison (piscicide) in water management programs. It is effectively synergized by PBO or MGK 264.

Mammalian Toxicity. Although rotenone is a potent cell toxin, mammals detoxify ingested rotenone efficiently via liver enzymes. As with pyrethrins, rotenone is more toxic by inhalation than by ingestion. Exposure to high doses may cause nausea, vomiting, muscle tremors, and rapid breathing. Very high doses may cause convulsions followed by death from respiratory paralysis and circulatory collapse. Direct contact with rotenone may be irritating to skin and mucous membranes. Treatment of poisoning is symptomatic.

Chronic exposure to rotenone may lead to liver and kidney damage. Although some rodent testing has shown that chronic dietary exposure to rotenone may induce tumor formation, the most recent EPA registration standard considers rotenone to be noncarcinogenic (see Hayes, 1982 and EPA report, 1989).

Rotenone is one of the more acutely toxic botanicals. As a matter of comparison, pure, unformulated rotenone is more toxic than pure carbaryl (Sevin) or malathion, two commonly used synthetic insecticides (see Table 1). In the form of a 1% dust, rotenone poses roughly the same acute hazard as the commonly available 5% Sevin dust.

Uses. Rotenone is a broad-spectrum contact and stomach poison that is particularly effective against leaf-feeding beetles and certain caterpillar pests. On a commercial level, rotenone has been used widely in the northeastern U.S. to control populations of the Colorado potato beetle that have become resistant to most other registered insecticides (Figure 2). Rotenone is also used extensively in fish management programs.

Several products containing rotenone in combination with pyrethrins, ryania, copper, and/or sulfur are registered for general garden and orchard insect and disease control. Rotenone is commonly sold as a 1% dust or a 5% powder (for spraying).

Rotenone degrades (oxidizes) quickly in air and sunlight; under warm, sunny conditions dust or spray residues on exposed plant surfaces provide some degree of protection for approximately 1 week. Degradation is greatly accelerated by mixing rotenone with alkaline materials such as soaps or lime.

SABADILLA

Source. Sabadilla is derived from the ripe seeds of *Schoenocaulon officinale*, a tropical lily plant which grows in Central and South America. Sabadilla is also known as cevadilla or caustic barley.

When sabadilla seeds are aged, heated, or treated with alkali, several insecticidal alkaloids are formed or activated. Alkaloids are physiologically active compounds that occur naturally in many plants. In chemical terms they are a heterogeneous class of cyclic compounds that contain nitrogen in their ring structures. Caffeine, nicotine, cocaine, quinine, and strychnine are some of the more familiar alkaloids. The alkaloids in sabadilla are known collectively as veratrine or as the veratrine alkaloids. They constitute 3 to 6% of the weight of aged, ripe sabadilla seeds. Of these alkaloids, cevadine and veratridine are the most active insecticidally.

European white hellebore (*Veratrum album*) also contains veratridine in its roots. Hellebore was once commonly used in Europe and the U.S. for insect control but is now unavailable commercially and is not registered by the EPA.

Mode of Action. In insects, sabadilla's toxic alkaloids affect nerve cell membrane action, causing loss of nerve function, paralysis, and death. Sabadilla kills insects of some species immediately, whereas others may survive in a state of paralysis for several days before dying. Sabadilla is effectively synergized by PBO or MGK 264.

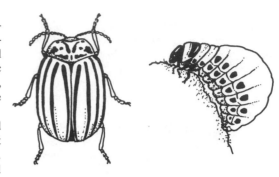

Figure 2 Colorado potato beetle, adult and larva. Rotenone is used on a commercial scale to control this major pest. In some areas of the U.S., the Colorado potato beetle has become resistant to most classes of synthetic chemical insecticides.

Mammalian Toxicity. Sabadilla, in the form of dusts made from ground seeds, is the least toxic of the registered botanicals (see Table 1). Purified veratrine alkaloids are quite toxic, however, and are considered on a par with the most toxic synthetic insecticides. Sabadilla can be severely irritating to skin and mucous membranes, and has a powerful sneeze-inducing effect when inhaled. Ingestion of small amounts may cause headaches, severe nausea, vomiting, diarrhea, cramps, and reduced circulation. Ingestion of very high doses may cause convulsions, cardiac paralysis, and respiratory failure. Sabadilla alkaloids can be absorbed through the skin or mucous membranes.

Uses. Sabadilla is mainly a broad-spectrum contact poison, but it also has some activity as a stomach poison. It is effective against certain "true bug" pests (order Hemiptera) such as harlequin bugs and squash bugs that are difficult to control with most other insecticides. Sabadilla is also highly toxic to honey bees, and care must be taken to avoid applying it when bees are present. The active alkaloids degrade rapidly in air and sunlight, and have little residual toxicity.

Sabadilla is registered by the EPA for use on squash, beans, cucumbers, melons, potatoes, turnips, mustard, collards, cabbage, broccoli, citrus, and peanuts. In citrus, sabadilla is applied with sugar as an insecticidal bait (stomach poison) for citrus thrips. For vegetable insect control, sabadilla is commonly applied as a contact insecticide in the form of a 20% dust or spray. *Several sabadilla products available for mail-order purchase are not registered for sale in specific states. See "Registration of Botanical Insecticides" for information on this problem.*

RYANIA

Source. Ryania comes from the woody stems of *Ryania speciosa*, a South American shrub. Powdered *Ryania* stem wood is combined with carriers to produce a dust or is extracted to produce a liquid concentrate. The most active compound in ryania is the alkaloid ryanodine, which constitutes approximately 0.2% of the dry weight of stem wood.

Mode of Action. Ryania is a slow-acting stomach poison. Although it does not produce rapid knockdown paralysis, it does cause insects to stop feeding soon after ingesting it. Little has been published concerning its exact mode of action in insect systems. Ryania is effectively synergized by PBO and is reported to be most effective in hot weather.

Mammalian Toxicity. Ryania is moderately toxic to mammals by ingestion and only slightly toxic by dermal exposure. Ingestion of large doses causes weakness, deep and slow respiration, vomiting, diarrhea, and tremors, sometimes followed by convulsions, coma, and death. Purified ryanodine is approximately 700 times more toxic than the crude ground or powdered wood and causes poisoning symptoms similar to those of synthetic organophosphate insecticides. (Depending on exposure, organophosphate poisoning symptoms may include sweating, headache, twitching, muscle cramps, mental confusion, tightness in chest, blurred vision, vomiting, evacuation of bowels and bladder, convulsions, respiratory collapse, coma, and death.)

Uses. Ryania currently is registered by the EPA for use on citrus, corn, walnuts, apples, and pears for the control of citrus thrips, European corn borer, and codling moth. Ryania has longer residual activity than most other botanicals and is therefore useful where the more quickly degrading compounds are ineffective. Ryania is also sold in mixtures containing rotenone and pyrethrins for use on a variety of vegetables and fruits. *Several products containing ryania and distributed through mail-order catalogs are not registered for sale in specific states. See "Registration of Botanical Insecticides."*

NICOTINE

Source. Nicotine is a simple alkaloid derived from tobacco, *Nicotiana tabacum*, and other *Nicotiana* species. Nicotine contitutes 2 to 8% of dried tobacco leaves. Insecticidal formulations generally contain nicotine in the form of 40% nicotine sulfate, and most are currently imported in small quantities from India.

Mode of Action. In both insects and mammals, nicotine is an extremely fast-acting nerve toxin. It competes with acetylcholine, the major neurotransmitter, by bonding to acetylcholine receptors at nerve synapses and causing uncontrolled nerve firing. This disruption of normal nerve impulse activity results in rapid failure of those body systems that depend on nervous input for proper functioning. In insects, the action of nicotine is fairly selective, and only certain types of insects are affected.

Mammalian Toxicity. Despite the fact that smokers regularly inhale small quantities of nicotine in tobacco smoke, nicotine in pure form is extremely toxic to mammals and is considered a Class I (most dangerous) poison (Figure 3). Nicotine is particularly hazardous because it penetrates skin, eyes, and mucous membranes readily; both inhalation and dermal contact may result in death. Ingestion is slightly less hazardous due to the effective detoxifying action of the liver.

Symptoms of nicotine poisoning are extreme nausea, vomiting, excess salivation, evacuation of bowels and bladder, mental confusion, tremors, convulsions, and finally death by respiratory failure and circulatory collapse. Poisoning occurs very rapidly and is often fatal. Treatment for nicotine poisoning is symptomatic, and only immediate treatment, including prolonged artificial respiration, may save a victim of nicotine poisoning.

Nicotine has been responsible for numerous serious poisonings and accidental deaths because of its rapid penetration of both skin and mucous

 KEEP OUT OF REACH OF CHILDREN DANGER POISON

PELIGRO

Figure 3 Nicotine is a Class I (extremely toxic) poison. The label on nicotine sulfate insecticides should display the signal words "DANGER POISON" along with the skull and crossbones symbol.

membranes and because of the concentrated form in which it is used.

Uses. Nicotine is used in greenhouses and gardens as a fumigant and contact poison to control soft-bodied sucking pests such as aphids, thrips, and mites. When nicotine sulfate is diluted with alkaline water or soap solutions, free nicotine alkaloid is liberated. Free nicotine is much more active than the sulfate form; it is fast-acting and degrades completely within 24 hours, leaving no toxic residues. Nonalkaline nicotine sulfate sprays liberate the free alkaloid more slowly (over 24 to 48 hours) and have limited use as stomach poisons for control of some leaf-eating pests. Certain roses and other ornamentals may be injured by nicotine sprays.

Tobacco teas are sometimes prepared by home gardeners for control of garden pests or for pests of houseplants. Although these teas are not as toxic as nicotine sulfate sprays, any nicotine solution that is toxic enough to kill insects is also toxic enough to be harmful to humans.

CITRUS OIL EXTRACTS: LIMONENE AND LINALOOL

Source. Crude citrus oils and the refined compounds d-limonene (hereafter referred to simply as limonene) and linalool are extracted from orange and other citrus fruit peels. Limonene, a terpene, constitutes about 90% of crude citrus oil, and is purified from the oil by steam distillation. Linalool, a terpene alcohol, is found in small quantities in citrus peel and in more than 200 other herbs, flowers, fruits, and woods.

Terpenes and terpene alcohols are among the major components of many plant volatiles or essential oils. Other components of essential oils are ketones, aldehydes, esters, and various alcohols. Essential oils are the volatile compounds responsible for most of the tastes and scents of plants. Many essential oils also have some physiological activity.

Mode of Action. The modes of action of limonene and linalool in insects are not fully understood. Limonene is thought to cause an increase in the spontaneous activity of sensory nerves. This heightened activity sends spurious information to motor nerves and results in twitching, lack of coordination, and convulsions. The central nervous system may also be affected, resulting in additional stimulation of motor nerves. Massive overstimulation of motor nerves leads to rapid knockdown paralysis. Adult fleas and other insects may recover from knockdown, however, unless limonene is synergized by PBO. Linalool is also synergized by PBO. Little has been published regarding the mode of action of linalool in insects.

Mammalian Toxicity. Both limonene and linalool were granted GRAS ("generally regarded as safe" — see the "Terms to Understand" section in the summary to this chapter) status by the U.S. Food and Drug Administration (FDA) in 1965, and are used extensively as flavorings and scents in foods, cosmetics, soaps, and perfumes. Both compounds are considered safe when used for these purposes because they have low oral and dermal toxicities (see Table 1). At higher concentrations, however, limonene and linalool are physiologically active and may be irritating or toxic to mammals.

When applied topically, limonene is irritating to skin, eyes, and mucous membranes. Both limonene and linalool may be allergenic. Limonene acts as a topical vasodilator and a skin sensitizer; it was also shown to promote tumor formation in mouse skin that had been previously sensitized to tumor initiation (see Roe and Field, 1965). Linalool is more active as a systemic toxin than as a skin irritant.

Both compounds affect the central nervous system, and moderate to high doses applied topically to cats and other laboratory animals cause tremors, excess salivation, lack of coordination, and muscle weakness. Even at the higher doses, however, these symptoms are temporary (lasting several hours to several days), and animals appear to recover fully. Some cats may experience minor tremors and excess salivation for up to 1 hour after applications of limonene or linalool at recommended rates.

Crude citrus peel oils and products prepared with the crude oils may be more toxic to animals than products containing purified limonene or linalool. Adequate research on the toxicity of crude citrus oils has not been conducted, and they are not recommended for use on animals.

Uses. Limonene and linalool are contact poisons and may also have some fumigant action against fleas. Both compounds are formulated as flea dips and shampoos (Figure 4). They also are included in some pet shampoos that do not directly claim to have insecticidal properties. These products are relatively new, and though showing promise, some questions persist concerning their toxicity. Citrus oil extracts have also been combined with insecticidal soap for use as contact poisons against aphids and mites; published evaluations of the effectiveness of these combinations are lacking. Linalool and limonene evaporate readily from treated surfaces and provide no residual control. Both compounds are most effective when synergized with PBO.

OTHER ESSENTIAL PLANT OILS: HERBAL REPELLENTS AND INSECTICIDES

Essential oils are volatile, odorous oils derived from plant sources. Although they are used mainly as

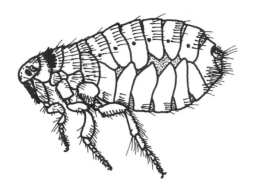

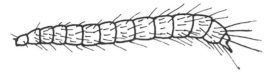

Figure 4 The cat flea, adult and larva. The citrus oil derivatives limonene and linalool are registered for use in controlling adult fleas on cats and dogs. Linalool also is registered for use in controlling all stages of fleas throughout the home.

flavorings and fragrances in foods, cosmetics, soaps, and perfumes, some of them also have insect repellent or insecticidal properties. Many essential oils have GRAS status; however, when applied topically at high concentrations they tend to be irritating to skin and mucous membranes. They are sometimes used as topical counterirritants to relieve or mask pain. Many of the essential oils that have low dermal toxicity may be toxic by ingestion.

The most common essential oils used as repellents are the oils of cedar, lavender, eucalyptus, pennyroyal, and citronella. They are used mostly on pets and humans to repel fleas and mosquitoes. With the exception of pennyroyal, these essential oils are thought to pose little risk to people or pets, though they should not be used above recommended rates. Some herbal pet products that contain essential oils recommend use daily or "as often as needed." These products should be used moderately and with careful observation of the pet to spot early signs of skin irritation or possible toxic effects.

Oil of pennyroyal contains pulegone, a potent toxin that can cause death in humans at doses as low as one tablespoon when ingested. At lower internal doses, it may cause abortion, liver damage, and renal failure. Although the dermal toxicity of pennyroyal is fairly low, some cats are susceptible to poisoning by topical application of oil of pennyroyal, possibly because they ingest it during grooming.

Citronella is sold mainly in the form of candles to be burned outdoors to repel mosquitoes from backyards or other small areas. It is also contained in some "natural" mosquito repellent lotions. Before the development of synthetic repellents, citronella was the most effective mosquito repellent available. Despite its wide usage, there is little scientific information available regarding its efficacy or mammalian toxicity.

NEEM

Source. Neem products are derived from the neem tree, *Azadirachta indica*, which grows in arid tropical and subtropical regions on several continents. The principal active compound in neem is azadirachtin, a bitter, complex chemical that is both a feeding deterrent and a growth regulator. Meliantriol, salannin, and many other minor components of neem are also active in various ways. Neem products include teas and dusts made from leaves and bark; extracts from whole fruits, seeds, or seed kernels; and an oil expressed from the seed kernel.

Mode of Action. Neem is a complex mixture of biologically active materials, and it is difficult to pinpoint the exact modes of action of various extracts or preparations. In insects, neem is most active as a feeding deterrent, but in various forms it also serves as a repellent, growth regulator, oviposition (egg deposition) suppressant, sterilant, or toxin.

As a repellent, neem prevents insects from initiating feeding. As a feeding deterrent, it causes insects to stop feeding, either immediately after the first "taste" (due to the presence of deterrent taste factors), or at some point soon after ingesting the food (due to secondary hormonal or physiological effects of the deterrent substance). As a growth regulator, neem is thought to disrupt normal development by interfering with insect hormone production or reception, thereby preventing insects from reaching reproductive maturity. Susceptibility to the various effects of neem differs by species.

Mammalian Toxicity. Neem has extremely low mammalian toxicity (see Table 1), and in most forms is nonirritating to skin and mucous membranes. The seed dust, however, may be extremely irritating to some people. In humans, neem has various pharmacological effects. Some of these effects are beneficial, such as lowering blood pressure and reducing inflammation and fever. Neem is also an antifungal agent. In addition, neem has antiulcer and strong spermicidal properties, depending on the type of extract. Neem is not mutagenic according to the Ames Test. It has been used in India and Asia for centuries for a multitude of practical and medicinal purposes.

REGISTRATION OF BOTANICAL INSECTICIDES

Laws governing pesticide sale and use require that any product sold to control pests must be registered (approved) by the EPA. In order to gain registration, the product must pass through several levels of testing for toxicity, carcinogenicity, mutagenicity, teratogenicity, environmental fate, and so forth. It also must be tested for safety to nontarget organisms. Natural as well as synthetic insecticides must be registered by the EPA before they may be sold and applied legally.

Under federal pesticide laws, state agencies are required to enforce the regulations that cover the safe use of EPA-registered pesticides. As a result, many states also require state registration before a pesticide may be sold and applied legally in that state. Registration at the state level provides both the records and the funding needed to carry out federally mandated registration enforcement. To register a pesticide at the state level, the manufacturer usually pays an annual fee for the company, plus an annual fee for each individual product that is registered. Although some states (most notably California) require additional testing, paying fees and supplying a minimal amount of background information meet the requirements of most state registration programs.

Several products containing botanical insecticides are not registered in state programs. This may result from the fact that many of the manufacturers of botanical insecticides are small companies that have limited sales. These companies have chosen not to afford the cost of registering their products with each state that requires such a step. Consequently, these unregistered botanical products cannot (or at least should not) be found on store shelves. Instead, they are available only by mail order from various distributors of natural or alternative pest management supplies in other states.

Although it is the responsibility of the manufacturers to register their products, it appears to be the responsibility of the mail-order distributors to alert customers to the fact that certain products are not registered in the customers' home states. Ideally, mail-order suppliers should refuse to ship unregistered products into those states. Unfortunately, many mail-order distributors and manufacturers are not aware of or do not acknowledge these responsibilities. They often will ship products illegally without alerting customers to the registration laws.

Where does this leave the consumer who wishes to purchase a botanical insecticide that is not registered in his or her state? Although it may seem that shipping alternative, "natural" pesticides is a harmless practice, adequate regulation of all pesticides depends on the existence of complete records of sales and use at both the state and federal levels. Requiring registration is the only means by which a state can monitor pesticides properly. Producers of all pesticides — whether synthetic or botanical — must support state registration programs.

Perhaps the most responsible approach is for consumers to write to their state regulatory agencies (often the state Department of Agriculture or state EPA) and to the manufacturer or distributor of the specific insecticide, stating a desire to see the product registered for sale and use in their state. By doing so, and by refusing to order the product until it is registered, consumers can encourage the proper regulation and sale of effective, environmentally sound insecticides.

Uses. Margosan-O, a neem seed extract, was registered briefly in the mid-1980s for outdoor control of gypsy moths and leafminers on trees and ornamental plants. A more stable formulation of Margosan-O received expanded EPA registration in 1989. In 1993, Azatin and Align were registered for use on ornamental and food crops, respectively.

The registration and availability of these and other neem products are expected to differ among individual states for at least a few years.

Azadirachtin controls whiteflies, thrips, mealybugs, and various caterpillar pests of ornamental plants, trees, and shrubs. It is formulated for use as a foliar spray or soil drench and acts as a feeding deterrent, growth regulator, or stomach poison, depending on the pest species.

INSECTICIDAL SOAPS

Source. Insecticidal soaps generally are not considered to be botanical insecticides, though the oils from which they are produced may be of plant origin. In chemical terms, insecticidal soaps (and all soaps in general) are made from the salts of fatty acids. Fatty acids are the principal components of the fats and oils found in animals and plants.

Numerous studies have been conducted to correlate insecticidal activity with the physical structure of fatty acids, and certain acids have been determined to be most insecticidal. Oleic acid, present in high quantities in olive oil and in lesser amounts in other vegetable oils, is especially effective. The common insecticidal soaps now available commercially contain potassium oleate (the potassium salt of oleic acid) as the active ingredient.

In this publication, the term insecticidal soap refers only to those products whose active insecticidal ingredient is the soap itself. Some insecticidal products contain soaps or shampoos in combination with organophosphates (for control of lice) or other kinds of insecticides (for control of fleas on pets). Such *insecticide-containing* soaps are not included in this discussion of *insecticidal* soaps.

Mode of Action. Despite many years of use, the manner in which insecticidal soaps work still remains somewhat unclear. Although the action of soaps involves some physical disruption of the insect cuticle (the outer body covering), additional toxic action is suspected. Some evidence indicates that soaps enter the insect's respiratory system and cause internal cell damage by breaking down cell membranes or disrupting cell metabolism. Soaps also exert some nonlethal developmental effects on immature insects.

Mammalian Toxicity. The mammalian toxicity of insecticidal soaps is basically the same as that of any soap or detergent. Ingestion causes vomiting and general gastric upset, but appears to have no serious systemic consequences. Insecticidal soap concentrates may contain ethanol (up to 30%), which causes intoxication at doses above several ounces; however, vomiting is likely to clear most of the alcohol from the system before it is absorbed into the bloodstream. Externally, soaps are irritating to eyes and mucous membranes and have drying effects on skin.

Some insecticidal soap products contain additional insecticidal compounds such as pyrethrins or citrus oil derivatives. These combination products have a higher toxicity than products containing only soap, and their additional toxic effects depend on the kinds of insecticides added.

Uses. Soaps are used as contact insecticides to control soft-bodied pests such as aphids, thrips, scales (crawler stage only), whiteflies, leafhopper nymphs, and mites. Soaps are effective only against those insects that come into direct contact with the spray before it has dried. Dried residues on plant surfaces are not insecticidal, and they degrade rapidly.

Soaps are not very effective against insects with heavier cuticles, such as adult beetles, bees, wasps, flies, or grasshoppers. In addition, highly mobile insects may escape soap spray applications by flying away. Mobility and a hardened cuticle protect the adult forms of most beneficial insects such as lady beetles, lacewings, or syrphid flies from the toxic action of soaps. The immature forms of these insects are flightless and soft-bodied, however, and may be more susceptible to injury from soaps.

Soaps are particularly useful for controlling pests of ornamental plants and houseplants, though they may be phytotoxic to some plant species. Plants that have pubescent (hairy) leaves may be more susceptible to soap injury than smooth-leafed plants because the hairs tend to hold the soap solution on the leaf surface where a lens effect may cause burning.

In addition to commercial insecticidal soaps, many common household soaps and detergents are insecticidal when applied to plants as a 1 to 2% aqueous solution. Dishwashing liquids and laundry detergents are designed to dissolve grease, however, and they may cause plant injury by dissolving the waxy cuticle on leaf surfaces. Also, detergents differ from soaps chemically and are sometimes more phytotoxic. Plant injury may possibly be avoided if the soap or detergent is rinsed from plant surfaces shortly after application. Commercial insecticidal soaps are less likely to dissolve plant waxes than household cleaning products.

CAUTION: Although homemade soap sprays may be fairly harmless, creating homemade poisons for pest control can be a dangerous practice. Some "recipes" for pesticides call for cleaning agents, fuel oils, polishes, solvents, and other materials that are toxic to plants and many animals (including humans). The fact that these chemicals are not generally considered to be pesticides does not mean that they are not toxic. Many of them are very toxic! *Readers are strongly urged not to prepare homemade pesticides from household chemicals.*

SUMMARY

Botanical insecticides and insecticidal soaps share a number of advantages. Foremost among these are their rapid degradation in the environment and their rapid action in insects. Rapid degradation is beneficial because it minimizes the risks posed by unwanted insecticide residues in food, water, and the environment in general. Rapid action in insects is advantageous because it serves to minimize the extent of feeding damage. On the other hand, rapid degradation imposes certain limitations on the use of botanicals. Where persistent residues are needed for continuous control, botanicals and soaps do not provide adequate protection.

The safety of botanical insecticides and insecticidal soaps in general is a topic that deserves careful consideration. Many of the insecticides discussed in this publication are attractive alternatives to synthetics because, in addition to their rapid degradation, they have low mammalian toxicities (high LD_{50} values). Pyrethrins, sabadilla, limonene, linalool, and the insecticidal soaps fall into this category.

It is very important, nevertheless, to note that all botanical insecticides are *not* safer than all synthetic insecticides simply because they are "natural." Ryania and rotenone are comparable in acute toxicity to commonly used synthetic insecticides such as malathion, carbaryl, diazinon, and permethrin. Nicotine is extremely toxic to mammals by both inhalation and dermal exposure. In addition, the chronic effects of repeated exposures to any insecticide — whether natural or synthetic — are not fully understood and are difficult to test realistically. Users are reminded to always read and follow pesticide label directions exactly and to wear proper protective clothing (long sleeves and long pants, shoes and socks, rubber or neoprene gloves, protective eye wear, and a respirator or dust mask) when applying any pesticide, natural or synthetic (Figure 5).

Finally, readers are encouraged to employ effective nonchemical control measures whenever possible. These may include crop rotation, sanitation, or other cultural controls; the release or conservation of beneficial insects; or the use of resistant crop varieties.

TERMS TO UNDERSTAND

Carcinogen: a substance or agent that produces or incites cancer (malignant tumors) in experimental animals or in humans.

Contact poison: a toxin that kills insects by contact; it does not have to be ingested to be effective. A contact poison either kills insects immediately (as they are touched by the wet spray) and then degrades quickly — as with pyrethrins or soaps — or has some residual action and kills insects over time (as they come into contact with the poison on plant surfaces) — as with the botanical insecticide sabadilla and the synthetic compounds carbaryl (Sevin) and permethrin, as well as many others.

GRAS: "Generally Regarded as Safe," a classification assigned by the FDA to certain compounds that have been used traditionally without apparent toxicity or are believed to be low enough in mammalian toxicity that certain testing requirements are waived during the pesticide registration process.

Hazard: the risk associated with the use of a formulated pesticide. Hazard depends on several factors, including the inherent toxicity of the active ingredients, the formulation (the form in which the product is used, such as liquid concentrate, wettable powder, ready-to-use spray), and the degree of exposure. Hazard, or risk, is sometimes defined as "toxicity times exposure."

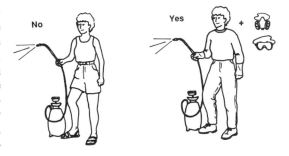

Figure 5 Applying pesticides of any sort is serious business requiring some degree of protection. Even though the active ingredients in most naturally occurring insecticides are relatively low in toxicity, they are often allergenic or irritating. The inert ingredients (solvents, carriers, and other additives) in formulated products also may be irritating to eyes, lungs, and skin. Proper protective clothing should be worn whenever mixing concentrates or applying sprays or dusts. Essential protection includes a loose-fitting, long-sleeved shirt and long pants; rubber boots or sturdy, closed shoes, and rubber gloves (not cloth or leather, as they can absorb toxins). Safety goggles and a cartridge respirator are recommended when mixing concentrates or when applying sprays or dusts in greenhouses or other enclosed areas where contact with fumes or airborne particles is likely.

Table 2 **Signal words**

Signal Word	Toxicity Category	Toxicity Rating	Oral LD_{50}
DANGER or DANGER-POISON	I	Extremely toxic	0 to 50
WARNING	II	Very toxic	50 to 500
CAUTION	III	Moderately toxic	500 to 5000
CAUTION	IV	Slightly toxic or practically nontoxic	>5000

LD_{50}: the abbreviation for "median lethal dose," the dose that kills 50% of the population of test animals. When dealing with pesticides, the LD_{50} values most commonly encountered are the mammalian oral and dermal LD_{50} values (measured in rats, mice, rabbits, etc.). These measurements provide an estimate of how toxic a substance might be to humans. LD_{50} values are usually reported in terms of milligrams of toxicant per kilogram of body weight of the test animal (mg/kg), so a lower LD_{50} indicates a more toxic substance.

Repellent: a volatile substance that keeps insects from alighting or feeding. Repellents are usually used to fend off mosquitoes, flies, fleas, ticks, and gnats (biting pests of humans and other animals).

Signal words: the words on a pesticide label that indicate the toxicity of the active ingredients. Although based on toxicity alone, signal words give an indication of potential hazard (see Table 2).

Stomach Poison: a toxin that must be ingested to be effective. A stomach poison usually has some residual activity, so it remains active until insects have a chance to feed on it. Most stomach poisons are used against plant-feeding insects or insects that will feed on baits. They are generally not effective against insects that suck plant sap. Rotenone and ryania are active as stomach poisons.

Synergist: in terms of pesticides, a material that is itself low in toxicity but increases the toxicity of the pesticide with which it is combined. Most synergists act by inhibiting the detoxification enzymes of the target pest, so that the pesticide has longer to act before being degraded in the insect's system.

Tolerance: the amount of pesticide residue permitted by the EPA or the FDA to be present in or on raw agricultural commodities, processed or semiprocessed food, or feed products.

Toxicity: the degree to which a substance is inherently poisonous (able to cause injury or death). The toxicity of pesticides is often expressed as an LD_{50} for the technical material (the pure active ingredient) in laboratory animals.

ACKNOWLEDGMENTS

The following reviewers contributed to this chapter: M.D. Atkins, Safer, Inc.; Whitney Cranshaw, Colorado State University; Martin Jacobson, USDA, Beltsville, Maryland; Donald Kuhlman, Cooperative Extension Service, University of Illinois; Diane Matthews-Gehringer, Rodale Research Center, Pennsylvania; Shirley McClellan, Illinois Natural History Survey; Frederick W. Plapp, Texas A&M University; John Williams, Necessary Trading Company.

REFERENCES

Casida, J. E., *Pyrethrum: The Natural Insecticide*, Academic Press, New York, 1973, 329 pp.

Farm Chemicals Handbook '89, 75th Anniversary Issue, Meister Publishing, 1989.

Grainge, M. and Ahmed, S., *Handbook of Plants with Pest-Control Properties*, John Wiley & Sons, New York, 1988, 470 pp.

Hayes, W. J., Pesticides derived from plants and other organisms, in *Pesticides Studied in Man*, Hayes, W. J., Ed., Williams & Wilkins, Baltimore, 1982, 75–111.

Hooser, S. B., Beasley, V. R., and Everitt, J. I., Effects of an insecticidal dip containing d-limonene in the cat, *J. Amer. Vet. Med. Assoc.*, 189(8), 905–908, 1986.

Jacobson, M., *Insecticides from Plants: A Review of the Literature, 1941–1953*, USDA Agric. Handbook 154, 1958, 299 pp.

Jacobson, M., *Insecticides from Plants: A Review of the Literature, 1954–1971*, USDA Agric. Handbook 461, 1975, 138 pp.

Jacobson, M., *Focus on Phytochemical Pesticides, Vol. I. The Neem Tree*, CRC Press, Boca Raton, FL, 1989, 178 pp.

Jacobson, M. and Crosby, D. G., *Naturally Occurring Insecticides*, Marcel Dekker, New York, 1971, 585 pp.

McIndoo, N. C. and Sievers, A. F., Plants tested for or reported to possess insecticidal properties, USDA Bulletin 1201, 1924.

Powers, K. A., Toxicological aspects of linalool: A review, *Vet. and Human Toxicol.*, 27(6), 484–486, 1985.

Roe, F. J. and Field, W. E., Chronic toxicity of essential oils and certain other products of natural origin, *Food and Cosmetic Toxicol.*, 3, 311–324, 1965.

Schmutterer, H., Ascher, K. R. S., and Rembold, H., *Natural Pesticides from the Neem Tree (Azadirachta indica* A. Juss.*)*, Eschborn, Rottach-Egern, FRG, 1982, 297 pp.

U.S. Environmental Protection Agency, *Guidance for the Reregistration of Pesticide Products Containing Rotenone as the Active Ingredient,* EPA Office of Pesticide Programs, Washington, DC, 1989, 116 pp.

Yamamoto, I., Mode of action of pyrethroids, nicotinoids, and rotenoids, *Ann. Rev. Entomol.*, 15, 257–272, 1970.

Chapter 49

Oils as Pesticides for Ornamental Plants

Warren T. Johnson, Department of Entomology, Cornell University, Ithaca, NY

Note: Certain portions of this chapter deal with research results on mixtures of certain oils and other materials. These are strictly scientific findings and are not to be construed as a recommendation for use. Only materials registered by the U.S. Environmental Protection Agency (EPA) may be legally used for pest control purposes.

CONTENTS

Introduction 557
Vegetable Oils 558
Petroleum Oils 560
Refining Petroleum Spray Oils 560
    Problems in Terminology 562
    Oil Dilutions 564
    Mechanisms of Toxicity (How Oils Work) 564
Range of Pests Controlled 569
Oils, Environmentally Friendly (?!) 573
    Safety to Humans and Other Mammals 573
    Safety to Insect Predators and Parasites 573
    Phytotoxicity 574
    Climatic Relationships 574
Application 575
    User Friendliness 575
Equipment 576
    Concentration and Formulation 576
    Dormant vs. Verdant 576
Outlook 577
    Control of Fungal Pathogens 577
    Adjuvants 577
    Combinations 578
    Arthropod Resistance 579
    Oil and IPM Principles 579
References 579

INTRODUCTION

This chapter describes the history, development, and current use of spray oils that have their origins from animal, vegetable, and petroleum sources. While spray oils are not new, major changes have taken place in the technology of refining specialty oils, and great strides have been made in understanding efficacy over a wide range of pests as well as the conditions under which phytotoxicity may occur. The EPA-registered spray oils are now as dependable and reliable as the synthetic organic and botanical pesticides. They are also useful adjuvants in aerial and ground applications of insecticides, fungicides, and herbicides. As adjuvants, they improve the efficacy of other pesticides and allow for reduced dosages of the active ingredients. To provide pest managers and sprayer operators the parameters for successful use of spray oils is another objective of this chapter.

0-87371-350-8/94/$0.00+$.50
© 1994 by CRC Press Inc.

The word "oil" comes from the Latin *oleum*, which translates as olive oil. Today, oil is typically defined as an organic liquid that is viscous at summer temperatures, combustible, and insoluble in water. Oils are of three distinct origins: vegetable, animal, and mineral (petroleum). Oils of any parentage may be toxic when applied to both plants and lower animals in low doses. Vegetable and mineral oils have lengthy traditions of use as pesticides; however, recorded use of petroleum oils goes back several centuries. Aboriginal people throughout ancient societies rubbed animal oils on their bodies to protect against biting insects. Animal oils are relative newcomers as products for plant protection. One of the earliest records in the U.S. (1800) is the use of whale oil for scale insect control. During the first half of the 20th century, fish and whale oils were utilized mostly in the form of sprayable soaps. Fish oil was commonly used for scale control in the Wenatchee apple-growing district of Washington during this time. In 1939, Marcovitch and Stanley wrote "It is well known that continuous spraying with summer mineral oil emulsions produce foliage injury and impair the vigor of the tree." Their answer to this problem was a 1% emulsion of wool grease.[1] Application to apple, peach, bean, tomato, and tobacco produced no foliage injury. Wool grease emulsion (2%) was effective against red spider mite. Both concentrations provided excellent sticking properties. While animal oils (fats) have insecticidal value, they probably have no place in modern pest management.

The mode of pesticidal action of all oils is primarily by slow suffocation (hypoxia).

VEGETABLE OILS

The early history of vegetable oils as insecticides has not been researched; the record is indeed meager. Speculation and circumstantial evidence allows us to believe that the early Greeks and Romans and other Mediterranean cultures used vegetable oil to control insects on plants but only after the oil became rancid, contaminated, or otherwise unfit for human consumption.

Serious consideration for the use of vegetable oils as insecticides came after soybeans became a major crop in the U.S. Prior to this, de Ong et al.,[2] in the mid-1920s, compared cottonseed, linseed, and castor seed oils with petroleum oils for their pesticidal and phytotoxic qualities. They found these oils to be more toxic to arthropods and attributed the toxicity to their fatty acid content causing chemical poisoning as well. Hill and Schoonhoven showed that triglycerides and oleic acid extracts alone were toxic to certain beetles.[3] These oils were also found to be more phytotoxic. Because of the phytotoxicity of raw vegetable oils, they were put back on the shelf and virtually ignored until the 1960s. The investigations in the 1960s confirmed previous work about efficacy on scales and other homopterans and provided a better understanding of the cause for phytotoxicity. The technology at that time for processing and refining raw vegetable oils was not sufficiently advanced to remove the phytotoxic elements economically.

Vegetable oils consist primarily of fatty acids and glycerides, and their chemistry is more complicated than that of the petroleum hydrocarbons used as spray oils. The term vegetable oil is not defined chemically, and these oils might best be called lipids, but they too are difficult to define. The oil extracted from each agricultural seed or grain commodity (corn, palm, soybean, etc.) has its own unique properties, such as solubility, viscosity, and boiling point, and these properties may hold the key to successful studies on pesticidal activity and phytotoxicity. Some lipids are highly phytotoxic.

As insecticides, oils must be made miscible for spray application or they may be treated with potassium alkali to make an insecticidal soap. Numerous fatty acids have been tested for insecticidal activity. Siegler and Popenoe found that coconut fatty acids provide excellent insecticidal activity,[4] and Dills and Menusan at Boyce Thompson Institute found capric and lauric acids to be effective insecticides.[5] While data are not present, empirical evidence would have us believe that some of the fatty acids pass through the integument of arthropods and interfere with cell and tissue functions as well as interfere with respiration.

The viscosity or flow properties are greater than those of most of the narrow-range mineral oils. This feature and the slow speed of evaporation may account for some of the phytotoxicity concerns with vegetable oils.

Purification and standardization is a refining process with few similarities to the refining process used for making mineral oils. The once-refined vegetable oils in common usage are refined chemically. The first process removes hydrates with phosphoric acid, followed by neutralization with sodium hydroxide, which also binds with other impurities. Following the chemical treatments two centrifugations are used, one of which mixes water with the product to remove all of the refining chemicals. This process results in once-refined oil. Table product vegetable oils undergo two other refining processes to bleach and destroy the color pigments. The final process is steam distillation at 500°F (2 to

Table 1 **Effect of Insecticidal oils on feeding and oviposition by leafminer (*Liriomyza trifolii*) on Chrysanthemum**

Chemical Spray	Mean Leaf Area (cm²)	Mean No. of Feeding Punctures	Mean No. of Eggs
Water	38	406	182
2% Sunspray 6E Plus	30	2	2
2% Ortho Volck Oil	33	4	2
2% Natur'l Oil	34	4	2

Note: Values are averages from 15 leaves: 3 leaves each from 5 plants. Sprayed plants were exposed to leafminers for 24 hours and then counted. Mean punctures and eggs are from a 100 cm² leaf area. Natur'l Oil is a seed-derived oil.

Table adapted from *Insecticide & Acaricide Tests, Vol. 13,* Entomological Society of America, 1988. With permission.

6 mmHg). At this point the oil is ready for the food market.[6,7]

In the late 1980s, George Butler and colleagues at the Western Cotton Research Laboratory in Phoenix[8,9] and chemists and entomologists associated with Stoller, Inc. in Houston found that once-refined soybean and cottonseed oils reduced the potential for phytotoxicity without decreasing efficacy. Work from these two laboratories has provided valuable information about efficacy on white fly (*Bemisia tabaci*), aphids (*Aphis gossypi, Myzus Persicae, Brevicoryne brassicae*), thrips (*Frankliniella* spp.), beet armyworm eggs and larva (*Spodoptera exiqua*), and eriophyoid mites and spider mites (eriophyoids and *Tetranychus* spp.). Efficacy not only included mortality of arthropods targeted but demonstrated repellent activity to several insect and mite species. Work at the Cotton Research Laboratory provides our most current information about a large number of vegetable oils including corn, coconut, palm, safflower, sunflower, peanut, cottonseed, and soybean and their effectiveness against arthropods largely on vegetable plants.

Only one vegetable oil, "Natur'l Spray Oil," is registered through EPA as a pesticide. Its active ingredient is soybean oil, and it is used to control insects and mites on citrus in Florida and perhaps other states. Thus no vegetable oil can legally be used on any ornamental crop or sprayed in the landscape by a commercial applicator except as noted in a later paragraph. Adequate data should soon be available for registration on certain woody ornamentals.

The best source of vegetable oils is the shelves of grocery markets. Name-brand cooking oils such as Wesson (soybean) and Crisco (corn) make good insecticides and miticides and do as well in this capacity as in the preparation of stir-fried veg-

etables for human consumption. When made miscible with water with a common dishwashing detergent (soft soap) like Dawn, Dove, or Ivory, and mixed with water to make a 2 to 3% mixture, you will have a pesticide that bears comparison to narrow-range superior horticultural mineral oil.*

In their studies, Butler and Henneberry made basic stock solutions of the above ingredients by adding 15 ml of dishwashing detergent to 237 ml of (cooking) oil.[9] Final spray solutions were made by adding 5 to 12.5 ml of the stock solution to 237 ml of water. These studies in Arizona do not account for the potential insecticidal activity of the detergents alone. All of the soft soaps named earlier have insecticidal properties and are discussed in another chapter.

Research in other laboratories[10] has provided data on the success of "Natur'l Oil" (a trade-name vegetable oil) as a repellent to the leafminer *Liriomyza trifolii* on chrysanthemum. The oil prevented feeding and egg laying by the adult flies (Table 1).

Studies by Messina and Renwick have provided evidence that peanut, coconut, and safflower oils applied to stored cowpeas deterred oviposition and caused high mortality to the eggs of the granary weevil, *Sitophilus granarius*.[11]

Work from the U.S. Department of Agriculture Aphis lab in Otis, ME, has produced data adequate to register soybean oil as a product that will kill gypsy moth eggs. It was registered in the summer of 1992 by the promoters of "Natur'l Oil" for spraying egg masses at a dose of 50% oil. Field tests provided evidence that 100% of the eggs would be killed when applied to runoff. It may be applied from September through April with equal effectiveness.[49]

Some entomologists and agricultural scientists recognize a bright potential for vegetable oils as

---

* This formulation does **not** comply with the laws and regulations that the EPA must enforce; it is thus illegal to use on a commercial basis.

natural and biodegradable insecticides. They also recognize that economics runs the machine of commerce. The cost for pesticide registration by current EPA regulations is staggering even though residual tolerance data are not required. Therefore, the potential for most of the vegetable oils to reach the market as pesticides, without major regulatory change, is not good.

PETROLEUM OILS

Petroleum or mineral oils do not fit the category "new-generation insecticides." They have been used as pesticides on plants, animals, and humans for centuries.[12] Before the time of Pliny (AD 23), the celebrated Roman author, mineral oils were used to control insects. In his 37-volume book *Natural History*, topics such as agriculture and forestry were covered, and he wrote about mineral oil as a means of insect pest management. Marco Polo (ca. 1300 AD) wrote about using oil to control mange of camels. By 1868, kerosene, refined from crude oil, but then often called coal oil, was emulsified with soap and applied to dormant fruit trees to control certain scale insects. Occasionally the "side effects" of these sprays were worse than the "disease." Smith wrote papers about crude petroleum as an insecticide.[13,14] He stated, as compared to kerosene, "it is practically harmless to dormant trees even when employed in liberal quantities." He described the crude he used to have a distinct advantage over kerosene in that the residue coating the bark lasts for months and is apparently fatal to all scale insects associated with apple trees.

REFINING PETROLEUM SPRAY OILS

Crude oil is a very complex and variable substance from which hundreds of useful products are refined. The basic chemical classes of petroleum crudes are paraffinic, naphthenic, aromatic, and asphaltic. It is primarily from paraffinic and, to a lesser extent, naphthenic stocks that horticultural oils are made.[12] Distillation is one of the refining processes. By controlled distillation of crude oil, refiners can produce methane gas described as a light end or fraction by the petroleum chemist. As the remainder of the crude boils at precise temperatures and cools in the distillation tower, liquid products of different molecular weight form. As the distillation temperature rises, the products become increasingly heavy until the final product, tar, is formed.

The early petroleum distillation products were developed for reasons other than agriculture or pest management use. It was likely that accident or trial and error brought about the use of kerosene for aphid and scale insect control. These early oils had many impurities such as sulfur, nitrogen, and oxygen atoms. They had to be removed as well as the aromatic compounds (the aromatic compounds are highly toxic to vegetation). A way had to be found to remove the carbon double bond compounds as well as the ring molecules (these are chemically active and phytotoxic). The ring molecules (mostly aromatics) are removed by the solvent extraction process (Figure 1). It was Gray and de Ong, about 1925,[2] who found that sulfuric acid would react with the unsaturated, hydrogen-deficient fractions (carbon double bond hydrocarbons) producing a sludge that floated to the surface of the treated batch. The sludge contained the impurities. To make the UR grade of 92% might require several acid treatments. Following the acid treatment the oil was again distilled to produce the correct molecular size range needed for spray oils. These cuts are made by precise distillation control to produce several grades of horticultural mineral oil. Thus the early production of lubricating class oils refined solely for horticultural purposes.

These processes did not adequately standardize the products of ten or more of the major refiners producing horticultural oils. In 1932, California established standards for different grades of spray oil for citrus largely through the work of R. H. Smith (see Table 2).[15] It was Chapman,[16,17] Riehl,[18,19] Jeppson, and others, in state agricultural experiment stations located in New York, Texas, Florida, and California, along with the cooperation of the oil companies, who developed standardized specifications for horticultural oils (termed superior oils in the East and supreme narrow-range oils in California).[20] The standards or properties provided minimums or ranges of acceptability and grades that provided products that could be used for dormant or verdant application. These properties are listed as unsulfonated residue (USR), viscosity, gravity, pour point, and distillation, all following standards of the American Society for Testing Materials (ASTM).

Unsulfonated residue is a measure of hydrogen saturation in the molecule; few carbon double bond compounds will be present. The minimum standard USR (=UR) is 92%. Most of today's spray oils exceed the minimum standard by several points, but with few exceptions, the higher USR products have not demonstrated a lesser degree of phytotoxicity. Published data show that high USR oils (above 95) are likely to be somewhat more injurious to foliage. Pharmaceutical mineral oil has a USR of 99.9%. Today the acid treatment is outdated, expensive, and the sludge would be a serious environ-

Steps in Refining Spray Oils

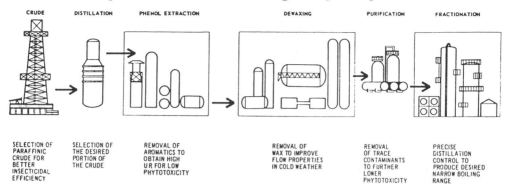

CRUDE	DISTILLATION	PHENOL EXTRACTION	DEWAXING	PURIFICATION	FRACTIONATION
SELECTION OF PARAFFINIC CRUDE FOR BETTER INSECTICIDAL EFFICIENCY	SELECTION OF THE DESIRED PORTION OF THE CRUDE	REMOVAL OF AROMATICS TO OBTAIN HIGH UR FOR LOW PHYTOTOXICITY	REMOVAL OF WAX TO IMPROVE FLOW PROPERTIES IN COLD WEATHER	REMOVAL OF TRACE CONTAMINANTS TO FURTHER LOWER PHYTOTOXICITY	PRECISE DISTILLATION CONTROL TO PRODUCE DESIRED NARROW BOILING RANGE

Figure 1   Spray-oil refining diagram, modified and used with permission of Exxon Company USA and *Farm Chemicals Journal.*

Table 2   **California grade standards (1932) for distinguishing types of petroleum spray oils used as insecticides and acaricides, based on unsulfonated residue (UR) and distillation properties**

Grade	Minimum UR, %	50% Point, °F	Percent Distilled at 636°F and Atmospheric Pressure
Heavy-medium	92	656	28–37
Medium	92	645	40–49
Light-medium	92	628	52–61
Light	90	617	64–79

Note: These oils are no longer recommended for use.

After Davidson, N. A., et al., *Managing insects and mites with spray oils,* Publ. 3347 IPM Education Publications, University of California, 1991, 47 pp.

mental liability. Other methods are now used to provide the USR standard (Figures 1 and 2).

Viscosity measures in seconds the flow rate through a standardized orifice and indicates the heaviness or thickness. Most horticultural oils range between 60 and 100 sec. To obtain spray oils with a specific viscosity, refiners in the past blended higher viscosity oils with oils of lower viscosity. Such a procedure interfered with standardization. Today, viscosity is redundant with narrow-range oils.

Pour point is the temperature at which the oil remains in the liquid state. All horticultural oils contain some wax even though dewaxing is a part of the refining process. At cold temperatures wax may cause oil to congeal and interfere in the spraying process. Acceptable pour points range from –5 to 20°F. The low pour point temperature assures flowability during early-season use in northern cli-

mates, but the specification has little practicality under field application conditions. It is important under certain storage and transport conditions.

Gravity measured by an API method indicates the hydrocarbon composition (molecular shape) and measures the degree of paraffinicity. The higher the API gravity the more paraffinic and lighter the oil. Typical gravity measurements for narrow-range oils are between 31 and 37. This specification will be subject to modification because of the uncertainty of crude oil sources. Any spray oil will have a mixture of paraffin and naphthene. Naphthenic base crudes refined using narrow distillation ranges may produce satisfactory spray oils,[11] but the higher the percentage of saturated naphthene ring compounds in the distillate, the greater the potential for phytotoxicity. Mineral oils for use on ornamental plants should be labeled as paraffinic. The hydrofining process (Figure 2) now makes it possible to use crude oils with less paraffins and at the same time control the molecular weight.

The *50% distillation point* (=boiling point [midpoint]) with the 10 to 90% distillation range is the most useful criterion for identifying the several grades of spray oils. The new oils are called narrow-range oils (NR) and are so designated because of the narrow separation temperatures providing more uniformity to molecular composition, while providing good pesticidal action and plant safety. The 10 to 90% distillation range is illustrated in Figure 4. There is also a range of ±8 points on the stated midpoint temperature, not because of the variability of the oil sample but as a reference to the sensitivity of the analytical method for calculating the midpoint. New technology and analytical methods using gas liquid chromatography (GLC) give greater accuracy to distillation specifications.[21] The distillation criterion provides an index of the relative speed of evaporation when applied to a plant. The lower the boiling point, the more rapid the evaporation. See Climatic Relationships for an expansion of these factors and the importance of the evaporation (dissipation) in efficacy and phytotoxicity.

Molecular weight is always provided in a technical description of a horticultural oil and indicates the size of the hydrocarbon molecules. The molecule size in spray oils is a function of the number of carbon atoms in the molecule. The range in spray oils is 16 atoms with a molecular weight of 226, to 32 carbon atoms weighing about 450 (Figure 3). Molecules with carbon numbers ranging from 20 to 26 have high pesticidal effectiveness with minlmal phytotoxicity. The molecular size may be shown on a tech sheet as the carbon or C number marked C 21-22 avg. Molecular weight of an oil product is controlled through distillation. It is through molecular weight analysis that a horticultural oil user (scientist or grower) can check the grade and quality of the product of a supplier who may have improperly advertised or labeled that product.[21]

Flash point specifications are now added to all of the newly relabeled products to provide important handling and storage safety information. As "nonreactive" oils the flash point in °F is relatively low (usually above 325°F) but *it will burn*. Horticultural oil is not an explosive substance, but once ignited it burns with intense heat.

Most of these refining specifications are not found on registered horticultural oil labels. The EPA and the Food and Drug Administration (FDA) recognize that mineral oil sprays have a high order of safety to humans and the environment. The health hazard is negligible, and there are no tolerance requirements. Old labels provide only a minimum of information about the product in the container; usually the ingredient statement will state petroleum distillate or paraffinic oil and give the percent active ingredient. Old labels will always provide the unsulfonated residue specification and may give a classification or viscosity. The trade name may provide an idea about its overall use. In addition to the above information, newly labeled products will provide distillation information and flash point. Directions for use provide most of the necessary details. There are many horticultural oil brand names, and some of these products are formulated by agricultural chemical companies that lack technical expertise and quality control. It may be important that the buyer know something about the reputation of the seller and vital that the user obtain a detailed product information sheet. All horticultural oils currently refined in the U.S. come from one of four oil-refining companies: Chevron, Exxon, Sun, or Unocal (Table 3). Most refine several grades of oil, and all provide technical information on all of their products to their distributors. None do direct marketing to the end-point user.

PROBLEMS IN TERMINOLOGY

With a lengthy history of petroleum oils used as insecticides we now have some terminology baggage that, in spite of all effort, creeps back into modern literature and Cooperative Extension recommendations. Some of these terms are now redundant, meaningless, or obsolete. Oils at one time were classified as dormant and summer. The dormant oils had the highest viscosity and the greatest use; the lighter summer oils were frequently used as adjuvant and later were used alone as sprays on actively growing plants. Today these terms provide no real meaning to a class or type of oils but designate an application season. Then came the terms "superior" and "supreme" oils. The superior oils, as coined and defined by Chapman, had to conform to clear specifications including a high content of paraffinic hydrocarbons.[16] All horticultural oils today are of this type but should now be called narrow-range oils. Both the terms superior and supreme, a western marketing designation, are included in the term narrow-range.

In the East, oil grades were often classified by their *viscosity* and sold as 60-, 70-, or 100-second oil. Today, reference to a spray oil by its viscosity only supplements the distillation temperature (dt) specification. It has no distinctive meaning to a spray contractor in describing NR oils.

"Narrow-range" is the current and most agreeable term for describing modern mineral spray

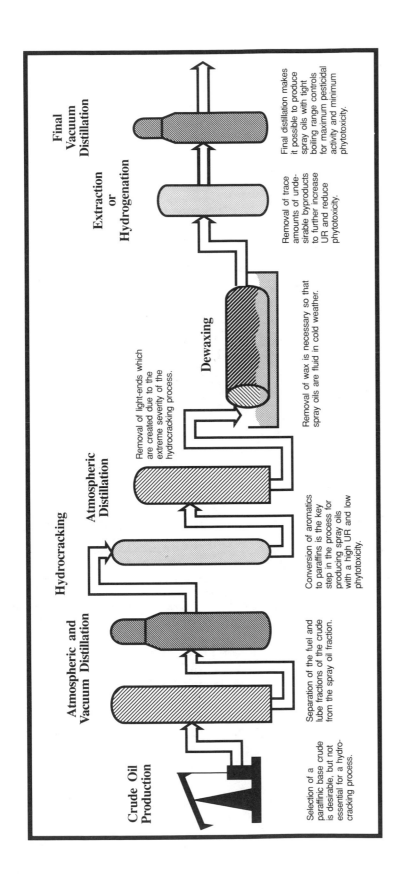

Figure 2 Process for producing "hydrocracked" spray oil.

oils.[19,21] The important property in NR distillation is the uniformity of the oil molecule size as identified by the 10 to 90% distillation range (Figure 4). The 10 to 90 range for oil A in Figure 4 is about 65°F. The small molecular size fractions that come off at lower temperatures reduce the efficacy while the larger molecules can be toxic. The object at this stage of refinement is to achieve as much molecule size uniformity as possible under present technical and economic constraints. There are several grades based upon the midpoint (50%) distillation temperature. Any oil with a midpoint dt above 440°F (about C-25) should be used only when the woody ornamental plant is in the dormant condition. Lower dt oils may be used for both dormant and verdant applications. When using such oils for dormant treatments a higher concentration may be necessary with a concomitant increase in cost.

OIL DILUTIONS

Pure oil, even the narrow-range variety, is toxic to temperate-zone plants by shutting off the transpiration and respiratory system, both in foliage and bark. To be useful, these oils must applied at controlled dosage levels. This is done by emulsification, of which there are two types. In one, the WO (water/oil) type, the water is the external phase and oil is the internal phase (see Figure 5). In the other, the phases are reversed, OW type. Because an emulsion can be diluted only by adding more of the external phase (water), the WO type of emulsion is the most universally employed for insecticidal purposes. There are many kinds of emulsifiers with several classifications, e.g., ionic and nonionic. In most cases the wetting agent used to make oil miscible in water is a trade secret of the formulator. Eddy wrote a detailed paper about miscible oil emulsifiers.[22] Martin made use of the oleic acid in vegetable oil as an emulsifier for mineral oil sprays.[23] Ideally, the WO mixture should break on contact with the target — the oil thinly coating the target, the water running off.

WO emulsions are cream in color. Creaming is due to the differences in specific gravity of the two phases. In the spray tank the two phases will separate over time without agitation. A promoter of one of the early superior oils claimed their emulsion phase would remain indefinitely without agitation. In some of the new narrow-range oils the emulsion will break if the spray material reaches 109°F. Agitation, however, will bring the emulsion back. The emulsifier is a nonactive inert ingredient. In most modern spray oils the inert ingredient is 2% or less.

MECHANISMS OF TOXICITY (HOW OILS WORK)

Horticultural oil will kill plants (on many plants a herbicide in concentrations exceeding 20% with the plant fully covered and act synergistically as an

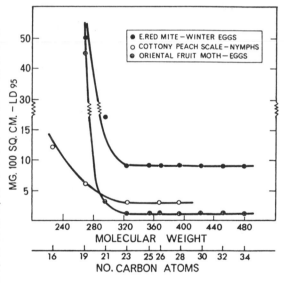

Figure 3 Correlations between pesticidal efficiency and molecular weight and of carbon atoms per molecule in a series of isoparaffins against three unrelated test species.

Table 3 **The range of most horticultural oil products made in 1993**

Product	Minimum UR, %	50% Dist. Point, °F	10–90% Dist. Range, °F	Minimum % Cp	Pour Point, °F	Viscosity	Molecular Weight	API Gravity
Gavicide Super 90	93	440	55	62	+5	86	352.0	33.0
Volck Supreme	≥99	476	85	68	+10	105	352.0	34.8
Orchex 796	92	440	68	60	+6	74	330.0	35.1
SunSpray Ultra-Fine™	92	414	65	NA	+10	68	305.0	32.0

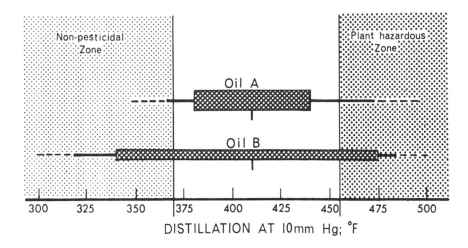

Figure 4 Showing the 10 to 90% distillation range (bar portion) of a narrow range (Oil A), a wide range (Oil B) horticultural oil, and the portion falling into two undesirable zones. The vertical line from bottom of bars shows the distillation midpoint. (*After Chapman, 1967.*)

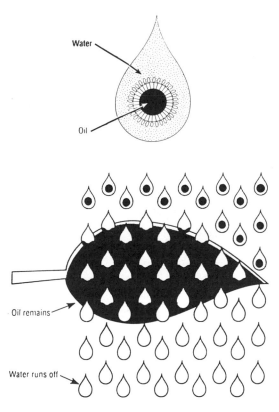

Figure 5 Oil remains on leaf; water drops away.

adjuvant with synthetic organic herbicides), certain fungi (e.g., mildews), arthropods, certain amphibians (e.g., salamanders), and fish. The exposed eggs of any oviparous animal may be in jeopardy from high oil concentrations. The primary toxic action

on both plant and animal is by interference with respiration. If oil through sprays covers the respiratory organs in higher plants (stomates in foliage, lenticels in bark) for an excessive period of time when the plant is not dormant, the plant or portion covered by oil will die of self-intoxication because it cannot dispose of or translocate waste products, the products of metabolism. By this same process other vulnerable organisms are killed as well.

The arthropods are very sensitive to anything that interferes with respiration over relatively short periods of time. This is the primary principle by which the new NR horticultural oils are so successful as arthropod pesticides. When oil membranes cover arthropod spiracles, respiration through the covered spiracles is prevented. Studies by Johnson and Baxendale[50] have demonstrated through microvideo pictures that a small droplet of dyed oil adhering to the spiracular opening appears to be sucked into the segment's tracheal system (Figure 6). The test animals were third-instar cabbage loopers, *Trichoplusia ni* (see Figure 7). Caterpillars with two or more abdominal spiracles so treated did not survive pupation. This study does not diminish the oil membrane cover theory, but adds information that implies an attempt by the test animal to clear the spiracle of the obstruction. There is no evidence, through the video, that oil penetrates the walls of the trachea. Studies with mosquito larvae suggest that petroleum oil or toxic components thereof probably penetrate the tracheal membrane.[23a]

It has also been suggested, with limited evidence, that light mineral oils also penetrate parts of the arthropod's integument. The exoskeleton has many points where movement between the

Table 4 Pests controlled with sprays containing Dendrol or Verdol (1940s through 1950s)

Column groups: all columns below fall under **Scales** — *Armored* (Obscura, Oyster, Putnam, S. Jose, Scurfy, Rose), *Soft* (Cottony Maple, Cottony Mapleleaf, Lecanium, Terrapin), *Other* (European Elm, Golden Oak, Mealybug).

Host Trees, Shrubs, House Plants	Armored						Soft				Other		
	Obscura	Oyster	Putnam	S. Jose	Scurfy	Rose	Cottony Maple	Cottony Mapleleaf	Lecanium	Terrapin	European Elm	Golden Oak	Mealybug
Ash		X	X										
Barberry													
Boxelder							X		X				
Buckeye		X		X	X								
Currant				X									
Dogwood		X		X	X								
Elm		X	X	X	X				X		X		
Euonymus													
Hackberry			X										
Hawthorne										X			
Lilac		X		X									
Linden			X	X									
Maple (Soft)			X				X	X		X			X
Mt. Ash				X									
Oak	X												
Pecan	X			X								X	
Poplar		X		X									
Quince				X	X								
Rose						X							
Sycamore													
Sweet Gum	X								X	X			
Tulip Tree					X								
Willow		X		X	X								
Arborvitae	Juniper												
Juniper	Juniper												
Pines	Pine needle												
Spruce	Pine needle			Spruce bud					X				
Fern	Fern							Hemispherical	X				

Table 5 Pests controlled with sprays containing Dendrol or Verdol (1940s through 1950s)

Host Trees, Shrubs, House Plants	Aphids	Adelgids	Psylla	Whitefly	Lacebug	Leaf-hopper	Thrips	Mealy-bug	Bagworm	Caterpillars	Leaf-roller	Leaf-miner	Shoot Moth	Mites
Ash											X			
Dogwood				O										
Elm	X										X			X
Hawthorn										†	X			X
Lilac											X	O		X
Maple (Soft)														X
Oak											X			X
Sycamore					X									X
Arborvitae			X						†			O		X
Juniper	O													
Pear			X											
Pine		X											O	X
Spruce		X					X							X
Azalea				O	O									
Red Cedar	O													
Flowering Crab	O							O		†				
Hackberry										†				
Rhododendron														
Viburnum	O													
Spiraea	O													
Carnation	O						O	O						X
Coleus	O			X				O						
Chrysanthemum	O			X			O							X
Cyclamen	O			X			O							X
Geranium	O							O						X
Rose	O					O	O		X					X
Sweet pea	O			X										
Currant	O			X										
Gooseberry	O													

Note: (X) oil alone; (†) with lead arsenate; and (O) with nicotine sulfate.

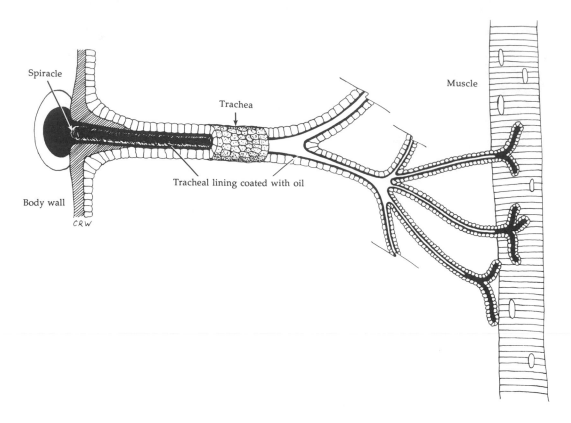

Figure 6 The tracheal system of Lepidoptera, as affected by horticultural oil spraying. Arrows point to spiracular openings.

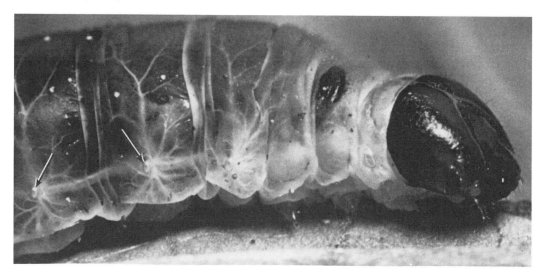

Figure 7 Micro-video photo of cabbage looper tested with dyed oil.

"armor" plates and appendage joints takes place. At these points there are thin articular membranes. It is through these flexible membranes that some investigators believe mineral oil penetrates to interfere with lipid chemistry and cell and tissue function.[24] Other thin cuticular areas are found at the base of setae, hairlike processes that extend through the body wall. There is evidence that immature stages and insects with soft, thin exoskeletons are more vulnerable to the actions of horticultural oil. Insect toxicologists of the past two decades have essentially ignored petroleum products and vegetable oils as pesticides, choosing to look upon them as carriers and adjuvants to enhance the synthetic and botanical pesticides.[25] Their work, however, has provided some insights regarding oil penetration through insect membranes. Matsumura states, "There is a threefold effect of oil as a carrier: (1) It gives the insecticide a chance to attach to the insect (attaches to the organic materials). (2) It breaks the epicuticular wax by dissolving it or else carries the insecticide through it. (3) It disrupts the internal protein organization of the cuticle. Also, the lighter the oil, the more insecticidal it is as a carrier."[26]

Evidence also shows that certain mobile insects avoid oil-treated surfaces, e.g., pear psylla. This has led to empirical evidence that a thin layer of oil on plant parts repels or produces negative tactile responses in several insect species. Such effects would be short-lived because light oils may evaporate in about 4 to 6 days. Bradley was among the first to enunciate some of the ways paraffinic oil impedes aphid transmission of virus disease.[27] Simons et al.[28] carried this direction of study further and produced evidence that a thin layer of oil on vegetable plants would prevent the feeding of sucking insects, primarily aphids, and prevent transmission of viral diseases. The purported action is by clogging the food tube in the sucking insects' stylets. The results of this work have led to a product called Stylet Oil, which contains 97% NR horticultural mineral oil. This product is effective when applied weekly at low concentrations.[28]

RANGE OF PESTS CONTROLLED

The North American use of oil products as pesticides began in the fruit-growing industry — apples in the North, citrus in the South — to protect these fruits from sucking arthropods, mainly including mites, scales, and aphids. All of the research efforts were allocated to the needs of these major food commodities. The "spin-off" came later for use on nut trees, woody ornamental crops, and landscape plants, ornamental greenhouse crops that include annual and perennial flowers and foliage plants, vegetable crops, and finally turf.

Oil has been used as a pesticide on woody ornamentals for over 60 years; the first was manufactured under the trade name "Volck" (Figure 8), named after the man who was largely responsible for its development. The Volck name continues to be used today.

Other oil-refining companies followed with their own versions of dormant and verdant use oils, with trade names like "Denrol" for dormant use and "Verdol" for summer use. Today the trade-name products are too numerous to name. Many agricultural experiment station researchers and university extension personnel acquired data for the promoters, but most of these data remained unpublished. We can assume that the promoters were satisfied with the data regarding efficacy and phytotoxicity because the results came out in product handbooks and advertising booklets. Such publications (Figure 9) described the complete range of uses. In the *Denrol and Verdol Spray Handbook* (circa 1950s), for example, the products were described as white-oil emulsions. These were probably refined to superior oil standards. Table 4 shows the list of pests and hosts for which control was claimed. The range of pest species is increasing from new trials on different pest species that attack all of the plant groupings listed above. The greatest potential for variety of insects and mites continues to be with ornamental plants, and some of the newest studies have been done on the pests of turfgrass. Baxendale and Johnson will soon publish data on the effect of horticultural mineral oils on brambles and other small fruits.

All of the soft-bodied phytophagous arthropods are likely to be susceptible to the toxic action of the new narrow-range horticultural mineral oils. The gross feeding sawfly larvae of conifers and the skeletonizing tenthredinid "slugs" of deciduous trees and shrubs are killed with 2 to 3% NR oils (414 to 435 dt). Most sawfly species conceal their eggs **in** plant tissues. There is no evidence so far of egg kill with the normal spray concentrations.

Larew and others report control of *Liriomyza trifoliia*, a dipterous leafminer common in greenhouse ornamental crops such as chrysanthemums. Control comes from the oil's repellent action against adult feeding as well as egg-laying, rather than any toxic action.[10]

Some caterpillar species are controlled, including some of the tortricids such as the fruit-tree leafroller, *Archips argyrospila*, oak webworm, *Archips fervidana*, and obliquebanded leafroller, *Choristoneura rosaceana*. Most of these tests were preliminary topograph laboratory trials. Other empirical reports include the young stages of forest and eastern tent caterpillars, oleander caterpillar, *Datana* species, and the oakleaf skeletonizer, *Bucculatrix ainsiella*. While some of these reports are not scientifically valid, it is reasonable to conclude that many free-feeding Lepidoptera, as well as the leafrollers and tiers, are vulnerable to sprays containing the new NR horticultural oils. Tests by Baxendale and Johnson[29] on the ermine moth *Yponomeuta multipunctella* showed that fourth-instar larvae

570

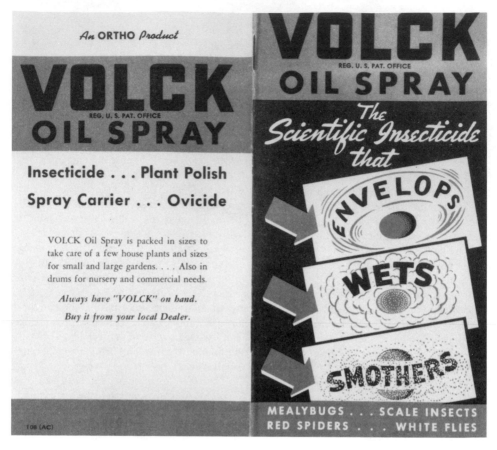

Figure 8 Volck is the oldest trade name horticultural oil in continuous use. The refining specifications of this product have been changed many times. The label dates to circa 1950, but patent rights go back to 1929.

sprayed with 2% 414 dt oil did not resume feeding. Some successfully constructed cocoons, but none completed pupal development.[29]

Labels and product leaflets from the old Standard Oil products have recommended oil with lead arsenate for control of yellownecked caterpillar, *Datana ministra*; cankerworms; bagworm, *Thyridopteryx ephemeraeformis*; and several leafroller species. When used alone, oil gave good control over eggs of many species of Lepidoptera.

The hairy and woolly caterpillars by reason of their covering are not affected by currently formulated spray solutions because the spray particles roll off. The eggs of the Lepidoptera not covered by scales, setae, and spume-like substances are highly vulnerable to present horticultural oil spray dosages. Several species of tortricids whose egg masses are placed on apple twigs and branches have been controlled by dormant treatments, and so were recommended by agricultural experiment stations for many years. From about 1935 through 1960 many

investigators wrote favorable reports about light mineral oil for control of the corn earworm.[30]

Some beetle larvae, particularly the leaf beetles, are likewise sensitive to oil sprays. Only empirical data are available for such insects as elm leaf beetle, *Pyrrhalta luteola*, and the asparagus beetle, *Crioceris asparagi*. The eggs of these species should be vulnerable, but in no case are adult beetles known to be killed by spray concentrations. Larvae of the willow leaf beetle are for the most part unaffected by 414 dt oil.

The weevil *Diaprepes abbreviatus* feeds on citrus foliage and lays its eggs in masses between leaves cemented together by the ovipositing female. Schroeder et al.[31] showed that 435 dt oil applied aerially would impair the bond between eggs and foliage, causing the eggs to detach and drop to the ground. When on the ground, ants were the major predator; eggs exposed to high soil temperatures were destroyed by desiccation. Oil residues on foliage acted as a feeding deterrent to adult weevils.[31]

Figure 9 Product handbook for commercial spray oils in the 1950s.

The pear thrips, *Taeniothrips inconsequens*, on pear and sugar maple may be controlled with spring applications of 414 dt oil at 2%. Only empirical data support this information. Work in the 1940s attests to the recommendations of those years that greenhouse thrips, *Heliothrips haemorrhoidalis*, and a species of *Frankliniella* are controllable with superior oil on carnations and rose, respectively. The privet thrips is highly susceptible to 414 dt oil applied to any of the several generations that develop during the late spring and summer in the Northeast.[32]

There is a large volume of literature about the effect of oils on the Homoptera and the Acari, namely the Tetranychidae, but little information about the phytophagous Heteroptera. The earliest efficacy data published about mirids are on the apple redbug, *Lygidea mendax*.[33] In the Northeast, the newly hatched nymphs were controlled with the then newly developed superior oil (1940s), but one investigator described problems. "These mirids are so fast that they escape being hit by sprays being applied to one side of the tree at a time." The strategy of the time was to have two sprayers operating so that both sides of the tree were sprayed simultaneously. The current New York State recommendation suggests application in the dormant to delayed dormant stages to kill the eggs deposited in the bark of the previous season's shoot.

Baxendale and Johnson found a 414 dt NR oil to be effective against nymphs of the honeylocust plantbug, *Diaphanocoris chlorionis*, but not the sycamore plantbug, *Plagiognathus albatus*.[29] Mirids may be most subject to the suffocating effect of oil in their overwintering egg stage as they are partially imbedded in twig tissue of host plants. The chinch bug (Lygaeidae) is primarily a pest of turfgrasses within the ornamentals industry. Baxendale[51] has had favorable results with 414 dt oil on the nymphal stage.

From the beginning of the fruit-growing industry in the U.S., the homopterans, namely the scale insects, were a plague on production and in many instances the life of the tree. It is here that the successful use of oil from crude petroleum to the new NR oils has been most evident. All of the fruit trees have their ornamental cultivars, and these have retained essentially the same susceptibility to these homopteran insects.

Published literature and labels of horticultural oils from the late 1800s to the present NR dormant and verdant oils name many species of armored scales (Diaspididae), soft scales (Coccidae), pit scales (Astrolecaniidae), Ericoccidae, Margarodidae, and Kermidae as subject to the lethal effects of oils in egg, crawler, and other immature stages. A wide range of aphid species is subject to suffocation. These include the single host species, e.g., honeysuckle aphid, *Hyadaphis tataricae*; two host aphids, e.g., rosy apple aphid, *Dysaphis plataginea*; multiple host aphids, e.g., *Myzus persicae*; and the woolly aphids. In most cases recommendations have been limited to dormant applications to control the overwintering egg. In areas where oleander grows, all stages of the oleander aphid, *Aphis nerii*, are easily controlled by 414 and 435 dt oil sprays. Most, if not all, of the Aphididae in all stages of development should be controlled or repelled by 414 or 435 dt oil sprays, provided the host will tolerate the oil. Existing data neither support nor refute this statement.

Adelgids are of concern on several conifers. Where tolerated by the plant species and applied at a vulnerable stage of the insect, oil provides excellent control. Currently several state extension services have recommendations for the control of the eastern spruce gall adelgid, Cooley spruce gall adelgid, pine bark adelgid, balsam woolly adelgid, and the hemlock woolly adelgid, *Adelges tsugae*.[34]

Ample data exist for the control of mealybugs under greenhouse conditions on such foliage plants as coleus and jade plant. Madsen has data supporting the use of oil on the grape mealybug.[35] In California, the eggs and crawlers of the obscure mealybug, *Pseudococcus obscurus*, are controlled by delayed dormant oil sprays on pear. In the East, the taxus mealybug, *Dysmiccocus wistariae*, was controlled under laboratory conditions, but the treatment failed when applied to *Taxus* under field conditions.

Psyllids may be either free-living on foliage or gall makers. Overwintering boxwood psyllid eggs (*Cacopsylla buxi*) tucked under bud scales are controlled by delayed dormant sprays. Pre-1960 oils were likely to cause bud and foliage damage on boxwood. NR 414 dt oils cause no foliar injury. Oil continues to be a dormant recommendation to control the eggs of pear psylla. The adult female overwinters, and most of her eggs are laid before buds break. NR (414 dt) oil is now used for summer generations of pear psylla but primarily in noncommercial plantings.[36] There are no current data on the effect of oil on the several psyllid gall makers on hackberry.

Early (1950s) records of oil recommendations for leafhopper control were limited to those leafhoppers on grape, of which there are several species. Successful control in the East required the addition of nicotine. Baxendale and Johnson[29] reported success with a honeylocust leafhopper, *Macropis fumipennis*, in early spring when using a 414 dt oil. Unpublished notes included laboratory data that 2% 414 dt oil will kill potato leafhopper nymphs (*Empoasca fabae*) with comments that in field trials the nymphs with their speed of movement and hiding behavior under leaves, are unlikely targets. Future successes for leafhopper control may rest on the degree of exposure permitted by foliage growth characteristics of the many ornamental host plants and with combination sprays that lay down other insecticides that provide residual action.

The early superior oils were effective for controlling unnamed species of whiteflies. Some of today's labels name the mulberry whitefly, *Tetraleurodes mori*, as controllable on azalea, hackberry, and dogwood. Current research supports the claim that oil will control sweetpotato whitefly, *Bemisia tabaci*, and greenhouse whitefly, *Trialeurodes vaporariorum*, in both greenhouse and nursery. Most of the whitefly pests of woody ornamentals continue to await efficacy trials.

The label on Verdol, a summer oil product of the old Standard Oil Company, claimed control over the pine spittlebug, *Aphrophora cribata*. There are no current data or label claims for spittlebugs. Judgments based on phytotoxicity studies and the assumption that results from the pine spittlebug might transfer to other species of spittlebugs make the outlook for desirable results worth the trial effort. We need environmentally friendly control chemicals on such spittlebugs as dogwood spittlebug, *Clastoptera protus*, and *C. arizonana* on *Acacia*.

Oil has been so successful for so long on the tetranychid and eriophyoid mites that evidence need not be covered here. Historically, it was the dormant applications to eggs (tetranychids) and overwintering adult females (eriophyoids) that sparked continuing studies on fruit trees.[37] When the verdant superior oils became available they found immediate use on deciduous fruit trees and on a wide range of woody ornamental plants. Labels only occasionally list specific mite species; most list only spider mites, leaving the implication "all spider mites." It was the common, multiple host mites such as southern red mite, twospotted mite, and the spruce mite that most growers and spray contractors were concerned about. The new NR oils have been tested against single-host species such as the sugar maple mite, *Oligonychus aceris*, the boxwood mite, *Eurytetranychus buxi*, the honeylocust spider mite,

Eotetranychus multidigituli, and others with equal success. It is a reasonable assumption that all of the recognized pest species of spider mites on both deciduous and conifer woody plants, perennial and annual flowering plants, as well as foliage plants are susceptible to NR oils and may provide acceptable control over eggs and all other growth stages of spider mites.

Eriophyoid mites are highly host specific and may be vagrants (rust), generally feeding on the underside of leaves, or gall makers. Most of the eriophyoids on ornamental plants are easy to control with appropriate timing and application techniques. Dormant or delayed dormant oil sprays provide highly effective control over the gall makers on maple, ash, cherry, mountain ash, beech, linden, and others. The vagrant mites of privet and pine are also easily controlled with verdant applications. Except for some of the eriophyoid mites of citrus it is rare to see specific mites named on labels or in recommendations from state Cooperative Extension services.

OILS, ENVIRONMENTALLY FRIENDLY (?!)

Is any chemical substance of animal, vegetable, or mineral origin categorically environmentally friendly? The question is probably more philosophical than scientific, and this is not the forum for establishing a definition for the term environmentally friendly. Comparisons can be made with biological and synthetic organic insecticides, and in this kind of analysis we can show distinct environmental advantages for the use of oils.

SAFETY TO HUMANS AND OTHER MAMMALS

Few if any substances labeled as insecticides are categorically safe to humans. Some of the vegetable oils are used as food, while the highly refined mineral oils (UR 99.9) may be used in medicines (e.g., laxatives) and as emollients (e.g., hand lotions). When used in spray concentrations (water and oil) the only known adverse results are dermal allergic reactions, which are uncommon. For spray concentrations there are no published reports of adverse effects to domestic or wild mammals. Any animal that respires through the skin, such as certain amphibians (salamanders) and earthworms, may be adversely affected, but they, like pests, require direct hits. Concentrates of mineral oil spilled in lakes or streams would be expected to kill or adversely affect some aquatic animals, including mosquito larvae and fish, in the immediate spill area, but only if they were to swim into the floating oil. The oil is not soluble ($<^1/_2$ ppm) in water. The emulsified oil concentrate when in water would be easily identified by a light creamy color over the contaminated area. Ingestion of minute quantities of oil by vertebrates would be inconsequential. The eggs of any terrestrial animal (reptile, bird) would likely be adversely affected with repeated or concentrated doses of oil sprays to the chorion or shell.

SAFETY TO INSECT PREDATORS AND PARASITES

Oil has a good but not totally clean record as it involves insect predators and parasites. All adult beetle predators appear to be unaffected by diluted oil sprays (1 to 3%), but this is not the case for the eggs and larvae. These stages, and particularly the exposed egg stages, do not tolerate oil. No LD_{50} data are available for any beetle larvae. Predatory mites appear to be equally susceptible to oil sprays as their phytophagous relatives. Adult neuropterans tolerate oil sprays, possibly a result of the umbrella effect provided by their wings. Predatory hemipterans in all stages appear to be unaffected by oil sprays.

Adult dipteran and the minute hymenopteran parasites frequently associated with scale insects and aphids are readily killed by direct hits with oil sprays. External and internal parasite larvae appear to be unaffected by the oil sprays that kill their hosts. This observation is empirical and requires additional data from planned experimentation.

While oil sprays kill some predators and parasites, they may safely be reintroduced soon (12 to 24 hours) after the oil application with no fear about the effect of the remaining oil residue. Natural movements into a treated area or eclosion of parasites from their dead hosts occurs following oil application without ill effects, and their presence at this time creates a greater survival pressure on host insects that may have survived.

A light coating of oil on actively growing plants reduces transpiration, and with plants growing under moisture stress conditions the oil provides a distinct benefit. Multiple applications of oil have resulted in remarkable growth increases explained by the conservation of water through reduced transpiration.[38]

Oils have also been used successful, particularly in Utah and the Northwest, for early-season frost protection. Oil applied to deciduous fruit trees before bud break slows development, allowing for a controlled delay of that phase of phenological development.

Oils are often considered biodegradable, but as a sprayed product on a plant, biodegradability is meaningless. The proper question would be, what are the evaporation products and should there be a concern about the volatiles as air pollutants?

PHYTOTOXICITY

Science has not prepared answers to these questions. Without regard for these answers, the volatiles from spray oils on plants would be a minute point source for such air pollutants.

Oils by their nature may be hazardous to plants through interference with respiration, by softening or dissolving the cuticular waxes or lipid membranes, or by penetration of the cuticle usually through stomata.[39]

When gas exchange through stomata is impaired for any reason, within about 48 hours the foliage will lose its normal color and become yellowish. With removal of short-term impairment the foliage returns to normal color. With increase or continuous stomatal dysfunction the symptom progresses, becoming increasingly yellow until the leaf ultimately drops prematurely. Most oil-caused phytotoxicity in foliage is related to stomatal dysfunction with the numerous ornamental plants showing a wide range of susceptibility: from no tolerance, e.g., eastern black walnut, to high tolerance, e.g., Scots pine. Most woody and herbaceous ornamentals fall into the middle ground. In this category foliar phytotoxicity becomes a function of species sensitivity, usually linked to early phenological development, oil dosage, and/or climatological conditions. Vegetative growth in the spring is so rapid that new foliage having escaped the bud scales is usually able to overcome the oil film that may have been deposited.

The leaf spotting symptom from oil usually begins as a small edema-like blister. The symptom is believed to be the result of dissolved cuticular waxes in a very localized area. As a result of symptom progression the edema blister may become a necrotic spot. Spray distribution may cause the leaf tip, and sometimes the margins, to turn brown or black. Leaf orientation with the tip facing downward collects a higher concentration of oil through gravity, and if drying conditions are poor, stomata in that region may remain clogged for a longer period and the oil may penetrate the stoma. In some cases cuticular waxes will be distorted or partially removed. This applies to both conifers and deciduous plants.

Injury by "dormant" sprays causes twig dieback, the result of clogged lenticels. Twigs up to $1^{1}/_{4}$ in. in diameter may be killed. Silver maple, *Acer saccharinum*, is especially sensitive to fall "dormant" sprays. (Needletip necrosis in conifers.)

High concentrations of oil (<50%) plus a bark-penetrating systemic insecticide applied as a banding treatment to thin or unsuberized bark may kill bark with a southern exposure. Heat from direct sun rays on the treated site may result in a concentration of heat adequate to kill or seriously damage the inner bark (Figure 10). Early symptoms are blisters and yellowish-orange blotches.

The flowers of some plants, e.g., magnolia and camellia, do not tolerate oil. The toxic action is through penetration of the cuticle.

With plants as well as animals, toxicity of any substance is dependent upon the dose. Dosage ranges have been established, but these must be adjusted in consideration of plant biology and climatological conditions. Biological conditions that lead to oil injury may involve plant growth stages, especially early phenological development, foliage anatomy, e.g., thick mats of trichomes (pubescence), thinness of cuticular waxes, dormancy status of aboveground parts (see Application-Timing), vigor, and the biological diversity of individuals that occurs in all species.

It is prudent not to generalize about oil sensitivity in plant taxonomic groups. Many shade-loving annual plants such as *Impatiens* are as tolerant to oil on their foliage as is privet. Garden variety roses are highly tolerant to oil, but the foliage of greenhouse roses may be seriously injured. Several investigators have prepared tables with several phytotoxic categories and made lists of plants that fit the categories. As we learn more about NR oils the list of tolerant plants increases as the sensitive lists get shorter and shorter. Only eastern black walnut and butternut consistently show foliage injury at verdant dosages. The dosage/phytotoxicity problem seems to be resolved with the Persian/Carpathian walnut in California.

Conifer trees and shrubs (Koster blue, Colorado blue spruce, Pfitzer juniper), whose primary ornamental attraction comes from foliage wax that reflects a bluish bloom, should not be sprayed with oil. The oil reacts with the epicuticular wax, changing the physical appearance. The needles will remain green for the remainder of their functioning life. Otherwise there is no damage.

CLIMATIC RELATIONSHIPS

There are limited degrees of freedom regarding the speed of evaporation built into the NR oils. If the oil evaporates too rapidly, pest control efficiency is impaired; if it does not evaporate rapidly enough, the plant will develop an adverse reaction. Temperature and relative humidity as well as dosage and dt play key roles in the speed of evaporation. A tree truly in the dormant condition may tolerate an oil film over its lenticels indefinitely. A verdant tree growing under luxuriant conditions and at optimum temperature may tolerate oil for as little as 1 day before showing symptoms of distress.

Figure 10 Silver maple shows bark damage from oil/insecticide treatment.

These are the two extremes. Pests associated with these plants are often in the same state of metabolic activity; the oil kills the pest before symptoms of stress occur on the host plant, provided the user has applied the proper dt oil and the correct dosage under the existing climatic conditions.

The predicted relative humidity and temperature over 24 to 48 hours must be considered in making dosage adjustments. More oil may be needed when the air is dry and hot, less when the air is wet and cool. These factors may account for success or failure especially when spraying near bodies of water or under semi-arid conditions where lands are being irrigated.

Temperature alone is of less concern now than previously believed. Old labels (1940s to 1970s) warned against applying oil to trees when the temperature was below 40°F or above 90°F. These cautions remain in effect for citrus in southern California. In Florida there are reports of palm and broadleaf evergreen defoliation when application temperature was near freezing and likewise when temperature was excessively high. A survey of arborists throughout the U.S. indicates no problems associated with temperature. Arborists in southern and western Texas routinely spray when the temperature is well above 90°F. Tippins[40] described the extreme temperatures reported to cause phyto-

toxic reactions as a "non-relationship" following his studies in Georgia. He produced no damage at 93°F or at 32°F.[40] While probably more academic than practical, low temperatures beginning at 50°F are said to change the viscosity and emulsion quality allowing for the oil to collect in large droplets. Large droplets of oil on foliage would cause damage to sensitive deciduous plants under verdant conditions and conifers under any climatic condition.

Cautions continue to appear on labels for those plants in low vigor or surviving under drought conditions. Much research is needed here for there are degrees of low vigor, and most pest-infested plants are in a low vigor condition. Davidson and colleagues[38] have produced evidence that 3% oil applied to drought-stressed trees and shrubs improves their growing conditions. There are empirical data to the contrary regarding drought stress that the writer accepts based upon the many degrees of drought stress. It seems reasonable that severely stressed plants would defoliate if another stress factor (oil spray) were added.

APPLICATION

The success or failure of any pesticide may ultimately rest on the abilities and dependability of the spray applicator — the person who holds the spray gun. That person must have a mechanical knowledge of the equipment, know limits of equipment capabilities, understand the instructions provided on the pesticide label, and understand the directions that may have been provided by an integrated pest management (IPM) coordinator. In some situations, e.g., treating for gypsy moth, the entire plant must be bathed in spray. Other situations may require the treatment of a limited area or parts of the plant. Such would be the control of white pine weevils, which feed only on pine and spruce leaders in the spring of the year. The application of oil for arthropod control has a special requirement. For the general foliage feeder (e.g., mites, sawfly larvae, aphids, etc.) and bark feeders (twigs, shoots branches), e.g., certain scales and adelgids, a full coverage spray is required.

USER FRIENDLINESS

The new NR oils are even more user friendly than the predecessors. The USR specification of spray oils is near the pharmaceutical grades, which attaches an added degree of safety. No pesticide is safer and more recognized as such by government regulatory agencies. It will burn, however, and must be stored and handled as a hazardous material. Health problems as presently understood are limited to those people with dermal petroleum allergies. The caution that must be emphasized relates to oils mixed with other insecticides or used as pesticide adjuvants. The oil allows for an increase in penetration of certain synthetic organic as well as botanical insecticides, e.g., nicotine, through the human skin. Protective clothing and a face shield must be used with such combinations.

EQUIPMENT

Spraying trees to control pests is a combination of mechanics, art, and science. The advent of low-volume (LV) spray equipment such as the rotomist blower, which uses air as the primary diluent, has resulted in a generation of sprayer operators who have never used a high-pressure hydraulic, high-volume machine. Successful use of oil as an insecticide/miticide requires high-volume equipment for the distribution of dilute spray (0.5 to 4% oil). The entire aboveground parts of a tree or shrub must be wetted down under all dormant applications and under most verdant applications. Generalizations, suggesting "spray to point of run-off" to establish the volume of spray applied are not adequate. Neither is the generalization "one gallon of spray per foot of tree height" or 800 gallons per acre under nursery conditions. Full coverage means that all sides of all exposed plant parts are wet with spray. Full coverage sprays are more labor-intensive with the bottom line translating to greater costs. Avoid the short-cut temptation especially when spraying several large trees growing close together. Spray them one at a time, wetting down all sides before going to another tree. This is an admonition to avoid double dosing portions of a tree. It is easy to do, especially under good drying conditions where water evaporates quickly.

Oil is most effective on arthropods when a thin oil film completely covers the animal. The target must be hit directly, and the size of some targets is measured in micrometers. Except for a few insects that object to the touch of oil or its odor, residue is of no pest control significance. Of course, this is not true when oil is used in combination with synthetic organic insecticides or certain botanicals. Aesthetically, oil provides leaves with a shiny lustre that gives the appearance of vigor.

Some airblast sprayers may be modified and used as high-volume sprayers. Aerial applications are not recommended.

CONCENTRATION AND FORMULATION

Oils are recommended for verdant use at concentrations as low as 0.5% (e.g., in the greenhouse) and as high as 4% on landscape or nursery plants in the dormant condition. Recommendations for summer outdoor use of 414 and 435 dt oils range from 1 to 3%. Labels and Cooperative Extension recommendations give IPM coordinators and applicators some leeway to account for plant sensitivity, weather conditions, and regional experience. Such dosage adjustments both up and down are important for best pest control and plant safety.

Most horticultural oils are formulated to contain about 98% of the active ingredient; some specialty products contain less. They are also formulated to allow the emulsion to break upon spray particle impact. Such formulations will also break in the tank reservoir or in spray hoses if the fluid temperature reaches 109°F. During a summer day, heat in the sprayer tank may reach the critical temperature and the emulsion break if there is no agitation. In the hose line there will be globules of separated oil, and if sprayed in this condition plant damage will occur usually in the form of spotted foliage, which in time will turn into necrotic spots (Figure 11). After a pause in sprayer use, particularly when moving from one job to another, make it a practice to pump the spray fluid from the hose back into the tank for remixing before commencing a new job.

DORMANT VS. VERDANT

"Timing means everything" were the words of a speaker at a recent pest management conference. There are distinct windows of opportunity for controlling plant pests. Some are wide windows, other very limited in their opening. There is both biological timing and timing limited by weather conditions. In biological timing, you must know when to look for your target. In some parts of the country the appearance of common pests is predicted by the Growing Degree Day technique. A prediction method even if by the calendar is important because the insect must be present and in a vulnerable stage before oil will be effective.

By implication, oil (verdant or dormant) may be effective and safely applied at any time of the year, but this is not true. For the health of the deciduous plant avoid autumn applications. During the transitional period of active growth to dormancy there is loss of tree leaves and a transfer to lenticels the total responsibility for respiration. Under field conditions there is no practical way to determine if the tree is fully dormant. Lack of foliage does not equate with dormancy! Serious injury has resulted from late fall applications particularly to silver maple (*Acer saccharinum*). When lenticels are clogged and daytime temperature reaches 60 to 65°F, metabolic activity continues in the shoots

Figure 11 Tiny dark spots indicate injury symptoms on Redbud from spray containing excessive amounts of oil.

and twigs, and gas exchange must not be restricted lest the toxic metabolic products kill the twig. Branches up to $1\frac{1}{2}$ in. in diameter have been killed; the symptoms are not apparent until the following spring. Early spring dormant applications of NR 435 to 440 dt oils have caused few phytotoxicity problems. Such timing has provided good efficacy for controlling early-season insects and has provided frost protection by slowing bud development.

Since the development and general availability of NR summer grade oils, verdant applications have caused fewer problems than the heavier dormant-grade oils. Light oils may be applied at 2-week intervals with up to three applications during the growing season according to label directions. Some investigators have applied up to six treatments at 2-week intervals with no phytotoxic symptoms.

After application to plant surfaces oil deposits evaporate, slowly on bark and much more rapidly on foliage. Rain and the sun's ultraviolet rays are resisted. Such qualities provide additional reasons for adding oil as an adjuvant to botanical, biological, and synthetic organic insecticides.

OUTLOOK

CONTROL OF FUNGAL PATHOGENS

Oil as a management tool for fungal plant pathogens is a rather recent innovation.[41] Sigatoka, a foliar disease of banana caused by *Cercospora musae*, has been controlled with horticultural oil since the 1940s. On banana, oil is considered to be fungistatic, providing a protective barrier that prevents the fungus spore from germinating or the germ tube from penetrating the cuticle.

In Florida, the disease greasy spot, caused by *Mycosphaerella citri,* is controlled by sprays of 435 dt oil. In all citrus growing areas of the world oil has been used not only to reduce phloem feeding insects, e.g., whiteflies, but also to remove caked sooty mold caused by several fungus species and genera. One of the common black sooty mold fungi in Florida is *Meliola palmicola*.

In 1990, Baxendale and Johnson reported successful control of several powdery mildews: *Microsphaera syrinqae* on common lilac, *Syringae vulgaris*; *Sphaerotheca pannosa* on rose; *Phyllactinia* spp. on *Aesculus* spp.; and *Erysiphe* spp. on English oak, *Quercus robur*.[29] Horst and Kawamoto used oil (Sunspray Ultrafine) to eradicate both powdery mildew, *S. pannosa* var. *rosae*, and the disease black spot caused by *Diplocarpon rosae* on several rose cultivars. Their best results came from the combined use of sodium bicarbonate, 0.5%, and horticultural oil (414 dt) at 1%. Both ingredients provide protection and eradication of both disease organisms, but they are most effective when used in combination (Figure 12).[42]

In 1976, Arneson published his experience with NR oil as an adjuvant to benomyl, a systemic fungicide, for control of apple scab, *Venturia inaequalis*. When one quart of oil was added in a tank mix and sprayed on apple trees, he obtained twice the usual effectiveness of benomyl.[43] He speculated that the oil rendered the apple leaf's cuticle more permeable to benomyl. Since Arneson's report was published, arborists and nurserymen have reported highly successful control of apple scab on ornamental crabapples.

ADJUVANTS

The use of vegetable or mineral oils, together with a surfactant to enhance the effect of herbicides, is well known, but references to such effects with insecticides are less abundant. An early reference is the work of Barber, who achieved "optimal" control of corn earworm with a combination of mineral oil and pyrethrins.[30] Unpublished work in the laboratory of Dr. I. D. Degheele at the University of Ghent, Belgium, and formulation specialists

Figure 12 Individual leaves of rose cultivar Samantha (A) treated with sodium bicarbonate plus oil which demonstrates protective properties on powdery mildew; (B) untreated with severe powdery mildew; and (C) treated with sodium bicarbonate plus oil, which demonstrates the eradicative properties on powdery mildew. *(Photos courtesy of Dr. K. Horst, Department of Plant Pathology, Cornell University.)*

at the ICI Specialty Chemical Laboratory, also in Belgium, have shown that the effects of oil enhancement of insecticides varied widely with different combinations.[44] The effectiveness of many commonly used insecticides, according to their tests, could be greatly increased; they claim up to three or four times in some cases. The most marked enhancement was obtained with hormonal insecticides that act through chitin-synthesis inhibition. At the proper concentration the oil/surfactant blend produced a remarkable synergistic effect. Work conducted by Schoones and Gilismee showed that the addition of 1% spray oil to methiodathion (Supracide) increased the toxicity to red scale, *Aonidiella aurantii*, on citrus by about 40 times.[45]

American formulators have mixed oils with ethion, parathion, malathion, Sevin (Sevin-4 oil for aerial application to conifer and hardwood forests and shade trees), Dursban,[46] pyrethroids, *Bacillus thuringiensis*, and others, all offering reduced dosage for equal or better pest control efficiency. This is both economically and environmentally advantageous.

Field tank mixes for application by certified applicators have been recommended through Cooperative Extension in a few states. The recommendation varies from 1 to 2 gal of NR grade oil plus about $^2/_3$ the labeled recommended dosage of synthetic organic insecticide. Oil added to foliar systemic insecticides enhances the penetration and often provides spectacular control.

It must be noted that not all insecticide/miticide uses are enhanced by oil as an adjuvant. Childers and Selhime found the lower rates of miticide in combination with NR mineral oil resulted in significant reduction in residual control of rust mite on citrus compared to standard rates alone.[47] Ochou believed that these differential effects could be explained by the differences in the solubility or miscibility of the insecticide with which they were combined. He found that the more oil-soluble pyrethroids had improved toxicity when combined with mineral oils, whereas the relatively less oil-soluble organophosphates and carbamates did not.[48]

COMBINATIONS

Mixes of oil with other compatible pesticides are of keen interest to farmers, spray contractors, and authorities and individuals concerned with environmental protection. With most of the natural and synthetic pesticides tested, the addition of oil improves effectiveness against the pest organism, residual life, and allows for a reduced dose of the traditional insect control chemicals, as well as a concomitant reduction in chemical costs.

The use of such mixtures may create a serious hazard to the applicator. If oil can synergize a nerve toxin insecticide and allow for penetration of the plant cuticle, the user should be alerted. Oil on human skin acts as an emollient, and the potential exists for carrying a toxin into the body either by reaction with skin-produced lipids or by direct penetration to susceptible tissues. There is also evidence that some synthetic organic insecticides are rendered less toxic to mammals as is the case for parathion. Human toxicological effects from many oil pesticide combinations have been scarcely studied. This is a hazard that demands attention.

ARTHROPOD RESISTANCE

Living entities put most of their energy into survival. Environmental pressures on these organisms provide us with fascinating survival mechanisms. It has been said that insects and mites cannot resist the toxic mode of action provided by oil. However, life being what it is, most organisms ultimately find a way to resist eradication. Already we have found what appears to be a new mechanism of survival among scale insects. The following account will illustrate how another avenue of resistance may be developing. In a Christmas tree plantation of Scots pine where oil had been used repeatedly in spring dormancy sprays for about 10 years against the pine needle scale, *Chionaspis pinifoliae*, it was found that the scale was not being controlled even when the dosage was increased to 6%. Upon close examination we found that scale covers (tests) were much thicker than those in nonsprayed plantations. The eggs under the thick scale covers were free of oil; no oil was evident anywhere except on the outer surface of the test.

From this we have made some assumptions that through natural selection the pine needle scale is developing a resistance to oil via a stronger, thicker, and more weather-resistant test. This same phenomenon may explain reports from the Midwest that oil is failing to control oystershell scale.

OIL AND IPM PRINCIPLES

Oils have now been rediscovered in the ornamentals industry, and they are making a favorable impact in integrated **plant** management. There is a potential for wider use of oil in early spring for frost protection and in summer for the reduction of transpiration (conservation of moisture) and short-term drought protection, especially in nursery production.

Horticultural oil was one of the first biorational/biocompatible insecticides. In lists ranking pesticides with human and mammal safety it ranks among the safest. Few if any chemicals with pesticide activity fit compatibly and effectively into three pest management categories: insect/mite, fungus, and weed control. Few pesticides protect biological control entities as well as oil, and few enhance botanical, hormonal, and semiochemicals that have pesticidal or repellent qualities. Likewise few pesticides can be produced as economically as oils.

These facts encourage the use of oils, and continued research will find even more ways to make oils even more environmentally friendly.

REFERENCES

1. **Marcovitch, S. and Stanley, W. W.,** Wool grease or degras as a substitute for mineral oil in sprays, *J. Econ. Entomol.*, 32, 154, 1939.
2. **De Ong, E. R., Knight, H., and Chamberlain, J. C.,** A preliminary study of petroleum oil as an insecticide for citrus trees, *Hilgardia*, 2, 351, 1927.
3. **Hill, J. and Schoonhoven, A. V.,** Effectiveness of vegetable oil fractions in controlling the Mexican bean weevil, *J. Econ. Entomol.*, 74, 478, 1981.
4. **Siegler, E. H. and Popenoe, C. H.,** The fatty acids as contact insecticides, *J. Econ. Entomol.*, 18, 292, 1925.
5. **Dills, L. E. and Menusan, H., Jr.,** A study of some fatty acids and their soaps as contact insecticides, *Contrib. Boyce Thompson Inst.*, 7, 63, 1935.
6. **Swern, D.,** *Bailey's Industrial Oil and Fat Products*, Vol. 7, 4th ed., John Wiley & Sons, New York, 1979, 841 pp.
7. **Anderson, A. J.,** *Refining of Oils and Fats for Edible Purposes*, 2nd ed., Pergamon Press, New York, 1962, 247 pp.
8. **Butler, G. D., Jr., Coudriet, D. L., and Henneberry, T. J.,** Sweet potato whitefly: host plant preference and repellent effects of plant derived oils on cotton, squash, lettuce and cantaloupe, *Southwest. Entomol.*, 14, 9, 1989.
9. **Butler, G. D., Jr. and Henneberry, T. J.,** Pest control on vegetables and cotton with household cooking oils and liquid detergents, *Southwest. Entomol.*, 15, 123, 1990.
10. **Larew, H. G.,** Oils and pests don't mix, *Greenhouse Grower*, March, 96, 1989.
11. **Messina, F. J. and Renwick, J. A.,** Effectiveness of oil in protecting stored cowpeas from the cowpea weevil (Coleoptera: Bruchidae), *J. Econ. Entomol.*, 76, 634, 1983.
12. **Baxendale, E. C., Baxendale, R. W., and Johnson, W. T.,** Petroleum Distillates for Pest Controls. A Bibliographic History of their Use and Development, Sun Refining and Marketing Co., Philadelphia, 1990, 56 pp.
13. **Smith, J. B.,** Crude petroleum as an insecticide, *N.J. Agric. Exp. Sta. Bull.*, 138, 1899.

14. **Smith, J. B.,** Crude petroleum versus the San Jose or pernicious scale, *N.J. Agric. Exp. Sta. Bull.*, 146, 1900.

15. **Smith, R. H.,** The tank mixture method of using oil sprays, *Calif. Agric. Exp. Sta. Bull.*, 527, 1932.

16. **Chapman, P. J.,** Dormant and semi-dormant sprays with special reference to the control of mites, *Proc. Penn. State Hortic. Soc.*, 88, 73, 1947.

17. **Chapman, P. J.,** Petroleum oils for the control of orchard pests, *N.Y. Agric. Exp. Sta. Bull.*, 814, 1967.

18. **Riehl, L. A. and Jeppson, L. R.,** Narrow-cut petroleum fractions of naphthenic and paraffinic composition for control of citrus red mite and citrus bud mite, *J. Econ. Entomol.*, 46, 1014, 1953.

19. **Riehl, L. A.,** Fundamental considerations and current development in the production and use of petroleum oils, *Proc. Int. Soc. Citriculture*, 2, 601, 1981.

20. **Davidson, N. A., Dibble, J. E., Flint, M. L., Marer, P. J., and Guye, A.,** Managing insects and mites with spray oils, Publ. 3347 IPM Education Publications, University of California, 1991, 47 pp.

21. **Furness, G. O., Walker, D. A., Johnson, P. G., and Riehl, L. A.,** High resolution g.l.c. specifications for plant spray oils, *Pestic. Sci.*, 18, 113, 1987.

22. **Eddy, C. O.,** Miscible oil emulsifiers and spreaders, *J. Econ. Entomol.*, 29, 722, 1936.

23. **Martin, H.,** The preparation of oil sprays. I. The use of oleic acid as emulsifier, *J.S.E. Agric. Coll. Wye Kent*, 28, 181, 1931.

23a. **Berlin, J. A. and Mix, D. W.,** Cellular and subcellular effects of petroleum hydrocarbons on mosquito larvae, *Ann. Entomol. Soc. Am.*, 66, 775, 1973.

24. **Ebeling, W.,** Permeability of insect cuticle, in *The Physiology of Insects*, Rockstein, M., Ed., Academic Press, New York, 1974, 271.

25. **Hesler, L. S. and Plapp, F. W., Jr.,** Uses of oils in insect control, *Southwest. Entomol. Suppl.*, 11, 1986.

26. **Matsumura, F.,** *Toxicology of Insecticides*, Plenum Press, New York, 1975, 503 pp.

27. **Bradley, R. H. E.,** Some ways in which a paraffin oil impedes aphid transmission of potato virus Y, *Can. J. Microbiol.*, 9, 369, 1963.

28. **Simons, J. N., McLean, D. L., and Kinsey, M. G.,** Effects of mineral oil on probing behavior and transmission of stylet-borne viruses by *Myzus persicae*, *J. Econ. Entomol.*, 70, 309, 1977.

29. **Baxendale, R. W. and Johnson, W. T.,** Efficacy of summer oil spray on thirteen commonly occurring insect pests, *J. Arboric.*, 16, 89, 1990.

30. **Barber, G. W.,** The use of insecticides in light mineral oil for corn earworm control, *J. Econ. Entomol.*, 32, 598, 1939.

31. **Schroeder, W. J., Sutton, R. A., and Selhime, A. G.,** Spray oil effects on *Diaprepes abbreviatus* on citrus in Florida, *J. Econ. Entomol.*, 70, 623, 1977.

32. **Moustafa, O. K. and El-Attal, Z. M.,** Enhancement of the efficiency of some insecticides against thrips and cotton leafworm by mineral oils, *J. Agric. Sci.*, 105, 63, 1985.

33. **Dean, R. W. and Chapman, P. J.,** Biology and control of the apple red bug, *N.Y. State Agric. Exp. Sta. Bull.*, 716, 1946.

34. **McClure, M. S.,** Biology and control of hemlock woolly adelgid, *Conn. Agric. Exp. Sta. Bull.*, 851, 9, 1987.

35. **Madsen, H. F.,** Behavior and control of the grape mealybug on pears, *J. Econ. Entomol.*, 55, 849, 1962.

36. **Madsen, H. F. and Westigard, P. H.,** Control of pear psylla with oils and oil-pyrethrins, *Calif. Agric.*, 17, 9, 1983.

37. **Smith, E. H. and Pearce, G. W.,** The mode of action of petroleum oils as ovicides, *J. Econ. Entomol.*, 41, 173, 1948.

38. **Davidson, J. A., Gill, S. A., and Raupp, M. J.,** Foliar and growth effects of repetitive summer horticultural oil sprays on trees and shrubs under drought stress, *J. Arboric.*, 16, 77, 1990.

39. **Ginsburg, J. M.,** Penetration of petroleum oils into plant tissues, *J. Agric. Res.*, 45(5), 469, 1931.

40. **Tippins, H. H.,** Nonrelationship between oil sprays and damage to ornamental plants during extreme temperatures, *J. Ga. Entomol. Soc.*, 9, 51, 1974.

41. **Martin, H. and Salmon, E. S.,** Fungicidal properties of certain spray fluids. VIII. The fungicidal properties of mineral tar and vegetable oils, *J. Agric. Sci.*, 21, 63, 1931.

42. **Horst, R. K. and Kawamoto, S. O.,** Effect of sodium bicarbonate and oil on the control of powdery mildew and black spot of rose, *Plant Dis.*, 76, 247, 1992.

43. **Arneson, P. A.,** Using oil as a fungicide adjuvant, *Am. Fruit Grower*, 96, 15, 1977.

44. **Anonymous,** Insecticide activity is enhanced by oil/surfactant adjuvants, *Int. Pest Control*, 28, 156, 1986.

45. **Schoones, J. and Gilismee, J. H.,** The toxicity of Methidathion and citrus spray oil to mature and immature stages of OP-resistant and susceptible red scale, *Aonidiella aurantii* (Mask.) (Hemiptera: Diaspididae), *J. Entomol. Soc. South Africa*, 45, 1, 1982.

46. **Stewart, D.,** Control of overwintering pests on dormant fruit trees with Lorsban and oil, *Down Earth*, 36, 2, 1990.

47. **Childers, C. C. and Selhime, A. G.,** Reduced efficacy of fenbutatin-oxide in combination with petroleum oil in controlling the citrus rust mite *Phyllocoptruta oleiivora*, *Fla. Entomol. Soc.*, 66, 310, 1983.

48. **Ochou, O. G.,** Plant Oils and Mineral Oils: Effects as Insecticide Additives and Direct Toxicity of *Heliothis virescens* (F.) and *Musca domestica*, M.S. thesis, Texas A&M University, College Station, TX, 1985.

49. **McLane, X.,** U.S. Department of Agriculture, personal communication.

50. **Johnson, W. T. and Baxendale, R. W.,** unpublished data, 1990.

51. **Baxendale, R. W.,** personal communication, 1992.

Grooming bare soil on a baseball diamond for
nonchemical weed control (photo
courtesy of Tim Rhay)

Overseeding a sports field (photo
courtesy of Tim Rhay)

A quality home lawn maintained with minimal pesticides through a
commercial IPM program (photo courtesy of Laurie Broccolo)

Chapter 50

Golf Course Turf Pest Monitoring Program in Monroe County, New York: 1991 Final Report

James D. Willmott and Maher Tawadros, *Cornell Cooperative Extension Service of Monroe County, Rochester, NY*

Jennifer Grant, *Cornell IPM Program, Geneva, NY*

CONTENTS

Background 585
Procedure 585
    Scouting 585
    Evaluation of Impact 585
Results 585
    Impact of Monitoring on Pesticides Use 585
    Diseases 586
    Insects 586
Discussion 587

BACKGROUND

This project is a continuation of efforts that were initiated in 1989. Monitoring was expanded from two clubs in 1990 to eleven in 1991. Six of these were scouted on a weekly basis, one was biweekly, and four were visited monthly. Funding has been provided by the New York State IPM Program and the participating golf courses.

PROCEDURE

SCOUTING

Turfgrass playing surfaces were scouted either weekly, biweekly, or monthly beginning in May and continuing through October 31. Greens, tees, and fairways were observed with each visit. Priority was given to greens and areas with a history of problems. Disease and insect occurrences were quantified, delineated, and documented during each visit. Samples from unknown problems were collected for clinical diagnosis at the Monroe County Cooperative Extension Lab. Diagnoses were usually completed within 24 hours. All information

was reported to the cooperating superintendents and the cooperative extension agent.

EVALUATION OF IMPACT

The impacts of monitoring on pest management practices were evaluated by comparing current and past pesticide-use records, and by cooperator responses to a questionnaire. Pesticide usage was summarized and evaluated by calculating acre treatments* (ATS) for greens, tees, and fairways for the period from May 1 until October 31. Cooperator comments and observations by IPM staff were also considered.

RESULTS

IMPACT OF MONITORING ON PESTICIDE USE

The following results are reported from five of the eleven participating golf courses. Each of these was scouted once a week.

Two of the five cooperators only applied pesticides on a curative basis: in response to the detection of infectious diseases or insect pests. For the

---

* Acre treatment = area treated × number of applications.

purpose of these results, "curative" will be further defined to include tolerance of symptoms at low levels. The other three cooperators treated primarily on a preventive basis, but in some cases made treatments for problems after they occurred.

DISEASES

Survey results from superintendents and scouting observations indicate that fungal disease pressure, in 1991, was relatively light in the greater Rochester area. However, an unusually hot, humid period in late spring did result in sporadic brown patch problems, and anthracnose was common on stressed *Poa annua* during the summer months. Additionally, one club which practiced a curative program, did have difficulty in managing dollar spot in late summer.

Clubs that practiced a curative program treated significantly less area with fungicides than those that treated on a preventive basis. Figure 1 shows the area treated with fungicides over the past seasons, beginning in 1988, for Club # 1 (a "curative" club). The total area treated, primarily for dollar spot, was 53 ATS, which is an increase of 38% from 1990. However, this is still only a third of the average area treated by clubs on a less tolerant, preventive program (see Figure 2). Club # 2, which also practiced a curative program, applied only 6.6 ATS of fungicides; 50% less than in 1990 (Figure 3). These treatments were made exclusively to greens in three applications. The "preventive" clubs made almost four times as many greens treatments

with 24 ATS of fungicides, made in eight applications.

In comparison with clubs # 1 and 2, pesticide use at the three courses treating mainly on a preventive basis is shown in Figure 2. Fungicide use in 1990 and 1991 (160 to 164 ATS) was down slightly from 1989 (188 ATS). As previously explained, these clubs have not adopted a tolerant attitude, whereby they would make treatments only in response to detected pest problems. Hopefully through the IPM program, these superintendents will shift toward a more curative approach, as they gain confidence in scouting results.

INSECTS

In general, most superintendents reported significant problems with insect pests in 1991. Field observations confirmed an early and damaging population of cutworms. Damage continued throughout the season and was widespread. Damage from the black turfgrass ataenius and Japanese beetle grubs was also heavy and widespread. The annual bluegrass weevil, while causing no damage, was detected on three golf courses. This pest has caused turf damage in other areas of the state and, in the future, should be monitored closely.

The average area treated for insect pests in 1991 was 43 ATS; down 9% from the 47 ATS in 1990. One club was planning to re-treat fairways for grubs, but decided not to based on scouting reports that failed to detect population levels above the damage threshold. Consequently, 30 ATS of insec-

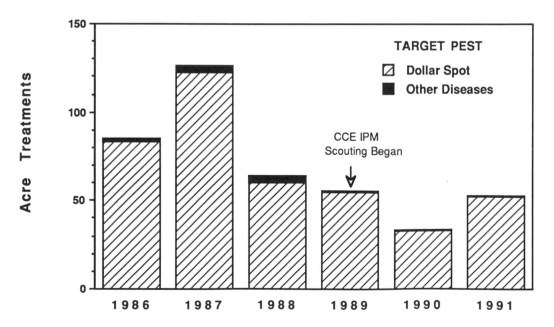

Figure 1 Total fungicides applied to tees and greens at Course #1.

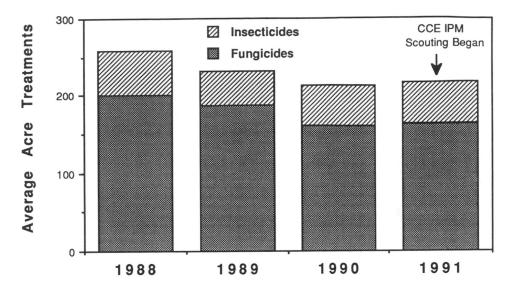

Figure 2 Pesticide applications at clubs on a preventive treatment program. Data taken from one club in 1988, two clubs in 1989, and three clubs in 1990 and 1991.

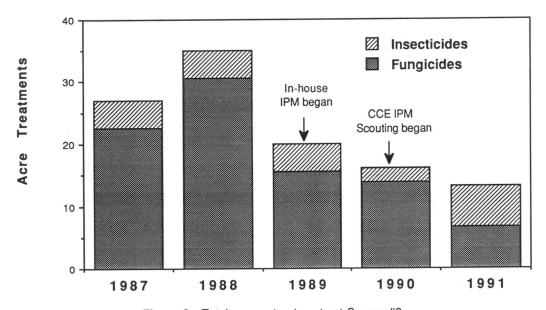

Figure 3 Total greens treatments at Course #2.

ticide treatments were avoided, resulting in a 24% reduction from 1990 at that club.

The greatest reduction in area treated was Club # 1, which applied 12 ATS in 1991; a 60% decrease from 1990. Pressure from fall white grubs was low to moderate at this club, necessitating 9.5 ATS of insecticide; black turfgrass ataenius grubs were not a problem, and only 0.5 ATS were applied for adult ataenius beetle control (Figure 4). Cutworms were the target of 2.5 additional ATS of insecticide at Course # 1.

DISCUSSION

The results indicate that timely monitoring can reduce pesticide inputs if managers are willing to tolerate symptoms and low levels of turf damage. Such restraint can only be expected if the superintendent has strong support from his or her club membership or owner. Additionally, superintendents need to have confidence in the scout's findings and to involve themselves and staff in the scouting process. Achieving this foundation takes

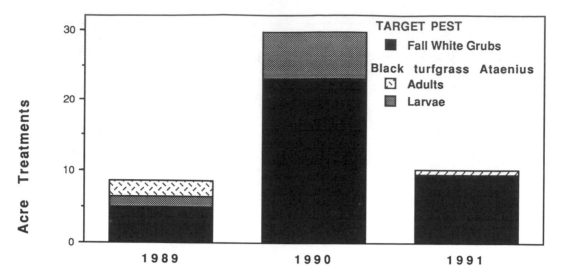

Figure 4 Insecticides applied for grub and beetle control at Course #1.

time, which is apparent in the results. The two clubs that have achieved the greatest degree of tolerance have been involved with IPM efforts for at least three seasons. Each of these superintendents demonstrate restraint in applying pesticides for problems. Their typical response to pest detection is to monitor it for further development. Treatments are made only when severity reaches unacceptable levels. Furthermore, treatments are made only to the areas threatened by the problem. Thus, the combination of tolerance and spot treating is necessary to reduce pesticide use.

Both clubs that practiced a curative program were pleased with the results. One club, which has reduced pesticide treatments to its lowest level ever — only 38% of the average course monitored, felt that playing surface quality was the best ever. In addition to monitoring, trees were removed to improve air flow, and the use of composted organic fertilizers was increased. These inputs may have helped to suppress disease occurrence.

The other club did feel that dollar spot problems were unacceptable in 1991, but generally the playing surfaces were equal to other clubs that averaged more than three times the area treated with fungicides and insecticides. Difficulties with dollar spot control may have resulted from waiting too long to treat; pathogen resistance to the fungicides; or fungicide failures for other unknown reasons.

The area treated in 1991, for clubs not practicing tolerance, was nearly equal to that of 1990. One club, however, reduced greens treatments by 36%. While monitoring did not seem to reduce pesticide inputs, pesticide records from these clubs suggest that superintendents were more accurately identi-

fying problems that occurred. Anthracnose was not recorded on spray records prior to 1991, but was common this season, and several superintendents stated that the scout helped them to recognize this problem.

The average area treated for insects in 1991 was down slightly (4%) from 1990. This may be more significant than it appears, since high populations of black turfgrass ataenius, Japanese beetle grubs, and cutworms were widespread. Several courses spot treated for these pests, including those who practice primarily a preventive program for infectious diseases. In one case, monitoring prevented an additional treatment of fairways, which is equal to 30 ATS. Another course reported a 75% reduction in the area treated for black turfgrass ataenius. Monitoring for insect pests allows for improved timing of, and spot treatment with pesticides.

In terms of cooperator satisfaction, the program has been very successful. All recognize that monitoring is a valuable tool to facilitate turf management decision making. The diagnostic lab has also been an integral part of the IPM program, working hand in hand with scouting observations, especially in disease diagnoses. Some superintendents are better able to practice tolerance, with regard to low levels of turf damage, than others. It is clear, however, that the areas treated with pesticides can be reduced significantly without jeopardizing playing surface quality.

Many cooperators do not feel free to risk any level of damage because of pressure from club membership for greater and greater playing surface quality. Developing support from golfers is essential for the practice of tolerance. Superintendents need reassurance that their jobs are not incessantly "on the line." IPM educational programs for club members, as well as increasing the awareness of the general public through greater media attention, should also improve superintendents' tolerance. One superintendent gave copies of his scouting report to his greens committee on a weekly basis, and found this a very useful communication tool.

Several other factors can also improve pest management and reduce pesticide use, i.e., proper calibration of application equipment. Also important is better designing of golf courses that considers site selection, resistant varieties, and local pest pressure and climatic conditions. Simple activities such as further deliberation of pesticide decisions, including communication with cooperative extension agents or other specialists, may also be helpful.

There are many obvious alternative turf management practices that could reduce reliance on pesticides. Ironically, however, measures to enhance turf health usually detract from what is perceived as a desirable and challenging playing surface. This is especially true of putting greens, where simply raising the cutting height or altering the fertility program could improve turf vigor and its ability to withstand greater pressure from infectious fungi, invasive weeds, and damaging insects. Such efforts will, however, slow green speeds — an unacceptable alternative for many players. Additional research is needed to develop new pest management alternatives. Some promising areas include: resistant varieties of bentgrass and other turf species; biological controls for weeds, diseases, and insects; selective suppression of *Poa annua* and improved fertility management.

Chapter 51

Avoid IPM Implementation Pitfalls

Tim Rhay, *Eugene Public Works Maintenance Division, Eugene, OR*

CONTENTS

Divorce Program Design and Operation from Partisan Controversy 591
Failure to Utilize Authentic IPM Methodology 592
Why the Confusion? 592
Segregation Is Not "Integrated" 592
Authentic IPM Lowers Costs 593
Don't Go "Cold Turkey" 593
Avoid Excessive, Unnecessary Documentation 594
Involve Field Staff in Program Design 594
The Benefits of IPM Are Reserved for Those Who Will Use It 595
References 595

Citizen concern about the use of pesticide by public agencies has increasingly led to calls for the use of alternative maintenance practices. Frequently, communities that have examined the options have accurately concluded that IPM provides the best means of developing viable vegetation and pest control strategies that are both environmentally sound and cost-effective. Unfortunately, many of those who have set out with good intentions to implement IPM programs have failed to achieve one or both of these objectives. Where such failures have taken place, one may hear that "IPM doesn't really work," or "IPM costs too much," neither of which is true of the historic or "authentic" methodology.

Why, then, the failures and inaccurate perceptions? In more than 10 years as the City of Eugene, Oregon's IPM program coordinator and a consultant to numerous other public agencies, I have observed that they are often the result of the "pitfalls" I will enumerate. These include: allowing the pesticide controversy to motivate and drive the program; failure to utilize authentic IPM methodology; imposing an arbitrary ban on pesticide use; developing overly ponderous written documentation, formal procedures, or both; and failure to involve field staff in program design.

DIVORCE PROGRAM DESIGN AND OPERATION FROM PARTISAN CONTROVERSY

Unworkable pseudo-IPM programs often grow out of the suspicion and hostility that seems to dominate the controversy between pesticide advocates and antichemical activists. Dwight Moody said, "The best way to show that a stick is crooked is not to argue about it, or spend your time denouncing it, but to lay a straight stick alongside it." His wisdom is usually lost on partisan activists, for whom argument and denunciation are a virtual way of life.

In this case, the straight stick is a functional IPM program that provides cost-effective vegetation and pest control without dependence on calendar-driven cover sprays, yielding substantial environmental benefits. Such a program is easy to defend to critics from either quarter, especially if they have nothing better to offer as a substitute. However, it is unusual for such programs to arise in a community that has already been polarized by the pro- vs. antipesticide struggle. And, once a nonauthentic "IPM" program, favoring the philosophical or political bias of either side, has been established, it becomes difficult or impossible to replace it with the actual/historic methods.

Your best opportunity to build a successful program, true to authentic IPM principles, is to do so before the controversy arises in your community, or at least before it has divided people into antagonistic, hostile factions. This means you must take the initiative. While you may not currently see the need to commit your agency to the type of effort this will require, especially if you are satisfied with your present vegetation and pest control programs, you will not regret doing so in the long run. As I will review below, IPM makes excellent sense from an economic and efficiency standpoint as well as environmentally. If the controversy never comes, it will still have been a good idea.

With the steady increase in environmental concern, it is not really a question of whether or not the pesticide controversy will eventually reach your community, but only a matter of when. Among the proverbs of wise King Solomon is the following: "A prudent man foresees evil and hides himself, but the simple pass on, and are punished."[1] Prepare yourself and your agency now.

If the controversy has already intensified in your community and is, to a large or small degree, the motivation for your efforts to implement IPM, it is critical that you understand that your success in doing so will be dependent upon your ability to divorce program design from the partisan controversy. To be successful, you must use authentic IPM, based firmly upon the historic model. Controversy-driven programs consistently fail to do so.

FAILURE TO UTILIZE AUTHENTIC IPM METHODOLOGY

IPM accurately refers to vegetation and pest control strategies similar to those recommended by U.S. Department of Agriculture entomologists W. D. Hunter and B. R. Coad in 1923, for the control of boll weevil infestations on cotton in the U.S. The planting of pest-tolerant cotton varieties and crop residue destruction were the primary means of control in that program, with insecticides considered supplementary and to be used only when monitoring identified weevil populations/damage above a predetermined level.[2]

A well-known organic gardening advocacy organization gives the following definition: "Integrated pest management (IPM) is a strategy for keeping plant damage within bounds by carefully monitoring crops, predicting trouble before it happens, and then selecting the appropriate controls — biological and cultural, or chemical controls as necessary."[3]

The University of California Statewide Integrated Pest Management Project fills in the details with the following: "Integrated pest management is an ecological approach to managing pests that often provides economical, long-term protection from pest damage or competition. Factors such as prior pest history, crop growth and development, weather, visual observations, pest monitoring information, and cultural practices are considered before control decisions are made. IPM programs emphasize prevention of weed competition and other types of pest damage by anticipating these problems whenever possible. Goals include conserving natural enemies and avoiding secondary pest problems through the use of nondisruptive and mutually compatible biological, cultural, and chemical methods."[4]

Another basic element of authentic IPM is the setting of economic and damage threshold levels. The economic threshold, a fundamental IPM concept, is the point at which the density of the pest requires a control measure (usually chemical) to prevent economic loss. The damage threshold is the lowest pest population density at which damage occurs.[5]

WHY THE CONFUSION?

Since IPM is so obviously widely and clearly understood, why are some confused as to what it really is, and why would people advocate something other than authentic IPM? In my own experience, the reasons are philosophical, political, or both. On the one hand are those opposed to any and all use of synthetic chemical pesticides. This faction typically tries to influence program design so that pesticide use is banned outright, very severely restricted, or so encumbered by process and regulation that it is rendered impractical. But such restrictions are not part of the authentic IPM methodology already reviewed in this chapter. Tampering with those proven methods typically results in the loss of one or more of the historic benefits of IPM and, not infrequently, reduces all of them.

On the other hand, advocates of traditional "cover spray" programs (blanket applications of pesticide made at scheduled intervals according to the calendar) sometimes try to inaccurately apply the IPM label to their methods, possibly as a way to insulate themselves from change. But such cover spray programs typically assume the need to treat, rather than setting realistic thresholds and basing treatment decisions on monitoring by knowledgeable field staff. Such an approach denies its users both the environmental advantages and financial savings of authentic IPM. Putting it bluntly, cover sprays often deposit large volumes of pesticide where no real "targets" (the problem pests or vegetation) are present. This is a waste of time and money in addition to being an unnecessary impact on the environment.

SEGREGATION IS NOT "INTEGRATED"

Another common misconception, or mislabeling, of a methodology that is not truly integrated as IPM can be illustrated by a hypothetical agency that manages, say, 400 acres. This agency submits a vegetation management plan stating that 100 acres will receive no treatment, 100 will be cultivated,

100 will be mowed, and 100 will be sprayed with herbicide. They declare their chosen strategy "integrated vegetation management." I submit, rather, that it is "site-segregated vegetation management," the range of control options available at any one site having been limited to only one.

In contrast, a truly integrated program would first identify the target pest (in this case, problematical vegetation) and set realistic tolerances for its presence, then develop a cultural program designed to encourage desirable vegetation while minimizing invasion by target species. Monitoring would reveal when and where additional control measures were required. If necessary, control action might involve a combined strategy of spot treatment with herbicide, posttreatment cultivation, and replanting with desirable species. Both examples are hypothetical, of course, but the difference in approach is obvious.

AUTHENTIC IPM LOWERS COSTS

From its beginnings, a major goal of IPM, in addition to efficacy, was cost-containment. IPM has been used, especially in agriculture, to achieve this same end ever since. Hence, when someone says to me, "IPM costs too much," I cannot help but be skeptical. Upon examination of such claims, I have always found that something other than the authentic methodology was being utilized, and incorrectly labeled IPM.

The substantial environmental advantages of IPM were recognized after its economic value, but they are just as real. Indeed, this is why the methodology is so well suited for the 1990s, when widespread environmental concern will apparently be accompanied by tight financial times. Enlisting the ecosystem as an ally instead of an enemy, setting realistic tolerances for pest populations, utilizing monitoring instead of assuming the need for treatment, and selecting control strategies that will be the least disruptive to predators and other nontarget organisms are obviously environmentally sound approaches. And, it is these same program features that provide the financial savings. Using authentic IPM, you really can have it both ways!

DON'T GO "COLD TURKEY"

Bans and moratoriums are not uncommon reactions to the pesticide controversy by public agencies. To administrators removed from in-field efficacy problems they may seem the bureaucratic course of least resistance, but they are rarely a good idea. Banning all use of pesticide, even the conservative and timely applications that are part of an IPM program, inevitably leads to drastically increased costs, an unacceptable deterioration in program efficacy or, most commonly, to both. These negative effects are unnecessary and, in the long term, can lead to pressure to reinstate the former cover spray program.

Indeed, some antichemical activist groups, having come to this realization, have replaced advocacy of bans with calls for the establishment of IPM. Others, however, have not, and continue to promote the fiction that workable alternatives exist for all present uses of pesticide.

I have worked in grounds management for more than 20 years, seeking workable alternatives to chemically dependent techniques throughout that time. Since 1980, I have coordinated the internationally recognized IPM program of the City of Eugene, and acted as an IPM consultant to numerous other agencies. Based on that experience I can tell you that, as much as I might wish such pronouncements were true, they simply are not.

A functional alternative to unworkable "cold turkey" bans is the technique I call "building-down." Briefly, the build-down approach tackles vegetation and pest control challenges sequentially, one (or a few) at a time. Workable alternative methods are incorporated gradually, as they are developed, and chemically dependent techniques are abandoned as they are no longer needed. In this way, efficacy is never compromised, but the agency moves inexorably farther and farther away from pesticide dependency.

In selecting where to begin, consider which of your existing management practices result in the highest volume of chemical application. For example, calendar-driven broadleaf weed control programs in turfgrass are high-volume applications. Setting realistic thresholds for broadleaf plants in various types or classifications of turf, employing cultural measures designed to give the grass the advantage over weeds, and converting to a threshold-driven application regimen will produce a significant reduction in application volume and, by extension, nontarget impact, without loss of truly necessary weed control.

We did this in Eugene, fully expecting that herbicide treatments would prove necessary on sports field turf fairly often because of high use levels and resultant stress. In practice, it has only been necessary to make a single herbicide application to parts of three fields in the last 12 years. Cultural measures have held the balance of more than 25 fields in an acceptable or "under threshold" condition.

A combination of thresholds and the use of water blasting and insecticidal soaps when treatment levels are reached has virtually eliminated the need for stronger chemical pesticides to control aphids anywhere in Eugene's park system. Efficacy has not diminished. It could be argued that it has improved. Our rose garden staff developed a disease-management strategy that eventually enabled them to forgo regular, preemptive fungicide treatments.

As positive as these results are, I do not feel we could have achieved them in "year one," or even now, if we had been faced with a comprehensive ban on pesticide use. In every case with which I am familiar, serious program efficacy problems have resulted wherever such a policy has been established without a prior effort to develop workable alternatives.

AVOID EXCESSIVE, UNNECESSARY DOCUMENTATION

Another feature of so-called IPM programs that are less effective than the historic model is a tendency toward overdocumentation. Of course, some documentation is required to implement an IPM program on an agencywide basis, especially if these methods have not been fully followed in the past. What I've termed overdocumentation involves the preparation of written materials or mandating of administrative procedures that are not truly necessary to effectively implement IPM in the field.

In my consulting work with other communities, two forms of excessive documentation or procedure have come up again and again. The first is production of voluminous "manuals" that attempt to anticipate all possible vegetation and pest control problems and prescribe treatment strategies in advance. Often, field staff are restricted to only those treatment options listed "in the book." While well intended, such manuals are rarely successful in anticipating every possible problem, often quickly made obsolete by improvements in technology, and less effective than direct training in teaching IPM methodology to field staff.

A loose-leaf manual, easily updated and available as a reference to all staff involved in vegetation or pest control, can be helpful, depending on the agency's size and communication system. Some use a computer file/network for this purpose. However, you should avoid setting a policy that only those treatments contained in your manual can be used. Such a policy inhibits the staff creativity that has led to many of our most successful innovations.

The other common error is establishment of mandatory, overly ponderous systems of documentation for monitoring data, pretreatment analysis, posttreatment evaluations, and the like. These are often coupled with review or approval by a nonstaff oversight board. Such administrative systems may so encumber the monitoring-analysis-treatment process that timely, needed control actions are interfered with or delayed. This can allow pest populations to build to unnecessarily large levels that may require extreme measures to control, thwarting the very goals that motivated the initial interest in IPM.

Avoid mistaking paper volume for progress, or assuming that the former will lead to the latter. If anything, an inverse relationship between the administrative/paper load and real progress in the field can be demonstrated by comparing successful IPM programs to those that have not produced the benefits inherent to the historic methodology.

INVOLVE FIELD STAFF IN PROGRAM DESIGN

In building a successful IPM program, your field staff is your most important resource. Those who have been responsible for vegetation and pest control "in the trenches," and will continue to be so, are an invaluable source of accurate information, essential to success. Often, this same group also is a major source of innovative control strategies.

Beware, again, of political or philosophical zealots who may seek to disenfranchise staff input because it is, or they perceive it to be, in conflict with their cherished notions of how the ideal system "should" function. Watch out, too, for societal biases which view the ideas of blue-collar workers, especially those in horticultural and agricultural vocations, as less valuable than those of managers or consultants with impressive academic credentials.

When the Eugene program began in 1980, our first priority was training all staff involved in vegetation and pest control in IPM theory and methodology. Next, we assigned field staff members to research and propose IPM strategies for dealing with our most frequently encountered pest and vegetation problems. With this foundation laid, our practice was to encourage staff to continue to apply what they had learned and build upon it. The result has been a sense of program ownership which, in turn, has led to a strong desire and motivation to see the program succeed. It has become routine, now, for staff to propose new control strategies that will enable us to further reduce pesticide use without adverse effects on program efficacy.

In contrast, I am familiar with agencies where so-called IPM programs have been established by administrative fiat with minimal to no staff input and, not infrequently, without addressing their expressed concerns. When combined with other pitfalls, especially overdocumentation and going cold turkey (which is not an atypical scenario) the result is alienation of the very group upon which program efficacy ultimately depends.

Predictably, field staffers in such an agency are unlikely to be willingly cooperative with the mandated program. Frustration in the field leads to coffee-break and lunch-time ridicule. Motivation is lacking. Enthusiasm never develops. The program, which was often ill-conceived and out of conformance with historic integrated practices in the first place, goes nowhere. Staff blames management and/or outside interference. Administrators and/or local activists blame the staff. Nothing positive results and, ultimately, it is the community that suffers.

THE BENEFITS OF IPM ARE RESERVED FOR THOSE WHO WILL USE IT

More than 10 years' experience has convinced me that IPM is the ideal pest and vegetation control strategy for the 1990s and beyond. I have yet to find a public agency resource management challenge to which it could not be applied with excellent results, yielding the parallel benefits of public safety, environmental protection, program efficacy, and cost efficiency. But I am even more convinced that these benefits will not be achieved by those who discard or tamper with the authentic, historic methodology. All such programs with which I am familiar have failed to do so.

Time and time again, the pitfalls discussed in this article have cost what could have been an effective program the results it should have achieved. Inevitably, this has come about because of the pro- vs. antipesticide controversy, with its hostility and suspicion, and the desire of one or both factions to bend reality to fit their preconceived philosophical position and/or exercise control.

My own bias is clear. The men and women asked to make the program work, to provide real results in the real world, also should control its design and direction. Those who cling rigidly to philosophical positions, but take no responsibility for demonstrating the workability or realism of their positions under actual field conditions, should be welcome to observe, suggest, or advise, but, under no circumstances to dictate or control.

IPM is the answer. It has provided environmentally sound, effective pest and vegetation control at reasonable cost for more than seven decades. It can do so for decades to come. Act now to bring your agency's methods in line with the authentic/historic model. Avoid these common pitfalls to effective implementation. If you have identified any of them in your program, have the courage to change it! You will not be sorry that you did.

REFERENCES

1. Proverbs 22:3, *The Holy Bible,* The New King James Version, Thomas Nelson Publishers, New York, 1983, 738.
2. **Flint, M. L. and van den Bosch, R.,** *Introduction to Integrated Pest Management*, Plenum Press, New York, 1981, 5.
3. **Yepsen, R. B., Ed.,** *The Encyclopedia of Natural Insect & Disease Control*, Rodale Press, Emmaus, PA, 1984, 189.
4. **Marer, P. J.,** *The Safe and Effective Use of Pesticides*, University of California, Statewide Integrated Pest Management Project, 1988, 67.
5. **Shurtleff, M. C., Fermanian, T. W., and Randell, R.,** *Controlling Turfgrass Pests*, Prentice-Hall, Englewood Cliffs, NJ, 1987, 361.

Chapter 52

An Award-Winning Management Plan

Laurie R. Broccolo, Broccolo Tree and Lawn Care, Henrietta, NY

As the environmental movement evolves, we as lawn and landscape managers must also evolve in our pest management techniques. A nonpesticide approach may sound good in theory, but in reality we must continually deal with insects, weeds, and diseases. How do we meet this challenge of eliminating or reducing pesticide usage and protecting the health of our trees and turf to the satisfaction of our customers and the public?

We must take a practical, common-sense approach in caring for our living environment. IPM is a comprehensive approach that includes cultural practices, with pesticides used as a tool when necessary. Understanding the theories of good cultural practices, such as proper mowing, pruning, watering, and fertilizing, is just the beginning. Taking the textbook approach into the field is the real challenge. Meeting the demands of customers (within their budgets), scheduling around weather conditions, and training and retaining employees, while juggling everyday business demands so that a profit is made, are a few of these challenges.

Communication is the key to a successful program. It involves **training customers**, the **public**, and **most importantly your employees.** The employee that physically cares for the plants is no longer just a laborer, but rather a manager of that plant's health. In addition, he/she also may be in the position to educate customers and the public.

Developing an ongoing training program that includes participation and **communication with employees** will keep the program interesting and help both new and old employees stay abreast of industry developments. Putting together a successful program takes commitment from management and employees to organize and follow through.

The first step in developing an effective communication program is to categorize the subjects into main topics i.e., safety, horticulture, company policies, and customer communications. Then outline subcategories within the main topics. Schedule a minimum of one formal meeting, approximately 30 minutes, per week. Also, allow 15 minutes at the beginning or end of each day for impromptu discussions. Be accessible!

Who has time to do all that? Delegate! This is a seasonal business. If you put some thought into your training program during the winter months, then implement it, you will find it becomes a part of your routine. Ask your employees for input. Involve them in the training. You may not have time to read all the trade magazines, but if you ask your employees what article, new product, or equipment they found of interest and have them share it with everyone at the meeting, then all of you will benefit. Send your employees to seminars and ask them to share what they learned. The more interest you show in educating your employees, the more effort your employees will put into participating in the training programs.

I have learned much more from communicating with employees than I could possibly absorb from reading all the trade magazines or attending all the trade association meetings and seminars. Two-way communication stimulates more ideas.

The formal meetings should include a subject from each of the main categories. Designate employees to discuss a different subject each week. Be prepared to add a subject that is timely and based on experiences during the week. Call an emergency meeting to address customer concerns or a safety problem that needs immediate attention. Programs that are flexible and involving employee participation are more meaningful to employees.

Resources must be readily available to all employees so they are encouraged to learn continually. For example:

- Take advantage of free publications such as trade magazines and vendors' newsletters.
- Join trade associations and seek cooperative extension advice.
- Develop a library by purchasing a few reference books and bring in your own. Encourage other employees to do the same.
- Post newsletters, magazine articles, and other

interesting information on a bulletin board. The bulletin board must be changed and updated weekly.

- To explore IPM, encourage everyone to bring in live (or dead) samples of insects, weeds, diseases, and symptoms. Preserve them in mounts and label them for future reference.
- Identify pests with the aid of reference books and discuss the best technique to control them.
- Pay attention to growing degree days for timing of applications.
- Explore all the alternatives for controls such as synthetic products, systemics, and biological controls.
- Include cultural practices such as: fertilizing birch trees to discourage bronze birch borer (*Agrilus anxius*) damage or mowing at $3^1/_2$-in. height to discourage weed seeds from germinating.

What about the cost of labor for all this training? The investment of training and educating employees is offset by less turnover of employees, which also means better retention of customers satisfied with the service by trained employees. Trained employees will work more efficiently and safely. Trained employees are the key to a good IPM program.

Record keeping is extremely important in a good IPM program. By keeping track of pest problems, you'll see patterns and develop a history by neighborhoods or soil type. This will help in training employees and planning control strategies based on the history. Knowing that shady areas of a lawn are not prone to insect or crabgrass problems will assist in targeting preventative or curative treatments. Mapping out problem areas associated with gypsy moths or Japanese beetles and grubs will make it easier for employees to scout an area and treat as needed.

A CASE HISTORY OF A LAWN AND TREE CARE COMPANY'S IPM PROGRAM

When I first tried to develop a practical IPM program, the employees and sales consultants had to be convinced that it could be done. They needed proof that there would be decent results and that customers would buy into an inspection program and actually pay for more consultation and fewer gallons of materials.

The first step was to look at equipment that would make an IPM program easier to carry out and to convince all customers of the benefit of this approach. By purchasing lawn equipment with a dual spray system, so we could fertilize and then inject pest control products at the applicator nozzle, we made this approach realistic. We set a policy against blanket weed applications. Most customers understood and liked this approach, but there were a few that were lost to competitors because of our refusal to treat without justification.

To test the market for tree care, we approached all customers who had expressed concerns about the use of pesticides. They had already committed to a set amount of dollars per year for the calendar approach of spraying. As long as their price did not increase and they were assured of satisfactory results, it was fairly easy to convert them to a monitoring program with treatment only as needed on their trees, shrubs, and lawns.

This approach could then be evaluated to see if the increased labor costs from monitoring would be compensated by the reduced cost of products. It was! In fact, the costs saved by using less product exceeded the cost of labor. However, since a highly educated employee is needed, this savings led to increased wages, which helped attract and retain qualified, trainable employees.

In addition, customers with concerns about pesticides liked this approach and stayed with the program instead of dropping the service.

The inspection program then became policy for all tree and shrub customers by 1987. Certain preventative treatments were automatic, such as dormant oil to trees and shrubs prone to aphid and mite attacks. Also, birch leaf miner and spruce gall treatments were automatic because once the damage is done it's too late to rectify. But all other blanket multipurpose applications by monthly schedules were eliminated and replaced with a monitor-and-treat program.

Because most customers were so accustomed to paying for gallons of materials, it was not as easy to convert them to the inspect-and-treat program. However, many were convinced to fertilize and prune as an addendum to the monitoring program.

The customers on a monitoring program for trees and shrubs expect much more communication. At the very least, leave written reports stating what was found and if anything was done. Customers also want recommendations as to other things they should be doing to enhance the health of the plants. If they are going to invest dollars in a program they must see something tangible. Consultations and written reports fill this need.

With the lawn care program, customers' lawns were fertilized at each visit for visible results. They could monitor the value of the program by the greenness of the lawn with few weeds.

With grub control however, there was much reluctance by sales employees and customers to limit the once-a-year August application. If we missed, damage repair costs would be much higher compared with retreating. By mapping out areas with a grub history, we easily implemented a preventative program. Not one lawn had a problem with grubs in the areas excluded from treatment.

Figure 1 Lawn and tree care technician Amy Campbell inspects the lawn and shows the customer that grub control is not needed. Communication is the key to a good IPM program. *(Photo courtesy of Broccolo Tree and Lawn Care.)*

After a couple of years it was time to cut back on grub applications based on an area's history and to treat each property individually. All lawns with a history of grubs in the neighborhood were then treated but only in the portion of the lawn more likely to have a grub problem. These areas included the south side of trees, south-facing slopes, or lawns that just were weak and thin. Problems were nonexistent, but a few needed extra treatment due to error in judgment or training of the employee.

The next step is to actually dig up square-foot samples in the areas prone to grubs and then treat only those lawns with spots of 4 to 5 grubs or more per square foot. Will the cost of scouting be justified by the savings of material and can this be done successfully to the satisfaction of the customer? Can we keep costs fairly competitive so the customer stays with you? Stay tuned!

Again we return to the importance of trained individuals who can do the labor but are also consultants that can communicate with the customer.

By having trained employees who are challenged and an integral part of the organization, you will then have more satisfied customers and retain both the customer and the employee. Customers will pay more for better service, and employees who are paid fairly and are satisfied with their work will continue the cycle.

Communication with the customer. The lawn/landscape and tree care industry is a service industry. In order to have a successful IPM program, the customers must have a good understanding of their role in the results. Budget affects how much a landscape manager can realistically accomplish. The relationship with the customer needs to be a partnership. Often the customer cannot afford or does not want to hire a professional for work he could perform himself. It's extremely important to communicate or train the customer how to properly mow the lawn or prune the shrubs.

Customers need to be educated about an IPM approach. Those who have had their properties treated yearly for certain pests based on a calendar approach will need convincing that spot treating is better. Many homeowners still believe that blanket spraying represents more value for the dollar. There is concern that many pests will be missed by spot treating and blanket spraying insures better control. Communication (Figure 1) will help them realize that blanket sprays of insecticides may actually cause more damage by eliminating natural predators and that continual blanket applications of weed controls will weaken the turfgrass.

It is always best to communicate in person or by telephone so that as new questions arise they can be quickly resolved. However, the opportunity for one-to-one conversation is rare. Employees must be ready to respond to customers or point out problems and solutions. If you can't communicate verbally, a handwritten note is the next best tool. Preprinted information sheets also can be very useful. For example, use a list that states reasons for proper mowing height, sharp mower blades, and leaving grass clippings. Using the list and then circling or underlining the problem/solution will personalize the written message. A newsletter can also be a good communication tool if done correctly. It should be educational and attractive with less emphasis on selling.

With so much competition from other professionals it is difficult to keep a customer long enough to show results through a good IPM program. Often you inherit problems from the mistakes of others and need up to 3 years to turn things around. Again, communicating this up front with the customers, gaining their confidence, and working closely with them will give you the opportunity to show results.

A perfect example is a customer who had employed a service for years to mow, fertilize, prune, design, plant, and maintain the property, but who just wasn't satisfied with the results. The lawn in particular just wasn't as green as the customer thought it should be. I was called for another opinion. Knowing that the lawn had been fertilized but wasn't responding, I suggested a soil test before making any recommendations. The results showed a very high pH. I suggested an approach that would include sulfur soil amendments along with soil aeration and a balanced fertilizer program, with pest treatments only as needed, as opposed to the blanket applications they had been receiving.

The customer decided to go along with these suggestions. In addition, the customer asked that the mowing also be done. This was the opportunity to show that by mowing higher with sharp blades and on a frequent schedule as the weather dictates — along with all of the fertilizer, soil amendments, aeration, and limited pest treatments — a good IPM program could be successful. The customer was very impressed with this approach and asked for a program for the trees and shrubs, which had been under another professional's care.

In evaluating the plants, it was obvious there were some insect and disease problems. The blue spruces had spruce gall adelgid and spruce spider mites, and showed very little growth and poor color. A soil test for the lawn showed high pH, so we decided to test the soil in the area surrounding the spruce. The birches hadn't put on much growth either and showed signs of stress, making them a prime target for borers. Junipers in the shady area

had blight; the pachysandra in the sun by the pool had leaf spot and mites and was chlorotic. Since the pachysandra and juniper problems were directly related to poor design, I advised the customer to remove them and consider a new design. The spruces and birches, however, could be turned around with a solid fertilization and pest management program. All the other plants would benefit from fertilizing and monitoring for pest problems. In addition to redesigning some of the landscape, the trees needed pruning. They would benefit from removing the grass around them and then making beds with mulch. After consulting with an expert in landscape design and construction, all of the suggestions have been implemented and the property is in excellent shape. The cultural practices continue, along with consultations with customers. Pesticides are used on a very limited basis.

Educating and **communicating with the public** takes time and commitment without immediate results. Like any training program it must be ongoing and organized. This is best handled through your trade associations, where you can find others who are like yourself, committed, and will share the responsibility and costs. Selecting two or three spokespersons to handle media inquiries will show consistency in the message. Contacting newspaper, television, and radio news directors will establish yourself as a media source for landscape related topics. Press releases should be sent for all community oriented programs. Those programs would include training courses for industry employees, because that shows a commitment to professionalism and career programs and Green Industry awareness programs that reach students and teachers of all levels. Awareness programs are also important for politicians.

The best way to communicate so that others understand is to relate to them through your own experiences. After 4 years of commitment by members of the New York State Lawn Care Association (NYSLCA), the message of IPM is getting through to the public. Dedication and perseverance do pay off. A look at the NYSLCA as a case history will illustrate how to communicate about IPM.

In 1986, Rochester, NY, was faced with a few one-sided antipesticide/antilawn care reports. Our neighbor city, Buffalo, had gone through some mass hysteria over lawn care companies and it seemed those activists were starting to infiltrate Rochester. At the Professional Lawn Care Association of America's summer seminar, some local managers and owners of lawn care companies expressed concern about how to respond to these negative reports. It was decided that we would form the NYSLCA to act as a clearing house for public concerns and also to organize educational meetings for those in the lawn care business. This would allow us to keep up with regulations and be the best professional industry.

A press release was sent announcing the formation of NYSLCA. Spokespersons were appointed and when reporters called a business, the owner could direct the calls to the spokespersons. The spokespersons were sent to a special training seminar by media expert Ford Rowan to learn how to handle the media questions on the controversial issue of pesticide use.

The television, newspapers, and radio were then given consistent messages about the materials used. We showed how companies were responding to concerns by choosing the least risky but effective products and employing trained individuals to apply materials.

Many school systems and other large institutions were made aware of the Pesticide Certification Program sponsored by NYSLCA and also began sending employees through this training. This annual program addresses safe use of pesticides, with IPM as the center of a good management program. Approximately 75 people attend this program each year. In addition, many of the companies have intensive in-house training programs.

Putting the Pesticide Certification Program together took a lot of planning and organizing. But once in place, it was easy to implement annually. The instructors are volunteers from the industry or teaching institutions. The curriculum focuses on personal and environmental safety with IPM as the approach.

Other programs to educate the public were started. Specifically, they were aimed at career awareness. As the lawn/tree care industry's need for educated employees with an IPM approach has grown, qualified individuals have been scarce. The image of this industry is one that is seasonal, labor intensive, and low paying.

In 1988, it was decided to invite the high schools and counselors to the New York State Turfgrass Association (NYSTA) trade conference held annually in Rochester. Literature was sent out and follow-up phone calls were made. There was very little response. When questioned as to why there was little interest, the counselors said they weren't going to send students to a program where the job opportunities were limited to mowing lawns for $5.00 per hour.

Those comments sparked more action on the part of NYSLCA members. It took a year of planning to sponsor a career awareness program for counselors. Announcements were printed and sent out. Follow-up phone calls were made to help increase participation. The program that was put together included slide shows, career videos, brochures, trade magazines, salary surveys, all of which are available through the trade associations. Speakers from the industry included equipment and product suppliers, landscape designers, arborists, landscape and lawn care managers, a cooperative extension agent, and a college professor.

This program was done with a lot of attention to detail and with many interested volunteers from the industry. Out of 60 counselors invited, about a dozen came. Since there were at least ten spokespersons from the industry, they had informal conversations with counselors during breaks and lunch. The counselors were treated like customers and were very impressed with what they saw. The counselors showed us ways to reach the students and other counselors. They also decided to sponsor their own Green Industry Awareness program and invited us to be the guest speakers the following fall.

Some of the programs resulting from the initial program include:

- Student shadow programs (an interested student is placed with a company for a day to see what a typical day may be like).
- Apprentice programs (a student works for one or more companies in various positions under direct supervision. The student receives school credit for a 10-week semester with no wages).
- Explorer programs through the Boy Scouts, which can include a variety of projects such as actually laying out a design, landscaping or cleaning up a community project, planting for Arbor Day, making wreaths, etc. All are ongoing projects to fulfill a certain time period.
- Visiting the schools to speak about our industry.
- Setting up displays and participating in job fairs.
- Participating in Industrial Management Programs which offers teachers night courses in exploring careers so they can then include the appropriate lessons in their curriculums.
- Sponsoring an annual field trip to the NYSTA trade conference in conjunction with a career presentation in which counselors can plan in advance and bring students to participate.

It has taken 3 years of commitment to get this far, but the more exposure received the more demand there is for the career awareness programs. Counselors and students now have a better understanding of the many career choices and paths available to them in the *Green Industry*.

In a similar fashion, local legislators and politicians have been invited to the career awareness program at the NYSTA trade conference with a special presentation to show them the economical and environmental impact and benefits of our industry. Legislators have a crucial impact on regulating our industry and must understand their actions so as to not impede the IPM approach.

Communication is the key for an IPM approach to become reality. The same effort put into obtaining customers is necessary to retain them. Retaining customers will only happen with satisfied, trained employees. Attracting employees and customers is dependent upon the perception of our Green Industry by the public.

As research develops better techniques for landscape management, the opportunity to implement an IPM approach becomes more of a reality. We must continually communicate with our employees, customers, and the public, verbally and through our actions so we are perceived as the environmentally responsible professionals that we are.

Chapter 53

A Lawn Care Alternative Service

Philip Catron, *NaturaLawn of America, Inc., Damascus, MD*

CONTENTS

Introduction 603
Implementation of an IPM Program into a Lawn Care Operation 604
Afterthoughts on Program Implementation 605
Deciding on a Control Strategy 607
    Evaluating Pest Pressure 607
        Tolerance Levels 607
        General Factors 607
        Agronomic Factors 608
Appendix 610
    Deciding on Whether or Not to Treat 610
    Facts About Choosing the Right Control 610

INTRODUCTION

Lawn care has many aspects about it that can create either a "friend or foe" image to the lawn care consumer. The homebuilding industry and homeowners in general have high expectations for turf and turf-care practices. These expectations include that lawns should grow when and where consumers want them to, but not necessarily where it is possible for a lawn to grow; lawns must look perfect during the entire year; and, finally, lawns must not be an obstacle to a homeowner's pursuit of leisure time and relaxation or professional and social activities.

Research has shown that many benefits are derived from lawns and turf landscaping. Other than the aesthetic benefits derived from lawns, the most important technical benefits are: turf's ability to control and minimize soil erosion, and to "intercept" excessive amounts of fertilizers and pesticides, tying them up and preventing their escape.

Without a program of application according to need, however, turf cannot always trap all the applied chemicals. Based on this realization, the lawn care industry must take a hard look at its general practices.

Historically, lawn care operations and programs were designed as a "one-treatment-fits-all" approach. This philosophy allowed lawn care service companies to blanket-apply pesticides over entire lawns without any regard to as to whether or not a weed or insect problem existed.

Turfgrass breeding efforts during the 1970s were designed to give the consumer the types of grasses that were aggressive growers and that had deep green coloration. In order to perform at their peak, these grasses often required a very high amount of nitrogen. This need for large quantities of nitrogen led to lawn care service companies applying additional applications with higher quantities of fertilizer than was necessary in the past. This increased level and number of fertilizer treatments often predisposed turf to insect and disease problems and thus encouraged and required additional pesticide treatments.

To justify cost increases, these turf industry service companies opted to use the less expensive, inorganic, and synthetic organic sources of fertilizers. Many of these replacement products contain highly soluble salt sources that can be detrimental to the microscopic plants and animals that live in soil. The end result of these excessive, unhealthy lawn care practices often resulted in a weakened root system for the lawn.

As the growth of the lawn care industry increased in the 1970s and 1980s, marketing and advertising campaigns convinced the general consumer that their lawn should be "picture perfect" green and absolutely weed and insect free. This promise of perfection would be provided for by the lawn care company and routine calendar treatments of fertilizers and chemicals.

This "green myth" was used extensively throughout the past 20 years with little regard to educating the lawn care consumer about lawn agronomics. In reality, lawns will always have some weeds and insects. Most lawns will always have some bare spots and tend to go brown during the summer or other stress periods, and thus the "green myth" is just that — a myth.

Our company's concept of natural, organic-based, environmentally safer lawn care is not a new idea. In fact, it is based on sound agronomic growing principles that were developed and proven many years before the advent of modern lawn care. Our company's program of lawn care is based on a three-pronged approach:

- The backbone of the program is the use of organic-based fertilizers derived from natural organic sources, not synthetic organics or inorganics. The choice of product based on formulation or delivery method does not correlate with source on toxicity. The lawn care industry must define terms accurately to earn credibility with the public.
- Our company has developed a strong IPM system for lawn care. Many lawn care companies state they practice an IPM system, but in reality, they practice see-and-spray (SAS) techniques. Remember, finding weeds or insects in a lawn doesn't necessarily mean it requires chemical treatment.
- Our company's third approach is to give preference to a biological weed or insect control material, if one has to be used, rather than a synthetic compound. Our company's development program included a lawn care service goal to be 100% biological within 3 years. After only 2 years in operation, the company was utilizing biological practices for 85% of its customers.

Through strong and continued customer education as to how lawn care can be accomplished more safely, the company's lawn care program for its customers has been able to reduce weed control material usage by over 85% and to reduce insect control usage by over 90% when compared to a typical chemical lawn care company. Through its lawn care treatment program, our company also has been able to reduce the amount of nitrogen and phosphorus entering into the soil system. These reductions in potential environmental pollutants have been accomplished without sacrificing the quality of the lawns grown by homeowners.

Lawn care consumers need a still higher level of awareness and education, not only regarding alternatives to the typical chemical lawn care service, but also to promote a better understanding and more realistic expectations of lawns and alternative lawn care services.

As the demand for alternative lawn care products and services increases, more research will be initiated to develop new biological tools necessary to maintain and nurture environmentally safer lawns and lawn care products.

IMPLEMENTATION OF AN IPM PROGRAM INTO A LAWN CARE OPERATION

Lawn care programs during the 1970s and 1980s were typically designed to apply repeated use of herbicides and insecticides on turf with little consideration given to the presence of pests, population numbers, and possible long-term damage to turf. Four to six calendar sprays were scheduled approximately 6 to 8 weeks apart, designed to apply pesticides over entire lawn areas. As mentioned earlier, this type of programming led to the creation of the myth that residential lawns could be perfectly weed and insect free. In response to continued public awareness on environmental concerns and oftentimes excessive use of pesticides by the lawn care industry, an alternative to the typical chemical lawn care programs of the 1970s and 1980s was designed and implemented by NaturaLawn, Inc., in the fall of 1987. The objectives of this alternative program were threefold.

The objectives were to adopt the concept of IPM to a lawn care situation in such a way that (1) customers could easily and readily understand; (2) technicians would be able to make operational decisions on a day-to-day and lawn-to-lawn basis; and (3) the results of the program in terms of lawn quality were either equal to or provided for a reasonable return on investment.

In order to fulfill the need for customer acceptance and understanding, a different approach toward marketing and advertising was developed so as to "educate" the customer as opposed to "selling" the customer. This was accomplished in part through the use of a direct mail, four-color piece depicting not the beautiful weed-free green lawn, but rather the lawns that had been environmentally abused by excessive use of pesticides either through improper homeowner applications or professional spray treatments. This pictorial was assisted by an accompanying dialogue which discussed the concept of IPM. Educational pieces were developed and distributed to the public. These pieces discussed (1) the need to preserve natural enemies by selective use of materials (biological or synthetic); (2) the introduction and encouraged use of newly developed turf varieties containing endophyte so as to have natural pest resistance; and (3) the importance of practicing proper cultural techniques when mowing and watering a lawn.

The second objective was met through the implementation of an extensive and continual training program that focuses on agronomic factors. This gives the technicians a foundation of knowledge with which to make decisions. These technicians' decision-making process for lawns was enhanced by two important factors: (1) all vehicles used for lawn treatments carry multiple choice selections — both biological and synthetic — and (2) technicians are financially rewarded for taking the necessary time to review properties to assist them in materials selection. The training of technicians was designed to be ongoing. Continued monitoring of lawns through soil testing, physical checks of lawns, collecting data on weather conditions, pest behavior/development, and the history of the lawn provided consistent feedback for proper decision making (see Table 1 for example of data collection).

The third criterion of better or equal results is to some extent subjective in nature. Industry ratios of service calls, cancellations, and net growth were used for comparative purposes to measure degrees of success or failure. With IPM programs implemented and functioning for over 48 months, results showed there were 76% fewer cancellations of lawn care customers in NaturaLawn's IPM program than industry averages generated. There also were 86% fewer service calls for NaturaLawn's IPM programs than for traditional chemical lawn care industry averages. New growth achieved in the first 26 months was comparable to the growth of a chemical company that had been in business a full 9 years. In addition, a significant reduction in pesticide use was achieved compared to a typical lawn care scenario. For acreage treated during the first 26-month period, a chemical lawn company's approach might have used in excess of 1500 gal of broadleaf herbicide and in excess of 600 gal of insecticide. With NaturaLawn's IPM approach, weed-control usage was cut by 86%, and insect-control usage was cut by 93%. This is a significant reduction in the amount of pesticides introduced into the urban environment.

The economics of implementing an IPM lawn care program is of major concern to most of the industry. While overall costs of materials went down because of decreased usage, there were increased costs in payroll, training, and recruitment areas. The net cost change was negligible, however. A significant increase in profit percentage over industry averages was realized due to a high degree of customer satisfaction that was shown by low cancellations and service calls.

The major results of this business program include (1) a high degree of customer acceptance; (2)

better trained technicians that make more rational pesticide treatment choices; (3) better-than-industry average in customer satisfaction; and (4) reduced pesticide usage (and cost) and better profit ratios.

Current activities conducted by NaturaLawn of America, Inc., include additional studies on biological weed-control materials, extended use of certain biological insect controls to ascertain geographical limitations of the organism, and a public relations effort to continue to further educate the general public as to the alternative approaches to chemical lawn care utilizing IPM.

AFTERTHOUGHTS ON PROGRAM IMPLEMENTATION

Many state regulations pertaining to their individual Pesticide Applicator's Law state that each licensee, certificate holder, permittee, or registered employee shall: "Consider all alternative pest control measures such as mechanical, cultural, and biological control."

Our industry should strongly encourage state agencies to reflect this regulation by including questions in the testing procedures and exams given to potential licensees in this area within the respective state's jurisdiction.

Inasmuch as the industry needs to be "encouraged" to use alternative approaches and biological materials when and where possible, we should support the use of a sign similar to any current pesticide notification sign being used in various states yet being different in color and/or shape so that the public would be aware of the use of a biological material rather than a synthetic pesticide.

In order to help reduce the amount of "cover" and "blanket" sprays by chemical lawn care companies we should consider supporting the use of a somewhat different sign (shape) that denoted a "spot" treatment was made to a particular problem area as opposed to an entire lawn or landscape. This could be limited further to applications made by nongasoline-powered sprayers (i.e., hand-can or backpacks).

There are new products on the market (parasitic nematodes) that are EPA-exempt from registration. Our industry should encourage regulators to look at these new materials and evaluate whether they are classified as "pesticides" under the definitions in regulations. We should recommend to the regulators that no sign posting be necessary when using these types of products, but should they be viewed as "traditional pesticides," these treatments be allowed to be posted with a sign different in color and/or shape to denote their "safer" and "softer" nature.

Table 1 **Temperature and weather record (copyright 1991 NaturaLawn Inc.)**

		1	2	3	4	5	6	7	8	9	10	11	12	13	14	15	16	17	18	19	20	21	22	23	24	25	26	27	28	29	30	31
January	Temperature																															
	Weather																															
February	Temperature																															
	Weather																															
March	Temperature																															
	Weather																															
April	Temperature																															
	Weather																															
May	Temperature																															
	Weather																															
June	Temperature																															
	Weather																															
July	Temperature																															
	Weather																															
August	Temperature																															
	Weather																															
September	Temperature																															
	Weather																															
October	Temperature																															
	Weather																															
November	Temperature																															
	Weather																															
December	Temperature																															
	Weather																															

Weather Codes
C - Cloudy	B - Bright
CB Part Cloud	R - Rain
	S - Snow

Consumers can get very confused over terminology. Our industry should strongly recommend that regulators look at the following words with their corresponding definitions and adopt them as standards:

Natural — when referenced with fertilizers and/or control materials, this term should signify that it is derived from plant or animal tissues, components, or sources.

Organic — when referenced to fertilizers should signify that it is derived from natural sources, not synthetic sources (i.e., urea). The public perceives organic as natural, not synthetic or manmade.

Organic-based — when referenced to fertilizers, this term should signify a minimum of 25% by weight of natural organic sources in the product.

To continue to allow producers and end users to use the term "organic" to signify synthesized products such as urea, sulfur-coated ureas, etc., simply because they contain carbon is ludicrous and deceptive.

DECIDING ON A CONTROL STRATEGY

EVALUATING PEST PRESSURE

The pressure from turf pests, whether they are weeds, disease, insects, or cultural, are indicators or symptoms of an underlying problem. The "reactive" turf manager often will treat symptoms without taking into account the various factors that lead to the symptoms' occurrence. The "responsive" turf manager will pull data from various sources (experiences, phenology charts, education, etc.) and attempt to determine the underlying factors causing the symptoms.

To simply "identify" the symptom (such as a disease lesion), and apply a fungicide for that disease is merely academic and will not lessen the possible chances of the problem reoccurring. To understand the reasons why the problem occurred, however, and make the necessary adjustments in your turf programs to minimize the problem's recurrence is the real challenge of every turf manager and the mark of someone who understands an IPM approach.

In this chapter, we will assume that a basic level of agronomic experience and expertise has been attained, and that the reader understands the typical jargon of turf management, can identify basic turf problems, and comprehend the way cultural and environmental conditions can influence the growth of turfgrass.

Tolerance Levels

In order to ascertain and evaluate pest pressures and make decisions on control procedures, the establishment of action thresholds or tolerance levels must be accomplished. By setting tolerance levels, the turf manager is able to make a series of decisions to answer the basic question, "Do I treat the problem or not?"

Generally speaking, there are four factors that are used in establishing tolerance levels:

1. Turf use
2. Location
3. Aesthetic value/replacement value
4. Playability and safety issues.

General Factors
Turf Use

Defining how the turf will be used is basic to establishing an "acceptable" level of turf damage. For example, is there active traffic (residential vs. commercial), will the turf be used all year long or for only show purposes? Identifying the primary and secondary uses of the turf area will assist the turf manager in determining what levels of damage may be acceptable throughout the season.

Turf Location

For turf use, the location of problem grass areas will have a significant bearing on whether treatment may be necessary. For example, turf on a golf course may be in use all season long, but fringe areas of fairways and certainly roughs will not be treated for problems in the same manner as tees and greens.

Residential and commercial turf are no different from golf courses in that certain locations (main entrance areas) have a higher perceived value than the backyard or rear parking lot areas.

Aesthetics and Replacement Value

The more valuable the turf in terms of cost of replacement, the lower the acceptable damage level from the turf manager's assessment. Using the golf course as an example again, the cost to repair or resod a green is very expensive when one considers not only the turf replacement and necessary labor costs, but also the lost playing time (revenues) and inconvenience to golfers.

Playability and Safety

One of the many benefits our society derives from turf is a certain level of enjoyment from its use, including the protection it gives us when it acts as a cushion to absorb falls. Pests that diminish or lower the grass area's playability and subsequent level of safety may pose a serious threat to this protection.

From the above brief discussion of the four general factors used to help evaluate pest pressures, the turf manager must then focus in on two more specific categories — the agronomic factors and the cultural factors that must be considered.

Agronomic Factors
Pest Identification

To correctly identify the specific turf pest that is causing the problem is of primary importance. While this is a very basic and logical fact, proper identification is the step that is most often done wrong.

For example, a big-eyed bug is frequently confused with a chinch bug. The mere presence of either does not necessarily indicate a problem. Nonetheless, it has become a far too common occurrence for the well-meaning turf manager to make an insecticide application in an attempt to control chinch bugs when in reality the "suspect" insect was the beneficial big eyed bug.

By properly identifying the pest problem (weed, insect, or disease), the turf manager will be better able to understand the pest's potential negative impact, since proper identification leads to knowledge of life cycles, turf preference, insect location, and susceptibility to control.

Type of Turf

Certain problems are more readily associated with specific turfgrass species. A solid understanding of the different strengths and weaknesses of the major turfgrasses assists the turf managers in evaluating pest pressure and the need to use controls.

Through the use of endophytic turfgrasses, levels of insect pressure can be greatly reduced, thus resulting in a lower need to apply control materials. Endophyte-enhanced turf can be introduced into existing stands of grass too, and this is a very desirable practice for the implementation of an IPM system. Any time a better suited turf species can be introduced into a grass area, the more positive the impact will be on the IPM program.

These improvements are often made in small increments such as:

- Replacement of powdery mildew-prone bluegrass in the shade with a more shade-tolerant turf.
- Introducing an improved ryegrass into a cool-season stand to reduce rust problems.
- Using improved tall fescues with endophyte to minimize insect pressure in a bluegrass area.
- Removal of turf in very difficult to manage areas and replacement with groundcovers.

The synergism of the interactions among all the "small" improvements is normally greater than one big improvement. Turf managers need to guard against looking for the "quick fix" and focus on a more long-term turf improvement program.

Turf Vigor

Many of the points already mentioned will have an impact on turfgrass vigor. The overall fertility program and cultural practices used on the turf will obviously impact vigor, too. The point that needs to be understood is that the more vigorous the turf, the better is its ability to withstand pest pressure. When turf can withstand pest pressure on its own, the turf manager need only continue to monitor the situation to evaluate when pest pressure is too heavy for the turf to withstand.

Calendar/Cultural Impact

Interacting at all times with the other agronomic factors is the seasonal aspect of the turf's growth.

There is a large amount of interplay among these factors in that the time of year will impact identification and effective treatment (i.e., whether the insect or disease or weed is at a susceptible stage or a protected stage in its cycle). The time of year will also impact the turf's vigor, and with the type of turf, it will affect its ability to resist damage.

When a turf manager begins the task of evaluating pest pressure so that threshold levels may be determined and available controls decided upon, he or she soon comes to the realization that this is a multifaceted area requiring a degree of educational background, experience, and common sense.

The final set of data used by the turf manager in assessing pest pressure deals with cultural factors: mowing, watering, and mechanical renovations.

By controlling or altering any one of these areas, damage to turf can usually be minimized and often avoided entirely. This becomes a balancing act, however. Should cultural practices be altered too much, the pressure from one pest may shift to another and damage can in fact be more severe.

Combining the *general*, *agronomic*, and *cultural* factors into the decision-making process enables the turf manager to follow a logical process in choosing control methods.

By referring to the decision-making flow chart (Figure 1) it can readily be seen how the various factors affect each other and where there are built-in safeguards to prevent or minimize the unnecessary use of pesticides.

As an additional guide for the turf manager, a step-by-step question and comment page (Appendix) can be used for assistance.

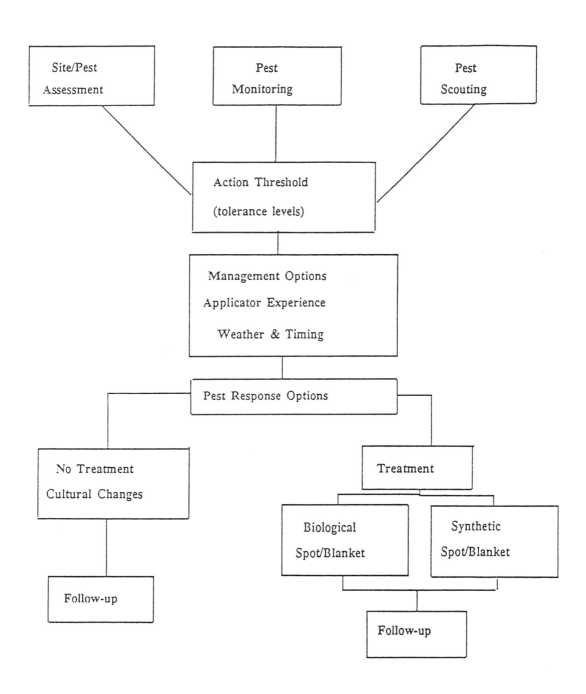

Figure 1 Decision-making flow chart (Copyright 1991 NaturaLawn, Inc.).

APPENDIX

DECIDING ON WHETHER OR NOT TO TREAT

(1) Identify the problem/pest — Are you dealing with a symptom or the true problem?

(2) Is control necessary?

(3) What are the alternatives?

FACTS ABOUT CHOOSING THE RIGHT CONTROL

(1) Must be effective against the pest.

(2) Label directions for intended use must be specific for the pest.

(3) Material should not cause unintentional damage.

(4) The control must cause the least damage to benefit organisms in the area (bees, worms, etc).

(5) Applied material must not persist or "move off" the treatment area to harm fish or other wildlife.

(6) Select the correct formulation for the type of machinery being used for application.

NOTE: The most "economical" choice for the long run often is not the cheapest!

Chapter 54

Maintenance of Infields and Other Bare Soil Areas

Tim Rhay, Eugene Public Works Maintenance Division, Eugene, OR

CONTENTS

Frequent, Shallow Cultivation 611
Soil Amendments 612
Concrete Underlayment 612
Selective Use of Low-Toxicity Herbicides 612
Meet the Challenge of the '90s 612

In spring, one's fancy often turns to … softball! While it may not yet rival romance, recreational softball enjoys a wide and increasing popularity with many sectors of the public. In numerous communities, demand for playable fields far outstrips supply. Gone, too, are the days when any vacant lot would do. Today's player wants a professional-looking facility with an infield surface that provides safe but solid footing and near-ideal playing conditions.

The responsibility of providing such fields frequently falls to school district and park agency grounds managers and their staffs. As public resources typically are limited and subject to multiple demands, proper field maintenance becomes a challenge to many, and a sizeable problem to some. Often, a significant herbicide spray program has been the way this challenge was met. But the public is increasingly concerned with routine, comprehensive herbicide applications to school and public sports facilities. So what's a responsible grounds manager to do?

The Eugene Public Works Maintenance Division has developed an infield maintenance program that emphasizes nonchemical techniques and has enabled us to provide more and better infields without substantially increasing costs. Too good to be true? Not at all. In fact, most agencies have, or can easily acquire, the means to duplicate our results.

FREQUENT, SHALLOW CULTIVATION

The foundational element in our program is the concept of frequent, shallow cultivation. Both frequency and depth are important. Like many, we used to use a tractor-mounted rotary tiller to culti-

vate our infields. Because of time and weather constraints, this was done only once each season. Tilling worked up the soil to a considerable depth. A bulky float was then transported from field to field and used to level the surface. The result was an infield that most considered too soft and slow for proper play, but one that rainfall or normal "wetting-down" for game preparation would all-too-quickly transform into a crusty, rock-hard surface that was equally unpleasant and unsuitable for play. The obvious solution was to cultivate more frequently, but not as deeply.

Our first attempt to acomplish these goals involved the use of an agricultural disk instead of the tiller. This lessened cultivation depth by half or better, and the time required to cultivate each field also was greatly reduced. Although now carried out more frequently and throughout the season, the total staff time devoted to cultivation did not significantly increase. However, it remained a two-step operation (a second staffer still followed the disk with the float to level the fields) and required two tractors, a trailer, and a flatbed truck, in addition to the staff.

We replaced this program with a single staffer and tractor by making use of a rotary harrow. This PTO-driven implement, also borrowed from agriculture, features pairs of counter-rotating tines that cultivate horizontally rather than digging vertically like a tiller. A trailing roller keeps the tines from penetrating too deeply and allows very shallow cultivation — at a depth of 2 in. and less. The roller also levels the cultivated soil. The float is no longer necessary.

Cultivation can now begin in the very early spring, as soon as the upper surface of the infields is dry, and can be done as required throughout the season by one or two staffers (we purchased a second rotary harrow this year), typically in one working day or less. Tilt-down brush attachments enable us to achieve a finished surface on the final pass, suitable for practice or informal games. We now rarely cultivate below 1 in. and can maintain infield quality with a 3- to 4-week cultivation cycle, depending on factors such as rainfall and how heavily a particular field is played. Further, since cultivation takes only about 30 minutes per field, additional service is no longer the problem it once was.

Of course, the reader should keep in mind that the cultivation depth and cycle information given above is based on the clay and clay-loam soils, weather patterns, and user levels found in Eugene. Regional differences in these variables may require adjustments.

SOIL AMENDMENTS

Another element of our infield management strategy is the incorporation of organic amendments into the infield soil. Our native soils, as mentioned, above, are high in clay content and tend to puddle and crust over when watered down for play. Incorporation of 20 to 30 yd^3 of sawdust per field greatly reduced this problem by enabling the soil to capture and distribute the moisture more efficiently. Sawdust-amended fields are also much less susceptible to compaction.

Spread the sawdust evenly over the playing surface of the infield and work it in with the rotary harrow, set for maximum penetration. This will incorporate it thoroughly into the upper 2 in. of the infield soil profile, which is adequate since most subsequent cultivation will occur above this depth.

Sawdust is readily available in Eugene, but other coarse, organic materials will work as well. Evaluate availability and price in your own area. You might wish to test your chosen material on one field before making wider use of it. Expect to add supplemental material after the initial application. With sawdust, our experience was that this was not necessary for 4 to 5 years, and we needed only half of the original volume to restore appropriate infield soil composition.

CONCRETE UNDERLAYMENT

A third way to reduce maintenance requirements and improve the level of service provided is to underlay fences, backstops, and bleacher sites with concrete. This one-time effort effectively eliminates the need for vegetation control, whether by chemical, manual, or mechanical means, and is very cost-efficient when evaluated from a "life-of-asset" standpoint.

SELECTIVE USE OF LOW-TOXICITY HERBICIDES

The integrated methodology developed in Eugene and described above will both improve the quality of infield surfaces and reduce the resource requirement for doing so. In particular, the need for herbicide application will be dramatically reduced. Some spot treatment may be necessary to deal with noxious perennial plants that do not respond to cultivation. In some cases, a comprehensive treatment may be needed to gain initial control of an area. In some climate zones, treatment may be needed only at the infield/outfield interface, to prevent opportunistic vegetation from creeping into the bare soil area. When such treatment is required, consider low-toxicity, granular preemergence materials which can be soil-incorporated during the dormant season, after the field is taken out of play.

Granular napropamide has proven useful for us in such applications, although trifluralin, oxadiazon, or other products should work as well. For postemergence work, newly available fatty acid-based herbicides may be useful for some types of vegetation. Others may require the use of foliar-applied, translocated materials such glyphosate. Consult local regulatory and reference sources before chosing herbicide materials.

MEET THE CHALLENGE OF THE '90S

The challenge of increasing demands for infield quality, doing more with the same resources, and less dependence on chemical herbicides can be met using an integrated approach stressing frequent, shallow cultivation, combined with the incorporation of organic soil amendments, the underlaying of fences, backstops, and bleachers with concrete, and the conservative, carefully timed use of well-chosen chemical controls. It is a strategy that is both effective and environmentally acceptable. The integrated methodology is the answer to the grounds maintenance challenges of the present decade — and the century to come.

Chapter 55

IPM in Municipal Parks Maintenance — A Case Study

Tim Rhay, Eugene Public Works Maintenance Division, Eugene, OR

CONTENTS

Ordered to Attend 613
Internal Staff Training 613
Field Staff Drives the Program 614
"Realism" Has Paid Off 614

Environmental concern is a longstanding community value in Eugene, OR. It is not surprising, then, to find a viable IPM program with more than 10 years of positive history behind it guiding vegetation and pest control activities of the Eugene Public Works Maintenance Division. What might not be expected is the history of how this came about.

ORDERED TO ATTEND

In 1980, another supervisor and I were ordered by the Superintendent to attend a seminar on IPM conducted by Dr. William Olkowski. As I've stated in the past (too many times to deny it now) I did not want to go to this presentation because, while I didn't know specifically what IPM was, I felt it would be some impractical, "no chemicals" pest control strategy, recently dreamed up by the radical fringe of the environmental movement, that would cost too much and wouldn't work under "real world" conditions. What I learned from Dr. Olkowski was that every one of these assumptions was, in fact, incorrect.

IPM was developed within, and for the use of, the agricultural industry. While ecologically based it is completely practical by nature and design. Originally formulated as a cost-containment strategy, it provides truly cost-efficient vegetation and pest control. "Real world" criteria drive the program, determining when and what action is taken. Finally, the methodology had at least a 57-year history of proven efficacy at the time I attended Dr. Olkowski's seminar. In short, IPM made perfect sense as a management tool to guide our division's vegetation and pest management operations.

Dr. Olkowski spent a second day with some of our staff in the field, seeing pests and vegetation problems first-hand and making suggestions. He also left us with some printed material. Our challenge was to effectively communicate and establish the IPM methodology as standard operating procedure throughout our organization. The approach we took to accomplish this, while it seemed obvious at the time, appears to have been somewhat unique.

INTERNAL STAFF TRAINING

An internal training program was organized for all staff engaged in pest or vegetation control activities. Prior to the first session, a half-dozen staff members were selected, based on experience and/or interest, to be briefed on IPM principles and to research how these might be applied to a specific weed or pest control challenge faced by the division. At the actual training session, following an initial review of the history and methodology of IPM, each of these "peer instructors," in turn, presented their specific proposed IPM strategies.

We purposely selected the six best-known or most problematical pests the division had to deal with. This, combined with the peer teacher technique, proved effective in "selling" the concept and providing staff with a useful (and useable) understanding of how it worked. Consequently, most of our staff viewed IPM as an asset and were enthusiastic about putting the methodology into practice in their own vegetation and pest control work.

Refresher training sessions brought the group together again, facilitated information sharing, and helped keep the momentum and enthusiasm high. These sessions also provided a ready means of training and indoctrinating new staff members. Most important, the field staff developed a strong sense of program ownership and pride in what the division had been able to accomplish. This has proven invaluable.

Periodic internal training continues to be a feature of our program. In addition, we require all staff involved in vegetation or pest control to obtain a state Public Pesticide Applicator's License, which obligates them to a cycle of state-certified refresher/update training or retesting. We've also established an electronic bulletin board for information sharing throughout the year.

FIELD STAFF DRIVES THE PROGRAM

Our efforts to enlist and involve the staff in the design as well as the implementation of our IPM program has been critical to our success. Many of our most useful alternative (nonchemical) vegetation and pest control techniques were developed or proposed by field staff, based on their observation and experience.

Some have gone even further. For example, our head gardener at the Hendricks Park Rhododendron Garden arranged a research effort in cooperation with the Oregon State University Department of Entomology and the Biosys Corporation of Palo Alto, CA. Their 2 years of experiments helped to establish the efficacy of garden sanitation and parasitic nematodes for the control of root weevils. Both of these techniques have now been incorporated into our IPM for this pest.

The Owen Municipal Rose Garden is another positive example of what motivated field staffers are able to accomplish. A rose display garden containing 4000 plants, Owen is very popular with the public and receives high visitation annually. Disease management is, of course, a considerable challenge. Working over several years and under a succession of supervisors, the head gardener and his assistant developed a strong cultural program and undertook some clever site modifications that enabled them to gain increasing control over conditions favoring disease. They also gave experienced thought to the reasonable and appropriate treatment thresholds for the diseases, varieties, and site. A modern irrigation system, installed in 1986, gave them the final tool they had lacked — much improved control of soil and foliage moisture levels. This spring, we began our fourth season without a significant fungicide treatment in the garden. It remains popular with the public and is frequently featured as a "background" for weather reports and public service announcements on all of the local television stations. The Owen Garden also was featured on the national television show, *Gardening in America.*

In the areas of shrub bed and turfgrass maintenance, too, staff input and willingness to try differ-

ent control strategies has brought us further that we originally imagined. Regular comprehensive pre- and postemergence herbicide applications in park plantings have been replaced by a multi-faceted cultural effort, careful monitoring of results and conditions, and judicious chemical programs where other measures are not effective. Adequate control in general-purpose and athletic turf areas has been achieved through cultural and mechanical measures. Even in aesthetic turf, where weed tolerances are lowest, cultural measures are the foundation of the program, and have been supplemented, in most cases, by only localized spot treatments with herbicide. Phenoxy herbicides have been phased out of our program in favor of other materials that provide better control of target species within program guidelines. Field staff's contribution to these successes has been invaluable.

"REALISM" HAS PAID OFF

An anonymous saying I read a few years after Eugene's IPM effort began has proved to be illustrative of the reason for our success:

> The idealist says the short run doesn't count. The cynic says the long run doesn't matter. The realist knows that what is done or left undone in the short run, determines the long run.

In the continuing controversy and debate over the use of pesticides in the public domain, it is easy to locate both cynics and idealists in profusion. Our program has been a success and has become an example because we were not driven either by a rigid refusal to examine and modify historic methods, nor a blind adherence to impractical philosophical or political biases. We chose to try IPM because it offered a method whereby we could be environmentally, operationally, and fiscally responsible concurrently — a method superior to those we had been using prior to that time, and to any alternative that has come to our attention in the more than 10 years since our program began.

Others can share our success by following a similar course. Be willing to educate yourself and your staff in this methodology and trust in the ability of experienced field personnel to effectively learn and apply it to whatever vegetation and pest control challenges your agency might face. You need not apologize to those who will not examine their present methods, or to those who demand much, but produce only controversy and ill will. While idealists and cynics struggle fruitlessly on, the realists will discover and develop viable pest control strategies for the 1990s and beyond.

Chapter 56

The Turfgrass Information File and TurfByte

Robert Emmons, Department of Plant Science, State University of New York, Cobleskill, NY

Peter Cookingham, Turfgrass Information Center, Michigan State University, East Lansing, MI

Duane Patton, O. M. Scott & Sons, Lawrence, KS

CONTENTS

The Turfgrass Information File 615
Using TGIF Online 616
Using TGIF Without a Computer 619
TurfByte 619
    The Bulletin Section 620
    The Message Section 620
    The Files Section 620
Other Computer Information Exchange Networks 620

It comes as no surprise that the most knowledgeable turfgrass managers develop the best IPM programs. There is an incredible amount of turf information available today. A turf manager's main problem is gaining access to this information in a timely manner.

Traditionally, turfgrass industry personnel have kept themselves updated by attending educational meetings and reading trade magazines. In recent years a powerful new method of information exchange has emerged. Modems and communications software enable data transmission between computers.

A modem converts computer language into phone language and vice versa. A computer user can go online (on a phone line) and obtain information from a database, or communicate with other turf professionals by typing a message and sending it to an electronic bulletin board.

The Turfgrass Information File is the primary online database serving the turfgrass industry. TurfByte is an electronic bulletin board used by golf course superintendents.

THE TURFGRASS INFORMATION FILE

In the early 1980s, the U.S. Golf Association (USGA) Turfgrass Research Committee identified the development of a turfgrass information computer database as a top priority. The USGA chose Michigan State University (MSU) as the site for the computerized bibliographic database. MSU was a logical choice because its libraries house the O.J. Noer Memorial Turfgrass Collection, which is the largest single concentration of published turf literature in one location in the world.

The online database that was developed is called the Turfgrass Information File (TGIF). Under construction since 1984, the early stages of database development were jointly financed by the USGA Turfgrass Research Program and the MSU libraries.

TGIF currently has over 26,000 records in it, and about 2500 new ones are added each year. To build the database, the MSU libraries collect periodicals, research and technical reports, theses and dissertations, trade and professional magazines, and books. All of the records relate to turfgrass culture, science, or management. The user can find a vast amount of information on bunker renovation, organic fertilizers, disease-resistant varieties, fairway mowing, and thousands of other topics.

This large collection of turf knowledge is available at a reasonable cost to people with a computer and a modem. To gain access a subscriber can use one of several software packages for communicating with the computer at MSU, including VuePort, which comes with the subscription package. For information on how to become a subscriber, call, write, or fax:

(800) 446-8443
Turfgrass Information Center
W-212 Main Library
East Lansing, MI 48824-1048
FAX: (517) 336-3693

Annual subscription fees are as low as $75 for an individual. There are also foundation, association, and corporate subscriptions. Subscribers receive two quarterly publications: the *Turfgrass Index*, which lists recent TGIF entries by author and subject, and *The Sward*, a newsletter for users.

USING TGIF ONLINE

TGIF can be searched by cultural practice, grass species, insect pest, pathogen, weed, pesticide, author, journal of publication, type of publication, etc. Search terms can be used alone or in combination with other terms. For example, if a user searched for information on *Poa annua* over 1200 articles would be found that mention this grass species. Combining "*Poa annua*" with the term "biological control" results in a listing of 14 records.

If a golf course superintendent wanted to learn about the effect of compost on red thread disease, he would enter "compost and red thread" on the search screen. He could then download the search results to his own computer or display the records found and read them before downloading. If there were too many items to review easily, the search could be limited to materials published more recently, say between 1988 and 1993. The following is an example of the information retrieved from TGIF by this search:

Record: 21681

Authors	Nelson, Eric B.; Craft, Cheryl
Affiliation	Department of Plant Pathology
Title	Biological control of red thread on perennial ryegrass
Journal Title	*1989–90 Cornell University Turfgrass Research Report 1990*, p.116
Keywords	Red thread; Lolium perenne; Biological control; Topdressing; Organic fertilizers
Abstract	"Our approach, as in the previous biocontrol studies, was to introduce individual microbial antagonists as well as complex mixtures of antagonists through top-dressing applications. Top dressing formulated with

sand: organic matter mixtures were applied to perennial ryegrass ('Palmer') to evaluate their effectiveness in suppressing Red Thread (*Laetisaria fuciformis*). A disease-suppressive microflora was introduced into top dressings as complex microbial mixtures found in composted organic wastes and organic fertilizers composed of various plant and animal meals. Top-dressing formulations consisted of 70% fine sand and 30% organic component (v:v) and were applied at the rate of 400 $cm^3/0.8m^2$ plot. One application was made and plots evaluated for disease severity approximately 1 month later. Cutting height was maintained at 1.5 in. throughout the experimental period. Of all the treatments evaluated, only Sustane (5-2-4) (poultry litter compost) was effective in suppressing red thread. A high level of suppression was observed at least 27 days after application. Diameter of infection centers was also reduced as compared with untreated plots and other compost treatments (data not shown)."

Notes	Tables; Lang: En.; O Ref.

Though searching for information in a database is always a bit disconcerting at first, TGIF is user-friendly. Subscribers receive a copy of the *Dial-Up User's Manual*, which is quite thorough. It also contains simplified mini-guides that explain how to sign on, search, download, use the electronic mail system, and sign off.

The TGIF simulator disk is very helpful. This training software enables a person interested in the database to learn how to use the TGIF without actually going online. The prospective user can practice signing on, searching, etc., without having to pay for telephone or database time.

In addition, to further simplify information retrieval, the TIC staff have already prepared searches on subjects that are "hot" topics in the turf industry. Nearly 100 "TOPICS" are presently available. Examples are: turf benefits, runoff, groundwater issues, green speed, protective covers, earthworms, clippings management, dollar spot, iron, effluent water use, *Poa annua* control, white grub control, and IPM.

These prepared searches are found by selecting the "Download Topics" choice on the main menu. If a turf professional chose to display or download the list of articles compiled for the IPM topic, the following are examples of what she would find:

Record: 25017

Authors	Redmond, C. T.; Georgis, R.
Title	Aspects of the use of entomo-pathogenic nematodes as IPM agents in the control of turfgrass insect pests
Source	*Journal of Nematology.* Vol. 23, No. 4, October 1991, p. 547.
Keywords	Gryllotalpidae; Pheromones; Integrated pest management

Record: 2477

Authors	Zontek, Stanley J.
Title	Post-emerge crabgrass and goose-grass control: Practical IPM
Source	*USGA Green Section Record.* Vol. 30, No. 4, July/August 1992, p. 17-18.
Keywords	Digitaria sanguinalis; Integrated pest management; Eleusine indica; Weed control

Record: 23175

Authors	Balogh;, James C.; Leslie, Anne R.; Walker, William J.; Kenna, Michael P.
Title	Development of integrated management systems for turfgrass
Source	*Golf Course Management & Construction: Environmental Issues*, 1992, p. 355-439.
Publisher	Boca Raton, FL: Lewis Publishers
Keywords	Integrated pest management; Turfgrass management systems; Biological control

Record: 22925

Authors	Eskelson, Dan
Title	Taking a holistic approach: Implementing IPM strategies
Source	*Golf Course Management.* Vol. 60, No. 2, February 1992, p. 68,70,72,74-75.
Keywords	Integrated pest management; Insect Control; Pest Control; Budgets

Record: 22801

Authors	Aldons, David
Title	Integrated pest management (IPM) for turfgrass systems
Source	*TurfCraft Aust..* Vol. Spring, No. 24, September/October 1991, p. 34, 36, 44.
Keywords	Integrated pest management; Pest control

Record: 21752

Authors	Skorulski, James E.
Title	Monitoring for improved golf course pest management results
Source	*USGA Green Section Record.* Vol. 29, No. 5, September/October 1991, p. 1-5.
Keywords	Pest control; Pest monitoring; Diagnosis; Integrated pest management; Scouting programs

Record: 21312

Authors	Carville, Jennifer
Title	Reducing golf course pesticide use; three examples. (III) A no-spray golf course in Squaw Valley California: Reality or hope?
Source	*Journal of Pesticide Reform.* Vol. 11, No. 3, Fall 1991, p. 9-11.
Keywords	Golf courses in the environment; Integrated pest management; Public involvement; Public relations; Groundwater monitoring; Pesticide use

Record: 21264

Authors	Giblin-Davis, Robin M.
Title	Potential for biological control of phytoparasitic nematodes in bermudagrass turf with isolates of the *Pasteuria penetras* group
Source	*Proceedings of the Florida State Horticultural Society.* Vol. 103, June 1991, p. 349-351.
Keywords	Cynodon; *Belonolaimus longicaudatus*; Biological control; Florida

Record: 21155

Authors	Burpee, Lee
Title	Biological turf disease control
Source	*Grounds Maintenance.* Vol. 26, No. 4, April 1991, p. 44,78,80.
Keywords	Disease control, Typhula blight; Take-all patch; Anthracnose; Spring dead spot; Red thread; Necrotic ring spot; Biological control

Record: 21108

Authors	Stack, Lois Berg
Title	Low chemical landscape management on a golf course
Source	*The Grass Roots.* Vol. 18, No. 4, July/August 1991, p. 19, 21, 23.
Keywords	Landscape design; Biological control; Integrated pest management; Ornamental plants; Trees

Record: 21039

Authors	Georgis, Ramon; Gaugler, Randy
Title	Predictability in biological control using entomopathogenic nematodes
Source	*Journal of Economic Entomology.* Vol. 84, No. 3, June 1991, p. 713-720.
Keywords	Nematoda; *Popillia japonica*; Biological control

Record: 20992

Authors	Anderson, Annette
Title	Integrated pest management (IPM)
Source	*Greenmaster.* Vol. 25, No. 3, May/June 1991, p. 24, 26, 28, 30, 34, 35.
Keywords	Integrated pest management; Instruments; Data logger; Pest control

Record: 20850

Authors	Frank, J. H.
Title	The biological control of mole crickets
Source	*Golf Course Management.* Vol. 59, No. 6, June 1991, p. 56-58.
Keywords	Talpidae; Biological control; *Scapteriscus acletus*; *Scapteriscus abbreviatus*; Integrated pest management

Record: 20567

Authors	Leslie, Anne R.
Title	An IPM program for turf
Source	*Grounds Maintenance.* Vol. 26, No. 3, March 1991, p. 84,86,116.
Keywords	Pest control; Integrated control, Integrated pest management

Record: 20418

Authors	Nelson, Eric B.; Craft, Cheryl M.
Title	Introduction and establishment of strains of *Enterobacter cloacae* in golf course turf for the biological control of dollar spot
Source	*Plant Disease.* Vol. 75, No. 5, May 1991, p. 510- 514.
Keywords	Dollar spot; *Enterobacter cloacae*; Biological control

Record: 20388

Authors	Boaz, Michael M.
Title	Soil modification and its role in integrated pest management
Source	*Golf Course Management.* Vol. 59, No. 2, February 1991, p. 133–136.
Keywords	Soil management; Integrated Pest Management

If the user was interested in the article on biological control of mole crickets, she could go to the search screen and call up the full TGIF record (20850) and read the abstract:

Record: 20850

Authors	Frank, J.H.
Affiliation	University of Florida
Title	The biological control of mole crickets
Journal Title	*Golf Course Management*
Date	Vol. 59, No. 6, June 1991, p. 56–58+
Call No.	SB 433 .A1 G5
Keywords	Talpidae; Biological control; *Scapteriscus acletus*; *Scapteriscus abbreviatus*; Integrated pest management
Abstract	Provides information on; the three species which have caused damage: (1) short-winged mole cricket (*Scapteriscus abbreviatus*), (2) southern mole cricket (*Scapteriscus acletus*) (3) tawny mole cricket (*Scapteriscus vicinus*). Tawny and short-winged mole crickets feed almost entirely on plants, and much of their damage is directly to the roots. Southern mole crickets include many other insects in their diet and do less damage directly to plants. But all three species tunnel actively in the ground and loosen and weaken plant roots, which is especially damaging during times of drought. Florida currently experiences more damage by Scapteriscus mole crickets than does any other state. Damage to turfgrass currently is estimated at $44 million annually in Florida. These levels of damage resulted in Florida's taking the lead in research of control methods. The three types of control are: (1) Chemical pesticides in the form of baits and sprays have been the principal means of controlling mole crickets. Chemicals, however, can only produce a temporary remedy, because mole crickets will reinvade the treated areas. (2) Native Natural Enemies — Many different animals kill mole crickets. Amphibians (frogs and toads), birds (egrets and sandhill cranes), mammals (armadillos and foxes), insects (tiger beetles, and assassin bugs) and spiders have been known to kill them. Perhaps as many

as one third of all mole cricket nymphs in the pastures are killed by such predators, but the number of mole crickets killed is simply not enough to prevent damage even in pastures, let alone turf. (3) Biological Control is the deliberate attempt to control pests with the use of natural enemies. Two major forms of biological control are classical (Inoculative) biocontrol and use of biopesticides. Classical biocontrol is used mainly against immigrant pest species (Scapteriscus mole crickets) and requires the search for and importation of specific natural enemies from the area of origin of the pest. Biopesticides are also biocontrol agents, but they generally are pathogens. They may be of native or foreign origin, but produce no permanent effect on the pest population. Biopesticides are applied the same way chemical pesticides are but can be safer for the environment. Other biocontrol agents used were Larra, Steinernema and Ormia. The University of Florida mole cricket program also discovered pathogens of mole crickets in South America. These include fungi, bacteria, protozoa and a virus. Two of the fungal pathogens seem promising in the laboratory, but field-release methods have not been developed. Integrated use of the above classical biocontrol agents with use of pesticides could employ either biopesticides or chemical pesticides. Chemical pesticides are available now. Biopesticides will require further research and development. In spring 1991, personnel at the University of Florida mole cricket program were highly optimistic about the impending success of the program in the near future.

Notes   Pictures, color, b/w; Lang: En.; O Ref.

Though the "full text" of most articles is not online, the information contained in an abstract is often very valuable. In the future, it is anticipated that increasing numbers of articles will be available "full text" in TGIF.

Though many university researchers use the database, the majority of subscribers are golf course superintendents and other "real world" practitioners throughout the industry. TGIF is a window into a vast treasury of knowledge. It has the potential to be a tremendous benefit to the turf industry.

Just as entire encyclopedias are now placed on a compact disk, eventually TGIF will be put on a single CD-ROM disk. By purchasing the compact disk users will have access to all of the information in the TGIF database without having to connect to the computer at Michigan State University. However, this new format won't become available until a substantial number of subscribers have the ability to read CD-ROM disks. Presently a majority of subscribers do not have CD-ROM drives.

USING TGIF WITHOUT A COMPUTER

Information from TGIF can also be accessed by dialing (800) 446-8443 and talking to the TIC staff. The TIC librarian will search the database, print out the search results, and mail the information to subscribers within 48 hours. Search requests can be faxed to a subscriber within 6 hours, if necessary.

TURFBYTE

TurfByte is an electronic bulletin board system for golf course superintendents. TurfByte evolved from an idea originally discussed by two superintendents, Jon Scott and Bill Spence, in the November 1987 issue of *Golf Course Management*. Duane Patton, another superintendent, with the help of Dale Gadd and Jon Scott, started TurfByte in 1988. Patton is the system operator, running the bulletin board from his home office in Lawrence, KS.

Bulletin boards are a convenient way for people with computers and modems to exchange information. This method of telecommunicating is becoming common in all professions. The purpose of TurfByte is to promote electronic conversations among turfgrass managers. They are encouraged to listen in and participate in any discussions that interest them.

For example, a superintendent can call up and comment that he is observing greater disease incidence after applying a plant growth regulator. He then asks if anyone else is experiencing the same problem. An electronic "bull" session results, with other superintendents offering their insights.

Turf managers go online and discuss weed control, what type of sprayer control system to buy, IPM techniques, and hundreds of other topics. The most attractive feature of the bulletin board is that users can join in at their convenience because these conversations take place over an extended period of time.

TurfByte has three main sections:

1. The Bulletin Section
2. The Message Section
3. The Files Section

THE BULLETIN SECTION

This section contains new information such as pesticide label changes or notification that a fungicide is contaminated with a herbicide. The user can also read or download articles on subjects such as hazard communication plans, lightning protection, wildflowers, *Poa annua*, computer usage, and topdressing.

THE MESSAGE SECTION

This is where electronic conversations occur. A message or question can be read by all who come online, or it can be left for a specific person. Private messages can only be seen by the sender, receiver, and system operator. Superintendents tired of playing "telephone tag" may find using TurfByte to be the easiest method of exchanging information with a colleague. By the spring of 1993, there were over 7000 messages on this section.

THE FILES SECTION

In this section TurfByte offers programs that can be downloaded to the caller's computer. Examples are programs for landscape plant management, record keeping, work scheduling, and resume writing. Some of the files are free, others are shareware. Shareware software is distributed without charge, but recipients are requested to send a payment to the author of the program if they use it regularly.

To connect with TurfByte, call (913)842-0618. The bulletin board is online 24 hours a day and is free. The only cost is the expense of the phone call. Duane Patton can be reached at (913)842-0146 for further information.

OTHER COMPUTER INFORMATION EXCHANGE NETWORKS

Many states have computerized information systems designed to serve Cooperative Extension agents and agronomists and horticulturists in the private sector. An example is CENET which is operated by Cornell Cooperative Extension in New York State. CENET provides information on pesticide regulations, pesticide label changes, turf insect, disease, and weed control, and fact sheets on cultural practices. In the IPM section there are predictions of tree and shrub pest activity based on degree-day accumulation.

Chapter 57

Enhancing Technology Transfer to the Homeowner

Laura Pottorff and William M. Brown, Jr., *Colorado State University, Golden, CO (Cooperative Extension) and Fort Collins, CO*

CONTENTS

Role of Master Gardeners in Homeowner Education 621
Role of the Master Gardener in Youth Education 622
Summary 623
References 626

Public concern about pesticide impact on human/animal health and the environment continues to increase.[1] As a result, educational and environmental advocacy organizations dealing with urban populations are searching for new ways to deliver IPM information to those who apply pesticides in maintaining landscapes and gardens.

The economics of the urban "Green Industry" are driven by urban citizens, whether they own homes or just enjoy the city parks and other public areas. Most citizens expect a perfect weed-, insect- and disease-free landscape. Garden centers and landscape maintenance services (public or private) exist to provide the perfect landscape the public wants. But this frequently happens at a cost, both economically and sometimes environmentally.

In agriculture, IPM practitioners use "economic thresholds." Unfortunately, in the urban setting it is difficult at best, and generally impossible, to define such economic thresholds. Homeowners, in their quest for the perfect lawn or blemish-free flower, revert to an "emotional threshold." The irony of this situation is that many people who expect perfection in the garden or landscape are frequently the very same persons most concerned about pesticide use and potential environmental or health consequences. They often hear a variety of conflicting information concerning pesticide use in the landscape and in their food and water supply. Whether these risks are real or perceived, a portion of the population has polarized to one extreme view or another concerning pesticide use. This is not a new issue and has been a point of controversy dating back to Rachel Carson's book, *Silent Spring,* published in the early 1960s.[2]

IPM offers a way to maintain an aesthetically pleasing landscape while minimizing environmental impact. But before the public will accept IPM, it must be educated. The more information the public receives regarding IPM the more often it will be able to make educated decisions regarding pest control and pesticide use. A result will be that more and more citizens will demand IPM landscape services and/or seek least-toxic pest control solutions from retail outlets.

How is IPM technology transferred to homeowners so they can make informed, educated decisions? There are many ways, including newsletters, educational bulletins, newspaper articles, and seminars. Often, however, average homeowners are not interested in pest control until actually faced with dead spots in the lawn or insects all over their rose or tomato plants. This is the "teachable moment," when a real impact can be made.

ROLE OF MASTER GARDENERS IN HOMEOWNER EDUCATION

Many state university Cooperative Extension programs throughout the country have been using the Master Gardener volunteer program to address this problem. The Master Gardener concept was initiated in King County, Washington, in 1972[3] to help Cooperative Extension professionals respond to horticultural questions from homeowners and others in the urban setting. Since then the program has been adopted by 45 states in the U.S.[4] Master Gardener volunteers receive in-depth horticultural training (including IPM) from university personnel and, in return, they volunteer their time, knowledge, and talent. Following are two examples of how Colorado State University Cooperative Extension uses

0-87371-350-8/94/$0.00+$.50

Master Gardener volunteers to transfer IPM technology to the homeowner.

The first contact the public has with the Master Gardener program is often through a phone call to their county cooperative extension office. In 1991, 678 Colorado Master Gardeners volunteered 23,309 hours and consulted with 66,304 individuals in 169 towns and cities in the state. These volunteers provided information, verbally and written, concerning everything from planting and care of various horticultural plants to household pest control.

The Master Gardener program in the seven-county Denver metropolitan area is very active in IPM technology transfer. Homeowners come to a plant diagnostic clinic when they are faced with a pest problem (or often what they think is a pest problem). Colorado's urban IPM programs target correct identification of pest and disease problems and subsequent recommendation of integrated environmentally sound control strategies.[5] Correct identification is one of the most important steps in IPM.

Master Gardeners help staff clinics. Volunteers receive approximately 20 hours of in-depth training in plant disease, insect, and abiotic (environmental, nutritional, and pesticide damage) problem diagnosis in addition to their basic Master Gardener training and volunteer experience. Volunteer duties include diagnosis of common disease, insect, and abiotic plant problems in the Jefferson County Cooperative Extension Plant Diagnostic Clinic and in other outreach activities at garden center, horticultural shows, open houses, and other locations. All pest or disease control recommendations follow IPM strategies. Master Gardeners are encouraged to explain to clientele what they should or should not expect in terms of control outcome. For example, melting out disease of turf is caused by a fungus that attacks turf that is stressed, usually because it has not been managed properly. Volunteers explain why the disease is present and how to manage the turf properly to obtain long-term improvement. If a fungicide is recommended, Master Gardeners emphasize that it must be used according to the label directions. They also explain that long-term control can only be achieved if the fungicide is used in conjunction with improved cultural practices that address problems of thatch buildup, soil compaction, and improper watering and fertilizer use.

In many other instances an effort is made to enhance the client's "emotional threshold" or tolerance of problems that only affect the appearance of plants. Plant problems that volunteers are unable to address are referred to the clinic supervisor (a postgraduate trained plant pathologist and urban IPM specialist).

Table 1 **Samples diagnosed by Master Gardener (M.G.) clinic volunteers over 3 years**

Site Profile	Year		
	1989–90	1990–91	1991–92
No. of volunteers	26	18	30
Total samples	996	1552	1783
M.G. diagnosis (%)	27	19	34

The Clinic Master Gardener Program began in Jefferson County in the mid-1980s.[2] The number of samples received by the clinic as well as the numbers of samples diagnosed by volunteers has increased over the past 3 years (Table 1). The goal in establishing this program was to help Cooperative Extension professionals in the Denver metro area meet the increased demand for homeowner plant diagnostic services. Prior to that time, all plant specimens were sent to a diagnostic facility at the state university. By using a trained volunteer force, increased homeowner demands have been met, allowing professional staff the opportunity to meet more complex homeowner problems and the demands of commercial urban agribusiness (Figure 1).

Use of Master Gardeners in plant problem diagnosis is very successful in Colorado. This is especially true when Master Gardeners take time to explain the biology of the pest, offer some IPM strategies to manage the pest, and outline what the client should or should not expect. This educational/service approach enables the homeowner to make an informed, environmentally sound decision on how or how not to proceed. Of the clients using the clinic in 1992 25% were surveyed to determine the effectiveness of this approach and the service received (Figure 2). The majority of clientele served were satisfied with the service and planned to, or already had, followed the IPM-based recommendations.

Not all private or public organizations dealing with urban horticulture and homeowners have access to a large volunteer base. However, it is clear that Master Gardeners taking time to explain IPM to homeowners, is leading to a greater understanding of pest control and more informed and environmentally conscious decisions.

ROLE OF THE MASTER GARDENER IN YOUTH EDUCATION

Another method of transferring IPM technology to the urban public is through children, who in turn share the information with parents and/or guardians. Master Gardeners also serve as volunteers in

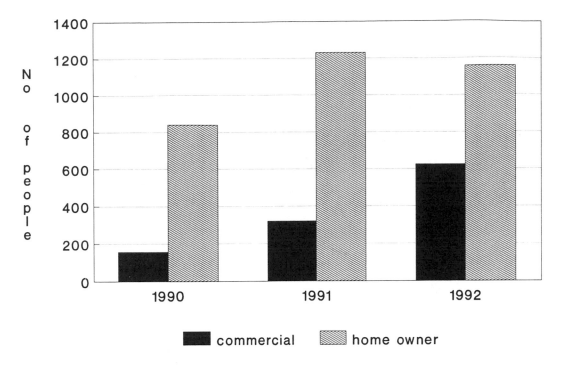

Figure 1 Number of clientele served during fiscal years 1990, 1991, and 1992.

an IPM awareness program that targets children with the "Bug Show."[6]

The Bug Show is an urban IPM educational program designed to compliment existing elementary science curriculum in the Kindergarten through third (K–3) grade levels. The program features three large-sized hand puppets (60 to 100 cm); Harry the Hornworm, Tom the Tomato, and Lucy the Ladybird Beetle (Figures 3 through 6). These puppets are used to teach children in K through 3 about the environment, some of the organisms that inhabit it, and how best to live with and without pesticides in an urban setting. Following a puppet show presentation each child receives a puzzle-coloring book to further the learning experience. The back of the book has a letter to the parents to allow the parent to become more aware of IPM through the child. The letter is routinely updated to address current public concerns relevant to food safety and environmental issues.

To evaluate the effectiveness of the Bug Show, enhanced IPM awareness was measured.[6] Children were asked preprogram questions about general IPM practices, plant growth, and general insect awareness. The puppet show concentrates on discussing beneficial organisms, pesticide safety and alternatives, and the consequences of pesticide misuse. The same preprogram questions were asked again as a postprogram test. The children's teachers were asked to evaluate child awareness before

and after the puppet show on a scale of 1 to 7 (1 being low, 7 being high). Teacher ratings of child awareness before vs. after the Bug Show presentation were then compared and significance determined by analysis of variance.

The Bug Show was evaluated 14 times for a total of 289 Denver area schoolchildren representing 5 schools or day care organizations between April and June 1991. The average rating of awareness before the Bug Show, again on a scale of 1 to 7, was 2.4, compared to an average rating of 5.25 after the show (Figure 7). This difference is highly significant.

The Bug Show is well received by teachers and students alike. It also provides an upbeat and meaningful educational activity to their classroom experience. The evaluation demonstrates that the puppet show clearly enhances child awareness of the IPM concept. It provides an opportunity to motivate young people at a very early age to explore and learn to appreciate the complexity of their environment, individually, as a group, and with parents. The ultimate goal is an increased awareness of the need to live in harmony with, and maintain the quality of, the urban environment.

SUMMARY

There is a critical need to educate homeowners, urban professionals, and the general public about

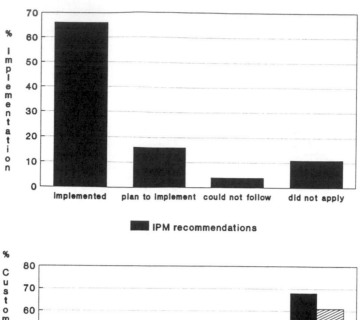

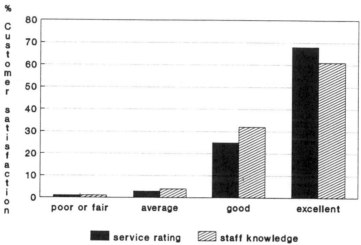

Figure 2 Results of clientele survey.

Figure 3 Harry the Hornworm.

Figure 4 Tom the Tomato.

Figure 5 Lucy the Ladybird Beetle.

Figure 6 The "Bug Show" being presented in a Denver classroom.

the safe use of pesticides and their alternatives. Using trained advance Master Gardeners to respond to homeowner plant health problems and to help educate children is an effective component of the Colorado State University Cooperative Extension Urban IPM program.

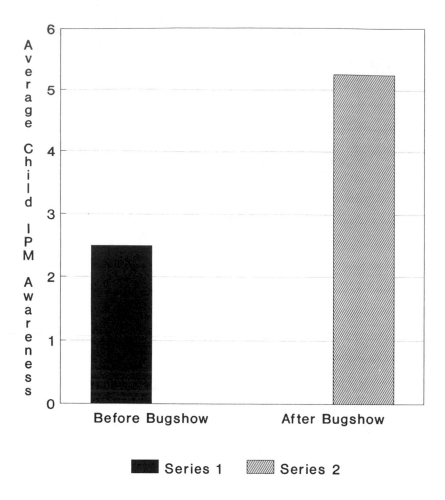

Figure 7 Evaluation of child IPM awareness. Significant at the 1% level.

REFERENCES

1. **Brown, W. M., Jr.,** The role of IPM in contemporary agriculture and environmental issues, in *Successful Implementation of Integrated Pest Management for Agricultural Crops*, Leslie, A. R. and Cuperus, G. W., Eds., CRC Press, Boca Raton, FL, 1993, 171.

2. **Brown, W., Cranshaw, W., and Rasmussen-Dykes, C.,** Urban integrated pest management education and implementation — implications for the future, in *Integrated Pest Management for Turfgrass and Ornamentals*, Leslie, A. R. and Metcalf, R. L., Eds., U.S. Environmental Protection Agency, Washington, DC, 1989, 57.

3. **Roberts, R.,** Master gardeners helping others to grow, *Family Food Garden,* 36, January 1982.

4. **Master Gardener Intl. Corp.,** Directory of Master Gardener Programs in the United States and Canada, ISSN 1054-9161, Master Gardener Intl., 1991.

5. **Pickett-Pottorff, L. S.,** Using master gardener volunteers in an extension plant clinic, *Phytopathology,* 82, 1145, 1992.

6. **Pickett, L. S. and Brown, W. M., Jr.,** The bug show — A Colorado youth urban IPM and pesticide awareness education progam, *Phytopathology,* 81, 1213, 1991.

INDEX

A

Acacia, 132
Acacia psyllid, 132
Acari, 60, 63, 64, 109
Acclaim, 307–311, see also Fenoxaprop
Acephate, 484, 485
Acer, 431, 461
Acizzia uncatoides, 132
Acremonium endophytes, 7, 521–526, 530–535
Acrostalagmus, 498
Actinolaimus, 499
Actinomycetes, 46, 48
Aculops fuchsiae, see *Fuschia* gall mite
Adalia bipunctata, see Two-spotted lady beetle
Adaptation zones, turfgrasses, 100–102
Adelgids, 20
    cultural practices and, 23, 134
    economic vs. aesthetic damage, 130
    oils and, 567, 572
Aeration, see also Cultivation and cultural practices
    golf course ponds, 177–178
    trees in urban environments, 207
Aesculus, 431, 461
Agasicles hygrophila, see Flea beetle
Aggregation pheromones, 337, 338
Agrilus anxius, see Bronze birch borer
Agrilus bilineatus, see Twolined chestnut borer
Agrobacterium radiobacter, 437
Agrobacterium tumefaciens, 437
Agrocin 84, 437
Agropyron repens, see Quackgrass
Agrostis alba, see Redtop
Agrostis canina, see Velvet bentgrass
Agrostis palustris, see Creeping bentgrass
Agrostis tenuis, see Colonial bentgrass
Airborne pesticide losses, 34
Air pollution, tree effects, 20, 209
Alachlor, 323
Alamo, see Propiconazole
Alarm pheromones, 337, 338
Albizzia, 431
Aldicarb, 36
Aleurothrixus floccosus (wooly white fly), 127
Alfalfa weevil, 445
Algae, golf course ponds, 170, 175–177, 179, 180
Allelochemicals, 11
Allelopathy, 19
Alliette, see Fosetyl-Al
Alligatorweed, 182, 278, 283
Allolobophora nocturna, 38, see also Earthworms
Alnus rhombifolia, 24

Alsophila pometaria, 343
Amaranthus hybridus, see Pigweed
Amblyseius, 444, 450
Amblyseius fallacis, 460–461
Amendments, see also Cultivation and cultural practices
    coarse-textured soils, 87
    fine-textured soils, 82
    weed control, 299
Amenity grasses, see Turfgrasses
Ammonium sulfate, 258
Amorbia looper, 459
Amphibians, 54
Andropogon gerardi, 320
Andropogon scoparius, 320
Andropogon virginicus, 320–321
Anguina, 492, 499
Anilazine
    earthworm toxicity, 261
    fate of in turfgrass environment, 33
    golf course use, 260
    nematode agent compatibility with, 486
Animals
    mole cricket control, 469
    wildflower-grass ecosystems, 319–321
Anisota senatoria, see Orangestriped oakworm
Annosus root rot, 23
Annual bluegrass (*Poa annua*), 11
    composts, disease suppression by, 414, 415
    fungal disease symptoms
        dollar spot, 250
        gray snow mold, 258
        yellow patch, 258
    growth and development, 94
    moist sites, 278
    most frequently encountered diseases, 251–252
    mowing regimes, 279
    nematodes, 262
    pesticide fate in turfgrass environment, 39
Annual bluegrass weevil, 586
    nematode agents, 483
    turf pests, 230
Annual lespedza, 278
Annuals
    bedding plants, 294, 311–312
    herbicide effectiveness, 298
    weed life cycles, 301
Annual sedge, 278
Antagonistic plants, 498
Anthracnose (*Colletotrichum graminicola*), 251
    cultural practices and, 405–406
    nutrition and, 389

trees
  cultural practices and, 23
  host resistance, 24
  turfgrasses
    models for in turfgrass management decisions,
      397–398
    symptoms and treatment, 391
Anthracnose (*Discula destructiva*), 431
Anthracnose (*Gnomonia platani*), 432
Anthracnose, tree, 23, 431–433
Antitranspirants, 437
Ants
  fire ants, 351–354
  insecticides and, 63
  turf pests, 60, 65, 220–223
Aonidiella aurantii, 460
Aonidiella citrina, 460
Aphelenchoides, 494
Aphelenchoides ritzemabosi, 492
Aphelenchus avenae, 498
Aphelinus mytilapsidis, 216
Aphidius matricariae, 458
Aphid lion, 447, 449
Aphid midge, 458
Aphidoletes aphidimyza, 458
Aphids, 445
  cultral practices for control, 134
  economic vs. aesthetic damage, 130
  fungal agents, *Beauveria bassiana,* 513
  natural enemies of, 458–459
  oils and, 567, 572, 579
  tree and shrub pests, 245
    cultural practices and, 23
    host resistance, 24
    pruning and, 21–22
Aphodius, 444
Aphytis lepidosaphes, 455
Aphytis melinus, 460
Apiosporina morbosa, 434
Apple, 238, 241, 437
Apple maggot, 337
Apple redbug, 571
Apple scab, 434, 437, 577
Application equipment
  calibration of, 143–161
    conversion tables, 161
    dry granular spreaders, 153–158
    factors affecting accuracy, 144–145
    frequency of, 144
    importance of, 144
    liquid application equipment, 146–153
    mixing and loading hemicals, 145–146
    safety guidelines, 159–160
    types of equipment, 144
    useful formulas, 160
  growth regulators, 271
  sewage sludge compost, 47–48

Apporectodea, 61, see also Earthworms
APTC, 303
Aquitards, 335–36
Aranae, 66, 67
Arborvitae, 342
Arbotect 20S, see Thiabendazole
Arbutus menziesii, see Madrone
Archips argyrospilia, see Fruit tree leafroller
Archips verasivoranus, see Uglynest caterpillar
Archops semiferans, see Oak leafroller
Argentine stem weevil (*Listronotus bonariensis*), 522
Argyresthia cupressella, see Cypress tip miner
Argyrotaenia velutinana, 340
Armillaria, 434–435
Armillaria mellea, 434
Armillaria root rot, 21, 23
Armyworm
  nematode agents, 483
  nematode host preference, 487
  sod production facilities, 123
  turfgrass susceptibility, 224
    St. Augustine grass, 110
    sod production facilities, 123
Arrowhead, 180
Artemisia artemisifolia, 322
Arthrobotrys, 497, 499
Arthrobotrys cynoides, 496
Arthrobotrys dactyloides, 496
Artichoke plume moth (*Platyptilia carduidactyla*), 341
Ascocalyx abietina, 437
Ascomycetes, 436
Ash (*Fraxinus*), 213–217, 241
Ash borer (*Podosesia syringae*), 20, 242
  economic vs. aesthetic damage, 130
  monitoring, 339
  pheromone lures for, 338
  tree and shrub pests, 243
Ash plant bug (*Tropidosteptes amoenus*), 246
Ash whitefly, 457
Ash yellows, 433
Asiatic garden beetle
  turf pests, 230–231
  white grub management, 354–359
Asparagus beetle, 570
Aspen serpentine leafminer (*Phyllocnistis poluliella,*) 244
Aspidiotus nerii, 460
Assassin bugs, 469
Aster, 322, 494
Aster pilosus, 323
Ataenius, 231, 509, see also White grub complex
Ataenius spretulus, 354–359
Athelia bombasina, 437
Athletic fields
  infield maintenance, 611–612
  turfgrass selection, 106
Atkinsonella, 529
Atrazine, 29, 36

Audubon Cooperative Sanctuary System, 56–57
Audubon Sustainable Resource Management Principles, 57
Augmentation, 446–447
Austrian pine, 433
Avocado looper, 459
Azadirachta indica, 498, 551, see also Neem
Azalea, 494, 572
Azalea lace bug (*Stephanitis pyrioides*), 134, 245–246
Azatin, 486

B

Bacillus
  compost, 412
  mode of action, 410
Bacillus popilliae, 196, 504, 508–509
  scarab larvae, 371, 375
Bacillus subtilis, 415
Bacillus thuringiensis, 446, 504, 578
  baculoviruses vs., 510, 511
  cankerworm control, 239
  and caterpillars, 459
  fall webworm control, 241
  nematode agent compatibility with, 484, 486
  ornamental and landscape plant insect pests, 506–508
  privatization of programs, 135
  resistance to, 127
  sod production, 123
  woody plant pest control, 133
Bacillus thuringiensis aizawa, 506
Bacillus thuringiensis israelensis, 182, 506
Bacillus thuringiensis japonensis, 506
Bacillus thuringiensis kurstaki, 133, 505–507
Bacillus thuringiensis san diego, 506–508
Bacillus thuringiensis ssp. *tenebrionis,* 506, 507
Bacillus thuringiensis tenebrionis, 506
Bacillus thuringiensis thuringiensis, 133
Backpack sprayers, 151–152
Bacteria, nematode gut, 470, 477
Bacterial canker, host resistance, 24
Bacterial controls
  golf course IPM program, 196
  mode of action, 410
  nematodes, 497–499
  ornamental and landscape plant pests, 505–508
  scarab larvae pathogens, 371
  turfgrass diseases, 390–391, 411
  weeds, 11
Baculoviruses, 509–511
Bagworm (*Thyridopteryx ephemeraeformis*), 129, 240, 342
  Bacillus thuringiensis, 506
  oils and, 567, 570
  pheromones, 343
  tree and shrub pests, 240
Bahiagrass (*Paspalum notatum*)
  fertilization of established lawns, 76, 77

fungal disease symptoms
  anthracnose, 391
  dollar spot, 392
  Rhizoctonia diseases, 391–392
  properties of, 388
Baking soda, 437
Balan, 306
Balansia, 529
Balsam woolly adelgid, 572
Bamboo, 299
Banana moth, 483
Banded ash borer, 338
Banks grass mite, 226–227
Banner, see Propiconazole
Banner 1.1 EC, see Propiconazole
Banvel 4E, see Dicamba
Bark composts, 414
Barnyardgrass, 283
Barricade, see Prodiamine
Basagran, 307-311
Basagran T/O, see Bentazon
Basidiomycetes, see also Dry spot; Fairy ring; Localized dry spot; Superficial fairy ring
Bathyplectes, 444
Bathyplectes anurus, 445
Bathyplectes curculionis, 445
Bats, 53, 54
Bayleton, see Triadimefon
Beauveria, 469
Beauveria bassiana
  black vine weevil control, 363
  ornamental and landscape plant pests, 513–515
Beauveria brongniartii, 371, 375, 513, 515
Beauveria popilliae, 9
Bedding plants, weed management plan
  annuals, 294
  perennial, 294–295
Beet armyworm
  Beauveria bassiana, 514
  nematode agents, 483
  oils and, 579
Beetles, see also White grub complex
  predators, 53
  as predators, 444
  turf pests, 230–235
    pesticide effects, 63, 67
    turfgrass ecosystem, 60
    white grub management, 354–359
Belonolaimus, 262, 492, 494, 497
  sewage sludge compost and, 48
  warm-season turfgrass infestations, 394
Beltsville aerated static pile method, 46
Bemisia tabaci, 462
Bendiocarb
  and earthworm populations, 63
  nematode agent comparison, 484, 485
  nematode agent compatibility with, 486

turfgrass ecosystems
    earthworm toxicity, 66
    and natural predators, 63
Beneficial invertebrates, see also Fungi, beneficial;
    Insects, beneficial; Nematodes, beneficial
  turfgrass systems, pesticide effects, 59–68
    acquired resistance and enhanced microbial
      degradation, 66–68
    on earthworm, 61–63
    earthworms and thatch, 60–61
    ecosystem, 59–60
    high-maintenance lawn care, cumulative effects,
      65–66
    on natural control of pest populations, 63–65
  sources of, 456–458
Benefin, 306
  landscape plantings, 296
  wildflower-grass plantings, 323
Benlate 50 WP, see Benomyl
Benomyl
  anthracnose, 391
  Bermudagrass decline, 391
  dollar spot, 392
  earthworm effects, 63, 66
  enhancement of fungal diseases by, 261
  golf course use, 260
  Helminthosporium leaf spot (melting-out), 393
  nematode agent compatibility with, 486
  spring dead spot, 393
Bensulide, 298, 302–303
  cumulative effects, 65
  landscape plantings, 296
Bentazon, 297, 302
Bentgrass
  fungal disease symptoms
    dollar spot, 250
    pink snow mold, 258
    stripe and flag smuts, 253
    yellow patch, 258
  golf course, 194
  nematodes, 262, 493
  pesticide interception, foliar, 30
  properties of, 104–106
  zoysiagrass collars, 109
Benzimidazoles, 261
Bermudagrass (*Cynodon*), 105, 299
  fertilization of established lawns, 76, 77
  fungal disease symptoms
    anthracnose, 391
    bermudagrass decline, 391
    dollar spot, 250, 392
    Helminthosporium leaf spot, 392–393
    Rhizoctonia diseases, 391–392
    rust, 393
    spring dead spot, 252, 393
  nematodes, 262, 493

pesticide interception, foliar, 30
properties of, 107–109
sewage sludge compost and, 47
sod production, 120, 122
white grub complex management, 358
zoysiagrass collars, 109
Bermudagrass decline
  symptoms and treatment, 391
Bermudagrass mite, 227
Bermudagrass scale, 227
Betasan, 307–311, see also Bensulide
Betula, 24
Biennial weeds, 301
Bifenthrin, 486
Big bluegrass (*Poa ampla*), 522
Big bluestem (*Andropogon gerardi*), 320
Big-eyed bug, 449, 450
Billbugs
  endophytes and, 522
  fungal agents, *Beauveria bassiana*, 513
  nematode agents, 483, 487
  pitfall traps, 334
  turfgrass
    lawn sustainability, 7
    monitoring, 332–333
    zoysiagrass, 109
  turf pests, 229–230
Binab T, 437
Bindweed, 299, 300
Biocon, 499
BioFix model, 339
Biological controls
  beneficial insects and mites, 443–452
  definitions, 455
  golf course, 181–182, 196
  insect and mite pests, commercial products, 455–464
  mole crickets, 467–474
  nematode agents, 477–487
  of nematodes, plant-parasitic, 491–500
    availability of agents, 499
    bacterial pathogens, 497–499
    fungal pathogens, 495–499
    host associations, 492–494
    other biocontrol agents, 497–498
    plant damage by, 492–494
    selection of agents, 498–499
    types of nematodes, 491–492
  optimizing, 462–464
  scarab larvae, 371
  turfgrass diseases, 409–423
    composts, 411–417
    microbial inoculants, 417–419
    suppressive soils, 417
    Typhula blight suppression with *T. phacorriza*,
      419–423
    warm-season, 390–391

Bio-Lure, 338
Biomass, turfgrass, 93
BioSafe, 478
Bipolaris, 252, see also Melting out
Birch leafminer (*Fenusa pusilla*), 133, 244
Birch skeletonizer (*Bucculatrix canadiensella*), 239
Birds
    biological control by, 455
    insect-eating, 53, 54
    mole cricket control, 469
    oystershell scale predators, 215–216
    wildflower-grass plantings and, 321
Birdsfoot trefoil, 278
Black cutworm, 483, 484
Black knot, 434
Black lady beetle, 457
Black light traps, 333–334
Black medic, 278, 283
Black scale, 23, 460
Black sooty mold, 577
Black spot of rose, 433
Black turfgrass ataenius, 354–359, 509, see also White
        grub complex
Black vine weevil (*Otiorhynchus sulcatus*), 239, 242, 462
    biorational controls, 133
    nematode agents, 483, 485
    nematode host preference, 487
Blight (*Phomopsis juniperovora*), 432
Blissus insularis, see Southern chinch bug
Blissus leucopterus hirtus, see Hairy chinch bug
Blotch leafminers, 244
Blue ash (*Fraxinus quadrangulata*), 214
Bluebirds, 53, 54
Blue fescue, 7, 9, 523, 525
Bluegrass billbug, 484
Bluegrass-red fescue turf ecosystem, 60
Bluegrass sod webworm, 226, 483
Blue green weevil, 483
Blue spruce aphid (*Elatobium abietinum*), 130
Bombardier beetle, 470–471
Boom sprayers, 145–150
Borers
    eucalyptus, 24
    trees
        cultural practices and, 23
        pruning and, 24
Botanical insecticides, 541–555
    advantages and disadvantages of, 542
    essential oils, 550–551
    nature of and synergism, 542–543
    neem, 551–553
    nicotine, 549–550
    pyrethrum and pyrethrins, 543–547
    rotenone, 547–548
    ryania, 549
    sabadilla, 548–549

Botryosphaeria, 3, 430
Botryosphaeria canker, 20
Botrytis cinerea, 437
Boxelder bug, 246
Boxwood leafminer (*Monoarthropalpus buxi*), 244
Boxwood mite, 572
Boxwood psyllid, 572
Bracon, 444
Bracted plantain, 278
Bradleygrass, 493
Brambles (*Rubus*), 321
Bravo, see Chlorothalonil
Broad mite, 460–461
Bromacil, 36
Bromus anomalus, 522
Bronze birch borer, (*Agrilus anxius*), 129, 130, 242
Broomsedge (*Andropogon virginicus*), 278, 283, 320–321
Brown black lady beetle, 462
Brown patch (*Rhizoctonia solani* and *R. zeae*), 251
    antagonism by *Typhula phacorrhiza,* 418
    biological control, 411, 418
    compost and, 413, 414, 416
    endophytes and, 525
    lawn grasses damaged by, 260
    models for in turfgrass management decisions, 401
    nutrition and, 406
    prediction of, 403
    symptoms and management, 254
    temperature and, 249
    turfgrass susceptibility
        tall fescue, 104
        zoysiagrass, 109
Brown rot
    predisposing factors, 20
    trees
        cultural practices and, 23
        host resistance, 24
Brown soft scale, 460
Brown thrush, 54
Bucculatrix ainsliella, see Oak skeletonizer
Bucculatrix canadiensella, see Birch skeletonizer
Buchloe dactyloides, see Buffalograss
Buckeye, 431
Buckhorn plantain, 284
Buffalograss (*Buchloe dactyloides*), 7, 299
    properties of, 111–112
    selection for environment, 108
Bug Chow, 446, 450
BugPro, 446, 450
Bulbs, 311
Bullheads, 179
Bull thistle, 279, 284
Bulrush, 178
Burdock, 279, 284
Burrowing sod webworm, 224–225
Butlerius, 499

C

Cabbage looper, 565, 568

Caddisflies, 53

Cadmium-based fungicides, 260

Calibration, defined, 143–144

California oakworm (*Phryganidia californicae*), 130

California poppy (*Eschscholzia californica*), 323

California red scale, 460

Calinsoga, 298

Caliper 90, see Simazine

Caliroa cerasi, see Pear sawfly

Calligrapha scalaria, 506

Callirhopathus bivasciatu, see Two-banded Japanese weevil

Cameraria hamadryadella, see Silatary oak leafminer

Canada thistle, 299

Canada thistle (*Cirsium arvense*), 321

Canker (*Septoria musiva*), 432

Cankerworms
  Bacillus thuringiensis, 506
  oil and, 570
  tree and shrub pests, 239

Captan, 436

Carabidae, 449
  insecticides and, 66
  turfgrass ecosystem, 60

Carbamates, 468, 543

Carbaryl, 456, 546
  and earthworm populations, 63
  fate of in turfgrass environment, 33, 38, 41
  nematode agent compatibility with, 486
  turfgrass ecosystems
    earthworm toxicity, 66
    enhanced microbial degradation, 68
    and natural predators, 63

Carbofuran, 62, 546

Cardinal, 54

Carp, 179, 182, 196

Carpenterworm (*Prionoxystus robiniae*), 133, 242

Carpetweed, 284

Casoron, see Diclobenil

Catbird, 54

Catenaria, 498

Caterpillars, see also specific genera and species
  natural enemies of, 459
  oils and, 567

Cattails, 178, 180, 181

Ceanothus, 132

Ceanothus stem gall moth (*Periploca ceanothiella*), 132

Cedar oil, 551

Celtis, 431

Centipedegrass (*Eremochloa ophiuroides*)
  fertilization of established lawns, 76, 77
  fungal disease symptoms
    anthracnose, 391

dollar spot, 250, 392
    Rhizoctonia diseases, 391–392
  nematodes, 262, 493
  properties of, 388
  sewage sludge compost and, 47
  St. Augustinegrass decline, 393–394

Cephalcia abietis, 134

Cephalosporium, 498

Ceratitis capitata, see Mediterranean fruit fly

Ceratocystis ulmi, 341

Cercospora leaf spot, 392

Cercospora rodmanii, 182

CGA 12223, 39

Chaetomium, 437

Chaitophorus polulicla, 134

Chalchid wasps, 216

Chamaecyparis, 431

Check-Mate, 338

Chemical controls, see also Pesticides
  biological controls and, 410–411
  golf course, 182, 196–197
  nematode agent compatibility with, 486
  turfgrasses, warm-season, 389–390

Chenopodium, see Lambsquarters

Cherry, 241

Chewings fescue (*Festuca rubra* var. *commutata*)
  endophytes, 7, 9, 522, 523, 525
  wildflower-grass mixes, 320

Chickadee, 54

Chickweed, 278, 279, 285, 289, 298

Chicory, 278, 279, 285

Chilochorus kuwanae, 131, 460

Chilochorus pinifolia, see Pine needle scale

Chilochorus stigma, 449, see also Twice-stabbed lady beetle

Chinch bug
  endophytes and, 522
  fungal agents, 513, 514
  turfgrass ecosystems
    lawn sustainability, 7
    pesticide effects, 64
    St. Augustinegrass, 110, 111
  turfgrass monitoring, 333, 334
  turf pests, 227–228

Chinese elm, 433

Chionaspis pinifolia, 455

Chionaspis pinifoliae, see Pine needle scale

Chitin, 49, 263, 496

Chitinase, 496

Chlopyrifos
  and earthworm populations, 63
  and enhanced microbial degradation, 68
  fate of in turfgrass environment, 41

Chlordane
  and earthworm populations, 62
  fate of in turfgrass environment, 32, 39, 41

golf course use, 194
mole cricket control, 468
Chloroneb
Pythium blight, 255
Pythium root rot, 393
snow mold, 258
turfgrasses, warm-season, 390
Chloropicrin, 119
Chlorothalonil
Beauveria bassiana and, 514
brown patch/Rhizoctonia blight, 255
Cercospora leaf spot, 392
dollar spot, 392
earthworm toxicity, 66, 261
enhancement of fungal diseases by, 261
fate of in turfgrass environment, 32, 33, 41
gray leaf spot, 392
leaf spot and melting out, 252
nematode agent compatibility with, 486
red thread and pink patch, 254
Rhizoctonia diseases, 392
rusts, 254
tree and shrub diseases, 436
yellow patch, 258
Chlorpyrifos, 546
cumulative effects, 65
earthworm toxicity, 66
fate of in turfgrass environment, 31, 32, 35, 39, 41
natural predator effects, 65, 67
nematode agent comparison, 484
nematode agent compatibility with, 486
Chlorthal dimethyl, 298, 303
landscape plantings, 296
nematode agent compatibility with, 486
Chlorthalonil, 391, 415
Chlosulfuron, 323
Chondrostereum purpureum, 437
Chordane, 64
Choristoneura fumiferana, see Spruce budworm
Christmas trees, 299
Chrysanthemum, 494
Chrysobothris femorata, see Flathead apple tree borer
Chrysomphalus dictyospermi, 460
Chrysoperla, 449–450
Chrysoperla carnea, 447, 449–450, 459
Chrysoperla rufalibris, 449–450, 459
Chrysoteuchia topiaria, 340
Chuck-will's widow, 54
Cicada killer wasps, 220–221
Cirsium, 322
Cirsium arvense, 321
Citronella, 551
Citrus, 457, 461
Citrus mealybug (*Planococcus citri*), 127
Citrus oils, 543
Citrus red mite (*Panonychus citri*), 127

Clarkus, 499
Clay, coarse-textured soils, 88
Clearwing borers
monitoring, 129
pheromones, 339, 341–342
wounding and, 134
Cleary's 3336, see Thiophanate methyl
Clethodim, 297
Clinodiplosis rhodoendri, see Rhododendron gall midge
Clover, 321
Coast live oak (*Quercus agrifolia*), 21
Coccinella septempunctata, 445, 447
Coccinellidae, 131, 447–449
Coccus hesperidum, 460
Cockchafer grub, 513
Cockspur thorn, 431
Codling moth (*Cydia pomonella*), 339
Coleomegilla maculata, see Spotted lady beetle
Coleophora, see El casebearer; Pistol casebearer
Collagenase, 496
Collembola
insecticides and, 63
turfgrass ecosystem, 60, 63
Colletotrichum graminicola, see Anthracnose
Colonial bentgrass (*Agrostis alba*), 105
Color, turfgrass, 93–94
Colorado blue spruce, 434
Colorado potato beetle, 506, 514
Comma leafminers, 244
Commercially available beneficial species, 456–462
Common damsel bug, 449
Common green lacewing, 443–444, 447
Common ragweed (*Artemisia artemisifolia*), 322
Common speedwell, 278
Compaction
fine-textured soils, 80–81
and weed types, 278
Competitive species
golf course pond management, 182
insects and mites, beneficial, 444
Compost, sludge, 45–49, 407, see also Cultivation and
cultural practices
and antagonistic microorganisms, 10
golf course applications, 194
turfgrass disease control, 411
Computer databases, 615–620
Container planting, 312
Convergent lady beetle, 447, 448, 458
Convolvulus on, 461
Cooley spruce gall aphid, 456, 572
Cool temperature brown patch, see Yellow patch
Copper hydroxide, 436, 486
Coreid bugs, 246
Cork oak (*Quercus suber*), 25
Corn poppy (*Papaver rhoeas*), 323
Corn speedwell, 278, 285

Cornus, 23
Cornus cousa, 431
Corythuca ciliata, see Sycamore lace bug
Corythuca cydoniae, see Hawthorn lace bug
Cotinus nitida, 354–359
Cotoneaster, 132
Cotton, 341
Cottontail rabbit, 321
Cottonwood borer, 514
Cottonwood leaf beetle, 506
Crabapple, 241
Crabgrass, 65, 276, 288
Crambus, see Sod webworm
Cranberry girdler (*Chrysoteuchia topiaria*), 340, 342,
    483, 513
Craneflies, 506
Crape myrtle (*Lagerstroemia indica*), 24
Crataegus, 431
Creeping bentgrass (*Agrostis palustris*), 00
    composts, disease suppression by, 414, 415
    endophytes, 523, 525–526
    fungal disease symptoms
        gray snow mold, 258
        Pythium blight, 255
    most frequently encountered diseases, 251–252
    nematodes, 262
    properties of, 101, 105–106
    Pythium blight, biological control, 418, 419
    root zone amendments, 413
Creeping red fescues
    endophytes, 7, 523
    sustainable turf, 13
Crickets, 53
Criconemella, 492–494, 497
Criconemoides, 494
Croesia semipurpurana, see Oak leaftier
Crown gall, 433
Crown rust, 254
    endophytes and, 525
    fungicides enhancing, 261
Crusade 5G, see Fonofos
Cryprolaemus montrouzieri, 447
Cryptocline twig blight, 20
Cryptolaemus montrouzieri, 459
Cryptostigmata, 60, 63
C-7, see *Coccinella septempunctata*
Cucumber, 461–462
Culms, 91
Cultivation and cultural practices
    fungicide use and, 259
    ornamental and landscape plants, 134, see also
        Pruning
        trees, 19–23
        in urban environment, 209–210
    soil and, 82
    turfgrasses
        cool-season, minimizing disease severity, 405–407

golf course, 193, 195–196, 198
    lawn turf sustainability, 6–8, 11–13
    sod production, 115–123, see also Turfgrasses,
        selection for environment
    warm-season, 387–389, 394
        nematode management, 394
    wildflower-grass mixes, 321–322
    weed control, 299, 311
Cunninghamella elegans, 496
Cupressaceae, 23
Cupressaceae, 132
Curvularia, 252
Cushiony cotton scale, 457
Cutflowers, 312
Cutless, see Flurprimidol
Cutworms
    nematode host preference, 487
    predators, 53
    turfgrass
        monitoring, 333
        St. Augustinegrass, 110
        sod production facilities, 123
    turf pests, 225
Cybocephalus nipponicus, 131
Cyclamen mite, 460–461
Cyclocephala borealis, 370
Cyclocephala lurida, 340
Cyclocephala (white grub complex), 354–359, 509
Cydia pomonella, 339, see also Codling moth
Cyfluthrin, 486
Cylindrocarpon, 497
Cynodon, see Bermudagrass
Cynodon dactylon, see Bermudagrass
Cynodon transvaalensis, 107
Cyperus, see Nutsedge
Cypress bark moth, 20
Cypress canker disease, 24
Cypress tip miner (*Argyresthia cupressella*), 130, 132
Cyst nematodes, warm-season turfgrass infestations, 394
Cystophage, 497
Cythion, 486
Cytospora canker, 23

D

2,4–D
    fate of in turfgrass environment, 31–34, 41
    Helminthosporium leaf spot (melting-out), 393
    nematode agent compatibility with, 486
    turfgrass ecosystem effects
        cumulative, 65
        earthworm toxicity, 66
Dacamine 4D, see 2,4–D
Daconil 2787, see Chlorothalonil
Dacthal, 33, 307–311, see also DCPA
Dacthal deacid, 32, 41
Dactylaria, 497

Dactylis glomerata, 320
Dactylon, 461
Dagger nematodes, see *Xiphinema*
Daisy fleabane (*Erigeron annuus*), 322
Damping off, 252, 255
Damsel bugs, 450
Dandelion, 285
Databases, 615–620
Datana mimistra, see Yellownecked caterpillar
Dazomet, 322–323
DBCP, 36
DCBA, 32
DCPA, 298, 303
    fate of in turfgrass environment, 29
    landscape plantings, 296
    wildflower-grass plantings, 323
Defense systems, see Host defenses
Delphastus pusillus, 462
Dendroctonus frontalis, see Southern pine beetle
Dendroctonus ponderosae, see Mountain pine beetle
Dendroctonus valens, see Red turpentine beetle
Deuteromycetes, 436
Devil's Pulpit Golf Course, 187–189
Devrinol, 307–311see also Napropamide
Diaphnocirus chlorionis, see Honeylocust plant bug
Diaspidid, 131
Diazinon
    and earthworm populations, 63
    fate of in turfgrass environment, 31, 33, 39
    nematode agent compatibility with, 486
    turfgrass ecosystem effects
        earthworm toxicity, 66
        cumulative, 65
        enhanced microbial degradation, 68
Dibromochloropropane, see DBCP
Dicamba
    fate of in turfgrass environment, 31–33, 41
    Helminthosporium leaf spot (melting-out), 393
    turfgrass ecosystem effects
        cumulative, 65
        earthworm toxicity, 66
Dicarboximide, 260
Dichloron, 34, 35
Diclobenil, 294, 297, 298
Dictyospermum scale, 460
Dienochlor, 486
Diflubenzurion
    nematode agent compatibility with, 486
    resistance to, 127
Dimethoate. 365
Dimethyl benzyl ammonium chloride, 486
Dioryctria, 339
Dioryctria ponderosae, 339
Dioryctria zimmermani, 339
Dipel, 508
Diplocarpon mespili, 431
Diplocarpon roseae, 433

Diplodia, 20, 24
Diquat, 297, 298
Discula destructiva, 433
Disease, defined, 387
Disease resistance, see Host defenses
Dithane M-45, see Mancozeb
Ditylenchus, 494
    fungal predators, 497
    Pasteuria susceptibility, 497
Ditylenchus myceliophagus, 499
Dogwood (*Cornus*), 180, 431, 572
Dogwood anthracnose, 433
Dogwood borer (*Synanthedon scitula*), 129, 134, 242
    pheromone lures for, 338
    tree and shrub pests, 243
Dogwood spittlebug, 572
Dogwood twig borer (*Oberea tripunctata*), 242
Dolichodorus, 497
Dollar spot (*Sclerotinia homeocarpa*), 251
    biological control, 411, 418
    compost and, 413, 414, 416
    cultural practices and, 405–406
    endophytes and, 523
    fungicides enhancing, 261
    golf course IPM program, 197
    lawn grasses damaged by, 260
    models for in turfgrass management decisions, 400–401
    nutrition and, 389
    preventative lawn treatment, 259
    sewage sludge compost and, 49
    symptoms and management, 250, 392
    zoysiagrass, 109
Dormancy
    and fungicide applications, 256
    turfgrass, 93–94, 96, 97
Dorylaimus, 499
Douglas fir, 433
Douglas-fir tussock moth (Orygia pseudotsugata, 238, 340, 342, 343, 510
Downy mildew, 436
2,4–DP, 32, 41
Dragonflies, 53
Drainage
    fine-textured soils, 82–83
    golf course siting and design, 167–170, 188
    trees in urban environments, 207
Drechslera, see Melting out
Drechslera dictyoides, 252
Drechslera or *Bipolaris,* see Leaf spot
Dreschslera, 252
Drop spreaders, 153
Drought stress
    endophytes and, 531–533
    sod production, 121
    trees in urban environments, 208, 209
    and weed types, 278
Dry spot (*Basidiomycetes*), 251

Duckweed, 179
Dursban, 578
Dursban 4E, see Chlorpyrifos
Dutch elm disease (*Ophiostoma ulmi*), 24, 126, 341
    cultural practices and, 23
    fungicides, 436
    sanitation pruning for, 134
    varieties resistant to, 432
Dwarf mistletoe, 20
Dyclomec, see Diclobenil
Dyslobus, 132

E

Earthworms, 38, 60–63, 66, 194, 261
Eastern spruce gall adelgid, 572
Eastern tent caterpillar (Malacosoma americanum), 133,
    240, 241, 569
Ectoparasitic nematodes, 262
Edge effect, 56
Education, public, 621–626
Egg parasitoids, 451
Elatobium abietinum, see Blue spruce aphid
Electronic monitors, turfgrass management decisions,
    401–402
Elm bark beetles, 24, 341
Elm calligrapha, 506
Elm casebearer(*Coleophora malivorella*), 240
Elm leaf beetle (*Xanthogaleruca luteola*), 129, 132, 133,
    239
    Bacillus thuringiensis, 506
    biorational controls, 134
    oil and, 570
    predisposing factors, 20
    tree and shrub pests, 239–240
Elm leafminer (*Fenusa ulmi*), 244
Elms, see also Dutch elm disease
    cankerworms, 238
    disease resistant varieties, 432–433
    pruning time, 24
    susceptibility to pests, 132
Elm yellows, 432–433
Elodea, 178
Embark, see Mefluidide
Encarsia, 444
Encarsia deserti, 462
Encarsia formosa, 448, 450–451, 462, 463
Encarsia parthenopea, 457
Encarsia tabacivora, 462
Encarsia transvena, 462
Endocronartium harknessi, 434
Endoparasitic nematodes, 262
Endophytes, turfgrass
    golf course IPM program, 196
    lawn sustainability, 7
    role in IPM, 521–526

    and stress tolerance, 529–534
        adaptive metabolic strategy, 533–534
        drought and flooding, 531–533
        nitrogen nutrition, 530–531
    tall fescue, 104
Endopiza viteana, 341
Endosulfan, 486
Endothenia albolinea, see Spruce needleminer
Endothia, 430
Endotoxin, 505-507
Energy partitioning, lawn turf, 6–7
English oak, 577
Enterobacter, 410
Enterobacter cloacae, 415
    vs. metalaxyl, 418, 419
    turfgrass disease control, 411
Entomopathogens, see Insects, beneficial
Environmental factors
    and biological controls, 410
    golf course design, 167, 189–190
    and nematode agent survival, 481–482
    oils and, 574–575
    pesticide fate in turfgrass environment, 36–37
    tree and shrub disease diagnosis, 429–430
    trees in urban environments, 205, 208, 209
    turfgrasses, see also Turfgrasses, selection for
        environment
        and disease outbreaks, 10
        and endophytes, 524
        growth and development, 94–95
        management, predictive models of disease outbreak,
            397–403
        turf sustainability, 7–8, 11–12
    turfgrasses, cool-season
        minimizing disease severity, 405–407
        zones of adaptation, 100–101
    turfgrass growth and development, 94–95
    white grub management, 355
    wildflower-grass mixes, 317–319
Environmental planning, 57
Environmental stress
    fungicide use and, 259
    trees, 19, 20
    turfgrasses
        endophytes and, 529–534
        sod production, 121
        and weed types, 278–279
Epichloe, 522, 529
Epilachna varivestis, 450
Eptam, 307–311, see also EPTC
EPTC, 296, 298
    landscape plantings, 296
    wildflower-grass plantings, 323
Equisetum, 299, 300
Eremochloa ophiuroides, see Centipedegrass
Eretmoceros californicus, 462

Erigeron annuus, 322

Erigeron canadensis, 322

Ermine moth, 569

Erosion, golf course siting and design, 167

Erwinia, 410

Erwinia amylovora, see Fireblight

Erwinia herbicola, 437

Erysiphe graminis, see Powdery mildew

Eschscholzia californica, 323

Esfenvalerate, 486

Essential oils, 543, 550–551

Ethazol, 390, 393

Ethion, 578

Ethoprop, 263

   and earthworm populations, 63, 66

   enhanced microbial degradation, 68

   nematode agent compatibility with, 486

   warm-season turfgrass treatment, 394

Ethylene dibromide, 194, see also Fumigation

Etridiazol, 255

Etridiazole, 486

Eucalyptus, 24

Eucalyptus long-horned borer, 24

Eucalyptus oil, 551

Eucosoma sonamana, 342

Euonymus, 132

Euonymus kisutschovicus, 132

Euonymus scale, 132, 246, 460

Euonymus shrubs, 25

European chafer

   fungal agents, *Metarhizium anisopliae,* 515

   life cycle, 374–375

   scarab larvae-turfgrass complex, 369–381

   stage-specific results of control agents,

      380–381

   turf pests, 220, 231

   white grub management, 354–359

European cockchafer, 515

European corn borer, 445, 449

European cranefly

   Bacillus thuringiensis, 506

   nematode agents, 483

   turf pests, 231–232

European elm bark beetle (*Scotylus multistriatus*), 242,

   341

European fruit lecanium, 127

European pine sawfly, 511

European pine shoot moth (*Rhyacionia buoliana*), 243

European red mite (*Panonychus ulmi*), 247

European spruce sawfly, 511

Eutypella canker, 433

Evapotranspiration, 35–36

Event, see Imidazolinones

Exhibit, 478

Exophiala, 497

Exotelia pinifoliella, see Pine needleminer

F

Face fly, 445

Fairy ring (*Basidiomycetes*), 251

   environmental stress and, 406

   lawn grasses damaged by, 260

   symptoms and management, 257, 392

Fall armyworm (*Spodoptera frugiperda*)

   endophytes and, 523, 524

   natural controls, pesticides and, 65

   turf pests, 225

Fall cankerworm (*Alsophila pometaria*), 343

Fall panicum, 286

Fall webworm (*Hyphantria cunea*), 240

   Bacillus thuringiensis, 506

   tree and shrub pests, 240–241

False cypress, 431

Fatty acids, 298

   insecticidal activity, 133

   landscape plantings, 297

   nematode agent compatibility with, 486

Fenamiphos, 263

   nematode agent compatibility with, 486

   warm-season turfgrass treatment, 394

Fenarimol, 269

   anthracnose, 391

   Bermudagrass decline, 391

   dollar spot, 392

   earthworm toxicity, 66

   phytotoxicity, 260

   Rhizoctonia diseases, 392

   spring dead spot, 252, 393

   summer patch, 256

   tree and shrub diseases, 436

Fenoxaprop, 302

   landscape plantings, 297

   wildflower-grass plantings, 324

Fenoxycarb, 486

Fenusa pusilla, see Birch leafminer

Fenusa ulmi, see Elm leafminer

Fern, nematode parasites, 494

Fertility, and weed species, 278

Fertilizers, see also Cultivation and cultural practices

   biological controls and, 410–411

   golf course siting and design considerations, 167, 168,

      170

   nematode agent compatibility with, 486

   soil tests and, 76, 77

   turfgrasses, 101

      cumulative effects, 65–66

      and earthworms, 60–63

Fescues

   endophytes, 7, 9

   fertilization of established lawns, 76, 77

   fungal disease symptoms

      dollar spot, 250

lawn, season of occurrence on, 260
golf course pond site, 180
nematodes and, 262
sewage sludge compost and, 49
sustainable turf, 13
wildflower-grass mixes, 320
Festuca, endophytes, 522, 525
Festuca arizonica, 522
Festuca arundinacea, see Tall fescue
Festuca elatior var *arundinacea,* 320
Festuca longifolia, see Hard fescue
Festuca obtusa, 522
Festuca rubra, see Red fescue
Festuca rubra var. *commutata,* see Chewings fescue
Field chickweed, 286
Field horsetail, 299, 300
Film-forming polymers, 437
Finale, see Glufosinate
Finches, 54
Fine fescues, see also Chewings fescue; Hard fescue;
  Red fescue
endophytes, 7, 9, 522, 523, 525
fertilization of established lawns, 76, 77
nematodes, 262
properties of, 106–107
most frequently encountered diseases, 251–252
symptoms of fungal disease
  lawn, season of occurrence on, 260
  melting out, 256
  Pythium and, 255
Fire ants (*Solenopsios invicta*), 351–354
fungal agents
  Beauveria bassiana, 513
  Metarhizium anisopliae, 515
turf pests, 221–222
Fireblight (*Erwinia amylovora*), 432–434, 437
predisposing factors, 20
trees
  cultural practices and, 23
  host resistance, 24
  pruning and, 24
Fish, 54
golf course ponds, 179–182
  IPM program, 196
  siting and design considerations, 168, 170
urban pesticide applications and, 216
Flag smut (*Urocystis agropyri*), 251, 253–254
Flathead apple tree borer (*Chrysobothris femorata*), 24,
  129, 242
Flathead borers, 24
Flavobacterium balustinum, 415
Flea beetle (*Agasicles hygrophila*), 182
Fletcher scale (*Parthenolecanium fletcheri*), 364–365
Flooding, endophytes and, 531–533
Floral lure traps, 334
Flotation sampling, 333
Flower flies, 450

Flowering crabapple, 432
Fluometuron, 37
Fluorazifop-p, 294, 298, 304, 307–311
  landscape plantings, 297
  wildflower-grass plantings, 324
Fluorprimidol, 269, 270
Flutalonil, 254, 255
Fluvalinate, 362
Flycatchers, 54
Foil, 508
Foliar blight, 398–400
Fomes annosus, 23
Fonophos, 63, 66
Food web, turfgrass ecosystem, 60
Forecasting models, 129, 397–403
Forest tent caterpillar (*Malacosoma disstria*), 241, 569
Formulations
herbicide, 312
and pesticide fate in turfgrass environment, 41
Fosetyl-Al, 415
nematode agent compatibility with, 486
Pythium blight, 255, 256
tree and shrub diseases, 436
turfgrasses, warm-season, 390
Fourlined plant bug (Poecilocapsus lineatus), 246
Fragaria, 461
Franklinella, 571
Fraxinus, Ash
Fraxinus excelsior, 214
Fraxinus pennsylvanica, 214
Fraxinus quadrangulata (blue ash), 214
Frit fly, 222
Froghoppers, 515
Fruit tree leafroller (*Archips argyrospilia*), 241, 506, 569
Fumigation
for fairy rings, 257
golf course, 194, 196
sod, 119
weed control, 299
wildflower-grass mixes, 323
Fungal associations, trees, 21
Fungal diseases
cultural practices and, 405–406
golf course IPM program, 196
models for in turfgrass management decisions, 397–491
nutrition and, 389
oil application in, 577
trees, cultural practices and, 23
turfgrass, see Turfgrass diseases, transition zone and
  northern regions
Fungi, beneficial
black vine weevil control, 363
endophytes, see Endophytes, turfgrasses
genera and modes of action, 410
golf course IPM program, 196
mole cricket control, 469
nematode control, 482, 495–499

ornamentals, 513–516
   trees and shrubs, 430
scarab larvae, 375
   stage-specific results in European chafer, 380–381
   stage-specific results in Japanese beetle, 376–379
scarab larvae pathogens, 371
sewage sludge compost and, 49
snow mold suppression, 258
tree and shrub treatment, 437
turfgrasses, warm-season, 390–391
turf insect control, 9
Typhula phacorrhiza, 419–423
vesicular-arbuscular mycorrhizae, 7
Fungicides
   biological controls and, 410–411
   biosphere targeting, 270, 271
   golf course use, 194
   nematode agent compatibility with, 486
   tree and shrub diseases, 435–437
   turfgrass
      factors associated with, 258–259
      on golf courses, 260–261
      in lawn care, 259–260
      warm-season, 389–390
   turfgrasses, warm-season, 391–393
Funginex, 436
Fungus gnats
   Bacillus thuringiensis, 506
   nematode agents, 483, 485
   nematode host preference, 487
Fusarium, 252, 497
   mode of action, 410
   sewage sludge compost and, 49
Fusarium blight (*Fusarium culmorum; fusarium poae*), 251
   necrotic ring spot vs., 252–253
   sod production considerations, 122, 123
   symptoms and management, 256
   thatch and, 407
Fusarium culmorum, see Fusarium blight
Fusarium heterosporum, 411
Fusarium nivale, 258
Fusarium poae, see Take-all patch
Fusarium wilt, trees resisant to, 431
Fuschia, 132
Fuscia gall mite (*Aculops fuchsiae*), 132
Fusilade, see Flurazifop-p

G

Gaeolaelaps, 499
Gaeumannomyces
   golf course IPM program, 197
   turfgrass disease control, 411
Gaeumannomyces graminis var *avenae,* see Gray
      snowmold
Galendromus occidentalis, 460–461
Galleria mellonella, 470

Gallery, 307–311, see also Isoxaben
Galls
   nematodes, 262
   tree and shrub pests, 238, 247
Galltrol-A, 437
Gambusia affinis (topminnow), 182
Garlon 3 A, see Trichlopyr
Geocoris, 449, 450
Geotextiles, 297
   flowerbeds, 294
   landscape plantings, 295, 296
   weed management plan, 297
German velvetgrass, 286
Giant foxtail, 286
Giant foxtail (*Setaria faberi*), 323, 324
Gilphinia hercyniae, 511
Gleditsia triacanthos f. *inermis* (honeylocust), 25
Gliocladium, 410, 497
Gliocladium virens, 411, 414, 418
Glufosinate, 297
Glycobius speciosus, see Sugar maple borer
Glyphosate, 269, 294, 298, 299
   landscape plantings, 297
   nematode agent compatibility with, 486
Glypta zozonae, 342
Gnatrol, 485
Gnomonia platani, 432
Goal, see Oxyfluorfen
Goldenrod, 299
Goldfish, 182
Golf course, 100
   fungicide use, 260–261
   minimizing environmental impact, 185–191
   monitoring programs, 585–589
   pest management strategies, 193–202
      biological controls, 196
      chemical controls, 196–197
      cultural controls, 195–196
      previous turf use, 194
      starting, 197–201
   pond construction, 173–183
      design considerations, 174–178
      drainage plan, 173
      IPM components, 178–183
   siting and design, 167–171
   turfgrass selection, 105
   weed scouting, 276–279
Goosegrass, 276, 278, 287
Granary weevil, 579
Grape, 321
Grapeberry moth (*Endopiza viteana*), 341
Grape mealybug, 459
Grapholitha molesta, 341
Grass-B-Gon, see Flurazifop-p
Grass grub (*Costelytra zealandica*), 509
Grasshopper Attack, 512
Grasslands, wildflower-grass mixes, 316–325

Gray leaf spot (*Pyricularia grisea*), 251, 389, 392

Gray snow mold, 251, see also Typhula blight

Gray willow leaf beetle (*Tricholochmaea decora*), 239

Greasy spot, 577

Greenbug (*Schizaphis graminum*)
    endophytes and, 524
    pesticide effects, 64
    turf pests, 228

Greenhouse thrips, oil and, 571

Greenhouse whitefly, 448, 450–451, 462, 463, 513

Green June beetle (*Cotinus nitida*), 232, 354–359, 515

Green lacewing, 449–450, 514

Green peach aphid, 445

Grey squirrel, 54

Grosbeaks, 54

Ground beetles, 449
    mole cricket control, 469
    turfgrass ecosystem, 60

Groundcover
    landscape plantings, 295–297
    weed management plan, 294
    wildflower-grass mixes, 316–325

Ground ivy, 287

Ground pearls, 232–233

Groundsel, 298

Groundwater
    fate of pesticides in turfgrass environment, 35–36
    golf course siting and design, 167, 170
    pesticide fate in turfgrass environment, 29, 32–33

Grow Gun, 435

Growth regulators, insect, 484, 486

Growth regulators, turfgrass, 267–273

Gryllotalpa orientalis, 467–474

Guignardia aesculi, 431, 434

Guppies, 182

Gymnosporangium, 432

Gypchek, 510

Gypsy moth (*Lymantria dispar*), 125, 126, 217, 445
    Bacillus thuringiensis, 133, 506
    biological control, 455
    control programs, 135
    evaluation of susceptibility to, 132
    insect predators, 449
    monitoring, 129
    ornamental plant pests, 239
    pheromones, 330, 342, 343
    predisposing factors, 20
    tree and shrub pests, 239

Gypsy moth baculovirus, 504, 510–511

H

Habitat creation and management, 55–56

Hackberry, 431, 572

Hairy chinch bug (*Blissus leucopterus hirtus*), 64

Hand gun sprayers, 145, 151–153

Hard fescue (*Festuca longifolia*)
    endophytes, 7, 9, 523
    wildflower-grass mixes, 320

Harposporium, 498

Harvester ants, 222

Hawthorn, 431

Hawthorn lace bug (*Corythuca cydoniae*), 132

Healall, 287

Heavy metals, sewage sludge compost, 46–47, 49

Helicotylenchus, 492– 494, 497
    endophytes and, 523
    grasses injured by, 262, 394

Helminthosporium diseases, 251
    cultural practices and, 405–406
    fungicide use and, 259, 261
    lawn grasses damaged by, 260
    nutrition and, 389
    symptoms and management, 250–252, 392–393
    thatch and, 407
    in warm-season turfgrasses, 392–393

Hemisarcoptes malus, 216

Hemispherical scale, 460

Hemlock adelgid, 20

Hemlock scale (*Fiorinia externa*), 134

Hemlock woolly adelgid, 134, 572

Henbit, 288

Hendersonia, 23

Heptachlor epoxide, 32, 41

Herbaceous perennials, 295–297

Herbicides
    Beauveria bassiana and, 514
    biological controls and, 410–411
    biosphere targeting, 270, 271
    container operations, 312
    and earthworm populations, 63
    fungal disease management, 252
    golf course design considerations, 174, 175
    growth regulators, 269
    Helminthosporium leaf spot (melting-out), 393
    landscape plantings, 296–297
    nematode agent compatibility with, 486
    sod production, 122
    turf classification, 272
    turf sustainability, 11
    weed control, 279–280
    weed management plan, 297–299
    wildflower-grass mixes, 322–324
    woody groundcovers, 294

Heterbasidion annosum, 437

Heterodera, 492–494, 499
    fungal antagonists, 496
    fungal parasites, 498
    fungal predators, 497
    Pasteuria susceptibility, 497

Heterorhabditis
    black vine weevil control, 363
    commercial products, 477–487

scarab larvae, 375
scarab larvae pathogens, 371
Heterorhabditis bacteriophora, 363, 478, 485
Heterorhabditis heliothidis, 363, 462
Heterorhabditis megidis, 478
Heterorjabites heliothidis, 133
Hippodamia convergens, 447, 448, 458
Hirsutella, 497, 499
Hirsutella thompsonii, 513
Holly (*Ilex opaca*), 244
Holly leafminer, 244
Homadaula anisocentra, see Mimosa webworm
Homoptera, 245
Honeybees, 514
Honeylocust (*Gleditsia triacanthos* f. *inermis*), 25
Honeylocust leafhopper, 572
Honeylocust plantbug (Diaphnocirus chlorionis), 246, 571
Honeylocust spider mite (Platyetranychus multidigitali),
      246–247, 572–573
Honeysuckle aphid, 572
Honeysuckle leaf blight, 434
Hoplolaimus, 262, 492, 497, 499
      sewage sludge compost and, 48
      warm-season turfgrass infestations, 394
Hoppers, cale insects, 245
Horn fly, 445
Horsechestnut, 431
Horsechestnut leaf blotch, 434
Horsetail, 299, 300
Horseweed (*Erigeron canadensis*), 322
Horticultural oil, 133
Host defenses, 217
      biological controls and, 410
      disease resistance
            improvement of, 9–10
            in trees, 20, 23–24
      endophytes and, 522–526
      trees and shrubs, 434–435
      wound hormones, 337
Hover flies, 450
Humidity, see Environmental factors;
Hummingbirds, 54
Hydrologic cycle, 35–36
Hydrophobic sands, 88
Hymenopteran parasitioids
      woody plant pest control, 131
      scale insects, 133
Hypericum, 25
Hyphantria cunea, see Fall webworm
Hypoaspis miles, 461, 499
Hypothalmicthys molitnx, 181

I

Ilex opaca, see Holly
Imidazolines, 269
Implementation pitfalls, 591–595

Indiangrass (*Sorghastrum nutans*), 320
Infield maintenance, 611–612
Injury
      defined, 387
      herbicide, 312
      trees, 20
            pruning and, 21–22
            in urban environment, 210
      woody landscape plants, 134
Insect attractants, 337–343
Insect-eating predators, 53–54
Insect growth regulators, 484
Insecticides
      Beauveria bassiana and, 514
      biological controls and, 410–411
      black vine weevil control, 362–363
      botanical, see Botanical insecticides
      and earthworm populations, 63
      earthworm toxicity, 261
      lawns, 5
      nematode agent compatibility with, 486
      ornamental trees, 126, 127
      rhododendron gall midge control, 364
      soaps and oils, see Soaps and oils
      and turfgrass ecosystems, 63
Insect pests
      endophytes and, 522–526
      fire ant management, 351–354
      golf course, 585–586
      mole crickets, 346–348
      ornamental and landscape plants
            biological control of, 503–515
            biological controls, commercial, 455–464
            chewing insects, 238–245
            gall insects, 238, 247
            sucking insects, 245–247
            trees, cultural practices and, 23
            woody ornamentals, 361–366
      turfgrass
            detection and sampling techniques, 331–336
            nuisance pests, 220–224
            sod production, 123
            subsurface or underground, 230–235
            surface and thatch pests, 224–230
            zoysiagrass, 109
      white grub complex, 354–359
Insects
      golf course ponds and, 179–181
      sampling
            mole crickets, 346–348
            pheromones and, 337–343
Insects, beneficial, 443–452
      competitors, 444
      mole cricket control, 469
      natural enemies of pest species, 443–444
      nematode genera attacked by, 499
      parasitoids, 444, 450–451

pesticides and, 22
predators, 443–444, 447–450
  lawns, 5, 8–9
  oystershell scale, 216
principles of biological control, 451–452
sources of, 456–458
types of biological control, 444–447
Insolibasidium deformans, 434
Integrated resource management, 53–57
Interspersion analysis, 56
Invade, 509
Inventory maintenance, sod, 123
Invertebrates, beneficial, see Beneficial invertebrates
Ipis beetle, 20
Iprodione
  brown patch—Rhizoctonia blight, 255
  Cercospora leaf spot, 392
  dollar spot, 392
  enhancement of fungal diseases by, 261
  fate of in turfgrass environment, 33
  leaf spot and melting out, 252
  nematode agent compatibility with, 486
  red thread and pink patch, 254
  Rhizoctonia diseases, 392
  snow mold, 258
  yellow patch, 258
Ips typographus, 340
Irrigation, see also Cultivation and cultural practices
  electronic controllers, 402
  golf course, 170
  and pesticide fate in turfgrass environment, 36–37
  trees, 26
  turfgrasses, 100, 389
Irritant sampling, 332–333
Isazophos
  and earthworm populations, 63, 66
  nematode agent comparison, 484
  nematode agent compatibility with, 486
  turfgrass ecosystem effects, 66, 68
  warm-season turfgrass treatment, 394
Isofenphos, 415
  fate of in turfgrass environment, 31, 32
  nematode agent comparison, 484, 485
  nematode agent compatibility with, 486
  turfgrass ecosystem effects
    earthworm toxicity, 63, 66
    enhanced microbial degradation, 68
    and natural predators, 63
Isoxaben, 296, 298, 304
Isozophos, 63
Istocheta aldrichi, 334

J

Japanese beetle (*Popillia japonica*), 6, 45, 217, 239, 445,
    see also White grub complex
  attractants, 338

Bacillus thuringiensis, 506
bacterial control, 504
endophytes and, 523
fungal agents, 9
  Beauveria bassiana, 513
  Metarhizium anisopliae, 515
golf course, 194, 196
large scale eradication programs, 127
life cycle, 372–373
natural controls, pesticide effects, 65
nematode agents, 480, 484, 485
pheromone traps, 339
scarab larvae-turfgrass complex, 369–381
Serratia, 509
stage-specific control agents, 376–379
trapping, 334, 340
turf pests, 233
Japanese honeysuckle (*Lonicera japonica*), 321
Japanese knotweed, 299
Japanese lawn cutworm, 483
Japanese lawngrass (*Zoysia jaonica*), 108, 109
Jeffrey pine, 456
Johnsongrass (*Sorgum halepense*), 321, 323
Juglans, 461
Juncus, 278
June beetle, 515, see also White grub complex
Junegrass (*Koeleria cristata*), 320
Juniper, 432
Juniperus chinensis, 132
Juniperus scopulorum, 132

K

Keifera lycopersicella, 341
Kentucky bluegrass (*Poa pratensis*)
  blending cultivars, 405
  composts, disease suppression by, 414
  earthworm populations, 61, 63
  endophytes, 523, 526
  fertilization
    earthworm effects, 61, 63
    of established lawns, 76, 77
  fungal disease symptoms
    dollar spot, 250
    leaf spot, 250
    melting out, 256
    necrotic ring spot, 252–253
    necrotic ring spot resistance, 253
    pink snow mold, 258
    red thread and pink patch, 253–254
    rusts, 254
    season of occurrence, 260
    stripe and flag smuts, 253
    summer patch, 256
  fungicide use, 259
  growth and development, 93, 97
  growth regulator application, 270

insect pests, 234

most frequently encountered diseases, 251–252

mowing height, 407

nematodes, 262

pesticide fate in turfgrass environment, 34, 39, 41

effects on natural predators, 63, 65, 67

foliar interception, 30

properties of, 102–103

Pythium and, 255

selection for environment, 106, 107

sewage sludge compost and, 49

sod production, 116, 120, 122, 123

sustainable turf, 5–7, 13

thatch degradation, earthworms and, 61

turf pests, 226

white grub complex management, 358

wildflower-grass mixes, 320

Kerb, see Pronamide

Key locations monitoring, insects, 331–332

Kildeer, 54

Kimberly ash (*Fraxinus excelsior*), 214

Kingbird, 54

Kinglets, 54

Kinoprene, 486

Kleenup, see Glyfosate

Kocide 101, see Copper hydroxide

Koeleria cristata, 320

Kudzu (Pueraria lobata), 321

L

Labronema, 499

Lacebug

oils and, 567

tree and shrub pests, 245–246

Lacewing larvae, 459

Lacewings, 444, 447

Lady beetles, 447–449, 457, 458, 462

Beauveria bassiana and, 514

as predators, 444

Laetisaria, 410, 411

Laetisaria arvalis, 418

Laetisaria fuciformis, see Red thread

Lagerstroemia indica, see Crape myrtle

Lambsquarters (*Chenopodium*), 288,

Lance nematodes, see *Hoplolaimus*

Landscape plants, see Ornamental and landscape plants;

Trees and shrubs; Trees

Lanzia, 250

Large crabgrass, 288

Large patch (*Rhizoctonia solani*), 251

Larra, 473

Larra bicolor, 467, 471, 472

Larra polita, 469–470

Lavender, oil of, 551

Lawn care, see also specific turfgrass topics

alternative service, 603–610

cumulative effects, 65–66

fertilization schedules, 76, 77

fungicide use, 258–259

pesticide effects on natural predators, 63

Lawrence linear pitfall trap, 334

Leaching

nutrient losses, 11–13

pesticides, in turfgrass environment, 35–41

Lead arsenate, 125

Leaf beetles, 515

Leaf blight (*Diplocarpon mespili*), 431

Leaf blotch (*Guignardia aesculi*), 431

Leaffooted bug, 246

Leafhoppers

insecticides and, 66

oils and, 567

trees, cultural practices and, 23

turf pests, 228

Leafminers

nematode agents, 483

nematode host preference, 487

oils and, 567, 569, 579

Leafrollers

oils and, 567

tree and shrub pests, 241

Leaf spot (*Drechslera* or *Bipolaris*), 251

endophytes and, 523

lawn grasses damaged by, 260

nutrition and, 389

prediction of, 403

symptoms and management, 250–252

turfgrass susceptibility

tall fescue susceptibility, 104

in warm-season turfgrasses, 392–393

in winter, 257

Leaftiers, 241

Leather jacket, 231–232

Lepidosaphes beckii, 455

Lepidosaphes ulmi, see Oystershell scale

Lepidosaphis beckii (purple scale), 127

Leptosphaeria, 197, see also Necrotic dead spot; Spring

dead spot

Leptosphaerulina blight (*Leptosphaerulina trifolii*), 251

Lescosan, see Bensulide

Lesion nematodes, see *Pratylenchus*

Lesser peachtree borer (*Synanthedon pictipes*), 338, 341

Life cycles

mole cricket, 346–347

optimizing biological control, 463

pheromones and, 129, 194, 337–343

white grub management, 355–356

Lilac borer (*Podosesia syringae*), 129, 242

economic vs. aesthetic damage, 130

monitoring, 339

pheromone lures for, 338

tree and shrub pests, 243

Lilac witches'-broom, 433

Lily ponds, 178
Limonene, 546, 550
Limonomyces roseipellis, see Pink patch
Linalool, 546, 550
Lipids
  insecticidal activity, 133
  landscape plantings, 297
  oil and, 578–579
Liriomyza trifolii (serpentine leaf miner), 120, 244, 569
Listronotus bonariensis, 522
Little bluestem (*Andropogon scoparius*), 320
Liverwort, 278
Localized dry spot (*Basidiomycetes*), 251
Locust borer (*Megacyllene robiniae*), 242
Lodgepole pine, 456
Lolium
  endophytes, 522
  nematode parasites, 493
London plane (*Platanus* x *acerifolia*), 24
Long-horned borer, 24
Longidorus, 492– 494, 497
Lonicera japonica, 321
Lophodermium needlecast, 433
Lumbricus terrestris, see Earthworms
Lygus lineolaris, see Tarnished plant bug
Lymantria dispar, see Gypsy moth

M

Macroposthonia, 262, see also Ring nematodes
Macrosiphum liriodendri, see Tuliptree aphid
Madrone (*Arbutus menziesii*), 20
Magnaporthe, 197
Magnaporthe poae, see Summer patch
Magnolia, 461
Magnaporthe, 197
Malacosoma americanum, see Eastern tent caterpillar
Malacosoma constrictum, see Pacific tent caterpillar
Malacosoma disstria, see Forest tent caterpillar
Maladera castanea, 354–359
Malathion, 546, 578
Maleic hydrazide, 268, 270, 271
Mallow, 288
Malus, 23, 129, 432
  disease susceptibility—resistance, 24
  Galendromus occidentalis on, 461
  insect pests, eastern tent caterpillar, 241
MAMA, 324
Management plans, 597–602
Mancozeb
  anthracnose, 391
  Beauveria bassiana and, 514
  brown patch—Rhizoctonia blight, 255
  Cercospora leaf spot, 392
  dollar spot, 392
  gray leaf spot, 392

leaf spot and melting out, 252
  Pythium blight, 256
  Pythium root rot, 393
  Rhizoctonia diseases, 392
  rusts, 254, 393
  and thatch accumulation, 261
  tree and shrub diseases, 436
Maneb
  dollar spot, 392
  enhancement of fungal diseases by, 261
  Rhizoctonia diseases, 392
  rust, 393
  and thatch accumulation, 261
Manilagrass (*Zoysia matrella*), 108, 109
Mantids, 444
Maple, 431, 433, 574
  insect pests, eastern tent caterpillar, 241
  pear thrips, 515
Maple twig borers (*Proteroteras*), 243
Margosan-O, 552
Marigolds, 498
Marshall ash (*Fraxinus pennsylvanica*), 214
Mascarenegrass (*Zoysia tenuifolia*), 108, 109
Masked chafer, 509
  Kentucky bluegrass injury, 358
  life cycles, 356
  milky spore disease, 509
  turf pests, 234–235
  white grub management, 354–359
Master gardeners, 621–626
Mastrus aciculatus, 342
May beetles, 233–234, 515
MCPP
  fate of in turfgrass environment, 33
  Helminthosporium leaf spot (melting-out), 393
  turfgrass ecosystem effects, 65
Meadow lark, 54
Mealybug, 245
  biological control, 457
  natural enemies of, 459–460
  oils and, 566, 567
Mealybug destroyer, 447
Mediterranean fruit fly (*Ceratitis capitata*), 127, 338, 445
Mefluidide, 268, 270, 271
Megacyllene robiniae, see Locust borer
Melampsora, 432
Melanapsis obscura (obscure scale), 129, 133
Melanoplus, 512
Melanotus phillipsii, see White blight
Meloidogyne, 492–494, 499, see also Root-knot
      nematodes
  fungal predators, 497, 498
  grasses injured by, 262
Meloidogyne javanica, 480
Melting out (*Bipolaris; Drechslera*), 251
  lawn grasses damaged by, 260

symptoms and management, 250–252, 256–257
thatch and, 407
in warm-season turfgrasses, 392–393
Menidia ardens, 181
Meria, 498
Mesoseiulus longipipes, 460–461
Metalaxyl
Beauveria bassiana and, 514
biological control agents (*Enterobacter cloacae*) vs., 418, 419
nematode agent compatibility with, 486
Pythium blight, 255, 256
Pythium root rot, 393
tree and shrub diseases, 436
Metam sodium, 119
Metaphycus helvolus, 460
Metarhizium, 469
Metarhizium anisopliae, 9, 371, 375, 513, 515
Metarhizium anisopliae var *anisopliae,* 363
Metarhizium flavoviride, 363
Metham sodium, 322–323
Methidathion, 486, 587, 578
Methoprene, 486
Methyl bromide
for fairy rings, 257
sod application, 119
wildflower-grass plantings, 322–323
Metolachlor, 304
fate of in turfgrass environment, 37
landscape plantings, 296
wildflower-grass plantings, 323
Metribuzin, 37, 323
Mexican bean beetle, 450
MGK 264, 543, 546
Mice, 455
Microbial associations, trees, 21
Microbial control, see also Biological controls
scarab larvae-turfgrass complex, insect pathogen interaction, 369–381
turfgrass disease control, 417–419
Microbial degradation
and pesticide fate in turfgrass environment, 42
pesticides, 66–68
pesticides in turfgrass systems, 66–68
Microbial ecology, trees, 21
Microclimates, trees in urban environments, 209
Microdochium nivale, see Pink snow mold
Micronutrients
sewage sludge compost, 47
soil tests, 76
Midge pod gall, 25
Midges, 180
Mimosa webworm (*Homadaula anisocentra*), 240
Mildew, trees
fog and, 24–25
host resistance, 24

Milfoil, 178
Milky diseases (*Bacillus popilliae*), 9, 196
in masked chafers, 234
ornamental and landscape plant insect pests, 508–509
scarab larvae, 375
stage-specific results in European chafer, 380–381
stage-specific results in Japanese beetle, 376–379
Mimosa, 431
Mimosa webworm (*Homadula aisocentra*), 133
Minute pirate bug, 449, 450, 461
Mississippi silversides, 181
Mites
beneficial, sources of, 456–458
natural enemies of, 460–461
oils and, 567, 579
oystershell scale predators, 216
tree and shrub pests, 247
turfgrass ecosystem, 60, 226–227, 229
evaluation of susceptibility to, 132
pesticide effects, 64
zoysiagrass, 109
turf pests
Miticides, nematode agent compatibility with, 486
Mixed plantings, 295–297
Mocap 10G, see Ethoprop
Mockingbird, 54
Models, forecasting, 129
Moellerodiscus, 250
Moisture, see Environmental factors
Mole cricket, 235
biological control, 467–474
fungal agents, *Metarhizium anisopliae,* 515
golf course, 194, 196, 198
monitoring, turfgrass sampling techniques, 333
nematode agents, 9, 483, 484
nematode host preference, 487
pitfall traps, 334
St. Augustinegrass, 110
zoysiagrass, 109
Monacrosporium, 497, 499
Monacrosporium ellipsosporum, 496
M-One, 508
Moneyworts, 278
Monitoring
golf course, 194–195
ponds, 180–181
turf pests, 585–589
insect attractants, 337–343
insects
mole cricket populations, 346–348
turfgrass, 331–336
optimizing biological control, 463–464
Monitor, 26
Monoarthropalpus buxi, see Boxwood leafminer
Monoculture, 405
Mononchoides, 499

Mononchus, 499
Monterey pine (*Pinus radiata*), 18
   causes of death, 21
   predisposing factors to pest problems, 20
Morning glory, 298
Mosquitoes, 179, 180
   golf course pond management, 182
   predators, 53
Mosquitofish (*Gambusia affinis*), 182
Moss, 278
Moth borers, 242–243
Moths, predators, 53
Mountain pine beetle (*Dendroctonus ponderosae*), 242
Mourningcloak butterfly (*Numphalis antiopa*), 238
Mouseear chickweed, 289
Mowing, see also Cultivation and cultural practices
   cool-season turfgrasses, minimizing disease severity,
      405–406
   and fungal diseases, 407
   sod production, 121–122
   and wildflower-grass mixes, 319
Mozambique mouthbrooder, 181
M-Pede, 543
M-Trak, 506
Mugo pine, 433
Mugwort, 289, 299, 300
Mulberry whitefly, 572
Mulches
   herbaceous perennial beds, 294–295
   landscape plantings, 296–297
   trees, 26
   weed control, 301, 312
   weed management plan, 296–297
Municipal parks, 613–614
Muscidifurax, 444
Mushrooms, 499
Mycoparasites, 410, 437, see also Fungi, beneficial
Mycoplasma-like organism, 433, see also Elm yellows
Mycorrhizae, 7, see also Endophytes, turfgrasses
Mycotol, 515
Mylonchulus, 499
Myriogenospora, 529
Myzocytium, 498

N

Nabis, 450
Nabis americoferus, 449
Nantucket pine tip moth (*Rhyacionia frustrana*), 129,
     132, 243, 338, 365
Napropamide, 298, 303
   fate of in turfgrass environment, 36, 37
   landscape plantings, 296
Natural predators, see also Fungi, beneficial; Nematodes,
    beneficial
   pesticides and, 63–65, 127
   woody plant pest control, 131

Necrotic ring spot (*Leptosphaeria korrai*), 251
   biological control, 411
   compost and, 413, 414
   lawn grasses damaged by, 260
   nutrition and, 406
   preventative lawn treatment, 259
   symptoms and management, 252–253
Nectria, 430
Nectria canker, 20
Nectria galligena, 437
Needleminers, 244
Neem, 498, 546, 551–553
Nemasys, 478
Nemasys-H, 478
Nematicides
   earthworm toxicity, 261
   golf course use, 194
Nematoctonus, 497, 498
Nematode pests, 251, 478
   biorational controls, 133
   cultural practices and, 405–406
   endophytes and, 523
   golf course, 182, 194, 196
   golf course pond management, 182
   scarab larvae, 375
   sewage sludge compost and, 48
   turfgrass, 261–263
   turfgrasses, warm-season, 394
Nematodes, beneficial, 10
   Beauveria bassiana and, 514
   biological control, 491–500
   black vine weevil control, 363
   commercial products, 477–487
      characteristics of, 477–482
      development of, 482–485
   golf course IPM program, 196
   host preference, 487
   mole cricket control, 467, 470
   nematode control by, 498, 499
   scarab larvae
      stage-specific results in European chafer, 380–381
      stage-specific results in Japanese beetle, 376–379
   scarab larvae pathogens, 371
   sewage sludge compost and, 48
   S. glaeri and, 480
   sources of, 456–458
   sustainable turf, 8–9
   woody plant pest control, 133, 462
Nematophthora, 497, 499
Nematus ventralus, see Willow sawfly
Nemocestes, 132
Neocletina bruchi, 181
Neocletina eichhorniae, 181
Neocurtilla, 467, 470
Neocurtilla hexadactyla, 467–469
Neodiprion lecontei, see Red-headed pine sawfly
Neoseiulus barkeri, 460–461

Neoseiulus californicus, 460–461
Neoseiulus cucumeris, 460–461
Net-blotch disease, 252, 260, 525
Network 8000 irrigation system, 402
New Zealand amber disease, 509
Nicotine, 543, 546, 549–550
Nighthawk, 54
Nigra scale, 460
Nigrospora blight (*Nigrospora sphaerica*), 251
Nigrospora sphaerica, see Nigrospora blight
Nimblewill, 289
Nitidulids, 131
Nitrogen, see Nutrition; Fertilizers
NoLo Bait, 512
Nontarget species
    Beauveria bassiana and, 514
    insecticides and, 66
Norbac, 437
Norosac, see Diclobenil
Northern masked chafer (*Cyclocephala borealis*), 354–
        359, 370, 506
Northern pine weevil (*Pissodes approximatus*), 242
Nosema fumiferanae, 512–513
Nosema locustae, 512
Nuclear polyhedrosis viruses, 510–511
Numphalis antiopa, see Mourningcloak butterfly
Nuthatches, 54
Nutrition, see also Cultivation and cultural practices;
        Fertilizers
    and fungal diseases, 406
    sewage sludge compost, 47
    soil tests, 76
    trees, 26, 207–208
    turfgrasses
        endophytes and, 530–531
        lawn turf utilization, 6–8
        and turf sustainability, 11–13
        warm-season, 389
    and weed types, 278–279
Nutsedge (*Cyperus*), 119, 196, 299

O

Oak borer, 338
Oak leafroller (*Archops semiferans*), 241
Oak leatier (*Croesia semipurpurana*), 241
Oak mite (*Oligonychus bicolor*), 246
Oak pit scale, 20
Oaks, 180
    branch dieback, 20
    insect pests, forest tent caterpillar, 241
    predisposing factors to pest problems, 20
    pruning time, 24
Oak skeletonizer (*Bucculatrix ainsliella*), 239
Oak webworm, 569
Oberea tripunctata, see Dogwood twig borer
Obliquebanded leafroller, 569

Obscure mealybug, 572
Obscure scale (*Melanapsis obscura*), 129, 133
Odontopus caleatus, see Yellow poplar weevil
Oftanol 5 G, see Isofenphos
Oils
    adjuvants, 577–578
    application of, 575–576
    arthropod resistance, 579
    climate and, 574–575
    combinations, 578
    equipment, 576–577
    fungal pathogen control, 577
    landscape plants, woody, 127
    and oystershell scale, 216
    petroleum, 560–569
        dilutions, 564
        mechanism of toxicity, 564–569
        terminology, 562–564
    phytotoxicity, 574–575
    range of pests controlled, 569–573
    safety, 573–574
    tree and shrub diseases, 437
    vegetable, 558–560
    woody plant pest control, 133
Oils, essential, 550–551
Old Marsh Golf Club, 190
Olea europea, 24
Oleander aphid, 572
Oleander scale, 460
Oligonychus bicolor, see Oak mite
Oligonychus ilicis, see Southern red mite
Oligonychus ununguis, see Spruce spider mite
Olive knot bacteria, 24
Onthophagus, 444
Ophiosphaerella herpotricha, see Spring dead sport
Ophiostoma ulmi, see Dutch elm disease
Orange lady beetle, 457
Orangestriped oakworm (*Anisota senatoria*), 130
Orchardgrass (*Dactylis glomerata*), 320
Organic amendments, and antagonistic microorganisms,
        10
Organic contaminants, sewage sludge compost, 48
Organophosphates, 263
    and natural predators, 63, 67
    nematode agent compatibility with, 486
Oribatid mites, 60
Oriental beetle, 370, see also White grub complex
    turf pests, 235
    white grub management, 354–359
Oriental fruit fly (*Dacus dorsalis*), 338
Oriental fruit moth (*Grapholitha molesta*), 341
Oriole, 54
Orius, 444, 461–462
Orius insidiosus, 449, 450, 461, 462
Orius tristicolor, 461
Ormia depleta, 467, 471, 473
Ornamec, see Fluorazifop-p

Ornamental and landscape plants, see also Trees and shrubs
  insect pests, 237–247, 361–366
    bacterial control, 505–508
    baculovirus control of, 509–511
    biological control, 503–516
    biological controls, commercial, 455–464
    fungal control agents, 513–515
    Nosema and, 512–513
  nematodes
    beneficial, 477–487
    plant parasitic, 491–500
  woody, 125–135
    biological controls, 131
    cultural practices, 134
    decision making, 129–131
    key pests and plants, 128–129
    monitoring, 129
    opportunities and needs, 135
    pesticides, 133–134
    programs for nurseries, 134–135
    reasons for IPM development for, 126–128
    resistant plant materials, 131
    special needs of trees, 17–26
Ortho garden fungicide, Captan
Orygia pseudotsugata, see Douglas-fir tussock moth
Oryzalin, 298, 305–311
  landscape plantings, 296
  nematode agent compatibility with, 486
Osmotic adjustment, endophytes and, 531
Otinem, 478
Otiorhynchus sulcatus, see Black vine weevil
Otiorhynchyus, 132
Oust, see Sulfonyl ureas
Oxadiazon, 304–305
Oxalis, 298
Oxazoidinedione, 486
Oxyfluorfen, 296–298, 305
Oystershell scale (*Lepidosaphes ulmi*), 213–217
Ozone, 134, 178
Pachysandra terminalis, 430

P

Pacific tent caterpillar (*Malacosoma constrictum*), 240
Paclobutrazol, 269, 270
Paecilomyces, 497, 499
Paecilomyces fumosoroseus, 363, 513
Paecilomyces lilacinus, 499
Paleacrita vernata, 343
Pales weevil, 515
Panicum mosaic virus, 110–111, 388, 389, 393–394
Panicum virgatum, 320
Panonychus citri (citrus red mite), 127
Panonychus ulmi, see European red mite
Papaver rhoeas, 323

Paralongidorus, 497
Paranguina, 492
Paranthene robiniae, 133
Parasaissetia nigra, 460
Parasites
  pest control, see Fungi, beneficial; Nematodes; beneficial
  pesticide effects, 63–65
Parasitoids
  insects and mites, beneficial, 444, 450–451
  mole cricket control, 469
  oystershell scale predators, 216
  pesticide applications and, 127
  pheromones and, 342
  woody plant pest control, 131
Parathion, 578
Paratrichodorus, 492
  endophytes and, 523
  grasses injured by, 262
  warm-season turfgrass infestations, 394
Paratylenchus, 262, 492, 493
Parthenocissus quinquefolia, see Virginia creeper
Parthenolecanium corni, 127
Parthenolecanium fletcheri, see Fletcher scale
Parthenolecanium quercifex, 133
Paspalum notatum, see Bahiagrass
Pasteuria, 497–499
Pasteuria penetrans, 498, 499
Patents, turfgrasses, 111, 112
PCNB
  biological control agents vs., 419, 421
  nematode agent compatibility with, 486
  Rhizoctonia diseases, 392
Pea aphid, 445
Peachtree borer (*Synanthedon exitiosa*), 242, 338, 341
Pear, 432, 437
Pearlwort, 278, 279
Pear sawfly (*Caliroa cerasi*), 239
Pear thrips, 515, 571
Pecan weevil. 515
Pectinophora gossypiella, 341
Pediasia, see Sod webworm
Pendimethalin, 298, 305
  earthworm toxicity, 66
  fate of in turfgrass environment, 32, 41
  landscape plantings, 296
Pendulum, see Pendimethalin
Penicillium, 410
Peniophora gigantea, 437
Pennant, 307–311, see also Metolachlor
Pennyroyal, oil of, 551
Pentachloronitrobenzene, 258
Penthaleus major, see Winter grain mite
Perennial ryegrass (*Lolium perenne*)
  composts, disease suppression by, 414
  endophytes, 9, 522–525, 529

fertilization of established lawns, 76, 77
fungal disease symptoms
   dollar spot, 250
   lawn, season of occurrence on, 260
   melting out, 256
   pink snow mold, 258
   Pythium blight, 255
   rusts, 254
   stripe and flag smuts, 253
   yellow patch, 258
fungicide use in lawn care, 259
growth and development, 97
most frequently encountered diseases, 251–252
mowing height, 407
nematodes, 262
properties of, 101, 106
sod production, 120, 123
Perennials, 295–297, 312, see also Ornamental and
   landscape plants
weed life cycles, 301
weed management plan, 294–295
wildflower-grass mixes, 316–325
Pergamascus, 499
Periploca ceanothiella, see Ceanothus stem gall moth
Persistence, 35
Pesticides
   Beauveria bassiana and, 514
   biological controls and, 410–411
   black vine weevil control, 362–363
   golf course use, 194
      IPM program, 198
      pond management, 178, 182
      siting and design considerations, 167, 168, 170,
        191
   large-scale applications of, 127–128
   monitoring program and, 585–586
   nematode agent compatibility, 484, 486
   optimizing biological control, 463
   ornamental and landscape plants
      tree and shrub diseases, 435–437
      woody, 133–134
   and oystershell scale, 216
   turfgrass environments, 29–42, 101
      foliar interception, 30–35
      leaching, 35–41
      persistence in, 35
      runoff, 41
   wildflower-grass mixes, 322–324
pH, see Environmental factors
pH, soil
   and disease development, 407
   and endophytes, 524
   fungicides and, 61
   lawn management, 5
   pesticides, cumulative effects, 65–66
   soil tests, 76

trees in urban environments, 207
   warm-season turfgrasses, 389
Phenacapsis pinifolia, 455
Phenological models, 129
Phenoxyherbicides, 252
Pherocon, 338
Pheromones, 129, 194, 337–343
   Japanese beetle, 372
   Nantucket pine tip moth control, 365
   turfgrass sampling, 334
Pheropsophus, 473
Pheropsophus aequinoctialis, 470–471
Phialophora radicicola, 411, 417
Phoebe, 54
Phoma, 497
Phomopsis juniperovora, 431
Photodegradation, pesticides, 34
Photosynthesis, 101
   air pollution and, 209
   endophytes and, 530, 532–533
   turf sustainability, 10–11
Photosynthesis, C-3, 5, 7, 10–11, 101
Photosynthesis, C-4, 5, 10–11
Phryganidia californicae, see California oakworm
Phyllocnistis ilicicola, see Holly leafminer
Phyllocnistis poluliella, see Aspen serpentine leafminer
Phyllonorycter blancardella, 340
Phyllophaga, 104, 356
Physical injury, Injury; Wound healing
Phythium aphanidarmatum, 410
Phytomyza ilicicola, see Holly leafminer
Phytophthora, 22, 436
Phytophthora cinnamomi, 433
Phytophthora root rot
   cultural practices and, 23
   landscape location and, 25
   predisposing factors, 20
Phytoseiulus, 444, 450
Phytoseiulus persimilis, 448, 450, 460–461, 463
Phytotoxicity
   fungicides, 260
   oils, 574
   sewage sludge compost, 49
Picea pingens, 434
Pigweed (*Amaranthus hybridus*), 322
Pine bark adelgid, 572
Pine needleminer (*Exotelia pinifoliella*), 244
Pine needle scale, 455, 579
   Chilochorus pinifolia, 133
   Chionapsis pinifoliae, 127, 129
   biological control by natural predators, 133
Pine-pine gall rust, 434
Pine shoot moths, 130
Pine tip moth (*Rhyacionia zozona*), 243–244, 342
Pink bollworm (*Pectinophora gossypiella*), 341
Pink patch (*Limonomyces roseipellis*), 251, 254

Pink snow mold (*Microdochium nivale*), 251
Pin nematodes, see *Paratylenchus*
Pinus, 132, 437
    pitch canker, 23
    pruning time, 24
Pinus mugo, 433
Pinus nigra, 433
Pinus radiata, see, Monterey pine
Pinus rigida, 433
Piperonyl butoxide, 546
Pissodes approximatus, see Northern pine weevil
Pissodes strobi (White pine weevil), 242
Pistacia chinensis, 25
Pistol casebearer (*Coleophora malivorella*), 240
Pitch canker disease, 23, 24
Pitch pine, 433
Pitfall traps, 334–335, 348
Plagiodera versicolor, 134
Planococcus citri (citrus mealybug), 127
Plant growth regulators, 267–273
Plant Health Care program, 26
Plastochron, 93
Platanus, 432
Platanus x *acerifolia,* 24, see also London plane
Platyetranychus multidigitali, see Honeylocust spider
    mite
Platyptilia carduidactyla, 341
Ploioderma needlecast, 433
Plovers, 54
Poa ampla, 522
Poa annua, see Annual bluegrass
Poa pratensis, see Kentucky bluegrass
Poast, see Sethoxydim
Podisus, 444
Podisus maculiventris, 450
Podosesia syringae, see also Ash borer; Lilac borer
Poecilocapsus lineatus, see Fourlined plant bug
Poison ivy, 321
Polymer films, 437
Ponds, golf course
    design of, 173–183
    IPM program, 196
Pondweed, 178, 180
Poor-will, 54
Popillia japonica, see Japanese beetle; White grub
    complex
Poplar
    biorational pesticides, 133
    defense strategies, 20–21
    insect pests, forest tent caterpillar, 241
Poplar borer (*Saperda calcarata*), 242
Population monitoring, mole cricket, 346–348
Populus, 20–21, 432, 514
Populus deltoides, 134
Pososesia, 130
Potamogeton, 178

Potassium chloride, 252
Potato beetle, *Bacillus thuringiensis,* 506
Potato leafhopper, 572
Powdery mildew, grasses
    environmental stress and, 406
    lawn grasses damaged by, 260
    symptoms and management, 254
Powdery mildew, shrubs and trees, 433, 437
    cultural practices and, 23
    oil application, 577
    pruning and, 21–22
    resistance to, 432
    sodium bicarbonate for, 437, 577, 578
Prairie dropseed (*Sporobolus heterolepis*), 320
Pratylenchus, 492–494, 497, 499, see also Lesion
    nematodes
    grasses injured by, 262
Praying mantids, 450
Precilia reticulata, 182
Precipitation, see also Environmental factors
    and pesticide fate in turfgrass environment, 36–37
    and wildflower-grass mixes, 319
Predators, see also Fungi, beneficial; Nematodes,
    beneficial
    attracting and maintaining, 446
    insects and mites, beneficial, 443–444, 447–450
    pesticide effects, 63–65
Predatory mite, 448
PredFeed, 446, 450
Prediction of outbreaks, 129
    monitoring and, 334
    in turfgrass management decisions, 397–403
Preen, see Trifluoralin
PreM 60 WDG, see Pendimethalin
Pre-San, see Bensulide
Primo, see Trinexepac-ethyl
Princep, see Simazine
Prionoxystus robiniae, see Carpenterworm
Prism, see Clethodim
Proact, 478
Prodiamine, 296, 298, 302
Pronamide, 298
    landscape plantings, 296, 297
    wildflower-grass plantings, 323
Propamocarb, 415
    Pythium blight, 255, 256
    Pythium root rot, 393
Propiconazole, 415
    anthracnose, 391
    Bermudagrass decline, 391
    dollar spot, 392
    earthworm toxicity, 66
    gray leaf spot, 392
    phytotoxicity, 260
    powdery mildew, 254
    Rhizoctonia diseases, 392

rust, 393
rusts, 254
smut diseases, stripe and flag smuts, 253
summer patch, 256
tree and shrub diseases, 436
turfgrass ecosystems
Prostrate knotweed, 278, 289
Prostrate spurge, 278, 290
Proteroteras, see Maple twig borers
Protozoa, insect pest control, 511–513
Pro-turf, see Bensulide
Proxol 80 WP, see Trochlorfon
Pruning, 21–22, 26, 435
benefits and problems, 23
sanitation, 134
timing, 24
in urban environment, 210
Prunus, 23, 342, 434
disease susceptibility/resistance, 24
Galendromus occidentalis on, 461
Pseudococcus maritimus, 459
Pseudomonas, 415, 417
mode of action, 410
turfgrass disease control, 411
Pseudomonas fluorescens, 437, 506, 507
Psyllids, oil and, 567, 572
Public education, 621–626
Public sector introductions of new species, 457–458
Puccinia, 196, see also Rust
Pueraria lobata, 321
Purple martins, 53
Purple scale (*Lepidosaphis beckii*), 127, 455
Purslane, 298
Pyracantha, 23, 132
Pyrenone, turfgrass sampling, 333
Pyrethrins and pyrethroids, 543–544, 578
and natural controls, 463
in oil, 577
and oystershell scale, 216
turfgrass sampling, 333
Pyrethrum, 543–544
Pyricularia grisea, see Gray leaf spot
Pyrrhalta luteola, see Elm leaf beetle
Pyrus, 432
Pyrus calleryana, 24
Pythium, 252
compost and, 49, 415
root dysfuction, 251
Pythium aphanidermatum, 398, see also Pythium blight
Pythium blight, 251, 259, 410, 415
biological control, 411, 418, 419
compost and, 413, 414
endophytes and, 523
environmental stress and, 406
fungicides enhancing, 261
golf course IPM program, 197

lawn grasses damaged by, 260
models for in turfgrass management decisions, 398–400
nutrition and, 389
symptoms and management, 255–256
temperature and, 249
turfgrass susceptibility
creeping bentgrass, 105
sod production considerations, 122
Pythium myriotylum, see Pythium blight
Pythium root diseases
fungicides, 436
Pythium root rot
biological control, 411
compost and, 414, 416
fungicides, 390
in warm-season turfgrasses, 393
Pythium torulosum, 415
Pythium ultimum, see Pythium blight

Q

Quackgrass (*Agropyron repens*), 290, 299, 321
Quadraspidotus juglansregiae, 133, see Walnut scale
Quadraspidotus perniciosus, 133, see also San Jose scale
Quercus, 461, see also Oaks
Quercus agrifolia, 21, 24
Quercus lobata, 24
Quercus rubor, 577
Quercus suber (cork oak), 25

R

Rabbit, 321
Rainfall, see Environmental factors; Environmental stress
Raxus cuspidata, 362
Redbanded leafroller (*Argyrotaenia velutinana*), 340
Redbud, 241
Red cedar, 432
Red fescue (*Festuca rubra*), 00
fungicide use in lawn care, 259
sewage sludge compost and, 49
subspecies, 106
sustainable turf, 13
Red-headed pine sawfly (*Neodiprion lecontei*), 238, 510, 511
Redhumped caterpillar (*Schizura concinna*), 238, 506
Red imported fire ants (*Solenopsios invicta*), 351–354, 515
Red scale, 578
Red sorrel, 290
Red thread (*Laetisaria fuciformis*), 251
biological control, 411
compost and, 413, 414, 416
fungicides enhancing, 261
lawn grasses damaged by, 260

nutrition and, 406
symptoms and management, 254
in winter, 257
Redtop (*Agrostis alba*), 105
Red turpentine beetle (*Dendroctonus valens*), 20, 242
Resmethrin, 546
Resource management, integrates, 53–57
Reward, see Diquat
Rhagoletis indifferens, 338
Rhagoletis pomonella, 337
Rhinoceros beetle, 515
Rhizoctonia, 416, 418
nutrition and, 389
turfgrass disease control, 411
Rhizoctonia blight (*Rhizoctonia solani* and *R. zeae*), 251
symptoms and management, 254–255
in warm-season turfgrasses, 391–392
Rhizoctonia cerealis, see Yellow patch
Rhizoctonia oryzae, 392, 401
Rhizoctonia solani, 252, 401, 415, see also Rhizoctonia
blight
Rhizoctonia zeae, 392, 401, see also Rhizoctonia blight
Rhizomes
sod production, 120
turfgrass, 91
Rhizotrogus majalis, 354–359
Rhodesgrass mealybug, 228–229
Rhododendron borer (*Synanthedon rhododendri*), 242,
338
Rhododendron catawbiense, see Rhododendron gall midge
Rhododendron lace bug (*Stephanitis rhododendri*), 245
Rhododendrons, 132
Rhyacionia, 339
Rhyacionia buoliana, see European pine shoot moth
Rhyacionia frustana, see Nantucket pine tip moth
Rhyacionia zozona, 342
Rhytisma, 434
Ring nematodes, see also *Macroposthonia*
sewage sludge compost and, 48
warm-season turfgrass infestations, 394
Ring spot, lawn grasses damaged by, 260
Rise, 494
Roadrunner, 54
Robin, 54
Rodolia cardinalis, 445
Ronamomermis culcivorax, 182
Ronstar, 307–311, see also Oxadiazon
Root-knot nematodes, 480, see also *Meloidogyne*
fungal antagonists, 496
warm-season turfgrass infestations, 394
Root rot, 21, 23, 25
golf course IPM program, 197
melting out and, 256
predisposing factors, 20
Roots
nematodes and, 262, 492–494

sustainability of turf, 7
trees, 19
pathogens, 433
in urban environments, 208–209
turfgrass, 91, 92–93
Root weevils, 132, 242, 513
Root-zone amendments, see Compost, sludge
Rose, 23, 433, 437
Rosy apple aphid, oils and, 572
Rotary spreaders, 153–154
Rotenone, 216, 546, 547–548
Rotylenchus, 497
Rotylenchus reniformis, 496
Roundheaded apple tree borer (*Saperda candida*), 242
Roundup, see Glyphosate
Rout, 296, 298, 307–311
Rove beetle, 448, 449, 499
pesticides and, 63, 67
turfgrass ecosystem, 60
Royal Slo-Gro, see Maleic hydrazide
Royal Woodbine Golf Course, 187–189
Rubigan, see Fenarimol
Rubus, 321, 461
Runoff
golf course siting and design considerations, 167, 174
pesticide fate in turfgrass environment, 32, 41
urban pesticide applications and, 216
Rushes, 180, 278
Russian wheat aphid, 445
Rust (*Gymnosporangium*), 431, 432
Rust (*Melampsora*), 432
Rust (*Puccinia*), 00, 251
cultural practices and, 405–406
lawn grasses damaged by, 260
nutrition and, 389, 406
turfgrass susceptibility
golf course IPM program, 196
sod production considerations, 122
symptoms and management, 254
in warm-season turfgrasses, 393
zoysiagrass, 109
Ryania, 543, 546, 549
Ryegrass
adaptation to high nitrogen, 278
endophytes, 522
nematode parasites, 493

S

Sabadilla, 543, 548–549
Safer, 543, 546
Safety
chemical application equipment, 159–160
urban pesticide applications and, 216
Saissetia coddeae, 460
Saissetia oleae, 460

Salinity
  coarse-textured soils, 88
  fine-textured soils, 81
  soil tests, 76–77
  trees in urban environments, 206, 208
Salix, 461, see also Willow
Sambucus, 461
Sameodes albiguttalis, 181–182
Sampling
  insects
    mole crickets, 346–348
    turfgrass, 331–336
  soil tests, 73–74
Sand, soil properties, 85–86, 88
Sanitation, trees and shrubs, 434
Sanitation pruning, 134
San Jose scale (*Quadraspidotus perniciosus*), 129, 338
Sanoplant, 478
Saperda calcarata, see Poplar borer
Saperda candida, see Roundheaded apple tree borer
Saprophytic fungi, 437
Sawflies
  biological control, 457
  biorational controls, 134
  cultural practices and, 23
  ornamental plant pests, 238, 239
Sawfly baculovirus, 511
Scab (*Venturia inaequalis*), 24, 432
Scale diseases
  biological control, 131, 133, 455–457
  economic vs. aesthetic damage, 130
  foreign origin of, 445
  hymenopteran parasitoids, 133
  monitoring, 129
  natural enemies of, 460
  oak pit scale, 20
  oils and, 566, 572
  trees and shrubs, 246
    cultural practices and, 23
    oystershell scale, 213–217
  turf pests, 227
Scapteriscus, see Mole crickets
Scapteriscus abbreviatus, 467–470
Scapteriscus borellii, 467–470
Scapteriscus didactylus, 467–470
Scapteriscus imitatus, 467–470
Scapteriscus vicinus, 467–470
Scarab beetles
  Bacillus thuringiensis, 506
  Metarhizium anisopliae and, 515
Scarabeid grubs, see White grub complex
Scarab larvae-turfgrass complex, see also White grub
      complex
  behavior of larvae in ecosystem, 370–371
  integration, 371–381
  Serratia and, 509

soil ecology, 371
Scentry, 338
Schizaphis graminum, see Greenbug
Schizura concinna, see Redhumped caterpillar
Sciopithes, 132
Scleroderris canker, 437
Sclerophthora macrospora, see Yellow tuft
Sclerotinia homeocarpa, see Dollar spot
Scolia dubia, 222–223
Scoliid wasps, turf pests, 222–223
Scolytus multistratus, see European elm bark beetle
Scots elm, 433
Scots pine, 433
Scotts OH2, 296, 298
Scotts TGR, see Paclobutrazol
Scotylus multistriatus, 341
Scotylus rugulosus, see Shothole borer
Scouting
  golf course, Monroe County NY, 585–589
  turfgrass weeds, 276–292
Seasonal cycles
  IPM treatment timing, 217
  tree pruning, timing of, 24
  turfgrass diseases, 249–258
  turfgrass growth, 95–97
  turf sustainability, 12
  weed life cycles, 301
Seattle street trees, 213–217
Sedges, 180
Seed bugs, 246
Seinura, 499
Semaspore, 512
Septoria musiva, 432
Sequoia pitch moth, 20, 24,134
Serpentine leaf miner (*Liriomyza trifolii*), 130, 244, 569
Serratia, 509
Serratia entomophila, 509
Serratia marcescens, 510
Setaria faberi, 323, 324
Sethoxydim, 269, 294, 298, 306
  landscape plantings, 297
  wildflower-grass plantings, 324
Sevin, 578
Sevin SL, see Carbaryl
Sewage, golf course siting and design, 167
Sewage sludge compost, 45–49, 407
Sex pheromones, 337, 338
Shade, tree culture, 23
Sharpshoot, 297, 307–311
Sheep fescue
  endophytes, 525
  wildflower-grass plantings, 320, 323
Shepherdspurse, 290
Shoot morphology, lawn turf, 7
Shoot moth, oils and, 567
Shoots, turfgrass, 91, 92

Shorttailed cricket, 235
Shothole borer (*Scotylus rugulosus*), 242
Shot hole damage, 239
Shrubs and trees, see Trees and shrubs
Siduron, 33
Sieridium, 23
Silt, 79–80
Silver carp, 181
Silver maple, 574
Simazine, 296, 323
Siphoninus phillyreae, 457
Sitka spruce weevil (*Pissodes strobi*), 242
Skeletonizing, 239
Skunks, 54
Slender creeping red fescue endophytes, 525
Sludge compost, see Compost, sludge
Smut, symptoms and management, 253–254
Snakes, 54
Snapshot, 296, 298
Snow mold diseases, symptoms and management, 257–258
Soap flush, insect sampling, 333
Soaps, 542–545
    landscape plants, woody, 127
    and oystershell scale, 216
    tree and shrub diseases, 437
    woody plant pest control, 133
Sodic, defined, 81
Sodium bicarbonate, 578
Sodium bicarbonate, as fungicide, 437
Sod production, 115–123
    establishment, 115–121
    growing, 121–123
    market preparation, 123–124
    pest management strategies, 117–118
    role and value of industry, 116–117
Sod webworms (*Crambus*; *Pediasia*)
    Bacillus thuringiensis, 507
    endophytes and, 522
    lawn sustainability, 7
    nematode host preference, 487
    pesticide effects, 65
    predator reduction, see 67
    sod production facilities, 123
    turfgrass
        monitoring, 333
        St. Augustinegrass, 110
        zoysiagrass, 109
    turf pests, 224–225
Soil conditions
    antagonistic microorganisms, 10
    golf course, 194, 198
    and nematode agent survival, 481
    and pesticide fate in turfgrass environment, 37–39
    trees, 19, 20
Soils
    and *Bacillus* thuringiensis, 507
    coarse-textured, 85–89

fine-textured, 79–83
fungicides and, 61
golf course
    ponds, 174
    siting and design considerations, 170
pesticide fate in turfgrass environment, 33, 36
scarab larvae-turfgrass complex, insect pathogen
    interaction, 369–381
sewage sludge compost, 45–49
testing, 73–77
trees in urban environments, 206–208
turfgrass disease control, 417
Soil survey
    detection techniques, 331–336
    mole cricket monitoring, 348
Solenopsios invicta (red imported fire ants), 351–354
Solitary oak leafminer (*Cameraria hamadryadella*), 244
Sonchus, 322
Sorghastrum nutans, 320
Sorgum halepense, 321
Sorosporella, 469
Southern bark beetles, 337
Southern blight
    biological control, 411
    sod production considerations, 122
Southern chinchbug (*Blissus insularis*), 00
    pesticide effects, 64
    St. Augustinegrass, 110, 111
Southern masked chafer (*Cyclocephala lurida*), 340,
    354–359
Southern pine beetle (*Dendroctonus frontalis*), 130
Southern red mite (*Oligonychus ilicis*), 247
Southern Weedgrass Control, see Pendimethalin
Spalangia, 444
Spanish Bay Golf Links, 190
Sparrows, 54
Speedwell, 291
Sphaeropsis, 430
Sphaerotheca fulginea, 437
Sphaerotheca phytophyla, 431
Spicaria, 498
Spicebush, 180
Spider mites, 246–247, 449
    natural enemies of, 460–461
    oils and, 579
    predators of, 450
    trees, 23
Spiders, 444, 449
    mole cricket control, 469
    pesticides and, 63, 67
Spikerush, 178
Spined soldier bug, 450
Spiral nematodes, see *Helicotylenchus*
Spittlebugs, 572
Spodoptera frugiperda, see Fall armyworm
Sporidesmium, 410
Sporobolus heterolepis, 320

Spot sampling, 332

Spotted lady beetle, 448, 449

Spotted tentiform leafminer (*Phyllonorycter blancardella*), 340

Sprayers, 145–163

Spreaders, 145, 153–158

Spring cankerworm (*Paleacrita vernata*), 343

Spring dead spot(*Leptosphaeria korrae; Ophiosphaerella herpotricha*), 251

    sod production considerations, 122

    symptoms and management, 252, 383

    in warm-season turfgrasses, 393

Spring tails, nematode control by, 499

Spring vetch (*Vicia sativa*), 323

Spruce bark beetle (*Ips typographus*), 340

Spruce budworm (*Choristoneura fumiferana*), 243, 506

Spruce gall adelgids, 130

Spruce gall aphid, 456

Spruce needleminer (*Endothenia albolinea*), 244

Spruce spider mite (*Oligonychus ununguis*), 247

Spurge, 276, 298

Staphylinidae, 449

    insecticides and, 66

    pesticide effects, 63, 67

    turfgrass ecosystem, 60, 63

St. Augustinegrass (*Stenotaphrum secundatum*), 110, 111

    fertilization of established lawns, 76, 77

    fungal disease symptoms

        Cercospora leaf spot, 392

        dollar spot, 250, 392

        gray leaf spot, 392

        Helminthosporium leaf spot, 392–393

        Rhizoctonia diseases, 391–392

        rust, 393

    nematodes and, 262, 493

    pesticide effects on natural predators, 64

    properties of, 109–111, 388

    sod production, 120

    soil pH for, 389

St. Augustinegrass decline, 111, 393–394

Steganosproium, 430

Steinernema, 477–487

Steinernema bibionis, 133

Steinernema carpocapsae, 9, 10, 462, 498

    black vine weevil control, 363

    commercial products, 478–482

Steinernema feltiae, 133, 363, 478

Steinernema glaseri, 9

    commercial products, 478–482

    scarab larvae pathogens, 371

Steinernema kraussei, 134

Steinernema riobravis, 478

Steinernema scapterisci, 9, 467, 473, 478

Stem borers, 483, 487

Stem rust, 254, 523, 525

Stenotaphrum secundatum, see St. Augustinegrass

Stephanitis pyrioides, see Azalea lace bug

Stephanitis rhododendri, see Rhododendron lace bug

Stethorus punctum, 449

Sting nematodes, see Belonolaimus

Stink bugs, 246

Stolons, 91, 129

Strawberry root weevil, 483

Stress tolerance, endophytes and, 531–533

Stripe smut (*Ustilago striiformis*), 251

    environmental stress and, 406

    lawn grasses damaged by, 260

    preventative lawn treatment, 259

    symptoms and management, 253–254

    thatch and, 407

Strong creeping red fescue endophytes, 525

Stubby root nematodes, see Paratrichodorus

Stunt nematodes, see *Tylenchorhynchus*

Stylet nematodes, see *Tylenchorhynchus*

Stylopage, 497

Subdue, see Metalaxyl

Sugar maple borer (*Glycobius speciosus*), 242

Sugar maple mite, 572

Sulfonyl ureas, 269

Sulfur-coated urea, 256, 407

Sulfur-containing fungicides, and thatch accumulation, 261

Sumagic, see Uniconazole

Summer patch (*Magnaporthe poae*), 251

    biological control, 411

    compost and, 413, 414

    cultural practices and, 405–406

    fungicides enhancing, 261

    lawn grasses damaged by, 260

    necrotic ring spot vs., 253

    preventative lawn treatment, 259

    symptoms and management, 256

    temperature and, 250

    thatch and, 407

Summit ash (*Fraxinus pennsylvanica*), 214

Sunspray 6E, 437

Superficial fairy ring (*Basidiomycetes*), 251

Supracide, 578

Surflan, see Oryzalin

Sustainability, lawn turf, 3–13

Swallows, 54

Swamp oak, 180

SWCT, 307–311

Sweet clover, 279

Sweet potato whitefly, 462

Swifts, 54

Swiss needlecast, 433

Switchgrass (*Panicum virgatum*), 180, **320**

Sycamore, 133

Sycamore anthracnose, 23

Sycamore lace bug (*Corythuca ciliata*), **245**

Sycamore plantbug, 571

Synanthedon culciformis, 133

Synanthedon exitiosa, 341, see also Peachtree borer

Synanthedon pictipes, 341
Synanthedon rhododendri, see Rhododendrom borer
Synanthedon scitula, see Dogwood borer
Syrphid fly, 444, 449, 450

T

Take-all patch (*Gaeumannomycers graminis* var *avenae*), 251
   biological control, 411, 418
   suppressive soils, 417
Talaromyces, 410
Tall fescue (*Festuca elatior* var *arundinacea*), 320
   composts, disease suppression by, 414
   endophytes, 7, 9, 522–525, 529
   fertilization of established lawns, 76, 77
   fungal disease symptoms
      lawn, season of occurrence on, 260
      net-blotch disease, 252
   fungicide use in lawn care, 259
   growth and development, 97
   growth regulator application, 270
   most frequently encountered diseases, 251–252
   nematodes, 262
   pesticide interception, foliar, 30
   properties of, 101, 103–104
   Pythium and, 255
   sod production, 120
   white grub complex management, 358
Tanagers, 54
Taphrina, 433
Tarnished plant bug (*Lygus lineolaris*), 246
Tar spot, 434
Tawny mole cricket (*Scapteriscus vicinus*), 346–348
Taxus, 363
Taxus cuspidata, 363
Taxus mealybug, 572
Taxus media, 362, 363
Teasel, 279, 291
Technology transfer, 621–626
Telar, see Sulfonyl ureas
Telonomus californicus, 342
Temperature, see Environmental factors
Tent caterpillars
   Bt control of, 505
   insect predators, 449
   oils and, 569
   tree and shrub pests, 241
Terbufos, 546
Tetranychus urticae, 448, see also Twospotted spider mite
Texas leafcutting ants, 220
Thatch
   and fungal diseases, 407
   fungicides and, 260
   pesticide effects, 60–61, 65–66

pesticide fate in, 31, 35, 41
sustainable turf, 13
turfgrasses, warm-season, 389
Thiabendazole, 436
Thiophanate methyl
   anthracnose, 391
   Bermudagrass decline, 391
   dollar spot, 392
   gray leaf spot, 392
   nematode agent compatibility with, 486
   tree and shrub diseases, 436
Thiophanates
   brown patch/Rhizoctonia blight, 255
   enhancement of fungal diseases by, 261
   summer patch, 256
   and thatch accumulation, 261
Thiram
   dollar spot, 392
   Rhizoctonia diseases, 392
   and thatch accumulation, 261
Thistles, 300, 322
Thornia, 499
Thridopteryx ephemeraeformis, 342
Thrips, 245
   natural enemies of, 461–462
   oils and, 567, 579
Thrushes, 54
Thymeleaf speedwell, 279
Thyridopteryx ephemeraeformis, see Bagworm
Thyronectria, 430
Thysanoptera, 245
Tifdwarf, 107
Tifgreen, 107
Tiger beetles, 469
Tiger mosquito, 445
Tilapia, 181, 182
Tilia cordata, 208
Tillage, see Cultivation and cultural practices
Tillers, 92, 93, 262
Timothy endophytes, 523
Tingidae, 245
Tipula, 506
Tomato pinworm (*Keifera lycopersicella*), 341
Topminnow (*Gambusia affinis*), 182
Topsin M, see Thiophanate methyl
Toxicity
   to earthworms, 66
   tree and shrub disease diagnosis, 429–430
Toxicodendron, 321
Toxic organic contaminants, sewage sludge compost, 48
Toxins, 513, 375
      stage-specific results in European chafer, 380–381
      stage-specific results in Japanese beetle, 376–379
Trail-marking pheromones, 338
Traps
   mole crickets, 347–348, 472

pheromone, 339–340
Trechispora alnicola, see Yellow ring
Trees
  special needs of, 17–26
    care of trees, 25–26
    development (life cycle), 19–22
    IPM programs, 18
    key plans in landscape, 22–25
    landscape settings, 17–18
    nature of trees, 18–19
  in urban environment, 205–211, 213–217
Trees and shrubs
  diseases of, 429–437
    biological control, 437–438
    diagnosis, 429–430
    established landscapes, 434–437
    new plantings, 430–433
  golf course siting and design considerations, 168–170
  insect pests, 237–247
  landscape plantings, 294–297
  wetland, 180
  woody landscape plants, 125–135
Treflan, 307–311, see also Trifluoralin
Triadimefon, 66, 269
  anthracnose, 391
  Bermudagrass decline, 391
  dollar spot, 392
  nematode agent compatibility with, 486
  phytotoxicity, 260
  powdery mildew, 254
  red thread and pink patch, 254
  Rhizoctonia diseases, 392
  rust, 393
  rusts, 254
  smut diseases, stripe and flag smuts, 253
  snow mold, 258
  spring dead spot, 393
  summer patch, 256
  tree and shrub diseases, 436
Trichlorfon
  and earthworm populations, 63, 66
  nematode agent compatibility with, 486
Trichlorpyr, 66
Trichlorpyridinol, 32, 41
Trichoderma
  Binab T, 437
  mode of action, 410
  turfgrass disease control, 411
Trichoderma hamatum, 414–416
Trichoderma harzanium, 411, 414, 416
Trichodorus, 492, 494, 497
Trichogramma, 444, 448, 451, 459
Trichogramma minutum, 459
Trichogramma platneri, 459
Trichogramma pretiosum, 459
Tricholochmaea decora, see Gray willow leaf beetle

Triclopyr, 486
Trifluralin, 298, 306
  landscape plantings, 296
  wildflower-grass plantings, 323
Trifolium, 321
Triforine, see Funginex
Trinexapac-ethyl, 269–272
Triumph 4E, see Isazophos
Tropical sod webworm, 483
Tropidosteptes amoenus, see Ash plant bug
Tufted titmouse, 54
Tuliptree aphid (*Macrosiphum liriodendri*), 130
Turcam 2.5G, see Bendiocarb
TurfByte, 619–620
Turfgrass
  biological controls, 409–423
  compost, sewage sludge, 45–49
  disease management in warm-season species,
    387–394
  disease models for management decisions, 397–403
  endophytes, see Endophytes, turfgrasses
  growth and development, 91–97
    ecosystem, 93–94
    environmental factors controlling, 94–95
    morphology, 91–93
    reproductive variation, 97
    seasonal cycles, 95–97
  growth regulators, 267–273
  insect detection and sampling techniques, 331–336
  insect pests, see Insect pests
  monitoring programs, 585–589
  nematode parasites, 491–500
  nematodes, beneficial, 477–487
  scarab larvae-turfgrass complex, insect pathogen
    interaction, 369–381
  selection for environment, 100–112
    cool-season species, 102–107
    warm-season species, 107–112
    zones of adaptation, see 100–102
  sod production, 115–123
  weed management, 276–292
Turfgrass, cool-season
  growth and development, environmental factors in, 94
  insect pests, 220
  minimizing disease severity, 405–407
  selection for environment, 102–107
  zones of adaptation, 101
Turfgrass, warm-season
  disease management in
    biological control, 390–391
    chemical control, 389–390
    cultural practices, 387–389
    diagnosis, 391–394
    nematode infestations, 394
  growth and development, environmental factors in,
    94–95

insect pests, 220
selection for environment, 107–112
sustainability of turf, 5
zones of adaptation, 101
Turfgrass diseases, transition zone and northern regions
  fungicide use
    factors associated with, 258–259
    on golf courses, 260–261
    in lawn care, 259–260
  nematodes, 261–263
  spring and fall, 250–254
  summer, 254–256
  temperature and, 249–250
  winter, 256–258
Turfgrass ecosystem
  dynamics of, 93–94
  pesticide use
    and beneficial invertebrates, 59–68
    pesticide fate in, 29–41
  sustainability, 3–13
    factors contributing to, 5–6
    grass adaptability, 6–11
    lawn ecosystem dynamics, 4–5
    managment considerations, 11–13
    value to environment, 3–4
Turfgrass endophytes, see Endophytes, turfgrass
Turfgrass Information and Pest Scouting (TIPS),
  194–195
Turfgrass Information File (TGIF), 615–619
Tussock moth, 449
Twice-stabbed lady beetle (*Chilochorus stigma*), 133,
  216, 448, 449
Twig blight, 431
Two-banded Japanese weevil (*Callirhopathus bivasciatu*),
  242
Twolined spittlebug, 229
Twolined chestnut borer (*Agrilus bilineatus*), 242
Two-spotted lady beetle, 448, 449
Two-spotted spider mite (*Tetranychus urticae*), 246, 247,
  448
Tylenchorhynchus, 262, 492, 494, 497, 499
Tylenchulus, 497, 499
Tylenorhynchus, 492
Typhula blight, see also Gray snow mold
  biological control, 411
  compost and, 413, 414, 416
  Typhula phacorrhiza and, 419–423
Typhula incarnata, see Brown patch
Typhula ishikariensis, 418, see also Brown patch
Typhula phacorrhiza, 258, 411, 418

U

Uglynest caterpillar (*Archips verasivoranus*), 240
Ulmus, 24, 432–433, see also Dutch elm disease; Elms
Unaspis euonymi, 131

Uniconazole, 269
Urban environments
  municipal parks, 613–614
  trees
    Seattle case study, 213–217
    survival of, 205–211
  turf value to, 3–4
Urea, sulfur-coated, 256, 407
Urocystis agropyri, see Flag smut
Ustilago striiformis, see Stripe smut

V

Valsa, 430, 434
Vantage, 307–311, see also Sethoxydim
Vapam, see Metam sodium
Vapor Gard, 437
Vector-MC, 478
Vector-WG, 478
Vedalia beetle, 445, 457
Velvet ants, 223
Velvet bentgrass (*Agrostis canina*), 105
Ventana Canyon Golf Course, 190
Venturia inaequalis, 432, 434
Verdol, 572
Vertalec, 515
Verticillium, 498, 499
  mode of action, 410
  nematode genera attacked, 497
Verticillium clamydosporium, 496
Verticillium dahliae, 431, 433
Verticillium lecanii, 513, 515–516
Verticillium wilt, 23, 431
Vesicular-arbuscular mycorrhizae, 8
Viburnum, 129
Vicia sativa, 323
Vinclozolin
  brown patch/Rhizoctonia blight, 255
  dollar spot, 392
  leaf spot and melting out, 252
Viral control agents, 509–511
Viral diseases, 110–111, 388, 393–394
Vireo, 54
Virginia creeper (*Parthenocissus quinquefolia*), 321
Vitis, 321, 461
Volutella pachysandricola, 430
VPM, see Metam sodium

W

Walnut, 23
Walnut scale (*Quadraspidotus juglansregiae*), 129
Warblers, 54
Wasps, 451
  mole cricket control, 467, 469–470
  oystershell scale predators, 216

parasitoid, 444
and scale parasites, 460
turf pests, 223
Water
coarse-textured soils, 86–88
golf course
ponds, 173–183
siting and design considerations, 167, 168
Waterfowl, golf course siting and design considerations, 169, 182
Water hyacinth, 182
Water lilies, 179
Watermilfoil, 178
Water quality
golf course design concerns, 173–183, 190
pesticide fate in turfgrass environment, 29
Water reqirements, see Environmental factors; Irrigation
Water stress, and weed types, 278
Water use, lawn turf, 6, 7–8
Wax moth, 470
Weather, see also Environmental factors
and chemical applications, 194
modeling disease outbreaks, 401–402
Weed management
for herbaceous ornamentals, 301–313
growing situations, 312
herbicide application, 312–313
herbicides, 302–311
options, 301
types of ornamentals, 311–312
landscape plantings, 293–300
life cycles, 301
turfgrass, 276–292
lawn adaptibility, 9–10
sod production system, 119, 122
sustainability, 10–11
wildflower-grass mixes, 321–324
Weed management plan, 293–300
installation and implementation, 300
planting type, 294–296
selecting options, 296–299
site assessment, 293–294
site preparation, 299–300
Western cherry fruit fly (*Rhagoletis indifferens*), 338
Western gall rust, 24
Western pine shoot borer (*Eucosoma sonamana*), 342
Wetlands, golf course design concerns, 190, 191
Wetland vegetation, 174, 175, 178–180
Wheast, 446, 450
Whippoorwill, 54
White amur carp, 196
White blight (*Melanotus phillipsii*), 251
Whitefly
natural enemies of, 462
oils and, 567, 579
White-footed mouse, 54

White grub complex
endophytes and, 7
fungal agents, *Metarhizium anisopliae,* 515
insect pests, 234
management, decision-making factors, 354–359
milky disease, 509
nematode agents, 483, 487
pesticides, cumulative effects, 65
scarab larvae-turfgrass complex, insect pathogen interaction, 369–381
turfgrass monitoring, 332–333
turfgrass susceptibility
St. Augustinegrass, 110
tall fescue, 104
zoysiagrass, 109
White heath aster (*Aster pilosus*), 323
White pine weevil (*Pissodes strobi*), 242
Whitetail deer, 321
White tall fescue, 97
Wild bees, 223–224
Wildflower-grass mixes, 316–325
fire and, 319
goals and species selection, 316–317
grasses for use in, 320–321
mowing, 319
natural grasslands, native species and, 317
pesticide role, 324–325
pests of, 321
rainfall, 317–319
soil disturbances, 319–320
weed control, 321–324
Wild garlic, 291
Wildlife
conservation programs, 57
golf course design concerns, 190
golf course pond management, 182
golf course ponds, 176
golf course siting and design considerations, 169, 170
habitat creation and management, 55–56
insect-eating, 53–54
integrated resource management, 54
Wild violet, 299
Willow (*Salix*), 180
biorational pesticides, 133
oystershell scale, 214
Willowbend Country Club, 190
Willow leaf beetle, 134
Willow sawfly (*Nematus ventralus*), 238
Wilt disease, cultural practices and, 23
Wilt Pruf, 437
Winged mole cricket (*Scapteriscus borellii*), 346–348
Winsome fly, 334
Winter grain mite (*Penthaleus major*), 64, 229
Witches'-broom (*Sphaerotheca phytophyla*), 431
Woodborers, 462
Woodchuck, 321

Woodpeckers, 54, 56
Woodthrush, 54
Woody ornamentals, see also Trees and shrubs
  insect control, 361–366
  insect pests, 237–247
  landscape plants, 125–135
  weed management plan, 294–297, 299
Woolly white fly (*Aleurothrixus floccosus*), 127
Wound healing
  lawn turf, 6
  trees, pruning and, 21–22
  woody landscape plants, 134
Wound hormones, 337
Xanthogaleria luteola, see Elm leaf beetle
Xanthomonas, 196
Xanthomonas campestris, 11
Xanthomonas maltophila, 415
Xenorhabdus, 470, 477, 478, 481
Xiphinema, 492, 494, 499
  fungal parasites, 498
  grasses injured by, 262
  Pasteuria susceptibility, 497
Xiphinema index, 496

X

XL, 296, 298, 306–311

Y

Yarrow, 292
Yellow nutsedge, 292
Yellow patch (*Rhizoctonia cerealis*), 251, 258
Yellow poplar weevil (*Odontopus caleatus*), 244
Yellow ring (*Trechispora alnicola*), 251

Yellow scale, 460
Yellow tuft (*Sclerophthora macrospora*), 251, 261
Yellow woodsorrel, 278, 292
Zelkova serrata, 132

Z

Zimmerman pine moth (*Dioryctria zimmermani*),
  339
Zones of adaptation, turfgrasses, 100–102
Zoysiagrass, 7
  fertilization of established lawns, 76, 77
  fungal disease symptoms
    dollar spot, 250, 392
    Helminthosporium leaf spot, 392–393
    lawn, season of occurrence on, 260
    Rhizoctonia diseases, 391–392
    rusts, 254, 393
  fungicide use in lawn care, 259
  growth and development, 94
  nematode parasites, 493
  nematodes, 262
  properties of, 108–109
  Pythium and, 255
  selection for environment, 107–108
  sewage sludge compost and, 47
  sod production, 120
  soil pH for, 389
Zoysiagrass rust, 254
Zoysia japonica (Japanese lawngrass), 108
Zoysia korenia, 108
Zoysia macrostaycha, 108, 109
Zoysia matrella (Manilagrass), 108
Zoysia sinica, 108, 109
Zoysia tenuifolia (Manilagrass), 108